DATE DUE

UNDERSTANDING DATA COMMUNICATIONS

UNDERSTANDING DATA COMMUNICATIONS

From Fundamentals to Networking

Third Edition

Gilbert Held

4-Degree Consulting

Macon, Georgia,

USA

JOHN WILEY & SONS, LTD

Chichester . New York . Weinheim . Brisbane . Singapore . Toronto

Baffins Lane, Chichester,
…019 1UD, England
…01243 779777
…(+44) 1243 779777
…e enquiries): cs-book@wiley.co.uk
…iley.co.uk
…iley.com

Other Wiley Editorial Offices

John Wiley & Sons, Inc., 605 Third Avenue,
New York, NY 10158-0012, USA

WILEY-VCH Verlag GmbH
Pappelallee 3, D-69469 Weinheim, Germany

John Wiley & Sons (Australia) Ltd, 33 Park Road, Milton,
Queensland 4064, Australia

John Wiley & Sons (Canada) Ltd, 22 Worcester Road,
Rexdale, Ontario M9W 1L1, Canada

John Wiley & Sons (Asia) Pte Ltd, 2 Clementi Loop #02-01,
Jin Xing Distripark, Singapore 129809

Library of Congress Cataloging-in-Publication Data

Held, Gilbert, 1943-
Understanding data communications: from fundamentals to networking / Gilbert Held.
p. cm.
ISBN 0-471-62745-3 (alk. paper)
1. Data transmission systems. 2. Computer networks. I. Title
TK5105 .H429 1997
004.6 — dc20 00-032094

British Library Cataloguing in Publication Data

A catalogue record for this book is available from the British Library

ISBN 0 471 627453

Typeset in $9\frac{1}{2}/11\frac{1}{2}$pt Bookman by Aarontype Ltd, Easton, Bristol
Printed and bound in Great Britain by Antony Rowe Ltd, Chippenham, Wiltshire.
This book is printed on acid-free paper responsibly manufactured from sustainable forestry, in which at least two trees are planted for each one used for paper production.

CONTENTS

PREFACE

Man's constant quest to communicate has resulted in a quantum leap in technology related to data communications. For the past quarter century the maximum obtainable transmission rate on many types of communications facilities has doubled every three to five years. During the past few years this growth rate has accelerated, with emerging technologies providing a transmission capability an order of magnitude or more above what were considered high operating rates just a year or two ago. Accompanying this growth and, in many cases, providing the impetus for the technological developments that made such growth possible are communications-dependent applications.

Today, data communications can be considered as the fiber that binds a modern society together. The past measurement of the strength of a nation, measured in the number of tons of steel manufactured per year, has essentially been replaced by the installed base of personal computers, workstations and other types of computational facilities, as well as the network structures that link those computers to one another. Unless stranded in a very remote location, you will use one or more communications facility almost every day of your life.

Due to the importance and, in many instances, our dependence upon communications, a detailed understanding of their evolution, technology and future directions is beneficial to most persons that work in a business, high technology, government or university environment, and provides a driving force for writing this book.

This book dates back to 1977 when the founding editor of *Data Communications* magazine, the renowned Harry Karp, asked me to develop a seminar to explain the characteristics, operation and utilization of data communications components which are the building blocks upon which networks are constructed. The resulting seminar, which I have continued to teach in both the United States and Europe, provided the basis for writing *Data Communications Networking Devices*, which has been blessed by reader demand to justify four editions. From teaching several data communications and computer courses at the university level, I became aware of many of the limitations of currently available books, including *Data Communications Networking Devices*. What my students desired was a comprehensive book that assumes no prior knowledge of communications and which presents concepts and theory, and relates practical experiences in a manner useful for persons involved or planning to work with data communications within an organization.

This new edition of *Understanding Data Communications* was written for both the student and the professional who wish to obtain a solid foundation concerning how data communications systems operate, why, where, and when certain types of equipment should be networked together, and the role of evolving communications technologies. In revising this book I continued to include and expand upon many basic communications concepts. History has a way of repeating itself and knowledge of how older communications systems operate that may not appear to be particularly important yesterday may be extremely useful tomorrow when attempting to understand the operation and utilization as well as limitations associated with a new technology. One key example of this is frequency division multiplexing, a technology considered relatively obsolete by the 1980s but which now forms the foundation for the operation of several types of high speed digital subscriber lines that represent a new generation of modem technology. Thus, while a major emphasis of this book is upon modern communications equipment and transmission systems, as an educator I felt it was important to include historical information and an overview of older technology that illustrates important concepts that are applicable for understanding modern technologies.

In developing this book I used a layered approach, building upon the knowledge presented in each prior chapter. This layered approach facilitates the utilization of this book as a one-semester course at a high undergraduate or at a first-year graduate level.

Throughout this book I have included numerous illustrations, tables and schematic diagrams to illustrate concepts, theory and practice. I believe this material will facilitate the use of this book long after a reader completes the course that it is used in, and will provide a reference for future endeavors in communications. Finally, at the end of each section I have included a comprehensive series of questions that cover many of the important concepts covered in the section. These questions can be used as a review mechanism prior to going forward in the book.

As both a professional author and an educator I highly value feedback. You can write to me through my publisher whose address is on page iv of this book, or you can communicate with me via email at gil_held@yahoo.com. Let me know if I committed the sin of omission and need to include other topics, if you feel I devoted too much space to a particular topic, or any other area you may wish to comment upon.

Gilbert Held
Macon, GA

ACKNOWLEDGEMENTS

The preparation of a manuscript that gives birth to a book requires the cooperation and assistance of many people. First and foremost, I must thank my family for enduring those long nights and for missing weekends while I drafted and redrafted the manuscript and reviewed proof pages.

As an 'old fashioned' author, I prefer the pen and paper to the modern convenience of the word processor. Although this may appear peculiar when writing on modern technology, my lifestyle of plane hopping, finding incompatible electrical outlets when traveling throughout the world and the extra weight of a portable computer makes pen and paper a most convenient mechanism of expression. Due to my method of writing I am indebted to Mrs Carol Ferrell who worked on the first edition of this book, and to Mrs Linda Hayes and Ms Junnie Heath, who worked on the second edition. Once again, I am indebted to the fine effort of Mrs Linda Hayes who also worked on the third edition of this book. Linda, as well as Junnie and Carol, were responsible for turning my handwritten manuscript revisions into the word processing files that were used for the creation of each edition of this book. Last but not least, one's publishing editor, editorial supervisor and copy editor are the critical link in converting the author's manuscript into a book. Thus, I would again like to thank Ian McIntosh and Ann-Marie Halligan for providing me with the opportunity to author three editions of this book, and Robert Hambrook and Sarah Lock and Sarah Corney for their fine efforts in moving my original and revised manuscripts through the production process.

1

COMMUNICATIONS IN A MODERN SOCIETY

The main objective of this chapter is to obtain an appreciation of the use of communications to enhance our daily work and recreation. To accomplish this, we will look at nine typical types of communication applications. Although an in-depth examination of many application areas will be deferred to later chapters in this book, the overview of communication applications presented in this chapter will illustrate our society's dependence upon the flow of timely and accurate information. Since there are many trade-offs involved in the design and operation of different communications systems, we will also focus our attention upon three key constraints and their effect upon different types of information flow in the second part of this chapter. Even though this is an introductory chapter it is important to understand the direction of technology as it relates to the field of data communications. Thus, the concluding section in this chapter will provide an overview of emerging trends and their potential effect upon your ability to communicate.

1.1 APPLICATIONS

The evolution of data communications has been nothing short of phenomenal. During a period slightly exceeding a century, the primitive telegraph has been replaced by a wide variety of networks that are the glue which binds our modern society together. As we perform our daily operations, it is most difficult to avoid coming into contact with an application that is not dependent upon data communications. Although we may take communications-related applications for granted, without the ability to communicate data, the banking, transportation and retail industries, as well as others, could not provide customers with an acceptable level of service. For other industries, such as publishing and finance, as well as many government agencies, the ability to rapidly communicate information is indispensable to their successful operation. Even the ability of countries to pursue policy is directly affected by communications. For example, in warfare the ability to successfully communicate can provide the margin which differentiates victory from defeat. This is vividly illustrated by the Gulf War, during which missiles with TV guidance, 'smart' bombs that could be directed down elevator shafts, and the ability to rapidly share intelligence gathered from the battlefield resulted in one of the most decisive military campaigns conducted in the history of warfare.

In this chapter we will examine a variety of applications that illustrate the important role of data communications in a modern society. This examination should provide readers with an insight into the ubiquitous nature of communications-dependent applications, as well as knowledge of some of the many industries that benefit from the ability to rapidly and accurately transmit information.

1.1.1 Data collection

Although many small firms still use manual time and attendance methods, the simple mechanical 'clock-punch' machine used in large industrial corporations and by companies with hundreds or thousands of employees is essentially only seen in movies of the 1960s or earlier. Today, most large organizations, as well as many firms with fewer than a hundred employees, use integrated data collection systems to track employee time and attendance data. Typically, employees insert their badges into a badge reader when they arrive at work or on the factory floor. Similarly, at break times, lunch and when they leave the premises, they insert their badges into a similar reader at the location where they 'clocked-in' or at another location.

Each badge reader recognizes and reads a magnetic strip on the badge, a series of vertical lines or perhaps hole punches that convey the unique identity of the employee. After reading the information, the badge reader transmits it to a computer center that may be located on the factory floor, in the same building or hundreds, or even thousands, of miles away.

Once the badge reader has transmitted the information it has read from the badge, the processing performed by the computer can range from simple time and attendance record keeping to the sophisticated alerting of management personnel to potential problems. Some problems that management might be alerted to include too few employees to perform a factory assembly function, excessive overtime or tardiness of employees. Within many organizations, the data collection facilities are integrated into the payroll system, relegating the use of time and attendance clerical employees to correcting such mistakes as forgetting to 'punch-in' or 'punch-out'.

A second pervasive example of data collection can be viewed by visiting many fast food retail chains. As you convey your order of a hamburger, large fries and shake to the clerk, you will probably notice that they press coded keys with symbols indicating each item on an electronic cash register type of device. Although that device functions as a cash register, totaling your purchase, adding applicable sales tax and computing change based upon your payment, it is also a data collection device more commonly known as a point-of-sale terminal. As the clerk presses a coded key, the information concerning the sale of each item is transmitted to a small computer system where it is recorded onto a diskette, cassette or other type of storage device. At the close of business or at a designated time, the computer system will print reports of the income received at each point of sale terminal to assist management in cash reconciliation as well as a summary report of items sold in the store. Taking the automation process a few steps further, some computer systems are programmed to automatically call a franchise distribution center or independent vendors. The computer will then electronically order such necessary supplies as hamburger wrappers, straws, napkins and cups, as well as meat patties and bags of french fries.

1.1.2 Transaction processing

Also known as inquiry-response, transaction processing is the key to customer support in the transportation and financial service industries where instant access to database information is required. Transaction processing differs from data collection in the fact that data transmitted to a computer in a transaction processing system can be used to immediately update a database. While this difference may appear trivial at first, it is the basis for ensuring that two persons do not purchase the same airline seat on the same flight, a bank customer does not charge an item beyond his or her credit limit, as well as other transactions dependent upon the immediate updating of information contained in a database.

Three of the more common uses of transaction processing include stock broker order entry systems, national credit card systems and automatic bank teller terminal operations. Although the actual execution of an order for securities varies based upon the market on which the security is traded and can be affected by other factors, in many instances an order to buy a security called into one stockbroker's office will be transmitted to a centralized market, where it is matched against an order to sell a security from a customer of a different security firm.

Today investors in securities have several methods they can use in addition to the traditional call to a registered representative. Some stock brokerage companies enable customers to bypass the registered representative and enter orders directly by punching keys on their telephone. Other companies established online Internet sites, enabling millions of investors to conduct electronic transactions.

In this book we will use the term Internet with a capital I to reference the global network of interconnected networks. In comparison, we will use the term internet to reference the connection of two or more public or private networks.

Figure 1.1 illustrates the initial or 'home' page of Waterhouse Securities, one of the pioneers of online brokerage accounts. Waterhouse, like many other brokerage and non-brokerage firms, established a presence on the Internet for electronic commerce. Their computer, referred to as a server, displays an initial screen referred to as a home page when accessed. In Figure 1.1 the Waterhouse home page is shown viewed through a Netscape browser, a software program that allows you to connect to literally an unlimited number of servers operated by an expanding universe of companies establishing a presence on the Internet. Note the $12 flat fee trading statement in the middle

Figure 1.1 Through the use of a browser you can access an online brokerage firm and perform different financial transactions at a fraction of the cost associated with the use of a traditional brokerage firm that requires you to use a broker

of the screen. A few years ago a typical purchase or sale of a few hundred shares of stock could result in a commission charge of over $100. Thus, online transaction processing is revolutionizing the manner by which consumers can perform a variety of tasks, ranging from purchasing stocks and airline tickets to validating bills for payment. Similar to the manner by which the refrigerator displaced the need for the iceman, electronic commerce can be expected to make many business obsolete.

Another popular example of transaction processing is the use of a credit card. Most major credit card companies have national and, in many instances, international credit authorization systems. When a customer makes a purchase in excess of a predefined amount, the merchant inserts the credit card into a terminal device and enters the amount of the purchase via a keyboard. Once the transmit key is pressed, the terminal transmits the credit card number and purchase amount to a computer system. First, the credit card number is electronically checked against cards reported lost or stolen, after which the amount of the purchase is added to the outstanding balance and compared to the maximum authorization limit for the credit card account. If the credit card is not lost or stolen and the authorization limit has not been exceeded, the transaction is accepted. If the transaction is rejected, the merchant may have to place a call to the credit card processing center to obtain additional information about the card.

Other stores that are part of a chain may use point-of-sale terminals that both authorize sales as well as transmit data, which is used by corporate market analysts to spot purchasing trends and to examine the relationship between the price of a product, its sales and geographical sales area. Some chain stores integrate their point-of-sale system with inventory control, using merchandise sale information transmitted with credit authorization data to track store sales and serve as a mechanism for the distribution of new merchandise to their stores.

While many readers have first-hand knowledge of the operation of bank teller terminals, for other readers their operation may be a slight mystery. In essence, a bank teller terminal can be considered to be a point-of-sale terminal that either dispenses information in the form of updating a passbook or dispenses cash and electronically updates one's account. The most conventional type of bank teller terminal simply dispenses information in the form of updating accounts and its operation depends upon the bank clerk who enters deposit or withdrawal information and accepts or dispenses cash. The second type of bank teller terminal, more formally known as an Automatic Teller Machine or ATM, dispenses predefined packets of cash, such as $10, $20, $50 or $100.

A person using an ATM first inserts his or her bank card and the machine reads and transmits magnetic coded information on the card to the bank's computer system. Assuming that the card was not reported lost or stolen and gobbled up by the machine, the computer will prompt the customer to enter his or her personal identification number, commonly referred to as a PIN. The PIN can be viewed as a secret number known only by the customer and his or her bank and serves to verify the identity of the person using the bank card. Thus, if the correct PIN associated with the card is not entered by the customer at the numeric keyboard of the ATM, the request for cash will not be granted. If the request is granted, after the cash is dispensed the customer's account is debited by the amount dispensed, with many banks adding a service fee which both pays for the facilities required to support the ATM system and contributes to their profit margin.

In addition to the previously mentioned transaction processing applications, other common examples of the use of this communications based technology include airline, hotel and automobile reservation systems. The key to the successful operation of each system is the ability of a terminal operator to query a database to determine the availability and cost of an airline trip, hotel room or a particular type of vehicle.

1.1.3 Conversational time sharing

The high cost of large scale computers resulted in the development of time sharing as a method to enable many users to share the computational power of a common facility. In a time sharing environment, each user obtains the use of a small fraction of time of the central processor known as a time slice. If the user's job is not completed during the allocated time slice, the job is queued by the operating system for service by subsequent assignments of time slices.

The development of interpretive languages, such as the Beginners All Purpose Symbolic Instruction Code (BASIC), as well as Formula Translation (FORTRAN), Common Business Oriented Language (COBOL) and Programming Language One (PL/I) compilers to operate under time sharing permits application programmers to develop and test their programs prior to placing them into a production environment. Since tens to hundreds, and in some cases thousands, of persons could create and execute programs concurrently on a time sharing system, their utilization made computing more economical than classical batch systems where one job must be completed prior to the start of the next job. Although the growth in the use of personal computers has considerably reduced the demand for time sharing, it is still an important computational facility in some large organizations.

Figure 1.2 illustrates a typical example of a modern time sharing application. In this example the main screen display for a version of IBM's OfficeVision calendar and electronic mail system is shown. This particular version of OfficeVision operates on an IBM mainframe computer which can support thousands of users.

Until the advent of personal computing, only time sharing extended the computational power of computers via communications facilities to terminals located on users' desks to provide 'desktop computational capability'. Even with the growth in the use of personal computers, there are many applications

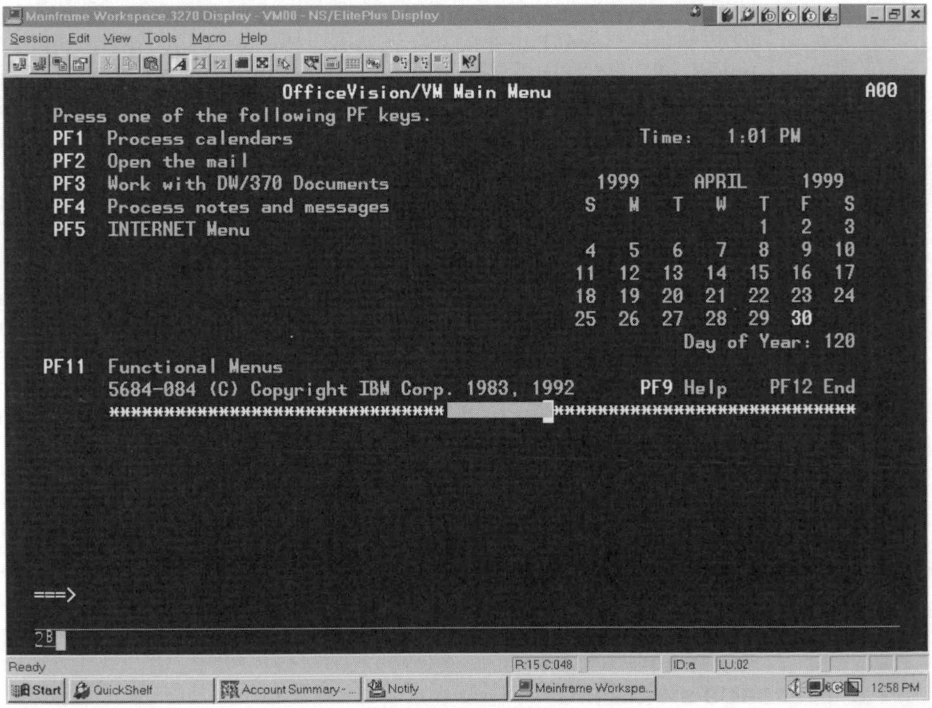

Figure 1.2 One example of a modern time sharing application is IBM's OfficeVision's calendar and electronic mail system. OfficeVision operates on several types of minicomputers and mainframes, with the latter supporting up to several thousand users

which, because of data storage capacity or processing power requirements, are restricted to operating in a time sharing environment. Due to this, the use of time sharing systems can be expected to coexist with personal computing for the foreseeable future.

1.1.4 Remote job entry

There are many types of data processing jobs, such as accounting and payroll, that require execution in a continuous manner. Organizations with diverse locations may prefer to use one or a few data processing centers to process payroll and accounting data.

To facilitate the timely processing of accounting and payroll data, most organizations that use centralized data processing centers employ remote batch transmission facilities. Typically, accounting and payroll data collected over a period of time at distributed locations are formed into a batch of records. At a predefined time or during a predefined time interval, the batched records are transmitted to the centralized data processing centers. There, the batched records received from the remote locations are combined and used as input to the organization's accounting and payroll programs.

During the 1960s and 1970s, physically large minicomputer based remote batch terminals were primarily employed to transmit batch data to centralized data processing locations. At these locations, mainframe computers were used to process the data received from the remote locations. By the late 1980s, many minicomputer-based remote batch terminals had been replaced by the use of personal computers to perform batch transmission applications. In the 1960s and 1970s, many batch terminal configurations included such peripheral devices as card readers, disk or magnetic tape storage units and high speed printers, while some terminals also supported interactive cathode ray tube terminals. By the use of interactive terminals, clerks could enter data throughout an accounting or payroll period. The data were then stored on disk or tape and transmitted to the central computer facility for processing. By the late 1980s, the tape and disks of many minicomputers had been replaced by the use of personal computer fixed disk and diskette on-line storage.

Some batch terminals include the capability to perform local data processing, executing small data processing jobs while transmitting larger jobs to the corporate mainframe. The results of those jobs, called system output (SYSOUT), as well as accounting reports, checks and other data, can be directed from the mainframe to the batch terminal via a communications facility where the data can be directly printed or stored on tape or disk for later printing. When first stored on tape or disk, the printing of the stored data is known as printer spooling.

1.1.5 Message switching

Message switching represents one of the earliest merging of communications and computer technologies. Beginning in the early 1950s, several computer manufacturers developed software specifically designed for message switching applications. Companies that purchased hardware and software obtained the capability to either develop an internal message switching facility for their organization or provide a commercial service that other organizations could subscribe to.

Early message switching systems required terminals to be permanently connected via a communications facility, precluding their use for other applications. Messages entered via a terminal were transmitted to a central computer facility where their heading was first examined. The message heading included information concerning the subscriber or subscribers that it was to be distributed to, the originator of the message and could include such optional information as its subject and priority. Depending upon the status of the destination subscriber's terminal, the message might be immediately switched to an output line routed to the destination terminal or stored on disk or tape. If the destination terminal was not busy servicing a previously transmitted message or sending data, the message might be switched directly to its destination. If the destination terminal was in use, the message would be stored. Then, when the destination terminal became available, the message would be retrieved from storage and forwarded to its destination. Due to this type of operation, message switching is commonly referred to as a store and forward system.

The use of message switching systems initially centered upon business and commercial activity. As the use of message switching increased, additional applications were developed, such as the electronic delivery of money orders that was well publicized by a series of television commercials. Although message switching as a technology has essentially been succeeded by electronic mail and value-added carrier services it provided a foundation for the movement of data between terminal users, Thus, it represents an underlying technology which formed the basis for the evolution of more modern technologies that today deliver electronic mail to tens of millions of persons each day.

1.1.6 Value-added carriers and electronic mail

The proliferation of the use of personal computers in the home and office during the late 1970s and early 1980s served as a driving force for the growth of value-added carriers and the introduction of a new type of message switching known as electronic mail. Essentially, value-added carriers can be considered as a new type of communications utility. By leasing communications lines from telephone companies and installing specialized computers, the value-added carriers developed extensive communications networks. The use of these networks was fostered by many companies connecting their computers to network nodes, permitting persons from their organization or other companies to access the carrier's network from hundreds of locations across the United States and via the entry of a code to be routed to the appropriate computer facility.

Although the initial use of value-added carriers was primarily by business, during the late 1970s many individuals began to subscribe to a variety of information retrieval services that provided financial, weather and text retrieval from selective databases. Some value-added carriers expanded into information utilities, adding their own computational facilities to their network to provide subscribers access to a variety of information services as well as the use of electronic mail facility. Other value-added carriers added electronic mail facilities for business users while providing a communications transportation facility for other users to access numerous commercial electronic mail services that were established during the 1980s.

One of the first, if not the first, commercially available electronic mail services was MCI's MCI Mail. MCI Mail was developed in the period prior to the expansion of the Internet for commercial use. At one time this text based electronic mail system was one of the most popular forms of electronic communications in use. Figure 1.3 illustrates the use of the HyperTerminal application bundled into Windows 95 and Windows 98 to access MCI Mail. Although Windows represents a graphic user interface (GUI), you cannot use the point and click capability of the operating system when working with MCI Mail. Instead, you must enter commands in the form of text, such as 'scan inbox' shown in the lower portion of Figure 1.3. In this example the command would result in the listing of five messages in the author's INBOX. To read each message would then require the entry of an appropriate 'read' command.

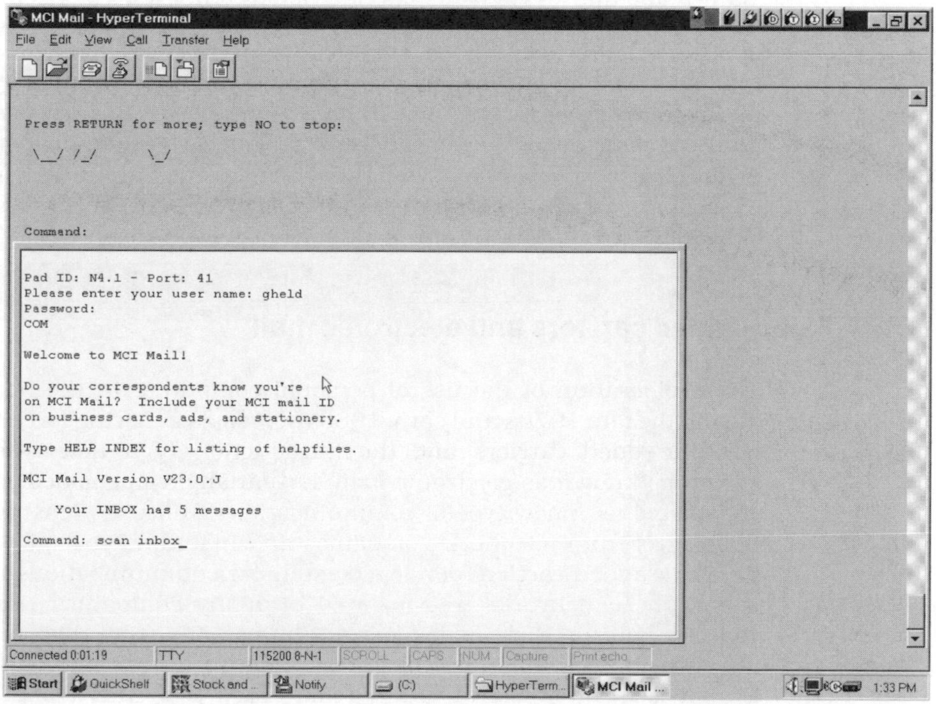

Figure 1.3 MCI Mail was one of the first commercial electronic mail systems. Although still used by this author, the popularity of the Internet where subscribers can perform many functions in addition to electronic mail has diminished the demand for systems strictly devoted to email

The growth in the use of the Internet makes the use of an Internet Service Provider (ISP) more attractive than the use of an electronic system restricted to mail delivery. Through an account with an ISP, both business and residential users can perform a number of functions in addition to sending and receiving electronic mail. MCIWorldCom, which represents the merger of MCI Communications and WorldCom, offers Internet access as well as numerous voice and data services, with the number of its Internet accounts now greatly exceeding its number of MCI Mail accounts, illustrating how advances in one area of communications can result in the rapid or gradual obsolescence of another area.

The use of a more modern electronic mail system is shown in Figures 1.4 and 1.5. Figure 1.4 illustrates the initial CompuServe mail center display. CompuServe was originally one of the earliest information utilities that provided subscribers access to shareware programs, news and weather information, and chat rooms in addition to electronic mail service. Figure 1.5 illustrates the point and click ease of use of CompuServe for creating an electronic message. After clicking on the icon labeled 'New' in Figure 1.4 the screen display was changed to the 'Create Mail' screen shown in Figure 1.5. Clicking on the button labeled 'Recipients' resulted in the display of the window labeled 'Message Recipients' shown in the middle of Figure 1.5. Note that by clicking on the rectangle labeled 'Address Book' a list of predefined names and addresses is displayed. Then another few clicks enables a person

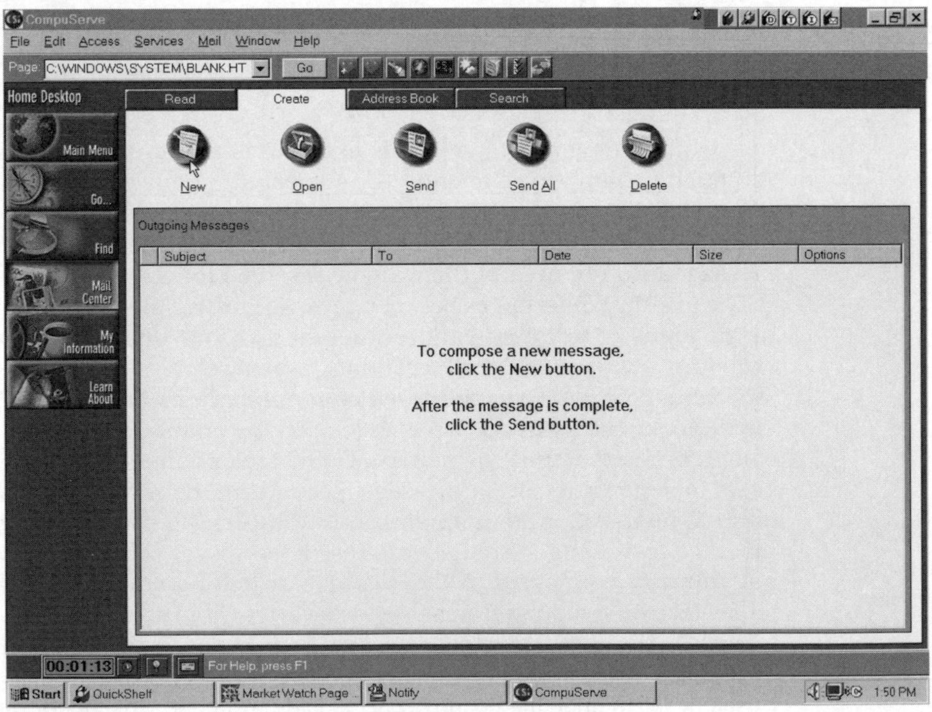

Figure 1.4 The CompuServe Mail Centre screen display provides users with a graphic user

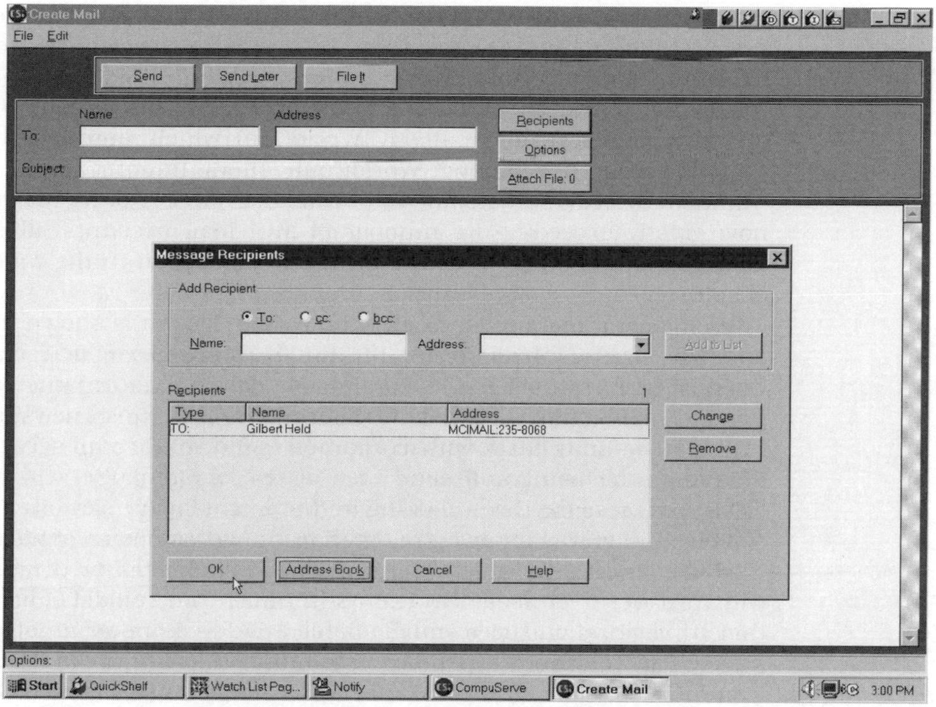

Figure 1.5 Using the graphic user interface of the CompuServe Mail Centre facilitates the creation of electronic mail messages

to select a recipient which in this example is the author's MCI Mail address. Although you still have to enter the subject and body of the message, through the use of Windows' cut and paste capability you could prepare your message using a word processor and either attach it as a file or copy and paste the message into the area of the window reserved for the body of the message.

The primary differences between message switching and electronic mail are in the areas of terminal connection and message delivery. Initially, message switching systems required terminals to be directly connected to the message switching computer via dedicated communications facilities. In comparison, electronic mail systems were developed to enable terminal and personal computer users to use the public switched telephone network on a temporary basis to send or receive a message, permitting the terminal or personal computer to be used for other applications. Concerning message delivery, initially message switching systems were restricted to delivering messages to terminals directly connected to the message switching computer. In comparison, most electronic mail systems provide a variety of message distribution options to include the conversion of an electronic message to hardcopy and its delivery by the postal service or via courier. Today, the use of electronic mail can range in scope from a corporation distributing new product announcements, to an individual bidding on a home or sending a birthday greeting to a friend or relative.

1.1.7 Office automation

Until the introduction of the microprocessor-based personal computer office, automation operations were highly centralized, with a mainframe or minicomputer typically used to provide computational resources to the employees of an organization. Those computational resources were usually limited to text processing and financial applications, and required the establishment of a communications infrastructure that could result in the transmission of information over hundreds or thousands of miles to perform relatively simple functions by today's computer environment, such as developing a mailing list or creating a form letter.

The use of a corporate mainframe for office automation functions represented perhaps the earliest example of client–server computing. Through the early 1980s dumb terminals without microprocessor based intelligence were used to communicate with corporate mainframe computers. The terminal, serving as a client, would send a request to the mainframe which functioned as a server, servicing the processing requirements of humerous clients. This type of client–server computing resulted in the development of hierarchical structured networks in which terminals were connected to control units which in turn were connected to the mainframe. The control unit can be viewed as a line sharing device which enabled two or more terminals to contend for access to relatively expensive communications lines and mainframe computer ports. Figure 1.6 illustrates an example of the mainframe-based client–server computing model which formed the basis for office automation through the mid-1980s.

During the 1980s the ubiquitous office typewriter was rapidly replaced by the personal computer. At first, a lack of application programs resulted in the PC being used as a dumb terminal in an office environment, with client–server computing continuing to resemble the illustration shown in Figure 1.6. In fact, the access to IBM's mainframe-based Office Vision calendaring and electronic mail system previously illustrated in Figure 1.2 occurred through the use of a PC acting as a dumb terminal. While some organizations continue to use mainframe centric computing, other organizations elected to distribute computing applications based upon the use of PCs.

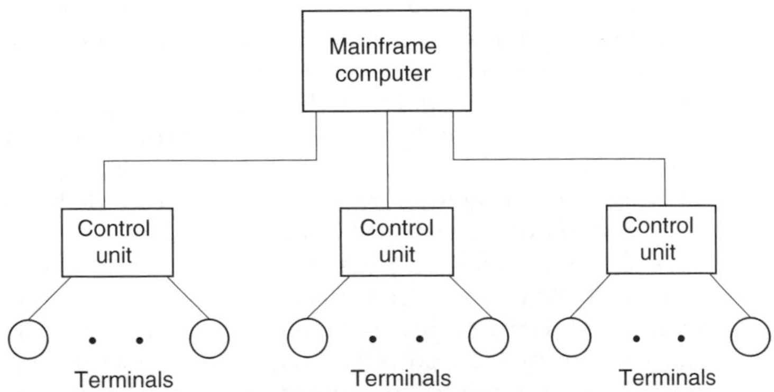

Figure 1.6 Mainframe-based client–server computing model

The rapid increase in the processing power of personal computers soon resulted in the development of a variety of office automation software to include word processing, electronic spreadsheets, visual presentations, database creation and retrieval and other programs. The expansion in the use of personal computers was accompanied by a requirement to share information between personal computer users. This requirement was primarily satisfied by the development of local area networks (LANs). Through the use of LANs small corporate departments within an organization, as well as companies that could not afford the expense associated with operating a mainframe computer, were able to establish their own client–server computing operations. In large corporations islands of workstations on individual LANs began to rapidly appear during the late 1980s, changing the corporate client–server model from a hierarchical mainframe centric model to a distributed computing environment with individual LANs connected to one another via specialized communications devices, as well as maintaining one or more connections to the corporate mainframe. This modern client–server model is illustrated in Figure 1.7.

In comparing the mainframe based client–server model illustrated in Figure 1.6 to the modern client–server model shown in Figure 1.7 the differences in potential network structures are apparent. The mainframe-based model communicated with dumb terminals, and it was difficult if not impossible to establish multiple routes for the transmission of information. In comparison, the modern client–server model is based upon the use of intelligent computers, as both workstations connected to a LAN as well as specialized communications devices that have routing capabilities. This makes it possible to use different topological structures to interconnect LANs as well as to support multiple communications paths between LANs.

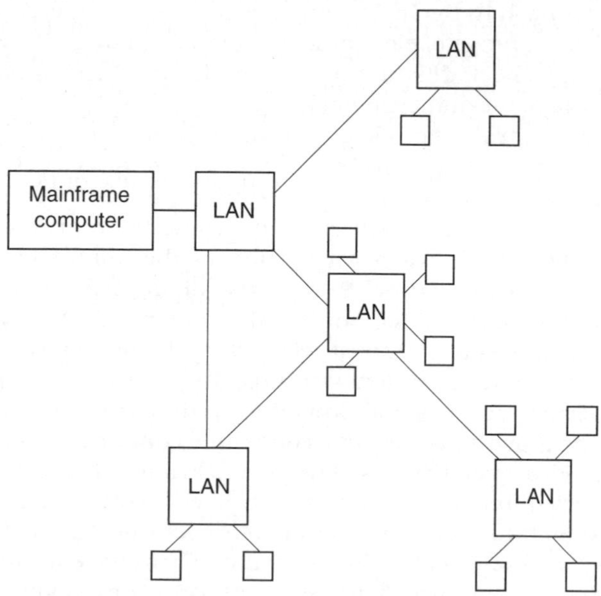

Figure 1.7 The modern client–server model

Although the centrally managed mainframe-based client–server model is easier to manage, its ability to adjust to organizational change is limited. In comparison, the modern client–server model is much more flexible in adjusting to a changing organizational structure, since LANs can easily be subdivided (a process known as segmentation) to accommodate growth or a changing user environment. Unfortunately, it is much more difficult to manage as an entity all of the LANs within an organization, a process commonly referred to as Enterprise network management, which can be viewed as the price paid for obtaining an increased flexibility to support the requirements of an organization. Readers should note that the process of downsizing or moving applications off the mainframe onto the corporate LAN results in a client–server model similar to the one illustrated in Figure 1.7, with the mainframe removed due to the effect of the downsizing effort.

In addition to being used in computers, the microprocessor has been incorporated into numerous office automation products which significantly improve worker productivity. Today pagers, inventory control scanners, and even the supermarket bar code reader are all based upon the use of microprocessors. Those small silicon chips interpret sequences of digital pulses to generate characters on a pager's display, convert the vertical lines scanned from a can of chicken soup into digits that a distant computer can use to determine the price of the product, and perform other operations that have significantly improved our lifestyles.

1.1.8 Electronic commerce

The growth in the Internet makes it possible for consumers and businesses to take advantage of electronic commerce opportunities. As a consumer you can literally check different merchants for product availability and price through simple point and click operations.

To illustrate the role of electronic commerce consider Figures 1.8 and 1.9. In Figure 1.8 I used my browser to access the Barnes & Noble World Wide Web home page. From this page I entered my name to check the price of books I authored. Because I gave away my complimentary copies of one book, I needed to order another copy. A portion of the simple electronic order process is illustrated in Figure 1.9. Note the dialog box named 'Security Information' displayed in the foreground of the screen. The Netscape browser is similar to other browsers in that it will automatically encrypt transmission to enable secure communications required to put the consumer's mind at rest when ordering products and providing credit card numbers over the Internet.

The growth in electronic commerce conducted over the Internet has literally exploded over the past few years. From a few sales of books, records and assorted items that may have reached \$100 million during 1996, by the new millennium electronic commerce over the Internet was estimated to have exceeded \$20 billion. Today you can purchase airline tickets, shop for a car, and buy insurance, flowers or perform your weekly food shopping, all literally at the click of a cursor.

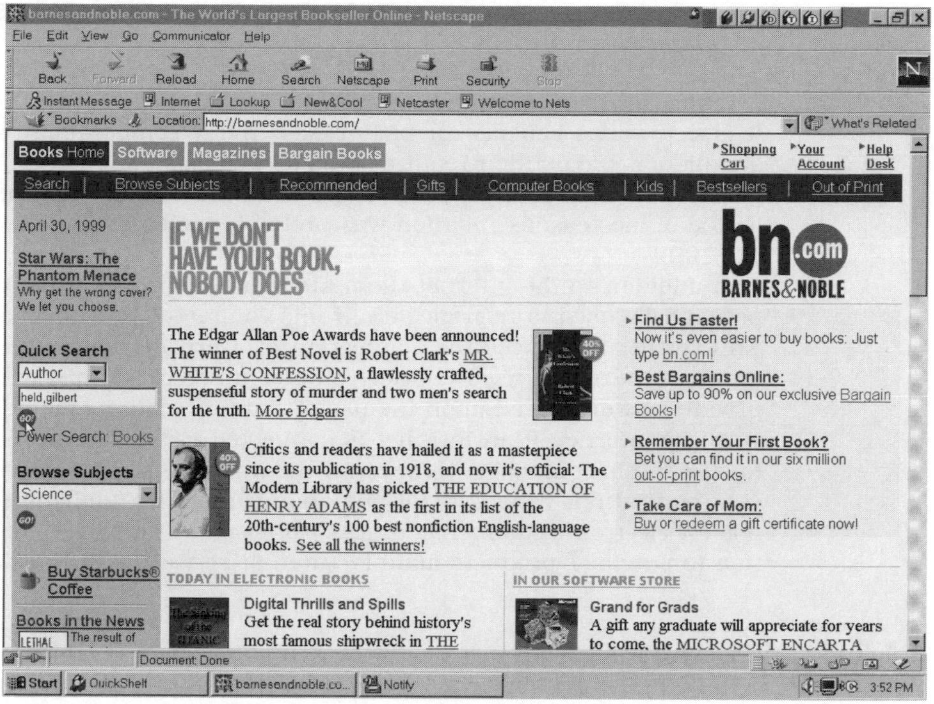

Figure 1.8 Using the Netscape browser to view the Barnes & Noble home page

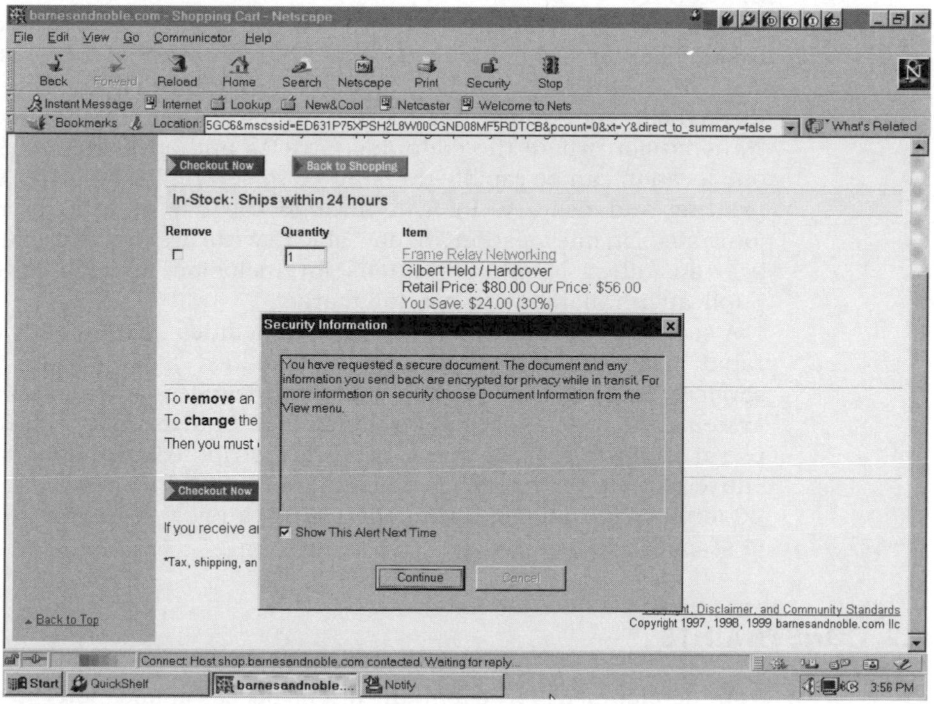

Figure 1.9 Ordering products over the Internet results in the browser encrypting information transmitted as a mechanism to protect credit card data

While electronic commerce provides a considerable benefit for consumers, it also provides benefits for businesses. Today companies run auctions for suppliers to bid on their requirements as well as allow potential employees to post their résumés. Thus, electronic commerce fosters competition, which is one of the reasons inflation was probably tamed during the latter part of the 1990s.

In addition to the Internet there are two other types of networks that are periodically used to reference electronic commerce: extranets and intranets. An extranet references the connection of a private network to the Internet and can indeed be used for electronic commerce. An intranet represents a private network based upon the use of communications methods associated with the Internet. If an intranet is connected to the Internet it can be used for electronic commerce. However, if the intranet is restricted to providing a communications capability for one organization, it is difficult to envision its use for electronic commerce, unless it is used by employees of the organization to purchase products manufactured or services sold internally.

1.1.9 Satellite transmission

One of the things many people take for granted is the ability to obtain a newspaper on the day of its publication. Without the use of satellite transmission, this minor event would be an impossibility in many areas of the world.

Today, satellite transmission and newspaper publications are closely linked to one another. Such publications as *USA Today*, *The New York Times* and *The Wall Street Journal* are printed simultaneously at several locations throughout the United States and overseas due to the use of satellite transmission where the editorials, articles and advertisements prepared at one location can be rapidly transmitted to several locations for simultaneous printing and delivery. In fact, through the use of satellite transmission, journalists in one location are now able to write articles and columns that can be transmitted to other locations for inclusion in different editions of a publication tailored for a specific market.

A second use of satellite transmission facilities which greatly enhances the rapid dissemination of news to include text and pictures involves wire services. Until the late 1970s, most wire services used message switching systems and facsimile transmission to distribute text and pictures. Today, the use of satellites permits wire services to distribute information to newspapers subscribing to their services much more rapidly. Pictures that required 10 to 20 minutes to transmit during the 1970s can now be transmitted in a matter of seconds.

1.2 CONSTRAINTS

The development of communications-based applications which are the foundation of our modern society involves many trade-offs in terms of the use of different types of communications facilities, types of terminal devices, hours of operation and other constraints. Four of the key constraints associated with the development of communications applications are throughput, response time, bandwidth and economics.

1.2.1 Throughput

Throughput is a measurement of the transmission of a quantity of data per unit of time, such as the number of records, blocks or print lines transmitted during a predefined interval. Throughput is normally associated with batch systems where the transmission of a large volume of data to a distant location occurs for processing, file updating or printing. As this is typically an extension of batch processing, and since it occurs remotely from a data center, the device that transmission is from or to is referred to as a remote batch or remote job entry device.

Although many readers may not realize it, every time you download or upload a program through a browser, use the file transfer protocol (ftp) or perform a similar operation, you are performing a batch transmission. Thus, your personal computer can function as a batch terminal.

In most batch transmission systems, a group of data representing a record, block or print line is transmitted as an entity. Its receipt at its destination must be acknowledged prior to the next grouping of data being transmitted. Figure 1.10 illustrates the operation of a batch transmission system by time, with the waiting time indicated by shaded areas. Since the throughput depends upon the time waiting for acknowledgements of previously transmitted data, one method used to increase throughput is to transmit more data prior to requiring an acknowledgement.

A second method to increase throughput can be obtained by acknowledging a group of blocks instead of on an individual basis. For example, acknowledging block n could signify that all blocks through block n were received correctly and the receiver now expects to receive block $n+1$. The number of blocks that can be outstanding prior to receiving an acknowledgement is

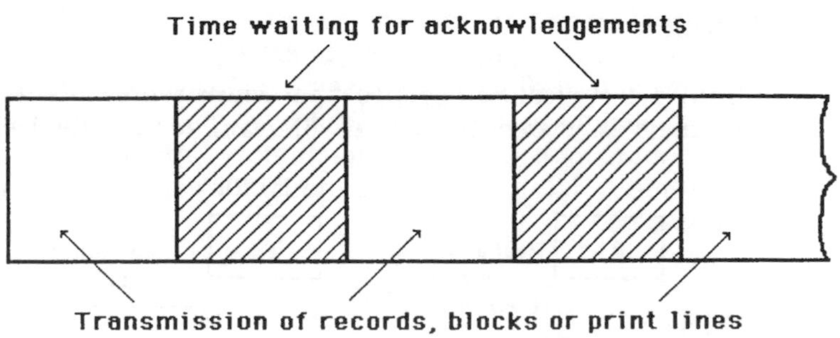

$$\text{THROUGHPUT} = \frac{\text{TOTAL RECORDS, BLOCKS OR PRINT LINES}}{\text{TOTAL TRANSMISSION TIME}}$$

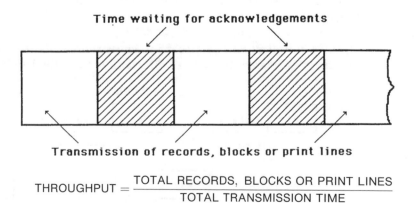

Figure 1.10 Batch transmission and throughput

referred to as a window. Later in this book we will examine the effect of different window size settings upon throughput.

1.2.2 Response time

Response time is associated with communications where two entities interact with one another, such as a terminal user entering queries into a computer system. Here each individual transaction or query elicits a response and the time to receive the response is of primary importance.

Response time can be defined as the time between a query being transmitted and the receipt of the first character of the response to the query. Figure 1.11 illustrates interactive transmission response time.

The optimum response time for an application is dependent upon the type of application. For example, a program that updates the inventory could have a slower response time than an employee badge reader or an airline reservation system. The reason for this is that an employee entering information from a bill of lading or other data which is used to update a firm's inventory would probably find a 5 or 10 s response time to be satisfactory. For a badge reader system where a large number of workers arrive and leave during a short period of time, queues would probably develop if the response time was similar. For airline reservation systems, many potential customers require a large amount of information concerning discount prices, alternative flights and time schedules. If the airline reservation clerk experiences a slow response time in scrolling through many screens of information to answer a customer query, the cumulative effect of a 5 s response time could result in the customer hanging up in disgust and calling a competitor. For other interactive communication applications, such as automated teller machines, competitive advertising has made slow response almost an issue involving the violation of a user's fundamental rights. In certain locations, it is quite common today to see banks battling against one another in advertisements over who has the fastest teller machines, yet another example of the use of communications to gain a competitive position.

1.2.3 Bandwidth

From a technical perspective bandwidth represents a range of contiguous frequencies, a concept that we will examine in some detail later in this book.

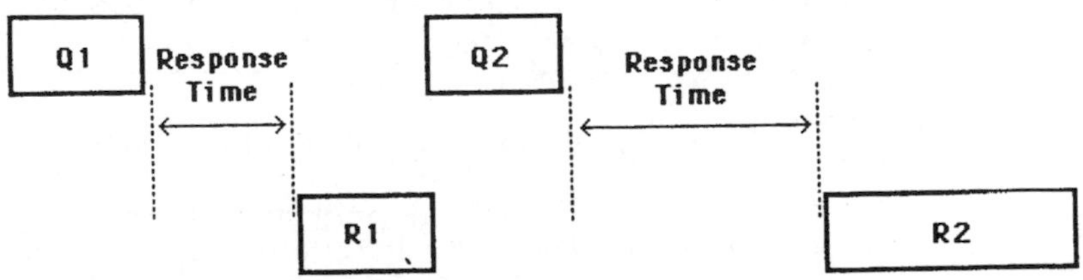

Figure 1.11 Interactive transmission response time

The range of frequencies is an important consideration for communications, since the maximum amount of data that can be transmitted per unit time is proportional to the bandwidth of transmission media. For example, fiber-optic cable, which has a relatively high bandwidth since it transports light, provides the ability to simultaneously transport thousands of telephone calls. In comparison, the twisted wire copper cable which forms the basis of most business and residential telephone service is limited to supporting only one or a few simultaneous telephone calls.

1.2.4 Economics

Similar to other technologies there are a range of economic trade-offs associated with the use of different types of communications. Some types of communications represent services for which users are billed on a per minute basis. Other types of communications involve leasing of a circuit for a fixed monthly fee regardless of use. Although a per minute service is less costly than a leased circuit when usage is minimal, as usage increases the situation could change and the leased line may be more economical.

While the preceding is an over-simplification of the economics associated with the use of communications, it illustrates an important concept. That concept is the fact that you should compare alternative means of communications as well as the cost of equipment required to support different communications methods. Doing so will provide you with the ability to select a cost-effective communications method required to satisfy your communications requirement.

1.3 EMERGING TRENDS

Through the 1970s communications was a highly regulated industry that provided customers with a limited choice of products and services. The divestiture of AT&T in the United States of its operating subsidiaries, the privatization of British Telecom and the sale of stock in other national communications carriers resulted in the emergence of a competitive market for communications services as well as a significant growth in the number of hardware and software vendors marketing communications products. In addition, telecommunications reform legislation in the United States and abroad are removing artificial barriers which limited the ability of local and long distance telephone companies and cable TV to compete with one another. Eventually, you can expect the distinctions between cable, local and long distance telephone services to diminish or even disappear.

In addition to changes in legislation, advances in technology are forming the basis for a profound change in the manner by which communications services are provided. The original communications infrastructure throughout the world was designed to transport voice. Although well-suited for carrying voice conversations, that infrastructure could not directly carry digital signals. The evolving conversion of the infrastructure of communications carriers to digital technology and the increased use of fiber-optic cable to interconnect buildings within cities and carrier offices in one city to offices in another city is having a profound effect upon the ability to merge voice, data and video, a process commonly referred to as multimedia.

The transport of voice requires an infrastructure that provides a minimal delay time. In comparison, the transport of images and data can tolerate relatively long delays. Recognizing the differences between optimum transmission methods, a technology known as Asynchronous Transmission Mode was developed to facilitate the merging of voice, data and video so that multimedia can be transported on local and wide area networks. At the same time 'fiber to the home' trials were in progress that extended fiber technology and its large bandwidth to residential customers, while the use of the Internet was being tested as a mechanism to transport digitized voice conversations.

In the first decade of the new century it is quite possible that products and services in limited use or not even presently offered will be commonly available as a result of advances in communications technology. Instead of visiting a library you will probably telecommunicate with your library and read a book on your home computer. Instead of simply listening to a person during a telephone conversation you will be able to see the person you are talking with. Similarly, research, business, finance and other functions can be expected to radically change as advances in communications unlock barriers and facilitate the interchange of information.

1.4 REVIEW QUESTIONS

1. Assume that your organization is considering the installation of badge readers to collect time and attendance data. Discuss how the time and attendance data can be used by management as well as serving as input for automation of other organizational functions.

2. Discuss the operation of a transaction processing system with respect to a database accessed by the system.

3. What effect do you expect electronic commerce to have upon the ability of persons to purchase securities, airline tickets, and other products?

4. Assume that you plan a trip that includes an airline flight from New York to San Francisco, the use of a rental car to drive to San Mateo and a week's stay at the San Mateo inn. Discuss the type of communications application you would probably use to plan your trip.

5. What is the function of a personal identification number (PIN) when entered into a bank automated teller machine (ATM)?

6. Why is time sharing considered as a predecessor to desktop computing obtained through the use of a personal computer?

7. What was a primary disadvantage of early message switching systems?

8. Why is message switching commonly referred to as a store and forward system?

9. What is client–server computing?

10. Discuss the differences between early and modern client–server models with respect to their operation and network infrastructure.

11. Describe an example of electronic commerce assisting the consumer and an example of how it can help a business.

12. What are three of the key constraints associated with the development of communications applications?

13. What does the term downsizing mean with respect to computer applications?

14. If the transmission of 5280 records required 2 minutes 80 seconds, what is the throughput?

15. Discuss the use of throughput and response time measurements with respect lo remote batch and interactive systems.

16. What is bandwidth and why is it an important consideration for transmission?

17. What is the term used to describe the merging of voice, data and video?

2

BASIC TELEGRAPH AND TELEPHONE OPERATIONS

The foundation of modern communications can be traced to the development of telegraph and telephone operations during the nineteenth century. The telegraph can be considered as the forefather of the automatic teleprinter and its use was based upon the development of an elementary code to convey information which is still in use today. The telephone has grown in use to the point where it is truly ubiquitous, with over 99.9% of homes and businesses in North America and Europe having one or more instruments. The development of telephone networks resulted in a structure used for the distribution of calls that remains in use over one hundred years after its initial development. Thus, both telegraph and telephone communications provided the foundation for modern communications, even though their operation and utilization have considerably changed over the past one hundred years.

In this chapter, we will first examine the evolution of communications from simple signaling by fire to early telegraph systems. In our examination of telegraph systems, we will focus attention upon the use of codes to convey information and two areas of technological development that were required to automate communications. This will be followed by an examination of the operation of the telephone, the routing of calls between telephone stations and the switching hierarchy established for the routing of long distance calls. From the information presented in this chapter, you will obtain an appreciation of the evolution of modern communications as well as why the operation and constraints of twentieth and twenty-first century communications can be traced to prior developments during the nineteenth century.

2.1 EVOLUTION OF COMMUNICATIONS

Man's method of communicating between diverse locations can be considered to form an index of our technological development. The first known methods of signaling were Greek and Roman signal fires which were limited in their information contents to the occurrence or non-occurrence of predefined events. In the mid 1600s, Portuguese explorers returning from Africa reported upon the use of jungle drums which transmitted messages between villages. Their use disseminated more information than fires, since the beat of the drum could be changed to convey different information. With the emergence of the Industrial Revolution, the requirement for timely and accurate mechanisms for information distribution grew, resulting in the development of machines that communicate with one another. In fact, much of our modern society is based upon the communication of messages whose information content is generated by or through the use of machines. Foremost among those machines are the telegraph and telephone, whose development can be considered as the foundation of modern communications systems.

2.2 TELEGRAPHY

Although Samuel F. B. Morse is credited by most persons as the man who invented the telegraph, in actuality the American physicist Dyer operated a single wire telegraph in 1828 based upon electrostatic electricity and which used litmus paper as a signal indicator. This telegraph operated over a distance of 10 km on a racecourse in Long Island and was in operation almost 16 years prior to the first telegraph line established to link two cities together.

Modern technology, which can be considered as the predecessor of other methods of electronic communications began in 1832 when Samuel Morse invented his telegraph alphabet, now known as the Morse code. By 1844, the first telegraph line had been constructed in the United States, linking Washington and Baltimore. On May 24, 1844, Morse transmitted the now famous phrase 'What hath God wrought' from the old Supreme Court Chamber in the United States Capitol to his partner Alfred Vale in Baltimore.

2.2.1 Operation

The Morse telegraph system is similar to all communications systems in that its operation requires a transmitter, a transmission medium and a receiver. The transmitter used in the first Morse telegraph system was the telegraph key, which was a switch with a knob or handle, which, when pressed down, resulted in the closure of an electrical circuit. The power for the circuit was provided by a battery or another source of direct current.

Morse's first telegraph receiver used wire coils wound around metal to form an electromagnet with a moving armature which was used to draw an inked line on a moving strip of paper. Morse soon observed that the noise of the receiver could be 'read' by a trained ear and modified the telegraph receiver. The modified receiver replaced the moving strip of paper with a thin piece of metal that would click on a contact due to the induced magnetism in the armature caused by the closure of the key at the transmitter. This type of receiver is also known as a Morse sounder.

Figure 2.1 illustrates the circuitry of a one-way telegraph system where the term simplex is used to denote the transmission of information in one direction. When the original Morse receiver was used to draw a line on a moving strip of paper, a mark was made on the paper when a pen attached to the armature was attracted to the coiled metal. Since a marking condition was caused by the closure of a key which resulted in current flowing through the resulting circuit, the term marking state has evolved to denote the flow of current in a line. Similarly, the opening of the telegraph key caused a break in the circuit which precluded the flow of current. This action caused the pen to be moved off the paper, resulting in a space. Hence, the term space or spacing state has evolved to denote a condition in which no current is flowing in a line. Although Morse didn't realize it, he had created a binary state machine. That is, a telegraph operates in one of two states – current flowing or current not flowing. As we will note later in this book, all modern communications systems are based upon binary operations. For example, the ability to communicate via a fiber optic cable is based upon the transmission of digitized voice conveyed as a series of light and absence of light pulses.

Since the telegraph system illustrated in Figure 2.1 was capable of transmitting in only one direction, it was soon modified to permit operators at each end of a telegraph line to communicate with one another. This modification resulted in the placement of a Morse sounder and key at each end of the circuit, as shown in Figure 2.2. In this configuration, the key at each station was provided with a switch to close the circuit when the station is receiving data. When neither end is transmitting, the line is in an idle state, both switches are closed, both sounders are operated and current is continually flowing in the resulting circuit.

When an operator has data to transmit, he or she first opens the key shorting switch, then depresses the key for varying short periods of time to produce the dots and dashes that make up the Morse code for each character to be transmitted. Since the sounder clicks when the operator presses the key, each operator hears the Morse code as he or she keys it. Once a message is completed, the operator shorts his or her key, enabling the operator at the opposite end of the line to begin transmission.

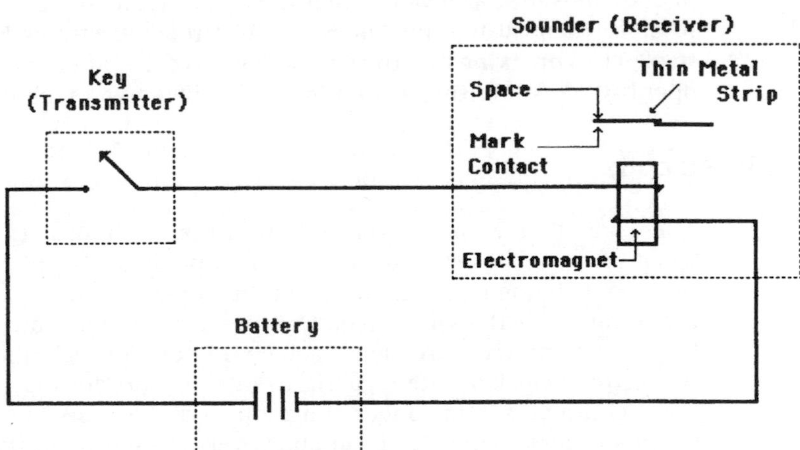

Figure 2.1 A simplex telegraph circuit. In a simplex (one way) telegraph circuit, the closure of the key causes a circuit to be formed, permitting current to flow. The flow of current around metal forms an electromagnet which causes the thin metal strip to strike the 'Mark' contact

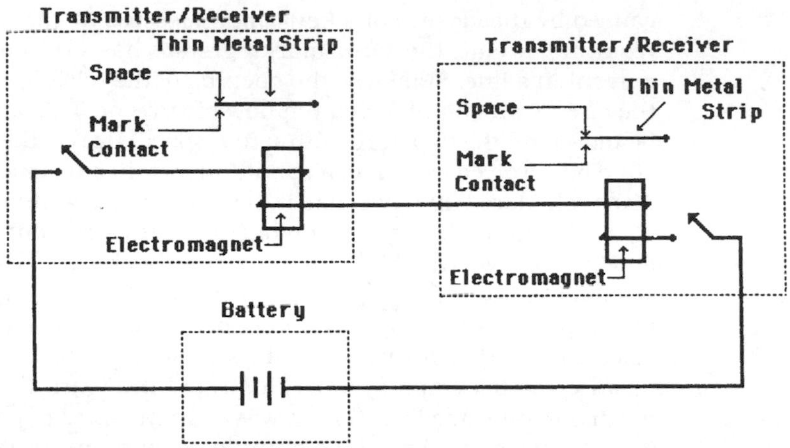

Figure 2.2 Half-duplex telegraph circuit. In a half-duplex telegraph circuit, both operators can transmit data, however, only one can do so at a time

The circuit illustrated in Figure 2.2 is called half-duplex. This type of circuit permits an operator to transmit and receive data, however, only one function can be performed at a time. A circuit which is capable of supporting the simultaneous transmission and reception of data is called full-duplex.

One obvious question you may have while examining the telegraph circuit illustrated in Figure 2.2 is how one operator can inform the other operator that he or she has data to send. If neither operator is transmitting data, the first operator to open his or her switch and begin keying could be considered to have seized control of the line. If the other operator desired to break-in, that operator could stop the transmission of the first operator by opening their key shorting switch. This would cause an open in the circuit, causing the sounders at both ends to stop. It would also serve as a signal to the transmitting operator to close their switch and listen for an urgent message. Since one operator, in effect, is breaking into the transmission of the other operator, the process of opening a key shorting switch is also known as a break-in operation.

2.2.2 Morse code

The code that Morse developed to transmit information resulted in the assignment of a series of short (dot) and long (dash) key depressions to represent characters. Legend has it that Morse visited a typesetter and counted the number of letters in each of the typesetter's letter drawers to obtain a basis for the assignment of a code to each character. Through his examination of the typesetter's stock of letters, Morse assigned short duration codes to frequently used characters and longer duration codes, consisting of more dots and dashes to less frequently used characters. Based upon this assignment, the letter E which is the most frequently occurring character in the English language is represented by a dot in Morse's code. The second most frequently occurring character, the letter T, is represented by a dash, and so on. Figure 2.3 lists the International Morse code for characters transmitted via telegraph.

Telegraph characters

	Morse		Morse
A	•-	T	-
B	-•••	U	••-
C	-•-•	V	•••-
D	-••	W	•--
E	•	X	-••-
F	••-•	Y	-•--
G	--•	Z	--••
H	••••	,	--••--
I	••	.	•-•-•-
J	•---	1	•----
K	-•-	2	••---
L	•-••	3	•••--
M	--	4	••••-
N	-•	5	•••••
O	---	6	-••••
P	•--•	7	--•••
Q	--•-	8	---••
R	•-•	9	----•
S	•••	0	-----

Figure 2.3 International Morse code. The dot (*) represents a short closure of the telegraph key, while the dash (–) represents a longer depression. A sequence of dots and dashes or a dot or dash by themselves are used to define unique characters

If you enjoy old movies and rented 'D-Day', you probably remember the signal sent to the French Resistance. With music in the background, the foreground sound of 'dot, dot, dot, dahh' represents the letter V in Morse. Not only was this a signal that the invasion was on, it also represented the goal of the Allied forces and represents perhaps the best known use of Morse code.

The first telegraph line which connected Washington, DC, to Baltimore was soon extended to New York. Within a few years, additional lines were installed throughout the United States and Europe. In the United States, the telegraph was initially used to convey a large volume of train dispatching information, resulting in a close collaboration between communications companies and railroads for the sharing of a 'right of way' that has been extended and expanded upon by other transportation companies. As communications evolved, several railroads and pipeline companies sold or leased the use of their 'right of way' to telephone companies. Those companies constructed microwave towers that at one time formed the backbone of the long distance telephone network. Beginning during the 1970s, the rights of way of railroads and pipeline operators were again used, this time for the construction of a fiber optic cable infrastructure that is now used for a majority of long distance communications in North America, Western Europe and Japan.

2.2.3 Morse code limitations

Although the telegraph revolutionized communications, until the early 1900s its use was limited to hand-keyed Morse code. This restricted the telegraph to a transmission rate between 30 and 60 words per minute when a pair of experienced operators were on each end of the line.

In attempting to automate telegraph operations, developers had to overcome two limitations associated with the Morse code. First, the code was variable in length, between one and six key elements – either dots, dashes or a combination of dots and dashes used to represent a character. This made it difficult to construct a machine to automatically recognize characters. A second limitation was the lack of a method to synchronize the transmitter of one machine with the receiver of another machine. Fortunately, due to the work of the Frenchman, Emil Baudot, and the American, Howard Krum, both limitations were overcome.

In 1874, Emil Baudot overcame the first limitation of the Morse code by devising a constant length, constant element code. In the resulting code, which was named as Baudot, the number of signal elements for each character was fixed at five, while the duration of each signal element was also fixed. This action simplified the decoding of data since each character has a uniform number of elements which defines its composition. In addition, since each element is of constant length, it is easier to recognize an element on the line instead of a dot or dash that would vary in duration between operators or even during the keying performed by one operator. The resulting five-level Baudot code can be considered as the forerunner of most modern day data processing codes, all of which are of constant length.

The second limitation of the Morse code which precluded its use for automatic operations was resolved in 1910 when Howard Krum devised a method for synchronizing the transmission and reception of characters. The method used by Krum involved appending a standard signal element to the beginning and end of each character. Since the first element was used to denote the start of a character while the second additional element denotes its end, Krum's method became known as start–stop synchronization. When combined with the use of the standard length Baudot code, the Krum synchronization method enabled the development of automatic telegraphy equipment.

2.2.4 Start–stop signaling and the Baudot code

The start–stop signaling method was developed from telegraphy where a closed circuit with current flowing indicates a marking condition and represents an idle state. To denote the occurrence of a character, the start element was defined as a line transition from a marking condition to a spacing condition. Since the start element is always followed by five data elements representing the code of a Baudot character, automatic telegraph equipment could be constructed to time six element durations (the start bit and five data elements) prior to encountering the stop element.

Unlike the start element which represents a line transition, the stop element may or may not indicate a transition. The stop element follows the last data element, which could be either at a marking or spacing condition. If the last data element was at a spacing condition, the stop pulse would represent a line transition. If the last data element was represented by a marking condition, the stop pulse would not represent a marking condition.

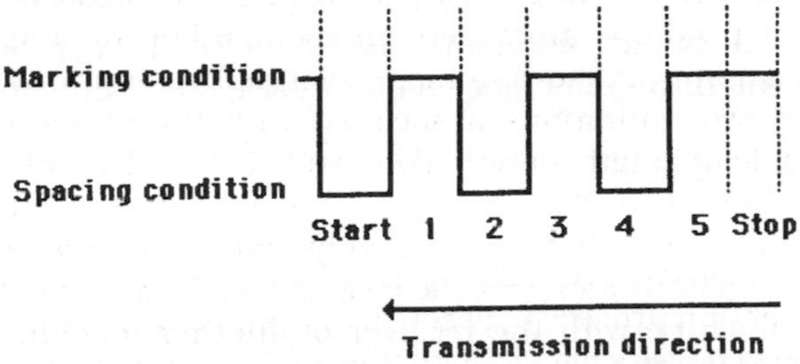

Figure 2.4 Start–stop synchronization. In start–stop synchronization, a start element causes a transition from a marking to a spacing condition. Since the stop element is always represented by a marking condition, the start element of the next character will always be represented by a line transition

The use of the stop element is one of forcing a line transition between transmitted characters. Since the stop element is a marking condition, the end of one character must always be followed by a line transition since the first pulse of the next character is represented by a spacing condition. The start–stop system thus enables synchronized transmission to begin anew on each character. Even though it takes time for an electrical signal to propagate down the transmission line which can cause a timing discrepancy, a start–stop system eliminates the accumulated timing discrepancy at the end of each character, permitting synchronization to be maintained even when the code element timing of the transmitter is not perfectly matched by the receiver.

Figure 2.4 illustrates start–stop synchronization for a five-code element character. Note that, initially, the line is held in the marking condition and the start pulse represents a transition from a marking to spacing condition which denotes the start of the character. If a Baudot code is used, the actual character is represented by five data elements, which in Figure 2.4 were arbitrarily assigned to different marking and spacing conditions.

Start–stop synchronization represents an important milestone in data communications. Through the use of start and stop transitions that prefix and terminate a sequence of bits that represent a character, it becomes possible for timing to be conveyed with each character. The modern use of start and stop bits to frame a character is referred to as asynchronous transmission.

2.2.5 Bits and codes

The operation of the telegraph involved the flow or absence of current, resulting in it representing a two-state or two-valued system. This type of system is easily represented by the 0 and 1 symbols of the binary system.

Each telegraph signal state, marking or spacing, could then be equated to a value of 0 or 1. Similarly, each individual signal element duration used to represent the start, stop and data elements could be represented as binary digits which are more commonly called bits.

Similar to the decimal system, the binary system uses positioned notation; however, each position has only two possible values instead of ten. Also, each position in a binary number represents a power of two. Thus, the number 1011 in binary, commonly noted as 1011_2, can be converted to its decimal equivalent by remembering that each position is a power of two as follows

$$1 \times 2^3 + 0 \times 2^2 + 1 \times 2^1 + 1 \times 2^0 = 11$$

As indicated in the preceding example, the weight of each position in a binary number is two times the weight of the one to its immediate right. To facilitate the conversion process between binary and decimal numbers, Table 2.1 contains a table of powers of two up to 2^8 and their equivalent binary number.

Until now we have referred to the five elements that make up a Baudot character as its data elements. Since each element can have one of two values, either marking or spacing, in effect each data element is a bit. Technically a bit is an acronym or mnemonic for the term 'binary digit'. A bit represents the smallest unit of information and its value can be either 0 or 1. When we refer to the Baudot code as a five-level code, we thus also mean that each character is defined by the value of five bits.

Since the Baudot code uses five bits to represent a character, one would logically assume that the code was limited to representing 2^5 or 32 unique characters. To overcome this limitation and obtain the ability to represent the 26 characters of the alphabet, the 10 decimal digits, punctuation marks and the space character, Baudot reserved two 5-bit combinations to select one of two subcodes known as letters (LTRS) and figures (FIGS). Since each five-bit code caused a shift into one of two subcodes, each character (LTRS and FIGS) is known as a shift-control character. The five-bit code which causes a shift into the letters subcode is thus called letters shift, while the five-bit code which selects the figure subcode is called figures shift.

Through the use of the letters shift and figures shift, Baudot extended his code to represent two character sets – 26 letters and 28 numerals, punctuation marks and the space character. The extension mechanism was accomplished by the use of two 'shift' characters: 'letters shift' and 'figures shift'. The transmission of a shift character informs the receiver that the characters which follow the shift character should be interpreted as characters from a symbol and numeric set or from the alphabetic set of characters.

The five-level Baudot code is illustrated in Table 2.2 for one particular terminal pallet arrangement. A transmission of all ones in bit positions 1 through

Table 2.1 Powers of two

Power of two	Positional weight	Binary number
2^0	1	0
2^1	2	10
2^2	4	100
2^3	8	1000
2^4	16	10000
2^5	32	100000
2^6	64	1000000
2^7	128	10000000
2^8	256	100000000

Table 2.2 Five-level Baudot code

Letters	Figures	Bit selection 1	2	3	4	5
Characters						
A	—	1	1	0	0	0
B	?	1	0	0	1	1
C	:	0	1	1	1	0
D	$	1	0	0	1	0
E	3	1	0	0	0	0
F	!	1	0	1	1	0
G	&	0	1	0	1	1
H		0	0	1	0	1
I	8	0	1	1	0	0
J	'	1	1	0	1	0
K	(	1	1	1	1	0
L	)	0	1	0	0	1
M	.	0	0	1	1	1
N	,	0	0	1	1	0
O	9	0	0	0	1	1
P	Ø	0	1	1	0	1
Q	1	1	1	1	0	1
R	4	0	1	0	1	0
S		1	0	1	0	0
T	5	0	0	0	1	0
U	7	1	1	1	0	0
V	;	0	1	1	1	1
W	2	1	1	0	0	1
X	/	1	0	1	1	1
Y	6	1	0	1	0	1
Z	"	1	0	0	0	1
Functions						
Carriage return	<	0	0	0	1	0
Line feed	=	0	1	0	0	0
Space		0	0	1	0	0
Letters shift		1	1	1	1	1
Figures shift		1	1	0	1	1

5 indicates a letter shift, and the characters following the transmission of that character are interpreted as letters. Similarly, the transmission of ones in bit positions 1, 2, 4 and 5 would indicate a figures shift, and the following characters would be interpreted as numerals or symbols based upon their code structure. Although the Baudot code is quite old in comparison to the age of personal computers, it is the transmission code used by the Telex network which is employed in the business community to send messages through the world.

To illustrate the use of the Baudot code for automated telegraphy, assume that the letter A is to be transmitted. Using a binary 1 to represent a marking condition and a binary 0 to represent a spacing condition, Figure 2.5 illustrates the composition of the character A in start–stop transmission to

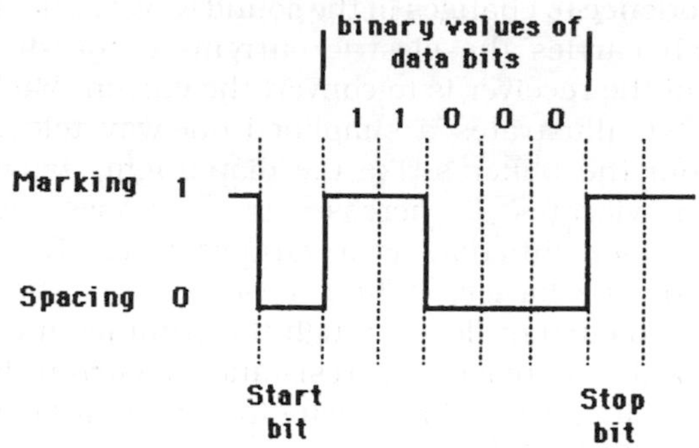

Figure 2.5 Transmitting an A in Baudot code using start–stop signaling. The transmission of a character in Baudot code results is represented by seven bits, including a start bit, five data bits and a stop bit

include its start and stop bits. In actuality, in many early automated telegraph systems, as well as later teletype systems, the stop bit was 1.5 or 2 bit durations in length. Today, most start–stop transmission systems use a stop bit whose duration is one bit interval.

2.3 TELEPHONY

Approximately 30 years after the installation of the first telegraph line in the United States, Alexander Graham Bell constructed a device which could vary the flow of electrical current in proportion to the sound which passed through it. Using a rudimentary diaphragm, Bell was able to convert sound waves generated by the twang of a clock spring into electric current and then to reconvert the current to sound. Further refinements to Bell's 'harmonic telegraph' continued until, on March 10 1876, Bell transmitted his now legendary sentence over a wire from his laboratory in Boston to his associate in an adjoining room. 'Mr Watson, come here, I want you!' Mr Watson, responded, as did the rest of the world, to this new invention.

In 1877, a telephone line was constructed between Boston and Somerville, MA. In the following year, the world's first local telephone exchange with 20 subscribers was opened in Newhaven, CT. By 1880, over 50 000 telephones had been installed in the United States and telephony was extended to Europe, with a seven-subscriber public telephone exchange being opened in London. Within 100 years of its invention, telephone networks blanketed most areas of the world with an excess of 500 million instruments in use.

2.3.1 Principle of operation

In common with all communication systems, telephone operations require a transmitter, transmission medium and receiver. In a telephone, the transmitter is used to convert sound waves into electric current which is varied in correspondence to changes in the sound waves. The transmission medium is a line which carries the electric current to the distant location, while the function of the receiver is to convert the current back into sound waves.

Figure 2.6 illustrates a simplified one-way telephone circuit. The sound waves from the talker strike the diaphragm, causing it to vibrate. As the diaphragm vibrates, it increases or decreases the density of the carbon granules which function as a variable level of resistance to current flow supplied by the battery. When the density of the granules decreases, the resistance to current flow through the granules increases. When the density of the granules increases, the resistance to current flow through the granules decreases. Since the primary winding of the induction coil is connected in series with the carbon granules and battery, the varying current also flows through the primary winding. This current creates a varying electromagnetic field across the primary winding, which induces a varying current through the secondary winding that is placed on the transmission line. Once placed on the transmission line, the current flows to the receiver at the distant end.

The receiver consists of three major parts – a magnet, iron diaphragm and voice coil. The voice coil is wound around the poles of a magnet which is positioned close to the diaphragm illustrated on the left-hand side of Figure 2.6.

The current which flows through the transmission line is applied to the voice coil of the receiver at the distant end, causing an electromagnetic field to develop around the coil which varies in conjunction with the intensity of sound applied to the transmitter. The fluctuating field causes the iron diaphragm to vibrate, which acts upon the air next to it. This action results in the reaction of the sound generated at the opposite end of the circuit.

To notify the party at the distant end of a call, each telephone is equipped with a ringer. The ringer originally consisted of one or two gongs that were struck by a clapper. In more modern telephones, the gongs and clapper of the ringer have been replaced by electronic circuitry that generates an equivalent alert signal.

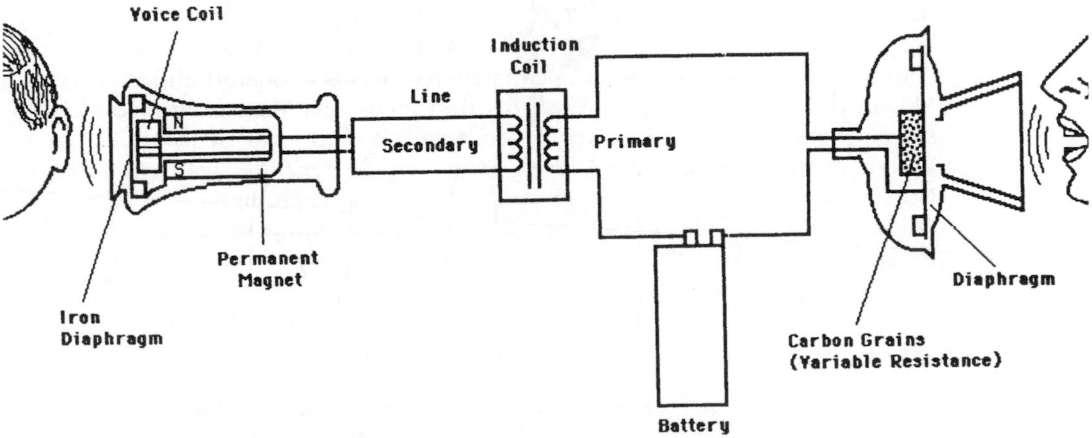

Figure 2.6 Simplified one-way telephone circuit

2.3.2 Sound wave conversion

The conversion of sound waves by a telephone into electrical current is based upon the characteristics of human speech. Thus, it is important to obtain an appreciation of the characteristics of human speech and its conversion to electricity as both sound and electricity are carried in a common waveform which plays an important role in the design of communication systems. Those waveforms in turn govern the characteristics of many transmission systems that will be discussed throughout this book.

As we talk our lungs and diaphragm expel air past our vocal cords. The resulting movement of air molecules creates a physical disturbance of the air which results in the generation of sound. Although most readers probably do not place much thought into the process of how speech is generated nor dwell over its technical components, the actual speech generation process is quite complex. As we talk the nasal cavity, lips, tongue and other parts of the body interact to produce different levels of air pressure. Periods of high pressure sound signals are referred to as condensations, whereas periods of low sound levels resulting from less dense air pressure are known as rarefractions. The continuous sequence of sound waves consists of a series of different sequences of condensations and rarefractions which form human speech. This is illustrated in Figure 2.7 in which the condensations and rarefractions generate a series of oscillating sound waves. As we talk, a complex set of waveforms are generated that can vary several thousand times per second.

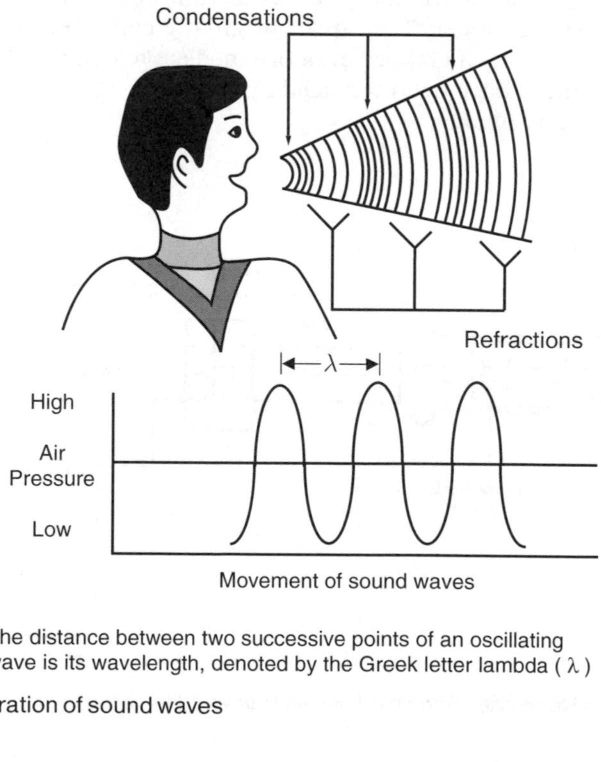

The distance between two successive points of an oscillating wave is its wavelength, denoted by the Greek letter lambda (λ)

Figure 2.7 Generation of sound waves

Those waves are converted by the telephone into a continuously variable electrical signal at one end and reconverted back into sound waves at the distant end. For simplicity, we will focus our attention upon the characteristics of a single wave which will enable us to discuss the general characteristics of acoustic (sound) and electrical waveforms.

The distance between two successive points of an oscillating wave is known as wavelength and is commonly referred to by the Greek letter lambda (λ).

In examining the lower portion of Figure 2.7 we note that the wavelength provides a measurement of the cycle of the wave between successive points. If you halve the wavelength the number of cycles will double. In comparison, doubling the wavelength would halve the number of cycles.

The number of cycles or oscillations of a wave per unit time is known as frequency (f) and is expressed in cycles per second (c.p.s.). However, in recognition of the work of the scientist Frederick Hertz, the term hertz (Hz) is used as a modern replacement or synonym for cycles per second.

Figure 2.8 illustrates a standard sine wave operating at one (top) and two (bottom) cycles per second. The time required for each signal to be transmitted over a distance of one wavelength is known as the period of the signal. That period represents the duration of the cycle and can be expressed as a function of the frequency. That is, if T denotes the period of the signal and f its frequency, then

$$T = 1/f$$

From the above we can represent the frequency as the reciprocal of the period. Thus

$$f = 1/T$$

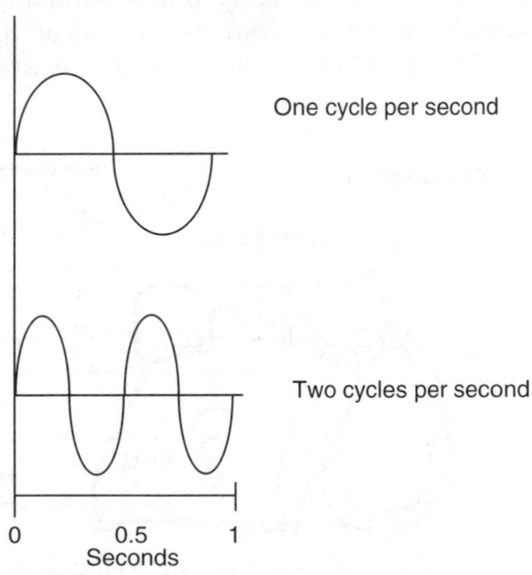

Figure 2.8 Oscillating sine waves at different frequencies

In examining Figure 2.8 note that the sine wave whose frequency is one cycle per second (1 Hz) has a signal period of 1/1 or 1 second. In comparison, the sine wave that has a frequency of two cycles per second (2 Hz) has a signal period of 1/2 or 0.5 seconds. Since the conversion of sound waves into electrical waves results in a similar electrical oscillating signal the range of frequencies produced by human speech is an important governing characteristic of the resulting electrical signal, In Chapter 3 we will investigate this relationship and the transmission of electrical signals representing human speech.

2.3.3 The basic telephone connection

Figure 2.9 illustrates the basic components of a standard telephone and its connection to a telephone company facility known as a central office. The handset houses the transmitter and receiver which are installed in the recesses located at each end of the handset shell. The pair of wires used to transmit a varying direct current are called tip and ring wires. As noted in Figure 2.9, the ringer is connected across the tip and ring pair of wires inside the telephone.

Among the many functions performed by a telephone company office is the generation of a ringing signal to a telephone. This signal is used to inform a subscriber of an incoming call. This is accomplished by the telephone company office supplying an alternating current (a.c.) ringing voltage to the telephone. This voltage causes a coil connected to a clapper to become energized and de-energized, resulting in the clapper oscillating back and forth, striking either one or two gongs located near the clapper. When the telephone line is not in use, the office supplies a constant negative voltage to the telephone on the ring lead. When the handset is lifted, the ring lead is shorted to the tip lead, resulting in a closed circuit which enables current to flow through the telephone to the telephone company office. The current flow is the mechanism by which equipment located at the office is informed that the phone is off-hook. When a conversation has been completed and the handset

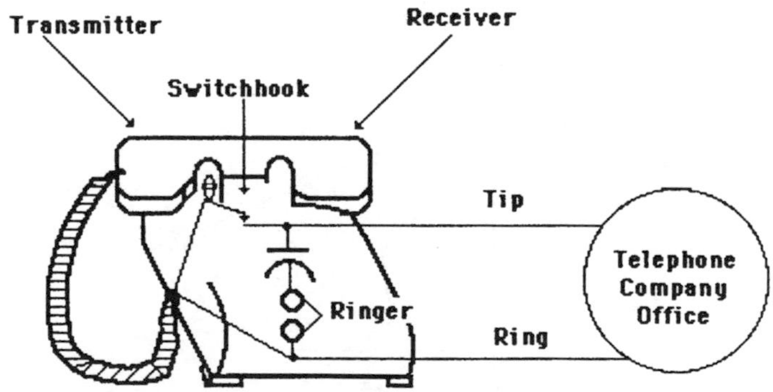

Figure 2.9 Basic telephone components. A two-wire circuit consisting of tip and ring wires connects a subscriber's telephone to a telephone company office

is placed back into its cradle, the tip and ring circuits are disconnected from one another. This causes a break in the current flow. This condition is known as on-hook and is used to inform equipment located at the telephone company office that the subscriber is disconnected from the line.

The modern telephone instrument uses a switch hook to enable and disable the flow of current based upon whether or not the handset is in the cradle depressing the switch hook. Two other components of the modern telephone that are not shown in Figure 2.9 are the dialer and slide tone. The dialer was originally a circular dial which you would turn to a digit position and release, generating a sequence of pulses. Hence this is referred to as pulse dialing. More modern telephones include a touch pad of keys which, when pressed, generate pairs of frequency tones and result in what is referred to as tone dialing. The slide tone enables a portion of speech to 'bleed' over to the receiver, allowing a person to hear themselves as they talk on the telephone.

2.3.4 Switchboards and central offices

Initially, the terms switchboard, exchange and central office were basically synonymous. Each facility, in essence, provided a mechanism to transfer calls between subscribers. To understand the necessity for a switching center, consider a telephone network which contains eight subscribers. To provide each subscriber with the ability to contact another subscriber, each telephone would require the ability to be connected to seven wires as illustrated in Figure 2.10.

To alleviate the complex wiring illustrated in Figure 2.10, telephone companies initially employed switchboards or exchanges. Each telephone was wired to the switchboard and the switchboard operator would use a connecting cord to patch the line from one subscriber to another as illustrated in Figure 2.11. The swltchboard or exchange was used to provide service to a specific locality, such as an area within a city, a town or a large building.

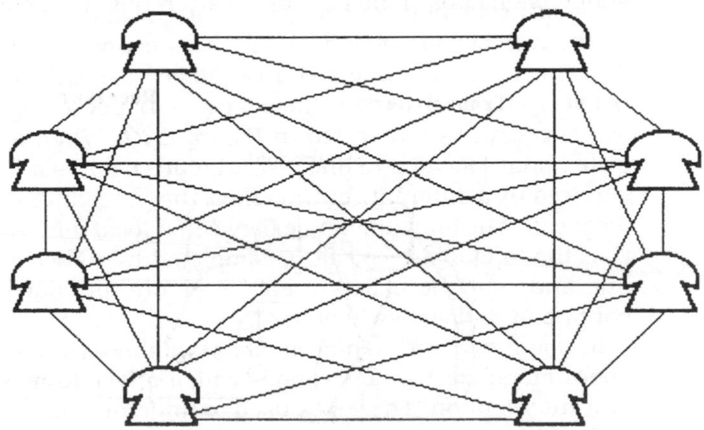

Figure 2.10 Telephone system without a switching center. Without a switching center, each telephone in a network must be cabled to every telephone in the network

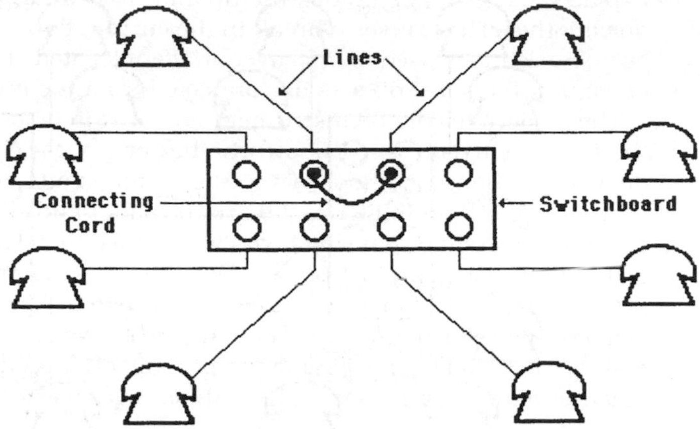

Figure 2.11 Using a switchboard. A switchboard provides a mechanism for connecting different subscribers to one another without requiring each subscriber to have a permanent connection to all other subscribers

Initially, all switchboards were manually operated, requiring operators to perform the required patching operations. A person at any telephone wishing to call another network subscriber would first ring the switchboard operator. After the operator answered the call, the call originator would supply the telephone number of the party he or she wished to talk to. Then, the operator would make the connection using the connecting cord and ring the destination party. Although switchboards are still in use today, they have evolved into sophisticated automatic electronic switching mechanisms, some of which are capable of supporting tens of thousands of subscribers.

In telephone company terminology, the locations where switching equipment is installed are called central offices. Since the probability of telephone subscribers connected to one central office simultaneously dialing telephones connected to a different central office is quite low, the telephone company interconnects their central offices with a small number of lines in comparison to the number of subscriber line connections supported by a central office. The lines which connect central offices are called trunks. The relationship between the subscriber lines and a trunk connecting two central office switchboards is illustrated in Figure 2.12. When a subscriber serviced by a switchboard located in one central office desires a connection to a subscriber serviced by a different central office, the procedure is similar to that used for a local call. The only difference between a local office and a long distance call is that the long distance call is completed by two or more operators instead of one and that the call is routed over one or more trunks instead of being confined to a local switchboard.

In modern central offices, the switchboard was replaced by automatic operationed electronic switches which alleviate most requirements for operator intervention. The trunk used to interconnect central offices is most likely to be fiber-optic cables or microwave towers, which provide a transmission capability for thousands of simultaneous telephone calls to be routed between central offices.

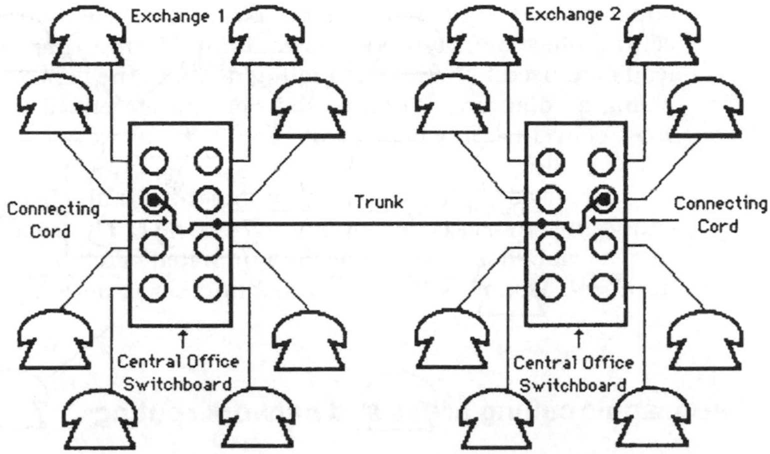

Figure 2.12 Connecting telephone company exchanges. A trunk is used to provide a connection between telephone company central offices or exchanges

2.3.5 Numbering plans

In the United States and many foreign locations, every telephone is assigned a seven-digit number. The first three digits are known as the central office prefix, or exchange, and identify the central office which services a particular telephone. The last four digits are called the line number and identify the subscriber loop to the central office. Although early central offices were limited to 10 000 subscribers, numbered from 0000 to 9999, modern central offices may have electronic switches capable of supporting 50 000 to 100 000 or more subscriber lines. Such central offices support several prefixes.

Figure 2.13 illustrates the routing of a local call from telephone number 744-2050 to telephone number 742-1031, assuming that one central office

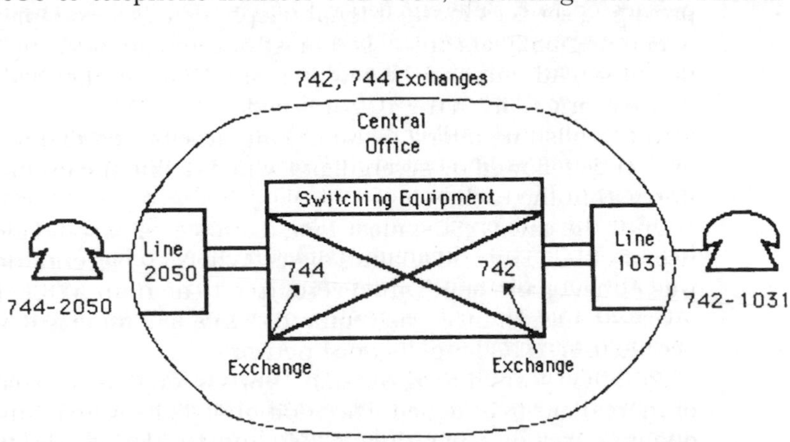

Figure 2.13 Automatic call processing by a central office. The switching equipment located at a telephone company central office performs the equivalent of an automated path between the calling and called parties

supports both 742 and 744 prefixes. Each telephone is connected to the central office by two wires known as tip (T) and ring (R) leads. The T and R leads are used to both send dialing digits to the central office as well as providing a communications path through the central office for the resulting voice conversation when the distant party answers the call.

The dialing digits passed from telephone number 744-2050 to the central office are used by the switching equipment at that office to make an automatic connection to the called number, 742-1031. The switching equipment can thus be viewed as performing an automated patch between the two telephone numbers, assuming that the called number is not busy.

2.3.6 Geographic calling areas and network routing

In North America, geographic areas were established which contain a number of central offices. Each of these geographic areas were assigned a three-digit identification code called the Numbering Plan Area (NPA), but which is better known as an area code. The North American numbering plan was created by AT&T Corporation in the 1940s as an internal numbering system to assist telephone operators route long distance calls. Later, the use of area codes was expanded to allow subscribers to make calls without operator assistance. Each area code is of the form *NXN*, where *N* can be any digit between 2 and 9, while *X* originally had to be assigned either the digit zero or 1.

Due to the significant growth in the use of fax machines, cellular telephones and second lines in residences, the ability of an area code to accommodate telephone number requests became severely taxed during the early 1990s. In 1995, the NPA was modified to provide for the creation of new area codes by allowing digits other than 0 or 1 as the middle digit in the area code. In actuality, new area code areas are being used to replace portions of older area code areas, providing an expansion capability for both the old and new area codes for providing services for additional telephones. For example, the area code 404 which originally encompassed most of the state of Georgia is now limited to downtown Atlanta, and the new area code 770 must be used to reach telephones in areas originally in the 404 area code.

In establishing a long distance call, at least ten digits are required for the area code followed by seven digits which define the exchange and subscriber line within the exchange.

Until the early 1980s, most long distance calls could be made through the use of either 10 or 11 dialing digits. When an eleventh digit was required, it was either a zero or 1, with a zero used to alert an AT&T toll operator that an operator's assistance was required, while a 1 was used to obtain automatic access to AT&T's long distance network.

For a few years during the mid 1980s a long distance call could require 20 or more digits to be dialed. The additional digits were required to access a long distance carrier other than AT&T and to identify the calling party to the carrier for billing purposes. Today, most long distance calls require 11 digits, since the use of a digit 1 prefix to the area code is used to access the primary long distance carrier selected by the subscriber.

As a call is routed between central offices to its ultimate destination, its actual path is dependent upon the volume of traffic flowing over telephone company facilities. During busy periods, such as Mother's Day, a call from New York City to Miami could be routed via St Louis and Dallas to its destination. If the same call was made at 2 p.m. on a normal Wednesday, it might be routed directly from New York City to Miami.

The central offices of all communications carriers are interconnected with one another either directly or indirectly and make up what is known as the public switched telephone network (PSTN). As calls are routed through the PSTN, they will be routed through a hierarchy of switching centers as illustrated in Figure 2.14. Each subscriber station (telephone) is connected via a station loop which is also known as the subscriber loop or local loop to their serving telephone company central office. As this office is at the bottom or end

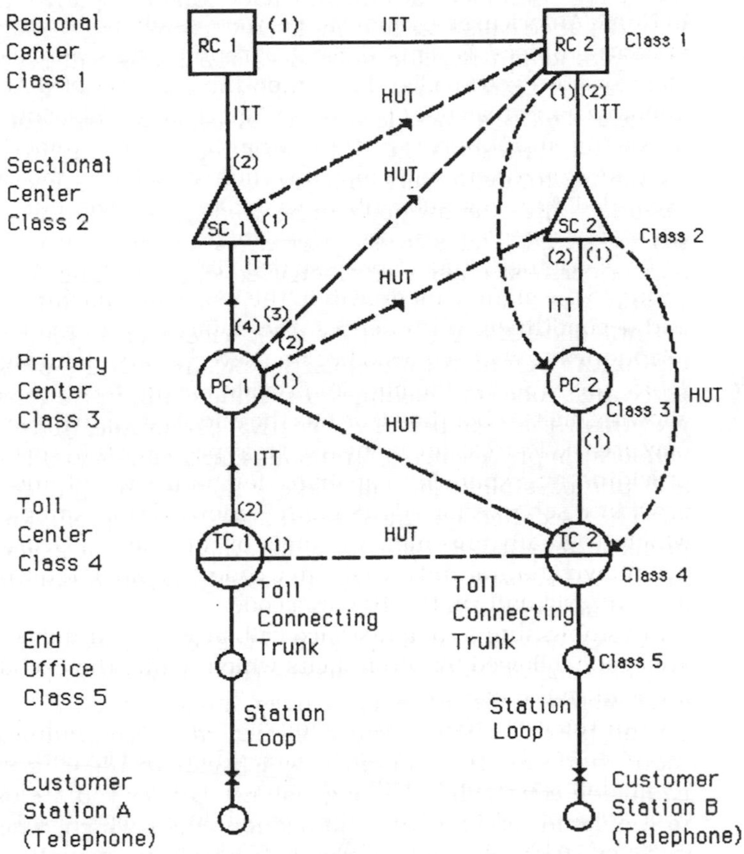

Figure 2.14 PSTN switching hierarchy.

1. Numbers in parentheses indicate order of choice of route at each center for calls originating at station A.
2. Dashed lines indicate high-usage paths.
3. ITT Intertoll Trunk.
4. HUT High Usage Intertoll Trunk.

of the hierarchy, it is also known as the end office. In telephone company terminology, the end office which directly services the subscriber is called a class 5 office.

In the United States and Canada, telephone companies established a hierarchy of switching centers to facilitate the routing of long distance calls through their network. This hierarchy has been adopted by other countries in a similar or modified structure for intracountry long distance telephone. communications.

Long distance calls from a class 5 office are first routed to a class 4 toll center. At that location, the call may be routed to another class 4 office that directly services the end office which, in turn, services the dialed number. Alternatively, the call could be routed higher up the PSTN switching hierarchy.

The actual path the call will take depends upon the volume of traffic flowing over the network as well as other factors to include the operational state of switching equipment at different telephone company offices. Although most long distance calls will be completed via a routing through class 4 or class 3 telephone company offices, on occasions calls will be routed up the PSTN switching hierarchy to a class 1 office.

The end office at most locations has a special demarcation point, referred to as a point of presence (POP). The POP represents an interface where the local telephone company servicing the end user or subscriber hands off long distance calls. The handoff can be predefined if the subscriber selects his or her preferred long distance carrier, or it can vary from call to call if the subscriber first dials a seven-digit code of the long distance company they wish to use.

Until mid-1998 the long distance access code you would dial to use an alternative communications carrier began with the prefix 10, followed by a three-digit code that identifies a specific communications carrier. Due to the growth in the number of alternative communications carriers, with some vendors simply resellers of transmission facilities of other carriers, there was a strong possibility that all of the available 999 carrier access codes would soon be used. Recognizing this situation, the 10 prefix was changed to a 1010 prefix in July 1998. By using a four-digit prefix it becomes possible for the prefix 1011 to be used when the need arises to obtain an additional 999 carrier access codes.

Carrier access codes are also referred to as dial-around codes. This is because dialing the carrier access code allows you to bypass your primary long distance carrier. Every carrier, including AT&T, MCI Worldcom, and Sprint, have a '1010' number. In fact, many communications carriers have more than one 1010 number, which can result in a significant degree of confusion concerning the long distance carrier you are using. To further add to the confusion, carrier identification codes which, when prefixed with the number 1010, become a carrier access code are actually four digits long, with the leading 0 dropped by most carriers.

Table 2.3 lists ten long distance companies that service middle Georgia and their access codes.

In many cities that have a heavy concentration of businesses that are heavy users of communications, several communications carriers now provide services which bypass the telephone company end office. Commonly referred

Table 2.3 Examples of long distance company access codes

Company	Access Code
AT&T	1010288
BTI	1010833
Cable & Wireless	1010488
Dial & Save	1010457
Frontier Communications	1010211
LCI International	1010462
LDDS Communications	1010488
MCI Worldcom	1010222
Spring	1010333
Touch 1 Long Distance	1010797

to as local loop bypass, carriers offering this service typically install fiber into several buildings within a city and route the fiber directly into their communications infrastructure. Not only does a local loop bypass commonly reduce the monthly phone charges of a business, but, in addition, it typically provides better transmission quality. The latter results from the telephone company's copper cable being replaced by fiber which has immunity to electrical interference.

2.3.7 The world numbering plan

To facilitate communications between persons located in different countries, a numbering plan was developed to enable persons in one country to directly dial a telephone number located in a second country. This numbering plan is called International Direct Distance Dialing (IDDD).

IDDD requires the use of a special prefix to gain access to the international network. In the United States and Canada, telephone subscribers can use 011 for a station-to-station call or 01 for an operator assisted call. Either prefix is followed by the country code and the telephone number of the called party. In dialing the telephone number, certain countries require the use of city codes to prefix the telephone number, while other countries use area codes similar to those used in the United States.

In making a call from overseas to the United States, a similar procedure is used; however, the international prefix will vary based upon your calling location. For example, in Africa the international prefix is 2, whereas it is 3 or 4 in Western Europe, 5 in Mexico, Central and South America, 6 in Southeast Asia, 8 in Japan and 9 in the Middle East.

2.4 REVIEW QUESTIONS

1. What are the three components common to all communications systems?

2. Define the terms marking and spacing with respect to the flow of current in a telegraph system.

3. Why is the telegraph considered to represent a binary state machine?

4. What is the term used to describe a transmission system capable of transmission in only one direction?

5. Discuss the difference between a half-duplex and full-duplex circuit with respect to their transmission support capability.

6. How does a telegraph operator inform a distant operator that they have data to send?

7. Discuss the method Morse used to assign codes to characters.

8. What is the International Morse code for HELLO?

9. What are the two key limitations associated with Morse code which precluded its use in automated telegraph operations?

10. Define a start element with respect to the signaling transition on a communications line. What is the significance of a start element?

11. What is the purpose of a stop element in start–stop signaling?

12. In start–stop communications, where is the timing information located when a character is transmitted?

13. What is the decimal value of the number $101\,1010_2$?

14. How did Baudot extend his five-level code to represent more than 32 characters?

15. Illustrate the marking and spacing conditions that define the character M in the Baudot code to include its start and stop bits.

16. What is the function of the diaphragm in a telephone transmitter?

17. If the period of a signal is 0.1 second what is its frequency in Hz?

18. What is the function of tip and ring wires?

19. Discuss the relationship between an off-hook and on-hook condition, the circuit formed by tip and ring wires and the current flow in that circuit.

20. What is the purpose of a telephone slide tone?

21. Name two types of telephone dialing.

22. What is the primary function of a switchboard?

23. Assume that you are dialing a long distance telephone number. If you enter the dialing digits in the format I XXX YYY ZZZZ, what do the 1, XXX, YYY and ZZZZ digits represent?

24. Where is a telephone company end office located in PSTN switching hierarchy?

25. What is a point of presence?

26. What are two advantages associated with local loop bypass?

27. Assume you want to use a dial-around carrier whose access code is 1010543 to dial the telephone number 2125551212. What string of digits do you dial?

3

BASIC CIRCUIT PARAMETERS: MEASUREMENT UNITS AND MEDIA OVERVIEW

In the first part of this chapter, you will be introduced to the basic circuit parameters of frequency and bandwidth. Building upon this information, several commonly used measurement units that express the quality of a circuit to include power ratios, the signal-to-noise ratio, the use of reference levels and the relationship between noise and power are explained. In the second portion of this chapter, an overview of several types of transmission media is presented. This overview includes an examination of the electrical properties that affect the transmission of both analog and digital signals on twisted pair wire, followed by an introduction to the use of coaxial cable, microwave and transmission of light energy on fiber-optic cable, topics that are covered in more detail later in this book.

In concluding this chapter we will obtain an overview of the concept of structured wiring, and discuss its use, a standard that classifies cable into different categories and the characteristics of some of those cable categories.

3.1 BASIC CIRCUIT PARAMETERS

3.1.1 Frequency and bandwidth

In Chapter 2 we examined the generation of speech resulting from the movement of air molecules and some elementary characteristics of sound waves. Now let us build upon those concepts to discuss the characteristics of circuits used to transmit speech.

Previously, we noted that frequency is a term used to refer to the number of periodic oscillations or waves that occur per unit time. Although frequency is usually measured in cycles per second, a synonymous term is hertz, abbreviated as Hz.

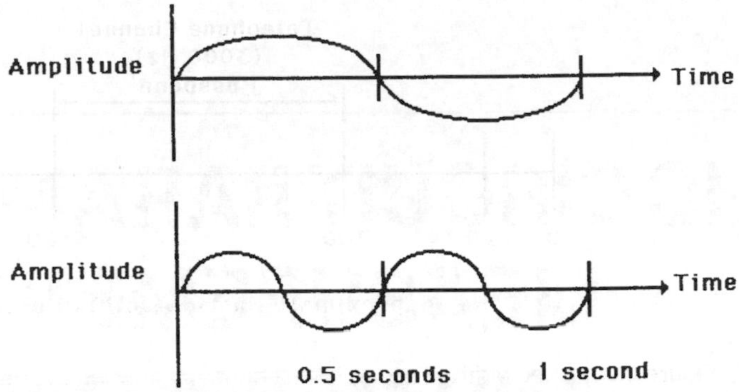

Figure 3.1 Frequency variance. The frequency of a periodic wave is inversely proportional to its duration

The top portion of Figure 3.1 illustrates a sine wave that has a frequency of one cycle in one second or 1 Hz. Note that a sine wave is a sinusoidal signal whose value changes over a 360° period and then repeats over the next 360° degree period. The maximum height of the signal occurs at 90° degrees, while its minimal value occurs at 270°. At 0°, 180° and 360° the height of a sine wave is zero.

The importance of the sine wave and other sinusoidal signals in communications is based upon their ability to be easily generated and modulated. The development of alternating current (ac) occurred by rotating a conductor through a magnetic loop. By varying the polarity of the loop in a magnetic field, a sinusoidal voltage was generated similar to the top portion of Figure 3.1. By varying (modulating) a sinusoidal signal the variance can be used to impress information onto the signal. By itself a sinusoidal signal conveys no information other than that there is continuity on the transmission path if the signal is received at the destination. However, by modulating the signal, different signal changes can be used to indicate different conditions, such as one type of change in a signal used to represent a '1' bit, while a different change in the signal is used to represent a '0' bit.

In the lower portion of Figure 3.1, the number of oscillations per period of time has been doubled, resulting in a frequency of 2 Hz. As indicated by Figure 3.1, the frequency of a periodic wave is inversely proportional to its period in seconds. The relationship between the frequency (f) in Hz and period (T) in seconds, therefore, is

$$f = \frac{1}{T}$$

Bandwidth is the measurement of the width of a range of frequencies and not the frequencies themselves. If the lowest frequency a circuit can transmit is f_1 and the highest is f_2, then the bandwidth of a circuit is $f_2 - f_1$.

Although an average person can hear a range of frequencies between approximately 20 and 20 000 Hz, the typical telephone circuit is restricted to passing approximately 3000 Hz. This block of frequencies that is passed actually represents the majority of frequencies required to make a conversation

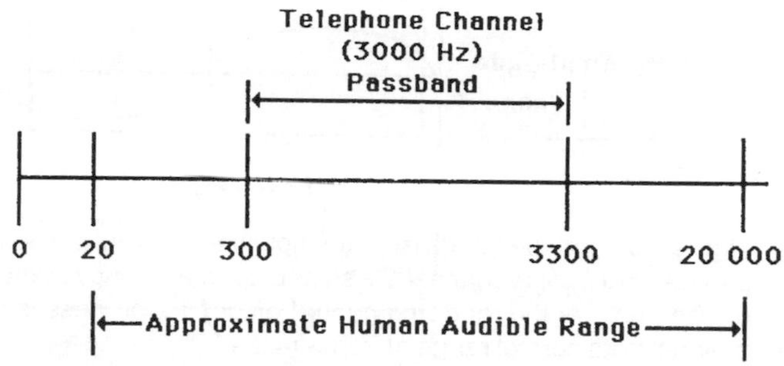

Figure 3.2 Bandwidth of a telephone channel. Although the human ear can hear a range of frequencies whose bandwidth is approximately 20 kHz, telephone companies limit the passband of the telephone channel to 3 kHz due to economic constraints

intelligible as well as those frequencies used in speech; that is, very rarely does one sing tenor notes nor speak at low frequencies below 300 Hz.

Figure 3.2 illustrates the relationship between the bandwidth of a telephone circuit and the audio spectrum heard by the human ear. The twisted-pair used to connect a telephone subscriber to an end office can actually support a range of frequencies up through 1 MHz. As we will note in the next section in this chapter, because of economics, filters are placed on subscriber lines that limit the range of frequencies that can be carried. Because those filters are located in telephone company equipment at end offices, this means that a subscriber line can actually transport approximately 1 MHz of frequencies. The ability to use approximately 1 MHz of frequency explains why a new generation of modems referred to as Digital Subscriber Line (DSL) devices are capable of transmitting data at rates as high as tens of millions of bits per second over the local subscriber line.

3.1.2 The telephone channel passband

The contiguous portion of the frequency spectrum which ranges from 300 to 3300 Hz is called the passband of a telephone channel. Here the term 'channel' is used to denote a portion of the frequencies that can be transmitted on a circuit that forms a subdivision of available bandwidth.

The passband of a telephone channel is formed by the use of low-pass and high-pass filters at telephone company offices. These filters are designed to pass all signals up to a predefined frequency or all signals under a predefined frequency, resulting in a contiguous range of frequencies that represents less than one-sixth of the human audible hearing range and a smaller fraction of the range of frequencies that twisted pair cable can actually pass.

The selection of a 3000 Hz passband was based upon economics. As mentioned in Chapter 2, the lines routed between telephone offices are called trunks. To maximize the use of trunks, long distance voice conversations that are required to be routed through one telephone company office to another office are first shifted up in frequency by a set amount prior to being placed onto the trunk. This process, which is called frequency division multiplexing

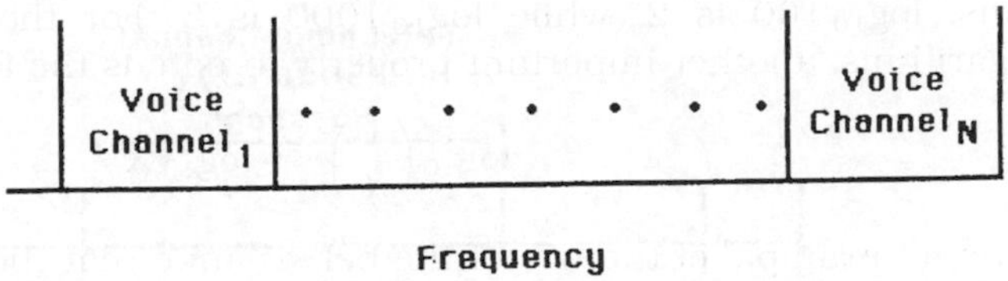

Figure 3.3 Frequency division multiplexing on trunks. The frequencies representing the passband of a voice channel are shifted up in frequency by multiples of 3 kHz for placement on the trunk. At the receiving central office, the process is reversed to restore the voice channel to its normal range of frequencies

(FDM), permits more than one voice conversation to be simultaneously routed over a trunk between telephone company offices, as illustrated in Figure 3.3. Since each telephone conversation occupies 3000 Hz instead of 20 000 or more Hz, the creation of a passband permits more simultaneous calls to be placed on a trunk.

Although a majority of the use of frequency division multiplexing by telephone companies was replaced during the 1960s by time division multiplexing, the selection of a 3000 Hz passband represents a restriction on the ability of subscribers to transmit data. As we will note later in this book, voice digitization and the use of multiplexing employed by telephone companies remains associated with the passband of a telephone channel.

3.2 MEASUREMENT UNITS

3.2.1 Power ratios

Due to the resistance to signal flow exhibited by cables, the power transmitted onto a circuit normally exceeds the power received. When amplifiers are used to boost the intensity of a signal, the received power can be equal to, less than or greater than the transmitted power. This potential variance between transmitted and received power resulted in the use of the bel to categorize the quality of transmission on a circuit.

Bel

The bel, named after the inventor of the telephone, uses logarithms to the base 10 to express the ratio of power transmitted to power received. The resulting gain or loss for a circuit is given by

$$B = \log_{10}\left(\frac{P_O}{P_I}\right)$$

where B is the power ratio in bels, P_O the output or received power and P_I the input or transmitted power.

In the preceding equation, the logarithm to the base 10 ($\log_{10}$) of a number is equivalent to how many times 10 is raised to a power to equal the number;

thus, $\log_{10} 100$ is 2, while $\log_{10} 1000$ is 3. For those not familiar with logarithms, another important property to note is the following relationship

$$\log_{10}\left(\frac{1}{X}\right) = -\log_{10} X$$

As an example of the use of the bel, assume that the received power on a channel was measured to be one-tenth that of the power transmitted. Then

$$B = \log_{10}\left(\frac{\frac{1}{10}}{1}\right) = \log_{10}\frac{1}{10}$$

Since $\log_{10}\frac{1}{10} = -\log_{10} 10$

$$B = -1$$

Decibel

The bel was used for many years to categorize the quality of transmission on a circuit since the response of the human ear to changes in volume is logarithmic. Gradually, the bel was replaced by the use of the decibel (dB), which is a more precise unit of measurement, as it represents one-tenth of a bel. The decibel, which also represents the gain or loss on a circuit, is given by

$$\text{dB} = 10\log_{10}\left(\frac{P_O}{P_I}\right)$$

Let us return to our previous example where the received power was measured to be one-tenth of the transmitted power. The power ratio in dB is then

$$\text{dB} = 10\log_{10}\left(\frac{\frac{1}{10}}{1}\right) = 10\log_{10}(\tfrac{1}{10})$$

Since

$$\log_{10}(\tfrac{1}{10}) = -\log_{10} 10$$

$$\text{dB} = -10\log_{10} 10 = -10$$

dB properties

In addition to the response of the human ear to changes in volume being logarithmic, the use of logarithms facilitates the calculation of gains and losses as a telephone call is routed through telephone company facilities. To illustrate the use of logarithmic properties, consider the circuit illustrated in Figure 3.4. The amplifier represents an electronic device that compensates for signal loss due to attenuation by rebuilding the signal. Unfortunately, the amplifier also rebuilds any impairments to the signal, boosting both the power level of the signal and any signal distortion. In Figure 3.4 an input of 10 dB is applied to a circuit that has a cable loss of 5 dB. If the amplifier provides a gain of 15 dB, what is the output?

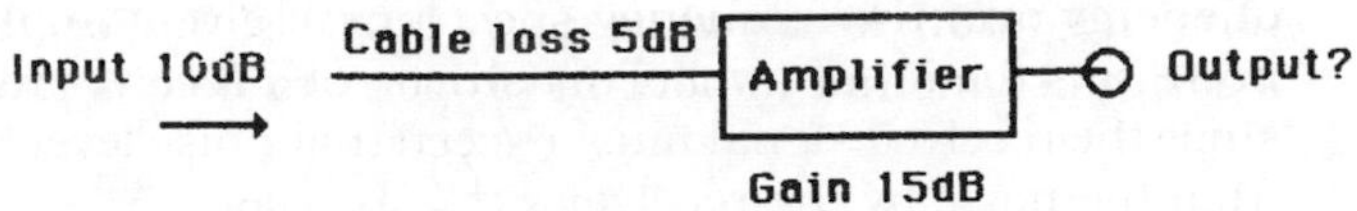

Figure 3.4 Properties of logarithms simplify gain and loss calculations. The properties of logarithms reduce gain and loss computations to the process of addition and subtraction. In this example, the output is the signal input less the cable loss plus the amplifier gain or $10 - 5 + 15 = 20$ dB

Due to the properties of logarithms, the computation of gains and losses is reduced to the process of addition and subtraction when expressed in dBs. The output is thus simply the input less the cable loss plus the amplifier gain. Then

$$\text{Output} = 10\,\text{dB} - 5\,\text{dB} + 15\,\text{dB} = 20\,\text{dB}$$

3.2.2 Signal-to-noise ratio

As electrons become excited due to the flow of a signal on a circuit, their vibration results in a degree of interference to the signal. This interference is called thermal or white noise and is one of two common types of noise that can affect the quality of a circuit. The second type of noise that commonly affects transmission is impulse noise, which results from lightning as well as from induced electromagnetic fields (EMF) caused by machinery.

Thermal noise

Thermal noise occurs in all transmission systems as a result of electron movement. Other causes of thermal noise include power line induction and cross-modulation from adjacent wire pairs. For readers with an FM radio set, you can hear thermal noise by tuning your set between stations in the evening.

Figure 3.5 illustrates the effect of thermal noise on an amplitude–frequency plot. Notice that thermal noise is characterized by a near uniform distribution

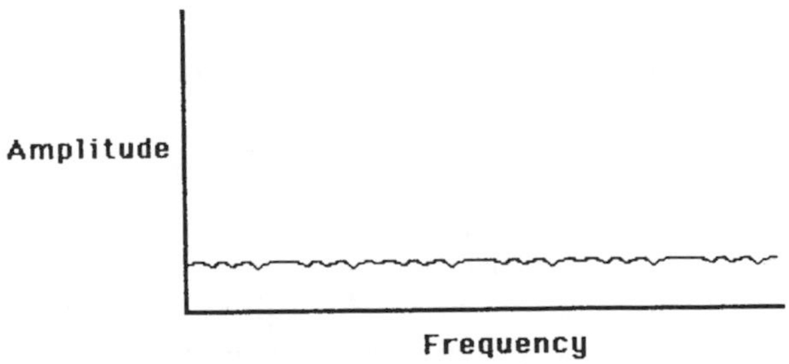

Figure 3.5 Thermal (white) noise. Thermal noise is characterized by a near uniform distribution of energy over the frequency spectrum

of energy over the frequency spectrum of given amplitude. In effect, thermal noise sets the lower level of sensitivity of a receiver in a transmission system since the received signal must exceed that noise level for it to be discriminated from the noise by the receiver.

Impulse noise

In comparison to thermal noise that is continuous, impulse noise is non-continuous. Impulse noise is formed from the effect of lightning and machinery disturbances, resulting in the creation of irregular pulses or spikes of relatively high amplitude and short duration, as illustrated in Figure 3.6. When occurring within the audio range, impulse noise will sound similar to a series of random clicks or bursts of static.

A ratio is used to categorize the quality of a circuit with respect to the intensity of a signal and noise on the circuit. This ratio is known as the signal-to-noise ratio (S/N) and is given by the signal power S divided by the noise power N on a circuit.

The S/N ratio is measured in dB. A high S/N ratio is desirable since it permits a receiver to discriminate a signal from the noise on a circuit. If, however, the signal power is increased too much, telephone company equipment, such as amplifiers, can be adversely affected. There are thus restrictions on the amount of signal power that can be placed on a circuit, and in the United States and many other countries equipment to be attached to the PSTN must include a built-in protection device to prevent excessive power from being placed onto the line.

The minimum limit for a received signal-to-noise power ratio for voice conversations can vary considerably based upon one's hearing capability. For data transmission, most communications carriers have a minimum limit for 24 dB for circuits to handle data. For reference purposes, Table 3.1 contains the relationship between dB and power or S/N ratios for 24 dB values.

In examining the entries in Table 3.1, there are several values that are significant and should be noted by the reader. First, consider a dB value of zero. Since the dB is defined as $10\log_{10}(P_O/P_I)$ this means that for a dB reading of zero, $\log_{10}(P_O/P_I)$ must be zero. This is only possible if $P_O = P_I$, which means that a dB value of zero means that the input power equals the

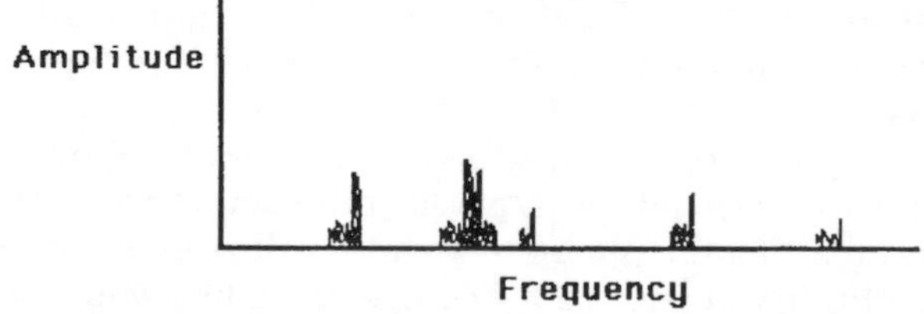

Figure 3.6 Impulse noise. Impulse noise occurs at random times and at random frequencies as a result of lightning and the electromagnetic fields created by machinery

Table 3.1 Relationship between dB and power measurements

dB	Power or S/N ratio
0	1.0 : 1
1	1.2 : 1
2	1.6 : 1
3	2.0 : 1
4	2.5 : 1
5	3.2 : 1
6	4.0 : 1
7	5.0 : 1
8	6.4 : 1
9	8.0 : 1
10	10.0 : 1
13	20.0 : 1
16	40.0 : 1
19	80.0 : 1
20	100.0 : 1
23	200.0 : 1
26	400.0 : 1
29	800.0 : 1
30	1 000.0 : 1
33	2 000.0 : 1
36	4 000.0 : 1
39	8 000.0 : 1
40	10 000.0 : 1
50	100 000.0 : 1

output power. Or, stated another way, a dB value of zero means that there is no gain or loss on the circuit.

A second important dB value is 3, which is equivalent to a power or S/N ratio of 2 : 1. Since a negative dB represents a power loss and a positive dB represents a power gain, −3 dB represents a 50% power loss, while +3 dB represents a gain of twice the input power.

3.2.3 Reference points

To define a standardized power relationship requires the use of a reference point. In the case of the decibel, 0 dB is equivalent to a 1 : 1 power ratio and serves as a reference point for dB measurements since it indicates no gain or loss.

In telephone operations the reference level of power is 0.001 W (1 mW). This level was selected because it represents the average amount of power generated in a telephone transmitter during a voice conversation.

To obtain a reference for comparing gains and losses in a circuit to denote its new power level, telephone company personnel use a 1 mW signal. This signal is applied to a circuit, and a meter is placed at intermediate points or at its distant end to determine a new power level based upon the gains or losses in the circuit up to a specific point or those occurring on an end-to-end basis.

For convenience, 1 mW of power is designated as being equal to 0 dB. To ensure that no one forgets that 1 mW is the reference level, the letter ‘m’ is attached to the power level. Thus

$$\text{dBm} = 10\log_{10}\left(\frac{\text{Signal power in milliwatts}}{1\text{ milliwatt}}\right)$$

In comparing dB (decibel) and dBm (decibel milliwatt), it is important to understand the difference between the two. dB is used to denote the gain or loss occurring on a circuit, whereas dBm is used to denote a new power level based upon a gain or loss.

Noise and power

The noise level at any point in a transmission system is the ratio of channel noise at that point to some predefined amount of noise chosen as a reference. This ratio is usually expressed in decibels above reference noise, which is abbreviated as dBrm.

Based upon the sampling of the hearing acuity of many persons, 0 dBrn (decibel reference noise) was set equal to −90 dBm, resulting in the ‘reference noise’ level illustrated in Figure 3.7, which also shows the relationship between noise and power.

One of the problems associated with measuring noise is the requirement to have a reference for standardization of noise measurements The resulting reference, which is a 1 mW signal transmitted at a frequency of 1004 Hz, results in a technician measuring both a test tone and noise.

To obtain a more precise measurement of noise, a technician will insert a ‘C-message’ filter between the meter and the line which serves to filter out the test tone, resulting in noise being received and measured. When this filter is used in noise measurements the units are called decibels above reference noise C-message weighted, which has the abbreviation dBmc.

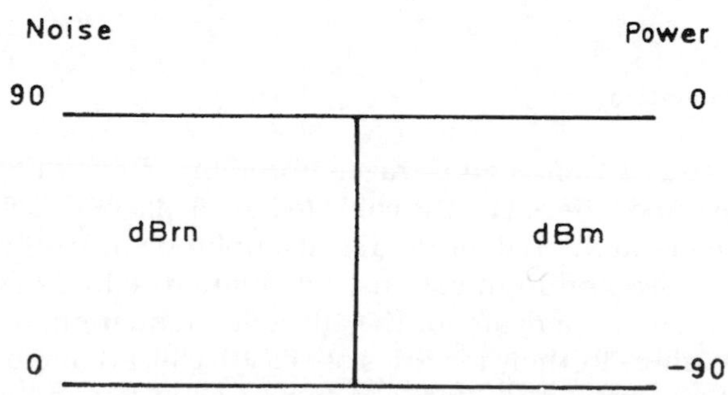

Figure 3.7 Noise and power relationship. 0 dBrn was set equal to −90 dBm based upon the measurement of the hearing acuity of many telephone subscribers

3.3 MEDIA OVERVIEW

3.3.1 Twisted-pair cable

The common twisted-pair cable, which many readers are familiar with from wiring a telephone extension in the home, forms the basis for many measurements used in the communications field. Although we may have only one telephone in our home, the twisted-pair cable also forms the basis for large diameter cables that connect up to 25 instruments to switchboards and automatic switching equipment in an office environment.

The twisted-pair cable consists of two insulated conductors that are twisted together to form a transmission line. The pairs are combined in a cable that typically contains 2, 4 or 25 pairs, with the cable wrapped with a protective jacket of plastic or similar material. Two- and four-pair cables are used in both the home and office. From the home, the cable pair will be routed to an end office, whereas in many offices the cable pair will be routed to a switchboard or to a wiring closet.

In an office environment the wiring closet functions as a wire distribution center, permitting telephones on a floor or within a particular geographic area to be cabled to a common point. From the wiring closet a 25-pair cable is typically used to connect up to 25 telephones to a switchboard or electronic switching device known as a private branch exchange (PBX). From the PBX, calls to other telephone numbers connected to the switch are routed through the switch and never leave the customers' premises. Calls destined to telephone numbers not serviced by the PBX are first routed via trunks that connect the organization's PBX to a telephone company office. From there, the call is routed over the public switched telephone network to its destination.

Signal degradation

Regardless of the method of call routing, there are numerous electrical properties associated with the twisted-pair cable that will affect the quality of an electrical signal moving through the cable. Four of the more prominent electrical properties that affect an electrical signal transmitted over a twisted-pair are attenuation, capacitance, crosstalk and delay distortion.

Attenuation

As an electrical signal travels through a cable, it becomes weaker due to the resistance offered by the cable to flow. This weakness is called attenuation and it refers to the reduction in the amplitude or height of a transmitted signal.

In voice communications, attenuation reduces the loudness of a conversation. To compensate for the effect of attenuation, telephone companies install amplifiers in their facilities at selective locations to boost signal levels. Figure 3.8 illustrates the effect of attenuation on an analog signal and the use of an amplifier to rebuild the signal level. When we speak of an analog signal, we are referring to a continuous signal as opposed to a digital signal which is discrete.

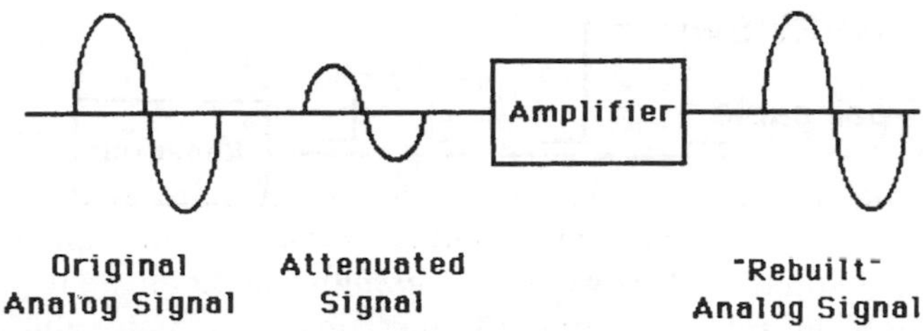

Figure 3.8 Attenuation of an analog signal. Amplifiers are used by the telephone company to boost the signal strength of an analog signal

A second cause of attenuation is the result of the creation of the telephone channel passband by the use of electrical filters. As explained at the beginning of this chapter, telephone companies use low and high pass filters to pass only a small subset of the 20 000 Hz bandwidth audible to the human ear. As a result of the use of electrical filters at approximately 300 and 3300 Hz and the fact that high frequencies attenuate more rapidly than low frequencies, the ideal passband of a telephone channel becomes skewed. The result is an increase in attenuation as frequencies increase as well as at both filter cut-off frequencies. Figure 3.9 illustrates the typical amplitude-frequency response across a voice channel.

Due to the us of low pass and high pass filters, a large degree of attenuation occurs both below 300 and above 3300 Hz. In addition, because high frequencies attenuate more rapidly than low frequencies, the amplitude frequency response is non-linear from 300 to 3300 Hz, with the amount of signal attenuation increasing as the frequency increases.

The effect of attenuation upon a digital signal is similar to its effect upon an analog signal. That is, attenuation of a digital signal reduces the height of the square waves that form the signal.

In place of amplifiers used on analog circuits digital repeaters, also called data regenerators, are used on a digital circuit to compensate for the loss in signal strength. The digital repeater as its name implies accepts a pulse and regenerates it at its original height and width, eliminating any previous

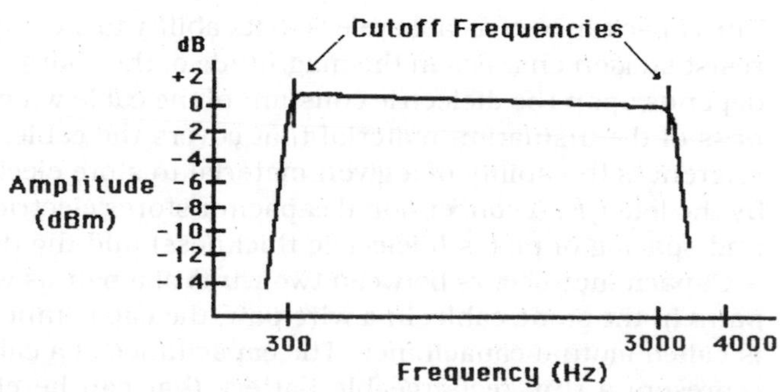

Figure 3.9 Typical amplitude–frequency response across a voice channel

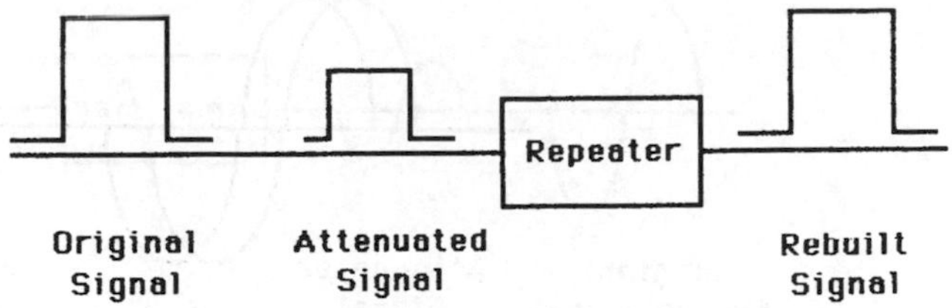

Figure 3.10 Effect of attenuation on digital signals. A digital repeater samples the line for the occurrence of a pulse and regenerates the pulse at its original height and width

distortion to the pulse. Figure 3.10 illustrates the attenuation of a digital signal and the use of a digital repeater to 'rebuild' the signals on a digital transmission medium.

Although both amplifiers and digital repeaters are designed to counteract the effect of attenuation, the resulting signals produced by each device differ in several respects. Amplifiers operate on analog signals and increase both the signal as well as any impairments to the signal, thus, a distorted analog signal will have its degree of distortion increased by an amplifier. In comparison, repeaters operate on digital signals and remove any distortion to the signal since the pulse is reconstructed to its original height and pulse by the repeater. The preceding comparison of amplifiers and digital repeaters provides the rationale for the growth in 'end-to-end' digital transmission systems during the 1980s that provide a higher signal quality than analog transmission.

When twisted-pair cable is used to transport digital data on local area networks (LANs) most cabling between a workstation and hub or concentrator cannot use a repeater. Owing to this, the digital signal cannot be rebuilt and the maximum length of individual LAN cables is affected by attenuation; it is typically limited to 100 m or less. Later in this chapter, in our discussion of structured wiring, we will examine the use of different types of twisted-pair cable for LAN applications and the attenuation limits by frequency for each type of cable.

Capacitance

The capacitance of a cable refers to its ability to store an electric charge and to resist sudden changes in the magnitude of the charge. Capacitance on a cable depends upon the dielectric constant of the cable which is related to the thickness of the insulation material that covers the cable. The dielectric constant references the ability of a given material to store electrons and is symbolized by the letter K. A conventional capacitor stores electricity based upon the area and spacing of plates (dielectric thickness) and the dielectric constant.

Capacitance occurs between two wires of a pair as well as between adjacent pairs in the same cable. In a wire pair, the capacitance between the two wires is called mutual capacitance. The capacitance of a cable can be considered to represent a tiny rechargeable battery that can be charged and discharged, with the amount of the charge commonly measured in microfarads (μF) or

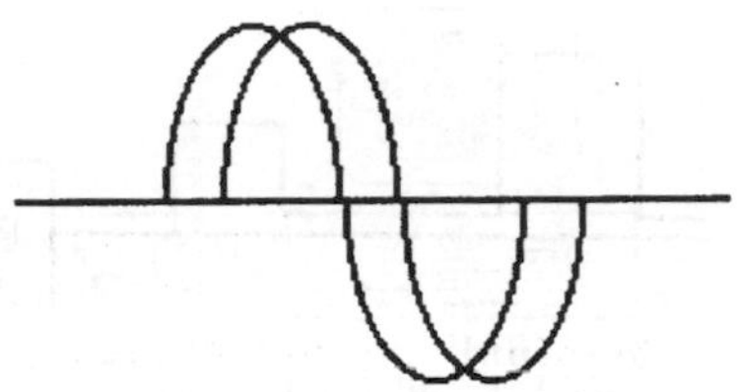

Figure 3.11 Effect of mutual capacitance on analog signals. Mutual capacitance results in the phase shift of analog signals

picofarads (F). The effect of mutual capacitance in voice communications is to shift the phase of the analog signal as illustrated in Figure 3.11. For most voice conversations, the resulting phase shift is normally not objectionable, since voice frequencies vary over a narrow range. When digital signaling occurs, the effect of mutual capacitance can be more severe. This is because the capacitance rounds and distorts the square shape of each digital pulse. When too much mutual capacitance is present or a combination of attenuation and capacitance occurs, the resulting signal distortion will result in data transmission errors. Figure 3.12 illustrates the effect of mutual capacitance upon digital signals.

Crosstalk

The presence on a cable of a signal which originated on a different cable is called crosstalk. Although telephone cables always exhibit a degree of crosstalk, since a signal on an 'excited' pair always induces a signal on a 'quiet' pair, in most instances the effect of crosstalk is negligible. When the effect of crosstalk becomes relatively large, its effect upon both voice and data can become considerable.

In general, crosstalk is proportional to the dielectric constant of a cable. A cable with a lower dielectric constant will have less capacitance than a cable with a higher dielectric constant. Since the amount of capacitance on a cable is proportional to its level of crosstalk, the level of crosstalk is proportional to the dielectric constant of the cable.

Near end crosstalk

A special type of crosstalk, referred to as near end crosstalk and abbreviated as NEXT, represents the biggest source of noise in twisted-pair cables

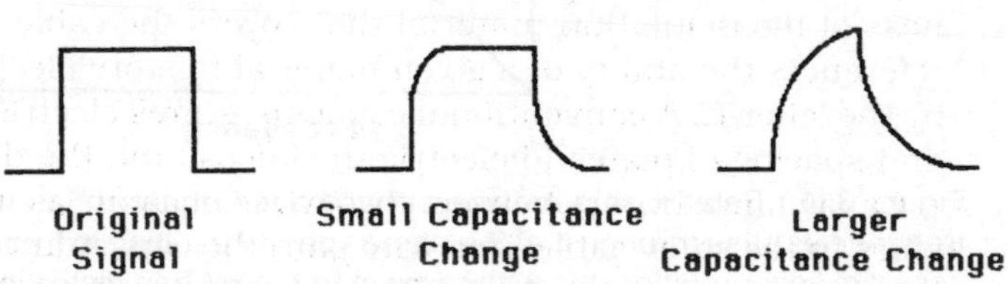

Figure 3.12 Effect of mutual capacitance on digital signals. As a result of mutual capacitance, the leading and trailing edges of a digital pulse are both rounded and distorted

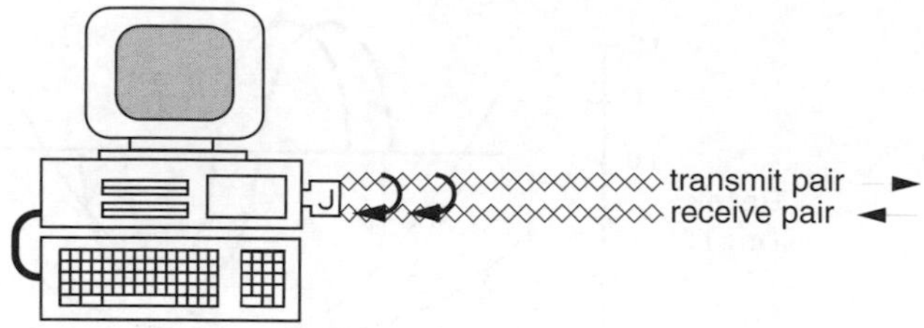

Figure 3.13 Near-end-crosstalk (NEXT) is generated by the coupling of the transmit signal onto the receiver pair

designed to transport digital data. NEXT occurs due to the electromagnetic coupling of a signal from a transmit pair of wires onto the receive pair. Since the transmit signal is strongest at its source, most of the crosstalk generated by a transmit signal occurs at the modular jack and in the first few feet of a cable. Figure 3.13 illustrates the generation of crosstalk by a computer connected to a LAN.

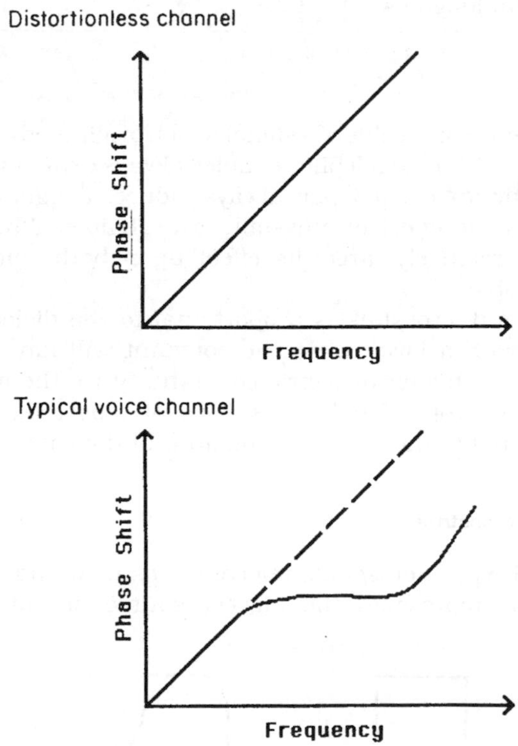

Figure 3.14 Relationship between phase and frequency. In a distortionless channel, all frequencies pass through it at the same speed, resulting in frequency and phase having a constant linear relationship with respect to time. When distortion occurs, the relationship becomes nonlinear with respect to time, causing some frequencies of a signal to reach the distant end of a channel before other frequencies

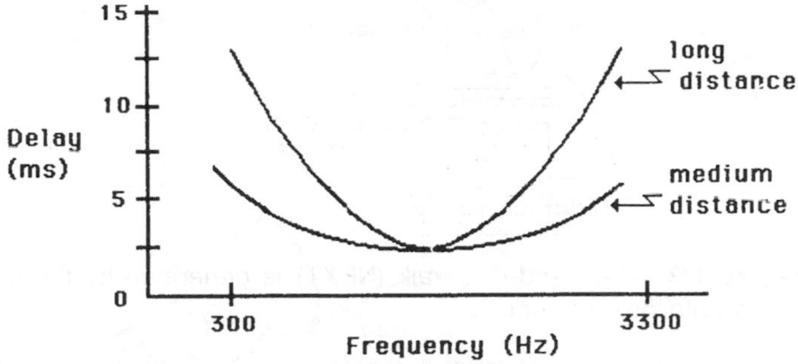

Figure 3.15 Typical signal delay curves. The delay in the propagation of a signal increases with respect to frequency approaching the passband cut-off frequencies and the distance the signal is transmitted

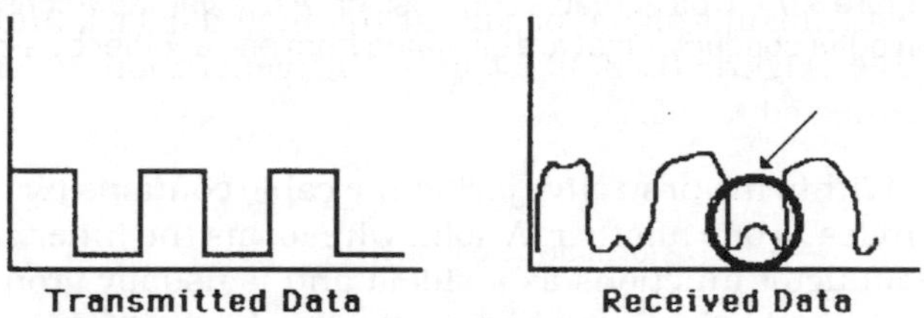

Figure 3.16 Delay distortion. The late arriving energy of one pulse can be misinterpreted as a new pulse, resulting in the occurrence of a digital error

Delay distortion

In a distortionless channel, all frequencies pass through it at the same speed. This results in the frequency and phase having a constant linear relationship with respect to time as illustrated in the top portion of Figure 3.14. When distortion occurs, the relationship between phase and frequency becomes nonlinear as illustrated in the lower portion of Figure 3.14.

Owing to the effect of filters at the edges of the passband as well as the nonlinear relationship between the harmonics of complex signals, such signals propagate through a transmission medium at different velocities. In general, both low and high frequencies require more time to propagate through a transmission medium than frequencies near the center of the passband. Figure 3.15 illustrates typical signal delay curves for medium (up to 300 miles) and long distance (3000 miles) transmission.

Although signal delay has a minimal effect upon voice communications, its effect upon digital transmission can be considerable. This is because a large degree of delay distortion can cause the late arrival of energy of one pulse to be misinterpreted as a new pulse, as illustrated in Figure 3.16.

3.3.2 Coaxial cable

A coaxial cable consists of an inner conductor and an outer conductor that are insulated from one another. The insulation is called the dielectric. Figure 3.17 illustrates the composition of a coaxial cable.

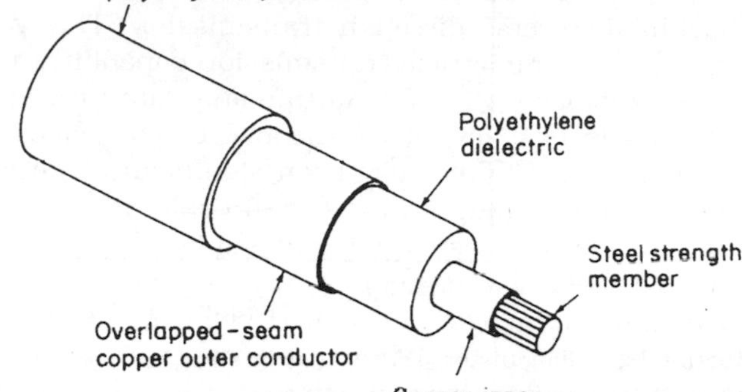

Figure 3.17 Coaxial cable composition. A coaxial cable consists of an inner conductor and an outer conductor that are insulated from one another by a dielectric

Within the protective jacket the cable contains two conductors in concentric circles to one another. A solid wire forms the inner conductor, while the outer conductor functions as a shield and is usually grounded.

Coaxial cables can transmit signals ranging in frequency from 1 kHz to 1 GHz per second with little loss, distortion or interference to or from outside signals.

There are many types of coaxial cable that include cable with multiple conductors. However, coax, a term used to denote this type of cable, can normally be classified into those used for baseband and those used for broadband transmission. Baseband refers to the transmission of one signal at a time, and baseband coax normally has a characteristic impedance of 50 Ω. By comparison, broadband refers to the ability to simultaneously transmit two or more signals on a cable by transmitting each signal at a different frequency. Broadband coax is the 75 Ω cable used with CATV systems.

Due to the larger bandwidth of coaxial cable than of the twisted pair, they have a larger data transmission capacity. Typical usage of coaxial cable includes community antenna television (CATV) and local area networks (LANs).

When used for CATV each television channel occupies a 6.0 MHz portion of coaxial cable bandwidth. For example, channel 2 has a video carrier of 55.25 MHz while channel 3 has a video carrier of 61.25 MHz. The actual frequencies used for other cable TV channels vary based upon the frequency band allocated for the off-the-air TV channel, even though a CATV company more than likely does not place your local channel X off-the-air station on channel X on their cable system. For example, channel 7, which represents a high band VHF (very high frequency) signal, uses a 175.25 MHz video carrier.

The relatively high bandwidth of coaxial cable permits CATV operators to offer high speed cable modem communications by allocating a 6 MHz TV channel for downstream communications. Because the original CATV infrastructure consisted of unidirectional amplifiers to support broadcast TV, cable operators must install bidirectional amplifiers to support two-way cable modem transmission. CATV operators that accomplished a system upgrade use a smaller bandwidth channel under the channel 2 video carrier signal for 'backhaul' reverse direction transmission. This results in a cable modem having an asymmetrical transmission capability, downloading at a data rate up to 36 Mbps while transmitting at a data rate typically up to 1 Mbps.

Over the past few years the use of coaxial cable to create the cabling infrastructure of LANs has been significantly surpassed by the twisted pair due to the advantages in cost and cable installation flexibility of the latter. Similarly to the twisted pair, the signal on a coaxial cable can be adversely affected. Electrical properties that are related to the quality of a signal on coaxial cable include impedance, capacitance, attenuation and delay distortion. The effect of the latter three properties upon coaxial cable is similar to their effect upon a twisted pair.

Impedance effect

Impedance defines the relationship of voltage and current in a coaxial cable. In general, the electrical requirements of hardware to which coaxial cable is connected dictate the impedance values for the interconnecting cables. Three common coaxial cables are designed to match 50, 75 and 90Ω impedances required by electronic hardware.

3.3.3 Microwave

Microwave is a line-of-sight transmission medium that provides the bandwidth and capacity of coaxial cable without requiring the laying of a physical cable. Since microwave is a line-of-sight transmission, its use requires the construction of towers which, due to the curvature of the Earth, are limited to distances of 30 miles or less from one another. Figure 3.18 illustrates the use of microwave towers, while Table 3.2 lists the frequency range of three of the more popular microwave transmission bands.

The large bandwidth of microwave communications permits communications carriers to simultaneously transmit thousands of calls between offices connected by microwave towers. As you travel around the globe the microwave

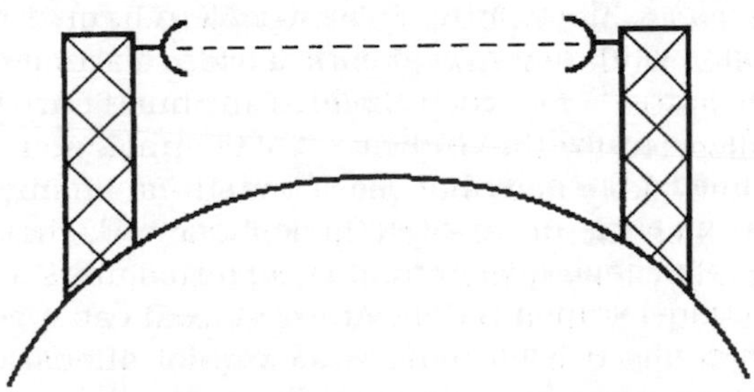

Figure 3.18 Microwave transmission. The curvature of the Earth limits the distance between stations to 30 miles

Table 3.2 Typical microwave transmission bands

3.7 to 4.2 GHz (4 GHz)
5.925 to 6.425 GHz (6 GHz)
10.7 to 11.7 GHz (11 GHz)

tower is a common sight, especially in small towns and cities where its height makes it stand out as the communications link of the town or city to the rest of the world.

The transmission of a microwave signal to a point above the Earth forms the basis for satellite communications. A communications satellite contains a number of transponders (transmitters–receivers) which function as a microwave relay in the sky. The satellite receives the microwave signal transmitted from an Earth station and rebroadcasts it back to Earth.

Unlike transmission via a cable, microwave transmission can be affected by sunspots and weather: heavy rain, thunderstorms and dust storms all have an adverse effect upon microwave transmission.

During the 1950s and 1960s microwave towers literally dotted the landscape of many countries, providing the most common method for supporting inter-city communications. During the 1970s communications carriers turned to the use of fiber optic cable as a replacement for microwave. Because light transmitted over fiber is immune to electrical disturbances, by the mid- to late 1980s the use of fiber optics resulted in a substantial reduction in the use of microwave towers.

3.3.4 Fiber-optic transmission

Once a laboratory and consumer product curiosity, optical transmission via fiber is now used for low cost, high data rate transmission. The major components of an optical system are similar to a conventional transmission system, requiring a transmitter, transmission medium and receiver. The transmitter, an electrical-to-optical (E/O) converter, receives electronic signals and converts them to a series of light pulses. The transmission medium is an optical fiber cable of plastic or glass. The receiver, an optical-to-electric (O/E) converter, changes the received light signals into corresponding electrical signals.

The bandwidth of a typical optical fiber can range up to 10 GHz. Normally a series of optical fibers, each shielded and bundled together, are routed between locations. To provide an indication of the transmission capability of an optical fiber, consider the potential use of two strands of fiber in a bundle. Each pair of strands can carry up to 24 000 telephone calls, with calls routed in different directions on each fiber strand in a pair as until recently they were used for unidirectional transmission. An equivalent capacity established through the use of metallic twisted pair would require 48 000 copper wires to establish the same number of telephone calls.

Although optical fiber has a relatively high bandwidth in comparison to copper, the growth in the use of the Internet and other applications began to tax that bandwidth. Beginning in the late 1990s a technique similar to FDM was used on many optical circuits. Based upon the subdivision of an optical fiber by wavelength, a technique referred to as wavelength division multiplexing (WDM) enables one optical fiber to transport up to 100 or possibly more simultaneous light sources, significantly increasing the transmission capacity of each optical fiber.

There are two general types of optical fiber cable used to transmit information: single-mode and multimode. A single-mode fiber has a core diameter of approximately 8 microns (μm) which is approximately the wavelength of light. This results in light flowing on one route through the fiber. In comparison, multimode fiber has a core diameter of approximately 62.5 microns, although 50, 85 and 100 micron cables are available. This diameter is large enough to permit light to travel on different paths, with each path considered a mode of propagation, hence the term multimode. Single-mode fiber is designed to transmit laser-generated light signals at distances up to 20 to 30 miles, This type of optical fiber is primarily used by communications carriers. In comparison, multimode was designed to carry relatively weak light emitting diode (LED) generated signals over relatively short distances, typically under a mile. Multimode fiber is primarily used within buildings and forms the infrastructure for optical LANs.

Optical fiber cable is typically specified using two numeric identifiers of the form x/y. Here x represents the diameter of the core, and y represents the cable diameter. Figure 3.19 compares the transmission of light in single and multimode fiber.

The large bandwidth of light permits a glass fiber with a diameter less than the tip of a pen to transmit the same number of voice conversations that would require the use of a copper cable several inches in diameter. Owing to this, communications carriers have installed tens of thousands of miles of fiber-optic cables to link most major metropolitan areas to one another.

Figure 3.20 illustrates the major portion of the frequency spectrum used for communications. The major omissions from Figure 3.20 are the visible light frequencies which range from 10^{12} Hz (infrared) to 10^{14} Hz (visible light) to 10^{16} Hz (ultraviolet). To place the frequencies in perspective, note that voice

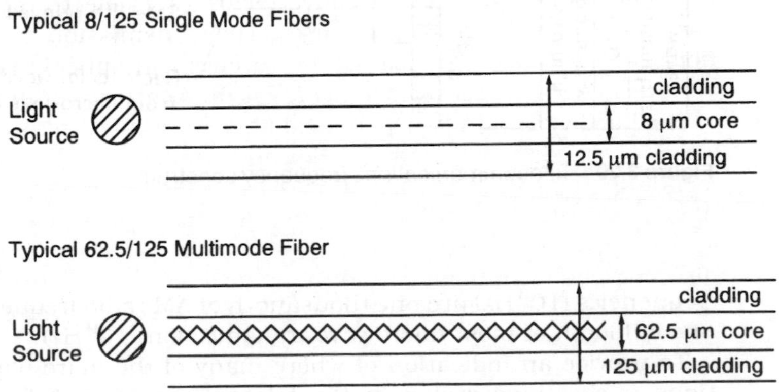

Figure 3.19 Comparing single and multimode fiber transmission

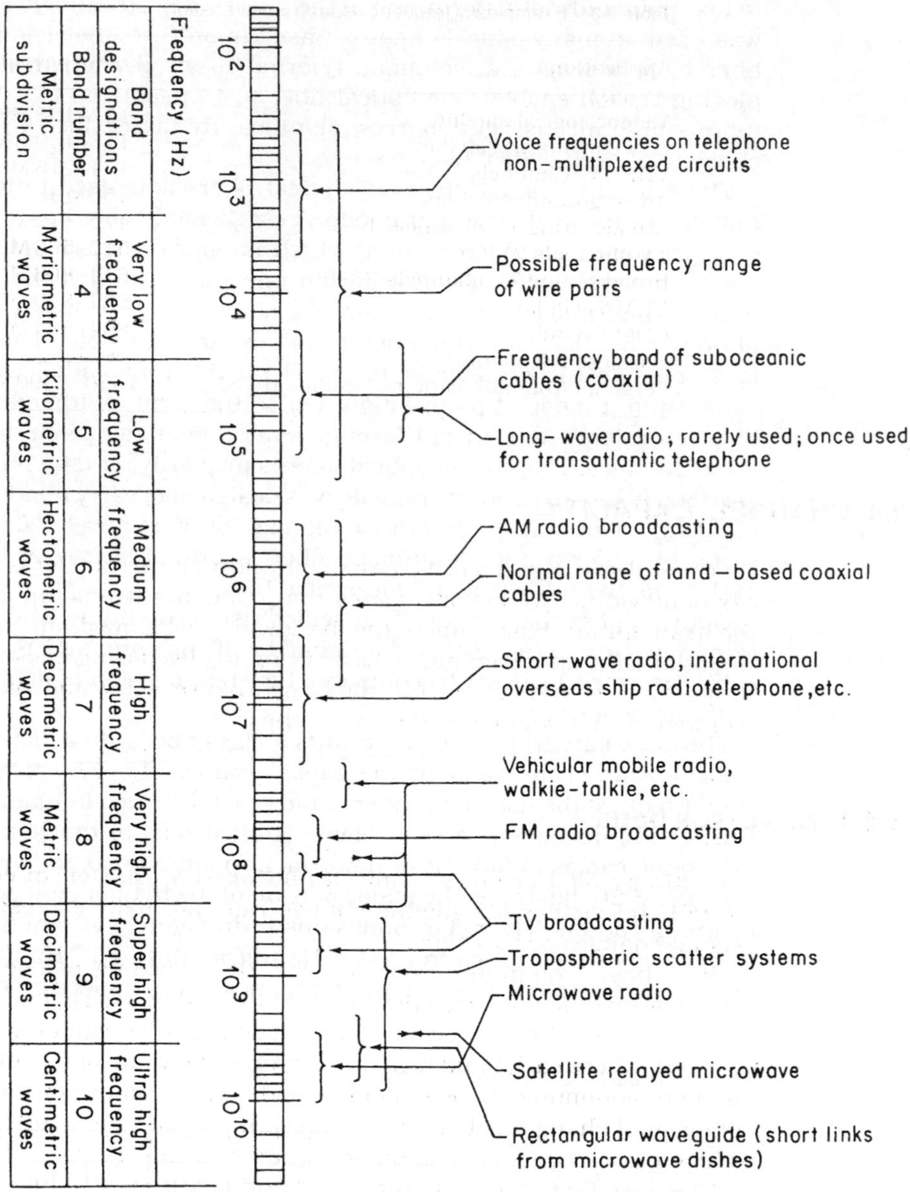

Figure 3.20 Telecommunications frequency spectrum

frequencies (10^3 Hz) are one-thousandth of AM radio frequencies (10^6 Hz) and one-millionth of most microwave transmission (10^9 Hz).

To provide an indication of where many of the more popular communications applications reside in the telecommunications frequency spectrum, Table 3.3 lists the frequency spectrum allocated to 11 activities.

Table 3.3 Examples of frequency spectrum allocation

Application	Frequency spectrum
Meteorological satellite	137–138 MHz
Low earth orbit satellite	148–149.9 MHz
VHF TV (Channels 7–13)	174–216 MHz
Air to ground telephone	450 MHz
Broadcasting TV channels 14–20	470–512 MHz
TV channels 21–36	512–560 MHz
Broadcasting TV channels 38–69	614–806 MHz
CDMA cellular	824–849 MHz
CDMA cellular	864–894 MHz
Cordless telephone	900 MHz
Personal Communications Systems (PCS)	1850–1990 MHz

3.4 CHANNEL CAPACITY

In general, the capacity of a channel for information transfer is proportional to its bandwidth. Two major theories that relate to the amount of data that can be transmitted based upon the bandwidth of a medium are the Nyquist Relationship and Shannon's Law. Prior to discussing these theories, it is important to understand the difference between bit and baud, due to the confusion that dominates the use of these terms.

3.4.1 Bit versus baud

The binary digit or bit is a unit of information transfer. In comparison, the term baud defines a signaling change rate, normally expressed in terms of signal changes per second.

In a basic communications system, the encoding of one bit per signal element results in equivalency between bit and baud. That is, an information transfer rate of X bits per second is carried by a signaling change rate of X baud, where each baud signal represents the value of one bit. Now suppose that our communications system is modified so that two bits are encoded into one signal change. This would result in the baud rate being half the bit rate, which obviously makes the bit and baud non-equivalent.

The encoding of two bits into one baud is known as dibit encoding, and is discussed in detail in Chapter 4 along with other coding schemes.

3.4.2 Nyquist relationship

In 1928, Harry Nyquist developed the relationship between the bandwidth and the baud rate on a channel as

$$B = 2W$$

where B is the baud rate and W the bandwidth in Hz.

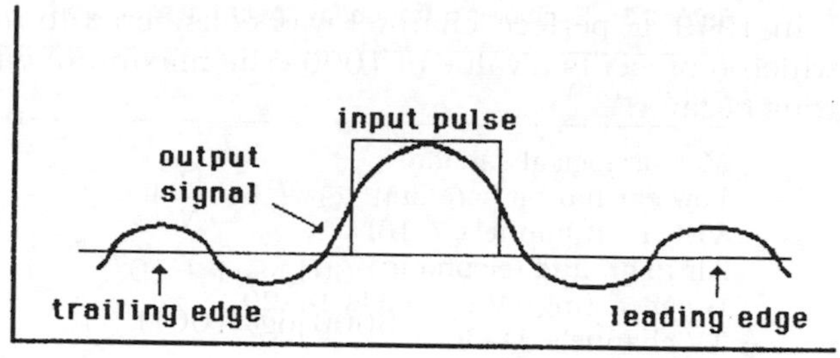

Figure 3.21 Pulse response through a band-limited channel. The bandwidth limitation of a channel causes the leading and trailing edges of a pulse to interfere with other pulses as the signal change exceeds twice the bandwidth of a channel. This condition is called intersymbol interference

The Nyquist relationship was based upon a problem known as intersymbol interference which is associated with band-limited channels. If a rectangular pulse is input to a band-limited channel, the bandwidth limitation of the channel results in a rounding of the corners of the pulse. This rounding results in the generation of an undesired signal in which the leading and trailing edges formed due to signal rounding can interfere with both previous and subsequent pulses. This signal interference is illustrated in Figure 3.21.

The Nyquist relationship states that the rate at which data can be transmitted prior to intersymbol interference occurring must be less than or equal to twice the bandwidth in Hz. Thus, an analog circuit with a bandwidth of 3000 Hz can only support baud rates at or under 6000 signaling elements per second.

Since an oscillating modulation technique such as amplitude, frequency or phase modulation halves the achievable signaling rate, a twisted pair telephone circuit supports a maximum signaling rate of 3000 baud.

3.4.3 Shannon's law

In 1948, Claude E. Shannon presented a paper concerning the relationship of coding to noise and calculated the theoretical maximum bit rate capacity of a channel of bandwidth W Hz. The relationship developed by Shannon is given by

$$C = W \log_2\left(1 + \frac{S}{N}\right)$$

where

C = capacity in bits per second,
W = bandwidth in Hz,
S = power of the transmitter,
N = power of thermal noise.

In 1949, a 'perfect' channel was considered to have a S/N ratio of 30 dB, which represents a value of 1000. The maximum data transmission capacity thus became

$$\begin{aligned} C &= W \log_2\left(1 + \frac{S}{N}\right) \\ &= 3000 \log_2(1 + 10^3) \\ &= 3000 \log_2(1001) \\ &= 3000 \times 10 \text{ (approximately)} \\ &= 30\,000 \text{ bps} \end{aligned}$$

It should be noted that Shannon's law is applicable to the transmission of digital data on an analog channel, requiring data to be modulated or converted to analog signals under the constraints associated with analog circuits. In comparison, it is possible to transmit data at much higher rates on a twisted pair. For example, when an analog signal takes advantage of the approximate 1 MHz bandwidth of twisted pair, a data rate of 1 Mbps or higher becomes obtainable.

It should be noted that when the transmission is in digital form at an operating note in the Mbps range, the transmission distance is limited, typically less than 100 m. This transmission characteristic forms the basis for the development of a twisted-pair cabling infrastructure commonly used to construct LANs.

3.5 STRUCTURED WIRING

The growth in the use of computers connected to local area networks, telephones, fax machines and other electronic devices requiring cabling resulted in the development of several commercial and industrial wiring standards. Those standards are similar in that they are based on specifications for similar wiring structures, some of which are the cabling infrastructures originally established by telephone companies when they wired buildings for telephone service.

3.5.1 The wiring closet

The wiring of a building for telephone service usually resulted in one or more rooms on each floor being dedicated as a wiring closet. From that closet individual or groups of twisted-pair cable were run to offices located on the floor. Wire closets in turn were connected to one another via backbone wiring.

3.5.2 The EIA/TIA-568 standard

Building on the telephone company wiring closet structure, the Electronics Industries Association (EIA) and the Telecommunications Industry Association (TIA) jointly developed a number of standards for intrabuilding wiring.

In 1991 the EIA/TIA published the EIA/TIA-568 Intrabuilding Wiring Standard which provides a guideline for the construction of a cabling system. That standard, as well as a 1992 revision and a subsequently issued standard, define the use of different types of twisted pair cable for different applications, where the applications are expressed genetically as a data rate and/or signaling rate. In addition, the standards define a cabling structure for horizontal and backbone wiring as well as for work area wiring from a communications outlet to station equipment.

Cable–pair categories

Under EIA/TIA-568 five levels or categories of unshielded twisted-pair (UTP) cable are defined. Each pair is constructed of four unshielded pairs of either 22 or 24 gauge solid conductors. The primary difference between each category of UTP is the number of twists per foot, with Category 5 cable having the highest density of twists in an attempt to reduce the potential of electromagnetic interference, since the twists reduce the antenna effect of the wire. Table 3.4 describes the type of communications supported by each cable category, with higher categories supporting the applications supported by all lower categories.

In addition to EIA/TIA cable categories being defined by the applications they support, they are also defined by the maximum attenuation and NEXT

Table 3.4 EIA/TIA cable categories

EIA/TIA cable category	Applications supported
Category 1	Voice and low speed data. Not suited for LAN applications.
Category 2	Data and LANs operating at 1 Mbps or less.
Category 3	Supports LANs at operating rates up to 10 Mbps and signaling rates up to 16 MHz.
Category 4	Supports LANs at operating rates up to 16 Mbps and signaling rates up to 20 MHz.
Category 5	Supports LANs at operating rates up to 100 Mbps and signaling rates up to 100 MHz.

Table 3.5 Attenuation and NEXT limits by frequency for 100 m cable

Frequency, MHZ	Category 3 Attenuation	NEXT	Category 4 Attenuation	NEXT	Category 5 Attenuation	NEXT
1.0	3.5	39.5	2.0	53.5	2.0	60.5
4.0	6.5	29.5	4.5	43.5	4.5	51.0
8.0	9.5	24.5	6.5	38.5	6.0	46.0
10.0	11.0	23.0	7.5	37.0	7.0	44.0
16.0	14.5	19.5	9.5	33.5	9.0	41.0
20.0	–	–	11.0	31.5	10.0	39.0
25.0	–	–	–	–	11.0	37.5
31.2	–	–	–	–	12.5	36.0
62.5	–	–	–	–	18.0	31.0
100.0	–	–	–	–	23.5	27.5

over frequency. Table 3.5 indicates the attenuation and NEXT limits for cable categories 3, 4 and 5 for a 100 m link between a wallplate and a wiring closet. Although we will defer a discussion of specific LANs until later in this book, you may wish to note that Category 3 cable supports conventional Ethernet, Category 4 cable supports Token-Ring and Category 5 cable supports Fast Ethernet and ATM.

The development of Gigabit Ethernet which operates at ten times the speed of Fast Ethernet represents a cabling problem, as its signaling rate exceeds 100 MHz. Through 1999 the primary method used to support Gigabit Ethernet was through the use of optical fiber cable. However, work has commenced upon an enhancement to category 5 cable referred to as cat5e, as well as upon category 6 cable. While no standard had been produced when this book revision was prepared, it is reasonable to expect further additions to the EIA/TIA standard to support higher LAN operating rates.

Cabling rules

Under the structured wiring concept a workstation is cabled to a communications outlet via a path cord whose length is 10 feet or less. The communications outlet is an RJ-45 jack wallplate which is cabled to a wiring closet. At the

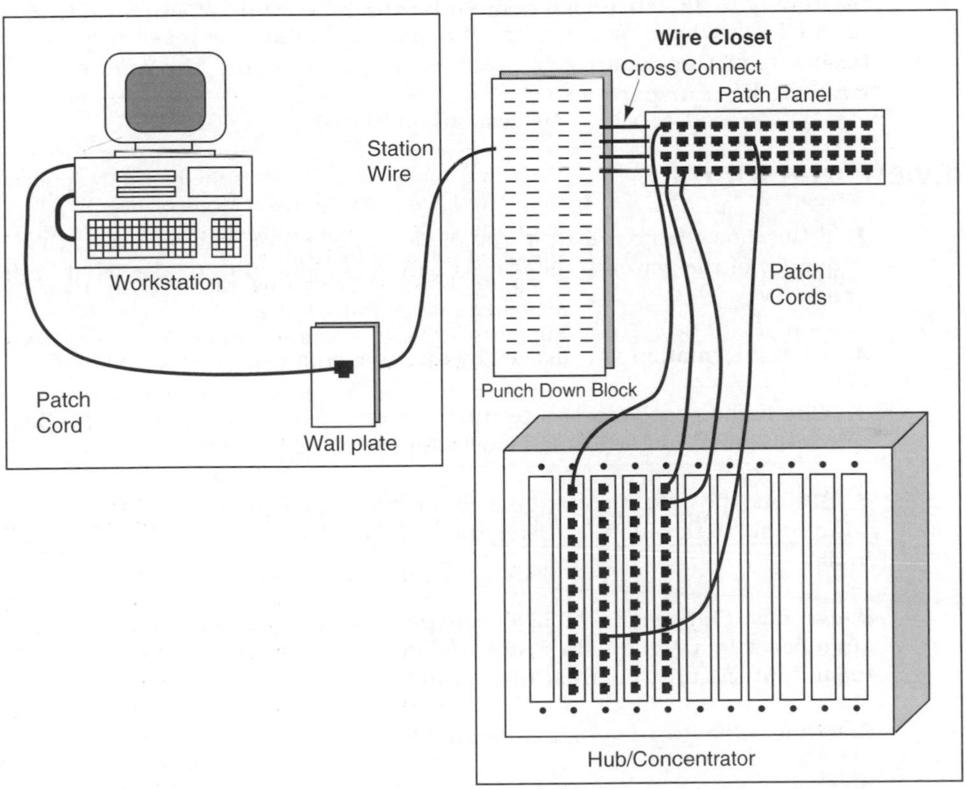

Figure 3.22 Horizontal wiring associated with a structured wiring system

wiring closet the cable is connected to a 25-pair punch down block. The block consists of multiple columns of 50 pins each, with the number of columns twice the number of pairs to be supported, since one column functions as the input to the block and the other side functions as its output. The punchdown block has its origins in the telephone industry where it was used as a mechanism to enable the individual pairs carried in a multi-pair cable to be distributed to specific wire pairs routed into a wiring closet. Rather than let a good idea fade, this wiring concept has been adapted for use with other types of building wiring to include LANs.

The length of the cable from the station to the punchdown block to include the patch cable is limited to 100 m. From the punchdown block individual twisted-pair cable are cross-connected to a patch panel. The patch panel in turn provides the mechanism to easily administer the connection of individual twisted-pair cable to different ports on a hub or concentrator used to form a particular type of local area network. Figure 3.22 illustrates the horizontal structed wiring associated with connecting a workstation to a LAN.

Under the EIA/TIA-568 standard horizontal cabling should run in a star topology from stations that can represent voice, data or LAN connections to a wiring closet now known as a telecommunications closet. The cable distance of 100 m is limited to 90 m from the punchdown block to the outlet, with 10 additional m allowed for all patch cables. The backbone cable, which is used to interconnect wiring closets can range in scope from multipair 100 Ω unshielded twisted pair to 62.5/125 μm optical fiber, with cabling constraints based upon the distance between wiring closets and the operating rate of traffic to be transported.

3.6 REVIEW QUESTIONS

1. If the time required for the completion of one sine wave is 0.1 s, what is the frequency of the wave? If the wave's period is reduced to 0.05 s, what is its new frequency?

2. What information is conveyed by an unmodulated sinusoidal signal?

3. Why is the passband of a telephone channel 3000 Hz even though the audio spectrum heard by the human ear is approximately 20 000 Hz?

4. Assume the output power on a channel was measured to be one-quarter of the power input to the channel. Using the data in Table 3.1, determine the power ratio in dB.

5. Suppose that a signal of 22 dB is applied to a cable that has a loss of 7 dB. If an amplifier with a 12 dB gain is inserted into the cable at its opposite end, what is the output signal strength at that location?

6. What is the function of an amplifier?

7. What is the difference between thermal noise and impulse noise? Which noise sets the lower level for receiver sensitivity?

8. What is the S/N ratio? Why is a high S/N ratio desirable but a very high S/N ratio undesirable.

9. Assume the reading on a dB meter inserted into a circuit showed zero. What would this indicate. If the dB meter reading was −10, what would this reading indicate?

10. What is the difference in the use of the dB and dBm?

11. Why was 0 dBrn set equal to −90 dBm?

12. What is the difference with respect to signal reconstruction obtained by the use of an amplifier and a digital repeater?

13. Discuss four causes of signal distortion on a twisted-pair cable.

14. Why does near end crosstalk (NEXT) diminish as the distance from the cable connector increases?

15. Explain the basic reason why a cable modem can operate at a data transmission rate in the millions of bits per second rate while a modem used on the switched telephone network is limited to an operating rate between 30 kbps and 56 kbps.

16. What is the relationship between microwave transmission and satellite communications?

17. What are two advantages associated with the use of transmission via optical fiber?

18. What is the difference between bit and baud with respect to transmission? Can they be equivalent? Can they be non-equivalent?

19. If the bandwidth of a channel is 100 Hz and two bits are encoded per signal change, what data transmission rate would be obtainable?

20. What are the core diameter and cladding diameter of a 62.5/125 μm fiber?

21. If a LAN encodes two bits into one baud and transmits at 22 Mbps can Category 4 cable be used? Explain your answer.

4

FUNDAMENTAL DATA TRANSMISSION CONCEPTS

The transmission of data is similar to the establishment of voice communications, requiring a transmitter, a receiver and a transmission medium to provide a path or link between the transmitter and the receiver. A second similarity between voice and data transmission is the requirement for the transmitter to encode information and the receiver to decode information back into its original form.

Although the transmissions of voice and data have many similarities, there are also important differences between the two methods of conveying information. In this chapter, we will focus our attention upon the fundamental concepts associated with the transmission of data, denoting the differences between voice and data transmission when applicable. In this chapter, as well as succeeding chapters in this book, we will interchangeably use the terms terminal and personal computer and refer to them collectively as terminals or terminal devices. The interchangeable use of these terms is in recognition of the exponential increase in the use of personal computers since the middle 1980s and a corresponding increase in communications between personal computers and other personal computers and large scale computers, where the personal computer, in effect, replaced most terminals. In certain instances we will however, focus our attention upon personal computers in order to denote certain hardware and software characteristics unique to such devices. In these instances we will use the term personal computer to explicitly refer to this terminal device.

4.1 ANALOG LINE CONNECTIONS

In this section, we will examine three basic types of twisted-pair based line connections commonly used to connect terminal devices to distance computers or other terminals: switched, leased and dedicated lines. The first two lines are based upon the plant facilities of communications carriers, and the third type of line can be installed by personnel within an organization.

Another term which warrants a degree of explanation is workstation. At one time the term workstation was exclusively used to denote a processing powerful desktop computer. Although this term is still used to refer to that type of computer, it is primarily used to represent a personal computer connected to a local area network. Since the focus of this book is on communications, unless otherwise noted, the term workstation will be used to denote a personal computer connected to a local area network.

4.1.1 The analog switched line

A switched line, often referred to as a dial-up line, permits contact with all parties having access to the public switched telephone network (PSTN). Because conversations represent analog signals, the PSTN was developed as an analog transmission facility. A second type of switched line which represents a digital transmission service will be discussed later in this chapter when we turn our attention to digital transmission facilities.

The use of the PSTN can be considered almost second nature to most persons; however, this author would be remiss if he did not gently explain its use for transmitting information. Thus, if the operator of a terminal device wants access to a computer, he or she dials the telephone number of a telephone which is connected to the computer. In using switched or dial-up transmission, telephone company switching centers establish a connection between the dialing party and the dialed party. After the connection has been set up, the terminal and the computer conduct their communications. When communications have been completed, the switching centers disconnect the path that was established for the connection and restore all paths used, so they become available for other connections.

The connection between a telephone subscriber and a telephone company serving office is called a local loop. The switched line local loop connection is obtained by the installation of a single pair of wires which enables transmission to occur in either direction. If, however, users on both ends of a connection simultaneously attempt to talk, their conversations will be superimposed on the wire path. Although the conversations will be heard, they will usually be distorted and probably could not be understood at the opposite end.

For transmission between telephone company offices, amplification of signals is required. This operation to boost signal strength is more easily implemented if the two directions of transmission are isolated from one another. Thus, transmission between telephone company offices is accomplished by the use of two pairs of wires, normally referred to as a four-wire circuit. This circuit permits one pair to be used for transmission in each direction.

As the switched call is routed through telephone company offices, the two-wire circuit is converted to a four-wire circuit until it reaches the office servicing the called party. The four-wire circuit between carrier offices is then converted back to a two-wire circuit that services the distant subscriber. Figure 4.1 illustrates the switched network wiring facilities that converts a two-wire subscriber loop into a four-wire transmission path.

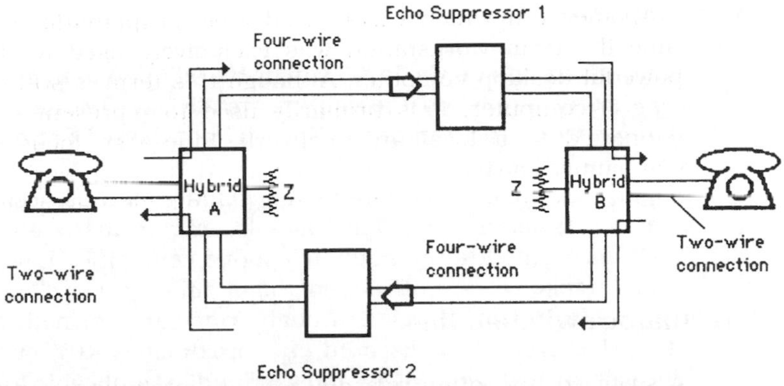

Figure 4.1 Two-wire to four-wire conversion. The two-wire to four-wire conversion results in an impedance (*Z*) mismatch. This causes some energy to be reflected back which is attenuated by the use of echo suppressors

The conversion of a two-wire to four-wire circuit is performed by hybrid circuits that couple the two directions of transmission. A hybrid circuit is similar to a transformer, coupling the energy on an incoming circuit onto an outgoing circuit. Unfortunately, not all of the energy is coupled from one circuit to another, since an impedance mismatch will occur between the two-wire and four-wire circuits. This impedance mismatch cannot be practically compensated for since each connection over the switched network can involve the establishment of a circuit using different local loops and trunks that have different impedance values. This mismatch causes a small portion of the energy of the transmitted signals to be reflected back to the subscriber. This signal reflection is known as an echo and can be caused by both the local hybrid (talker echo) and the distant hybrid (listener echo).

Since an echo can be very annoying to voice conversations, telephone companies incorporated echo suppressors into their facilities, locating them at each conversion point to block reflected energy.

Echo suppressors operate on four-wire circuits by comparing the speech volume on each pair and inserting a high level of attenuation on the path with the lowest power level. Thus, if A is speaking to B, that portion of A's signals that is reflected back by B's hybrid is blocked by echo suppressor 2. Similarly, that portion of B's signals that is reflected back by A's hybrid is blocked by echo suppressor 1. If both A and B speak at once, the echo suppressors would attenuate reflections in both directions. While this situation has minimal effect upon a voice conversation, its effect upon certain types of data transmission can be considerable, especially when it is desired to simultaneously transmit data in both directions. To alleviate the effect of echo suppressors, certain types of communications equipment include built-in circuitry called an echo suppressor tone disabler. After an initial period of time for circuit establishment, the circuitry will place a tone on the line at a predefined frequency which will disable the echo suppressors, permitting transmission to occur simultaneously in both directions. To be able to accomplish this, echo suppressors are manufactured to recognize a tone at a predefined frequency as a signal to cease operations.

Dial network cost elements

The cost of a call on the PSTN can be based upon many factors, which include the time of day when the call was made, the distance between called and calling parties, the duration of the call and whether or not operator assistance was required in placing the call.

Direct dial calls made from a residence or business telephone without operator assistance are billed at a lower rate than calls requiring operator assistance. In addition, many telephone companies have three categories of rates: 'weekday', 'evening' and 'night and weekend'. Calls made between 5 p.m. and 11 p.m. Monday through Friday are normally billed at an evening rate, which reflects a discount of approximately 40% over the weekday rate. The last rate category, night and weekend, is applicable to calls made between 11 p.m. and 8 a.m. on weekdays as well as anytime on weekends and holidays. Calls during this rate period are usually discounted 50% or more from the weekday rate.

Figure 4.2 illustrates an example of an AT&T rate discount chart. Many communications carriers have similar rate discount charts based upon the establishment of day, evening and night and weekend rate periods.

Table 4.1 contains a sample PSTN rate table which is included for illustration but which should not be used by readers to determine the actual cost of a PSTN call. This is due to the fact that the cost of intrastate calls by state and interstate calls vary. In addition, the cost of using different communications carriers to place a call between similar locations will typically vary from vendor to vendor. Thus you should obtain current schedules from the vendors they plan to use in order to determine or project the cost of using PSTN facilities.

During the mid-1990s several communications carriers began to offer subscribers flat-rate pricing plans. Initially such plans were commonly applicable to hours other than 8 a.m. through 5 p.m. Monday through Friday. However, by

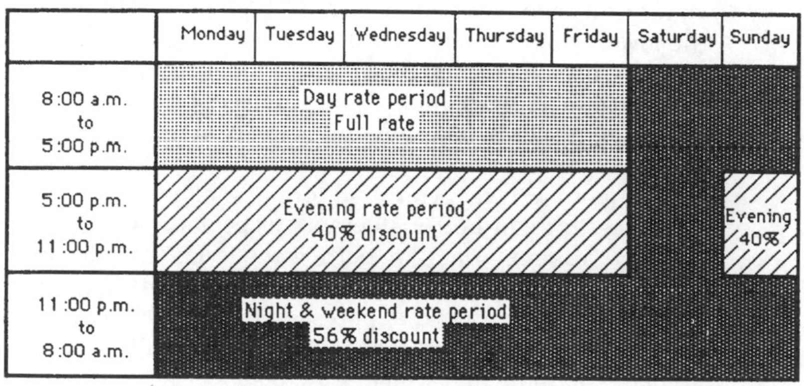

Monday
Tuesday
Wednesday
Thursday
Friday
Saturday
Sunday
8:00 a.m. to 5:00 p.m.
5:00 p.m. to 11:00 p.m.
11:00 p.m. to 8:00 a.m.
Day rate period
Full rate
Evening rate period
40% discount
Evening
40%
Night & weekend rate period
56% discount

Figure 4.2 AT&T rate discount chart. Most communications carriers have a discount rate schedule for switched network calls based upon day, evening and night and weekend rate periods

Table 4.1 Sample PSTN rate table (cost per minute in cents)

	Rate category					
	Weekend		Evening		Night and weekend	
Mileage between locations	First minute	Each additional minute	First minute	Each additional minute	First minute	Each additional minute
1–100	0.31	0.19	0.23	0.15	0.15	0.10
101–200	0.35	0.23	0.26	0.18	0.17	0.12
201–400	0.48	0.30	0.36	0.23	0.24	0.15

1999 almost all communications carriers offered calling plans that provided a flat per minute usage rate regardless of the time of the call.

4.1.2 Analog leased line

A leased line is commonly called a private line and is obtained from a communications company to provide a transmission medium between two facilities which could be in separate buildings in one city or in distant cities. In addition to a one-time installation charge, the communications carrier will normally bill the user on a monthly basis for the leased line, with the cost of the line usually based upon the distance between the locations to be connected.

One of the key differences between the PSTN and leased lines is the ability to have the communications carrier tune an analog leased line. The communications carrier can tune an analog leased line since its routing is fixed. In comparison, calls over the switched network vary in routing based upon traffic on the network which precludes the tuning of switched network calls. The tuning process applied to analog leased lines is called conditioning and it results in the communications carrier adding fixed equalizers to correct distortions due to attenuation and envelope delay, or providing minimum defined signal to C-notched noise ratio and intermodulation distortion. The use of equalizers results in what is known as C-type conditioning, while the tuning of an analog line to provide a minimum signal to C-notched noise ratio and intermodulation distortion is known as D-type conditioning.

C-type conditioning

A communications carrier provides C-type conditioning at circuit installation time by adding and adjusting equalizers to provide an approximate signal equal to the opposite signal attenuation and envelope delay of the circuit's characteristics.

Figure 4.3 illustrates the effect of the use of attenuation and delay equalizers upon the channel characteristics of an analog leased line. Note that the equalizer provides a variable gain which compensates for the differences in

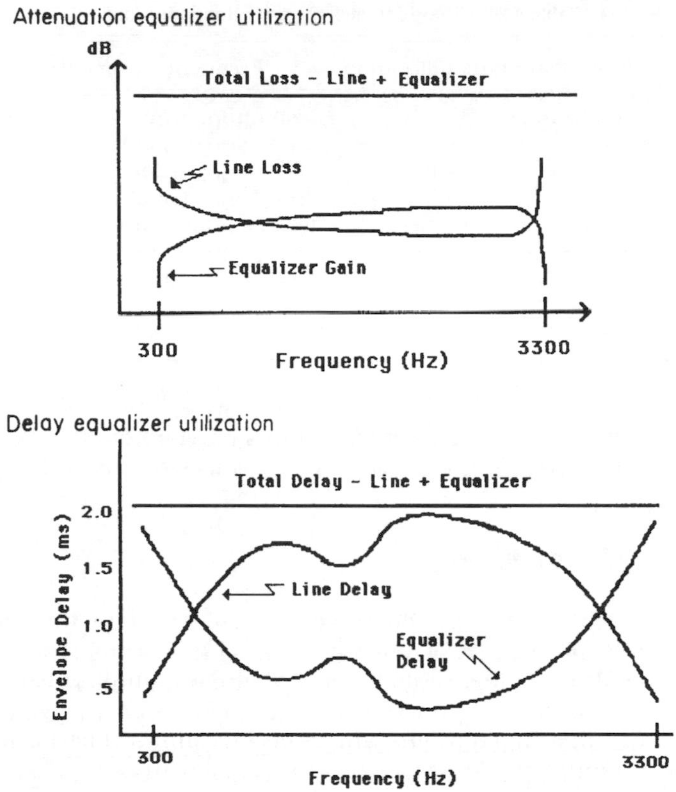

Figure 4.3 Effect of attenuation and delay equalizers used in C-type conditioning. C-type conditioning results in the use of attenuation and delay equalizers to provide near uniform loss and delay across the passband

attenuation between high and low frequencies as well as the increased attenuation at the edges of the passband, resulting in a uniform or near uniform level of attenuation across the passband. By making the signal loss near uniform, it becomes more practical to use amplifiers to boost overall signal power without adding to the distortion previously encountered by a signal.

Although all communications circuits will exhibit a degree of delay, it is important to flatten the delay time across the passband. Doing so reduces the potential of intersymbol interference.

Delay times can be made to approach a linear value through the use of a delay equalizer. Here the delay equalizer introduces a delay that is approximately the inverse of that exhibited by the channel, resulting in a relatively flat delay across the passband as illustrated in the lower portion of Figure 4.3.

To illustrate the need for delay equalization, consider a modem that uses tones at 1020 Hz and 2040 Hz to represent a binary 1 and binary 0, respectively. Without the use of a delay equalizer it becomes possible for a high frequency tone to be delayed more than a low frequency tone. If your modem transmitted tones at 2040 Hz and 1020 Hz to represent the sequence '10', it is entirely possible that the tone at 2040 Hz reaches the receiving

Table 4.2 C-type conditioning and attenuation loss

Channel/conditioning	Frequency range (Hz)	Loss limits (dB)
BASIC	300–500	−3 to +12
	500–2500	−2 to +8
	2500–3000	−3 to +12
C1	300–1000	−3 to +12
	1000–2400	−1 to +3
	2400–2700	−2 to +6
	2700–3000	−3 to +12
C2	300–500	−2 to +6
	500–2800	−1 to +3
	2800–3000	−2 to +6
C4	300–500	−2 to +6
	500–3000	−2 to +6
	3000–3200	−2 to +6
C5	300–500	−1 to +3
	500–2800	−0.5 to +1.5
	2800–3000	−1 to +3

modem after the tone transmitted at 1020 Hz, resulting in the receive modem interpreting the sequence '10' as '01.'

Although the ultimate goal of C-type conditioning is a near uniform delay and line loss across the passband, in actuality, communications carrier conditioning only guarantees certain limits of performance. Table 4.2 lists the attenuation loss for a basic unconditioned channel and four types of C-conditioning offered by U.S. carriers. Table 4.3 lists similar information for envelope delay distortion. In examining the entries in Table 4.2, note that the

Table 4.3 C-type conditioning and envelope distortion delay

Channel/conditioning	Frequency range (Hz)	Delay limits (μs)
BASIC	800–2600	1750
	800–1000	1750
	1000–2400	1000
	2400–2600	1750
C2	500–600	3000
	600–1000	1500
	1000–2600	500
	2600–2800	3000
C4	500–600	3000
	600–800	1500
	800–1000	500
	1000–2600	300
	2600–2800	500
	2800–3000	3000
C5	500–600	600
	600–1000	300
	1000–2600	100
	2600–2800	600

Table 4.4 D1-type conditioning parameters

	Parameter limits (dB)
Signal to C-notched noise ratio	28
Signal to second-order modulation products	35
Signal to third-order modulation products	40

loss limits for attenuation loss vary from negative to positive. In interpreting the loss limits, a negative value means less loss or a gain, while a positive value reflects signal loss. Both loss limits and delay limits are expressed in a range of tolerances for different types of C-type conditioning. Note also that these ranges decrease as the type of conditioning increases, with C5 conditioning providing the tightest range of values.

D-type conditioning

D-type conditioning is offered by some communications carriers for the control of signal to C-notched noise ratio and intermodulation distortion. Normally, D-type conditioning is applied to circuits that will support high speed transmission at 19.2 kbps or above. C-type and D-type conditioning can be combined on the same circuit to minimize the effect of line parameters on information flow. Table 4.4 lists examples of currently available D1-type conditioning parameters.

4.1.3 Dedicated line

A dedicated line is similar to a leased line in that the terminal is always connected to the device on the distant end, transmission always occurs on the same path and, if required, the line can be easily tuned to increase transmission performance.

The key difference between a dedicated and a leased line is that a dedicated line refers to a transmission medium internal to a user's facility, where the customer has the right of way for cable laying, whereas a leased line provides an interconnection between separate facilities. The term facility is usually employed to denote a building, office or industrial plant. Dedicated lines are also referenced as direct connect lines and normally link a terminal or business machine on a direct path through the facility to another terminal or computer located at that facility. The dedicated line can be a wire conductor installed by the employees of a company or by the computer manufacturer's personnel, or it can be a local line installed by the telephone company.

Normally, the only cost associated with a dedicated line in addition to its installation cost is the cost of the cable required to connect the devices that are to communicate with one another.

4.1.4 Switched network vs leased line economics

A leased line is billed on a monthly basis regardless of its level of utilization. In comparison, a switched network service, such as the use of the switched telephone network, is billed on a usage basis.

To illustrate an example of the manner by which you would compare the usage of the switched network with the use of a leased line, let's assume you are investigating an application that is estimated to require two hours of communications per day. Let's further assume there are 22 business days in the month, your long distance communications carrier's rate plan permits calls to be made any time during the day for 10 cents per minute, and a leased line used to connect the two locations would have a monthly cost of $300.

Because our estimate involves two hours of communications per business day and there are 22 business days per month, this results in 44 hours of communications per month. Based upon the use of a flat rate dialing plan that charges 10 cents per minute, the cost of using the switched network becomes:

$$44 \text{ hours/month} \times 60 \text{ min/hr} \times 10¢\text{/min} = \$264.00\text{/month}$$

If we compare the cost of anticipated switched network usage to the cost of a leased line, it is apparent that the cost of the switched network is more economical. However, what happens if we expect daily usage to increase to three hours per day?

Based upon communicating three hours per day, the total usage of the switched network becomes 3 hours/day × 22 days/month, or 66 hours per month. Continuing our use of a flat rate calling plan of 10¢/min, the monthly cost of the switched network becomes:

$$66 \text{ hours/month} \times 60 \text{ min/hr} \times 10¢\text{/min} = \$396.00\text{/month}$$

Note that the cost of using the switched network now exceeds the cost of a leased line. The preceding computations illustrate several key communications concepts. First, as usage of the switched network increases, it is possible to reach a point where the monthly cost of a leased line is more economical. Secondly, because the monthly cost of a leased line is fixed regardless of usage, many times a decision to install a leased line should be made even when it appears that the cost of using the switched network may be more economical. This is because if we are slightly wrong in our estimate of switched network usage, a leased line could be more economical and its expense is fixed, making its cost easier for budgeting purposes. In addition, if estimated usage is on the low side but the cost of using the switched network and a leased line are within close proximity of one another, we should probably select the usage of a leased line. This is because its use caps our communications cost.

To illustrate the use of rate tables, let's examine a second comparison of the use of the PSTN versus a leased line. For this example, assume that a personal computer located 50 miles from a mainframe has a requirement to communicate between 8 a.m. and 5 p.m. with the mainframe once each business day for a period of 30 minutes. Using the data in Table 4.1, each call would

Table 4.5 Line selection guide

Line type	Distance between transmission points	Speed of transmission	Use for transmission
Dedicated (direct connect)	Local	Limited by conductor	Short or long duration
Switched (dial-up)	Limited by telephone access availability	Normally less than 56 000 bps	Short duration transmission
Leased (private)	Limited by telephone company availability	Limited by type of facility	Long duration or short duration calls

cost $0.31 \times 1 + 0.19 \times 29$ or \$5.82. Assuming there are 22 working days each month, the monthly PSTN cost for communications between the PC and the mainframe would be $\$5.82 \times 22$ or \$128.04. If the monthly cost of a leased line between the two locations was \$250, it is obviously less expensive to use the PSTN for communications.

Suppose the communications application lengthened in duration to 2 hours per day. Then, from Table 4.1, the cost per call would become $0.31 \times 1 + 0.19 \times 119$ or \$22.92. Again assuming 22 working days per month, the monthly PSTN charge would increase to \$504.24, making the leased line more economical.

If data communications requirements to a computer involve occasional random contact from a number of terminals at different locations and each call is of short duration, dial-up service is thus normally employed. If a large amount of transmission occurs between a computer and a few terminals, leased lines are usually installed between the terminal and the computer.

Since an analog leased line is fixed as to its routing, it can be conditioned to obtain a uniform level of attenuation and delay distortion, as well as provide a minimum signal-to-noise ratio which will reduce errors in transmission. Normally, switched circuits are used for transmission at speeds up to 56 000 bps.

Some of the limiting factors involved in determining the type of line to use for transmission between terminal devices and computers are listed in Table 4.5.

4.2 TYPES OF SERVICE AND TRANSMISSION DEVICES

Digital devices which include terminals, mainframe computers and personal computers transmit data as unipolar digital signals, as indicated in Figure 4.4(a). When the distance between a terminal device and a computer is relatively short, the transmission of digital information between the two devices may be obtained by cabling the devices directly together. As the distance between the two devices increases, the pulses of the digital signals become distorted because of the resistance, inductance and capacitance of the cable used as a transmission medium. These impairments to signal flow result in attenuation and delay distortion that alter the shape of the received signal. At a certain distance between the two devices the pulses of the digital data will

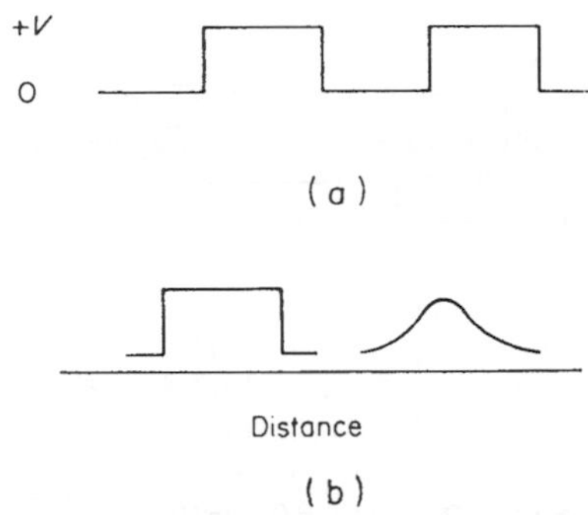

Figure 4.4 (a) Digital signaling. Digital devices to include terminals and computers transmit data as unipolar data signals. (b) Digital signal distortion. As the distance between the transmitter and receiver increases, digital signals become distorted because of the resistance, inductance and capacitance of the cable used as a transmission medium

distort, such that they are unrecognizable by the receiver, as illustrated in Figure 4.4(b). To extend the transmission distance between devices, specialized communications equipment must be employed, with the type of equipment used dependent upon the type of transmission medium employed.

4.2.1 Digital repeaters

Although digital signaling is considered to be a modern technology, its roots go back to early telegraph systems. In such systems, current flowed in the form of pulses to operate a sounder to emit the 'clicks' corresponding to the dots and dashes of Morse code. Since signals could not travel over long distances without being subjected to distortion, repeaters were employed to regenerate or recreate the transmitted signals. The modern digital repeater is a device that essentially scans the line looking for the occurrence of a pulse and then regenerates the pulse into its original form. Another name for the repeater is thus a data regenerator. As illustrated in Figure 4.5, a repeater extends the communications distance between terminal devices to include personal computers and mainframe computers.

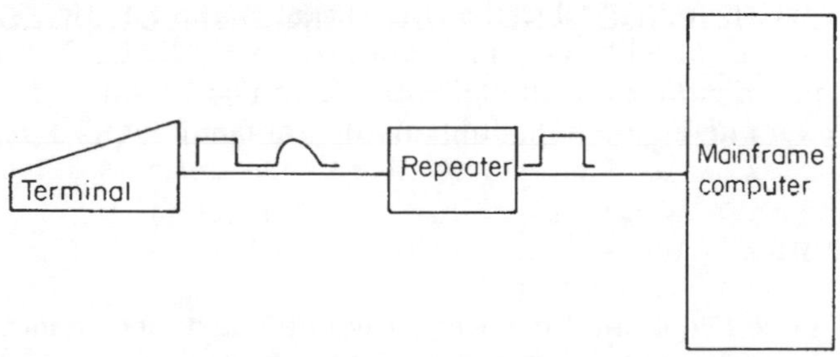

Figure 4.5 Transmitting data in digital format. To transmit data over long distances in digital format requires repeaters to be placed on the line to reconstruct the digital signals

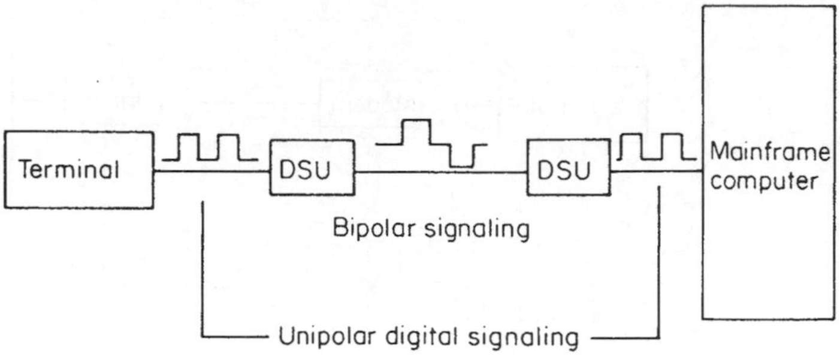

Figure 4.6 Transmitting data on a digital network. To transmit data on a digital network, the unipolar digital signals of terminal devices and computers must be converted into a bipolar signal

Modern digital transmission services permit data to be transmitted from source to destination in a modified digital form, alleviating the necessity of converting the signal into an analog form. Since repeaters are used throughout a digital network, this network provides users with a lower error rate than obtained through the use of an analog service. In comparison, amplifiers used in an analog network amplify the total signal, including any signal distortion.

Unipolar and bipolar signaling

In unipolar signaling, each binary one is represented by a positive voltage of a predefined duration based upon the data transmission rate, with lower rates represented by a wider duration pulse than higher data rates. In bipolar signaling, each binary one is represented by alternating voltage polarities of predefined duration; that is, a positive voltage used to represent the nth binary one will be followed by a negative voltage to represent the $(n + 1)$th binary one. This signaling technique eliminates dc voltage buildup, permitting repeaters to be spaced further apart from one another.

Since unipolar signaling results in a dc voltage buildup when transmitting over a long distance, digital networks require unipolar signals to be converted into a modified bipolar format for transmission on this type of network. This requires the installation at each end of the circuit of a device known as a data service unit (DSU) in the United States and a network terminating unit (NTU) in the United Kingdom. The utilization of DSUs for the transmission of data on a digital network is illustrated in Figure 4.6. Later in this chapter we will examine digital signaling and digital facilities in more detail.

4.2.2 Modems

Since telephone lines were originally designed to carry analog voice signals, the digital signals transmitted from a terminal to another digital device must be converted into a signal that is acceptable for transmission by the telephone line. To effect transmission between distant points, a data set or

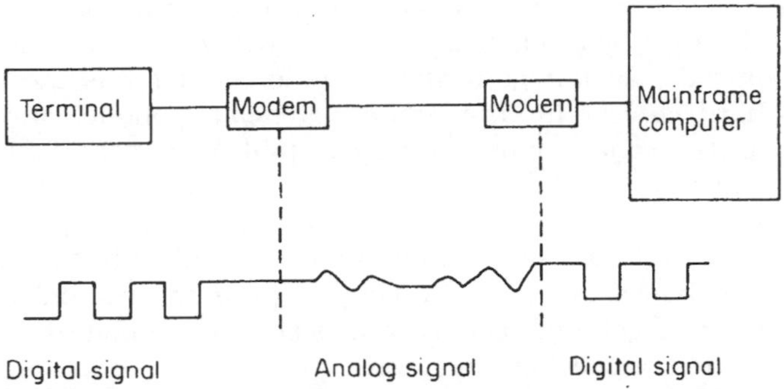

Figure 4.7 Signal conversion performed by modems. A modem converts (modulates) the digital signal produced by a terminal into an analog tone for transmission over an analog facility

modem is used. A modem is a contraction of the compound term 'modulator–demodulator' and is an electronic device used to convert the digital signals generated by computers and terminal devices into analog tones for transmission over telephone network analog facilities. At the receiving end, a similar device accepts the transmitted tones, reconverts them to digital signals and delivers these signals to the connected device.

In its simplest form of operation, a modem places a steady tone on the line which is called a carrier. The carrier frequency is normally near the center of the passband, as that is the area which is usually least affected by impairments. By itself the carrier tone conveys no information. To impress information on the carrier the modem modulates the carrier, altering its amplitude, frequency, phase or some combination of the three characteristics of an analog signal.

Signal conversion

Signal conversion performed by modems is illustrated in Figure 4.7. This illustration shows the interrelationship of terminals, mainframe computers and transmission lines when an analog transmission service is used. Both leased lines and switched lines employ analog service; therefore, modems can be used for transmission of data over both types of analog line connection. Although an analog transmission medium used to provide a transmission path between modems can be a direct connect, a leased or a switched line, modems are connected (hard-wired) to direct connect and leased lines, whereas they are interfaced to a switched facility. A terminal user can communicate with one distant location on a leased line, but he or she can communicate with many devices when there is access to a switched line.

4.2.3 Acoustic couplers

Although popular with data terminal users in the 1970s, today only a very small percentage of persons use acoustic couplers for communications. The acoustic coupler is a modem whose connection to the telephone line is obtained by acoustically coupling the telephone headset to the coupler. The primary advantage of the acoustic coupler was the fact that it requires no hard-wired connection to the switched telephone network, enabling terminals and personal computers to be portable with respect to their data transmission capability. Owing to the growth in modular telephone jacks, modems that interface the switched telephone network via a plug, in effect, are portable devices. Since some hotels and older office buildings still have hard-wired telephones, the acoustic coupler permits terminal and personal computer users to communicate regardless of the method used to connect a telephone set to the telephone network.

Signal conversion

The acoustic coupler converts signals generated by a terminal device into a series of audible tones, which are then passed to the mouthpiece or transmitter of the telephone and in turn onto the switched telephone network. Information transmitted from the device at the other end of the data link is converted into audible tones at the earpiece of the telephone connected to the terminal's acoustic coupler. The coupler then converts those tones into the appropriate electrical signals recognized by the attached terminal. The interrelationship of terminals, acoustic couplers, modems and analog transmission media is illustrated in Figure 4.8.

In examining Figure 4.8, note that a circle subdivided into four equal parts by two intersecting lines is used as the symbol to denote the switched telephone network or PSTN. This symbol will be used in the remainder of the book to illustrate communications occurring over this type of line connection.

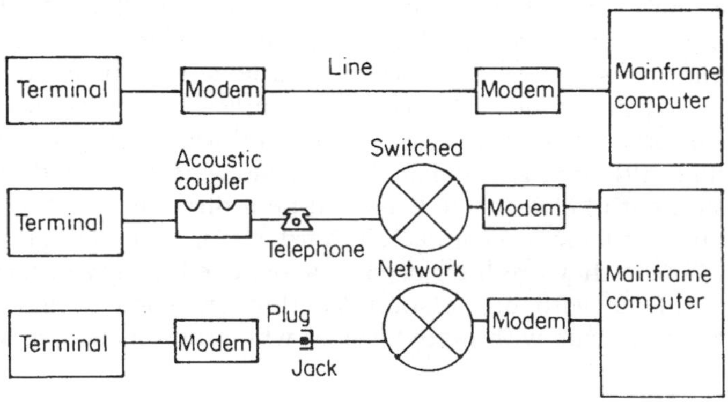

Figure 4.8 Interrelationship of terminals, modems, acoustic couplers, computers and analog transmission mediums. When using modems on an analog transmission medium, the line can be dedicated, leased or switched facility. Terminal devices can use modems or acoustic couplers to transmit via the switched network

4.2.4 Analog facilities

Several types of analog switched facility are offered by communications carriers. Each type of facility has its own set of characteristics and rate structure. Normally, for extensive communications requirements, an analytic study is conducted to determine which type or types of service should be utilized to provide an optimum cost-effective service for the user. The common types of analog switched facilities are direct distance dialing, wide area telephone service and foreign exchange service.

Direct distance dialing (DDD)

Direct distance dialing (DDD) permits the user to dial directly any telephone connected to the public switched telephone network. The dialed telephone may be connected to another terminal device or mainframe computer. The charge for this service, in addition to installation costs, may be a fixed monthly fee if no long distance calls are made, a message unit rate based upon the number and duration of local calls, or a fixed fee plus any long distance charges incurred. Depending on the time of day that a long distance call is initiated and its destination (intrastate or interstate), discounts from normal long distance tolls are available for selected calls made without operator assistance.

Wide area telephone service (WATS)

Introduced by AT&T for interstate use in 1961, wide area telephone service (WATS) is now offered by several communications carriers, including MCI and Sprint in the United States and British Telecom in England, with service at the later location referred to as Freephone. Its scope of coverage has been extended from the continental United States to Hawaii, Alaska, Puerto Rico, the US Virgin Islands and Europe, as well as selected Pacific and Asian countries.

Wide area telephone service (WATS) may be obtained in two different forms, each designed for a particular type of communications requirement. Outward WATS is used when a specific location requires placing a large number of outgoing calls to geographically distributed locations. Inward WATS service provides the reverse capability, permitting a number of geographically distributed locations to communicate with a common facility. Calls on WATS are initiated in the same manner as a call placed on the public switched telephone network. Instead of being charged on an individual call basis, however, the user of WATS facilities pays a flat rate per block of communications hours per month occurring during weekday, evening and night and weekend time periods.

Access line

A voice-band trunk called an access line is provided to the WATS user. This line links the facility to a telephone company central office. Other than cost considerations and certain geographical calling restrictions which are functions of the service area of the WATS line, the WATS subscriber may place as many calls as desired on this trunk if the service is outward WATS or receive as many calls as desired if the service is inward. Inward WATS, the well known '800' area code, permits remotely located personnel to call your facility toll-free from the service area provided by the particular inward WATS-type of service selected.

The popularity of the inward WATS service resulted in the use of all available 800 area code numbers a few years ago. This resulted in the addition of the 888 area code, followed shortly thereafter by the use of the 887 area code for WATS service in an attempt to satisfy consumer demand for this bulk rate discount service.

The charge for WATS is a function of the service area. This can be intrastate WATS, a group of states bordering the user's state where the user's main facility is located, a grouping of distant states, or International WATS which extends the inbound 800 service to the United States from selected overseas locations.

Another service very similar to WATS is AT&T's 800 READYLINE^{SM} service. This service is essentially similar to WATS; however, calls can originate or be directed to an existing telephone in place of the access line required for WATS service.

During the late 1990s AT&T began to refer to WATS service as 'Toll-Free Service' and announced the addition of several routing plans that enable toll free calls to be dynamically routed to different organizational call centers based upon predefined times. For example, a national organization that has call centers located east and west of the Mississippi could arrange for calls originating on the east coast to be routed to the call center located in Los Angeles after 5 p.m. to provide an additional three hours of coverage without having to hire additional employees.

Figure 4.9 illustrates an example of how a communications carrier could provide different levels of WATS service to a subscriber located in Georgia. In this example customers in Georgia can select one of four service areas, with service area one representing adjacent states. If this service area is selected and a user in Georgia requires inward WATS service, he or she will pay for toll-free calls originating in the states surrounding Georgia: Florida, Alabama, Mississippi, Tennessee, Kentucky South Carolina and North Carolina. Similarly, if outward WATS service is selected for service area one, a person in Georgia connected to the WATS access line will be able to dial all telephones in the states previously mentioned. The states comprising a service area vary, based upon the state in which the WATS access line is installed. Thus, the states in service area one when an access line is in New York would obviously differ from the states in a WATS service area one when the access line is in Georgia.

In general, since WATS is a service based on volume usage, its cost per hour is less than the cost associated with the use of the PSTN for long distance calls. One common application for the use of WATS facilities is to install one or more inward WATS access lines at a data processing center. Then terminal and personal computer users distributed over a wide geographical area can use the inward WATS facilities to access the computers at the data processing center.

Figure 4.9 AT&T WATS service area one access line located in Georgia

Since International 800 service enables employees and customers of US companies to call toll-free from foreign locations, this service may experience a considerable amount of data communications usage This usage can be expected to include applications requiring access to such databases as hotel and travel reservation information, as well as order entry and catalog sales data updating. Persons traveling overseas with portable personal computers, as well as office personnel using terminals and personal computers in foreign countries, who desire to access computational facilities and information utilities located in the United States represent common International 800 service users. Due to the business advantage of WATS its concept has been implemented in several foreign countries, with inward WATS in the United Kingdom marketed under the term Freefone.

Foreign exchange

Foreign exchange (FX) service may provide a method of transmission from a group of terminal devices remotely located from a central computer facility at less than the cost of direct distance dialing. An FX line can be viewed as a mixed analog switched and leasing line. To use an FX line, a user dials a local number which is answered if the FX line is not in use. From the FX, the information is transmitted via a dedicated voice line to a permanent connection in the switching office of a communications carrier near the facility with which communication is desired. A line from the local switching office which terminates at the user's home office is included in the basic FX service. This is illustrated in Figure 4.10.

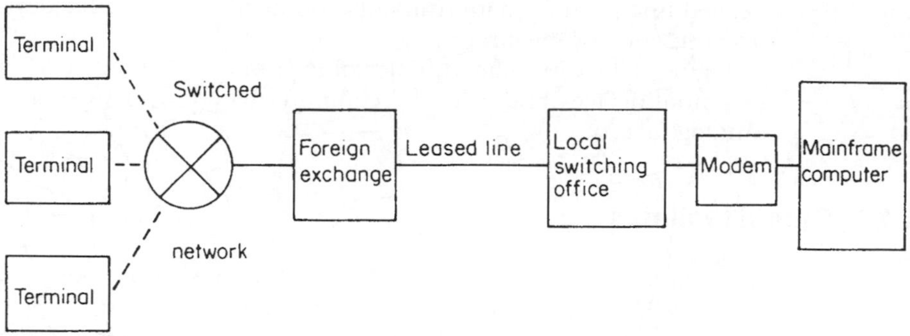

Figure 4.10 Foreign exchange (FX) service. A foreign exchange line permits many terminal devices to use the facility on a scheduled or on a contention basis

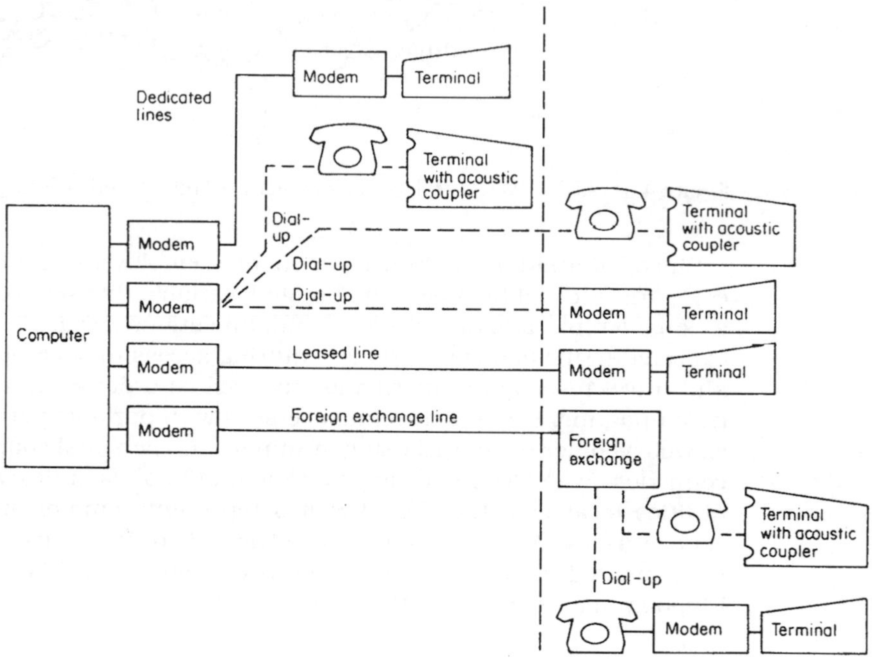

Figure 4.11 Terminal-to-computer connections via analog mediums. A mixture of dedicated, dial-up, leased and foreign exchange lines can be employed to connect local and remote terminals to a central computer

The use of an FX line permits the elimination of long distance charges that would be incurred by users directly dialing the distant computer facility. Since only one person at a time may use the FX line, normally only groups of users whose usage can be scheduled are suitable for FX utilization.

With respect to data communications, the major difference between an FX line and a leased line is that any terminal dialing the FX line provides the second modem required for the transmission of data over the line, whereas a leased line used for data transmission normally has a fixed modem attached at both ends of the circuit.

Figure 4.11 illustrates the possible connections between remotely located terminal devices and a central computer where transmission occurs over an analog facility.

4.2.5 Digital facilities

In addition to analog service, numerous digital service offerings were implemented by communications carriers over the last decade. Using a digital service, data is transmitted from source to destination in digital form without the necessity of converting the signal into an analog form for transmission over analog facilities, as is the case when modems or acoustic couplers are interfaced to analog facilities.

To understand the ability of digital transmission facilities to transport data requires an understanding of digital signaling techniques. Those techniques provide a mechanism to transport data end-to-end in modified digital form from locations hundreds or thousands of miles apart.

4.2.6 Digital signaling

Digital signaling techniques have evolved from use in early telegraph systems to provide communications for different types of modern technology, ranging in scope from the data transfer between a terminal and modem to the flow of data on a LAN and the transport of information on high speed wide area network digital communications lines. Instead of one signaling technique, numerous techniques are used, with each technique developed to satisfy different communications requirements. In this section we will focus our attention on digital signaling used on wide area network transmission facilities, deferring a discussion of LAN signaling until later in this book.

Unipolar non-return to zero

Unipolar non-return to zero (NRZ) is a simple type of digital signaling which was originally used in telegraphy. Today, unipolar non-return to zero signaling is used in computers as well as in the common RS-232/V.24 interface between data terminal equipment and data communications equipment.

Figure 4.12 illustrates an example of unipolar non-return to zero signaling. In this signaling scheme, a dc current or voltage represents a mark, and the

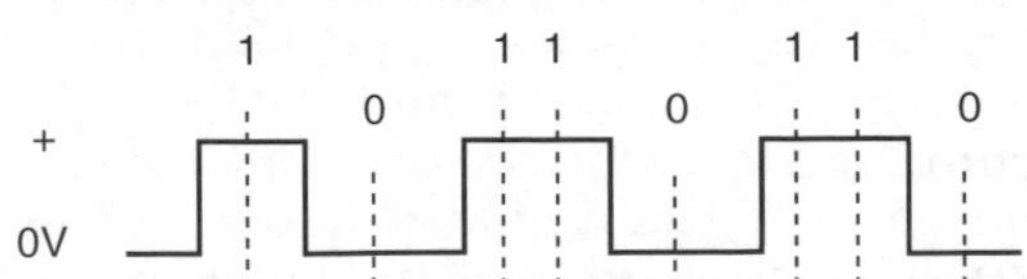

Figure 4.12 Unipolar non-return to zero signaling

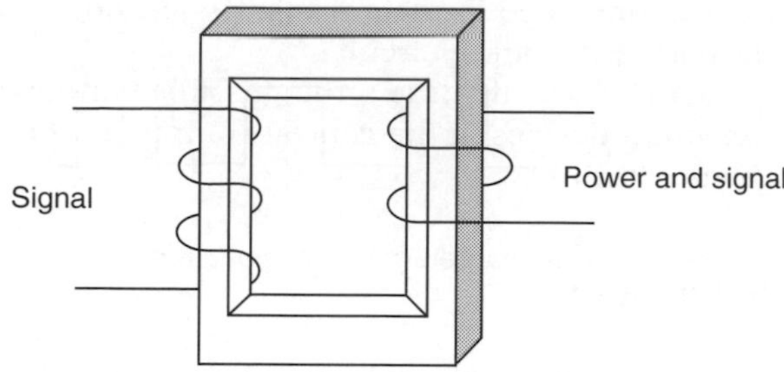
Signal
Power and signal
A signaling technique which does not produce residual dc enables the use of transformer coupling to separate the signal from power

Figure 4.13 Transformer coupling

absence of current or voltage represents a space. Since voltage or current does not return to zero between adjacent set bits, this signaling technique is called non-return to zero. Because voltage or current is only varied from 0 to some positive level the pulses are unipolar, hence the name unipolar non-return to zero.

There are several problems associated with unipolar non-return to zero signaling which make it unsuitable for use as a signaling mechanism on wide area network digital transmission facilities. Those problems include the need to sample the signal and the fact that it provides residual dc voltage buildup.

Since two or more repeated marks or spaces can stay at the same voltage or current level, sampling is required to distinguish one bit value from another. The ability to sample requires clocking circuitry which drives up the cost of communications. A second problem related to the fact that a sequence of marks or set bits can occur is that this condition results in the presence of residual dc levels. Residual dc requires the direct attachment of transmission components while the absence of residual dc permits ac coupling via the use of a transformer. When communications carriers engineered their early digital networks they were based on the use of copper conductors, as fiber optics did not exist. Communications carriers, like persons, attempt to do things in an economical manner. Rather than install a separate line to power repeaters they examined the possibility of carrying both power and data on a common line, removing the data from the power at the distant end, as illustrated in Figure 4.13. To accomplish this required transformer coupling at the distant end, which is only possible if residual dc is eliminated. Thus, communications carriers began to search for an alternative signaling method.

Unipolar return to zero

One of the first alternative signaling methods examined was unipolar return to zero (RTZ). With this signaling technique which is illustrated in Figure 4.14,

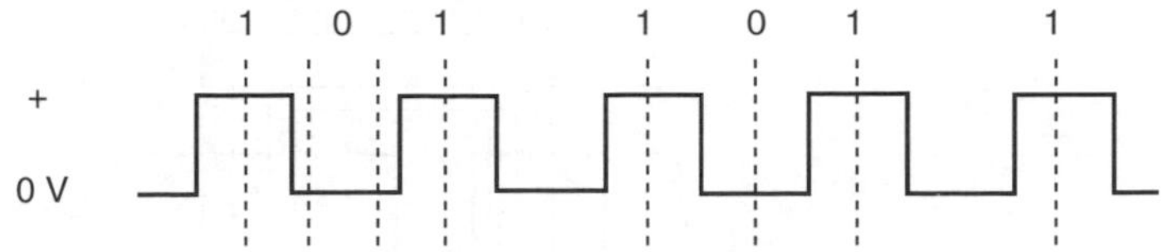

Figure 4.14 Unipolar return to zero signaling

the current or voltage always returns to zero after every '1' bit. Although this signal is easier to sample since each mark has a pulse rise, it still results in residual dc buildup, and was unsuitable for use as the signaling mechanism on communications carrier digital transmission facilities.

Bipolar return to zero

After examining a variety of signaling methods, communications carriers focused their attention on a technique known as bipolar return to zero. Under bipolar signaling alternating polarity pulses are used to represent marks, and a zero level pulse is used to represent a space. In the bipolar return to zero signaling method the bipolar signal returns to zero after each mark. Figure 4.15 illustrates an example of bipolar return to zero signaling.

The key advantage of bipolar return to zero signaling is the fact that it precludes dc voltage buildup on the line. This enables both power and data to be carried on the same line, enabling repeaters to be powered by a common line. In addition, repeaters can be placed relatively far apart in comparison to other signaling techniques, which reduces the cost of developing the digital transmission infrastructure.

On modern wide area network digital transmission facilities a modified form of bipolar return to zero signaling is employed. That modification involves the placement of pulses in their bit interval so that they occupy 50% of the interval, with the pulse centered at the center of the interval. This positioning eliminates high frequency components of the signal that can interfere with other transmissions, and it results in a bipolar pulse known as a 50% duty cycle alternate mark inversion (AMI). An example of this pulse is illustrated in Figure 4.16. Now that we have a general appreciation for the type of digital signaling used on digital transmission facilities and the rationale for the use of that type of signaling, let us discuss some of the types of digital transmission facilities available for use.

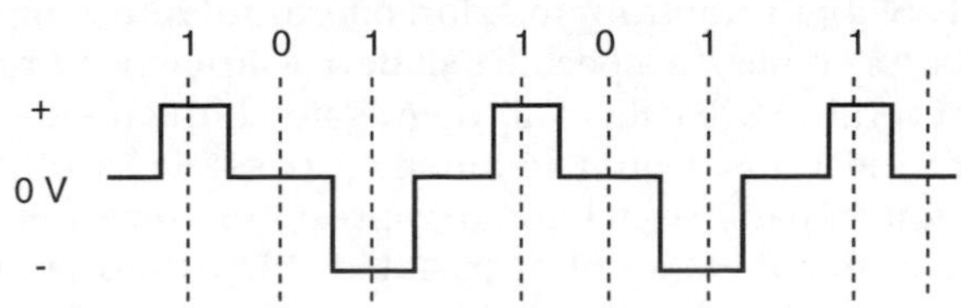

Figure 4.15 Bipolar return to zero

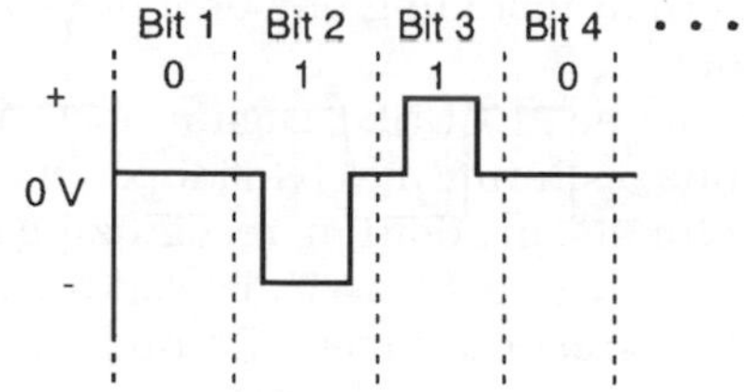

Figure 4.16 Bipolar (AMI) RTZ 50% duty cycle

4.2.7 Representative AT&T digital offerings

In the United States, AT&T offers several digital transmission facilities under the ACCUNET℠ Digital Service service mark. The Dataphon® Digital Service was the charter member of the ACCUNET family and is deployed in 100 major metropolitan cities in the United States as well as having an interconnection with Canada's digital network as well as digital networks in other countries. Dataphone Digital Service operates at synchronous data transfer rates of 2.4, 4.8, 9.6, 19.2 and 56 kilobits per second (kbps), providing users of this service with dedicated, two-way simultaneous transmission capability.

Originally all AT&T digital offerings were leased line services, where a digital leased line is similar to a leased analog line in that it is dedicated to full time use by one customer. In the late 1980s, AT&T introduced its Accunet Switched 56 service, a dial-up 56 kbps digital data transmission service. This service was originally promoted to enable users to maintain a dial-up backup for previously installed 56 kbps AT&T Dataphone Digital Services leased lines. Since then AT&T and other communications carriers have considerably expanded the types of digital transmission facilities available for commercial usage. Today AT&T offers switched 56 kbps, switched 384 kbps and even switched T1 service, the latter a technology originally developed to enable 24 digitized voice conversations to be transported on a common digital circuit operating at 1.544 Mbps.

Another type of switched service that has gained in popularity due to its expanded support by telephone companies during the late 1990s is Integrated Services Digital Network (ISDN). A basic rate ISDN connection provides two bearer (B) channels, each operating at 64 kbps, and a data (D) channel operating at 16 kbps over one telephone line, with the D channel used primarily for call control. Each B channel can support a digitized voice conversation or the transmission of data, and the two B channels can be aggregated to provide a 128 kbps data transmission operating rate. One popular business use of ISDN is to employ a specialized device known as an inverse multiplexer to dial several ISDN calls and aggregate the transmission over two or more ISDN connections. Doing so makes it possible to achieve an $n \times 64$ kbps transmission capability, where n represents the number of ISDN B-channel calls. One popular use of aggregated ISDN is to support videoconferencing. If an organization requires videoconferencing only on a periodic basis, it is often less expensive to use a switched digital transmission service billed on a usage basis than installing one or more leased lines between videoconferencing locations.

Under the name ACCUNET Digital Spectrum of Services AT&T offers leased digital circuits operating at 1.544 Mbps and 44 Mbps, commonly referred to as T1 and T3 circuits. In addition, recognizing that many firms cannot justify the cost associated with a T1 or T3 leased line, AT&T markets fractions of those leased lines as fractional T1 (FTl) and fractional T3 (FT3) service. Other communications carriers such as MCI Worldcom and Sprint provide equivalent transmission facilities. For example, under the tradename Clearline Sprint provides FT1 service in increments of 64 kbps up to a full T1.

European offerings

In Europe, a number of countries established digital transmission facilities. One example of such offerings is British Telecom's KiloStream service. KiloStream provides synchronous data transmission at 2.4, 4.8, 9.6, 48 and 64 kbps and is very similar to AT&T's Dataphone Digital Service. Each KiloStream circuit is terminated by British Telecom with a network terminating unit (NTU), which is the digital equivalent of the modem required on an analog circuit. In comparison, Dataphone Digital Service users can terminate their digital facilities with either a data service unit or a channel service unit. The European equivalent of a the North American T1 circuit is an E1 line which operates at 2048 Mbps. Later in this book we will examine both types of digital circuit in considerable detail.

Data service units (DSUs)

A data service unit (DSU) provides a standard interface to a digital transmission servlce and handles such functions as signal translation, regeneration, reformatting and timing. The low speed DSU is designed to operate at one of five speeds: 2.4, 4.8, 9.6, 19.2 and 56 kbps. The transmitting portion of the DSU processes the customer's signal into bipolar pulses suitable for transmission over the digital facility. The receiving portion of the DSU is used both to extract timing information and to regenerate mark and space information from the received bipolar signal. A second interface arrangement is called a channel service unit (CSU) and was originally provided by the communication carrier to those customers who wish to perform the signal processing to and from the bipolar line, as well as to retime and regenerate the incoming line signals through the utilization of their own equipment. At data rates below 56/64 kbps most communications manufacturers incorporate both DSU and CSU functions into a common device now generally referred to as a DSU, even though that term is technically incorrect. At data rates above 64 kbps the DSU function is commonly built into data terminal equipment devices and separate CSUs are required to serve as an interface between the digital line and the equipment used to transmit and receive data.

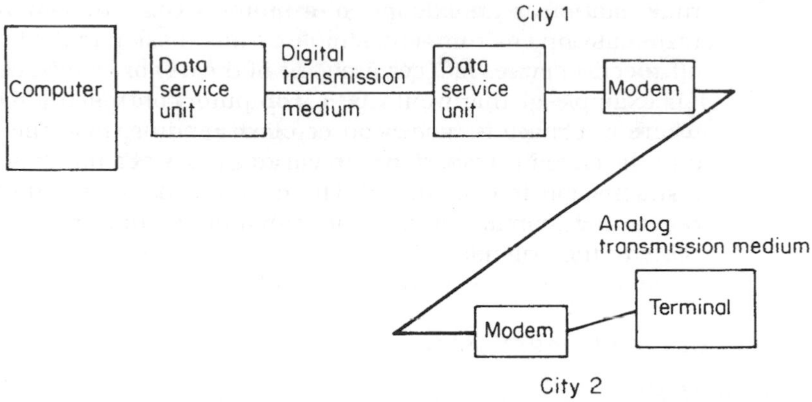

Figure 4.17 Analog extension to access a digital service. Although data are transmitted in digital form from the computer to city 1, they must be modulated by the modem at that location for transmission over the analog extension

As data are transmitted over digital facilities, the signal is regenerated by the communications carrier numerous times prior to its arrival at its destination. In general, digital service gives data communications users improved performance and reliability when compared to analog service, owing to the nature of digital transmission and the design of digital networks. This improved performance and reliability is due to the fact that digital signals are regenerated, whereas, when analog signals are amplified, any distortion to the analog signal is also being amplified.

Although digital service is offered in many locations, for those locations outside the serving area of a digital facility the user will have to employ analog devices as an extension in order to interface to the digital facility. The utilization of digital service via an analog extension is illustrated in Figure 4.17. As depicted in Figure 4.17, if the closest city to the terminal located in city 2 that offers digital service is city 1, then to use digital service to communicate with the computer an analog extension must be installed between the terminal location and city 1. In such cases, the performance, reliability and possible cost advantages of using digital service may be completely dissipated.

4.3 TRANSMISSION MODE

One method of characterizing lines, terminal devices, computers and modems is by their transmission or communications mode. The three classes of transmission modes are simplex, half-duplex and full-duplex.

4.3.1 Simplex transmission

Simplex transmission is that transmission which occurs in one direction only, disallowing the receiver of information a means of responding to the transmission. A home AM radio which receives a signal transmitted from a radio station is an example of a simplex communications mode. In a data transmission environment, simplex transmission might be used to turn on or off specific devices at a certain time of the day or when a certain event occurs. An example of this would be a computer-controlled environmental system where a furnace is turned on or off depending upon the thermostat setting and the current temperature in various parts of a building. Normally, simplex transmission is not utilized where human–machine interaction is required due to the inability to turn the transmitter around so that the receiver can reply to the originator.

4.3.2 Half-duplex transmission

Half-duplex transmission permits transmission in either direction; however, transmission can occur in only one direction at a time. Half-duplex transmission is used in citizen band (CB) radio transmission where the operator can either transmit or receive but cannot perform both functions at the same time on the same channel. When the operator has completed a transmission, the other party must be advised that he or she has finished transmitting and is ready to receive by saying the term 'over'. Then the other operator can begin transmission.

When data are transmitted over the telephone network, the transmitter and the receiver of the modem or acoustic coupler must be appropriately turned on and off as the direction of the transmission varies. Both simplex and half-duplex transmission require two wires to complete an electrical circuit. The top of Figure 4.18 illustrates a half-duplex modem interconnection and the lower portion of that illustration shows a typical sequence of events in the terminal's sign-on process to access a computer. In the sign-on process, the user might first transmit the word NEWUSER to inform the computer that

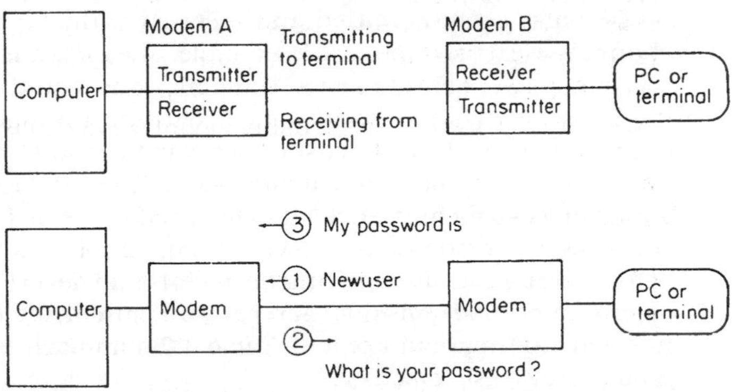

Figure 4.18 Half-duplex transmission. Top: Control signals from the computer and terminal operate the transmitter and receiver sections of the attached modems. When the transmitter of modem A is operating, the receiver of modem B operates; when the transmitter of modem B operates, the receiver of modem A operates. Only one transmitter, however, operates at any one time in the half-duplex mode of transmission. Bottom: during the sign-on sequence, transmission is turned around several times

a new user wishes to establish a connection. The computer responds by asking for the user's password, which is then furnished by the user.

In the top portion of Figure 4.18, when data is transmitted from a computer to a terminal, control signals are sent from the computer to modem A which turns on the modem A transmitter and causes the modem B receiver to respond. When data are transmitted from the terminal to the computer, the modem B receiver is disabled and its transmitter is turned on while the modem A transmitter is disabled and its receiver becomes active. The time necessary to effect these changes is called transmission turnaround time, and during this interval transmission is temporarily halted, which affects data throughput. Half-duplex transmission can occur on either a two-wire or four-wire circuit. The switched network is a two-wire circuit, whereas leased lines can be obtained as either two-wire or four- wire links. The terms two-wire and four-wire refer to the physical number of wires leaving the subscriber's premises and routed to the communication carrier's end office. A four-wire service provides two separate signal paths and is essentially a pair of two-wire links which can be used for transmission in both directions simultaneously. This type of transmission is called the full-duplex mode.

4.3.3 Full-duplex transmission

Although you would normally expect full-duplex transmission to be accomplished over a four-wire connection that provides two two-wire paths, full-duplex transmission can also occur on a two-wire connection. This is accomplished by the use of modems that subdivide the frequency bandwidth of the two-wire connection into two distinct channels, permitting simultaneous data flow in both directions on a two-wire circuit. This technique will be examined and explained in more detail in Chapter 8, when the operating characteristics of modems are examined in detail.

Full-duplex transmission is often used when large amounts of alternate traffic must be transmitted and received within a fixed time period. If two channels were used in our CB example, one for transmission and another for reception, two simultaneous transmissions could be effected. While full-duplex transmission provides more efficient throughput, this efficiency was originally negated by the cost of two-way lines and more complex equipment required by this mode of transmission. The development of low cost digital signal processor chips enabled high speed modems to operate in a full-duplex transmission mode on a two-wire circuit through a technique referred to as echo cancellation, This technique will also be described when we discuss the operation of modems in Chapter 8. In Figure 4.19, the three types of transmission modes are illustrated, Table 4.6 summarizes the three transmission modes previously discussed.

Note that the column that includes the term 'CCITT' in Table 4.6 refers to the Consultative Committee on International Telephone and Telegraph. The CCITT operates as part of the International Telecommunications Union (ITU), which is a United Nations agency. Until the mid-1980s CCITT modem standards were primarily followed in Europe and resulted in compatibility problems when persons attempted to communicate between continents.

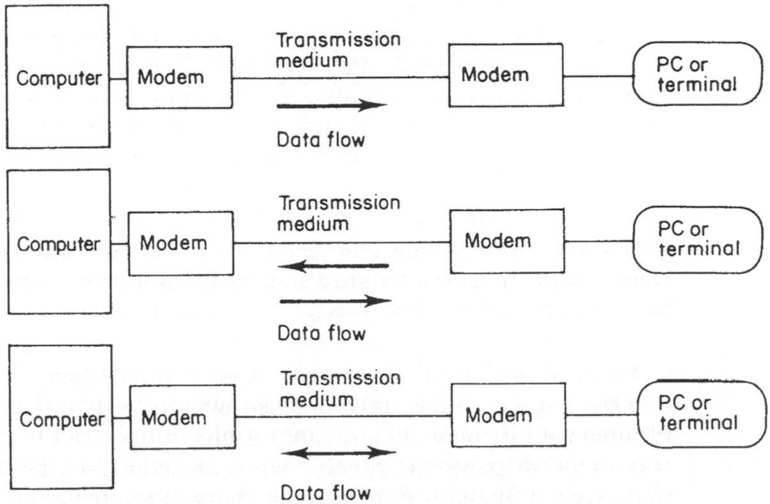

Figure 4.19 Transmission modes. Top: Simplex transmission is in one direction only, transmission cannot reverse direction. Center: Half-duplex transmission permits transmission in both directions but only one way at a time. Bottom: full-duplex transmission permits transmission in both directions simultaneously

Table 4.6 Transmission mode comparison

Symbol	ANSI	US telecommunications industry	ITU/ CCITT	Historical physical line requirement
←	One-way only	Simplex		Two-wire
⇆	Two-way alternate	Half-duplex (HDX)	Simplex	Two-wire
↔	Two-way simultaneous	Full-duplex (FDX)	Duplex	Two-wire

Fortunately, most modems today are manufactured to CCITT (now ITU) standards, which has to a large degree eliminated the compatibility problems that can occur between US and European-manufactured modems.

4.3.4 Terminal and mainframe computer operating modes

When referring solely to terminal operations, the terms half-duplex and full-duplex operation take on meanings different from the communications mode of the transmission medium. Vendors commonly use the term 'half-duplex' to denote that the terminal device is in a local copy mode of operation. This means that each time a character is pressed on the keyboard it is printed or displayed on the local terminal as well as transmitted. A terminal device operated in a half-duplex mode would thus have each character printed or displayed on its monitor as it is transmitted.

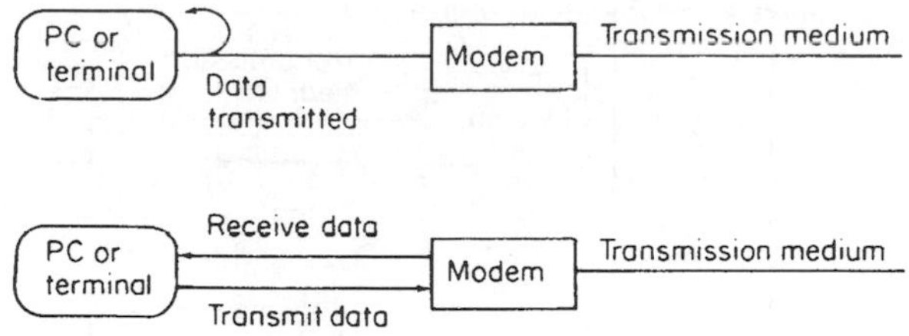

Figure 4.20 Terminal operation modes. Top: the term half-duplex terminal operation implies that data transmitted are also printed on the local terminal. This is known as local copy. Bottom: the term full-duplex terminal operation implies that no local copy is provided

When we say that a terminal is in a full-duplex mode of operation this means that each character pressed on the keyboard is transmitted but not immediately displayed or printed. Here the device on the distant end of the transmission path must 'echo' the character back to the originator, which, upon receipt displays or prints the character. A terminal in a full-duplex mode of operation would thus only print or display the characters pressed on the keyboard after the character is echoed back by the device at the other end of the line.

Figure 4.20 illustrates the terms full- and half-duplex as they apply to terminal devices. Note that although most conventional terminals have a switch to control the duplex setting of the device, personal computer users normally obtain their duplex setting via the software program they are using. Thus, the term 'echo on' during the initialization of a communications software program would refer to the process of displaying each character on the user's screen as it is transmitted.

When we refer to half- and full-duplex with respect to mainframe computer systems we are normally referring to whether or not they echo received characters back to the originator. A half-duplex computer system does not echo characters back, while a full- duplex computer system echoes each character it receives.

Different character displays

When considering the operating mode of a terminal device, the transmission medium and the operating mode of the computer on the distant end of the transmission path as an entity, three things could occur in response to each character pressed on a keyboard. Assuming that a transmission medium is employed that can be used for either half- or full-duplex communications, a terminal device could print or display no character for each character transmitted, one character for each character transmitted, or two characters for each character transmitted. Here the resulting character printed or displayed would be dependent upon the operating mode of the terminal device and the host computer you are connected to, as indicated in Table 4.7.

To understand the character display column in Table 4.7, let us examine the two-character display result caused by the terminal device operating in a half-duplex mode while the distant computer operates on a full-duplex mode.

Table 4.7 Operating mode and character display

	Operating mode	
Terminal device	Host computer	Character display
Half-duplex	Half-duplex	1 character
Half-duplex	Full-duplex	2 characters
Full-duplex	Half-duplex	No characters
Full-duplex	Full-duplex	1 character

When the terminal is in a half-duplex mode it echoes each transmitted character onto its printer or display. At the other end of the communications path, if the computer is in a full-duplex mode of operation it will echo the received character back to the terminal, causing a second copy of the transmitted character to be printed or displayed. Two characters should then appear on your printer or display for each character one transmits. To alleviate this situation, you would change the transmission mode of the terminal to full-duplex. This would normally be accomplished by turning 'echo' off during the initialization of a communications software program, if using a personal computer; or you would turn a switch to half-duplex if operating a conventional terminal.

4.4 TRANSMISSION TECHNIQUES

Data can be transmitted either asynchronously or synchronously. Asynchronous transmission is commonly referred to as a start–stop transmission where one character at a time is transmitted or received. Start and stop bits are used to separate characters and synchronize the receiver with the transmitter, thus providing a method of reducing the possibility that data becomes garbled.

Most devices designed for human–machine interaction that are teletype compatible transmit data asynchronously. By teletype compatible, we refer to terminals and personal computers that operate similar to the Teletype™ terminal manufactured by Western Electric, a subsidiary of AT&T. Various versions of this once popular terminal were manufactured for over 30 years and at one time it had an installed base of approximately one million such terminals in operation worldwide. As characters are depressed on the device's keyboard they are transmitted to the computer, with idle time occurring between the transmission of characters. This is illustrated in the top of Figure 4.21.

4.4.1 Asynchronous transmission

In asynchronous transmission, each character to be transmitted is encoded into a series of pulses. The transmission of the character is started by a start pulse equal in length to a code pulse. The encoded character (series of pulses) is followed by a stop pulse that may be equal to or longer than the code pulse, depending upon the transmission code used.

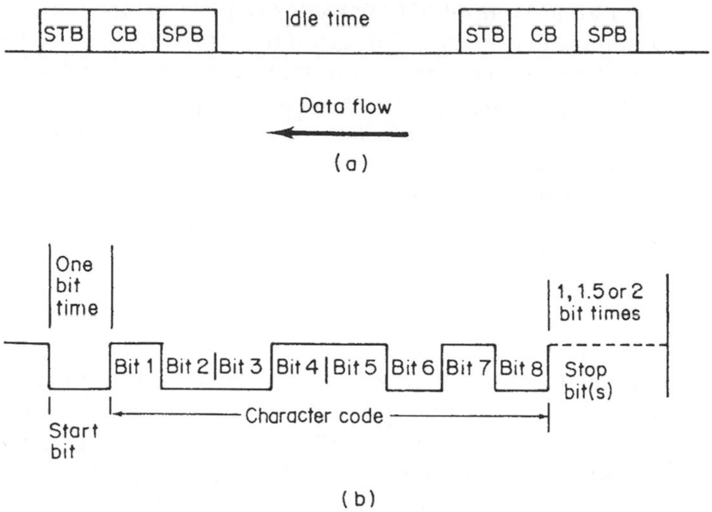

Figure 4.21 Asynchronous (start–stop) transmission. (al Transmission of many characters. STB = start bit; CB = character bits; SPB = stop bit(s); idle time is between character transmission. (b) Transmission of one 8-bit character

The start bit represented a transition from a mark to a space. Since in an idle condition when no data are transmitted the line is held in a marking condition, the start bit serves as an indicator to the receiving device that a character of data follows. Similarly, the stop bit causes the line to be placed back into its previous 'marking' condition, signifying to the receiver that the data character is complete as well as enabling a start bit to provide a marking to spacing line transition.

As illustrated in the lower portion of Figure 4.21, on of an 8-bit character requires either 10 or 11 bits, depending upon the length of the stop bit. In actuality the eighth bit may be used as a parity bit for error detection and correction purposes. The use of the parity bit is described in detail in Chapter 10. In the start–stop mode of transmission, transmission starts anew on each character and stops after each character. This is indicated in the upper portion of Figure 4.21. Since synchronization starts anew with each character, any timing discrepancy is cleared at the end of each character, and synchronization is maintained on a character-by-character basis. Asynchronous transmission is normally used for transmission at speeds up to 56 000 bps over the switched telephone network, and data rates up to 64 kbps or higher are possible over a direct connect cable whose distance is limited to 50 feet.

The term 'asynchronous TTY' or 'TTY compatible' refers to the asynchronous start–stop protocol employed originally by Teletype[1] terminals and is the protocol in which data are transmitted on a line-by-line basis between a terminal device and a mainframe computer. In comparison, more modern terminals and personal computers with cathode ray tube (CRT) displays are usually designed to transfer data on a full screen basis.

Personal computer users only require an asynchronous communications adapter and a software program that transmits and receives data on a line-by-line basis to connect to a mainframe that supports asynchronous TTY compatible terminals. Here the software program that transmits and receives data on a line-by-line basis is normally referred to as a TTY emulator program and it is the most common type of communications program written for use with personal computers. Since a personal computer includes a video display onto which characters and graphics can be positioned, the PC can be used to emulate a full screen addressable terminal. Thus, with appropriate software or a combination of hardware and software you can use a personal computer as a replacement for proprietary terminals manufactured to operate with a specific type of mainframe computer as well as to perform such local processing as spreadsheet analysis and word processing functions.

4.4.2 Synchronous transmission

A second type of transmission involves sending a grouping of characters in a continuous bit stream. This type of transmission is referred to as synchronous or bit-stream synchronization.

In the synchronous mode of transmission, modems or other communications devices located at each end of the transmission medium normally provide a timing signal or clock that is used to establish the data transmission rate and enable the devices attached to the modems to identify the appropriate bits in each character as they are being transmitted or received. In some instances, timing may be provided by the terminal device itself or a communication component, such as a multiplexer or front-end processor channel. No matter what timing source is used, prior to beginning the transmission of data the transmitting and receiving devices must establish synchronization among themselves. In order to keep the receiving clock in step with the transmitting clock for the duration of a stream of bits that may represent a large number of consecutive characters, the transmission of the data is preceded by the transmission of one or more special characters. These special synchronization or 'syn' characters are at the same code level (number of bits per character) as the coded information to be transmitted. They have a unique configuration of zero and one bits which are, however, interpreted as the syn character. Once a group of syn characters is transmitted, the receiver recognizes and synchronizes itself onto a stream of those syn characters.

After synchronization has been achieved, then actual data transmission can proceed. Synchronous transmission is illustrated in Figure 4.22. In synchronous transmission, characters are grouped or blocked into groups of characters, requiring a buffer or memory area so characters can be grouped together. In addition to having a buffer area, more complex circuitry is required for synchronous transmission since the receiving device must remain in phase with the transmitter for the duration of the transmitted block of information. Synchronous transmission is normally used for data transmission rates in excess of 2000 bps. The major characteristics of asynchronous and synchronous transmission are denoted in Table 4.8.

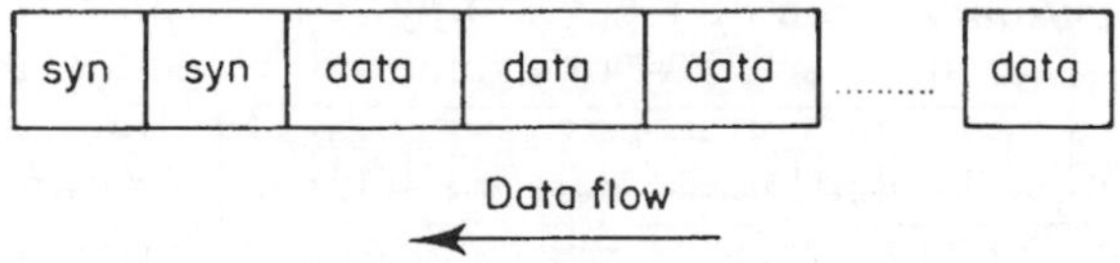

Figure 4.22 Synchronous transmission. In synchronous transmission, one or more syn characters are transmitted to establish clocking prior to the transmission of data

Table 4.8 Transmission technique characteristics

Asynchronous
1. Each character is prefixed by a start bit and followed by one or more stop bits.
2. Idle time (period of inactivity) can exist between transmitted characters.
3. Bits within a character are transmitted at prescribed time intervals.
4. Timing is established independently in the computer and terminal.
5. Transmission speeds normally do not exceed 56 000 bps over analog switched facilities.

Synchronous
1. Syn characters prefix transmitted data.
2. Syn characters are transmitted between blocks of data to maintain line synchronization.
3. No gaps exist between characters.
4. Timing is established and maintained by the transmitting and receiving modems, the terminal, or other devices.
5. Terminals must have buffers.
6. Transmission speeds are normally in excess of 2000 bps.

In examining the entries in Table 4.8 note that the ability to transmit data at 56 000 bps over analog switched network facilities is based upon only one end of a point-to-point communications facility having an analog to digital conversion. This means that the other end of the transmission facility has a direct digital interface to a communications carrier's digital network, reducing the potential errors associated with analog to digital conversion as an analog waveform that can have an infinite number of heights is converted into a digital signal and vice versa. A special type of modem referred to as a V.90 modem must be used to achieve 56 kbps transmission and the operation of this modem is described later in this book. By enhancing the signal to noise ratio, the use of a V.90 modem provides users with the ability to overcome Shannon's Law which restricts communications over a conventional analog telephone channel to approximately 30 000 bps.

4.5 TYPES OF TRANSMISSION

The two types of data transmission that one can consider are serial and parallel. For serial transmission the bits which comprise a character are transmitted in sequence over one line; whereas, in parallel transmission characters are transmitted serially but the bits that represent the character are transmitted in parallel. If a character consists of eight bits, then parallel transmission would require a minimum of eight lines. Additional lines may be

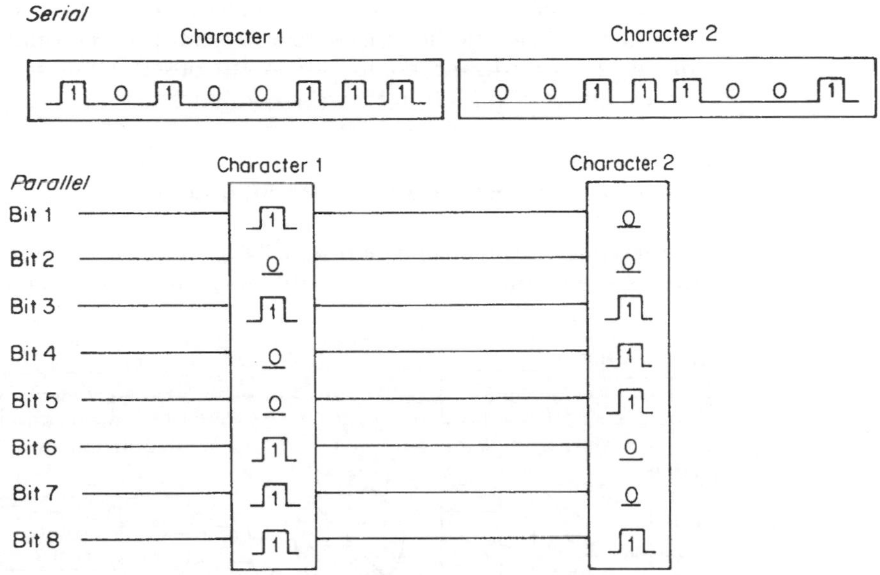

Figure 4.23 Types of data transmission. In serial transmission, the bits that comprise the character to be transmitted are sent in sequence over one line.
In parallel transmission, the characters are transmitted serially but the bits that represent the character are transmitted in parallel

necessary for control signals or for the transmission of a parity bit. Although parallel transmission is used extensively in computer-to-peripheral unit transmission, it is not normally employed other than in dedicated data transmission usage owing to the cost of the extra circuits required.

A typical use of parallel transmission is the in-plant connection of badge readers and similar devices to a computer in that facility. Parallel transmission may also reduce the cost of terminal circuitry since the terminal does not have to convert the internal character representation to a serial data stream for transmission. The cost of the transmission medium and interface will, however, increase because of the additional number of conductors required. Since the total character can be transmitted at the same moment in time using parallel transmission, higher data transfer rates can be obtained than are possible with serial transmission facilities. For this reason, most local facility communications between computers and their peripheral devices are accomplished using parallel transmission. In comparison, communications between personal computers and other computers normally occur serially, since this requires only one line to interconnect two devices that need to communicate with one another. Figure 4.23 illustrates serial and parallel transmission.

4.6 WIDE AREA NETWORK TRANSMISSION STRUCTURES

The geographical distribution of personal computers and terminal devices and the distance between each device and the device it transmits to are important parameters that must be considered in developing a network configuration. The method used to interconnect personal computers and terminals to mainframe computers or to other devices is known as line structure and it results in a network configuration.

4.6.1 Mainframe computer-based network structure

The two types of line structure primarily used in mainframe-based computer networks are point-to-point and multipoint, the latter also commonly referred

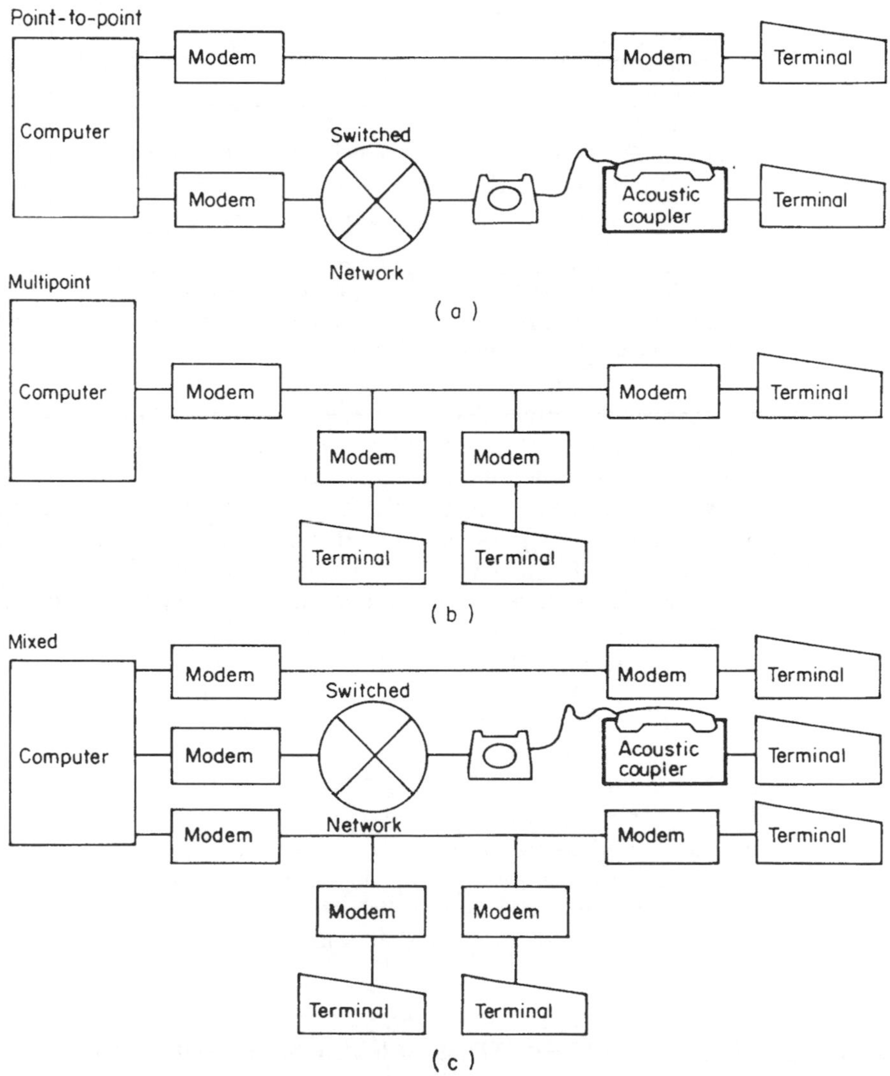

Figure 4.24 Wide area network line structures. Top: point-to-point line structure. Center: multipoint (multidrop) line structure. Bottom: mixed network line structure

to as multidrop lines. Communications lines that only connect two points are point-to-point lines. An example of this line structure is depicted at the top of Figure 4.24. As illustrated, each terminal device transmits and receives data to and from a computer via an individual connection that links a terminal to the computer. The point-to-point connection can utilize a dedicated circuit or a leased line, or can be obtained via a connection initiated over the switched (dial-up) telephone network.

When two or more terminal locations share portions of a common line, the line is a multipoint or multidrop line. Although no two devices on such a line can transmit data at the same time, two or more devices may receive a message at the same time. The number of devices receiving such a message is dependent upon the addresses assigned to the message recipients. In some systems a 'broadcast' address permits all devices connected to the same multidrop line to receive a message at the same time. When multidrop lines are employed, overall line costs may be reduced since common portions of the line are shared for use by all devices connected to that line.

To prevent data transmitted from one device from interfering with data transmitted from another device, a line discipline or control must be established for such a link. This discipline controls transmission so no two devices transmit data at the same time. A multidrop line structure is depicted in the second portion of Figure 4.24. For multidrop line linking n devices to a mainframe computer, $n+1$ modems are required, one for each device as well as one located at the computer facility.

Both point-to-point and multipoint lines may be intermixed in developing a network and transmission can be either in the full- or half-duplex mode. This mixed line structure is shown in the lower portion of Figure 4.24.

4.6.2 LAN network structure

There are three types of line structure associated with the construction of most local area networks. On an individual LAN basis, the bus, ring and star represent common network topologies. The top portion of Figure 4.25 illustrates a bus structured LAN, and the lower portion of that illustration shows ring and star structured LANs.

Access to a bus-based LAN occurs on a contention basis, with a workstation first 'listening' to determine if the cable is in use. If it is not in use, the workstation can transmit data, otherwise it must wait for a period of inactivity to gain access to the bus.

On a ring-based LAN a token circulates which is used to control access of workstations to the ring. If a token is not in use a workstation can acquire it, adding data which are then transported to another workstation. If the token is in use the workstation must wait until a free token becomes available for use.

A third type of LAN topology is represented by a star based configuration as illustrated in the lower portion of Figure 4.25. In a star-based topology a switch at the center of the star enables each workstation to have immediate access to the LAN; however, if the destination of the data is in use the switch may either block or queue the data from being forwarded to its destination.

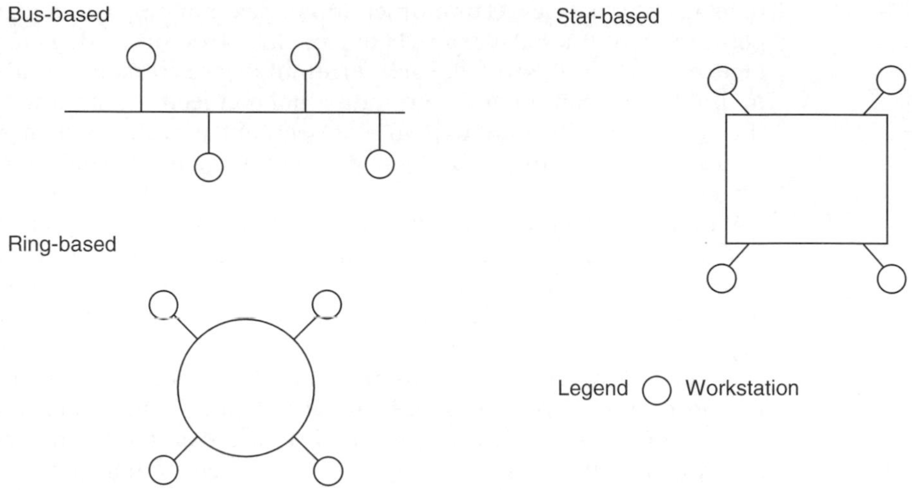

Figure 4.25 Common LAN network structures

A fourth type of LAN network structure that warrants discussion is an inverted tree, with the root or origin at the top of the network while branches and leaves fan out downward similar to an inverted tree. This network structure represents the structure used by cable television and one type of Ethernet network. Because cable modems in effect operate as an Ethernet local area network, they represent a LAN network that is gaining in popularity and that will be described in detail later in this book.

4.6.3 LAN internetworking structure

The interconnection of locally and geographically separated LANs is commonly referred to as internetworking, A variety of network structures can be developed to interconnect LANs. Two common network structures include an H or rotated H and a mesh.

The top portion of Figure 4.26 illustrates the H structure formed by bridging LANs together. A simple bridge examines the source addresses of information packets on one side to build a table of known addresses. If an information packet contains a destination address that is not contained in the table associated with source addresses known on one side of the device then the packet is forwarded onto the other side.

A mesh structure is developed through the use of a more complex network device known as a router. The use of routers to develop a mesh structured network is illustrated in the lower portion of Figure 4.26. Routers can make dynamic decisions concerning the use of different paths for transmission. For example, data from a workstation located on LAN D destined for LAN A might flow directly to the router connected to LAN A, or they could be routed on paths D-B-A or D-C-A. Later in this book we will examine the operation of bridges and routers in detail.

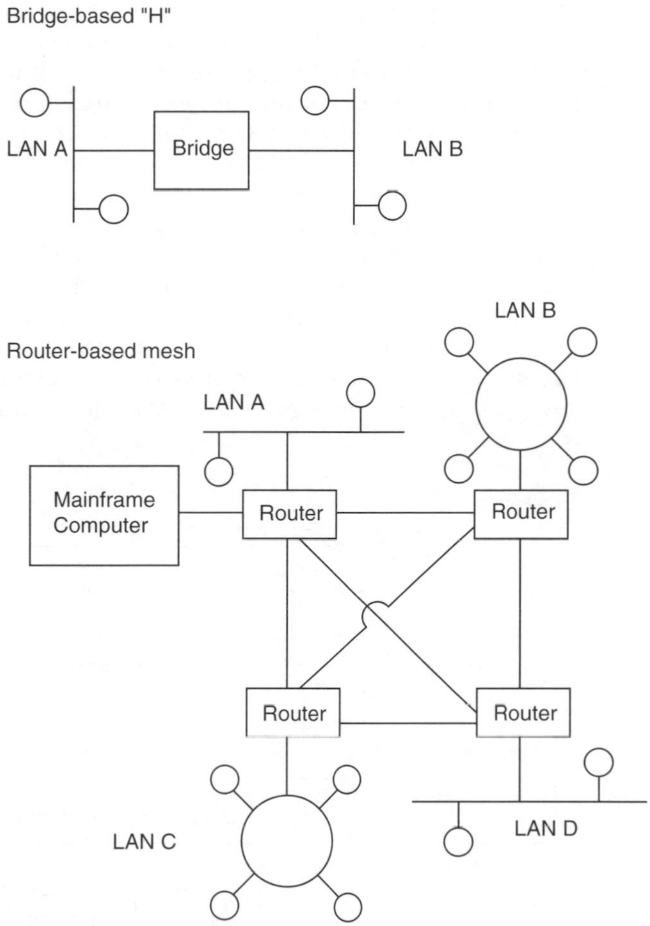

Figure 4.26 Common LAN internetworking structures

4.7 LINE DISCIPLINE

When several devices share the use of a common, multipoint communications line, only one device may transmit at any one time, although one or more devices may receive information simultaneously. To prevent two or more devices from transmitting at the same time, a technique known as 'poll and select' is utilized as the method of line discipline for multidrop lines. To utilize poll and select, each device on the line must have a unique address of one or more characters as well as circuitry to recognize a message sent from the computer to that address. When the computer polls a line, in effect it asks each device in a predefined sequence if it has data to transmit. If the device has no data to transmit, it informs the computer of this fact and the computer continues its polling sequence until it encounters a device on the line that has data to send. Then the computer acts on that data transfer.

As the computer polls each device, the other devices in the line must wait until they are polled before they can be serviced. Conversely, transmission of data from the computer to each device on a multidrop line is accomplished by the computer selecting the device address to which that data is to be transferred, informing the device that data are to be transferred to it, and then transmitting data to the selected device. Polling and selecting can be used to service both asynchronous or synchronous operating terminal devices connected to independent multidrop lines. Due to the control overhead of polling and selecting, synchronous high-speed devices are normally serviced in this type of environment. By the use of signals and procedures, polling and selecting line control ensures the orderly and efficient utilization of multidrop lines. An example of a computer polling the second personal computer or terminal on a multipoint line and then receiving data from that device is shown at the top of Figure 4.27. At the bottom of that illustration, the

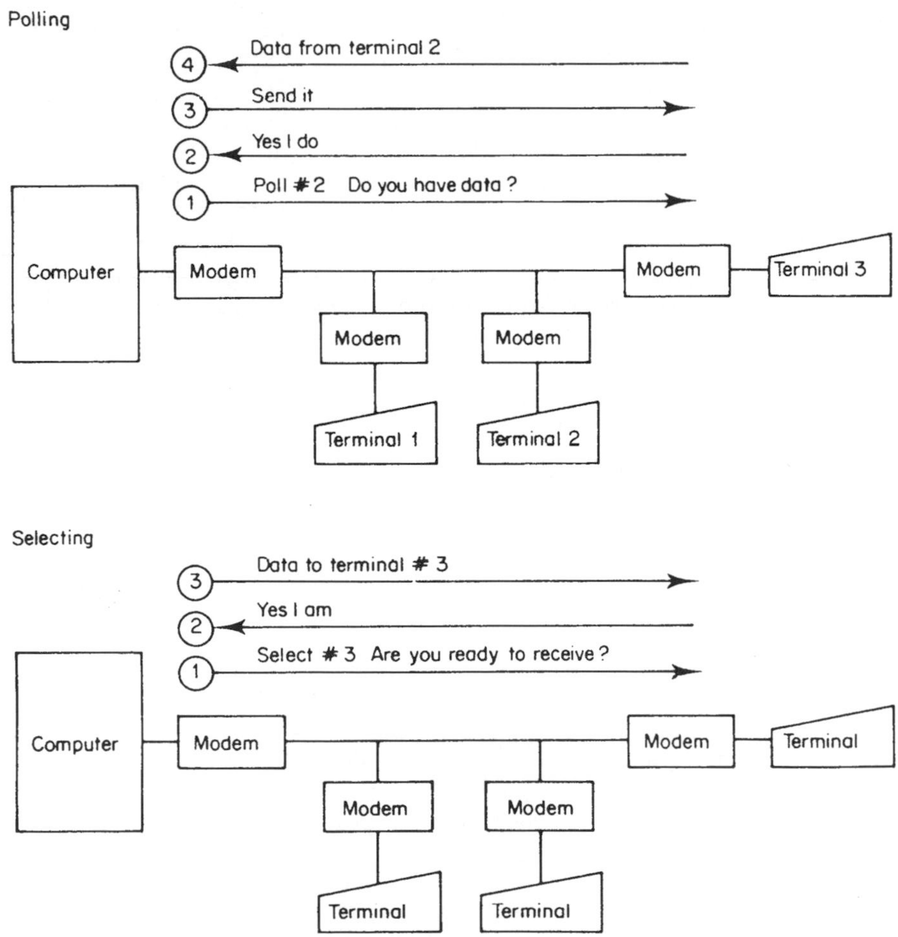

Figure 4.27 Poll and select line discipline. Poll and select is a line discipline which permits several devices to use a common line facility in an orderly manner

computer first selects the third terminal on the line and then transfers a block of data to that device.

When terminals transmit data on a point-to-point line to a computer or another terminal, the transmission of that data occurs at the discretion of the terminal operator. This method of line control is known as 'non-poll-and-select' or 'free- wheeling' transmission.

In a LAN environment line discipline is commonly referred to as the LAN's access method. Contention and token passing represent two commonly used LAN access methods.

4.8 TRANSMISSION RATE

Many factors can affect the transmission rate at which data is transferred. The types of modems, data service units or acoustic couplers used, as well as the line discipline and the type of computer channel to which a terminal is connected via a transmission medium, play governing roles that affect transmission rates; however, the transmission medium itself is a most important factor in determining transmission rates.

Data transmission services offered by communications carriers such as AT&T are based on their available plant facilities. Depending upon terminal and computer locations, two types of transmission services may be available. The first type of service, analog transmission, is most readily available and can be employed on switched or leased telephone lines. Digital transmission is available in most large cities, and analog extensions are required to connect to this service from non-digital service locations as previously illustrated in Figure 4.17. Within each type of service several grades of transmission are available for consideration.

4.8.1 Analog service

In general, analog service offers the user three grades of transmission: narrowband, voice-band and wideband. The data transmission rates possible on each of these grades of service is dependent upon the bandwidth and electrical properties of each type of circuit offered within each grade of service. Basically, transmission speed is a function of the bandwidth of the communications line: the greater the bandwidth, the higher the possible speed of transmission.

Narrowband facilities are obtained by the carrier subdividing a voice-band circuit or by grouping a number of transmissions from different users onto a single portion of a circuit by time. Transmission rates obtained on narrowband facilities range between 45 and 300 bps. Teletype™ terminals that connect to message switching systems are the primary example of the use of narrowband facilities.

While narrowband facilities have a bandwidth in the range of 200 to 400 Hz, voice-band facilities have a bandwidth in the range of 3000 Hz. Data transmission speeds obtainable on voice-band facilities are differentiated by the type of voice-band facility utilized: switched dial-up transmission or transmission via a leased line. For transmission over the switched telephone network, maximum data transmission is normally between 28 800 and 33 600 bps, with a 56 000 bps data rate obtainable when transmission occurs through modern electronic switches and one end of a point-to-point connection is directly interfaced to a digital transmission facility. Although leased analog lines can be conditioned, their use involves two analog to digital conversions, limiting their maximum data transmission rate to 33.6 kbps via a 3000 Hz channel.

Although low data speeds can be transmitted on both narrowband and voice-band circuits, do not confuse the two since a low-data speed on a voice circuit is transmission at a rate far less than the maximum permitted by that type of circuit. On the other hand, a low rate on a narrowband facility is at or near the maximum transmission rate permitted by that type of circuit.

Facilities which have a higher bandwidth than voice-band are termed wideband or group-band facilities since they provide a wider bandwidth through the grouping of a number of voice-band circuits. Wideband facilities are available only on leased lines and permit transmission rates in excess of 33 600 bps. Transmission rates on wideband facilities vary with the offerings of communications carriers. Speeds normally available include 40.8, 50 and 230.4 kbps.

For direct connect circuits, transmission rates are a function of the distance between the terminal and the computer as well as the gauge of the conductor used.

4.8.2 Digital service

In the area of digital service, numerous offerings are currently available for user consideration. Digital data service offered by AT&T as DATAPHONE'® digital service (DDS) provides interstate, full-duplex, point-to-point and multipoint leased line, as well as synchronous digital transmission at speeds of 2400, 4800, 9600, 19 200 and 56 000 bps.

Table 4.9 Common transmission facilities

Facility	Transmission speed (bps)	Use
Analog		
Narrowband	45–300	Message switching
Voice-band	less than 4800–56 600	Time sharing; remote job entry; information
Switched	up to 56 000	utility access; file transfer
Leased	up to 33 600	Computer-to-computer; remote job entry;
Wideband	over 33 600	tape-to-tape transmission; high-speed terminal to high-speed terminal
Digital		
Switched	56 kbps, 384 kbps, 1.544 Mbps	LAN internetworking, videoconferencing, computer-to-computer
Leased line	2.4, 4.8, 9.6, 19.2 kbps	remote job entry; computer-to-computer
	56 kbps, n56/64 kbps	high speed fax; LAN internetworking
	1.544 Mps	integrated voice and data
	44 Mbps	integrated voice and data

Under the Accunet Spectrum of Digital Services AT&T offers digital switched transmission at 56 kbps, 384 kbps and 1.544 Mbps. In addition, you can obtain a variety of fractional and full T1 and T3 digital transmission lines from AT&T, MCI, Sprint and other communications carriers. Table 4.9 lists the main analog and digital facilities, the range of transmission speeds over those facilities and the general use of such facilities. The entry $n56/64$ represents fractional service with n varying from 1 to 12 on some carrier offerings, whereas other carriers support n varying up to 24, the latter providing a full T1 service. When we discuss the T1 carrier later in this book we will also discuss the reason that the fractional T1 operating rate increments at either 56 or 64 kbps intervals, and why 24×64 kbps provides an operating rate 8000 bps less than the T1 operating rate of 1.544 Mbps.

4.9 TRANSMISSION CODES

Data within a computer are structured according to the architecture of the computer. The internal representation of data in a computer is seldom suitable for transmission to devices other than the peripheral units attached to the computer. In most cases, to effect data transmission, internal computer data must be redesigned or translated into a suitable transmission code. This transmission code creates a correspondence between the bit encoding of data for transmission or internal device representation and printed symbols. The end result of the translation is usually dictated by the character set used by the remote terminal. Frequently available terminal codes include: Baudot, which is a five-level (5 bits per character) code; binary-coded decimal (BCD), which is a six-level code; American Standard Code for Information Interchange (ASCII), which is normally a seven-level code; and the extended binary-coded decimal interchange code (EBCDIC), which is an eight-level code.

In addition to information being encoded into a certain number of bits based upon the transmission code used, the unique configuration of those bits to represent certain control characters can be considered as a code that can be used to effect line discipline. These control characters may be used to indicate the acknowledgement of the receipt of a block of data without errors (ACK), the start of the message (SOH), or the end of a message (EOT), with the number of permissible code characters standardized according to the code employed. With the growth of computer-to-computer data transmission, a large amount of processing can be avoided by transferring the data in the format used by the computer for internal processing. Such transmission is known as binary mode transmission, transparent data transfer, code-independent transmission, or native mode transmission.

4.9.1 Morse code

One of the most commonly known codes, the Morse code, is not practical for utilization in a computer communications environment. As previously discussed in Chapter 2, this code consists of a series of dots and dashes, which, while easy for the human ear to decode, are of unequal length and are not practical for data transmission implementation. In addition, since each character in the Morse code is not prefixed with a start bit and terminated with a stop bit, it was initially not possible to construct a machine to automatically translate received Morse transmissions into their appropriate characters.

4.9.2 Baudot code

The Baudot code, which is a five-level (5 bits per character) code, was the first code to provide a mechanism for encoding characters by an equal number of bits, in this case five.

Through the work of Alfred Kahn, a method was developed to permit the synchronization of transmitted characters to a receiver, thus permitting the development of automatic transmission and reception equipment. Refer to Chapter 2 for information concerning the character composition of data represented in the Baudot code.

4.9.3 BCD code

The development of computer systems required the implementation of coding systems to convert alphanumeric characters into binary notation and the binary notation of computers into alphanumeric characters. The BCD system was one of the earliest codes used to convert data to a computer-acceptable form. This coding technique permits decimal numeric information to be represented by four binary bits and permits an alphanumeric character set to be represented through the use of six bits of information. This code is illustrated in Table 4.10. An advantage of this code is that two-decimal digits can be stored in an 8-bit computer word and manipulated with appropriate computer instructions. Although only 36 characters are shown for illustrative purposes, a BCD code is capable of containing a set of 2^6 or 64 different characters.

4.9.4 Extended binary-coded decimal interchange code (EBCDIC)

In addition to transmitting letters, numerals and punctuation marks, a considerable number of control characters may be required to promote line discipline. These control characters may be used to switch on and off devices which are connected to the communications line, control the actual transmission of data, manipulate message formats and perform additional functions. Thus, an extended character set is usually required for data communications. One such character set is EBCDIC code. The extended binary-coded decimal interchange code (EBCDIC) is an extension of the BCD system and uses 8 bits for character representation. This code permits 2^8 or

Table 4.10 Binary-coded decimal system

Bit position						
b_6	b_5	b_4	b_3	b_2	b_1	Character
0	0	0	0	0	1	A
0	0	0	0	1	0	B
0	0	0	0	1	1	C
0	0	0	1	0	0	D
0	0	0	1	0	1	E
0	0	0	1	1	0	F
0	0	0	1	1	1	G
0	0	1	0	0	0	H
0	0	1	0	0	1	I
0	1	0	0	0	1	J
0	1	0	0	1	0	K
0	1	0	0	1	1	L
0	1	0	1	0	0	M
0	1	0	1	0	1	N
0	1	0	1	1	0	O
0	1	0	1	1	1	P
0	1	1	0	0	0	Q
0	1	1	0	0	1	R
1	0	0	0	1	0	S
1	0	0	0	1	1	T
1	0	0	1	0	0	U
1	0	0	1	0	1	V
1	0	0	1	1	0	W
1	0	0	1	1	1	X
1	0	1	0	0	0	Y
1	0	1	0	0	1	Z
1	1	0	0	0	0	0
1	1	0	0	0	1	1
1	1	0	0	1	0	2
1	1	0	0	1	1	3
1	1	0	0	1	1	4
1	1	0	1	0	1	5
1	1	0	1	1	0	6
1	1	0	1	1	1	7
1	1	1	0	0	0	8
1	1	1	0	0	1	9

256 unique characters to be represented, although currently a lower number is assigned meanings. This code is primarily used for transmission by byte-oriented computers, where a byte is a grouping of eight consecutive binary digits operated on as a unit by the computer. The use of this code by computers may alleviate the necessity of the computer performing code conversion if the connected terminals operate with the same character set.

Several subsets of EBCDIC exist that have been tailored for use with certain devices. As an example, IBM 3270 type terminal products would not use a paper feed and its character representation is omitted in the EBCDIC character subset used to operate that type of device as indicated in Table 4.11.

Table 4.11 EBCDIC code implemented for the IBM 3270 information display system

		00				01				10				11				← Bits 0, 1
	Hex 1	00	01	10	11	00	01	10	11	00	01	10	11	00	01	10	11	← 2, 3
Bits 4567	↓	0	1	2	3	4	5	6	7	8	9	A	B	C	D	E	F	← Hex 0
0000	0	NUL	DLE			SP	&	-									0	
0001	1	SOH	SBA					/		a	j			A	J		1	
0010	2	STX	EUA		SYN					b	k	s		B	K	S	2	
0011	3	ETX	IC							c	l	t		C	L	T	3	
0100	4									d	m	u		D	M	U	4	
0101	5	PT	NL							e	n	v		E	N	V	5	
0110	6			ETB						f	o	w		F	O	W	6	
0111	7			ESC	EOT					g	p	x		G	P	X	7	
1000	8									h	q	y		H	Q	Y	8	
1001	9		EM							i	r	z		I	R	Z	9	
1010	A					¢	!	¦	:									
1011	B					.	$	,	#									
1100	C		DUP		RA	<	*	%	@									
1101	D		SF	ENQ	NAK	(	)	—	'									
1110	E		FM			+	;	>	=									
1111	F		ITB		SUB	|	¬	?	"									

4.9.5 ASCII code

As a result of the proliferation of data transmission codes, several attempts to develop standardized codes for data transmission have been made. One such code is the American Standard Code for Information Interchange (ASCII). This seven-level code is based upon a 7-bit code developed by the International Standards Organization (ISO), and permits 128 possible combinations or character assignments to include 96 graphic characters that are printable or displayable and 32 control characters to include device control and information transfer control characters. Table 4.12 lists the ASCII character set and Table 4.13 lists the ASCII control characters by position and their meaning. A discussion covering the operation and utilization of communications control characters is deferred until Chapter 11. In this chapter the use of communications control characters to develop methods of controlling the flow of data between terminal devices is covered. The primary difference between the ASCII character set listed in Table 4.12 and other versions of the CCITT (now known as the ITU) International Alphabet Number 5 is the currency

Table 4.12 The ASCII character set

Bits												
b_7					0	0	0	0	1	1	1	1
b_6					0	0	1	1	0	0	1	1
b_5					0	1	0	1	0	1	0	1
b_4	b_3	b_2	b_1	COLUMN / ROW	0	1	2	3	4	5	6	7
0	0	0	0	0	NUL	DLE	SP	0	@	P	\	p
0	0	0	1	1	SOH	DC1	!	1	A	Q	a	q
0	0	1	0	2	STX	DC2	"	2	B	R	b	r
0	0	1	1	3	ETX	DC3			C	S	c	s
0	1	0	0	4	EOT	DC4	$	4	D	T	d	t
0	1	0	1	5	ENQ	NAK	%	5	E	U	e	u
0	1	1	0	6	ACK	SYN	&	6	F	V	f	v
0	1	1	1	7	BEL	ETB	/	7	G	W	g	w
1	0	0	0	8	BS	CAN	(	8	H	X	h	x
1	0	0	1	9	HT	EM	)	9	I	Y	i	y
1	0	1	0	10	LF	SUB	*	:	J	Z	j	z
1	0	1	1	11	VT	ESC	+	;	K	[	k	{
1	1	0	0	12	FF	FS	,	<	L	\	l	|
1	1	0	1	13	CR	GS	–	=	M	]	m	}
1	1	1	0	14	SO	RS	.	>	N	∧	n	~
1	1	1	1	15	SI	US	/	?	O	—	o	DEL

symbol. Although the bit sequence 0100010 is used to generate the dollar ($) currency symbol in the United States, in the United Kingdom that bit sequence results in the generation of the pound sign (£). Similarly, this bit sequence generates other currency symbols when the International Alphabet Number 5 is used in other countries.

Code conversion

A frequent problem in data communications is that of code conversion. Consider what must be done to enable a computer with an EBCDIC character set to transmit and receive information from a terminal with an ASCII character set. When that terminal transmits a character, that character is encoded according to the ASCII character code. Upon receipt of that character,

Table 4.13 ASCII control characters

Column/Row	Control character	Mnemonic and meaning	
0/0	^@	NUL	Null (CC)
0/1	^A	SOH	Start of Heading (CC)
0/2	^B	STX	Start of Text (CC)
0/3	^C	ETX	End of Text (CC)
0/4	^D	EOT	End of Transmission (CC)
0/5	^E	ENQ	Enquiry (CC)
0/6	^F	ACK	Acknowledgement (CC)
0/7	^G	BEL	Bell
0/8	^H	BS	Backspace (FE)
0/9	^I	HT	Horizontal Tabulation (FE)
0/10	^J	LF	Line Feed (FE)
0/11	^K	VT	Vertical Tabulation (FE)
0/12	^L	FF	Form Feed (FE)
0/13	^M	CR	Carriage Return (FE)
0/14	^N	SO	Shift Out
0/15	^O	SI	Shift In
1/0	^P	DLE	Date Link Escape (CC)
1/1	^Q	DC1	Device Control 1
1/2	^R	DC2	Device Control 2
1/3	^S	DC3	Device Control 3
1/4	^T	DC4	Device Control 4
1/5	^U	NAK	Negative Acknowledge (CC)
1/6	^V	SYN	Synchronous Idle (CC)
1/7	^W	ETB	End of Transmission Block (CC)
1/8	^X	CAN	Cancel
1/9	^Y	EM	End of Medium
1/10	^Z	SUB	Substitute
1/11	^[	ESC	Escape
1/12	^/	FS	File Separator (IS)
1/13	^]	GS	Group Separator (IS)
1/14	~	RS	Record Separator (IS)
1/15	.	US	Unit Separator (IS)
7/15	^_	DEL	Delete

the computer must convert the bits of information of the ASCII character into an equivalent EBCDIC character. Conversely, when data are to be transmitted to the terminal, they must be converted from EBCDIC to ASCII so the terminal will be able to decode and act according to the information in the character that the terminal is built to interpret.

One of the most frequent applications of code conversion occurs when personal computers are used to communicate with IBM mainframe computers.

Normally, ASCII to EBCDIC code conversion is implemented when an IBM PC or compatible personal computer is required to operate as a 3270 type terminal. This type of terminal is typically connected to an IBM or IBM compatible mainframe computer and the terminal's replacement by an IBM PC requires the PC's ASCII coded data to be translated into EBCDIC. There are many ways to obtain this conversion, including emulation boards that are

Table 4.14 ASCII and EBCDIC digits comparison

ASCII				
Dec	Oct	Hex	EBCDIC	Digit
048	060	30	F0	0
049	061	31	F1	1
050	062	32	F2	2
051	063	33	F3	3
052	064	34	F4	4
053	065	35	F5	5
054	066	36	F6	6
055	067	37	F7	7
056	070	38	F8	8
057	071	39	F9	9

inserted into the system unit of a PC and protocol converters that are connected between the PC and the mainframe computer. Later in this book, we will explore these and other methods that enable the PC to communicate with mainframe computers that transmit data coded in EBCDIC.

Table 4.14 lists the ASCII and EBCDIC code character values for the ten digits for comparison purposes. In examining the difference between ASCII and EBCDIC coded digits, note that each EBCDIC coded digit has a value precisely Hex C0 (decimal 192) higher than its ASCII equivalent. Although this might appear to make code conversion a simple process of adding or subtracting a fixed quantity depending upon which way the code conversion takes place, in reality many of the same ASCII and EBCDIC coded characters differ by varying quantities. As an example, the slash (/) character is Hex 2F in ASCII and Hex 61 in EBCDIC, a difference of Hex 92 (decimal 146). In comparison, other characters such as the carriage return and form feed have the same coded value in ASCII and EBCDIC, while other characters are displaced by different amounts in these two codes. Thus code conversion is typically performed as a table lookup process, with two buffer areas used to convert between codes in each of the two conversion directions. One buffer area might thus have the ASCII character set in hex order in one field of a two-field buffer area, with the equivalent EBCDIC hex values in a second field in the buffer area. Then, upon receipt of an ASCII character its hex value is obtained and matched to the equivalent value in the first field of the buffer area, with the value of the second field containing the equivalent EBCDIC Hex value which is then extracted to perform the code conversion.

Modified ASCII

Members of the IBM PC series and compatible computers use a modified ASCII character set which is represented as an eight-level code. The first 128 characters in the character set, ASCII values 0 through 127, correspond to the

ASCII character set listed in Table 4.12, and thc ncxt 128 characters can be vicwcd as an extension of that character set since they require an 8-bit representation.

Caution is advised when transferring IBM PC files, since characters with ASCII values greater than 127 will be received in error when they are transmitted using 7 data bits. This is because the ASCII values of these characters will be truncated to values in the range 0 to 127 when transmitted with 7 bits from their actual range of 0 to 255. To alleviate this problem from occurring, you can initialize your communications software for 8-bit data transfer; however, the receiving device must also be capable of supporting 8-bit ASCII data.

Although conventional ASCII files can be transmitted in a 7-bit format, many word processing and computer programs contain text graphics represented by ASCII characters whose values exceed 127. In addition, EXE and COM files which are produced by assemblers and compilers contain binary data that must also be transmitted in 8-bit ASCII to be accurately received. While most communications programs can transmit 7- or 8-bit ASCII data, some older programs may not be able to transmit binary files accurately. This is due to the fact that communications programs that use the control Z character (ASCII SUB) to identify the end of a file transfer will misinterpret a group of 8 bits in the EXE or COM file being transmitted when they have the same 8-bit format as a control Z, and upon detection prematurely close the file. To avoid this situation, you should obtain a communications software program that transfers files by blocks of bits or converts data into hexadecimal or octal ASCII equivalent prior to transmission if this type of data transfer is required.

Another communications problem that you may encounter occurs when attempting to transmit files using some electronic mail services that are limited to transferring 7-bit ASCII. If you created a file using a word processor which employs 8-bit ASCII codes to indicate special character settings, such as bold and underline text, attempting to transmit the file via a 7-bit ASCII electronic mail service will result in the proverbial gobbledegook being received at the distant end. Instead, you should first save the file as a text file prior to transmission. Although you will lose any embedded 8-bit ASCII control codes, you will be able to transmit the document over an electronic mail system limited to the transfer of 7-bit ASCII.

4.10 REVIEW QUESTIONS

1. Discuss the function of each of the major elements of a transmission system.

2. What is the relationship between each of the three basic types of line connections and the use of that line connection for short or long duration data transmission sessions?

3. What is the function of a hybrid installed at a telephone company's central office?

4. What is the function of an attenuation equalizer?

5. Assume a flat rate plan results in the cost of a long distance call becoming 12 cents per minute. If a leased line costs $400 per month, how many minutes per month must occur over the switched telephone network between two locations prior to the use of a leased line becoming more economical?

6. Discuss the relationship of modems and data service units to digital and analog transmission systems. Why are these devices required and what general functions do they perform?

7. Name four types of analog facilities offered by communications carriers and discuss the utilization of each facility for the transmission of data between terminal devices and computer systems.

8. Draw the bit sequence 101101 encoded as a unipolar non-return to zero, unipolar return to zero and a bipolar return to zero signal.

9. What are the advantages associated with removing the dc residual voltage from a signal?

10. What is the advantage associated with using a 50% duty cycle to encode digital data?

11. Why is transformer coupling used on digital transmission facilities where transmission occurs over metallic conductors?

12. What is the function of an analog extension?

13. What is the difference between simplex, half-duplex and full-duplex transmission?

14. Discuss the relationship between the modes of operation of terminals and computers with respect to the printing and display of characters on a terminal in response to pressing a key on the terminal's keyboard.

15. In asynchronous transmission how does a receiving device determine the presence of a start bit?

16. What is the difference between asynchronous and synchronous transmission with respect to the timing of the data flow?

17. How can a V.90 modem provide a data transmission rate over the approximate 30 kbps maximum resulting from Shannon's Law?

18 What is the difference between serial and parallel transmission? Why do most communications systems use serial transmission?

19. Discuss the difference in terminal requirements with respect to point-to-point and multidrop line usage.

20. Draw and label three common LAN network topologies.

21. Discuss the difference between a linc disciplinc and an access method.

22. Why is the Morse code basically unsuitable for transmission by terminal devices?

23. What are the bit compositions of the ASCII characters A and a?

24. Assume that your terminal is set to a half-duplex mode of operation and that you access a full-duplex computer system. How many characters will be displayed or printed on your terminal each time you press a character on the keyboard? Why is this number not one?

25. Assume you just used a word processor to create a document that you wish to transmit to an associate using an electronic mail system limited to supporting 7-bit ASCII. What should you do to the file which contains the document you wish to transmit?

5

TERMINALS, WORKSTATIONS AND WAN AND LAN NETWORKING OVERVIEW

The purpose of this chapter is twofold – to introduce you to the operation and utilization of terminal devices and to provide you with an overview of local and wide area networking.

Most data communications applications involve the transmission of information between terminal devices and workstations or from terminal devices and workstations to different types of servers, such as fileservers and mainframe computer systems. While at first it might appear that referring to a mainframe as a server is odd, a bit of thought suggests this is logical. After all, when conventional terminals were first connected to mainframes they were clients of the mainframe. Even when PCs communicate with a mainframe, client–server processing occurs. Thus, regardless of whether or not a mainframe is attached to a local area network, access to the computer by terminals and other computers can be considered to represent client–server communications.

Due to the importance of terminals and the constraints they place upon communications, we will first examine this category of communications equipment in this chapter. This examination will include several categories of terminals to include devices commonly referred to as workstations and will form the basis for the overview of wide area networking presented later in this chapter.

Once we complete our examination of terminals, we will turn our attention to obtaining an appreciation for the operation and functionality of different LAN components. Although the title of this chapter refers to workstations which are indeed an important LAN component, we will also focus our attention on the operation of other devices on which local area networks are constructed and which provide network interconnectivity. In concluding this chapter we will obtain an appreciation of the operation and utilization of several types of equipment used to construct wide area networks, to include the rationale for using such equipment.

5.1 TERMINALS

Terminals were originally categorized as interactive or remote batch, the latter also commonly referred to as a remote job entry device. An interactive terminal is typically used to transmit relatively short queries or to provide an operator with the ability to respond to computer-generated screen displays by entering data into defined fields prior to transmitting the filled in screen. Once the data have been received, the destination device will respond to the transmission in a relatively short period of time, typically measured in seconds. In comparison, remote batch terminals provide the operator with the ability to group or batch a series of jobs that can range in scope from programs developed for execution on a large computer to queries that are also structured for execution against a database maintained on a large computer system.

The introduction of the personal computer altered the previous distinction. That is, the ability to use the hard drive of a PC as a storage mechanism enables the personal computer to transmit and receive files that can range up to a gigabyte or more in size. Although the distinction between interactive and batch terminals was essentially eliminated by the PC, it is an important topic to note as many software programs were developed to turn a personal computer into a specific type of terminal by emulating the features of a terminal. As you might expect, such software is referred to as terminal emulation software.

5.1.1 Interactive terminal classification

One method commonly used to classify interactive terminals is based on their configuration. This method of terminal classification has its origins with teletype terminals in which those terminals could be configured as a receive only (RO), keyboard send–receive (KSR) or automatic send–receive (ASR) device. The largest manufacturer of terminals in the United States during the earlier part of the twentieth century was the Teletype Corporation, then a subsidiary of AT&T. Although the term teletype has evolved to refer to any low-speed terminal with a serial interface, this usage is actually improper since the term is a trademark of AT&T. In spite of this, the term teletype transmission is commonly used to refer to a start–stop asynchronous terminal and the term teletype compatible is used to refer to a start–stop, and asynchronous terminal that transmits and receives data on a line-by-line basis.

Receive only (RO)

A receive only (RO) terminal consists of a stand-alone printer with a serial communications interface but lacking a keyboard. Originally developed to simply receive messages transmitted on message switching systems developed in the 1930s, a limited number of RO terminals are still in use today. One example of the use of RO terminals is the weather bureau meteorological service network. This network broadcasts weather information from a central computer system onto multidrop lines routed throughout the country. At each drop a RO terminal prints the weather forecast for the region in which the terminal is located. Since no response is required to the forecast there is no need to incorporate a keyboard into the terminal.

Keyboard send–receive (KSR)

Originally, keyboard send–receive (KSR) terminals included a printer, serial communications interface, and keyboard. This permitted the terminal operator to both originate a message from the keyboard as well as to print a received message. Some KSR terminals use a cathode ray tube (CRT) display screen for receiving data and may or may not have a printer attached to the device.

Automatic send–receive (ASR)

The third interactive terminal classification is automatic send–receive (ASR). An ASR terminal consists of a printer, serial communications interface, keyboard and auxiliary storage. Here the auxiliary storage permits messages to be composed 'off-line' with the terminal not attached via a communications facility to its intended transmission destination. Once a message is composed the terminal can be used to obtain a connection to its intended transmission destination. Then the previously created and stored message can be transmitted at the operating rate of the terminal instead of the typing rate of the operator.

The first ASR devices were teletype terminals which used paper tape for auxiliary storage. More modern ASR devices incorporate cassette or diskette storage. In fact, the personal computer can be considered to be an automatic send–receive terminal.

5.1.2 Terminal evolution

The evolution of interactive terminals corresponds to changes in most of the components upon which the manufacture of those devices is based. Due to this, we will examine the basic components common to most interactive terminals and review their product development which formed the basis to the development of a wide range of terminal devices.

Basic terminal components

Figure 5.1 illustrates the basic components of interactive terminals in schematic form. Until the development of the microprocessor, all terminals

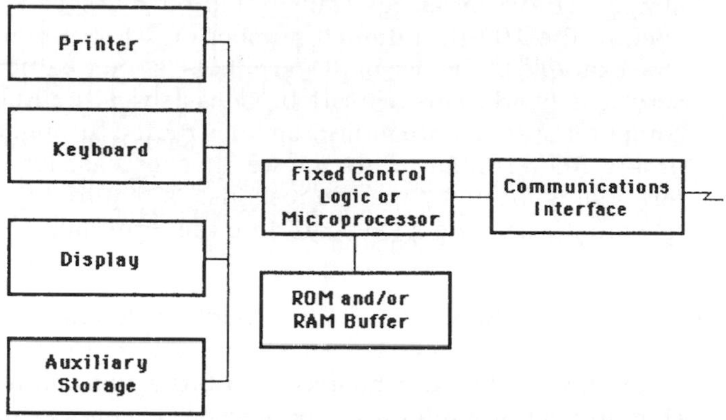

Figure 5.1 Basic components of interactive terminals. The construction of most interactive terminals is based upon the use of many common components

employed fixed control logic circuitry which governed the operation of the device. This circuitry could not be altered, resulting in the term fixed logic being used to represent the operation of such terminals.

Due to the use of fixed logic, it was costly to incorporate more than a limited number of functions into terminals with this method of control. Such terminals are often called dumb terminals to denote their lack of functionality as well as intelligence to perform user requested functions built into other terminals in the form of software that is executed by the terminal.

The incorporation of microprocessor technology into terminals permitted manufacturers to develop a variety of operator selectable functions. Since the microprocessor is in essence a computer on a chip, its use in a terminal provided vendors with the ability to add intelligence to terminals. This resulted in the term smart terminal being applied to microprocessor-based terminals. Among the key functions added to smart terminals are data editing and formatted data entry, both of which are described later in this section.

Communications interface

The communications interface functions as a sophisticated parallel to serial converter. As characters are pressed on the keyboard of an asynchronous terminal, the resulting *n*-level code that represents each character is passed to special circuitry. For asynchronous transmission this circuitry frames the character with a start and stop bit prior to presenting the data bit by bit on to the serial communications line. In modern terminals a special computer chip called a UART (universal asynchronous receiver transmitter) performs both character framing and the presentation of the character on a bit by bit basis onto the line.

To correctly receive data the communications interface samples the line, usually at a rate four to eight times faster than the bit rate. This sampling rate provides the mechanism whereby each bit interval is recognized as well as the transition in the state between bits, such as a marking condition followed by a

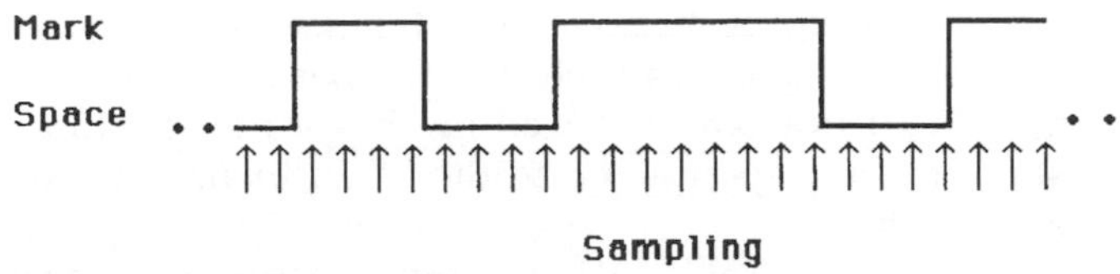

Figure 5.2 Sampling detects state transitions. Circuitry in the communications interface samples the line typically four to eight times faster than the bit rate, permitting the state of each bit period to be recognized

spacing condition or a spacing condition followed by a marking condition. Figure 5.2 illustrates how the sampling circuitry can detect the transitions between marking and spacing conditions which define the value of a particular bit interval.

As the bits are detected, the control circuitry or UART will first strip the start bit and then pass the data bits into a buffer or temporary storage area. When *n* data bits that form a character are assembled in the buffer, they are passed in parallel to circuitry which either displays or prints the character that corresponds to the binary value that represents the character. Finally, the stop bit is discarded or stripped, since it is not used internally within the terminal.

Printers and displays

A serial or character printer, whose name resulted from the fact that they print one character at a time, was the display mechanism first used with terminals. The first serial printers were 'fully formed' impact printers in which different types of mechanisms, including a daisy wheel, type-ball or rotating cylinder which formed characters from a single piece of type, were used to strike a ribbon to produce a printed image. The editing capability provided by the terminal was minimal, typically permitting the operator to delete a previously entered character or the current line, since the terminal transmitted and received data on a line by line basis. A second type of impact printer which grew in popularity during the 1970s and 1980s to where it has virtually replaced fully formed printers is the dot matrix printer.

The dot matrix printer employs a matrix of pins in its print head. The first dot matrix printers used a rectangular matrix of dots, typically 7 dots high by 5 dots wide or 9 dots high by 7 dots wide. The pins in the matrix are selectively 'fired' to form each character. Printing of characters results from the movement of the print head containing a column of 7 or 9 pins across the paper, with the printer selectively firing the pins at 5 or 7 successive intervals to form each character.

Until the mid-1980s, the matrix of pins used to form characters resulted in a considerable amount of white space between dots. This space made the dot matrix pattern easily discernible to the eye and limited the use of printed output produced by this type of printer to draft correspondence. By the mid-1980s, advances in print head technology resulted in the inclusion of more pins on some print heads. The additional pins were used to considerably reduce the space between pin impacts in forming a character. Other dot matrix developments included two-pass printing in which the first pass of printing a line was followed by the printer feeding the paper upward by a slight amount, perhaps 1/256th of an inch, prior to the line being printed a second time.

One result of placing more pins on the print head and using a two-pass printing technique was. a higher quality print. Since this print resembled the letter quality print of a full impact printer, it became known as near letter quality (NLQ).

The firing of additional pins to form a better print image required additional time, resulting in NLQ printing being slower than conventional dot matrix printing. Thus, most modern dot matrix printers have two or more user selectable print modes – draft and NLQ – with the draft print mode providing a considerably faster print rate than the NLQ print mode.

A second major category of printers employs non-impact technology to form characters. Non-impact printers include thermal matrix ink jet and laser devices. The thermal matrix printer forms characters by applying a voltage to pins in a matrix, causing the pins to be heated. The heated pins interact with heat-sensitive paper used in these printers, resulting in the formation of characters. The ink jet printer has a nozzle consisting of a matrix of holes out of which ink is squirted to form characters. Thus, both thermal matrix and ink jet printers are based upon dot matrix technology. In comparison, the laser printer uses a rotating drum and a small amount of current to generate a magnetic field, which results in toner from a cartridge adhering to distinct locations on paper passing around the drum. Laser printers have resolutions between 300 and 1200 dots per inch (dpi) and through software can form characters in almost any shape.

For comparison purposes, nine major characteristics of two types of impact and three types of non-impact printer are presented in Table 5.1. Note that due to the large variety of printers manufactured by different vendors, the entries in Table 5.1 are general characteristics representative of the majority of devices based on a particular printing technology. Thus, although there are exceptions to most entries, those exceptions, such as a higher maximum print speed, result in an increase in the cost of the printer.

The key limitation associated with the use of printers for both input and output is the elementary editing capability provided by this type of terminal device. Data entered from the keyboard are either printed and transmitted as each character is pressed or stored in a buffer area. The buffer storage area contained in most ASR and KSR terminals is only capable of holding one line of data or 72 to 80 characters depending upon the type of terminal. By using the backspace key to eliminate a previously entered character, an operator can perform elementary editing. Once the carriage return key has been pressed, however, the entire line is transmitted, resulting in an operator having to re-enter the line with any changes he or she desires to correct a previously entered line. As an alternative to the use of the backspace key, an operator can simultaneously press the control (Ctrl) key and an alphabetic key, canceling the present line and removing its contents from buffer storage. This action causes a carriage return and line feed to be automatically generated, permitting the operator to begin his or her data entry anew.

Table 5.1 Printer characteristic comparison

Characteristic	Impact full character	Impact dot matrix	Non-impact thermal	Non-impact ink jet	Non-impact laser
Noise level	Generally noisy	Generally noisy	Quiet	Quiet	Very quiet
Maximum print speed	10–120 cps	60–240 cps	10–200 cps	80–200 cps	4–10 pages per minute
Character formation	Typewriter like	Depends upon dot density; 7 × 7, 9 × 7, 18 × 9 and 24 × 9	Depends upon dot density; 5 × 7 typical	Depends upon dot density; 12 × 9 typical	Depends upon software
Legibility	Good–excellent	Good–excellent	Fair	Good–excellent	Excellent
Copies	Usually 46	Usually 46	Original only	Original only	Original only
Paper	Ordinary	Ordinary	Treated	Ordinary	Ordinary
Paper feed	Friction, pin	Friction, pin	Friction	Friction, pin	Friction
Forms	Can be preprinted	Can be preprinted	Difficult to use	Can be preprinted	Can be preprinted
Line width (character positions)	Usually 4132	Usually 4132	Usually 480	Usually 4132	Usually 4132

CRT

The development of the cathode ray tube (CRT) is based on television technology, incorporating a moving electron beam to excite a phosphor coating on the inside of a glass tube. This results in predefined points within a matrix of points becoming illuminated, forming predefined characters. Since its appearance in 1965, the CRT terminal has revolutionized the field of data communications, rapidly replacing most teletype devices.

The main components of a CRT terminal include a keyboard for data input, a picture tube onto which both keyboard input and received data are displayed, control circuits that govern the operation of the terminal and a metal or plastic housing. Some CRT terminals have a keyboard built into a common housing, while other terminals have a stand-alone keyboard cabled to the housing containing the picture tube and control logic.

The display area of most CRTs is 80 columns by 24 or 25 rows. Although the first generation of CRT terminals used fixed logic and can be classified as dumb devices in most instances, they provide much more capability than obtainable through the use of conventional teletype terminals. This additional capability was primarily in the areas of display features and editing.

Display features of CRT terminals typically include reverse video obtained by the display of a negative image of data, blinking of data, underlining data and programmable levels of brightness in which the intensity of a display area is altered to correspond to the desired level of brightness.

The display capability of a CRT depends on its control logic which is designed to recognize codes which specify the display function to be performed. A CRT designed to display characters may require two characters to be transmitted to it for each character to be displayed. The first character is the character that is

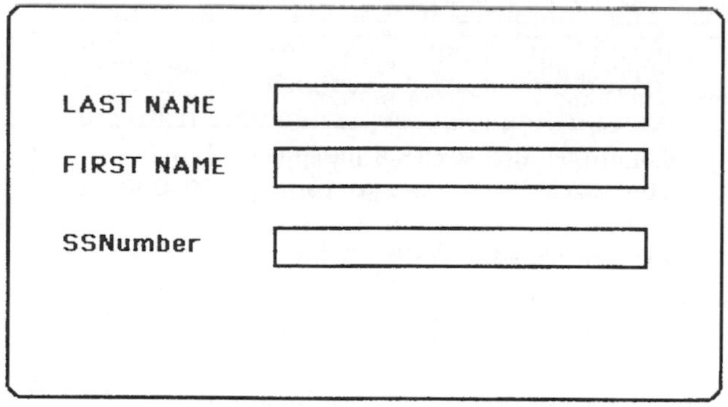

Figure 5.3 Protected field utilization. The use of protected fields (shown as rectangles) can be used to reduce the probability of operator keyboard mistakes occurring

displayed, and the second character, commonly called the character display attribute, defines the display feature associated with the character.

The editing features of CRT terminals are also dependent on the terminal's control logic. Most CRT terminals include four cursor control keys and Insert and Delete keys. The cursor control keys permit the operator to move a blinking symbol called a cursor which denotes the position of the next keyboard entry up one line, down one line, to the left one character position or to the right one character position each time the applicable cursor key is pressed. The Insert key permits characters to be entered at the cursor position, shifting all characters to the right of the cursor one position to the right as each character is entered. Pressing the Delete key causes the characters at the cursor to be deleted, and all characters to the right of the deleted character to be shifted to the left one position each time the Delete key is pressed.

Another significant capability incorporated into many first generation CRTs is the ability to support protected format display and data entry. A protected format display results from the terminal's logic receiving a code that defines the row and column where the field commences, its length in characters and the type of protection assigned to the field. Typical protected field formats include all numeric, all alphabetic and alphanumeric.

The use of protected fields facilitates 'fill in the blank' data entry operations. As an example, consider the CRT screen illustrated in Figure 5.3. This screen contains three field labels and three protected fields which are illustrated by rectangles. By assigning alphabetic protection to the last and first name fields and numeric protection to the SSNumber field, some potential operator keyboard entry mistakes are eliminated. As an example, an operator cannot enter a numeric value into an alphabetic field nor a letter of the alphabet into a numeric field.

Limitations

One of the problems associated with fixed logic CRT terminals is their support of only one set of video attribute codes to include codes used to support protected fields. This means that the terminal must communicate with a specific computer on which application programs have been developed that generate appropriate codes to obtain the desired CRT display capabilities. A second limitation of fixed logic CRT terminals involves the addressability of screen positions for displaying data transmitted to the terminal.

Terminal designers incorporated logic into CRT terminals to recognize codes related to screen positioning. This allowed, as an example, a short code to define the vertical and horizontal location where character display should commence and avoided the necessity of transmitting many lines of null characters to position the cursor to a desired position on the screen.

If a computer system's software is compatible with the positioning codes built into a CRT terminal, the transmission efficiency of the computer to the terminal can be substantially increased. Unfortunately, the manufacturers who developed many first generation CRT terminals were also computer system manufacturers. Since computer systems designed in the 1960s and 1970s were incompatible with one another, the development of CRT terminals to operate with a specific computer system basically ensured that that terminal was not compatible with computer systems manufactured by other vendors. The only exception to this incompatibility is CRT terminals that operate on a line by line display basis and recognize a standard character code, such as ASCII. Thus, the term 'ASCII display terminal' commonly refers to a CRT terminal that transmits and receives data on a line by line basis asynchronously. A second common term used to refer to this type of terminal is 'teletype compatible display', since the CRT terminal in essence functions as a teletype terminal. Both terms are synonymous and should be compared to the terms 'full screen ASCII' and 'full screen display' which are used to refer to CRT terminals that recognize screen position codes which enable a transmission to the terminal to display data at any location on the screen.

CRT terminals that emulate teletype line by line display capability are normally classified as dumb terminals. When control logic is added to CRT terminals to provide full screen positioning capability as well as the display and editing features previously discussed, the terminal is known as a smart terminal. Both dumb and smart terminals are restricted to recognizing one particular set of predefined codes, and their utilization is either limited to line by line 'teletype compatible' operations or for communications with one type of vendor's computer system.

The role of the microprocessor

The invention of the microprocessor by Intel Corporation resulted in the well-known growth of personal computing. Perhaps equally as important but less formally recognized, the microprocessor has also revolutionized communications technology.

In the area of CRT terminals the use of microprocessors resulted in the development of intelligent or programmable devices. Some of the earliest intelligent CRT terminals incorporated cassette and cartridge storage. This permitted the operator to insert a cassette or a cartridge containing read-only memory (ROM) whose contents were read under the control of the microprocessor. The data read into the terminal's memory area (RAM) represents program instructions which are then executed by the microprocessor.

One of the most common functions performed by microprocessors incorporated into CRT terminals is terminal emulation. The process of terminal emulation can be defined as the coding of software and/or firmware which enables one terminal to mimic the operation of another terminal. Software which converted received Baudot code to ASCII for display and keyboard-entered ASCII characters to their equivalent Baudot code for transmission would thus represent a Baudot terminal emulation capability for an ASCII terminal.

Currently, the majority of terminals used by businesses and home users are ASCII devices. The IBM PC series and compatible products manufactured by hundreds of vendors can be considered to have an ASCII terminal capability. In fact, the personal computer has resulting in a significant contraction in the market for conventional ASCII CRT terminals. This is because, in addition to functioning as one of several types of full screen terminals with appropriate emulation software, the personal computer can also perform local processing. This permits operators to compose memorandums, use spreadsheets to perform calculations, write programs in a programming language, as well as use the personal computer as a terminal. In addition, diskette and fixed disk storage provides operators with the ability to store and retrieve correspondence and programs, which significantly improves the efficiency of most office operations.

The first generation of ASCII terminals were dumb devices that were primarily used for switched network operations. This required the terminal operator to dial the telephone number of a telephone line connected to an auto-answer modem which in turn was connected to a computer port. Figure 5.4 illustrates the typical utilization of ASCII terminals in accessing computers over the public switched telephone network.

The type of communications connection illustrated in Figure 5.4 is called a point-to-point connection. This type of communication requires the use of a business line, a pair of modems and a computer port for each terminal which will simultaneously 'communicate with the computer. For business organizations with many terminals installed at a common location, the cost associated with providing one business line, a pair of modems and a computer port for each terminal user could become prohibitive. A method was thus required to economize on the communications equipment and facilities required to support a large number of terminals at different locations or many terminals installed at a common location. One method developed to overcome the

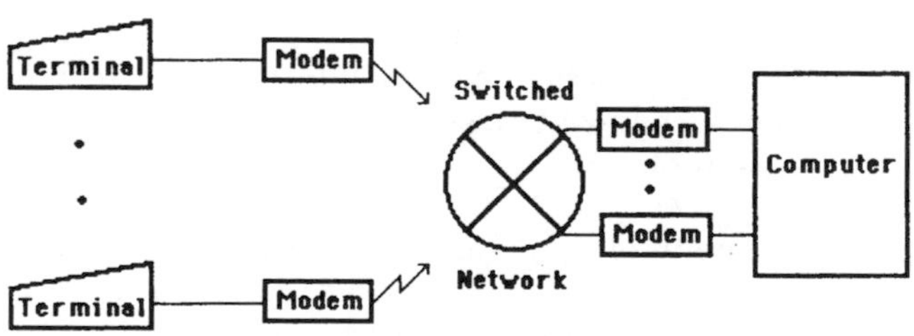

Figure 5.4 Point-to-point dial-up communications. Point-to-point dial-up communications requires the use of a business line, pair of modems and a computer port for each terminal that will simultaneously communicate with the computer

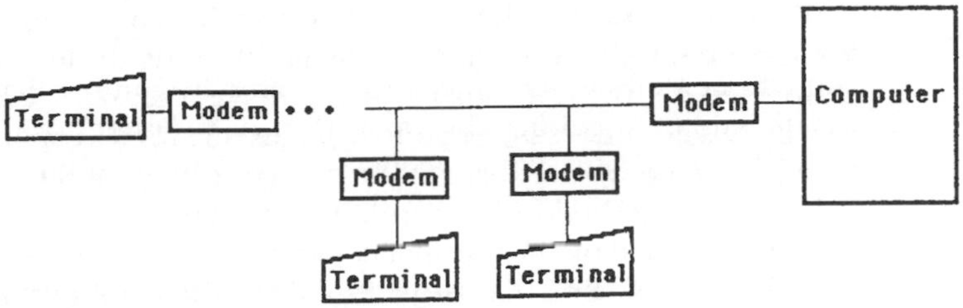

Sharing a control unit

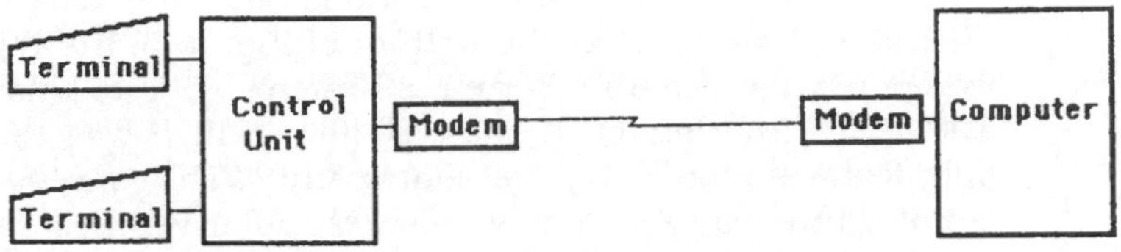

Figure 5.5 Memory and addressability allow facility sharing

previously mentioned constraint of each ASCII terminal requiring the use of separate communications lines and equipment was the development of synchronous terminals that were addressable and which contained memory that could be used to buffer keyboard input.

Addressability enabled terminals at geographically dispersed locations to share the use of a common line, as illustrated in the top portion of Figure 5.5. Since each terminal was addressable they could be controlled by software transmitted from a single computer port. The software could address each terminal in a specific order to determine if the terminal had data to transmit. Then, if the terminal had data to transmit and responded to the computer's inquiry positively, the computer would tell the terminal it could transmit once the computer was ready to receive. This process is known as polling and selecting.

Since one terminal operator could be entering data while the computer was occupied with receiving data from another terminal, a buffer area was required to temporarily store data. Originally, this buffer area consisted of RAM memory equal in byte capacity to the number of characters that could be displayed on the CRT screen. Since then the memory capacity of terminals has significantly increased, with many synchronous terminals becoming capable of storing a minimum of several screens of data. This permits the terminal operator to store two or more 'pages' of data and select any screen by number or by scrolling.

When the terminal operator presses his or her transmit key, the terminal is placed in a 'locked' condition until the information on the screen is transmitted to the computer. To denote this condition, the first generation of synchronous terminals included a 'Busy' button. This button was illuminated when the operator pressed the transmit key and would stay illuminated until the computer accepted the contents of the screen transmitted by the terminal.

When the computer has data to transmit to a specific terminal, it first polls the terminal using the address assigned to the device. Although all terminals connected to the multidrop line receive the poll, only the terminal that recognizes its address responds. If the terminal responds that it is ready to receive, the computer selects the device to receive data.

A second method of economizing on communications facilities and hardware is obtained by the use of a control unit. As illustrated in the lower portion of Figure 5.5, a control unit is useful for those situations where a number of terminals are colocated and require access to a common computer. In this type of communications environment, the computer polls the terminals through the control unit, with the control unit performing many processing functions to improve communications efficiency. The communications configuration illustrated at the bottom of Figure 5.5 is representative of many computer manufacturer terminal systems. Due to this we will examine the IBM 3270 Information Display System, which was the first series of communications products to use control units. The concept behind the IBM 3270 architecture was adopted by most computer vendors and today represents a popular method used to provide full screen terminal access to persons employed in business, industry and government.

The IBM 3270 Information Display System

The IBM 3270 Information Display System describes a collection of products ranging from display stations with keyboards and printers that communicate with mainframe computers through several types of cluster controllers, as well as a member of the IBM PC series known as the 3270 PC.

First introduced in 1971, the IBM 3270 Information Display System was designed to extend the processing power of the mainframe computer to locations remote from the computer room. Controllers, which are called control units by IBM, were made available to economize on the number of lines required to link display stations to mainframe computers. Typically, a number of display stations are connected to a control unit on individual cables and the control unit, in turn is connected to the mainframe via a single cable. Both local and remote control units are offered, with the key differences between the two pertaining to the method of attachment to the mainframe computer and the use of intermediate devices between the control unit and the mainframe.

Local control units are usually attached to a channel on the mainframe, whereas remote control units are connected to the mainframe's front-end processor, which is also known as a communications controller in the IBM environment and which is described later in this chapter. Since a local control unit is within a limited distance of the mainframe, no intermediate communications devices, such as modems, are required to connect a local control unit to the mainframe. In comparison, a remote control unit can be located in another building or in a different city and normally requires the utilization of intermediate communications devices, such as a pair of modems or DSUs for communications to occur between the control unit and the communications controller. The relationship of local and remote control units to

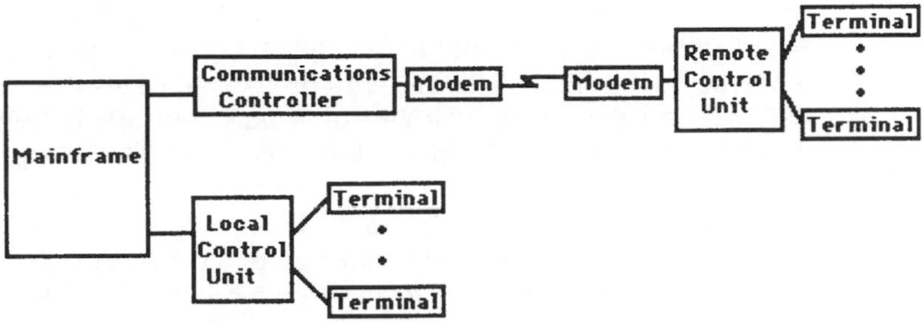

Figure 5.6 Relationship of 3270 information display products. In a 3270 environment, the terminals are also known as display stations

display stations, mainframes and a communications controller is illustrated in Figure 5.6.

Control unit operation

The control unit polls each connected display station to ascertain if the station has data stored in its transmit buffer. If the station has data in its buffer, it will transmit it to the control unit when it is polled. The control unit then formats the data with the display station's address, adds the control unit's address and other pertinent information and transmits it in a synchronous data format to the communications controller or to the I/O channel on the mainframe, depending on the method used to connect the control unit to the mainframe.

3270 protocols

Two different protocols are supported by IBM to connect 3270 devices to a mainframe. The original protocol used with 3270 devices which is in limited use today is the byte-oriented bisynchronous protocol, often referred to as 3270 bisync or BSC. In the late 1970s, IBM introduced its Systems Network Architecture, which is basically an architecture that permits distributed systems to be interconnected based upon a series of conventions which includes a bit-oriented protocol for data transmission known as Synchronous Data Link Control, or SDLC. Communications between an IBM mainframe and the control units attached to the communications controller will thus be either BSC or SDLC, depending on the type of control units obtained and the configuration of the communications controller which is controlled by software. Both BSC and SDLC protocols are discussed in Chapter 11 of this book.

Types of control unit

Control units marketed support up to 8, 16, 32 or 64 attached devices, depending upon the model. The IBM 3276 control unit supports up to eight devices while the IBM 3274 control unit can support 16, 32 or 64 attached devices. Older control units, such as the 3271, 3272 and 3275 have largely been replaced by the 3274 and 3276 and only operate bisynchronously, whereas certain models of the 3274 are 'soft' devices that can be programmed with a diskette to operate with the originally developed bisynchronous protocol or with the newer synchronous data link control (SDLC) protocol.

Devices to include display stations and printers are normally attached to each control unit via coaxial cable. Thus, under this design philosophy, every display station must first be connected to a control unit prior to being able to access a mainframe application written for a 3270 type terminal. This method of connection excluded the utilization of dial-up terminals from accessing 3270 type applications and resulted in numerous third-party vendors marketing devices to permit lower cost ASCII terminals to be attached to 3270 networks. In late 1986, IBM introduced a new controller known as the 3174 Subsystem Control Unit. This controller can be used to connect terminals via standard coaxial cable, shielded twisted-pair wire and telephone type twisted-pair wire. Other key features of this controller include an optional protocol converter which can support up to 24 asynchronous ports and the ability of this controller to be attached to IBM's Token-Ring Local Area Network.

Recent releases of the 3174 are referred to as a Network Processor and include support for Ethernet and Token-Ring LAN connections as well as the ability to connect to X.25 and frame relay packet networks. A single terminal connected to a 3174 Network Processor can establish up to five simultaneous sessions, providing connectivity to mainframes, AS/400 minicomputers, and TCP/IP devices. This capability enables a terminal user to click on different windows on their screen to work with different applications.

Terminal displays

Although IBM has essentially replaced its series of 3270 terminal displays with PCs with appropriate graphics adapters and software, it is important to obtain an appreciation of the different models of terminal in the 3270 Information Display System. This is because PCs used as a specific type of 3270 terminal must be obtained with appropriate hardware and software to emulate the terminal.

IBM 3270 terminals fall into three display classes: monochrome, color and gas plasma. Members of the monochrome display class include the 3278, 3178, 3180, 3191 and 3193 type terminals. The 3278 is a large, bulky terminal that easily covers a significant portion of one's desk and was replaced by the 3178, 3180, 3191 and 3193 display stations which are lighter, more compact and less expensive versions of the 3278. The 3279 color display station was similarly replaced by the 3179 and 3194 which are cheaper and more compact color display terminals. The last class of display stations is the gas plasma display, consisting of the 3270 flat panel display.

The physical dimensions of a 3270 screen may vary by class and model within the class. As an example, the 3178 and 3278 Model 2 display stations have a screen size of 24 rows by 80 columns, while the 3278 Model 3 has a screen size of 32 rows by 80 columns and the 3278 Model 4 has a screen size of 43 rows by 80 columns.

Table 5.2 IBM display stations

Model number	Display type	Screen (inches)
3178	monochrome	12
3179	color	14
3180	monochrome	15
3191	monochrome	12
3193	monochrome	15
3194	color	14

Table 5.3 3270 terminal field characteristics

Field characteristic	Result
Highlighted	Field displayed at a brighter intensity than normal intensity field.
Nondisplay	Field does not display any data typed into it.
Protected	Field does not accept any input.
Unprotected	Field accepts any data typed into it.
Numeric-only	Field accepts only numbers as input.
Autoskip	Field sends the cursor to the next unprotected field after it is filled with data.
Underscoring	Causes characters to be underlined.
Blinking	Causes characters in field to blink.

To use a personal computer to emulate a specific type of 3270 display at a minimum requires an appropriate graphics display adapter, as the adapter governs the ability of a computer to support different screen resolutions. In addition, appropriate software must be obtained to map the terminal field characteristics and keys supported by 3270 terminals to equivalent field characteristics and keys associated with a personal computer.

Table 5.2 lists the family of terminals marketed for use with the IBM 3270 Information Display System. Note that the 3179 and 3180 display stations can also be used with IBM System/36 and System/38 mini- computers.

Each 3270 screen consists of fields that are defined by the application program connected to the display station. Attributes sent by the application program further define the characteristics of each field as indicated in Table 5.3. As a minimum, any technique used to enable a personal computer to function as a 3270 display station requires the PC to obtain the field attributes listed in Table 5.3.

3270 keyboard functions

In comparison with the keyboard of a member of the IBM PC series and compatible computers, a 3270 display station contains approximately 40 additional keys, which, when pressed, perform functions unique to the 3270 terminal environment. A list of the more common 3270 keys which differ from the keys on an IBM PC keyboard is given in Table 5.4.

Table 5.4 Common 3270 keys differing from an IBM PC keyboard

Key(s)	Function
CLEAR	Erases screen except for characters in message area, repositioning cursor to row 1, column 1.
PA1	Transmits a code to the application program which interpreted as a break signal. Thus, in TSO or CMS the PA1 key would terminate the current command.
PA2	Transmits a code to the application program that is often interpreted as a request to redisplay the screen or to clear the screen and display additional information.
PFnn	24 program function keys on a 3270 terminal are defined by the application program in use.
TAB	Moves the cursor to the next unprotected field.
BACKTAB	Moves the cursor to the previous unprotected field.
RESET	Disables the insert mode.
ERASEEOF	Deletes everything from the cursor to the end of the input field.
NEWLINE	Advances the cursor to the first unprotected field on the next line.
FASTRIGHT	Moves the cursor to the right two characters at a time.
FASTLEFT	Moves the cursor to the left two characters at a time.
ERASE INPUT	Clears all the input fields on the screen.
HOME	Moves the cursor to the first unprotected field on the screen.

Since most, if not all, of the 3270 keyboard functions may be required to successfully use a 3270 application program, the codes generated from pressing keys on a personal computer keyboard must be converted into appropriate codes that represent 3270 keyboard functions to enable a PC to be used as a 3270 terminal. Due to the lesser number of keys on a personal computer keyboard, a common approach to most emulation techniques is to use a two- or three-key sequence on the PC keyboard to represent many of the keys unique to a 3270 keyboard.

Emulation considerations

In addition to converting keys on an IBM PC keyboard to 3270 keyboard functions, 3270 emulation requires the PC's screen to function as a 3270 display screen. The 3270 display terminal operates by displaying an entire screen of data in one operation and then waits for the operator to signal that he or she is ready to proceed with the next screen of information. This operation mode is known as 'full-screen' operation and is exactly the opposite of teletype (TTY) emulation where a terminal operates 'on a line-by-line' basis.

A key advantage of full-screen editing is the ability of the operator to move the cursor to any position on the screen to edit or change data. To use an IBM PC as a 3270 display station, the transmission codes thus used to position the 3270 screen and effect field attributes must be converted to equivalent codes recognizable by the PC.

Remote batch terminals

The first generation of remote batch terminals developed in the late 1960s were based on the use of minicomputers for processing power. The minicomputer supported such peripheral devices as a card reader and card punch, card reader/punch, disk driver, magnetic tape unit and high speed printer. The RBT enabled operators at locations remote from an organization's central computer facility to transmit batched jobs represented by previously keypunched cards that were either read from a card reader, or if previously transferred to tape or disk, read from tape or disk storage. Once a job was executed on the mainframe computer it was placed in an output queue called SYSOUT (system output). The remote batch terminal could then 'pull' SYSOUT, directing the output from the mainframe to the RBT's printer or to disk or tape storage, from which it could be printed.

In addition to providing the ability to transmit batched jobs to the mainframe and retrieve SYSOUT, most RBTs include a local processing capability. This processing capability enables an RBT to execute programs locally. Some RBTs include a foreground/background processing capability, permitting its communications program to be executed as a foreground task receiving priority processing while it executes local processing jobs as a background task. RBT terminals are synchronous devices whose data rates normally exceed 4800 bps. With the growth in the popularity of personal computers, a second generation of remote batch terminals was developed during the early 1980s based upon the use of this class of computers. Several vendors including remote batch terminal manufacturers as well as third parties developed adapter cards designed for insertion into the system unit of personal computers. Each adapter card contains a synchronous serial port as well as ROM modules, which, when supplemented by software, emulate the functions previously performed by minicomputer-based remote batch terminals.

5.2 WORKSTATIONS AND OTHER LAN COMPONENTS

The term workstation is used to represent a computer that functions as a participant on a local area network. To obtain the ability to become a participant on a LAN requires a combination of hardware and software as well as the installation of a cabling infrastructure.

5.2.1 Network interface card

A network interface card (NIC), which is also commonly referred to as a network adapter card, is normally installed in the system unit of a personal computer and provides an interface between the computer on the media in the form of a LAN cable. In addition to providing a physical interface to the LAN, the adapter card contains instructions, usually in the form of read only memory, which perform network access control functions, as well as the framing of data for transmission onto the network and the removal of framing from data received from the network.

To illustrate an example of access control performed by a network adapter, let us assume that the LAN is contention-based. Then, the adapter card will 'listen' to the LAN prior to attempting to transmit data. If no activity is heard on the LAN the adapter card will transmit data onto the media, whereas the presence of activity will result in the adapter deferring transmission and returning to a listening state.

One of the problems associated with an access methodology based upon contention is the fact that two or more workstations that listen and have data to transmit will do so. As you might expect, this will result in the collision of data, and a network adapter which supports a contention access scheme will contain collision detection circuitry. Then, once a collision has been detected each adapter card will employ circuitry which results in the generation of a random time interval to be used prior to attempting to retransmit. This method of LAN access is referred to as Carrier Sense Multiple Access with Collision Detection (CSMA/CD) and represents the access method used on Ethernet LANs. Later in this book we will examine in detail the access methods and data flow associated with several types of local area networks to include Ethernet.

5.2.2 Hubs

Most local area networks developed during the 1970s were based on the use of coaxial cable as a transmission media. The resulting network topology was a long cable into which workstations were connected and the backbone cable formed a bus structure, hence the term bus-based LAN.

The development of twisted pair cable that could transport high speed LAN traffic resulted in the introduction of hubs that were used as wiring centers to support the attachment of individual workstations. Figure 5.7 illustrates the use of a hub to connect eight workstations to a LAN. By cabling one hub to another you obtain the ability to expand the number of workstations connected to the LAN. To accomplish this most hubs are now manufactured as stackable, allowing one hub to be easily placed above another inside a wiring closet.

There are two major types of hub: those that function as repeaters and those that dynamically switch data between ports based on the destination address carried within a data unit. As you might expect, the latter type of hub is referred to as a switching hub.

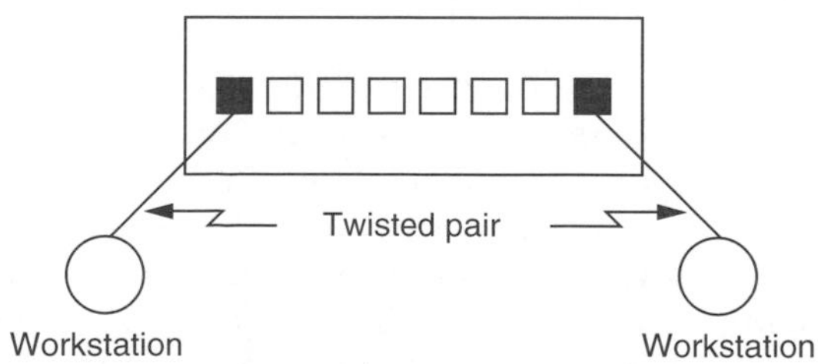

Figure 5.7 Using a hub as a wiring center

The earliest type of hub was the repeating hub, which is still commonly used. This type of hub receives information on one port and broadcasts the data onto all other ports, functioning as a large repeater. Since each data unit contains the source address of the transmitting workstation and the destination address for the data, only the workstation with the destination address of the data will read the data. Because the use of the media is shared among all workstations, another name for this type of hub as well as the network formed from its use is 'shared-media'.

The operating rate of a shared-media hub is shared among users. For example, a 10 Mbps network with ten active users results in each user obtaining an average transmission capability of 10 Mbps/10 users, or 1 Mbps per user.

A key limitation of a hub functioning as a repeater is its data throughput capability. For example, consider a 10 port hub used to form a 10 Mbps LAN. On a per port basis each workstation on the average obtains a data transfer capability of 10 Mbps/10 or 1 Mbps. Although this may still seem high, as LANs expand the average data transfer capability correspondingly decreases. Thus, interconnecting ten 10 port hubs to form a 100 station LAN would result in an average data transfer capability of 10 Mbps/100 or 100 kbps.

To overcome the throughput constraints associated with repeating hubs, vendors developed switching hubs. Since an n port switch can support up to $n/2$ simultaneous connections a switching hub can significantly increase the average throughput of workstations connected to a LAN. Thus, a 10 port switching hub used to create a 10 Mbps LAN would enable a throughput of up to 50 Mbps, since it could support up to 10/2 or five simultaneous connections between LAN devices, each operating at 10 Mbps. However, if only two servers were connected to individual switch ports, its practical data transfer capability would be limited to 20 Mbps. This is because most LAN communications involve client–server interaction.

5.2.3 File server

A file server represents a special type of workstation where LAN management functions reside in the form of software which performs network operating system functions. Those functions include the creation and control of LAN user accounts, application program sharing among network users, controlling network printing and other network related functions.

The network operating system on a file server requires a mechanism to operate in tandem with workstations on a LAN, responding to network-related requests issued by workstation users. To accomplish this, one or more software modules are normally loaded onto each LAN workstation. Those modules perform two key functions. First, they function as a filter, examining commands entered at the workstation to determine if they are local or network-related commands. Local commands are passed to the workstation's operating system for execution on the workstation. In comparison, a network-related command must be transported to the file server for execution. To accomplish this function a second software module converts the network-related request into a format for transmission that the file server can interpret. Figure 5.8 illustrates the relationship of the data flow between a workstation and a file server.

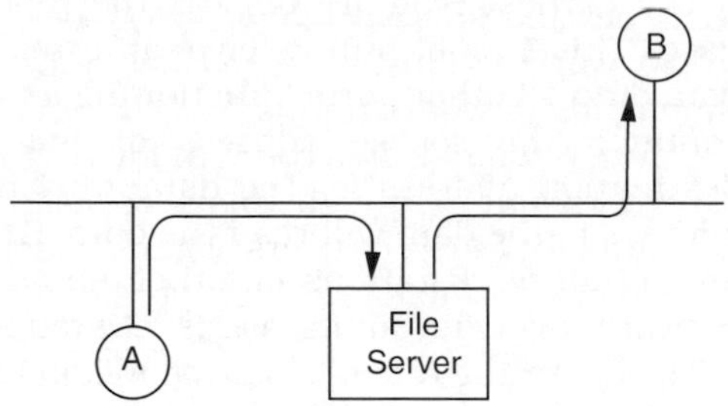

Workstation A transmitting a message to workstation B on a server-based network first transmits the message to the server. The server then transmits the message to workstation B

Figure 5.8 Dataflow on a server-based LAN

The use of a network operating system which depends on one or more file servers is known as a server-based LAN. All communications from workstations are routed through the server and this type of processing is known as client–server processing where each workstation is a client of a server. Another type of LAN which avoids the use of a server is a peer-to-peer LAN. In a peer-to-peer LAN data flow from one workstation to another until they reach

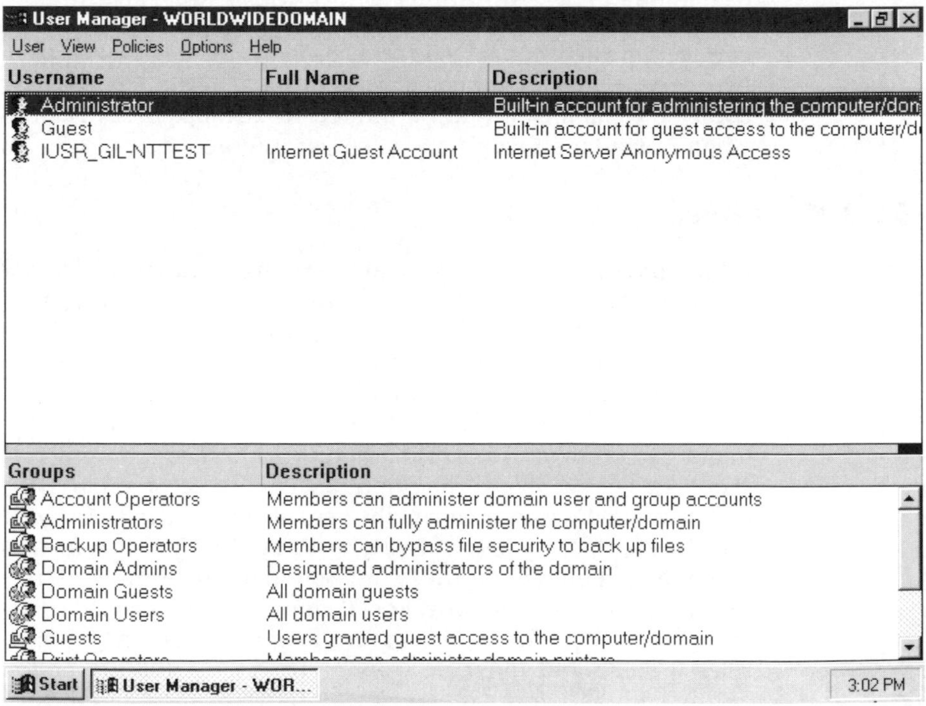

Figure 5.9 The Windows NT/Windows 2000 Server User Manager display provides network administrators with the ability to enable users to access different resources through the establishment of user accounts

their destination. Normally peer-to-peer LANs are limited to supporting 10 to 20 workstations, whereas server-based LANs can support hundreds to thousands of workstations.

Figure 5.9 illustrates the Windows NT/Windows 2000 Server User Manager screen display. Through this screen a network administrator can establish accounts for different users. Then, by assigning a user to one or more groups, it becomes possible to assign predefined access rights to different resources.

5.2.4 Print server

A print server was originally a personal computer which operated special LAN software, which enabled it to service network printing requests. Since it was based on a PC platform the print server can support multiple printers. Although print servers based on the use of personal computers are still commonly used, newer print servers consist of a miniature network card and software in read only memory and a parallel port packaged in a housing no bigger than a pack of cigarettes. This type of print server eliminates the necessity to dedicate a computer to network printing; however, most miniature print servers are limited to supporting one printer.

On a file server based LAN print queues are established on the server which are associated with printers attached to print servers. When a workstation transmits a network printing job the print job first flows into a print queue on a file server. The file server then transmits the print job to the appropriate print server which directs the job to an appropriate printer connected to the print server. Figure 5.10 illustrates the flow of a print job on a file server based LAN.

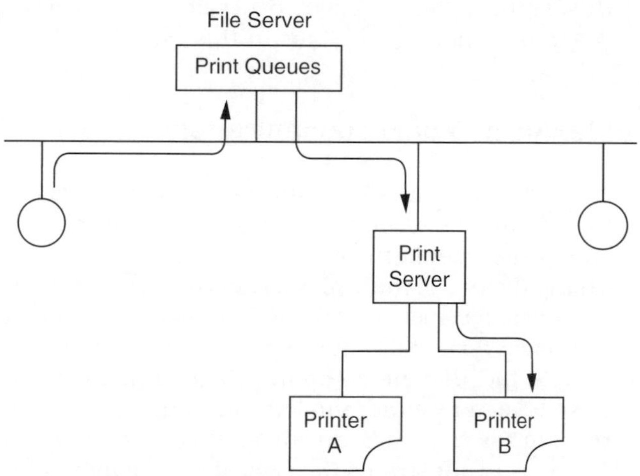

A print job directed to printer B first flows into a print queue on the file server. The file server transmits the print job from the queue to the print server which directs the job to the requested printer.

Figure 5.10 Transmitting a print job on a server-based LAN

5.2.5 Other types of servers

Three additional popular types of servers are communications servers, database servers, and Web servers. A communications server has a number of connections to the public switched telephone network and/or to a packet network as well as a connection to a LAN. This enables persons beyond the boundary of the LAN to access the network via remote communications. A database server represents a special type of file server which structures the contents of files so that data can be rapidly retrieved. The third type of popular server is a Web server, which supports HyperText Transmission Protocol (HTTP) queries in the form of Uniform Resource Locator (URL) addresses and returns Web pages. When we discuss the Internet later in this book, we will turn our attention to Web servers.

It is important to note that most LANs contain several types of servers that have a predefined relationship to one another. For example, remote users may access a LAN via the use of a communications server in attempting to display a particular Web page by accessing a Web server. If the user needs to search for information or place an electronic order, his or her request may be forwarded from the Web server to the database server and the results sent back to the Web server for transmission to the user, with the functions of the communications server and database server normally transparent to the user.

5.3 WIDE AREA NETWORKING OVERVIEW

In this section we will focus our attention on obtaining an overview of the operation, and the utilization of some more commonly employed wide area network communications hardware products. Each of the products covered in this section can be used to develop large corporate networks, and will be covered in more detail later in this book.

5.3.1 Multiplexing and data concentration

As an economy measure multiplexers and data concentrators were developed to combine the serial data stream of two or more digital data sources onto one composite communications line. Figure 5.11 illustrates how economics can justify the use of multiplexers. In the top portion of Figure 5.11, two terminals located at a remote location with respect to a computer center are shown. Note that each terminal requires the use of a separate communications facility to access the distant computer. The transmission devices represented by a boxed T can be either modems or data service units, depending upon whether transmission occurs on an analog or digital facility. The lower portion of Figure 5.11 illustrates the typical employment of multiplexers, which are also referred to as data concentrators, since they 'concentrate' data from several sources onto one line.

In comparing the top and bottom portions of Figure 5.11, note that the use of a pair of multiplexers permits many terminals to share the use of a

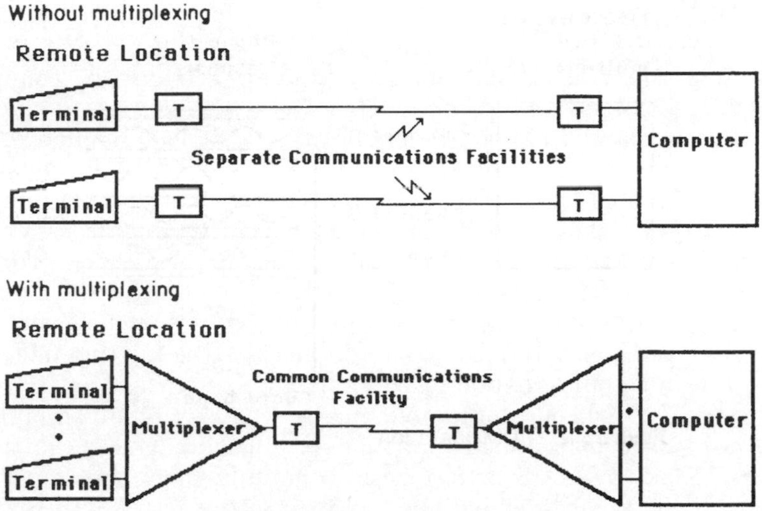

Figure 5.11 Economics justify multiplexing

common communications facility and one pair of modems or data service units. In comparison, without multiplexing, each terminal would require the use of a separate communications facility and pair of modems or digital service units.

The time division multiplexer (TDM), statistical time division multiplexer (STDM) and intelligent time division multiplexer (ITDM) can be used on both analog and digital facilities. In comparison, the frequency division multiplexer (FDM) can only be used on analog facilities, while the T1 multiplexer can only be used on a digital T1 facility.

Frequency division multiplexer (FDM)

The frequency division multiplexer (FDM) operates by subdividing the 3000 Hz bandwidth of an analog leased line into subchannels or derived bands. As data enter, the FDM tones are placed on the line in a predefined subchannel to correspond to the bit composition; that is, a tone is placed on the line in the subchannel at one frequency to correspond to a binary 1 while a second tone at another frequency is placed on the line to correspond to a binary 0.

Figure 5.12 illustrates the operation of a four-channel FDM. Since each digital data stream is converted into a series of tones at predefined frequencies, no modem is required when this multiplexer is used. Here a device known as a channel set which converts the digital data stream into tones and the received tones into digital data replaces the modem.

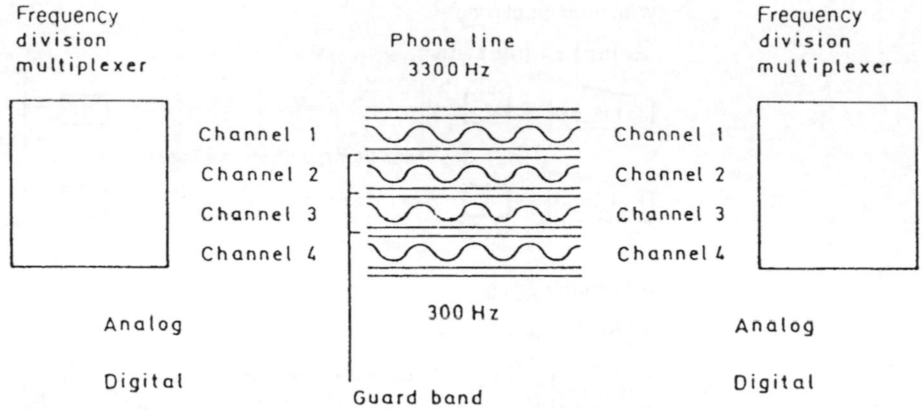

Figure 5.12 FDM operation

Time division multiplexer (TDM)

The time division multiplexer (TDM), as its name implies, separates the high-speed composite line into intervals of time. The TDM assigns a time slot to each low-speed channel whether or not a data source connected to the channel is active. If the data source is active at a particular point in time, the TDM places either a bit or a character into the time slot, depending on whether the TDM performs bit interleaving or character interleaving. If the data source is not active, the TDM places an idle bit or character into the time slot. At the opposite end of the high-speed data link another multiplexer reconstructs the data for each channel based on its position in the sequence of time slots. This process is called demultiplexing.

The top of Figure 5.13 illustrates the use of TDMs to multiplex four data sources onto one high-speed line. The lower portion of Figure 5.13 illustrates how the TDMs interleave bits or characters into time slots on the high-speed line at one end of the communications link (multiplexing) and remove the bits or characters from the high-speed line (demultiplexing), sending them to their appropriate channels.

Statistical (STDM) and intelligent time division multiplexers (ITDM)

Both statistical and intelligent time division multiplexers (STDM and ITDM) are more efficient than TDMs due to their use of a technique commonly referred to as the dynamic allocation of bandwidth. In this technique, a microprocessor in the multiplexer only places data into a time slot when a device connected to the multiplexer is both active and transmitting. Since a time slot can now hold data from any channel, the dynamic allocation of bandwidth requires that an address indicating from what channel the data originated be included in the slot to enable demultiplexing to occur correctly.

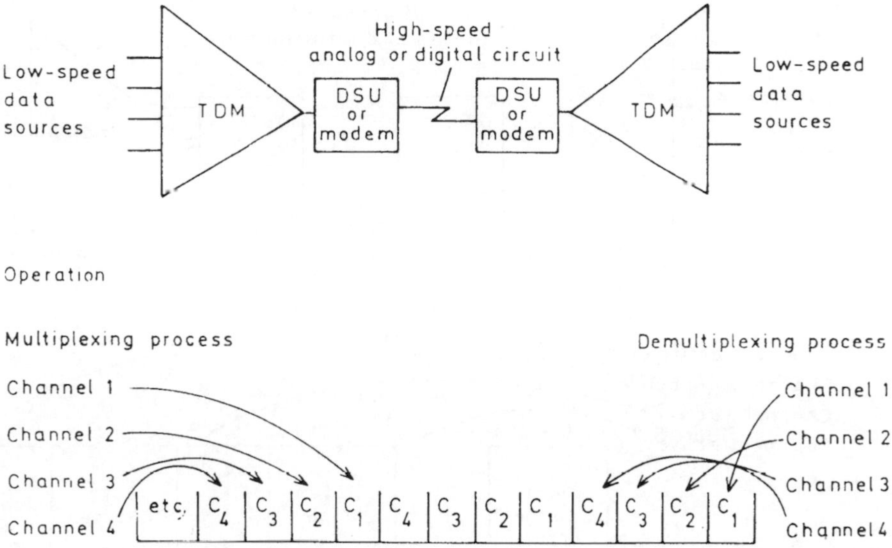

Figure 5.13 Time division muitiplexing

The primary advantage of STDMs and ITDMs over conventional TDMs is their greater efficiency in transmitting and receiving data on the high-speed line connecting two multiplexers. Although STDMs and ITDMs both employ the dynamic allocation of bandwidth, ITDMs also perform data compression. Since this reduces the amount of data that has to be transmitted, it results in the ITDM being the more efficient device. STDMs and ITDMs service between one and a half and four times as many data sources as a TDM.

If all or a majority of the data sources connected to an STDM or an ITDM became active at the same time, the buffer in these multiplexers would overflow, causing data to be lost. This situation would occur because the composite data rate of active channels would exceed the operating rate of the high-speed line, causing the buffers in the multiplexers to fill and eventually overflow. To prevent this situation from occurring, both STDMs and ITDMs incorporate one or more techniques that inhibit data transmission into the multiplexer when the data in its buffer reach a predefined level. Then, after data have been transferred from the buffer onto the composite high-speed line the buffer occupancy is lowered until another predefined level is reached. This lower level then becomes a trigger mechanism for the multiplexer to enable previously inhibited data sources to resume transmission.

The process of inhibiting and enabling data transmission is known as flow control. The most common method of flow control is obtained by lowering and raising the Clear to Send control signal. Another common method of flow control is obtained by having the multiplexer transmit the XOFF and XON characters to devices that recognize those characters as a signal to stop and resume transmission. Figure 5.14 illustrates the statistical multiplexing process.

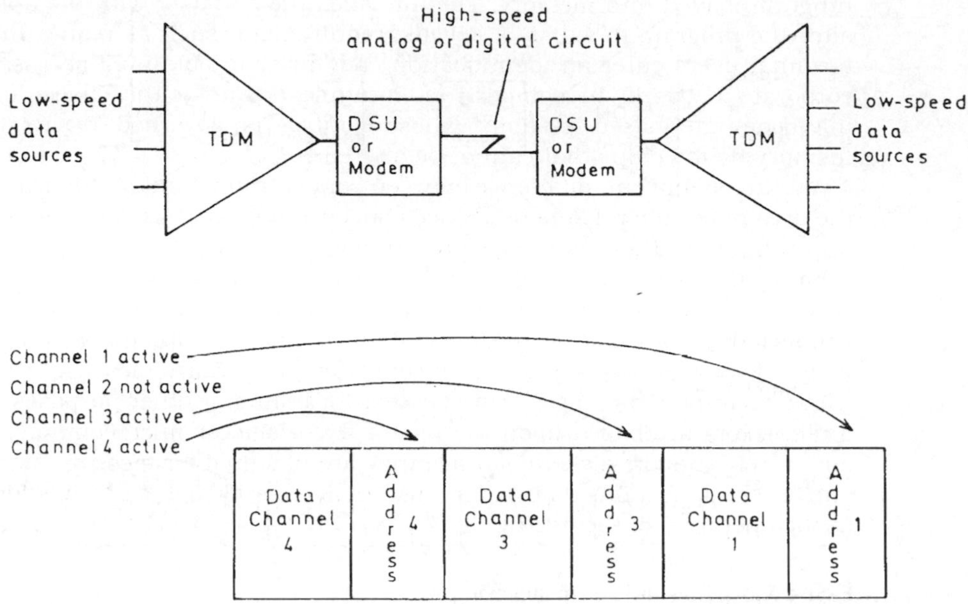

Figure 5.14 Statistical multiplexing

T1 multiplexer

The T1 multiplexer is a data concentration device specifically designed to operate at the T1 data rate of 1.544 Mbps in North America and 2.048 Mbps in Europe. Due to the high data rate that this devicc supports, it is used to multiplex both voice and data as well as video onto a single channel.

Most T1 multiplexers are marketed with several optional voice digitization modules that end-users can select. The most common module digitizes voice based upon a technique known as pulse code modulation (PCM) into a 64 kbps data stream. If used only to support digitized voice resulting from PCM, a T1 multiplexer would enable 24 voice conversations to be multiplexed onto one T1 line.

Many T1 multiplexers include modules that permit voice to be digitized at 32 kbps, while some multiplexers support modules that digitize voice at 16 kbps. With these modules users could multiplex 48 or 72 voice conversations onto one T1 circuit through the use of T1 multiplexers. Chapter 12 provides additional information covering the operation and utilization of the multiplexers briefly described in this section.

Concentrator

Although multiplexers are considered to be data concentrators, another similar device which predates most multiplexers is the concentrator. The concentrator is a minicomputer-based system which is controlled via a program loaded into memory. This provides the end-user with the ability to alter the program to satisfy a specific requirement, such as requesting that terminal users enter an identification code prior to obtaining access to the corporate network. In comparison, multiplexers are either fixed logic or microcomputer-based systems whose control is governed by read only memory (ROM) chips that cannot be altered.

Another significant difference between concentrators and multiplexers is in the area of peripheral support. Since concentrators are based on the use of a minicomputer, they can typically support magnetic tape systems, large capacity fixed disks and high speed printers. This enables a concentrator to service peak loads that exceed the capacity of the high speed circuit that it is connected to by temporarily storing data to disk. In addition, by the use of magnetic tape or disk storage, a concentrator can perform message switching where records of messages must be kept for billing or other purposes. Since multiplexers are based upon the use of fixed logic or microcomputers that only rarely support diskette storage they are ill suited for message switching and other applications that require historical record keeping or a modification in the operation of the device.

5.3.2 Front-end processor

The requirements for communications processing differ significantly from the requirements of conventional data processing. In communications processing, a large amount of processing time is spent performing serial-to-parallel and parallel-to-serial conversion, sampling pulse widths and determining character compositions. In comparison, conventional data processing systems primarily perform parallel operations, conducting additions, subtractions and comparing values. Thus, the architecture required for efficient communications processing differs from the architecture required to perform efficient data processing.

To relieve the burden of communications processing from mainframes, front-end processors were developed. These computers are designed to efficiently process communications-oriented data, as well as to prepare such data for further processing by the mainframe computer.

5.3.3 Network configurations

Figure 5.15 represents a 'typical' corporate communications network. Note that the data flow to and from a central point. This type of network is a hub- or star-based network. Other common network configurations (topology) include ring networks, bus-based networks and mesh networks, all of which are discussed later in this book.

At location A, a control unit is used to service a large number of co-located terminals. This control unit is connected via a pair of modems and an analog leased line to the front-end processor at the computer center. Terminals at locations B and C are assumed to require a minimal amount of access to the computer center and are shown to dial modems connected to the front-end processor via the use of the PSTN. At location D, a large number of terminals

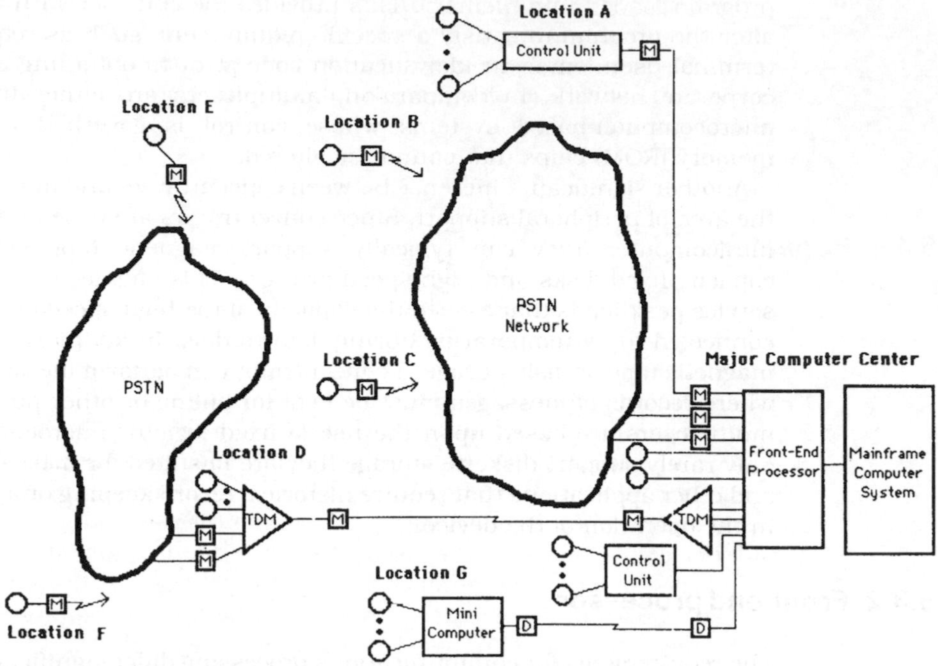

Figure 5.15 'Typical' corporate communications network. ○ terminal, [M] modem, [D] data service unit

are located that are serviced by a time division multiplexer. This multiplexer is shown attached to the PSTN via several business lines that are connected to modems, permitting terminal users at locations E and F to dial the multiplexer at location D instead of dialing long distance to the computer center. At location G, a minicomputer is shown connected via data service units and a digital transmission facility. If the minicomputer transmits and receives database information with the mainframe computer system, it is distributing the workload among computers. This type of application is commonly referred to as distributed processing. At the major computer center, a cluster of terminals is serviced by a local control unit, while other terminals are directly connected via cables to individual ports on the front-end processor.

The network illustrated in Figure 5.15 is a 'typical' corporate mainframe computer based communications network only in the sense that it shows the diversity of possible network structures based upon the use of different facilities and equipment. In actuality, due to the difference in organizational requirements, including the use of computers and terminal locations to be serviced, most networks differ from one another.

5.4 LOCAL AREA NETWORKING OVERVIEW

Although there are distinct differences between local and wide area networking equipment, data flow and networking techniques, there are also certain similarities between the two. In this section we will first focus our attention on obtaining an overview of equipment used to interconnect LANs. Once this has been accomplished we will turn our attention to the use of equipment which enables workstations on LANs to obtain access to mainframe computers.

5.4.1 Repeaters

As noted earlier in this book, digital signals are affected by noise, attenuation and other impairments which limit the distance over which they can be transmitted prior to the signal becoming unrecognizable. To extend the transmission distance of LANs, repeaters are used.

A repeater looks for a pulse rise and then ignores the pulse and regenerates a new pulse, creating an exact duplicate of the original signal.

Figure 5.16 illustrates the use of a repeater to extend the transmission distance on an Ethernet bus-based LAN. From a theoretical basis, an unlimited number of repeaters can be used to extend the distance of a LAN. From a practical standpoint the number of repeaters that can be used is limited, due to other constraints which limit the distance between workstations as a result of the method of signaling used on different networks.

In Figure 5.16, which illustrated the use of a repeater, each portion of the interconnected LAN was labeled as a segment. The term segment is used to represent a collection of workstations and other network devices that are treated as an entity. Through the use of a repeater all data transmitted on segment A are repeated or duplicated onto segment B, and vice versa. Thus, a repeater operates at the physical layer and does not concern itself with addresses.

5.4.2 Bridges

A bridge is similar to a repeater in that it is used to connect two LAN segments. However, unlike a repeater which simply duplicates all traffic on

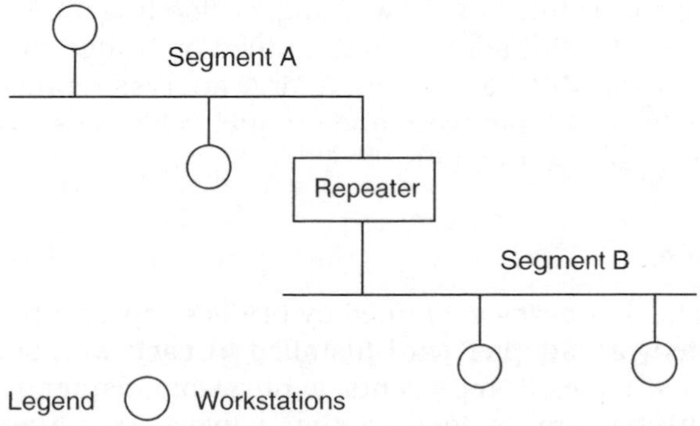

Figure 5.16 Using a repeater to extend a LAN

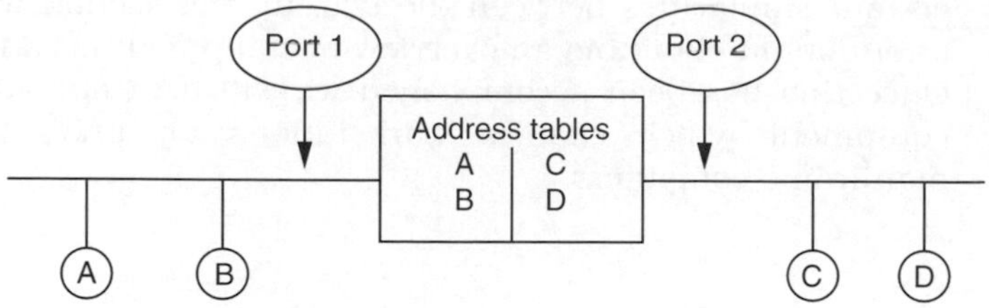

Bridges "learn" the addresses of workstations connected to each port by examining the source address in each transmission

Figure 5.17 Bridge address table construction

one segment to the other, a bridge reads addresses and makes decisions concerning the transfer of information from one segment to another based on those addresses.

Figure 5.17 illustrates the construction of address tables for a two-port bridge used to connect two LAN segments. For simplicity only two workstations are shown connected to each LAN segment.

As data are transmitted on each LAN segment both the source and destination addresses are included in the transmission. A bridge 'learns' the addresses of workstations connected to each port by examining the source address in each transmission, using those addresses to construct a table associated with a port. At the same time the bridge examines the destination address in each transmission, and compares that address to entries in its address table. If the address is in its address table, this indicates that the destination is on the LAN segment connected to the port. Thus, the bridge will not forward the data to the LAN segment connected to its other port. If the destination address is not in the address table this indicates that the destination is not on the current LAN segment. Thus, the bridge will forward the data onto the LAN segment connected to its second port. The examination of data is referred to as the bridge's filtering rate, and the transfer of data from one LAN to another is known as its forwarding rate. If a bridge does not have an entry for a destination address in its address table, it will transmit data onto every port other than the port on which it was received. This operation is referred to as flooding. Thus, a bridge performs filtering, forwarding, or flooding on data.

The actual addresses examined by bridges are referred to as media access control (MAC) addresses. A MAC address represents an address 'burnt into' a network adapter card and uniquely identifies a computer using the adapter to gain access to a LAN.

5.4.3 Routers

The addresses examined by bridges represent identifiers associated with the network adapter card installed in each workstation. Although each address is unique, it represents a physical adapter card and does not provide a mechanism to identify the network that the workstation containing the adapter card is located on.

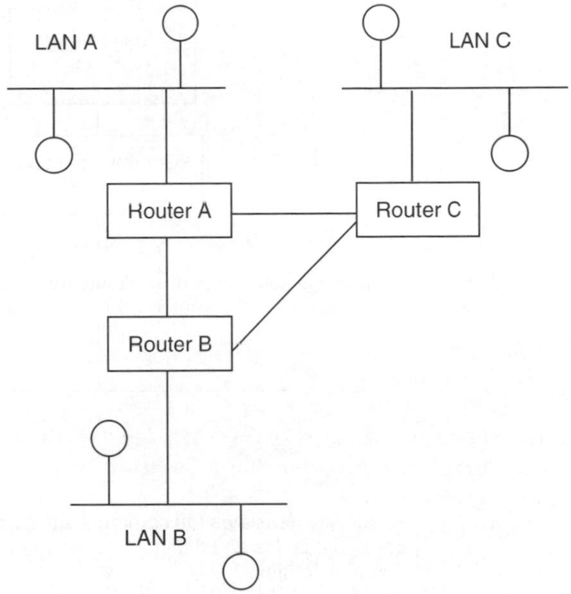

A router supports the addressing of network segments

Figure 5.18 Using routers

A router supports the addressing of network segments. This addressing is normally performed by a network administrator and it enables a logical network address to be associated with a group of workstations. This in turn enables a router to examine network addresses and make routing decisions based on those addresses. Figure 5.18 illustrates the use of three routers to interconnect three geographically separated local area networks. Although remote bridges could be used in place of routers, routers can be programmed to make intelligent decisions concerning the path to be used to forward data. For example, if the circuit connecting routers A and C were to fail, data from LAN A could be forwarded to LAN C via router B to router C. In addition to using logical rather than physical addresses, routers also differ from bridges in their use of routing algorithms. Various routing algorithms have been developed to calculate the best path through an internetwork based upon the time to reach a destination, the number of routers (hops) required to reach a destination or other metrics. Most routing algorithms are dynamic, and enable a new path to be established between two locations if a path in use becomes inoperative.

5.4.4 Gateways

A gateway represents a special type of networking device which provides a variety of translating or conversion functions. Although primarily used to enable workstations on a LAN to communicate with a mainframe computer,

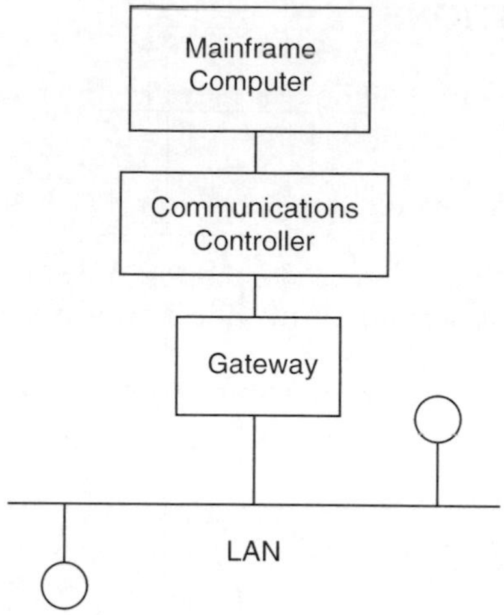

A gateway translates the communications protocol, data code, keystroke codes and data format between LAN workstations and a mainframe

Figure 5.19 Using a gateway to provide LAN workstations with connectivity to a mainframe computer

gateways can also be used to connect LANs to packet switching networks and similar applications that require translations of data.

To illustrate the general functions performed by a gateway, let us assume that we wish to provide workstations on an Ethernet LAN with access to application programs operating on an IBM mainframe computer.

Figure 5.19 illustrates the use of a gateway to connect an Ethernet LAN to an IBM SNA network. In this example it was assumed that the Ethernet LAN was located in the same building as the mainframe, resulting in the gateway being directly cabled to a port on the communications controller. Otherwise, the gateway would be connected via a leased analog or digital circuit, requiring the use of a pair of modems or DSUs.

The IBM communications controller transmits and receives data using the SDLC communications protocol. Thus, one function performed by a gateway is to translate the LAN transmission protocol into SDLC and vice versa. Since IBM mainframes use EBCDIC whereas workstations use extended ASCII, another function performed by a gateway is code conversion.

As previously noted in this chapter, IBM 3270 terminals include approximately 40 additional keys that are not found on a personal computer keyboard that are used to perform special predefined mainframe application recognized functions. To function as an IBM terminal to access mainframe applications each workstation will use emulation software which, for example, enables certain key combinations on a PC to represent unique 3270 keys. Thus, another function performed by a gateway is to translate emulation software codes that represent 3270 keys into appropriate EBCDIC code that actually represents the desired 3270 key.

5.5 REVIEW QUESTIONS

1. Discuss the difference between interactive and remote batch terminals with respect to their utilization.

2. Discuss the difference between RO, KSR and ASR terminals.

3. What is a 'teletype' compatible terminal? Why is the use of the word 'teletype' in the first portion of this question actually improper?

4. Discuss the relationship between fixed control logic, microprocessors and dumb and intelligent terminals.

5. How does a terminal recognize received data bits to include the transition in the state between bits?

6. Discuss two techniques used to increase the print quality of a dot matrix printer.

7. Why are ASR and KSR teletype terminals limited to providing an elementary editing capability?

8. What is the purpose of a character display attribute?

9. What does the term 'teletype-compatible display' signify? Contrast the meaning of that term to 'full-screen ASCII'.

10. Discuss the importance of addressability and memory with respect to synchronous terminals transmitting on a multidrop line.

11. What is the difference between local and remote control units in an IBM 3270 Information Display System?

12. Discuss some of the functions that must be performed by a personal computer emulating a 3270 terminal.

13. How can a PC obtain the display capability of a specific type of 3270 terminal?

14. Discuss two functions performed by a network interface card.

15. What is the difference between a repeating hub and a switching hub?

16. Assume 20 active users are connected to a 10 Mbps shared-media LAN. What is the average transmission rate per network user?

17. Assume that a 12 port switching hub operates at 10 Mbps. What is the maximum throughput through this switch?

18. Assume the 12 port switch in question 17 has two servers connected to different switch ports. What is the maximum client–server data transfer capability of the switching hub?

19. What is the difference between a client–server and a peer-to-peer LAN?

20. Discuss the role of four types of LAN servers.

21. Compare and contrast FDM and TDM operations.

22. What is the significance of the use of an idle bit or character in TDM operations?

23. What is flow control? Why is it important in STDM and ITDM operations?

24. Discuss the differences in hardware between a concentrator and multiplexer with respect to the communications applications they can support.

25. How does a repeater operate?

26. What is the difference between a repeater and a bridge with respect to data transmitted on a LAN?

27. What do the terms filtering, forwarding and flooding refer to with respect to bridges?

28. Where do the addresses that bridges use as a decision criterion come from?

29. Discuss the general operation of a router with respect to the use of network addresses.

30. Discuss several functions performed by a gateway used to connect a LAN to an IBM mainframe.

6

REPRESENTATIVE STANDARDS ORGANIZATIONS: THE OSI REFERENCE MODEL

Standards can be viewed as the 'glue' that binds hardware and software from different vendors so they can operate together. Thus, the importance of standards and the work of standards organizations have proved essential for the growth of worldwide communications. In the United States and many other countries, national standards organizations have defined physical and operational characteristics that enable vendors to manufacture equipment compatible with transmission line facilities provided by communications carriers as well as equipment produced by other vendors. At the international level, standards organizations promulgated several series of communications-related recommendations. These recommendations, while not mandatory, have become highly influential on a worldwide basis for the development of equipment and facilities and have been adopted by hundreds of public companies and communications carriers.

In addition to national and international standards, a series of *de facto* standards has evolved through the licensing of technology among companies. Such *de facto* standards, as an example, have facilitated the development of communications software for use on personal computers. Today, consumers can purchase communications software that can control modems manufactured by hundreds of vendors, since most modems are now constructed to respond to a uniform set of control codes.

In this chapter we will first focus our attention on national and international standards bodies as well as discuss the importance of de facto standards. Due to the importance of the Open Systems Interconnection (OSI) reference model in data communications, we will conclude this chapter with an examination of that model. Although this examination will only provide you with an overview of the seven layers of the OSI model, it will provide a foundation for a detailed investigation of several layers of that model presented in later chapters in this book.

6.1 NATIONAL STANDARDS ORGANIZATIONS

In our discussion of national standards organizations we will define the term national to mean 'within a country' as opposed to a governmental entity. Doing so enables us to examine the activities and influence of more organizations in this section.

Table 6.1 lists ten representative national standards organizations – four whose activities are primarily focused upon within the United States and six whose activities are directed within specific foreign countries. Although each of the organizations listed in Table 6.1 is an independent entity, their level of coordination and cooperation with one another and international organizations is very high. In fact, many standards adopted by one organization have been adopted either intact or in modified form by other organizations.

One standardization not listed in Table 6.1 which deserves mention and which will be discussed in detail later in this chapter is the International Organization for Standardization (ISO). The ISO can be considered as the head organization of national standardization bodies whose efforts are directed on attempting to harmonize national standards. The result of its efforts are published as ISO standards and include such subjects as international stationery sizes, the metric system of units, computer protocols, programming languages and a variety of communications standards.

6.1.1 American National Standards Institute (ANSI)

The American National Standards Institute (ANSI) is the principal standards-forming body in the United States. This non-profit, non-governmental organization is supported by approximately 1000 professional societies,

Table 6.1 Representative national standards organizations

United States
American National Standards Institute
Electronic Industries Association
Federal Information Processing Standards
Institute of Electrical and Electronic Engineers
Foreign
British Standards Institution
Canadian Standards Association
Dansk Standard (Denmark)
Deutsches Institut fuer Normung (Germany)
Nederlands Normalisatie-instituut (Netherlands)
Norges Standardiseringsforbund (Norway)

Table 6.2 ANSI X3S3 data communications technical committee

Task group	Responsibility
X3S31	Planning
X3S32	Glossary
X3S33	Transmission format
X3S34	Control procedures
X3S35	System performance
X3S36	Signaling speeds
X3S37	Public data networks

companies and trade organizations. ANSI represents the United States in the International Organization for Standardization (ISO) and develops voluntary standards that are designed to benefit the consumers and manufacturers.

ANSI standards are developed through the work of approximately 300 Standards Committees, and from the efforts of associated groups, such as the Electronic Industries Association (EIA). Recognizing the importance of the computer industry, ANSI established its X3 Standards Committee in 1960. Established to investigate computer industry related standards, the X3 Standards committee has grown to 25 technical committees to include the X3S3 Data Communications Technical Committee which is composed of seven task groups. Table 6.2 lists the current task groups of the ANSI X3S3 Data Communications Technical Committee and their responsibilities.

Foremost among ANSI standards are X3.4 which defines the ASCII code and X3.1 which defines data signaling rates for synchronous data transmission on the switched telephone network.

Table 6.3 lists representative ANSI data communications standards publications by number and title. Many of these standards were adopted by the US Government as Federal Information Processing Standards (FIPS). Table 6.6 at the end of this section includes a list of addresses of standards organizations mentioned in this chapter.

6.1.2 Electronic Industries Association (EIA)

The Electronic Industries Association (EIA) is a trade organization headquartered in Washington, DC, which represents most major US electronics industry manufacturers. Since its founding in 1924, the EIA Engineering Department has published over 400 documents related to standards.

In the area of communications, the EIA established its Technical Committee TR-30 in 1962. The primary emphasis of this committee is on the development and maintenance of interface standards governing the attachment of data terminal equipment (DTE); such as terminals and computer ports, to data communications equipment (DCE), such as modems and data service units.

TR-30 committee standards activities include the development of the ubiquitous RS-232 interface standard, which describes the operation of a 25-pin conductor which is the most commonly used physical interface for connecting DTE to DCE.

Table 6.3 Representative ANSI publications

Publication number	Publication title
X3.1	Synchronous signaling rates for data transmission
X3.4	Code for information interchange
X3.15	Bit sequencing for the American National Standard Code for information interchange in serial-by-bit data transmission
X3.16	Character structure and character parity sense for serial-by-bit data communications information interchange
X3.24	Signal quality at interface between data processing technical equipment for synchronous data transmission
X3.25	Character structure and character parity sense for parallel-by-bit communications in American National Standard Code for information interchange
X3.28	Procedures for the use of the communications control characters for American National Standard Code for information interchange in specific data communications links
X3.36	Synchronous high-speed signaling rates between data terminal equipment and data communication equipment
X3.41	Code extension techniques for use with 7-bit coded character set of American National Standard Code for information interchange
X3.44	Determination of the performance of data communications systems
X3.57	Structure for formatting message headings for information interchange for data communication system control
X3.66	American National Standard for advanced data communication control procedures (ADCCP)
X3.79	Determination of performance of data communications systems that use bit-oriented control procedures
X3.92	Data encryption algorithm

Two other commonly known EIA standards and one emerging standard are RS366A, RS-449 and RS-530. RS-366 describes the interface used to connect terminal devices to automatic calling units; RS-449 was originally intended to replace the RS-232 interface due to its ability to extend the cabling distance between devices, and RS-530 may eventually evolve as a replacement for both RS-232 and RS-449, as it eliminates many objections to RS-449 that inhibited its adoption. Refer to Chapter 7 for detailed information covering the previously mentioned EIA standards.

The TR-30 committee works closely with both ANSI Technical Committee X3S3 and with groups within the Consultative Committee for International Telephone and Telegraph (CCITT) which was renamed the International Telecommunications Union (it has its Telecommunications Standardization Sector (ITU-T)). In fact, the ITU-T V.24 standard is basically identical to the EIA RS-232 standard, resulting in hundreds of communications vendors designing RS-232/V.24 compatible equipment.

As a result of the widespread acceptance of the RS-232/V.24 interface standard, a cable containing up to 25 conductors with a predefined set of connectors can be used to cable most DTEs to DCEs. Even though there are exceptions to this interface standard, this standard has greatly facilitated the manufacture of communications products, such as terminals, computer ports, modems and data service units that are physically compatible with one another and which can be easily cabled to one another.

Another important EIA standard resulted from the joint efforts of the EIA and the Telecommunications Industry Association (TIA). Known as EIA/TIA-568, this standard defines structured wiring within a building, and it was examined earlier in this book.

6.1.3 Federal Information Processing Standards (FIPS)

As a result of US Public Law 89-306 (the Brooks Act) which directed the Secretary of the Department of Commerce to make recommendations to the President concerning uniform data processing standards, that agency developed a computer standardization program. Since Public Law 89-306 did not cover telecommunications, the enactment of Public Law 99-500, known as the Brooks Act Amendment, expanded the definition of automatic data processing (ADP) to include certain aspects related to telecommunications. FIPS, an acronym for Federal Information Processing Standards, are the indirect result of Public Law 89-306 and is the term applied to standards developed under the US Government's computer standardization program.

FIPS specifications are drafted by the National Institute of Standards and Technology (NIST), formerly known as the National Bureau of Standards (NBS). Approximately 80 FIPS standards have been adopted, ranging in scope from the ASCII code to the Hollerith punched card code and such computer languages as COBOL and FORTRAN. Most Federal Information Processing Standards (FIPS) have an ANSI national counterpart. The key difference between FIPS and its ANSI counterpart is that applicable FIPS standards must be met for the procurement, management and operation of ADP and telecommunications equipment by Federal agencies, whereas commercial organizations in the private sector can choose whether or not to obtain equipment that complies with appropriate ANSI standards.

6.1.4 Institute of Electrical and Electronic Engineers (IEEE)

The Institute of Electrical and Electronic Engineers (IEEE) is a US-based engineering society that is very active in the development of data communications standards. In fact, the most prominent developer of local area networking standards is the IEEE, whose subcommittee 802 began its work in 1980 prior to the establishment of a viable market for the technology.

The IEEE Project 802 efforts have been concentrated on the physical interface of equipment and the procedures and functions required to establish, maintain and release connections among network devices to include defining data formats, error control procedures and other control activities governing the flow of information. This focus of the IEEE actually represents the lowest two layers of the ISO model, physical and data link, which are discussed later in this chapter.

6.1.5 British Standards Institution (BSI)

The British Standards Institution (BSI) is the national standards body of the United Kingdom. In addition to drafting and promulgating of British National Standards, BSI is responsible for representing the United Kingdom at ISO and other international bodies. BSI responsibilities include ensuring that British standards are in technical agreement with relevant international standards, resulting in, as an example, the widespread use of the V.24/RS-232 physical interface in the United Kingdom.

6.1.6 Canadian Standards Association (CSA)

The Canadian Standards Association (CSA) is a private, non-profit-making organization which produces standards and certifies products for compliance with their standards. CSA functions similarly to a combined US ANSI and Underwriters Laboratory, the latter also a private organization which is well known for testing electrical equipment ranging from ovens and toasters to modems and computers.

In many instances, CSA standards are the same as international standards, with many ISO and CCITT standards adopted as CSA standards. In other instances, CSA standards represent modified international standards.

6.2 INTERNATIONAL STANDARDS ORGANIZATIONS

In this section we will look at some of the functions and operation of two international standards organizations: the Consultative Committee for International Telephone and Telegraph (CCITT) which, as previously mentioned, is now known as the ITU and the International Standards Organization (ISO). The ITU can be considered as a governmental body as it functions under the auspices of an agency of the United Nations. Although the ISO is a non-governmental agency, its work in the field of data communications is well recognized.

6.2.1 International Telecommunications Union (ITU)

The International Telecommunications Union (ITU) is a specialized agency of the United Nations, headquartered in Geneva, Switzerland. The telecommunications section of the ITU (ITU-T) is tasked with direct responsibility for developed data communications standards and consists of 15 Study Groups, each tasked with a specific area of responsibility.

The work of the ITU-T is performed on a four-year cycle which is known as a Study Period. At the conclusion of each Study Period, a Plenary Session occurs. During the Plenary Session, the work of the ITU-T during the previous four years is reviewed, proposed recommendations are considered for adoption and items to be investigated during the next four-year cycle are considered.

The ITU-T Tenth Plenary Session met in 1992 and its eleventh session occurred in 1996. Although approval of recommended standards is not intended to be mandatory, ITU-T recommendations have the effect of law in some Western European countries and many of its recommendations have been adopted by both communications carriers and vendors in the United States.

ITU-T recommendations

Recommendations promulgated by the ITU-T are designed to serve as a guide for technical, operating and tariff questions related to data and telecommunications. ITU-T recommendations are designated according to the letters of the alphabet, from Series A to Series Z, with technical standards included in Series G to Z. In the field of data communications, the most well known ITU-T recommendations include Series I which pertains to Integrated Services Digital Network (ISDN) transmission, Series Q which describes ISDN switching and signaling systems, Series V which covers facilities and transmission systems used for data transmission over the PSTN and leased telephone circuits, the DTE-DCE interface and modem operations and Series X which covers data communications networks to include Open Systems Interconnection (OSI).

One emerging series of ITU-T recommendations that can be expected to become relatively well known in the next few years is the G.922 recommendations. The G.922 recommendations define standards for different types of digital subscriber lines to include splitterless G.lite which is covered later in this book.

The ITU-T V-Series

To provide a general indication of the scope of ITU-T recommendations, Table 6.4 lists those promulgated for the V-Series at the time this book was prepared. In examining the entries in Table 6.4, note that ITU-T Recommendation V.3 is actually a slightly modified ANSI X3.4 standard, the ASCII code. For international use, the V.3 recommendation specifies national currency symbols in place of the dollar sign ($) as well as a few other minor differences. You should also note that certain ITU-T recommendations, such as V.21, V.22 and V.23, among others, while similar to AT&T Bell modems, may or may not provide operational compatibility with modems manufactured to Bell specifications. Refer to Chapter 8 for detailed information concerning modem operations and compatibility issues.

6.2.2 International Standards Organization (ISO)

The International Standards Organization (ISO) is a non-governmental entity that has consultative status within the UN Economic and Social Council. The goal of the ISO is to 'promote the development of standards in the world with a view to facilitating international exchange of goods and services'.

Table 6.4 ITU-T V-series recommendations

General

V.1 —Equivalence between binary notation symbols and the significant conditions of a two-condition code.
V.2 —Power levels for data transmission over telephone lines.
V.3 —International Alphabet No. 5
V.4 —General structure of signals of International Alphabet No. 5 code for data transmission over public telephone networks.
V.5 —Standardization of data signaling rates for synchronous data transmission in the general switched telephone network.
V.6 —Standardization of data signaling rates for synchronous data transmission on leased telephone-type circuits.
V.7 —Definitions of terms concerning data communication over the telephone network

Interface and voice-band moderns

V.10—Electrical characteristics for unbalanced double-current interchange circuits for general use with integrated circuit equipment in the field of data communications. Electrically similar to RS-423.
V.11—Electrical characteristics of balanced double-current interchange circuits for general use with integrated circuit equipment in the field of data communications. Electrically similar to RS-422.
V.15—Use of acoustic coupling for data transmission.
V.16—Medical analogue data transmission modems.
V.19—Modems for parallel data transmission using telephone signaling frequencies.
V.20—Parallel data transmission modems standardized for universal use in the general switched telephone network.
V.21—300 bps duplex modem standardized for use in the general switched telephone network. Similar to the Bell 103.
V.22—1200 bps duplex modem standardized for use on the general switched telephone network and on leased circuits. Similar to the Bell 212.
V.22bis—2400 bps full-duplex two-wire.
V.23—600/1200 baud modem standardized for use in the general switched telephone network. Similar to the Bell 202.
V.24—List of definitions for interchange circuits between data terminal equipment and data circuit-terminating equipment. Similar to and operationally compatible with RS-232.
V.25—Automatic calling and/or answering equipment on the general switched telephone network, including disabling of echo suppressors on manually established calls. RS-366 parallel interface.
V.25bis—serial RS-232 interface.
V.26—2400 bps modem standardized for use on four-wire leased telephone-type circuits. Similar to the Bell 201B.
V.26bis—2400/1200 bps modem standardized for use in the general switched telephone network. Similar to the Bell 201C.
V.26ter—2400 bps modem that uses echo cancellation techniques suitable for application in the general switched telephone network.
V.27—4800 bps modem with manual equalizer standardized for use on leased telephone-type circuits. Similar to the Bell 208A.
V.27bis—4800/2400 bps modem with automatic equalizer standardized for use on leased telephone-type circuits.
V.27ter—4800/2400 bps modem standardized for use in general switched telephone network. Similar to the Bell 208B.
V.28—Electrical characteristics for unbalanced double-current interchange circuits (defined by V.24; similar to and operational with RS-232).
V.29—9600 bps modem standardized for use on point-to-point four-wire leased telephone-type circuits. Similar to the Bell 209.

Table 6.4 *(continued)*

V.31—Electrical characteristics for single-current interchange circuits controlled by contact closure.
V.32—Family of 4800/9600 bps modems operating full-duplex over two-wire facilities.
V.33—14.4 kbps modem standardized for use on point-to-point four-wire leased telephone- type circuits.
V.34—28.8 kbps modems standardized for operating full-duplex over two-wire facilities.
V.34bis—33.6 kbps modem standardization for operating full-duplex over two-wire facilities.
V.35—Data transmission at 48 kbps using 60–108 kHz group band circuits. CCITT balanced interface specification for data transmission at 48 kbps, using 60–108 kHz group band circuits. Usually implemented on a 34-pin M block type connector (M 34) used to interface to a high-speed digital carrier such as DDS.
V.36—Modems for synchronous data transmission using 60–108 kHz group band circuits.
V.37—Synchronous data transmission at a data signalling rate higher than 72 kbps using 60–108 kHz group band circuits.
V.90—56 kbps modem standardization for operating downstream and 33.6 kbps upstream over two-wire facilities.

Error control
V.40—Error indication with electromechanical equipment.
V.41—Code-independent error control system.
V.42—Error detection and correction for modems.

Data compression
V.42bis—Data compression for modems.

Transmission quality and maintenance
V.50—Standard limits for transmission quality of data transmission.
V.51—Organization of the maintenance of international telephone-type circuits used for data transmission.
V.52—Characteristics of distortion and error-rate measuring apparatus for data transmission.
V.53—Limits for the maintenance of telephone-type circuits used for data transmission.
V.54—Loop test devices for modems.
V.55—Specification for an impulse noise measuring instrument for telephone-type circuits.
V.56—Comparative tests of modems for use over telephone-type circuits.

The membership of the ISO consists of the national standards organizations of most countries, with approximately 100 countries currently participating in its work.

Perhaps the most notable achievement of the ISO in the field of communications is its development of the seven-layer Open Systems Interconnection (OSI) Reference Model. This model is discussed in detail in Section 6.4.

6.3 *DE FACTO* STANDARDS

Prior to the breakup of AT&T, a process referred to as divestiture, US telephone interface definitions were the exclusive domain of AT&T and its research subsidiary, Bell Laboratories. Other vendors which developed equipment for use on the AT&T network had to construct their equipment to those interface definitions. In addition, since AT&T originally had a monopoly on equipment that could be connected to the switched telephone network, on liberalization of that policy third-party vendors had to design communications equipment, such as modems, that was compatible with the majority of equipment in use.

As some third-party vendor products gained market acceptance over other products, vendor licensing of technology resulted in the development of *de facto* standards. Another area responsible for the development of a large number of *de facto* standards is the Internet community. In this section we will examine both vendor and Internet *de facto* standards.

In discussing *de facto* standards it is important to note that their use is not normally regulated by a legal authority. Thus, unlike a *de jure* standard that has the rule of law behind its adoption, the use of a *de facto* standard can be considered for the most part optional.

6.3.1 AT&T compatibility

Since AT&T originally had a monopoly on equipment connected to its network, when third-party vendors were allowed to manufacture products for use on AT&T facilities they designed most of their products to be compatible with AT&T equipment. This resulted in the operational characteristics of many AT&T products becoming *de facto* standards. In spite of the breakup of AT&T, this vendor still defines format and interface specifications for many facilities that third-party vendors must adhere to for their product to be successfully used with such facilities. AT&T, like standards organizations, publishes a variety of communications reference publications that define the operational characteristics of its facilities and equipment.

AT&T's catalog of technical documents is contained in two publications. AT&T's '*Publication 10000*' lists over 140 publications and includes a synopsis of their contents, date of publication and cost. Formally known as the *Catalog of Communications Technical Publications, Publication 10000* includes several order forms as well as a toll-free 800 number for persons who wish to call in their order. AT&T's *Publication 10000A*, which was issued as an addendum to *Publication 10000* lists new and revised technical reference releases as well as technical references deleted from *Publication 10000* and the reason for each deletion. In addition, *Publication 10000A* includes a supplemental list of select codes for publications listed in *Publication 10000*. The select code is the document's ordering code, which must be entered on the AT&T order form. Readers should obtain both documents from AT&T to simplify future orders.

Publication 10000, Publication 10000A and the publications listed therein can be obtained by writing or calling AT&T Technologies at the address or telephone numbers listed in Table 6.6.

Table 6.5 is an extract of some of the AT&T technical publications listed in *Publication 10000* and *Publication 10000A*. As can be seen, a wide diversity of publications can be ordered directly from AT&T.

Table 6.5 Selected AT&T publications

Publication number	Publication title
CB142	The Extended Superframe Format Interface Specification
CB143	Digital Access and System Technical Reference and Compatibility Specification
PUB 41449	Integrated Services Digital Network (ISDN) Primary Rate Interface Specification
PUB 41457	SKYNET Digital Service
PUB 48502	Network Circuit Access Test Set Functional Criteria
PUB 52411	ACCUNET T1.5 Service Description and Interface Specification
PUB 54010	X.25 Interface Specification and Packet Switching Capabilities
PUB 54012	X.75 Interface Specifications and Packet Switching Capability
PUB 54070	Bit Compression Multiplexing
PUB 54075	56 kbps Subrate Data Multiplexing

6.3.2 Cross-licensed technology

As the deregulation of the US telephone industry progressed, hundreds of vendors developed products for the resulting market. Many vendors cross-licensed technology, such as the command set which defines the operation of intelligent modems. Thus, another area of *de facto* standards developed based upon consumer acceptance of commercial products. In certain cases, *de facto* standards have evolved into *de jure* standards with their adoption by one or more standard-making organizations.

6.3.3 Bellcore/Telcordia Technologies

A third area of *de facto* standards is Bellcore. Upon divestiture in 1984, AT&T formed Bell Communications Research Inc. (Bellcore) with its seven regional Bell holding companies. Bellcore provides technical and research support to these holding companies in much the same way that Bell Laboratories supports AT&T. Bellcore maintains common standards for the nation's telephone systems, ensures a smoothly operating telephone network and coordinates telecommunications operations during national emergencies. With approximately 7000 employees, hundreds of research projects and an annual budget of nearly $1 billion, Bellcore is among the largest research and engineering organizations in the United States.

In 1997 Bellcore was purchased by Science Applications International Corporation and in 1999 the research organization was renamed Telcordia Technologies. While continuing to provide research for carrier customers, Telcordia is also at the forefront in developing products that combine, via a common packet data network architecture, voice, data, Internet, cable TV, and emerging electronic commerce and multimedia applications.

Like AT&T, Telcordia Technologies publishes a catalog of technical publications called *Catalog 10000*. The most recent catalog lists approximately

Table 6.6 Communications reference publications sources

ANSI 1430 Broadway New York, NY 10018, USA (212) 354-3300	CCITT General Secretariat International Telecommunications Union Place des Nations 1211 Geneva 20, Switzerland
AT&T Technologies Customer Information Center Indianapolis, IN 46219, USA (800) 432-6600 (317) 352-8557	IEEE 345 East 47th St. New York, NY 10017, USA (212) 705-7900
Telcordia Technologies Information Operations Center 60 New England Ave. Piscataway, NJ 08854, USA (201) 981-5600	US Department of Commerce National Technical Information Service 5285 Port Royal Rd. Springfield, VA 22161, USA (703) 487-4650
EIA Standard Sales 2001 Eye St. NW Washington, DC 20006, USA (202) 457-4966	

500 publications that vary in scope from compatibility guides, which list the interface specifications that must be adhered to by manufacturers building equipment for connection to telephone company central offices, to a variety of technical references. *Catalog 10000* and the publications listed therein can be ordered directly from Telcordia Technologies by mail or telephone. The address of Telcordia Technologies is contained in Table 6.6.

6.3.4 Internet standards

The Internet can be considered as a collection of interconnected networks that use the Transmission Control Protocol/Internet Protocol (TCP/IP) protocol suite. The Internet has its roots in experimental packet switching work sponsored by the U.S. Department of Defense Advanced Research Projects Agency (ARPA) which resulted in the development of the ARPANet. That network was responsible for the development of file transfer, electronic mail and remote terminal access to computers which became applications incorporated into the TCP/IP protocol suite. The efforts of ARPA during the 1960s and 1970s were taken over by the Internet Activities Board (IAB) whose name was changed to the Internet Architecture Board in 1992. The IAB is responsible for the development of Internet protocols, to include deciding if and when a protocol should become an Internet standard.

Although the IAB is responsible for setting the general direction concerning the standardization of protocols the actual effort is carried out by the Internet Engineering Task Force (IETF). The IETF is responsible for the development of documents called Requests For Comments (RFCs) which are normally issued

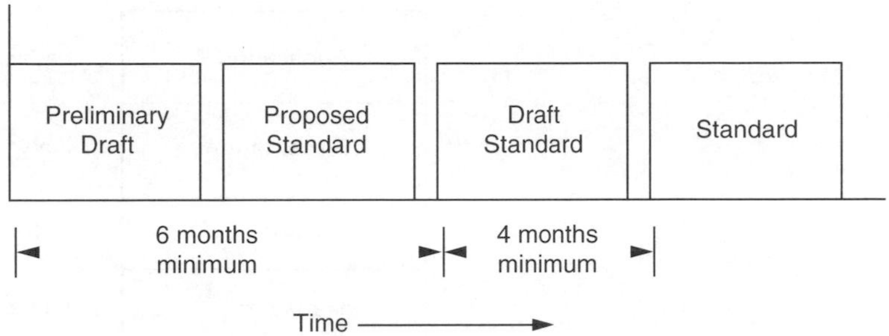

Figure 6.1 Internet standards time track

as a preliminary draft. After a period allowed for comments, the RFC will be published as a proposed standard or, if circumstances warrant, it may be dropped from consideration. If favorable comments occur concerning the proposed standard it can be promoted to a draft standard after a minimum period of six months. After a review period of at least four months the draft standard can be recommended for adoption as a standard by the Internet Engineering Steering Group (IESG).

The IESG consists of the chairperson of the IETF and other members of that group, and it performs an oversight and coordinating function for the IETF. Although the IESG must recommend the adoption of an RFC as a standard the IAB is responsible for the final decision concerning its adoption. Figure 6.1 illustrates the time track for the development of an Internet standard. RFCs cover a variety of topics, ranging from TCP/IP applications to the Simple Network Management Protocol (SNMP) and the composition of databases of network management information. Over 2500 RFCs were in existence when this book was published, and some of those RFCs will be described in detail later in this book.

A second Internet organization that warrants a brief note is the Internet Assigned Numbers Authority (IANA). IANA is responsible for assigning unique identifiers for different fields in protocols as well as performing other related activities. One of the key functions of IANA is registering the use of 'port numbers' for use in the Transmission Control Protocol (TCP) and User Datagram Protocol (UDP), both of which will be covered later in this book

6.4 THE OSI REFERENCE MODEL

The International Standards Organization (ISO) established a framework for standardizing communications systems called the Open System Interconnection (OSI) Reference Model. The OSI architecture defines the communications process as a set of seven layers, with specific functions isolated to and associated with each layer. Each layer, as illustrated in Figure 6.2 covers low layer processes, effectively isolating them from higher layer functions. In this way, each layer performs a set of functions necessary to provide a set of services to the layer above it.

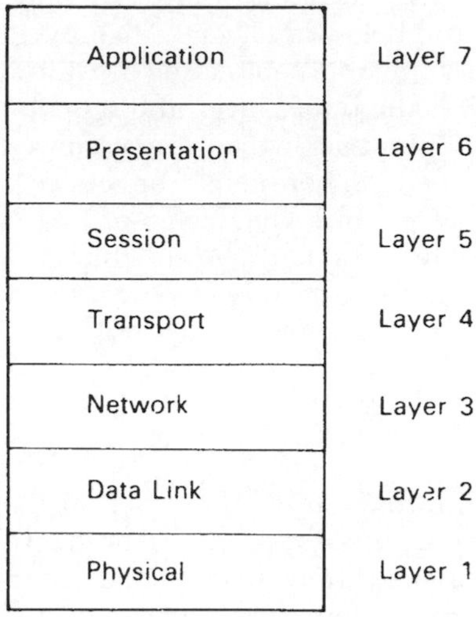

Figure 6.2 OSI reference model

Layer isolation permits the characteristics of a given layer to change, without impacting on the remainder of the model, provided that the supporting services remain the same. One major advantage of this layered approach is that users can mix and match OSI conforming communications products to tailor their communications system to satisfy a particular networking requirement.

The OSI Reference Model, while not completely viable with current network architectures, offers the potential to directly interconnect networks based upon the use of different vendor equipment. This interconnectivity potential will be of substantial benefit to both users and vendors. For users, interconnectivity will remove the shackles that in many instances tie them to a particular vendor. For vendors, the ability to easily interconnect their products will provide them with access to a larger market. The importance of the OSI model is such that it has been adopted by the ITU-T as Recommendation X.200.

6.4.1 Layered architecture

As previously discussed, the OSI Reference Model is based on the establishment of a layered, or partitioned, architecture. This partitioning effort can be considered as being derived from the scientific process whereby complex problems are subdivided into functional tasks that are easier to implement on an aggregate individual basis than as a whole.

As a result of the application of a partitioning approach to communications network architecture, the communications process was subdivided into seven distinct partitions, called layers. Each layer consists of a set of functions designed to provide a defined series of services which relate to the mission of that layer. For example, the functions associated with the physical connection of equipment to a network are referred to as the physical layer.

With the exception of layers 1 and 7, each layer is bounded by the layers above and below it. Layer 1, the physical layer, can be considered to be bound below by the interconnecting medium over which transmission flows, whereas layer 7 is the upper layer and has no upper boundary. Within each layer is a group of functions which can be viewed as providing a set of defined services to the layer which bounds it from above, resulting in layer n using the services of layer $n - 1$. Thus, the design of a layered architecture enables the characteristics of a particular layer to change without affecting the rest of the system, assuming that the services provided by the layer do not change.

6.4.2 OSI layers

An understanding of the OSI layers is best obtained by first examining a possible network structure that illustrates the components of a typical network. Figure 6.3 illustrates a network structure which is only typical in the sense that it will be used for a discussion of the components upon which networks are constructed.

The circles in Figure 6.3 represents nodes, which are points where data enters or exits a network or is switched between two paths. Nodes are connected to other nodes via communications paths within the network, where the communications paths can be established on any type of communications media, such as cable, microwave or radio.

From a physical perspective, a node can be based on the use of one of several types of computer to include a personal computer, minicomputer or mainframe computer or specialized computer, such as a front-end processor. Connections to network nodes into a network can occur via the use of terminals directly connected to computers, terminals connected to a node via the use of one or more intermediate communications devices or via paths linking one network to another network.

The routes between two nodes, such as C–E–A, C–D–A, C–A and C–B–A which could be used to route data between nodes A and C are information paths. Due to the variability in the flow of information through a network, the shortest path between nodes may not be available for use or may represent a

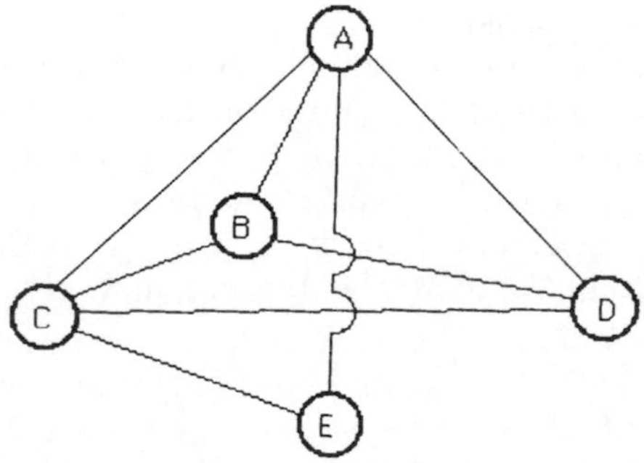

Figure 6.3 Network components. The path between a source and destination node established on a temporary basis is called a logical connection (* node, lines represent paths)

non-efficient path with respect to other paths constructed through intermediate nodes between a source and destination node. A temporary connection established to link two nodes whose route is based on such parameters as current network activity is known as a logical connection. This logical connection represents the use of physical facilities to include paths and node switching capability on a temporary basis.

The major functions of each of the seven OSI layers are described in the following seven sections.

Layer 1: the physical layer

At the lowest or most basic level, the physical layer (level 1) is a set of rules that specifies the electrical and physical connection between devices. This level specifies the cable connections and the electrical rules necessary to transfer data between devices. Typically, the physical link corresponds to established interface standards, such as RS-232.

Layer 2: the data link layer

The next layer, which is known as the data link layer (level 2), denotes how a device gains access to the medium specified in the physical layer; it also defines data formats, to include the framing of data within transmitted messages, error control procedures and other link control activities. From defining data formats to include procedures to correct transmission effors, this layer becomes responsible for the reliable delivery of information. Data link control protocols such as binary synchronous communications (BSC) and high-level data link control (HDLC) reside in this layer.

Since the development of OSI layers was originally targeted toward wide area networking, its applicability to local area networks required a degree of modification. Under the IEEE 802 standards, the data link layer was originally divided into two sublayers: logical link control (LLC) and media access control (MAC). Later, the requirement to develop new signaling methods to support recently developed high speed LANs operating at 100 Mbps or higher required another layer subdivision involving the physical layer that will be discussed later in this book.

The LLC layer is responsible for generating and interpreting commands that control the flow of data and perform recovery operations in the event of errors. In comparison, the MAC layer is responsible for providing access to the local area network, which enables a station on the network to transmit information. Later in this chapter, we will examine the IEEE 802 subdivision of the data link layer into LLC and MAC sublayers, as well as the operation of each layer.

Layer 3: the network layer

The network layer (level 3) is responsible for arranging a logical connection between a source and destination on the network to include the selection and management of a route for the flow of information between source and destination based on the available data paths in the network. Services provided by this layer are associated with the movement of data through a network, to include addressing, routing, switching, sequencing and flow control procedures. In a complex network the source and destination may not be directly connected by a single path, but instead require a path to be established that consists of many subpaths. Thus, routing data through the network onto the correct paths is an important feature of this layer.

Several protocols have been defined for layer 3 to include the CCITT X.25 packet switching protocol and the CCITT X.75 gateway protocol. X.25 governs the flow of information through a packet network, whereas X.75 governs the flow of information between packet networks. In the TCP/IP protocol suite the Internet Protocol (IP) represents a network layer protocol.

Layer 4: the transport layer

The transport layer (level 4) is responsible for guaranteeing that the transfer of information occurs correctly after a route has been established through the network by the network level protocol. Thus, the primary function of this layer is to control the communications session between network nodes once a path has been established by the network control layer. Error control, sequence checking, and other end-to-end data reliability factors are the primary concern of this layer, and they enable the transport layer to provide a reliable end-to-end data transfer capability.

Examples of transport layer protocols include the Transmission Control Protocol (TCP) and the User Datagram Protocol (UDP). As we will note later in this book, although both TCP and UDP are both transport layer protocols there are significant differences between the two. TCP is a connection-oriented protocol that provides a reliable end-to-end data transfer capability. In comparison, UDP is a connectionless, unreliable protocol which depends upon higher layers in the protocol suite for reliability, if desired.

Layer 5: the session layer

The session layer (level 5) provides a set of rules for establishing and terminating data streams between nodes in a network. The services that this session layer can provide include establishing and terminating node connections, message flow control, dialogue control and end-to-end data control.

Layer 6: the presentation layer

The presentation layer (level 6) services are concerned with data transformation, formatting and syntax. One of the primary functions performed by the presentation layer is the conversion of transmitted data into a display format appropriate for a receiving device. This can include any necessary conversion

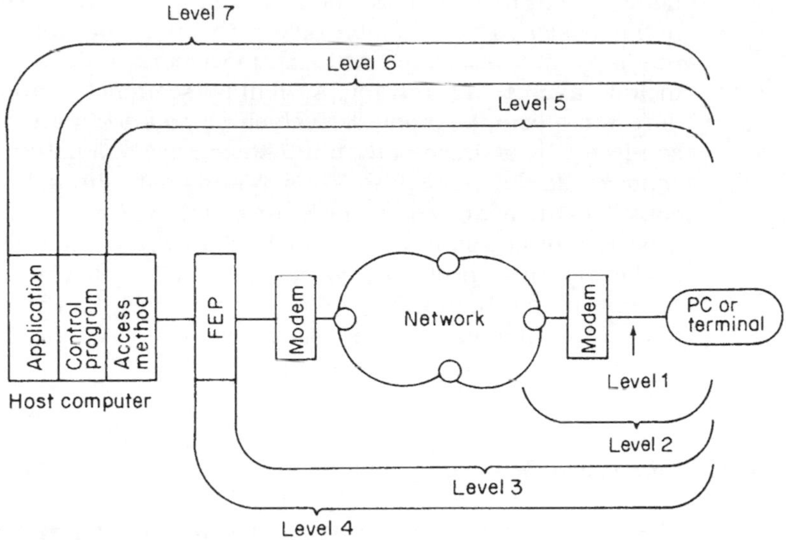

Figure 6.4 OSI model schematic

between different data codes. Data encryption/decryption and data compression and decompression are examples of the data transformation that could be handled by this layer.

Layer 7: the application layer

Finally, the application layer (level 7) acts as a window through which the application gains access to all of the services provided by the model. Examples of functions performed at this level include file transfers, resource sharing and database access. Although the first four layers are fairly well defined, the top three layers may vary considerably, depending on the network used. Figure 6.4 illustrates the OSI model in schematic format, showing the various levels of the model with respect to a terminal accessing an application on a host computer system.

6.4.3 Data flow

As data flow within an OSI network, each layer appends appropriate heading information to frames of information flowing within the network while removing the heading information added by a lower layer. In this manner, layer (n) interacts with layer ($n - 1$) as data flows through an OSI network.

Figure 6.5 illustrates the appending and removal of frame header information as data flow through a network constructed according to the ISO Reference Model. Since each higher level removes the header appended by a lower level, the frame traversing the network arrives in its original form at its destination.

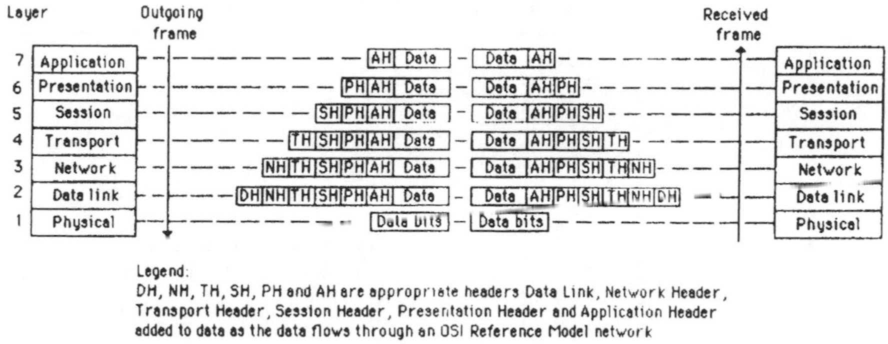

Figure 6.5 Data flow within an OSI reference model network. DH, NH, TH, SH, PH and AH are appropriate headers data link, network header, transport header, session header, presentation header and application header added to data as the data flows through an OSI reference model network

As you will surmise from the above illustrations, the ISO Reference Model is designed to simplify the construction of data networks. This simplification is due to the eventual standardization of methods and procedures to append appropriate heading information to frames flowing through a network, permitting data to be routed to its appropriate destination following a uniform procedure.

6.5 IEEE 802 STANDARDS

The Institute of Electrical and Electronic Engineers (IEEE) Committee 802 was formed at the beginning of the 1980s to develop standards for emerging technologies. The IEEE fostered the development of local area networking equipment from different vendors that can work together. In addition, IEEE LAN standards provided a common design goal for vendors to access a relatively larger market than if proprietary equipment were developed. This, in turn, enabled economies of scale to lower the cost of products developed for larger markets.

6.5.1 802 committees

Table 6.7 lists a portion of the organization of IEEE 802 committees involved in local area networks. In examining the lists of committees in Table 6.7 it is apparent that the IEEE early on noted that a number of different systems would be required to satisfy the requirements of a diverse end-user population. Accordingly, the IEEE adopted the CSMA/CD, Token Bus, and Token Ring as standards 802.3, 802.4 and 802.5, respectively.

The IEEE Committee 802 published draft standards for CSMA/CD and Token Bus local area networks in I982. Standard 802.3, which describes a baseband CSMA/CD network similar to Ethernet, was published in 1983.

Table 6.7 IEEE Series 802 committees

802.1	High Level Interface
802.2	Logical Link Control
802.3	CSMA/CD
802.4	Token-Passing Bus
802.5	Token-Passing Ring
802.6	Metropolitan Area Networks
802.7	Broadband Technical Advisory Group
802.8	Fiber Optic Technical Advisory Group
802.9	Integrated Voice and Data Networks
802.10	Network Security
802.11	Wireless LANs
802.12	100VG-AnyLAN

Since then, several addenda to the 802.3 standard have been adopted to govern the operation of CSMA/CD on different types of media. Those addenda include: 10BASE-2 which defines a 10 Mbps baseband network operating on thin coaxial cable; 1BASE-5, which defines a 1 Mbps baseband network operating on twisted-pair; 10BASE-T, which defines a 10 Mbps baseband network operating on twisted-pair; and 10BROAD-36, which defines a broadband 10 Mbps network that operates on thick coaxial cable.

The next standard published by the IEEE was 802.4, which describes a token-passing bus-oriented network for both baseband and broadband transmission. This standard is similar to the Manufacturing Automation Protocol (MAP) standard developed by General Motors.

The third LAN standard published by the IEEE was based upon IBM's specifications for its Token-Ring network. Known as the 802.5 standard, it defines the operation of token-ring networks on shielded twisted-pair cable at data rates of 1 and 4 Mbps. That standard was modified to acknowledge three IBM enhancements to Token-Ring network operations. These enhancements include the 16 Mbps operating rate, the ability to release a token early on a 16 Mbps network, and a bridge routing protocol known as source routing.

In late 1992 Grand Junction proposed to the IEEE a method for operating Ethernet at 100 Mbps. At approximately the same time, AT&T and Hewlett-Packard proposed a different method to the IEEE which was originally named 100BaseVG, with VG referencing voice grade twisted pair cable. IBM joined AT&T and HP, adding support for Token-Ring to 100BaseVG, resulting in the proposed standard having its name changed to 100VG-AnyLAN in recognition of its ability to support either Ethernet or Token-Ring. Due to the merits associated with each proposed standard, the IEEE approved both in 1995. 100Base-T, also commonly referred to as Fast Ethernet, was approved as an update to 802.3. The specification for 100VG-AnyLAN was approved as 802.12.

Other Ethernet-related standards include Fast Ethernet which is denoted as 802.3μ and was published as an addendum to the 802.3 standard in 1995, and the Gigabit Ethernet standard which was published in 1998. The latter did not define 1 Gbps transmission over copper pair wire, which required a new standard referred to as 802.3ab which was published in 1999.

6.5.2 Data link subdivision

One of the more interesting facets of IEEE 802 standards is the subdivision of the ISO Open System Interconnection Model's data link layer into a minimum of two sublayers: logical link control and medium access control. Figure 6.6 illustrates the relationship between IEEE 802 local area network standards and the first three layers of the OSI Reference Model.

The separation of the data link layer into two entities provides a mechanism for regulating access to the medium that is independent of the method for establishing, maintaining and terminating the logical link between workstations. The method of regulating access to the medium is defined by the medium access control portion of each local area network standard. This enables the logical link control standard to be applicable to each type of network.

Medium access control

The medium access control sublayer is responsible for controlling access to the network. To accomplish this, it must ensure that two or more stations do not attempt to transmit data onto the network simultaneously. For Ethernet networks, this is accomplished through the use of the CSMA/CD access protocol. Under the CSMA/CD access protocol a station on a network first 'listens' to determine if another station is transmitting. If the station does not hear anything, it will transmit a unit of information referred to as a frame. If two stations both 'listen', hear no activity, and simultaneously or near-simultaneously transmit data, a collision will occur. Thus, each station

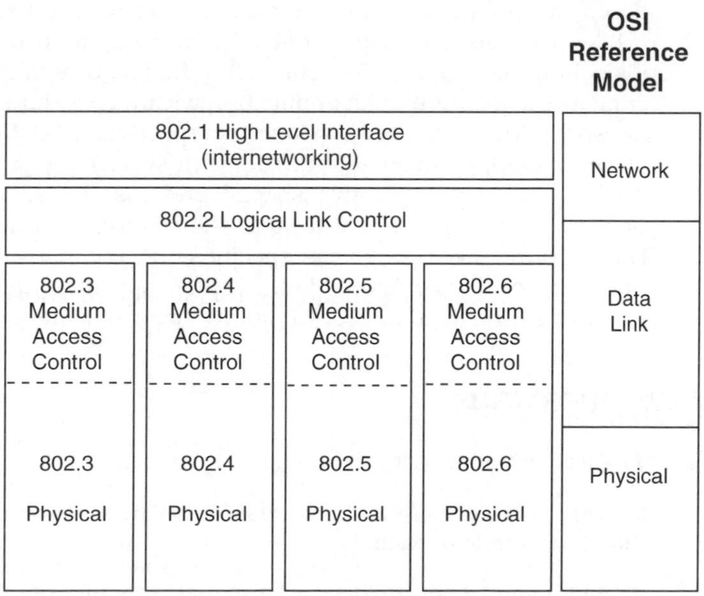

Figure 6.6 Relationship between IEEE standards and the OSI Reference Model

connected to a CSMA/CD network has a collision detection capability. Upon detecting a collision, a station uses a random time to delay an attempted retransmission to reduce the possibility of a subsequent collision.

In addition to network access control, the MAC sublayer is responsible for the orderly movement of data onto and off of the network. To accomplish this, the MAC sublayer is responsible for MAC addressing, frame type recognition, frame control, frame copying, and similar frame-related functions.

The MAC address represents the physical address of each station connected to the network. That address can belong to a single station, can represent a predefined group of stations (group address), or can represent all stations on the network (broadcast address). Through MAC addresses, the physical source and destination of frames are identified.

Frame type recognition enables the type and format of a frame to be recognized. To ensure that frames can be processed accurately, frame control prefixes each frame with a preamble, which consists of a predefined sequence of bits. In addition, a frame check sequence (FCS) is computed by applying an algorithm to the contents of the frame; the results of the operation are placed into the frame. This enables a receiving station to perform a similar operation. Then, if the locally computed FCS matches the FCS carried in the frame, the frame is considered to have arrived without error.

Logical link control

Logical link control frames are used to provide a link between network layer protocols and media access control. This linkage is accomplished through the use of Service Access Points (SAPs), which operate in much the same way as a mailbox. That is, both network layer protocols and logical link control have access to SAPs and can leave messages for each other in them.

Like a mailbox in a post office, each SAP has a distinct address. For the logical link control, a SAP represents the location of a network layer process, such as the location of an application within a workstation as viewed from the network. From the network layer perspective, a SAP represents the place to leave messages concerning the network services requested by an application.

LLC frames contain two special address fields, known as the Destination Services Access Point and the Source Services Access Point. The Destination Services Access specifies the receiving network layer process. The Source Service Access Point (SSAP) specifies the sending network layer process. Both DSAP and SSAP addresses are assigned by the IEEE.

6.6 REVIEW QUESTIONS

1. Discuss the importance of having standards.

2. Discuss the difference between national, international and *de facto* standards. Cite an example of each.

3. Why would it be in the best interest of a manufacturer to build a product compatible with appropriate standards, such as the RS-232/V.24 standard?

4. Discuss the applicability of FIPS and ANSI standards with respect to federal agencies and private sector firms.

5. Why can it take up to four years or more for the ITU-T to adopt a recommendation?

6. Name two sources of *de facto* communications standards.

7. Discuss the role of the IETF and the IANA with respect to Internet standards.

8. What is a Request for Comment (RFC)?

9. What is the purpose of layer isolation in the OSI Reference Model?

10. What are the functions of nodes and paths in a network?

11. What is the difference between a physical and a logical network connection?

12. Discuss the seven OSI layers and the functions performed by each layer.

13. Do all transport layer protocols operate in the same manner?

14. What is the function of the Medium Access Control Sublayer?

15. What is the function of the Logical Link Control Sublayer?

7

THE PHYSICAL LAYER, CABLES, CONNECTORS, PLUGS AND JACKS

As discussed in Chapter 6, the physical layer is the lowest layer of the ISO Reference Model. This layer can be considered to represent the specifications required to satisfy the electrical and mechanical interface necessary to establish a communications path. Standards for the physical layer are concerned with connector types, connector pin-outs and electrical signaling, to include bit synchronization and the identification of each signal element as a binary one or binary zero. This results in the physical layer providing those services necessary for establishing, maintaining and disconnecting the physical circuits that form a communications path.

Since one part of communications equipment utilization involves connecting data terminal equipment (DTE) to data communications equipment (DCE), the physical interface is commonly thought of as involving such standards as RS-232, RS-449, ITU-T V.24 and ITU-T X.21. Another less recognized aspect of the physical layer is the method whereby communications equipment is attached to communications carrier facilities.

In this chapter we will first focus attention upon the DTE/DCE interface, examining several popular standards and emerging standards. This examination will include the signal characteristics of several interface standards, including the interchange circuits defined by the standard and their operation and utilization. Since the RS-232/V.24 interface is by far the most popularly employed physical interface, we will examine the cable used for this interface in the second portion of this chapter. This examination will provide the foundation for illustrating the fabrication of several types of null modem cables as well as the presentation of other cabling tricks.

In addition to the RS-232/V.24 interface, the first section in this chapter also covers RS-442, RS-449, V.35, RS-366-A, X.21 and X.20, X.21 bis, RS-530, the High Speed Serial Interface (HSSI), the High Performance Parallel Interface (HIPPI), and two emerging interfaces that can be expected to gain in popularity due to their high data transfer capability. Those two interfaces are the

Universal Serial Bus (USB) and the IEEE 1394 interface, the latter referred to as FireWire. The multitude of interfaces and standards presented in the first section in this chapter should be examined based upon your need for information. While some persons may never require information about a particular interface or standard, other persons may require such information on a periodic or daily basis. Thus, in addition to functioning as a textbook, the material in this chapter is similar to the material in other chapters in that it also provides a considerable amount of reference material.

Because communications, in most instances, are dependent on the use of facilities provided by a common carrier, in the last section of this chapter we will discuss the interface between customer equipment and carrier facilities. In this section we will examine the use of plugs and connectors to include the purposes of different types of jacks.

7.1 DTE/DCE INTERFACES

In the world of data communications, equipment that includes terminals and computer ports is referred to as data terminal equipment or DTEs. In comparison, modems and other communications devices are referred to as data communications equipment or DCEs. The physical, electrical and logical rules for the exchange of data between DTEs and DCEs are specified by an interface standard; the most commonly used is the EIA RS-232-C and RS-232-D standards which are very similar to the ITU-T V.24 standard used in Europe and other locations outside of North America.

The term EIA refers to the Electronic Industries Association, which is a national body that represents a large percentage of the manufacturers in the US electronics industry. The EIA's work in the area of standards has become widely recognized and many of its standards were adopted by other standard bodies. RS-232-C is a recommended standard (RS) published by the EIA in 1969, with the number 232 referring to the identification number of one particular communications standard and the suffix C designating the revision to that standard.

In the late 1970s it was intended that the RS-232-C standard would be gradually replaced by a set of three standards: RS-449, RS-422 and RS-423. These standards were designed to permit higher data rates than were obtainable under RS-232-C as well as to provide users with added functionality. Although the EIA and several government agencies heavily promoted the RS-449 standard, its adoption by vendors has been limited. Recognizing the fact that the universal adoption of RS-449 and its associated standards was basically impossible, the EIA issued RS-232-D (Revision D) in January, 1987, and RS-232-E (Revision E) in July, 1991, as well as a new standard known as RS-530.

Other DTE/DCE interfaces that warrant attention are the RS-366-A and the CCITT (ITU-T) X.20, X.21 and V.35 standards. The RS-366-A interface governs the attachment of DTEs to a special type of DCE called an automatic calling unit. The CCITT (ITU-T) X.20 and X.21 standards govern the attachment of DTE to DCE for asynchronous and synchronous operation on public networks, respectively. The CCITT (ITU-T) V.35 standard governs high-speed data transmission, typically at 48 kbps and above, with a limit occurring at approximately 6 Mbps.

Two additional standards that we will examine are the High Speed Serial Interface (HSSI) and the High Performance Parallel Interface (HIPPI). HSSI provides support for serial operating rates up to 52 Mbps, whereas for extremely high bandwidth requirements, such as extending the channel on a supercomputer, HIPPI supports maximum data rates of either 800 Mbps or 1.6 Gbps on a parallel interface.

Two emerging standards we will examine are the Universal Serial Bus (USB) and the IEEE 1394 interface, with the latter referred to as FireWire. Both interfaces were developed to provide PCs and other types of computers with an easy-to-use high speed interface for connecting digital cameras, keyboards, tape and floppy drives, and other peripheral devices without having to open a PC's system unit cover, set jumpers, or adjust configurations via software. Today several modem manufacturers are introducing products based upon these new standards, and other communications products are expected to be introduced in the near future that will support one or both interface standards.

Figure 7.1 provides a comparison of the maximum operating rate of the eight interfaces that we will discuss in this section. Although RS-232 is

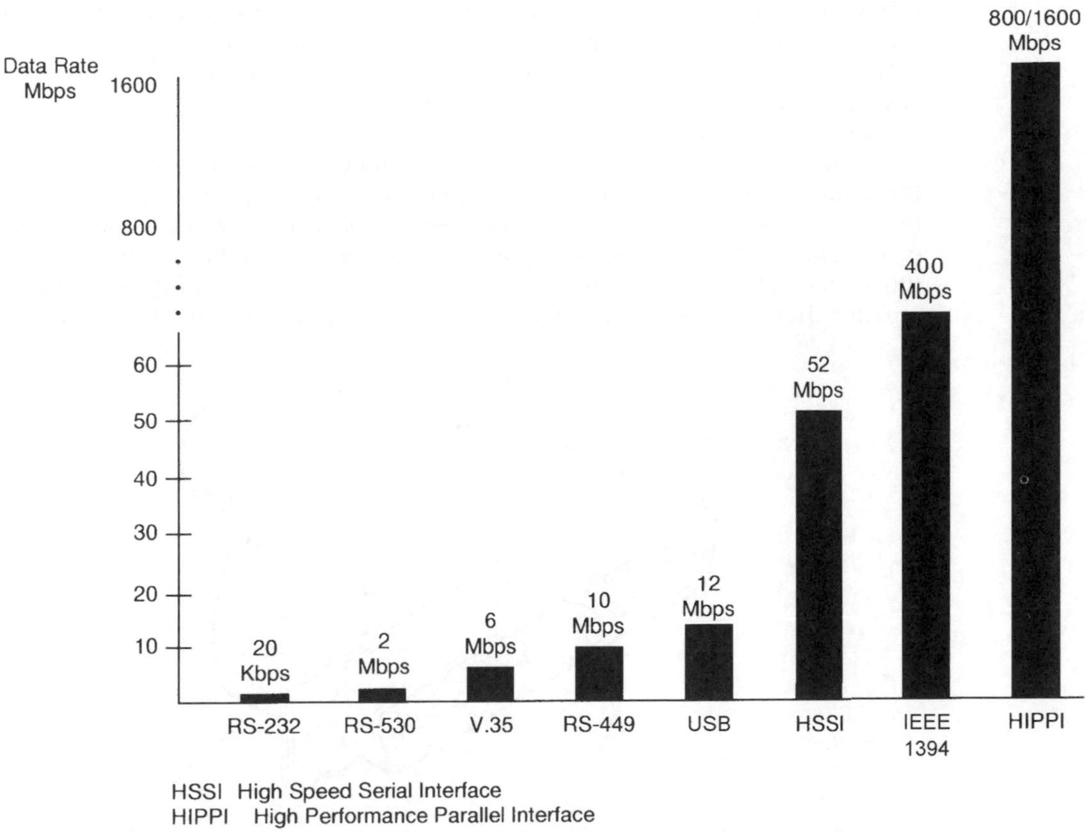

Figure 7.1 Maximum interface data rates

'officially' limited to approximately 20 kbps, that limit is for a maximum cable length of 50 feet, which explains why that interface can still be used to connect terminal devices to include personal computers to modems operating at data rates up to 56 kbps. However, when the compression feature built into modems operating at 56 kbps is enabled, most vendors recommend that you set the interface speed between the DTE and DCE to four times the modem operating rate to permit the modem to effectively perform compression. Doing so results in the inability to use a cable length anywhere near the maximum 50 foot RS-232 cable length.

In examining the entries in Figure 7.1, it should be noted that a second version of USB was being considered when this book revision was prepared. That revision could boost the operating rate of the interface to between 120 Mbps and 240 Mbps.

7.1.1 Connector overview

RS-232 and the ITU-T V.24 standard as well as RS-530 formally specify the use of a D-shaped 25-pin interface connector similar to the connector illustrated in Figure 7.2. A cable containing up to 25 individual conductors is fastened to the narrow part of the connector, and the individual conductors are soldered to predefined pin connections inside the connector.

ITU-T X.20 and X.21 compatible equipment use a 15-position subminiature D connector that is both smaller and has 10 less predefined pins than the 25-position D connector illustrated in Figure 7.2. ITU-T V.35 compatible equipment uses a 34-position 'M' series connector similar to the connector illustrated in Figure 7.3. RS-449 compatible equipment uses a 37-position D connector and may optionally use a 9-position D connector, while RS-366 compatible equipment uses a 25-position D connector similar to the connector illustrated in Figure 7.2. Although the HSSI connector has 50 pins, it is actually smaller than the 32-position V.35 connector. Another interesting feature of

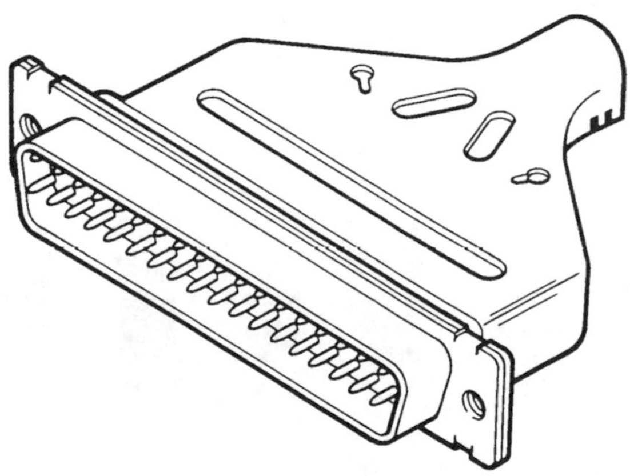

Figure 7.2 The D connector. The 25-pin D connector is used by RS-232, RS-366, RS-530 and V.24 equipment

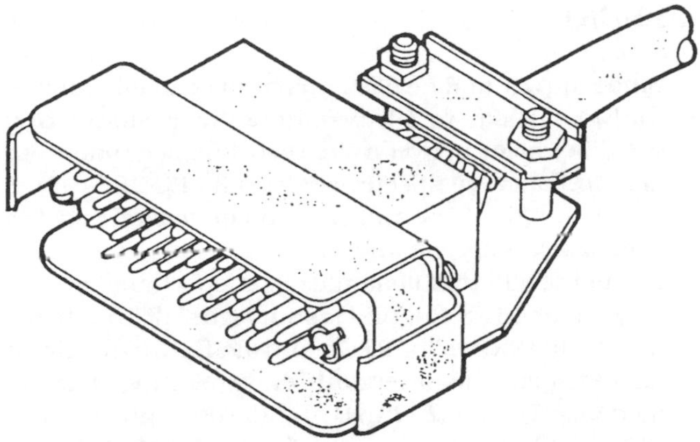

Figure 7.3 The V-35M series connector. The V.35 connector has 34 pins

HSSI connectors is their genders. The cable connectors are specified as male, where both DTE and DCE connectors are specified as receptacles. This specification minimizes the need for male–male and female–female adapters commonly required to mate equipment and cables based on the use of other interface standards.

If you recently purchased a PC or notebook computer you probably noticed two rectangular slots on the back or side of your computer. Those rectangles represent USB ports into which a USB cable with a rectangular plug is inserted, simplifying connectivity in a manner similar to plugging a telephone cord into a wall socket.

In comparison to RS-232-D and RS-530, the RS-232-C standard only referred to the connector in an appendix and stated that it was not part of the standard. In spite of this omission, the use of a 25-pin D-shaped connector with RS-232-C is considered as a *de facto* standard. Although a *de facto* standard, many RS-232-C devices, in fact, use other types of connectors.

Perhaps the most common exception to the 25-pin connector resulted from the manufacture of a serial–parallel adapter card for use in the IBM PC AT and compatible personal computers. The serial RS-232-C port on that card uses a 9-pin connector, which resulted in the development of a viable market for 9-pin to 25-pin converters consisting of a 9-pin and 25-pin connector on opposite ends of a short cable whose interchange circuits are routed in a specific manner to provide a required level of compatibility.

The major differences between RS-232-D and RS-232-C is that the new revision supports the testing of both local and remote communications equipment by the addition of signals to support this function and modified the use of the protective ground conductor to provide a shielding capability. A more recent revision, RS-232-E, added a specification for a smaller alternative 26-pin connector and slightly modified the functionality of a few of the interface pins. Since RS-232-C and RS-232-D are by far the most commonly supported serial interfaces, we will first focus our attention upon these interfaces. Once this has been accomplished we will discuss the newest revision to this popular interface, RS-232-E.

7.1.2 RS-232-C/D

Since the use of RS-232-C has been basically universal since its publication by the EIA in 1969, we will examine both revisions C and D in this section, denoting the differences between the revisions when appropriate. When both revisions are similar, we will refer to them as RS-232-C/D. In general, devices built to either standard as well as the equivalent ITU-T V.24 recommendation are compatible with one another. There are some slight differences that can occur due to the addition of signals to support modem testing under RS-232-D.

Since the RS-232-C/D standards define the most popular method of interfacing between DTEs and DCEs in the United States, they govern, as an example, the interconnection of most terminal devices to stand-alone modems. The RS-232-C/D standards apply to series data transfers between a DTE and DCE in the range from 0 to 19 200 bits per second. Although the standards also limit the cable length between the DTE and DCE to 50 feet, since the pulse width of digital data is inversely proportional to the data rate, one can normally exceed this 50 foot limitation at lower data rates, as wider pulses are less susceptible to distortion than narrower pulses. When a cable length in excess of 50 feet is required, it is highly recommended that low-capacitance shielded cable be used and tested prior to going on-line, to ensure that the signal quality is acceptable. This type of cable is discussed later in this chapter.

Another part of the RS-232-D standard specifies the cable heads that serve as connectors to the DTEs and DCEs. Here the connector is known as a DB-25 connector and each end of the cable is equipped with this 'male' connector that is designed to be inserted into the DB-25 female connectors normally built into modems. Figure 7.4 illustrates the RS-232-C/D interface between a terminal and a stand-alone modem.

Signal characteristics

The RS-232-C/D interface specifies 25 interchange circuits or conductors that govern the data flow between the DTE and DCE. Although you can purchase a 25-conductor cable, normally a smaller number of conductors is required. For asynchronous transmission, normally 9 to 12 conductors are required, while synchronous transmission typically requires 12 to 16 conductors; with the number of conductors required a function of the operational characteristics of the devices to be connected to one another.

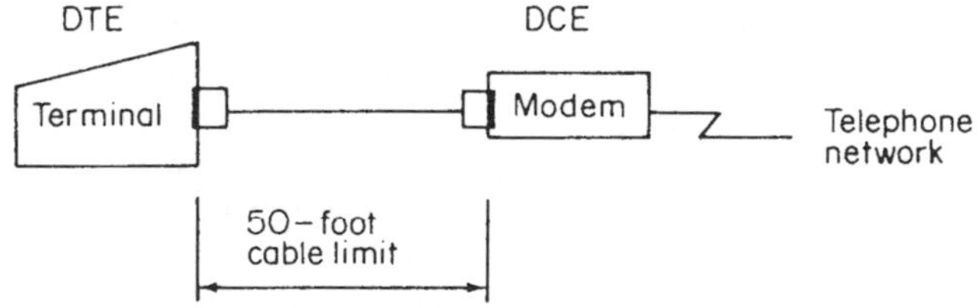

Figure 7.4 The RS-232-C/D physical interface standard cables are typically 6, 10 or 12 ft in length with 'male' connectors on each end

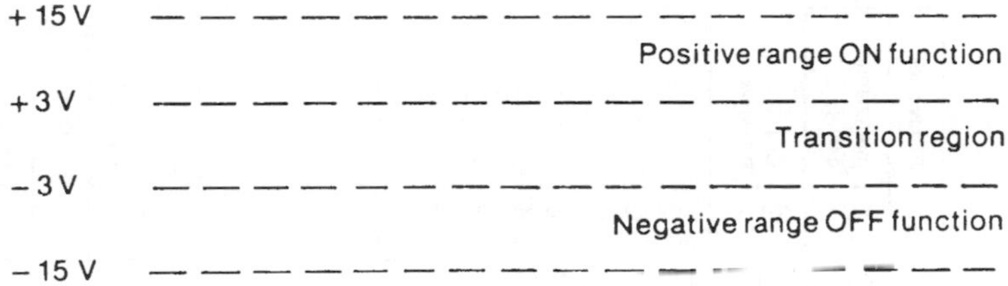

Figure 7.5 Interchange circuit voltage ranges

Table 7.1 Interchange circuit comparison

	Interchange circuit voltage	
	Negative	Positive
Binary state	1	0
Signal condition	Mark	Space
Function	OFF	ON

The signal on each of the interchange circuits occurs based upon a predefined voltage transition occurring as illustrated in Figure 7.5. A signal is considered to be ON when the voltage (V) on the interchange circuit is between +3 V and +15 V. In comparison, a voltage between −3 V and −15 V causes the interchange circuit to be placed in the OFF condition. The voltage range from +3 V to −3 V is a transition region that has no effect upon the condition of the circuit.

Although the RS-232 and V.24 standards are similar to one another, the latter differs with respect to the actual electrical specification of the interface. The ITU-T V.24 recommendation is primarily concerned with how the interchange circuits operate. Thus, another recommendation, known as V.28, actually defines the electrical specifications of the interface that can be used with the V.24 standard.

Table 7.1 provides a comparison between the interchange circuit voltage, its binary state, signal condition and function. As a binary 1 is normally represented by a positive voltage with a terminal device, this means that data signals are effectively inverted for transmission

Since the physical implementation of the RS-232-C/D standard is based on the conductors used to interface a DTE to a DCE, we will examine the functions of each of the interchange circuits. Prior to discussing these circuits, an explanation of RS-232 terminology is warranted since there are three ways in which one can refer to the circuits in this interface.

Circuit–conductor reference

The most commonly used method to refer to the RS-232-C/D circuits is by specifying the number of the pin in the connector which the circuit uses. A second method used to refer to the RS-232-C/D circuits is by the two- or

PIN Number	Interchange circuit	CCITT equivalent	Description	Gnd	Data		Control		Timing		Testing	
					From DCE	To DCE	From DCE	To DCE	From DCE	To DCE	From DCE	To DCE
1	AA	101	Protective Ground (Shield)	X								
7	AB	102	Signal Ground/Common Return	X								
2	BA	103	Transmitted Data			X						
3	BB	104	Received Data		X							
4	CA	105	Request to Send					X				
5	CB	106	Clear to Send				X					
6	CC	107	Data Set Ready (DCE Ready)				X					
20	CD	108.2	Data Terminal Ready (DTE Ready)					X				
22	CE	125	Ring Indicator				X					
8	CF	109	Received Line Signal Detector				X					
21	(RL)/ CG	110	(Remote Loopback)/Signal Quality Detector				X					
23	CH	111	Data Signal Rate Selector (DTE)					X				
23	CI	112	Data Signal Rate Selector (DCE)				X					
24	DA	113	Transmitter Signal Element Timing (DTE)							X		
15	DB	114	Transmitter Signal Element Timing (DCE)						X			
17	DD	115	Receiver Signal Element Timing (DCE)						X			
14	SBA	118	Secondary Transmitted Data			X						
16	SBB	119	Secondary Received Data		X							
19	SCA	120	Secondary Request to Send					X				
13	SCB	121	Secondary Clear to Send				X					
12	SCF	122	Secondary Received Line Signal Detector				X					
8	—	—	Reserved for Testing									X
9	—	—	Reserved for Testing								X	
18	(LL)		(Local Loopback)									X
25	(TM)		(Test Mode)								X	

Figure 7.6 RS-232-C/D and CCITT V.24 (now referred to as ITU-T) interchange circuits by category. RS 232-D additions – changes to RS-232-C indicated in parentheses

three-letter designation used by the standards to label the circuits. The first letter in the designator is used to group the circuits into one of six circuit categories as indicated by the second column labeled 'interchange circuit' in Figure 7.6. As an example of the use of this method, the two ground circuits have the letter A as the first letter in the circuit designator and the signal ground circuit is called 'AB', since it is the second circuit in the 'A' ground category. Since these designators are rather cryptic, they are not commonly used.

A third method used is to describe the circuits by their functions. Thus, pin 2 which is the transmit data circuit can be easily referred to as transmit data. Many persons created acronyms for the descriptions which are easier to remember than the RS-232 pin number or interchange circuit designator. For example, transmit data are referred to as 'TD', which is easier to remember than any of the RS-232 designators previously discussed.

Although the list of circuits in Figure 7.6 may appear overwhelming at first glance, in most instances only a subset of the 25 conductors are employed. To better understand this interface standard, we will first examine those interchange circuits required to connect an asynchronously operated terminal device to an asynchronous modem. We can then expand on our knowledge of these interchange circuits by examining the functions of the remaining circuits, to include those additional circuits that would be used to correct a synchronously operated terminal to a synchronous modem.

Asynchronous operations

Figure 7.7 illustrates the signals that are required to connect an asynchronous terminal device to a low-speed asynchronous modem. Note that although a 25-conductor cable can be used to cable the terminal to the modem, only ten conductors are actually required to support most asynchronous modems. Thus, a 10-conductor cable could be used to connect an asynchronous modem to an asynchronously operated terminal device, which could result in a significant reduction in cable costs when cabling many DTEs to DCEs.

By reading the modem vendor's specification sheet you can easily determine the number of conductors required to cable DTEs to DCEs. Although

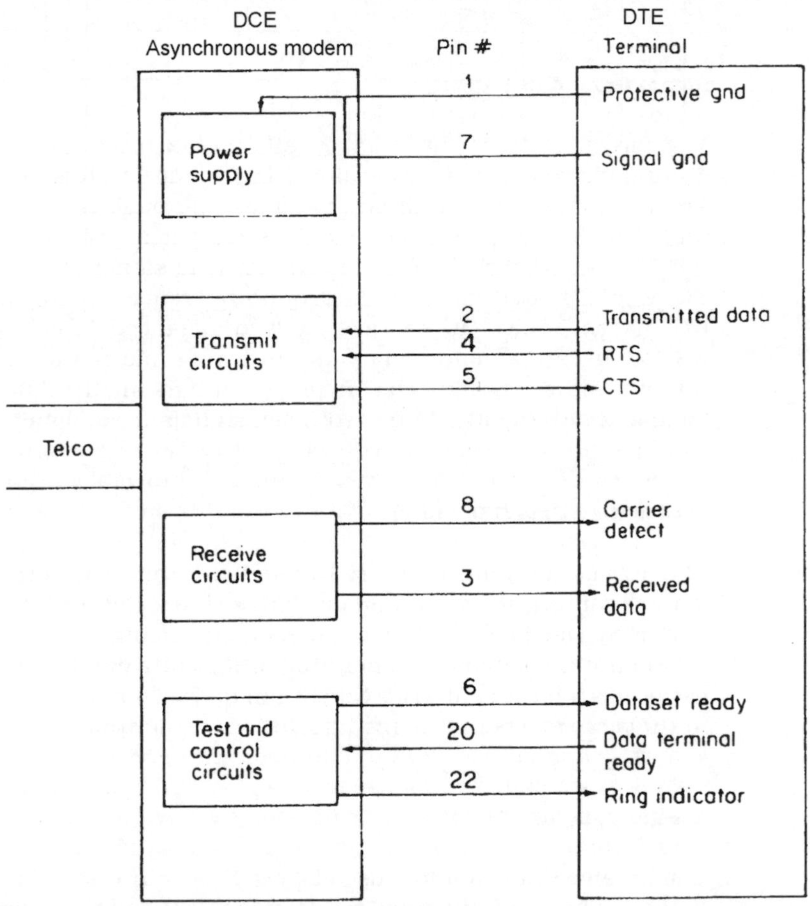

Figure 7.7 A typical DTE–DCE interface supporting asynchronous modems. In this example pin 7 can be tied to pin 1, resulting in the use of a common nine-conductor cable

most cables have straight-through conductors, in certain instances the conductor pins at one end of a cable may require reversal or two conductors may be connected onto a common pin, a process called strapping. In fact, many times only one conductor will be used for both protective ground and signal ground, with the common conductor cabled to pins 1 and 7 at both ends of the cable. In such instances a 9-conductor cable could be employed to satisfy the cabling requirements illustrated in Figure 7.7. With this in mind, let us review the functions of the ten circuits illustrated in Figure 7.7.

Protective Ground (GND, Pin 1)

This interchange circuit is normally electrically bonded to the equipment's frame. In some instances, it can be further connected to external grounds as required by applicable regulations. Under RS-232-D this conductor's use is modified to provide shielding for protection against electromagnetic or other interference occurring in high-noise environments.

Signal Ground (SG, Pin 7)

The circuit must be included in all RS-232 interfaces, as it establishes a ground reference for all other lines. The voltage on this circuit is set to 0 V to provide a reference for all other signals. Although the conductors for pins 1 and 7 can be independent of one another, typical practice is to 'strap' pin 7 to pin 1 at the modem. This is known as tying signal ground to frame ground. Since RS-232-C/D use a single ground reference circuit the standard results in what is known as an electrically unbalanced interface. In comparison, RS-422 uses differential signaling in which information is conveyed by the relative voltage level in two conductors, enhancing the data rate and distance for that standard. The latter is known as balanced signaling.

Transmitted Data (TD, Pin 2)

The signals on this circuit are transmitted from data terminal equipment to data communications equipment or, as illustrated in Figure 7.7, a terminal device to the modem. When no data are being transmitted the terminal maintains this circuit in a marking or logical 1 condition. This is the circuit over which the actual series bit stream of data flows from the terminal device to the modem where it is modulated for transmission.

Receive Data (RD, Pin 3)

The Receive Data circuit is used by the DCE to transfer data to the DTE. After data have been demodulated by a modem, they are transferred to the attached terminal over this interchange circuit. When the modem is not sending data to the terminal, this circuit is held in the marking condition.

Request to Send (RTS, Pin 4)

The signal on this circuit is sent from the DTE (terminal) to the DCE (modem) to prepare the modem for transmission. Prior to actually sending data, the terminal must receive a Clear to Send signal from the modem on pin 5.

Clear to Send (CTS, Pin 5)

This interchange circuit is used by the DCE (modem) to send a signal to the attached DTE (terminal), indicating that the modem is ready to transmit. By turning this circuit OFF, the modem informs the terminal that it is not ready to receive data. The modem raises the CTS signal after the terminal initiates a Request to Send (RTS) signal.

Carrier Detect (CD, Pin 8)

Commonly referred to as a Received Line Signal Detector (RLSD), a signal on this circuit is used to indicate to the DTE terminal) that the DCE (modem) is receiving a carrier signal from a remote modem. The presence of this signal is also used to illustrate the carrier detect light-emitting diode (LED) indicator on modems equipped with that display indicator. If this light indicator should go out during a communications session, it indicates that the session has terminated owing to a loss of carrier, and software that samples for this condition will display the message 'carrier lost' or a similar message to indicate that this condition has occurred.

Data Set Ready (DSR, Pin 6)

Signals on this interchange circuit flow from the DCE to the DTE and are used to indicate the status of the data set connected to the terminal. When this circuit is in the ON (logic 0) condition, it serves as a signal to the terminal that the modem is connected to the telephone line and is ready to transmit data. Since the RS-232 standard specifies that the DSR signal is ON when the modem is connected to the communications channel and not in any test condition, a modem using a self-testing feature or automatic dialing capability would pass this signal to the terminal after the self-test had been completed or after the telephone number of a remote location was successfully dialed. Under RS-232-D this signal was renamed DCE Ready.

Data Terminal Ready (DTR, Pin 20)

The signal on this circuit flows from the DTE to the DCE and is used to control the modem's connection to-the telephone line. An ON condition on this circuit prepares the modem to be connected to the telephone line, after which the connection can be established by manual or automatic dialing. If the signal on this circuit is placed in an OFF condition, it causes the modem to drop any telephone connection in progress, providing a mechanism for the terminal device to control the line connection. Under RS-232-D this signal was renamed DTE Ready.

Ring Indicator (RI, Pin 22)

The signal on this interchange circuit flows from the DCE to the DCE and indicates to the terminal device that a ringing signal is being received on the communications channel. This circuit is used by an auto-answer modem to 'wake-up' the attached terminal device. Since a telephone rings for one second and then pauses for four seconds prior to ringing again, this line becomes active for one second every five seconds when an incoming call occurs.

The Ring Indicator circuit is turned ON by the DCE when it detects the ON phase of a ring cycle. Depending on how the DCE is optioned, it may either keep the RI signal high until the DTE turns DTR low or the DCE may turn the RI signal ON and OFF to correspond to the telephone ring sequence.

If the DTE is ready to accept the call its DTR lead will either be high, which is known as a Hot-DTR state, or be placed into an ON condition in response to the RI signal turning ON. Once the RI and DTR circuits are both ON, the DCE will actually answer the incoming call and place a carrier tone onto the line.

If a computer port connected to a modem is not in a Hot-DTR state the first ring causes the modem to turn ON its RI circuit momentarily, alerting the computer port to the incoming call. As the computer port turns on its DTR circuit the modem's RI circuit will be turned off as it cycles in tandem with the telephone company ringing signal. The DCE must thus wait for the next ON phase of the ring cycle to answer the call, explaining why many modems may require two rings to answer a call.

Control signal timing relationship

The actual relationship of RS-232 control signals varies in time, based upon the operational characteristics of devices connected as well as the strapping option settings of those devices. Figure 7.8 illustrates a common timing relationship of control signals between a computer port and a modem.

At the top of Figure 7.8 it is assumed that the data set is powered, resulting in the Data Set Ready (DSR) control signal being high or in the ON state. Next, two Ring Indicator (RI) signals are passed to the computer port, resulting in the computer responding by raising its Data Terminal Ready (DTR) signal. The DTR signal in conjunction with the second Ring Indicator (RI) signal results in the modem answering the call, presenting the Carrier Detect (CD) signal to the computer port. Assuming that the computer is programmed to transmit a sign-on message, it will raise its Request to Send (RTS) signal. The modem will respond by raising its Clear to Send (CTS) signal if it is ready to transmit, which enables the computer port to begin the actual transmission of data.

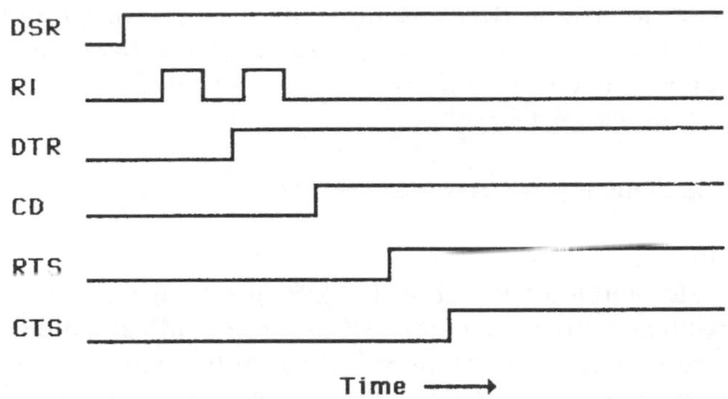

Figure 7.8 An example of a possible control signal timing relationship. The state of the control signals varies in time, based upon the operational characteristics of devices connected as well as the strapping option settings of those devices

Synchronous operations

One major difference between asynchronous and synchronous modems is the timing signals required for synchronous transmission.

Timing signals

When a synchronous modem is used, it puts out a square wave on pin 15 at a frequency equal to the modem's bit rate. This timing signal serves as a clock from which the terminal would synchronize its transmission of data onto pin 2 to the modem. Pin 15 is thus referred to as the transmit clock as well as its formal designation of Transmission Signal Element Timing (DCE), with DCE referring to the fact that the communications device supplies the timing.

Whenever a synchronous modem receives a signal from the telephone line it puts out a square wave on pin 17 to the terminal at a frequency equal to the modem's bit rate, while the actual data is passed to the terminal on pin 3. Since pin 17 provides receiver clocking, it is known as the 'receiver clock' as well as its more formal designation of Receiver Signal Element Timing.

In certain cases a terminal device such as a computer port can provide timing signals to the DCE. In such situations the DTE will provide a clocking signal to the DCE on pin 24 while the formal designator of transmitter signal element timing (DTE) is used to refer to this signal.

The process whereby a synchronous modem or any other type of synchronous device generates timing is known as internal timing. If a synchronous modem or any other type of synchronous DCE is configured to receive timing from an attached DTE, such as a computer port or terminal, the DCE must be set to external timing when the DTE is set to internal timing.

Intelligent operations

There are three interchange circuits that can be employed to change the operation of the attached communications device. One circuit can be used to first determine that a deterioration in the quality of a circuit has occurred, and the other two circuits can be employed to change the transmission rate to reflect the circuit quality.

Signal quality detector (CG, Pin 21)

Signals on this circuit are transmitted from the DCE (modem) to the attached DTE (terminal) whenever there is a high probability of an error in the received data due to the quality of the circuit falling below a predefined level. This circuit is maintained in an ON condition when the signal quality is acceptable and turned to an OFF condition when there is a high probability of an error. Under RS-232-D this circuit can also be used to indicate that a remote loopback is in effect.

Data signal rate selector (CH/CI, Pin 23)

When an intelligent terminal device such as a computer port receives an OFF condition on pin 21 for a predefined period of time, it must be programmed to change the data rate of the attached modem, assuming that the modem is capable of operating at dual data rates. This can be accomplished by the terminal device providing an ON condition on pin 23 to select the higher data signaling rate or range of rates, while an OFF condition would select the lower data signaling range or range of rates. When the data terminal equipment selects the operating rate, the signal on pin 23 flows from the DTE to the DCE and the circuit is known as circuit CH. If the data communications equipment is used to select the data rate of the terminal device, the signal on pin 23 flows from the DCE to the DTE and the circuit is known as circuit CI.

Secondary circuits

In certain instances a synchronous modem will be designed with the capability to transmit data on a secondary channel simultaneously with transmission occurring on the primary channel. In such cases the data rate of the secondary channel is normally a fraction of the data rate of the primary channel.

To control the data flow on the secondary channel the RS-232-C/D standards employ five interchange circuits. Pins 14 and 16 are equivalent to the circuits on pins 2 and 3, except that they are used to transmit and receive data on the secondary channel. Similarly, pins 19, 13 and 12 perform the same functions as pins 4, 5 and 8 used for controlling the flow of information on the primary data channel.

In comparing the interchange circuits previously described to the connector illustrated in Figure 7.9, note that the location of each interchange circuit is explicitly defined by the pin number assigned to the circuit. In fact, the RS-232-D connector is designed with two rows of pins, with the top row containing 13 while the bottom row contains 12. Each pin has an explicit

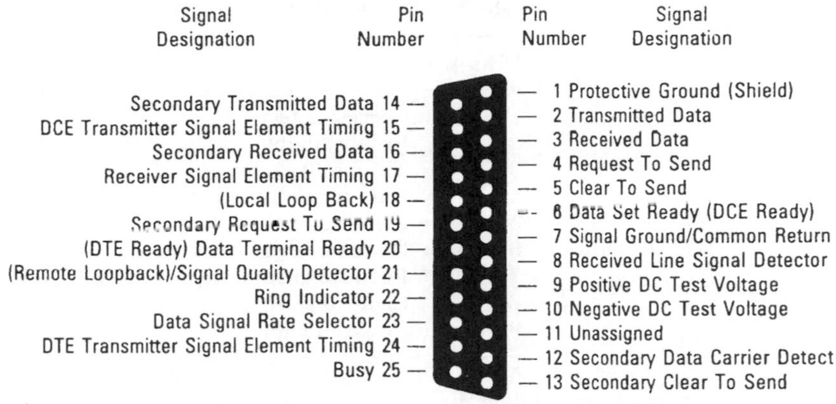

Figure 7.9 RS-232 interface on D connector

signal designation that corresponds to a numbering assignment that goes from left to right across the top row and then left to right across the bottom row of the connector. For ease of illustration the assignment of the interchange circuits to each of the pins in the D connector is presented in Figure 7.9 by rotating the connector 90° clockwise. In this illustration, RS-232-D conductor changes from RS-232-C are denoted in parentheses.

Test circuits

RS-232-D adds three circuits for testing that were not part of the earlier RS-232-C standard. The DTE can request the DCE to enter Remote Loopback (RL, pin 21) or Local Loopback (LL, pin 18) mode by placing either circuit in the ON condition. The DCE, if built to comply with RS-232-D, will respond by turning the TEST MODE (TM, pin 25) circuit ON and performing the appropriate test.

Connector conversion

In Table 7.2 the reader will find a list of the corresponding pins between a DB-9 connector used on an IBM PC AT serial port and a standard RS-232 DB-25 connector. Data in this table can be used to develop an appropriate DB-9 to DB-25 converter, enabling users to attach a DB-25 connector modem, as an example, to a DB-9 connector serial port on an IBM PC AT or compatible personal computer. As an example of the use of Table 7.2, the conductor for Carrier Detect would be wired to connect pin 1 at the DB-9 connector to pin 8 at the DB-25 connector.

RS-232-E

The latest revision to RS-232, RS-232-E (Revision E), resulted in several minor changes to the operation of some interface circuits and the specification of an alternate interface connector (Alt A). Although none of the changes

Table 7.2 DB-9 to DB-25 pin correspondence

DB-9		DB-25
1	Carrier detect	8
2	Receive data	3
3	Transmitted data	2
4	Data terminal ready	20
5	Signal ground	7
6	Data set ready	6
7	Request to send	4
8	Clear to send	5
9	Ring indicator	22

were designed to create compatibility problems with prior versions of RS-232, the use of the alternative physical interface can only be accomplished if a Revision E device is cabled via an adapter to an earlier version of that interface or if a dual connector cable is used. The 26-pin Alt A connector is about half the size of the 25-pin version and was designed to support hardware in which connector space is at a premium, such as laptop and notebook computers. Pin 26 is only contained on the Alt A connector and presently is functionless.

In addition to specifying an alternative interface connector RS-232-E slightly modified the functionality of certain pins or interchange circuits. First, pin 4 (Request to Send) is defined as Ready for receiving when the DTE enables that circuit. Next, pin 18 which was used for Local Loopback will now generate a Busy Out when enabled. A third modification is the use of Clear to Send for hardware flow control, a function used by countless vendors over the past two decades but only now formally recognized by the standard. The term flow control represents the orderly control of the flow of data. By toggling the state of the Clear to Send signal DCE equipment can regulate the flow of information from DTE equipment, a topic that we will discuss in detail later in this book.

RS-232/V.24 limitations

There are several limitations associated with the RS-232 standard and the V.28 recommendation which defines the electrical specification of the interface that can be used with the V.24 standard. Foremost among these limitations are data rate and cabling distance.

RS-232/V.24 is limited to a maximum data rate of 19.2 kbps at a cabling distance of 50 feet. In actuality, speeds below 19.2 kbps allow greater transmission distances, and for cable lengths of only a few feet a data rate up to approximately 100 kbps becomes possible.

7.1.3 Differential signaling

Over long cabling distances the cumulative cable capacitance and resistance combine to cause a significant amount of signal distortion. At some cabling distance, which decreases in an inverse relationship to the data rate, the signal cannot be recognized. To overcome the cabling distance and speed limitations associated with RS-232 a different method of signaling was thus devised. This signaling technique, known as differential signaling, results in information being conveyed by the relative voltage levels in two wires. Instead of using one driver to produce a larger voltage swing as RS-232 does, differential signaling uses two drivers to split a signal into two parts.

Figure 7.10 illustrates differential signaling as specified by the RS-422 interface standard. To transmit a logical '1' the A driver output is driven more positive while the B driver output is more negative. Similarly, to transmit a logical '0' the A driver output is driven more negative while the B driver output is driven more positive. At the receiver a comparator is used to examine the relative voltage levels on the two signal wires.

With the use of two wires, RS-449 specifies a mark or space by the difference between the voltages on the two wires. The difference is only 0.4 V under RS-422, whereas it is 6 V (+3 V and −3 V) under RS-232/V.24. Thus, if the difference signal between the two wires is positive and greater than 0.2 V, the receiver will read a mark. Similarly, if the difference signal is negative and more negative than −0.2 V, the receiver will read a space. In addition to requiring a lower voltage shift, source and load impedance of RS-232, which is approximately 5 kΩ, is reduced to 100 Ω by the use of differential signaling. Another benefit of this signaling method is the effect of noise on signal distortion. Since the presence of noise results in the same voltage being imposed on both wires, there is no change in the voltage difference between the signal wires. The combination of lower signaling levels and impedances coupled with voltage difference immunity to noise thus permits differential signaling to drive longer cable distances at higher speeds. In Figure 7.11 the reader will find a plot of cable distances versus signaling rate for the RS-422 standard.

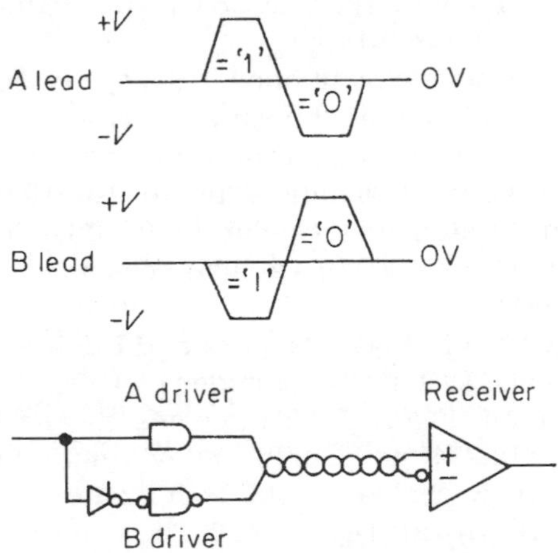

Figure 7.10 Differential signaling. The RS-422 specifies balanced differential signaling since the sum of the currents in the differential signaling wires is zero

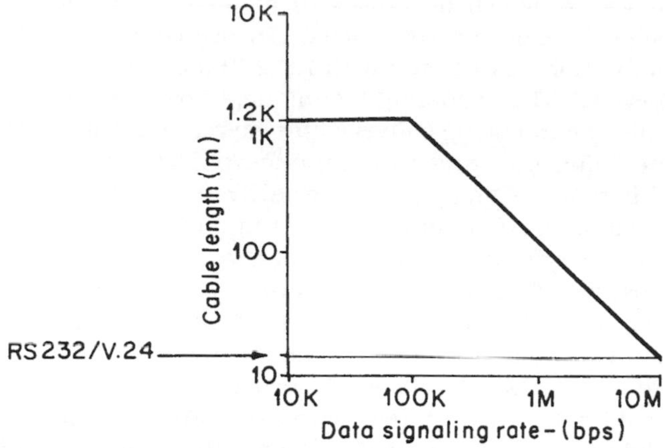

Figure 7.11 RS-422 cable distance versus signaling rate

For comparison purposes, the RS-232/V.24 cable distance versus signaling rate is shown in Figure 7.11. As indicated, RS-422 offers significant advantages over RS-232 with respect to both cabling distance and data signaling rate.

7.1.4 RS-449

RS-449 was introduced in 1977 as an eventual replacement for RS-232-C. This interface specification calls for the use of a 37-pin connector as well as an optional 9-pin connector for devices using a secondary channel. Unlike RS-232, RS-449 does not specify voltage levels. Two additional specifications known as RS-442-A and RS-423 cover voltage levels for a specific range of data speeds. RS-442-A and its counterpart, the ITU-T X.27 (V.11), define the voltage levels for data rates from 20 kbps to 10 Mbps; whereas RS-423-A and its ITU-T counterpart, X.26 (V.10), define the voltage levels for data rates between 0 and 20 kbps.

As previously mentioned, RS-442 (as well as its ITU-T counterpart) defines the use of differential balanced signaling. RS-422 is designed for twisted-pair telephone wire transmission ranging from 10 Mbps at distances up to 40 feet to 100 kbps at distances up to 4000 feet. RS-423 defines the use of unbalanced signaling similar to RS-232. This standard supports data rates ranging from 100 kbps at distances up to 40 feet, to 10 kbps at distances up to 200 feet.

The use of RS-422, RS-423 and RS-449 permits the cable distance between DTEs and DCEs to be extended to 4000 feet in comparison to RS-232's 50 foot limitation. Figure 7.12 indicates the RS-449 interchange circuits. In comparing RS-449 to RS-232, you will note the addition of ten circuits which are either new control or status indicators and the deletion of three functions formerly provided by RS-232. The most significant functions added by RS-449 are local and remote loopback signals. These circuits enable the operation of diagnostic features built into communications equipment via DTE control, permitting as an example, the loopback of the device to the DTE and its

RS-232 Destination	Circuit mnemonic	Circuit name	Circuit direction	Circuit type	
Signal ground	SG	Signal ground	—	Common	
—	SC	Send common	to DCE		
—	RC	Receive common	from DCE		
—	IS	Terminal in service	to DCE	Control	
Ring Indicator	IC	Incoming call	from DCE		
Data terminal ready	TR	Terminal ready	to DCE		
Data set ready	DM	Data mode	from DCE		
Transmit data	SD	Send data	to DCE	Data	Primary channel
Receive data	RD	Receive data	from DCE		
Transmit timing (DTE)	TT	Terminal timing	to DCE	Timing	
Transmit timing (DCE)	ST	Send timing	from DCE		
Receive timing	RT	Receive timing	from DCE		
Request to send	RS	Request to send	to DCE	Control	
Clear to send	CS	Clear to send	from DCE		
Receive signal detector	RR	Receiver ready	from DCE		
Signal quality detector	SO	Signal quality	from DCE		
—	NS	New signal	to DCE		
—	SF	Signal frequency	to DCE		
Data receive selector (DTE)	SR	Signaling rate selector	to DCE		
Data rate selector (DCE)	SI	Signaling rate indicator	from DCE		
Secondary transmit data	SSD	Secondary send data	to DCE	Data	Secondary channel
Secondary receive data	SRD	Secondary receive data	from DCE		
Secondary request to send	SRS	Secondary request to send	to DCE	Control	
Secondary clear to send	SCS	Secondary clear to send	from DCE		
Secondary received signal detector	SRR	Secondary receiver ready	from DCE		
Local loopback (D/E)	LL	Local loopback	to DCE	Control	
—	RL	Remote loopback	to DCE		
—	TM	Test mode	from DCE		
—	SS	Select standby	to DCE	Control	
—	SB	Standby indicator	from DCE		

Figure 7.12 RS-449 interchange circuits

placement into a test mode of operation. With the introduction of RS-232-D a local loopback function was supported. Thus, the column labeled RS-232 Destination with the row entry Local Loopback indicates that that circuit is only applicable to revisions D and E of that standard by the entries D/E in parentheses after the circuit name.

Although a considerable number of articles have been written describing the use of RS-449, its complexity has served as a constraint in implementing this standard by communications equipment vendors. Other constraints limiting its acceptance include the cost and size of the 37-pin connector arrangement and the necessity of using another connector for secondary operations. By late 1999, less than a few percent of all communications devices were designed to operate with this interface. Due to the failure of RS-449 to obtain commercial acceptance, the EIA issued RS-530 in March, 1987. This new standard is intended to gradually replace RS-449.

7.1.5 V.35

The V.35 standard was developed to support high-speed transmission, typically 48, 56 and 64 kbps. Originally the V.35 interface was designed into adapter boards inserted into mainframe computers to support 48 kbps transmission on analog wideband facilities. Today, the V.35 standard is the prevalent interface to 56 kbps common carrier digital transmission facilities in the United States and 48 kbps common carrier digital transmission

V.35		Direction		RS-232	
Pin	Name	DCE DTE	Function	Pin	Name
A	FG		Frame GND	1	AA
B	SG		Signal GND	7	AB
C	RTS	→	Request to send	4	CA
D	CTS	←	Clear to send	5	CB
E	DSR	→	Data set ready	6	CC
F	RLSD	→	Received line signal	8	CF
H	DTR	←	Data terminal ready	20	CD
J	RI	→	Ring indicator	22	CE
R, T	RD	→	Receive data	3	BB
U, X	SGR	→	Receive clock	17	DD
P, S	SD	←	Send data	2	BA
U, W	SCTE	←	Send clock (EXT)	24	DA
Y, a	SCT	→	Send clock	15	DB
m	TST	→	Reserved for test (D/E)	25	TM

Figure 7.13 V.35/RS-232 signal correspondence

facilities in the UK. In addition, the V.35 standard can support data transfer operations at operating rates up to approximately 6 Mbps. This has resulted in the V.35 interface being commonly employed in videoconferencing equipment, routers and other popularly used communications devices.

The V.35 electrical signaling characteristics are a combination of an unbalanced voltage and a balanced current. Although control signals are electrically unbalanced and compatible with RS-232 and CCITT (ITU-T) V.28, data and clock interchange circuits are driven by balanced drivers using differential signaling similar to RS-422 and V.11. V.35 uses a 34-pin connector specified in ISO 2593, similar to the connector illustrated in Figure 7.13.

Figure 7.13 illustrates the correspondence between RS-232 and V.35. Note that the V.35 pin pairs tied together by a brace are differential signaling circuits that use a wire pair.

7.1.6 RS-366-A

The RS-336-A interface is employed to connect terminal devices to automatic calling units. This interface standard uses the same type 25-pin connector as RS-232, however, the pin assignments are different. A similar interface to RS-366-A is the ITU-T V.25 recommendation, which is also designed for use with automatic calling units. Figures 7.14 and 7.15 illustrate the RS-366-A and V.25 interfaces. Note that for both interfaces each actual digit to be dialed is transmitted as parallel binary information over circuits 14 through 17. The pulse on pin 14 represents the value 2^0, whereas the pulse on pins 15 through 17 represent the values 2^1, 2^2 and 2^3, respectively. Thus, to indicate to the automatic calling unit that it should dial the digit 9, circuits 14 and 17 would become active.

Originally, automatic calling units provided the only mechanism to automate communications dialing over the PSTN. This enabled the use of RS-366 automatic dialing equipment under computer control to re-establish communications via the PSTN if a leased line became inoperative, a process called dial-backup. Another common use of automatic calling units is to poll remote terminals from a centrally located computer in the evening when rates are lower. Under software control the central computer would dial each remote terminal and request the transmission of the day's transactions for processing. Due to the development and wide acceptance of the use of intelligent modems with automatic dialing capability the use of automatic calling units has greatly diminished.

Until recently, only intelligent asynchronous modems had an automatic dialing capability, restricting the use of automatic calling units to mainframe computers that required a method to originate synchronous data transfers over the PSTN. In such situations a special adapter was required to be installed in the communications controller of the mainframe, which controlled the operation of the automatic calling unit. The introduction of synchronous modems with automatic dialing capability has significantly diminished the requirement for automatic calling units, since their use eliminates the

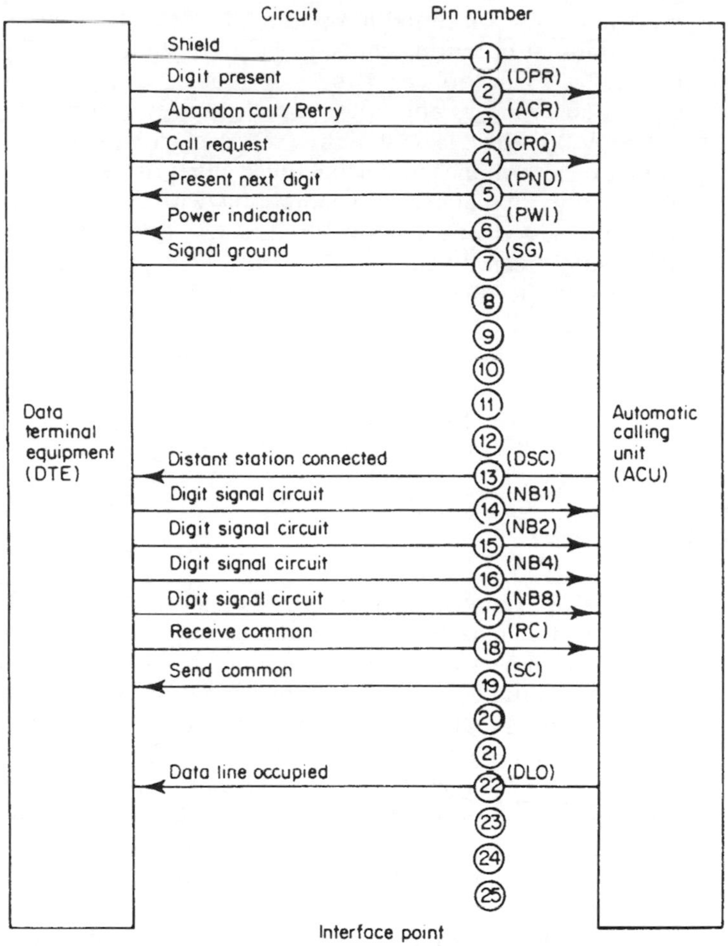

Figure 7.14 RS-366-A interface

requirement to install an expensive adapter in the communications controller as well as the cost of the automatic calling unit. The operation and utilization of intelligent modems is discussed in Chapter 11.

7.1.7 X.21 and X.20

Interface standards X.21 and X.20 were developed to accommodate the growing use of public data networks. X.21 is designed to allow synchronous devices to access a public data network and X.20 provides a similar capability for asynchronous devices.

Instead of assigning functions to specific pins on a connector like RS-232, RS-449 and V.35, X.21 assigns coded character strings to each function to

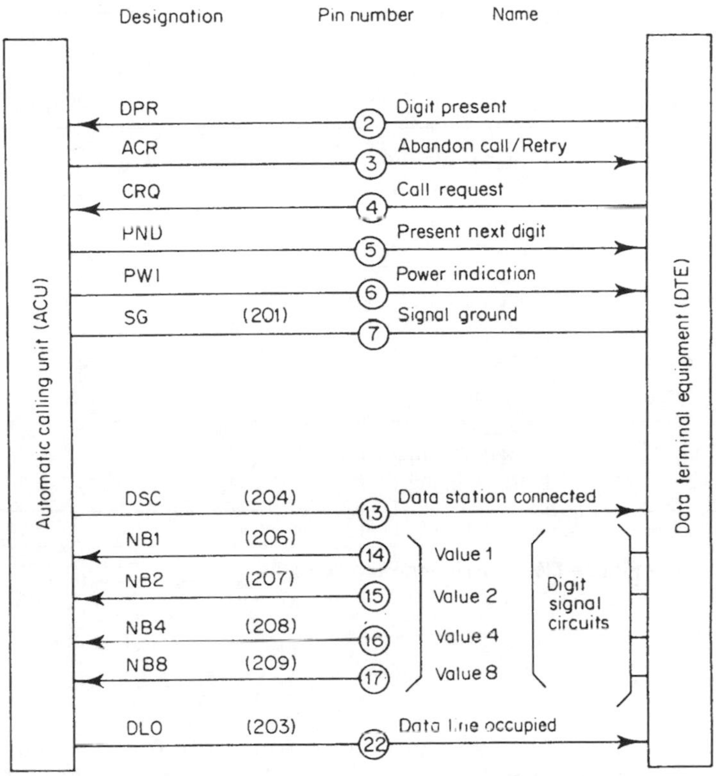

Figure 7.15 V.25 interface

establish connections through a public data network. For example, a dial tone is represented to a computer as a continuous sequence of ASCII '+' characters on the X.21 receive circuit. The computer can then dial a stored number by transmitting it as a series of ASCII characters on the X.21 transmit circuit. Once the call dialing process is complete, the computer will receive call progress signals from the modem on the receive circuit indicating such conditions as number busy and call in progress.

The X.21 interface specifies the use of the balanced signaling characteristics of CCITT X.27 (RS-422) for the network side of the interface and either X.27 or X.26 (RS-423) for the terminal equipment side. This specification enables terminal equipment to be designed for several applications. Unlike RS-232 and V.24, X.21 standard specifies the use of a 15-pin connector. The X.20 interface uses the same 15-pin connector as X.21, however since it supports asynchronous transmission it only needs transmitted data, receive data and ground signals.

Figure 7.16 illustrates the X.21 interchange circuits by circuit type. As indicated, X.21 specifies four categories of interchange circuits: ground, data transfer, control and timing. The operation of the circuits in each category is described in the four following sections.

Interchange circuit	Name	Direction		Circuit Type
		to DCE	from DCE	
G	Signal ground or common return			
Ga	DTE common return	×		Ground
Gb	DCE common return		×	
T	Transmit	×		Data
R	Receive		×	transfer
C	Control	×		
I	Indication		×	Control
S	Signal element timing		×	
B	Byte timing		×	Timing

Figure 7.16 X.21 interchange circuits

Ground signals

Circuit G, signal ground or common return is used to connect the zero volt reference of the transmitter and receiver ends of the circuit. If X.26 differential signaling is used the G circuit is split into two. The Ga circuit is used as the DTE common return and it is connected to ground at the DTE. The Gb circuit is used as the DCE common return and is connected to ground at the DCE.

Data transfer circuits

Circuit T is the transmit circuit used by the DTE to transmit data to the DCE. Circuit R is the receive circuit which is used by the DCE to transmit data to the DTE.

Control circuits

The X.21 specification has two control circuits: control and indication.

Circuit C is the control circuit used by the DTE to indicate to the DCE the state of the interface. During the data transfer phase in which coding flows over the transmit circuit, circuit C remains in the ON condition.

Circuit I, which is the indication circuit, is used by the DCE to indicate the state of the interface to the DTE. When circuit I is ON the representation of the signal occurs in coded form over the receive circuit. Thus, during the data transfer phase circuit I is always ON.

Timing circuits

There are two timing circuits specified by X.21: signal element timing and byte timing.

Circuit S, which is signal element timing, is generated by the DCE and it controls the timing of data on the transmit and receive circuits. In providing a clocking signal, circuit S turns ON and OFF for nominally equal periods of time. The second timing circuit, circuit B, which is byte timing, provides the DTE with 8-bit timing information for synchronous transmission generated by the DCE. Circuit B turns OFF whenever circuit S is ON, indicating the last bit of an 8-bit byte. At other times within the period of the 8-bit byte circuit B remains ON. This circuit is not mandatory in X.21 and is only used occasionally.

Limitations

The X.21 standard has not gained wide acceptance for several reasons, including the popularity of the RS-232/V.24 standard and the cost of implementing X.21. With respect to cost, X.21 transmits and interprets coded character strings. This requires more intelligence to be built into the interface, adding to the cost of the interface. Due to the preceding limitations, the ITU-T defined an alternate interface for public data network access known as X.21 bis, where bis is the Latin term for secondary.

7.1.8 X.21 bis

The X.21 bis recommendation is both physically and functionally equivalent to the V.24 standard, which is compatible with RS-232. The X.21 bis recommendation is designed as an interim interface for X.25 network access and will gradually be replaced by the X.21 standard as more equipment is manufactured to meet the X.21 specification. The X.21 bis connector is the common DB-25 connector used by RS-232 and V.24. The connector pins for X.21 bis are defined in an ISO specification called DIS 2110.

7.1.9 RS-530

Like RS-232, RS-530 uses the near universal 25-pin D-shaped interface connector. Although this standard is intended to replace RS-449, both RS-422 and RS-423 standards specify the electrical characteristics of the interface and will continue in existence. These standards are referenced by the RS-530 standard.

Similar to RS-449, RS-530 provides equipment meeting this specification with the ability to transmit at data rates above the RS-232 limit of 19.2 kbps. This is accomplished by the standard originally specifying the utilization of balanced signals in place of several secondary signals and the Ring Indicator signal included in RS-232. As previously mentioned in our discussion of differential signaling, this balanced signaling technique is accomplished by using two wires with opposite polarities for each signal to minimize distortion.

The RS-530 specification was first outlined in March, 1987, and was officially released in April, 1988. A revision known as RS-530-A was approved in May, 1992. By supporting data rates up to 2 Mbps and using the standard 'D' type 25-pin connector, RS-530 offers the potential to achieve a high level of adoption during the new millennium. One significant change resulting from Revision A is the specification of an alternative 26-position interface connector (Alt A) which is the same optional connector as specified in Revision E to RS-232. Another significant change resulting from Revision A was the addition of support for Ring Indicator which enables the interface to support switched network applications.

Figure 7.17 summarizes the RS-530-A interchange circuits and compares those circuits to both RS-232 and RS-449. Note that RS-530 has maintained the standard RS-232 circuit structure for data, clock and control, all of which are balanced signals based upon the RS-442 standard. RS-530 has also adopted the three test circuits specified in RS-232-D: local loop, remote loop and test mode. Like RS-232-D, these three circuits are single ended. Finally, RS-535 has maintained pin 1 as frame ground or shield and pin 7 as signal ground.

The original RS-530 interface specified balanced circuits for DCE Ready and DTE ready, using pins 22 and 23 for one pair of each signal, respectively. Under Revision A, Ring Indicator support was added through the use of pin 22 while Signal Common support was added through the use of pin 23.

Based upon the RS-530 pin assignments contained in Figure 7.17, the interchange circuits for the D connector specified by the standard are illustrated in Figure 7.18. For comparison purposes, the reader can compare Figure 7.18 to Figure 7.8 to see the similarity between RS-530 and RS-232 D-connector interfaces.

7.1.10 High Speed Serial Interface

The High Speed Serial Interface (HSSI) was jointly developed by T3Plus Networking Inc. and Cisco Systems Inc. as a mechanism to satisfy the growing demands of high-speed data transmission applications. Although the development of HSSI dates to 1989, its developers were forward-looking, recognizing that the practical use of T3 and Synchronous Optical Network (SONET) terminating products would require equipment to transfer information well beyond the capability of the V.35 and RS 422/449 interfaces. The result of their efforts was HSSI, which is a full duplex synchronous serial interface capable of transmitting and receiving information at data rates up to 52 Mbps between a DTE and a DCE. This standard was ratified by the American National Standards Institute (ANSI) in July 1992. ANSI document SP2795 defines the electrical specifications for the interface at data rates up to 52 Mbps, and document SP2796 specifies the operation of the DTE-DCE interface circuits.

Designation	RS-232				RS-449	
Shield			1 1	1		
Transmitted data	BA BA	(A) (B)	2 14	2 —	4 22	Send Data
Received data	BB BB	(A) (B)	3 16	3 —	6 24	Received data
Request to send	CA CA	A) (B)	4 19	4 —	7 25	Request to send
Clear to send	CB CB	(A) (B)	5 13	5 —	9 27	Clear to send
DCE ready	CC	(A)	6	6 —	11 29	Data mode
DTE ready	CD	(A)	20	20 —	12 30	Terminal ready
Signal ground	AB		7	7	19	Signal ground
Received line signal detector	CF CF	(A) (B)	8 10	8 —	13 31	Receiver ready
Transmit signal element timing (DCE source)	DB DB	(A) (B)	15 12	15 —	5 23	Send timing
Receive signal element timing (DCE source)	DD DD	(A) (B)	17 9	17 —	8 26	Receive timing
Local loopback	LL		18	—	10	Local loopback
Remote loopback	RL		21	—	14	
Transmit signal element timing (DTE source)	DA DA	(A) (B)	24 11	24 —	17 35	Terminal timing
Test mode	TM		25	—	18	Test mode
Ring indicator (Revision A)	CD		22	22	15	Incoming call
Signal common (Revision A)	AC		23	—	20	Receive common

Figure 7.17 Pin comparison – RS-530-A to RS-232 and RS-449

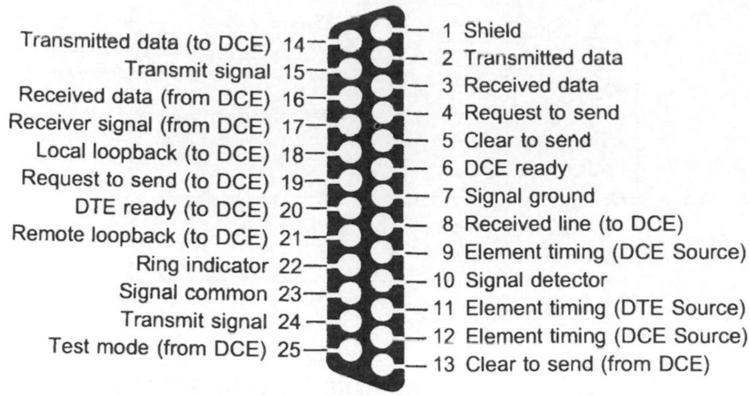

Figure 7.18 RS-530-A interface on D connector

Rationale for development

The rationale for the development of HSSI was not only due to the operating limit of 6 Mbps for V.35 and 10 Mbps for RS-422/449, but, in addition, the problems that occur when those standards are extended to higher operating rates. Several manufacturers developed proprietary methods to increase the data transfer rate of those standards; however, doing so resulted in an increase in radio frequency interference (RFI) which in some instances resulted in the disruption of the operation of other nearby equipment and cable connections.

HSSI eliminates potential RFI problems while obtaining a high speed data transfer capability through the use of emitter-coupled logic (ECL). ECL is faster than complimentary metal–oxide semiconductor (CMOS) or transistor-to-transistor logic (TTL) commonly used in other interfaces, while generating a lower amount of noise. To accomplish this ECL has a voltage swing of 0.8 between defining 0 and 1 bits, which is considerably smaller than the voltage swings in CMOS and TTL logic.

Signal definitions

HSSI can be viewed as a simple V.35 type interface based on the use of emitter-coupled logic for transmission levels, with twelve signals currently defined. Figure 7.19 illustrates the twelve HSSI currently defined interchange circuits, with the normal dataflow indicated by arrowheads on each circuit.

Under HSSI signaling the DCE manages clocking, similar to the V.35 and RS-499 standards. The DCE generates Receive Timing (RT) and Send Timing (ST) signals from the network clock. In comparison, the DTE returns the clocking signal as Terminal Timing (TT) with data on circuit SD (Send Data) to ensure that data are in phase with timing.

HSSI signaling was designed to support continuous as well as gapped, or discontinuous clocking. The latter is associated with the DS3 signal used for

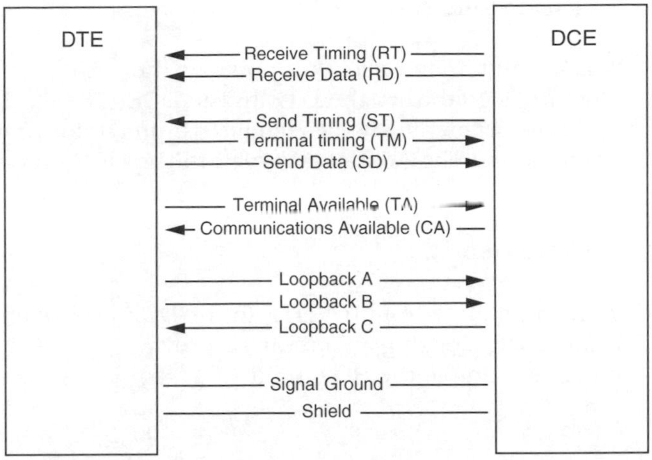

Figure 7.19 HSSI signaling between DTE and DCE

T3 transmission at 44.736 Mbps. Under DS3 signaling every eighty-fifth frame is a control frame, requiring the DCE clock to run for 84 contiguous pulses and then miss one pulse or gap over it to correctly achieve the DS3 frequency. We can obtain an appreciation of the operation of HSSI by examining the operation and functionality of each of the twelve signals currently supported by the interface.

Receive Timing (RT)

The Receive Timing circuit presents the DCE clocking obtained from the network to the attached terminal device. As previously discussed, RT is a gapped clock and it has a maximum bit rate of 52 Mbps. The clocking signal on RT provides the timing information necessary for the DTE to receive data on circuit RD.

Receive Data (RD)

Data received by the DCE from an attached communications circuit are transferred on the RD circuit to the DTE.

Send Timing (ST)

The Send Timing circuit transports a gapped clocking signal with a maximum bit rate of 52 Mbps from the DCE to the DTE. This clock provides transmit signal element timing information to the DTE which is returned via the Terminal Timing (TT) circuit.

Terminal Timing (TT)

The Terminal Timing circuit provides the path for the echo of the Send Timing clocking signal from the DTE to the DCE. The clocking signal on this circuit provides transmit signal element timing information to the DCE which is used for sampling data forwarded to the DCE on circuit SD.

Send Data (SD)

The flow of data from the DTE to the DCE occurs on circuit SD. As previously mentioned, clocking on circuit TT provides the DCE with the timing signals to correctly sample the SD circuit.

Terminal Available (TA)

Terminal Available can be considered as the functional equivalent of Request to Send (RTS); however, unlike TRS, TA is asserted by the DTE independently of DCE when the DTE is ready to both send and receive data. Actual data transmission will not occur until the DCE has asserted a Communications Available (CA) signal.

Communications Available (CA)

The Communications Available (CA) signal can be considered as functionally similar to the Clear to Send (CTS) signal. However, the CA signal is asserted by the DCE independently of the TA signal whenever the DCE is prepared to both transmit and receive data to and from the DCE. The assertion of voltage on circuit CA indicates that the DCE has a functional data communications channel; however, transmission will not occur until the TA signal is asserted by the DTE.

Through the elimination of the Data Set Ready (DSR) and Data Terminal Ready (DTR) signals commonly used in other interfaces, the HSSI interface becomes relatively simple to implement. This in turn simplifies the DTE-to-DCE data exchange by eliminating the complex handshaking procedures required when using other interfaces.

Loopback circuits

Through the use of three loopback circuits HSSI, provides an expanded diagnostic testing capability that can be extremely valuable when attempting to isolate transmission problems. Circuits LA and LB are asserted by the DTE to inform the near or far end DCE to initiate one of three diagnostic loopback modes: loopback at the remote DCE line, loopback at the local DCE line, or loopback at the remote DTE. The third loopback circuit, LC, is optional and it is used to request the local DTE to provide a loopback path to the DCE. When

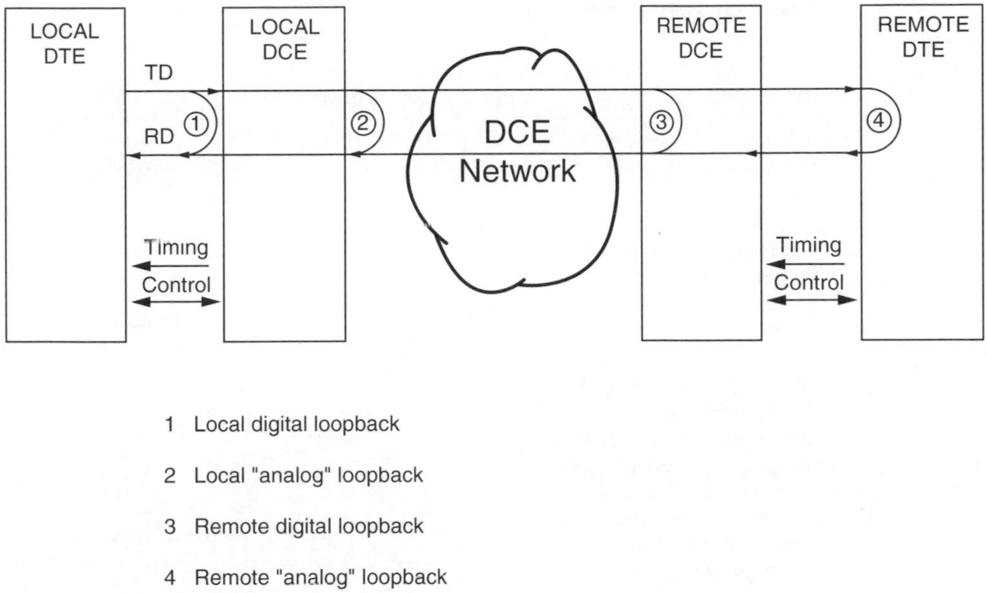

Figure 7.20 HSSI supports four loopbacks

the LC circuit is asserted the DTE would set TT = RT and SD = RD, enabling testing of the DCE to DTE interface independent of the DTE. The ST circuit would not be used as it cannot be relied on as a valid clocking source.

Figure 7.20 illustrates the four possible HSSI loopbacks with respect to the local DTE. Although data flow end-to-end in digital form, the term analog used to refer to two loopbacks is a carryover from modem loopback terminology, and it indicates that data are converted from unipolar to bipolar in the same manner as certain modem loopbacks convert digital to analog to test the modulator of a modem.

Signal Ground (SG)

Signal Ground is used to ensure that transmit signal levels remain within the common input range of receivers. The SG circuit is grounded at both ends.

Shield (SH)

The shield is used to limit electromagnetic interference. To accomplish this the shield encapsulates the HSSI cable.

Pin assignments

HSSI employs a 50-pin plug connector and receptacle. The connectors are mated to a cable consisting of 25 twisted pairs of 28 AWG cable and are limited

Table 7.3 HSSI pin assignments

Signal name	Signal direction	Pin no. + side	Pin no. – side
SG: Signal Ground	N/A	1	26
RT: Receive Timing	←	2	27
CA: DCE Available	←	3	28
RD: Receive Data	←	4	29
LC: Loopback circuit C (optional)	←	5	30
ST: Send Timing	←	6	31
SG: Signal Ground	N/A	7	32
TA: DTE Available	→	8	33
TT: Terminal Timing	→	9	34
LA: Loopback circuit A	→	10	35
SD: Send Data	→	11	36
LB: Loopback circuit B	→	12	37
SG: Signal Ground	N/A	13	38
Reserved for future use	→	14–18	39–43
SG: Signal Ground	N/A	19	44
Reserved for future use	←	20–24	45–49
SG: Signal Ground	N/A	25	50

Note: Pin pairs 18 and 43, and 20 and 45 to 24 and 49 are reserved for future use.

to a 50 foot length. The 25 twisted pairs are encapsulated in a polyvinylchloride (PVC) jacket. Table 7.3 lists the pin assignments. Note that the signal direction is indicated with respect to the DCE.

Applications

Since its initial development in 1989 HSSI has been incorporated into a large number of products designed to support high speed serial communications. In addition, its relatively simple interface has resulted in its use at data rates that would normally require the use of a V.35 or RS-499 interface. For example, many routers, multiplexers, inverse multiplexers and Channel Service Units operating at 1.544 Mbps can now be obtained with HSSI as well as products designed to operate at T3 (44.736 Mbps) and SONET Synchronous Transmission Service Level 1 (STS-1) (51.84 Mbps).

7.1.11 High Performance Parallel Interface

The High Performance Parallel Interface (HIPPI) represents an ANSI switched network standard which was originally developed to support direct communications between mainframes, supercomputers and directly attached storage devices. A series of ANSI standards currently define the physical layer operation of HIPPI as well as data framing, disk and tape connections and link encapsulation. Table 7.4 lists ANSI HIPPI related standards and their areas of standardization.

Table 7.4 ANSI HIPPI related standards

ANSI standard	Area covered
X3.183- 1991	Physical layer
X3.222- 1993	Switch control
X3.218- 1993	Link encapsulation
X3.210- 1992	Framing protocol
ANSI/ISO 9318-3	Disk connections
ANSI/ISO 9318-4	Tape connection

Transmission distance

Although its name implies the use of a parallel interface a number of extensions to HIPPI resulted in its ability to support a 300 m serial interface over multimode copper as well as parallel transmission via a 50 pair shielded twisted pair wiring group for relatively short distances. In its basic mode of operation, a HIPPI-based network consists of two computers connected via a pair of 50-pair copper cables to HIPPI channels on each device. Each 50-pair cable supports transmission in one direction, resulting in the pair of 50-pair cables providing a full duplex transmission facility. That transmission facility can extend up to 25 m and operate at either 800 Mbps or 1.6 Gbps, the latter accomplished by doubling the data path.

Through the use of one or more HIPPI switches you can develop an extended HIPPI network. That network can use copper cable between switches which permits cabling runs up to 200 m in length or a fiber extender can be used to support extending the distance between switches up to 10 km. The fiber extender functions as a parallel to serial converter as well as an electrical to optical converter to support serial light transmission between switches.

Figure 7.21 illustrates the creation of a HIPPI-based network on a college campus to connect a research laboratory to an administrative file server. HIPPI

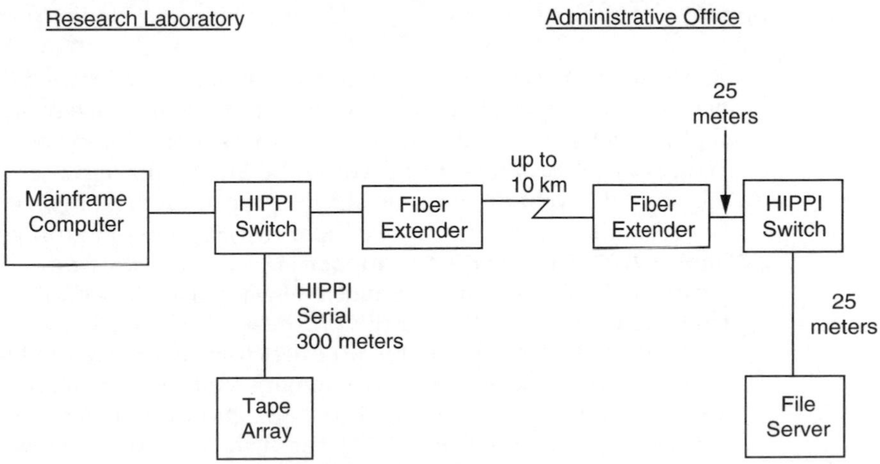

Figure 7.21 Creating an HIPPI-based network

interfaces are now available for a wide range of products to include personal computers, routers and gateways as well as mainframes and supercomputers.

Operation

HIPPI operates by framing data to be transmitted as well as by using messages to control data transfer operations. A HIPPI connection is set up through the use of three messages. A Request message is used by the data originator to request the establishment of a connection. A Connect message is returned by the desired destination to inform the requestor that a connection was established. The third message is Ready, which is issued by the destination to inform the originator that it is ready to accept data.

7.1.12 Universal Serial Bus

The Universal Serial Bus (USB) represents one of two relatively new interfaces developed to enhance the connection of external devices to personal computers. This interface is designed for 'plug and play' operations, allowing users to connect devices via a cable to a PC with the operating system performing all required configuration settings.

The USB supports both asynchronous and isochronous data transfer at 12 Mbps. Isochronous data transfer provides a guaranteed data transfer capability at a predetermined rate needed to support a particular USB device. A pending revision to the standard is designed to increase data transfer to between 120 Mbps and 140 Mbps. Both the current and pending revision of USB permit up to 127 devices to share a common bus, thus, the actual data transfer to a specific device on a USB depends upon the activity of other devices connected to the bus. However, in general, the USB provides a significant enhanced throughput in comparison to conventional serial and parallel ports.

Device support

The USB is designed to facilitate the connection of a wide range of external devices to a computer. Those devices can range in scope from keyboards, mice and joysticks to digital cameras, scanners, faxes, modems, printers, storage devices and USB hubs. Concerning the latter, through the use of a USB hub you can daisychain multiple devices together to share the use of several USB-compliant devices. An example of USB daisy-chaining is illustrated in Figure 7.22. Note that USB supports access to and from a PC and multiple peripheral devices cimultaneously, allowing users to perate numerous devices from a single connection to the PC.

The goal of USB is to enable products to be connected to PCs via a common interface. Because some USB compatible devices reserve a portion of the 12 Mbps operating rate of the bus, the actual number of devices that can be connected to a computer's USB port will more than likely be less than its theoretical maximum of 127. Concerning USB ports, most modern PCs include two in the form of rectangular slots, typically located on the rear of the

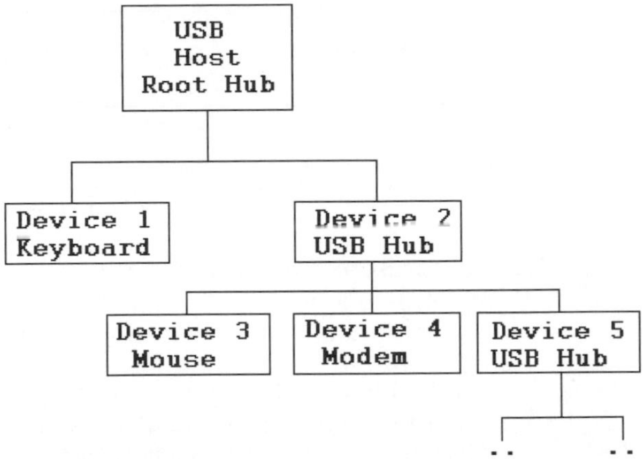

Figure 7.22 USB daisy-chaining

Table 7.5 Maximum data transfer rates

Port/bus	Maximum data transfer rate
Serial port	115 kbps
Parallel port (standard)	115 kbytes/s
ECP/EPP parallel port	3 Mbytes/s
IDE	3.3–16.7 Mbytes/s
SSCI-1	5 Mbytes/s
SSCI-2	10 Mbytes/s
SSCI-3	20 Mbytes/s
IEEE 1394	400 Mbps
UltraIDE	33 Mbytes/s
USB	12 Mbps
USB 2.0	120/240 Mbps

system unit. Even modern notebooks have at least one USB port, which facilitates peripheral connections. Because USB is designed to distribute electrical power to many peripherals, chip sets supporting this interface automatically sense the power requirements of different devices as they are connected to the daisy-chain and deliver appropriate power to each device, eliminating the necessity to have individual power supplies for each peripheral.

Table 7.5 provides a list of common PC interfaces/buses and their data transfer capabilities. Note that the IEEE 1394 standard is also known as FireWire and will be described and discussed in the next section in this chapter. Also note from the entries in the table that while the existing USB standard is hard pressed to support most types of hard drives, the USB 2.0 specification will extend the performance of the bus by 10 to 20 times over its existing data transfer capability.

Dataflow

The USB transfers data in packets, using a synchronization byte (hex 01) followed by a packet identifier (PID) field, data field, and cyclic redundancy

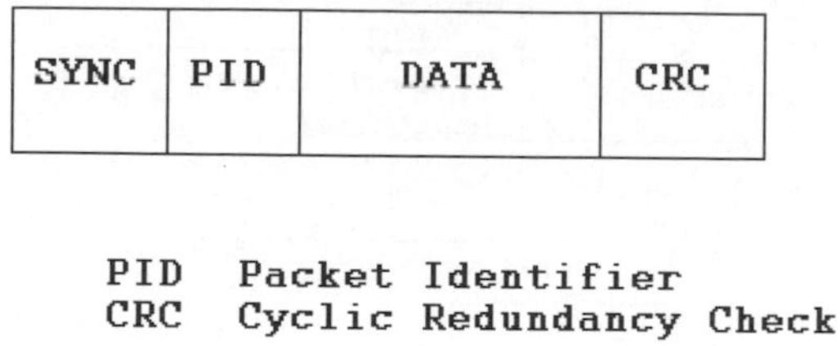

Figure 7.23 USB packet composition

check (CRC) field. Figure 7.23 illustrates the general format of a USB packet. The Packet Identifier can contain control, data or handshaking information. Control packets include IN, OUT, SOF and SETUP. Handshake PIDs include ACK, NAK and STALL.

The plug and play design of the USB results in the transmission of handshaking and data packets between USB-compatible equipment, which enables the host to dynamically support different devices. For example, if you plug a USB device into a USB hub port, the hub will inform the host of the vendor identification of the USB device, its product ID, and the assigned address. The host receives the USB device descriptor and records the assigned address. Once this is accomplished, the host informs the hub to activate the port the USB device was plugged into. For persons familiar with the setting of jumpers and dip switches, plug and play USB devices alleviate the need for the time-consuming process associated with manipulating those controls.

7.1.13 IEEE 1394 (FireWire)

The IEEE 1394 serial bus represents a more complex protocol than the USB; however, it also provides support for operating rates of 100 Mbps, 200 Mbps and 400 Mbps. Commonly referred to as FireWire, the IEEE 1394 bus was originally developed by Apple Computer as a replacement for the SCSI bus.

FireWire can be viewed as a complement to USB. USB's original specification is well suited for low-speed devices, such as scanners, modems and printers. In comparison, FireWire is more suitable for higher speed data transfers required to support disk drives and digital video cameras. In addition, it may be used in the future to support LAN connections without requiring PC users to install network adapter cards in their computer system units.

Similar to USB, FireWire represents a daisy-chain bus. However, it also supports the use of a tree, star or combination topology. In doing so, it supports a maximum of 63 devices, while USB supports 127 devices. For just about all real-world applications the maximum number of supported devices will never be reached, since from a practical perspective most consumers commonly have no more than three or four devices they would place on a chain. Table 7.6 provides a comparison of the operational characteristics of the USB and the IEEE 1394/FireWire serial buses.

Table 7.6 USB and IEEE 1394/FireWire

Optional characteristic	USB	IEEE 1394/FireWire
Maximum number of devices	63	127
Hot-swap capability	yes	yes
Maximum cable length between devices	4.5 m	5 m
Data transfer rate (Mbps)	100	12.5
	200	120*
	400	240*

* Version 2.0

The protocol stack

The IEEE 1394/FireWire interface operates based upon a three-layer protocol stack. An overview of that protocol stack is illustrated in Figure 7.24.

Physical layer

The physical layer, which represents the lowest layer in the stack, defines the transmission media as well as the electrical and signal characteristics. Functions defined at the physical layer include the connectors, connection slate, signal levels, signal encoding, and arbitration required for devices to access the bus. Although most literature, including this book, commonly refers to the operating rate of the IEEE 1394 specification as 100 Mbps, 200 Mbps and 400 Mbps, in actuality the operating rates are 98.304 Mbps, 196.608 Mbps and 393.216 Mbps. Most literature usually references a rounded operating rate. In addition, because communications are automatically configured to the highest operating rate, some literature simply defines the rate of this serial interface as 400 Mbps, which is not correct.

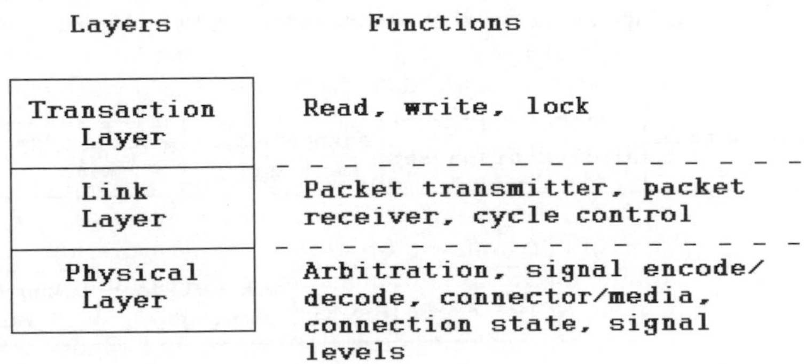

Layers
Functions
Transaction Layer
Read, write, lock
Link Layer
Packet transmitter, packet receiver, cycle control
Physical Layer
Arbitration, signal encode/decode, connector/media, connection state, signal levels

Figure 7.24 The IEEE 1394 protocol stack

Link layer

The link layer defines the manner by which packets are used to transport data. When data is transmitted asynchronously, a variable amount of data and several bytes of transaction layer information are transmitted as a packet to an explicit address, with an acknowledgement returned for each packet. Device addressing supported by the IEEE 1394 bus is 64 bits, with 10 bits for node identifiers and 48 bits for memory addresses.

Under isochronous transmission a variable amount of data is transferred as a sequence of fixed length packets at regular intervals. This method of transmission does not require acknowledgement to be returned.

Bus access

The use of asynchronous transmission supports situations where there are no fixed data transfer requirements. Because the IEEE 1394 bus can be arranged in a tree structure, the node at the top of the tree can function as a central arbiter and process requests for bus access. This allows bus access requests to be processed on a first-come, first-served basis, with a device with the highest priority gaining access to the bus when simultaneous requests occur. The previously described arbitration method is supplemented by two additional functions – fair arbitration and urgent arbitration. Under fair arbitration bus time is organized into fairness intervals and each node can compete for bus access. Once a node gains access it is inhibited from competing for fair access during the interval period. Thus, this technique prevents high-priority devices from monopolizing the bus. Both fair and urgent arbitration are supported by asynchronous transmission, with fair arbitration the default.

Figure 7.25 illustrates an example of IEEE 1394 asynchronous transmission. The process of transmitting a packet is referred to as a sub-action and the gap between packets is referred to as a sub-action gap.

The transmission of a packet results in five intervals of time. First, an arbitration sequence occurs during which signals are exchanged that enable a device to gain control of the bus. Next, a header is formed which contains source and destination identifiers, packet type information, parameter information for the packet, and a CRC. Once a packet is transmitted, an

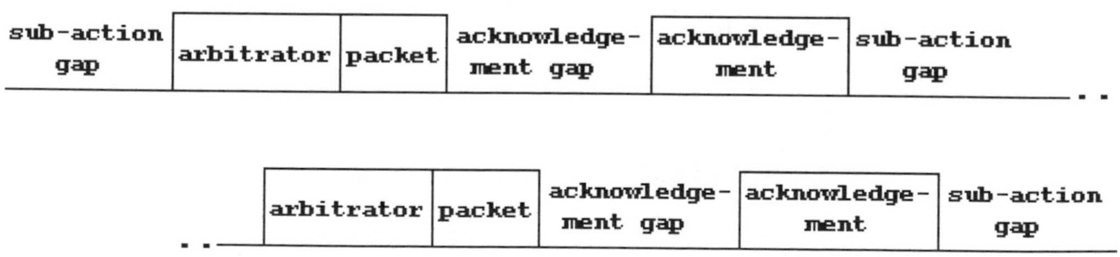

Figure 7.25 IEEE 1394 asynchronous transmission

acknowledgement gap occurs which represents the delay between the time the destination receives and decodes a packet and the time it generates an acknowledgement. The acknowledgement packet that contains a code that indicates the recipient has taken action follows this.

The fifth interval represents the sub-action gap. That gap represents an enforced idle period that ensures other nodes on the bus do not commence arbitrating prior to the acknowledgement being received. Because the acknowledgement node is in control of the bus when an acknowledgement is sent, a request-response interaction between two nodes can occur without a new arbitration sequence. Thus, it becomes possible for the packet in the sub-action 2 response shown in Figure 7.25 to directly follow the sub-action 1 request acknowledgement.

Isochronous operations

The second type of transmission supported by the IEEE 1394 bus is for isochronous data transfer. This method of data transfer guarantees delivery within a specified latency with a guaranteed data rate.

Because both asynchronous and isochronous data transfers are supported, it becomes possible for mixed traffic to flow over the IEEE 1394 bus. To accommodate this situation one node is designated as a cycle master and periodically issues a signal to all other nodes that an isochronous cycle has begun. During the isochronous cycle, only isochronous packets can be transmitted, with each isochronous data source arbitrating for bus access.

Figure 7.26 illustrates an example of IEEE 1394 isochronous transmission. Note that each isochronous packet is unacknowledged, permitting other isochronous data sources to immediately arbitrate for the bus after the prior isochronous packet is transmitted.

The isochronous gap illustrated in Figure 7.26 represents a small period of time between the transmission of one packet and the arbitration period for the next packet. Once the isochronous cycle is completed, a sub-action gap occurs. This gap represents a period of time longer than the isochronous gap and informs asynchronous devices that they can now compete for access to the bus until the next isochronous cycle begins.

Transaction layer

The top of the IEEE 1394 protocol stack is the transaction layer. This layer supports read, write, and lock transactions. When a read transaction occurs, data at a particular address is transferred back to the requester. The resulting

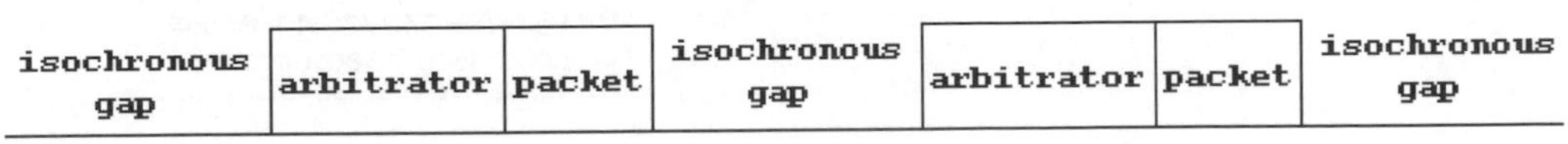

Figure 7.26 IEEE 1394 isochronous transmission

packet contains the destination address and the data length. In comparison, a write transaction results in data transferred from a requester to an address within one or more responders. The contents of the request packet include the destination address, the write data and the data length for the write.

A third type of transaction is a lock transaction. During a lock transaction data is transferred from a requester to a responder, processed within the responder, and transferred back to the requester. The request packet contains the address of the location in the target node to be lock accessed, the type of lock access to be performed, the write data, and the length of data to be written.

Both the USB and IEEE 1394 bus can be considered as potential successors to the serial and parallel ports included with just about all PCs. While the use of the USB and IEEE 1394 bus will not make the use of the serial and parallel port obsolete overnight, you can reasonably expect a gradual migration of manufacturers towards developing products that support each bus.

7.2 CABLES AND CONNECTORS

A variety of cables and connectors can be employed in data transmission systems. In this section we will focus our attention on several types of cable and connectors as well as several cabling tricks based upon our previous examination of the operation of RS-232/V.24 interchange circuits.

7.2.1 Twisted-pair cable

The most commonly employed data communications cable is the twisted-pair cable. This cable can usually be obtained with 4, 7, 9, 12, 16 or 25 conductors, where each conductor is insulated from another by a PVC shield.

For EIA RS-232 and ITU-T V.24 applications, those standards specify a maximum cabling distance of 50 feet between DTE and DCE equipment for data rates ranging from 0 to 19 200 bps; and normal industry practice is to use male connectors at the cable ends which mate with female connectors normally built into such devices as terminals and modems. Figure 7.27 illustrates the typical cabling practice employed to connect a DTE to a DCE.

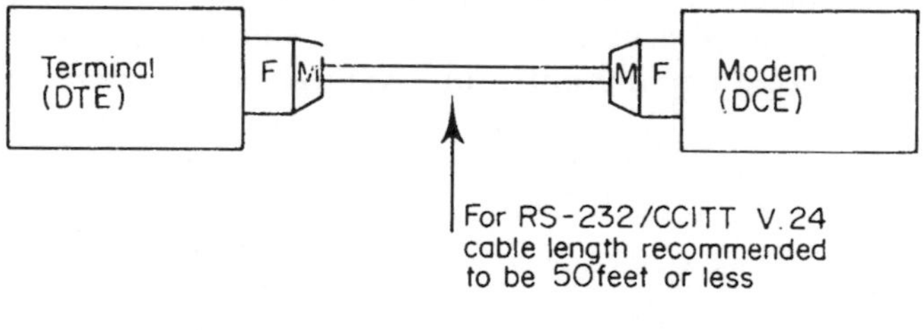

Figure 7.27 DTE to DCE cabling

7.2.2 Low-capacitance shielded cable

In certain environments where electromagnetic interference and radio frequency emissions could be harmful to data transmission, you should consider the utilization of low-capacitance shielded cable in place of conventional twisted-pair cable. Low-capacitance shielded cable includes a thin wrapper of lead foil that is wrapped around the twisted-pair conductors contained in the cable, thereby providing a degree of immunity to electrical interference that can be caused by machinery, fluorescent ballasts and other devices.

7.2.3 Ribbon cable

Since an outer layer of PVC houses the individual conductors in a twisted-pair cable, the cable is rigid with respect to its ability to be easily bent. Ribbon or flat cable consists of individually insulated conductors that are insulated and positioned in a precise geometric arrangement that results in a rectangular rather than a round cross-section. Since ribbon cable can be easily bent and folded, it is practical for those situations where you must install a cable that must follow the contour of a particular surface.

7.2.4 The RS-232 null modem

No discussion of cabling would be complete without a description of a null modem, which is also referred to as a modem eliminator. A null modem is a special cable that is designed to eliminate the requirement for modems when interconnecting two collocated data terminal equipment devices. One example of this would be a requirement to transfer data between two collocated personal computers that do not have modems and use different types of diskettes, such as an IBM PC which uses a $5\frac{1}{4}$ inch diskette and an IBM PS/2 which uses a $3\frac{1}{2}$ inch diskette. In this situation, the interconnection of the two computers via a null modem cable would permit programs and data to be transferred between each personal computer in spite of the media incompatibility of the two computers. Since DTEs transmit data on pin 2 and receive data on pin 3, you could never connect two such devices together with a conventional cable as the data transmitted from one device would never be received by the other.

In order for two DTEs to communicate with one another, a connector on pin 2 of one device must be wired to connector pin 3 on the other device. Figure 7.28 illustrates an example of the wiring diagram of a null modem cable, showing how pins 2 and 3 are cross-connected as well as the configuration of the control circuit pins on this type of cable.

Since a terminal will raise or apply a positive voltage in the 9 to 12 V range to turn on a control signal, you can safely divide this voltage to provide up to three different signals without going below the signal threshold of 3 V previously illustrated in Figure 7.4. In examining Figure 7.28, we note the following control signal interactions are caused by the pin cabling.

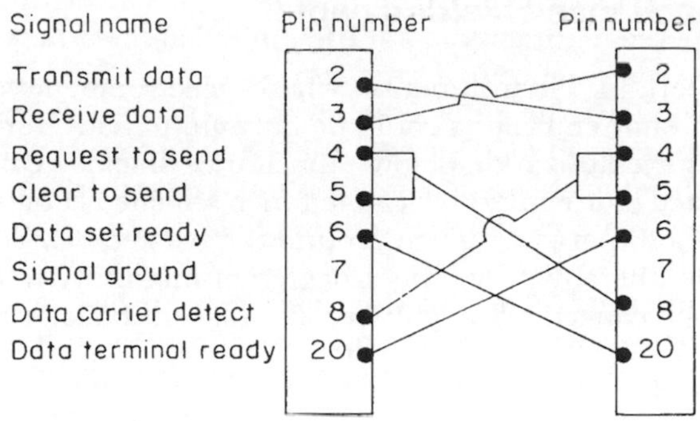

Figure 7.28 Null modem cable

(1) Data Terminal Ready (DTR, pin 20) raises Data Set Ready (DSR, pin 6) at the other end of the cable. This makes the remote DTE think a modem is connected to the other end and powered ON.

(2) Request to Send (RTS, pin 4) raises Data Carrier Detect (CD, pin 8) on the other end and signals Clear to Send (CTS, pin 5) at the original end of the cable. This makes the DTE believe that an attached modem received a carrier signal and is ready to modulate data.

(3) Once the handshaking of control signals is completed, we can transmit data onto one end of the cable (TD, pin 2) which becomes receive data (RD, pin 3) at the other end.

The cable configuration illustrated in Figure 7.28 will work for most data terminal equipment interconnections; however, there are a few exceptions. The most common exception is when a terminal device is to be cabled to a port on a mainframe computer that operates as a 'ring-start' port. This means that the computer port must obtain a Ring Indicator (RI, pin 22) signal. In this situation,

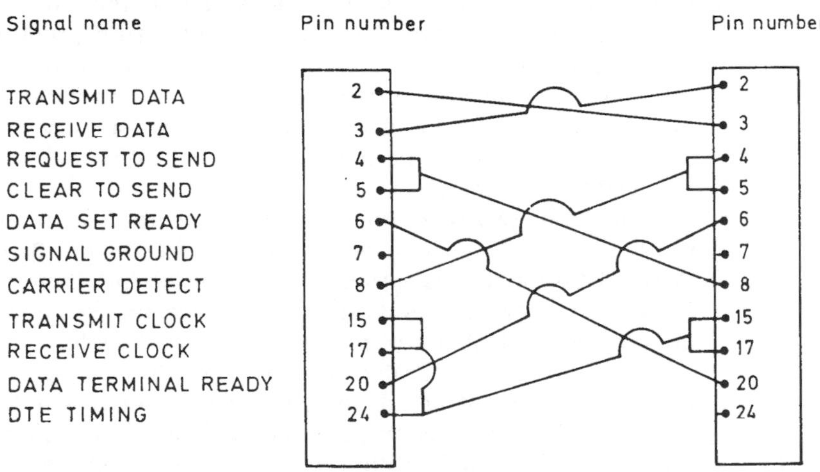

Figure 7.29 Synchronous null modem cable

the null modem must be modified so that Data Set Ready (DSR, pin 6) is jumpered to Ring Indicator (RI, pin 22) at the other end of the cable to initiate a connect sequence to a 'ring-start' system.

Owing to the omission of transmit and receive clocks, the previously described null modem can only be used for asynchronous transmission. For synchronous transmission you must either drive a clocking device at one end of the cable or employ another technique. Here you would use a modem eliminator which differs from a null modem by providing a clocking signal to the interface. If a clocking source is to be used, DTE Timing (pin 24) is normally selected to develop a synchronous null modem cable. In developing this cable, pin 24 is strapped to pins 15 and 17 at each end of the cable as illustrated in Figure 7.29. DTE Timing then provides transmit and receive clocking signals at both ends of the cable.

7.2.5 RS-232 cabling tricks

A general purpose 3-conductor cable can be used when there is no requirement for hardware flow control and a modem will not be controlled. Here the term flow control refers to the process that causes a delay in the flow of data between DTE and DCE, or two DCEs or two DTEs resulting from the changing of control circuit states. Figure 7.30 illustrates the use of a 3-conductor cable for DTE to DCE and DTE to DTE or DCE to DCE connections. When this situation occurs it becomes possible to use a 9-conductor cable with three D-shaped connectors at each end, with each connected to three conductors on the cable connector. Doing so eliminates the necessity of installing three separate cables.

Figure 7.31 illustrates a 5-conductor cable that can be installed between a DTE and DCE (modem) when asynchronous control signals are required. Similar to the use of 9-conductor cable to derive three 3-conductor connections, standard 12-conductor cable can be used to derive two 5-conductor connections.

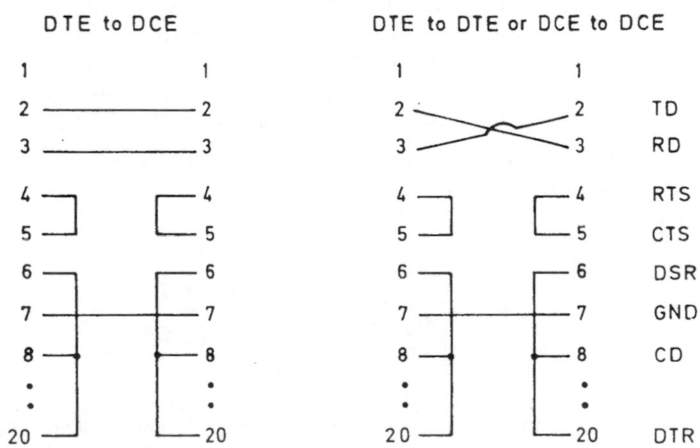

Figure 7.30 General purpose 3-conductor cable

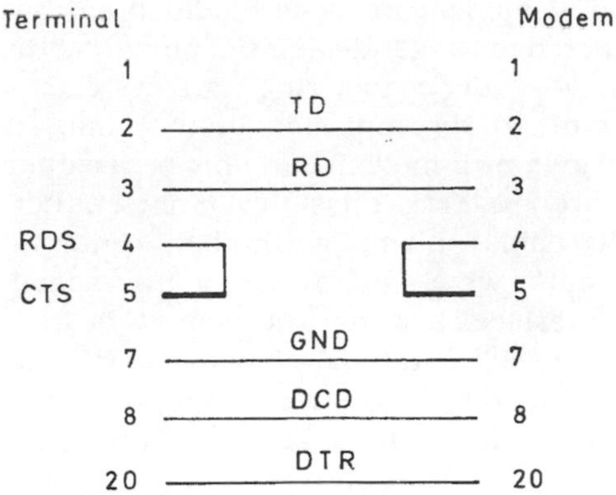

Figure 7.31 General purpose 5-conductor cable

7.3 PLUGS AND JACKS

Modern data communications equipment is connected to telephone company facilities by a plug and jack arrangement as illustrated in Figure 7.32. Although the connection appears to be, and in fact is, simplistic, the number of connection arrangements and differences in the types of jacks offered by telephone companies usually ensure that the specification of an appropriate jack can be a complex task. Fortunately, most modems and other communications devices include explicit instructions covering the type of jack the equipment must be connected to as well as providing the purchaser with information that must be furnished to the telephone company in order to legally connect the device to the telephone company line.

Most communications devices designed for operation on the PSTN interface the telephone company network via the use of an RJ11C permissive or an RJ45S programmable jack.

Figure 7.33 illustrates the conductors in the RJ11 and RJ45 modular plugs. The RJ11 plug is primarily used on two-wire dial lines. This plug is used in both the home and in an office to connect a single instrument telephone to the PSTN. In addition, the RJ11 also serves as an optional connector for four-wire private lines. Although the RJ11 connector is fastened to a cable containing four or six stranded-copper conductors, only two wires in the cable are used for switched network applications. When connected to a four-wire leased line, four conductors are used.

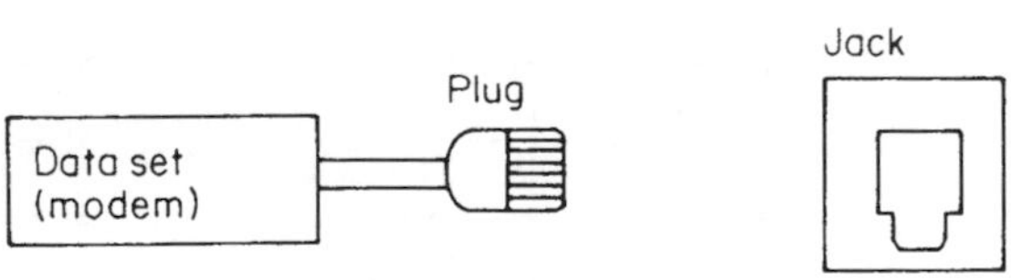

Figure 7.32 Connection to telephone company facilities. Data communications equipment can be connected to telephone company facilities by plugging the device into a telephone company jack

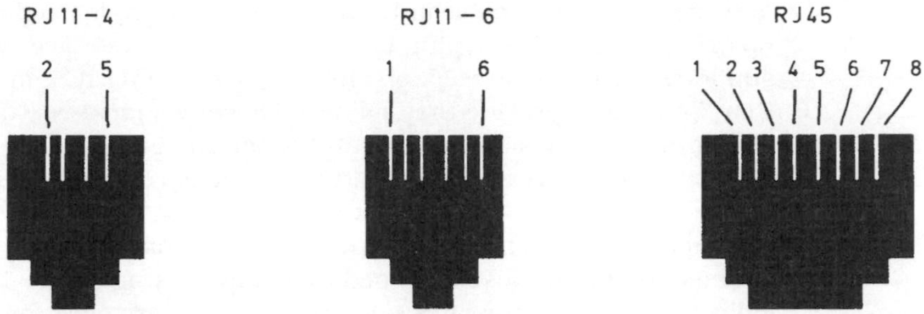

Figure 7.33 RJ11 and RJ45 modular plugs

The development of the RJ11 connector can be traced to the evolution of the switchboard. The plugs used with switchboards had a point known as the 'tip' which was colored red, while the adjacent sleeve known as the ring was colored green.

The original color coding used with switchboard plugs was carried over to telephone wiring. If you examine a four-wire (two-pair) telephone cable, you will note that the wires are colored yellow, green, red and black. The green wire is the tip of the circuit while the red wire is the ring. The yellow and black wires can be used to supply power to the light in a telephone or used to control a secondary telephone using the same four-wire conductor cable.

The most common types of telephone cable used for telephone installation are four-wire and six-wire conductors. Normally, a four-wire conductor is used in a residence that requires one telephone line. A six-wire conductor is used in either a residential or business location that requires two telephone lines and can also be used to provide three telephone lines from one jack. Table 7.7 compares the color identification of the conductors in four-wire and six-wire telephone cable.

During the late 1970s, telephone companies replaced the use of multiprong plugs by the introduction of modular plugs which in turn are connected to modular jacks.

The RJ11C plus was designed for use with any type of telephone equipment that requires a single telephone line. Thus, regardless of the use of either

Table 7.7 Color identification of telephone cables

Four-wire		Six-wire	
Pair	Color	Pair	Color
1	Yellow	1	Blue
	Green		Yellow
2	Red	2	Green
	Black		Red
		3	Black
			White

four-wire or six-wire cable only two wires in the cable need be connected to an RJ11C jack. The RJ11 plug can also be used to service an instrument that supports two or three telephone lines; however, RJ14C and RJ25C jacks must then be used to provide that service. These two jacks are only used for voice. For data transmission both 4- and 6-conductor plugs are available for use, with conductors 1, 2, 5 and 6 in the jack normally reserved for use by the telephone company. Then, conductor 4 functions as the ring circuit while conductor 5 functions as the tip to the telephone company network.

The RJ45 plug is also designed to support a single line although it contains eight positions. In this plug, positions 4 and 5 are used for ring and tip and a programmable resistor on position 8 in the jack is used to control the transmit level of the device connected to the switched network.

The RJ45 plug and jack connectors are also used in some communications products to provide an RS-232 DTE-DCE interface via twisted-pair telephone wire. In certain cases an RJ45 to DB25 adapter may be needed. This adapter will, as an example, permit the cabling of a cable terminated with an RJ45 plug to a DB25 connector or a DB25 connector cable end to a RJ45 socket. RJ45 connectors typically support the Transmitted Data, Received Data, Data Terminal Ready (DTE Ready), Data Set Ready (DCE Ready), Data Detect (Received Line Signal Detector), Request to Send, Clear to Send and Signal Ground circuits.

The physical size of the plugs used to wire equipment to each jack as well as the size of each of the previously discussed jacks are the same. The only difference between jacks is in the number of wires cabled to the jack and the number of contacts in the jack which are used to pass telephone wire signals.

7.3.1 Connecting arrangements

There are three connecting arrangements that can be used to connect data communications equipment to telephone facilities. The object of these arrangements is to ensure that the signal received at the telephone company central office does not exceed -12 dBm.

Permissive arrangement

The permissive arrangement is used when you desire to connect a modem to an organization's switchboard, such as a private branch exchange (PBX). When a permissive arrangement is employed, the output signal from the modem is fixed at a maximum of -9 dBm and the plug that is attached to the data set cable can be connected to three types of telephone company jacks as illustrated in Figure 7.34. The RJ11 jack can be obtained as a surface mounting (RJ11C) for desk sets or as a wall-mounted (RJ11W) unit; however, the RJ41S and RJ45S are available only for surface mounting.

Since permissive jacks use the same six-pin capacity miniature jack used for standard voice telephone installations, this arrangement provides for good mobility of terminals and modems.

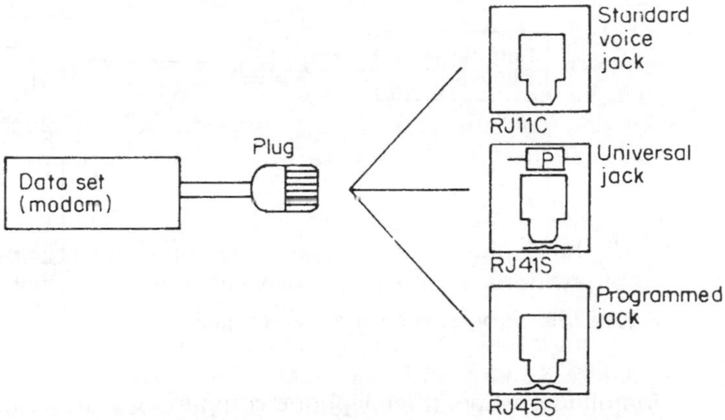

Figure 7.34 Permissive arrangement jack options

Fixed loss loop arrangement

Under the fixed loss loop arrangement the output signal from the modem is fixed at a maximum of −4 dBm and the line between the subscriber's location and the telephone company central office is set to 8 dBm of attenuation by a pad located within the telephone company provided jack. As illustrated in Figure 7.35, the only jack that can be used under the fixed loss loop arrangement is the RJ41S. This jack has a switch labeled FLL-PROG, which must be placed in the FLL position under this arrangement. Since the modem output is limited to −4 dBm, the 8 dB attenuation of the pad ensures that the transmitted signal reaches the telephone company office at −12 dBm. As the pad to the jack reduces the receiver signal-to-noise ratio by 8 dB, this type of arrangement is more susceptible to impulse noise and should only be used if one cannot use either of the two other arrangements.

Programmable arrangement

Under the programmable arrangement configuration a level setting resistor inside the standard jack provided by the telephone company is used to set the transmit level within a range between 0 and −12 dBm. Since the line from the user is directly routed to the local telephone company central office at

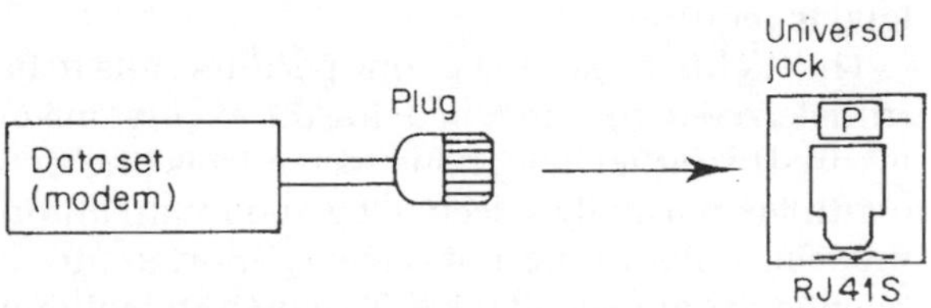

Figure 7.35 Fixed loss loop arrangement

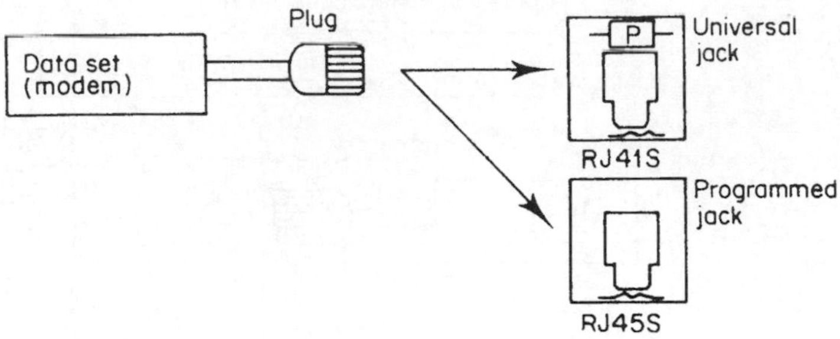

Figure 7.36 Programmed arrangement jack

installation time, the telephone company will measure the loop loss and set the value of the resistor based upon the loss measurement. As the resistor automatically adjusts the transmitted output of the modem so the signal reaches the telephone company office at −12 dBm, the modem will always transmit at its maximum allowable level. As this is a different line interface in comparison to permissive or fixed loss data sets, the data set must be designed to operate with the programmability feature of the jack.

Either the RJ41S universal jack or the RJ45S programmed jack can be used with the programmed arrangement as illustrated in Figure 7.36. The RJ41S jack is installed by the telephone company with both the resistor and pad for programmed and fixed loss loop arrangements. By setting the switch to PROG, the programmed arrangement will be set. Since the RJ45S jack can operate in either the permissive or programmed arrangement without a switch, it is usually preferred, as it eliminates the possibility of an inadvertent switch reset.

7.3.2 Telephone options

Prior to the use of modular jacks, telephones were hardwired to the switched telephone network. Even with the growth in the use of modular connecting arrangements, there are still many locations where telephones are still connected the 'old fashioned way'. Those telephone sets require the selection of specific options to be used with communications equipment. As part of the ordering procedure you must specify a series of specific options that are listed in Table 7.8.

When the telephone set is optioned for telephone set controls the line, calls are originated or answered with the telephone by lifting the handset off-hook. To enable control of the line to be passed to a modem or data set an 'exclusion key' is required.

The exclusion key telephone permits calls to be manually answered and then transferred to the modem using the exclusion key. The exclusion key telephone is wired for either 'telephone set controls line' or 'data set controls line'. Data set control is normally selected if you have an automatic call or automatic answer modem, since this permits calls to be originated or answered without taking the telephone handset off-hook. To use the telephone for voice communications the handset must be raised and the exclusion key placed in an upward location

Table 7.8 Telephone ordering options

Decision		Description
A	1	Telephone set controls line
	2	Data set controls line
B	3	No aural monitoring
	4	Aural monitoring provided
C	5	Touchtone dialing
	6	Rotary dialing
D	7	Switchhook indicator
	8	Mode indicator

The telephone set control of the line option is used with manual answer or manual originate modems or automatic answer or originate modems that will be operated manually. To connect the modem to the line the telephone must be off-hook and the exclusion key placed in an upward position. To use the telephone for voice communications the telephone must be off-hook while the exclusion key is placed in the downward position.

When the data set that controls the line option is selected, calls can be automatically originated or answered by the data equipment without lifting the telephone handset.

Aural monitoring enables the telephone set to monitor call progress tones as well as voice answer back messages, without requiring the user to switch from data to voice.

Users can select option B3 if aural monitoring is not required, while option B4 should be selected if it is required. Option C5 should be selected if touchtone dialing is to be used, while option C6 should be specified for rotary dial telephones. Under option D7, the exclusion key will be bypassed, resulting in the lifting of the telephone handset causing the closure of the switchhook contact in the telephone. In comparison, option D8 results in the exclusion key contacts being wired in series with the switchhook contacts, indicating to the user whether he or she is in voice or data mode.

7.3.3 Ordering the business line

Ordering a business line to transmit data over the switched telephone network currently requires you to provide the telephone company with four items of information. First, you must supply the telephone company with the Federal Communications Commission (FCC) registration number of the device to be connected to the switched telephone network . This 14-character number can be obtained from the vendor, who must first register their device for operation on the switched network prior to making it available for use on that network.

Next, you must provide the ringer equivalence number of the data set to be connected to the switched network. This is a three-character number, such as 0.4A, and it represents a unitless quotient formed in accordance with certain circuit parameters. Finally, you must provide the jack numbers and arrangement to be used as well as the telephone options if you intend to use a handset.

7.3.4 LAN connectivity

The migration of LAN cabling infrastructure to the use of twisted pair cable has been accompanied by a significant increase in the use of the modular RJ-45 connector. Today most network adapter cards, hubs and concentrators are manufactured to accept the use of the RJ-45 connector.

One of the first networks to use the RJ-45 connector was 10BASE-T, which represents a 10 Mbps version of Ethernet designed for operation over unshielded twisted pair (UTP). Since UTP cable previously installed in commercial buildings commonly contains either three or four wire pairs, the RJ-45 connector or jack was selected to be used with 10BASE-T, even through this version of Ethernet only supports the use of four pins. Table 7.9 compares the 10BASE-T pin numbers to the RJ-45 jack numbers and indicates the signal names used with 10BASE-T UTP cable.

Although 10BASE-T only uses four of the eight RJ-45 pins, other versions of Ethernet and different LANs use the additional pins in the connector. For example, a full duplex version of Ethernet requires the use of eight pins. Thus, the original selection of the RJ-45 connector has proven to be a wise choice.

Ethernet is a term that actually references a series of local area networks that use the same access protocol (CSMA/CD) but can use different types of cable. Early versions of Ethernet operated over either thick or thin coaxial cable. The attachment of an Ethernet workstation to a thick coaxial cable is accomplished through the use of a short cable which connects the workstation to a device known as a transceiver. This connection is accomplished through the use of a DB-15 connector.

The attachment of an Ethernet workstation to a thin coaxial cable requires the use of a T connector on the coax. The T connector is then cabled to a BNC connector on the network adapter card installed in the workstation.

Recognizing the fact that a workstation could be connected to a thick or thin coaxial cable or a twisted-pair based network manufacturers would have to support three separate types of adapter cards based upon different connectors required. Rather than face this inventory nightmare, most Ethernet network adapter card manufacturers now incorporate all three connectors on the cards they produce. Figure 7.37 illustrates an example of an Ethernet multiple media network interface card. Not only does this type of card simplify the

Table 7.9 10BASE-Twiring

10BASE-T pin #	RJ-45 pin #	10BASE-T signal name
1	1	Transmit Data +
2	2	Transmit Data −
3	3	Receive Data +
—	4	Not used
—	5	Not used
6	6	Receive Data −
—	7	Not used
—	8	Not used

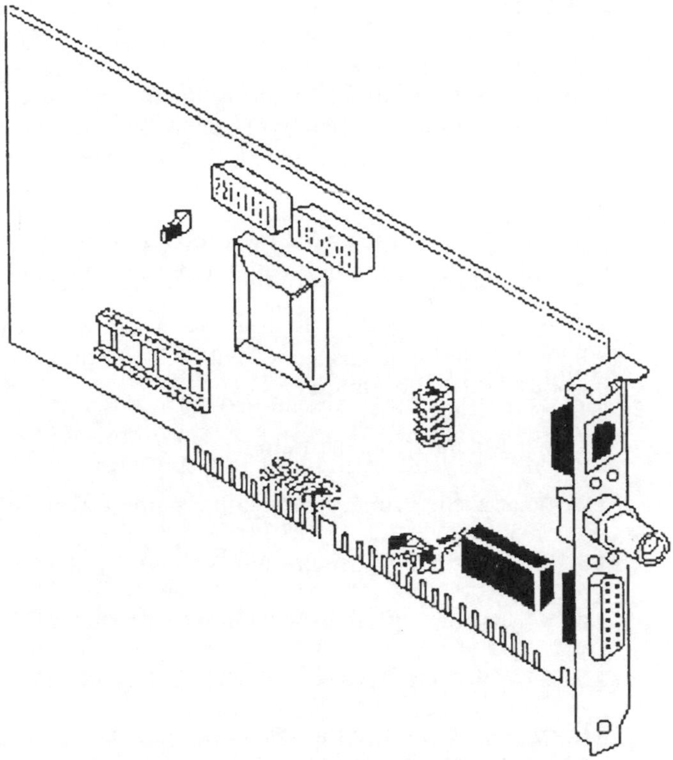

Figure 7.37 Multiple media network interface card. Some multiple media network interface cards, such as the one illustrated, support the direct attachment to UTP and thin coaxial cable, while the DB-15 connector permits the card to be cabled to a transceiver connected to thick coaxial cable.

manufacturer's inventory but, in addition, it provides end-users with the ability to easily migrate from one wiring infrastructure to another without having to replace network interface cards.

7.4 REVIEW QUESTIONS

1. What is the primary difference between RS-232-C and RS-232-D with respect to interchange circuits and connectors?

2. What is the difference in connector requirements between RS-232, V.24, V.35 and X.20 standards?

3. What is the purpose of the Alt A interface connector specified under RS-232-E?

4. What interface is built into many modems operating at 28.8 kbps to overcome the data rate limitation at RS-232?

5. Discuss the relationship between the voltage to represent a binary one in a terminal and the RS-232 signal characteristics that represent a binary one.

6. What are three methods commonly used to refer to RS-232 circuits? Which method do you feel is most popular in industry? Why?

7. What is the purpose of the Ring Indicator signal? Why do some modems require two rings prior to answering a call?

8. What is the difference between internal and external timing?

9. Using the data presented in Table 7.2 construct a wiring diagram to indicate the cabling required to fabricate a DB-9 to DB-25 adapter.

10. What are two key limitations associated with RS-232? Describe how differential signaling associated with RS-449 and RS-530 alleviate a considerable portion of those limitations

11. What is balanced signaling?

12. What is the primary application using a V.35 interface?

13. What is the purpose of the RS-366-A interface?

14. Why is the X.21 interface more costly than an RS-232/V.24 interface?

15. What is the purpose of the X.21 bis interface?

16. Why does the RS-530 standard offer the potential to achieve a high level of adoption during the 1990's?

17. What are two key differences between the original RS-530 standard and its Revision A?

18. Discuss the advantages associated with emitter-coupled logic (ECL) used in HSSI.

19. Compare and contrast the data rates supported by RS-232, RS-530, V.35 and HSSI.

20. Describe two advantages associated with the USB.

21. Do you feel the maximum number of devices that can be connected to a USB port is a viable constraint? Why?

22. Compare and contrast the IEEE 1394 bus to the USB.

23. What are the two types of transmission supported by the USB and the IEEE 1394 serial buses?

24. What is a key advantage associated with the use of low-capacitance shielded cable?

25. What is a null modem? Why are pins 2 and 3 reversed on that cable?

8

BASIC TRANSMISSION DEVICES: LINE DRIVERS, MODEMS AND SERVICE UNITS

In this chapter we will focus our attention on extending the transmission distances between data terminal equipment (DTE) and data communications equipment (DCE). The line driver is the first transmission device that is covered in this chapter, as it is normally used to extend the DTE to DCE transmission distance within a building.

After examining the operation and utilization of line drivers, we will turn our attention to basic modem operations, examining the modulation process and methods used by modem designers to pack more information into each signal change. In doing so we will also examine several types of common modems to obtain information about modem compatibility issues and the reason why the current generation of analog devices reached the maximum transmission rate obtainable over the public switched telephone network.

Due to the incorporation of microprocessors into modems providing intelligence that enables them to respond to commands, the third section in this chapter is focused upon intelligent modems. In this section we will examine basic and extended modem command sets and the manner by which software can be used to control modem operations.

Although conventional analog modems use under 4 kHz of bandwidth, it is possible to use approximately 1 MHz of bandwidth when transmitting data over twisted-pair wire. Recognizing this potential resulted in the development of a new series of analog modems referred to as digital subscriber line (DSL) devices. DSL modems represent one of two new types of next generation modems covered in the fourth section of this chapter, with the other type of modem being used on the cable TV infrastructure. Based upon each modem requiring the use of a broad frequency spectrum, we will collectively refer to them as broadband modems in this book. Because knowledge of the telephone company and cable TV infrastructure is essential for understanding the operation of these modems we will also review the manner by which telephone and cable television service is delivered in the fourth section in this chapter. Once this is accomplished, we will conclude this chapter by obtaining an appreciation for the operation and utilization of service units that enable transmission on digital facilities and that are often referred to as digital modems.

Recognizing the growing importance of digital transmission facilities, the last section in this chapter is focused on what many persons refer to as digital modems. Although those devices do modulate data as they convert unipolar into bipolar digital signals, they also perform other functions and are more commonly referred to as service units.

8.1 LINE DRIVERS

Without considering such unconventional techniques as laser transmission through fiber-optic bundles, there are four basic means of providing a data link between a terminal and a computer. These methods are listed in Table 8.1. It is interesting to observe that each of the methods listed in Table 8.1 provides for progressively greater distances of data transmission while incurring progressively greater costs to the users. In this chapter the limitations and cost advantages of each of these methods will be examined in detail.

8.1.1 Direct connection

The first and most economical method of providing a data circuit is to connect a terminal directly to a computer through the utilization of a wire conductor. Surprisingly, many installations limit such direct connections to 50 ft in accordance with terminal and computer manufacturers' specifications. These specifications are based on RS-232 and V.24 standards. If the maximum 50 ft standard is exceeded, manufacturers may not support the interface, yet terminals have been operated in a reliable manner at distances in excess of 1000 ft from a computer over standard data cables. This contradiction between operational demonstrations and usage and standard limitations is easily explained.

If you examine both the RS-232 and V.24 standards, such standards limit direct connections to 50 ft of cable for data rates up to and including 19.2 kbps. Since the data rate is inversely proportional to the width of the data pulses transmitted, taking capacitance and resistance into account, it stands to reason that slower operating terminals can be located further away than 50 ft from the computer without incurring any appreciable loss in signal quality. This also explains why you can connect, for example, a 56 kbps modem to a serial RS-232 computer port and set the data transfer speed to 115 kbps when data compression is enabled. Simply stated, the longer the

Table 8.1 Terminal-to-computer connection methods

Direct connection of terminals through the use of wire conductors
Connection of terminals through the utilization of line drivers
Connection of terminals through the utilization of limited distance modems
Connection of terminals through the utilization of modems or data service units

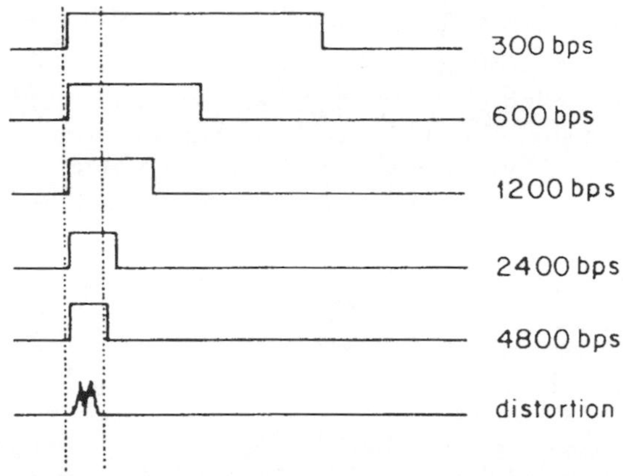

Figure 8.1 Distortion and data rate

cable length the weaker the transmitted signal at its reception point and the slower the pulse rise time. As transmission speed is increased, the time between pulses is shortened until the original pulse may no longer be recognized at its destination. This becomes more obvious when you consider that a set amount of distortion will effect a smaller (less wide) pulse than a wider pulse as illustrated in Figure 8.1.

In Figure 8.2, the relationship between transmission speed and cable length is illustrated for distances up to approximately 3400 ft and speeds up to 40 000 bps. This figure portrays the theoretical limits of data transfer speeds over an RS-232 serial connection based on the use of 22 American wire gauge (AWG) conductors. Many factors can have an effect on the relationship between transmission speed and cabling distance, including noise, distortion introduced owing to the routing of the cable, and the temperature of the surrounding area where the cable is installed. The ballast of a fluorescent fixture, for instance, can cause considerable distortion of a signal transmitted over a relatively short distance.

The pulse width is inversely proportional to the data rate, resulting in a small disturbance (in time) having a greater effect on transmission at high data rates than low data rates.

The diameter of the wire itself will affect the total signal loss. If the cross-sectional area of a given length of wire is increased, the resistance of the wire to current flow is reduced. Table 8.2 shows the relationship between the dimensions and resistances of several types of commercially available copper wire denoted by gauge numbers. By increasing the gauge from 22 to 19, the resistance of the wire is reduced by approximately one-half.

In examining the entries in Table 8.2, note that the gauge number has a reverse relationship to the diameter of a wire conductor, that is, the smaller the wire gauge the greater the diameter of the wire. Since there is less resistance to current flow in a wire as its diameter increases (column 3 of Table 8.2) transmission distance can be increased by the use of a smaller wire gauge conductor. Thus, the speed and cable length relationship illustrated in Figure 8.2 is not entirely accurate as it only represents the use of one

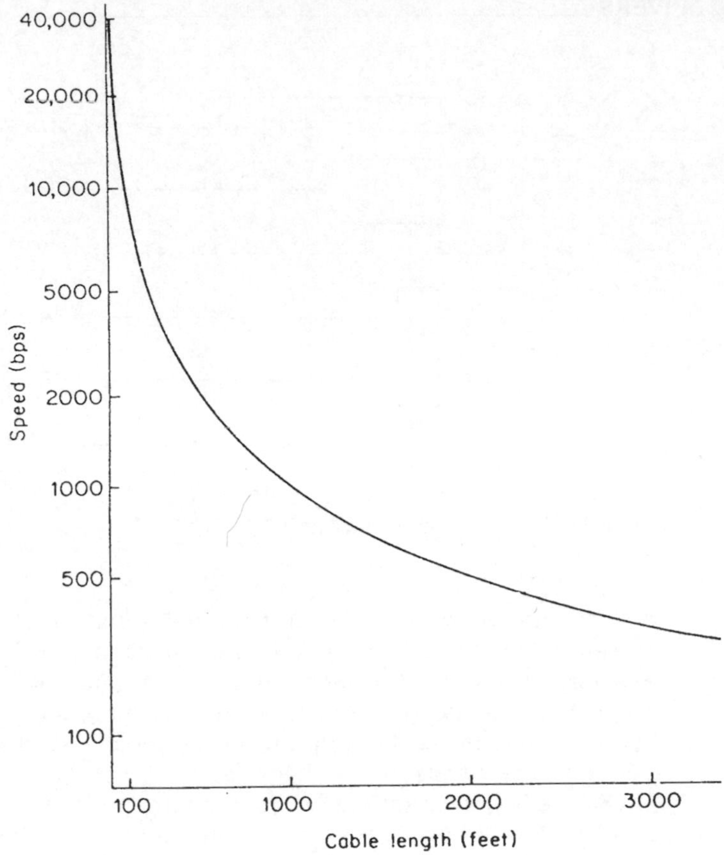

Figure 8.2 Speed and cable length relationship on an RS-232 serial connection

Table 8.2 Relationship between wire diameter and resistance

Gauge number	Diameter (inch)	Resistance (Ω/1000 ft) at 70 8F
10	0.102	1.02
11	0.091	1.29
12	0.081	1.62
13	0.072	2.04
14	0.064	2.57
15	0.057	3.24
16	0.051	4.10
17	0.045	5.15
18	0.040	6.51
19	0.036	8.21
20	0.032	10.30
21	0.028	13.00
22	0.025	16.50
23	0.024	20.70
24	0.020	26.20
25	0.018	33.00
26	0.016	41.80
27	0.014	52.40
28	0.013	66.60
29	0.011	82.80

particular type of wire conductor. In actuality, Figure 8.2 can be modified to show a series of curves similar to the curve drawn on that illustration. Curves above the drawn curve would represent smaller gauge wire conductors as they have greater diameter, permitting the transmission distance to be extended. Curves drawn below the curve in Figure 8.2 would represent larger gauge wire conductors as they have a smaller diameter, resulting in a reduced transmission distance.

Another method which can be utilized to extend the length of a direct wire connection is limiting the number of signals transmitted over the data link. After the connect sequence or handshaking is accomplished, only two signals leads are required for asynchronous data transfer: the transmitted data and receive data leads. With some minor engineering at both ends of the data link and an available dc voltage source, the remaining signals can be held continuously high, permitting the use of a simple paired cable to complete the data link.

Cable length can be further extended by the use of commercially available low-capacitance shielded cable. The shield consists of a thin wrapping of lead foil around the insulated wires and is quite effective in reducing the overall capacitance of the data cable with a very modest increase in price over standard unshielded cables. The use of low-capacitance shielded cables is strongly recommended when several cables must be routed through the same limited-diameter conduit. Once the practical limitation of cable length has been reached, signal attenuation and line distortion can become significant and either reduce the quality of data transmission or prevent its occurrence. One method of further extending the direct interface distance between terminals and a computer is by incorporating a line driver into the cable connection.

8.1.2 Using line drivers

As the name implies, a line driver is a device which performs the function of extending the distance or drive over which a signal can be transmitted down a line. A single line driver, depending on manufacturer and transmission speed, can adequately drive signals over distances ranging from hundreds of feet up to a mile. One manufacturer introduced a line driver capable of transferring signals at a speed of 100 kbps at a distance of 5000 ft and a 1 Mbps signal over a distance of 500 ft using a typical multipair cable.

A multitude of names have been given to the various brands of line drivers to include local data distribution units and modem eliminator drivers. For the purpose of this discussion a line driver is a device inserted into a digital transmission line in order to extend the signal distance.

Operation

A line driver can be considered as an interface converter, as it changes an interface, such as the RS-232/V.24 (actually V.28) signal levels to a low voltage, low impedance format for transmission on two-wire and four-wire twisted

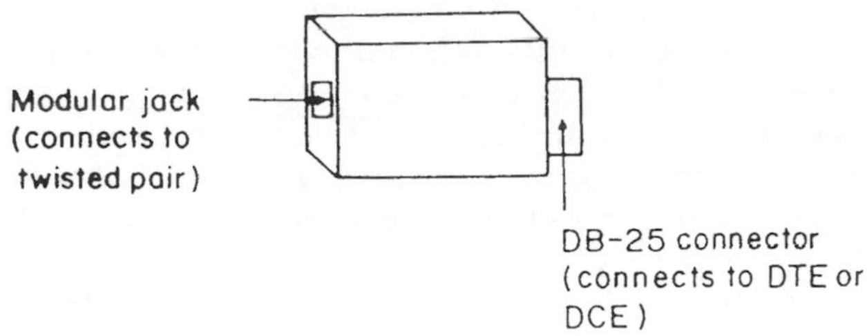

Figure 8.3 Interface powered line driver

pair installed within a building. In general, the use of line drivers is restricted to lines that are not routed through telephone company facilities. This is because those lines have filters that reduce the frequency band to 3000 Hz based on the original design of the telephone network. Since the line driver uses a wider frequency spectrum it is generally used on intra-building wiring that is not routed to a telephone company office, where filters would reduce the band-width of the line.

Figure 8.3 illustrates the physical shape of a typical line driver for use with an RS-232 25-pin interface. This line driver derives its power from the data and control signals on the RS-232 interface and has a built-in modular jack for connection to a two-twisted pair modular cable. Some line drivers include a switch for DTE/DCE interface selection, since a DTE transmits on pin 2 and receives on pin 3, whereas a DCE transmits on pin 3 and receives on pin 2. Other line drivers without a DTE/DCE interface selection switch must be purchased specifically for use with a DTE or DCE.

Types of line driver

There are several types of line drives, with each type based on the use of different signaling technology. One type of line driver that warrants discussion is actually an interface converter, changing the RS-232 signal into an RS-422 balanced signal. This type of interface converter can provide transmission of data up to distances of 4000 ft at a data rate up to 100 kbps. Unlike the previously described line driver that permits longer distance communications to occur over inexpensive twisted pair, the interface converter requires the use of a conventional RS-232 cable which links the RS-422 signaling output between a pair of interface converters.

A third type of line driver can be used by itself, unlike the previously mentioned line drivers that operate in pairs. This type of line driver is frequently called a modem eliminator or data regenerator, as it regenerates all RS-232 signals at the point where it is inserted into the cable.

In Figure 8.4 the distinction between single- and multiple-line drivers functioning as data regenerators is illustrated and contrasted with limited distance modems, which will be explored later in this chapter. The primary distinction between line driver and limited distance modems is that two identical units must be used as limited distance modems to pass data in

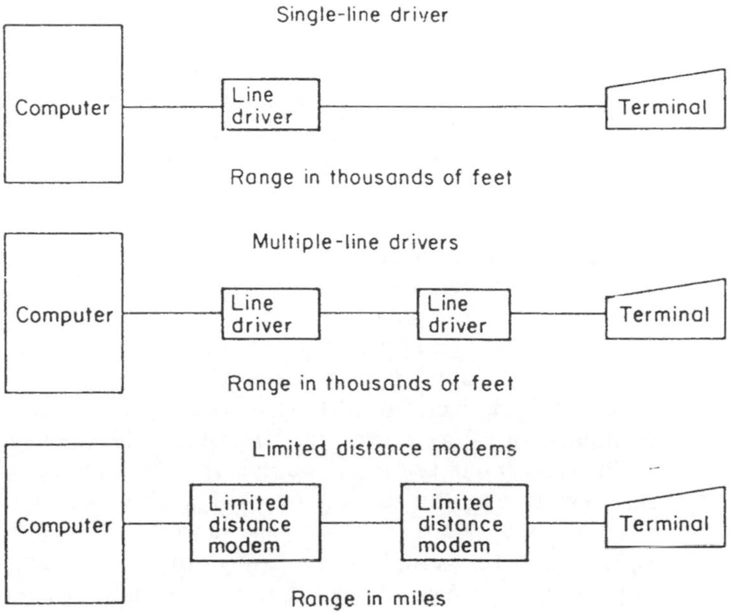

Figure 8.4 Line drivers and limited distance modems. When line drivers functioning as modem eliminators are used, the signal remains in its digital form for the entire transmission. Distances can be extended by the addition of one or more line drivers which serve as digital repeaters. For longer distances, limited distances modems can be utilized where the transmitted data is converted into an analog signal and then reconverted back into its original signal by the modem

analog form over a conductor, whereas a data regenerator line driver serves as a repeater to amplify and reshape digital signals.

Although some models of line drivers can theoretically have an infinite number of repeaters installed along a digital path, the cost of the additional units as well as the extra cabling and power requirements must be considered. Generally, the use of more than two line drives functioning as data regenerators in a single digital circuit makes the use of limited distance modems a more attractive alternative.

Applications

The characteristics of line drivers functioning as modem eliminators become important when considering their incorporation into a data link. If the EIA RS-232 signals are accepted, amplified, regenerated and passed over the same leads, they can be used as repeaters. If, however, the line driver also serves as a modem eliminator by providing a synchronous clock, inserting RTS–CTS delays and reversing transmit and receive signals, care should be taken when attempting to use them as repeaters. If this type of line driver is used and strapping options are provided for the RTS–CTS delay and such desirable features as an internal–external clock, it is a simple matter to convert it to a repeater by setting the delay to zero, setting the clock to external, and using a short pigtail cable to reverse the signals. One manufacturer offers a single stand-alone device that performs the function of a pair of two synchronous modems along with a less expensive remote cable extender option which serves as a matching line driver similar to that previously described.

In Figure 8.5, a typical application where line drivers would be installed is illustrated. In this office building a mainframe computer system is located in the basement. The three terminals to be connected to the computer are located on different floors of the building. A remote terminal located on the second floor of the building is only 100 ft from the computer and is directly connected by the use of a low-capacitance shielded cable. The second terminal is located on the eighth floor, approximately 500 ft from the computer and is connected by the use of line drivers to extend the signal transmission range. A third terminal, located on the 30th floor, uses a pair of limited-distance modems for transmission, since a large number of line drivers would be cost-prohibitive.

The replacement of many mainframes by LAN based client–server computing resulted in the development of a new series of line drivers or data regenerators. These relatively new devices are designed to operate with the signaling method employed on LANs which, although digital, differs considerably from the unipolar digital signaling method used on the RS-232, V.35 and similar interfaces. When we examine LAN signaling methods later in this book we will also examine the operation of data regenerators designed to extend LAN transmission.

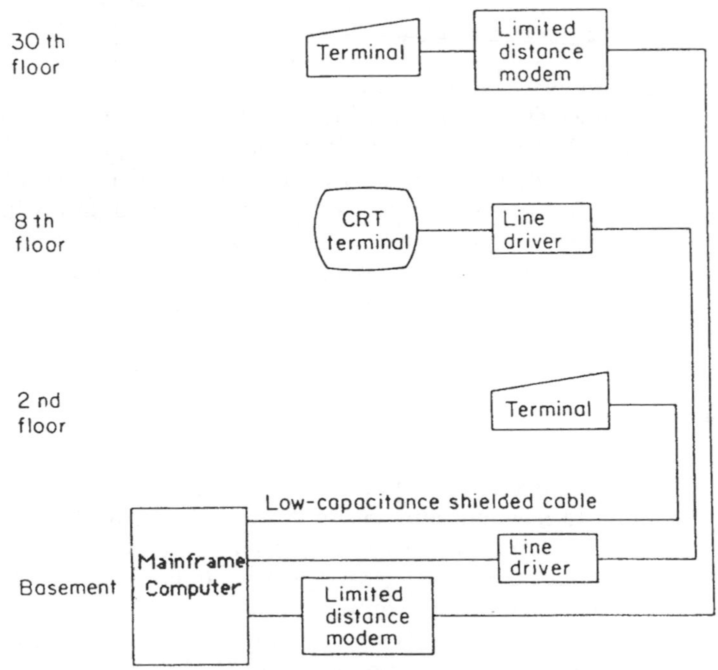

Figure 8.5 A variety of methods can be used to connect a terminal to a computer located in the same building

8.2 MODEM OPERATIONS

Today, despite the introduction of a number of all-digital transmission facilities by several communications carriers, the analog telephone system remains the primary facility utilized for data communications. Since terminals and computers produce digital pulses, whereas telephone circuits are designed to transmit analog signals which fall within the audio spectrum used in human speech, a device to interface the digital data pulses of terminals and computers with the analog tones carried on telephone circuits becomes necessary to transmit data over such circuits. Such a device is called a modem, which derives its meaning from a contraction of the two main functions of such a unit: modulation and demodulation. Although modem is the term most frequently used for such a device that performs modulation and demodulation, 'data set' is another common term whose use is synonymous in meaning.

In its most basic form a modem consists of a power supply, a transmitter, and a receiver. The power supply provides the voltage necessary to operate the modem's circuitry. In the transmitter, a modulator and amplifier, as well as filtering, waveshaping, and signal control circuitry convert digital direct current pulses; these pulses, originated by a computer or terminal, are converted into an analog wave-shaped signal which can be transmitted over a telephone line. The receiver contains a demodulator and associated circuitry which reverse the process by converting the analog telephone signal back into a series of digital pulses that is acceptable to the computer or terminal device. This signal conversion is illustrated in Figure 8.6.

8.2.1 The modulation process

The modulation process alters the characteristics of a carrier signal. By itself, a carrier is a repeating signal that conveys no information. However, when the

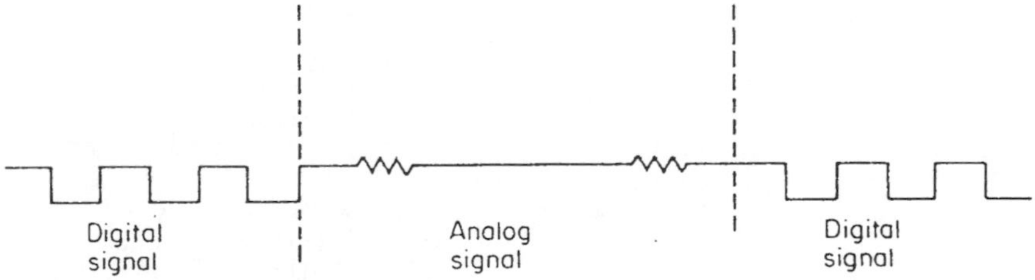

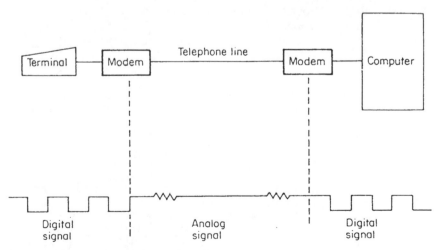

Figure 8.6 Signal conversion performed by modems. A modem converts a digital signal to an analog tone (modulation) and reconverts the analog tone (demodulation) into its original digital signal

carrier is changed by the modulation process information is impressed on the signal. For analog signals, the carrier is normally a sine wave, represented by

$$a = A\sin(2\pi ft + \phi)$$

where a = instantaneous value of voltage at time t, A = maximum amplitude, f = frequency, and ϕ = phase.

Thus, the carrier's characteristics that can be altered are the carrier's amplitude for amplitude modulation (AM), the carrier's frequency for frequency modulation (FM), and the carrier's phase angle for phase modulation (ϕM).

Amplitude modulation

The simplest method of employing amplitude modulation is to vary the magnitude of the signal from a zero level to represent a binary zero to a fixed peak-to-peak voltage to represent a binary one. Figure 8.7 illustrates the use of amplitude modulation to encode a digital data stream into an appropriate series of analog signals. Although pure amplitude modulation is normally used for very low data rates, it is also employed in conjunction with phase modulation to obtain a method of modulating high-speed digital data sources.

Frequency modulation

Frequency modulation refers to how frequently a signal repeats itself at a given amplitude. One of the earliest uses of frequency modulation was in the design of low-speed acoustic couplers and modems, where the transmitter shifted from one frequency to another as the input digital data changed from a binary one to a binary zero or from a zero to a one. This shifting in frequency is known as frequency shift keying (FSK) and is primarily used by modems operating at data rates up to 300 bps in a full-duplex mode of operation and up to 1200 bps in a half-duplex mode of operation. Figure 8.8 illustrates FSK frequency modulation.

Phase modulation

Phase modulation is the process of varying the carrier signal with respect to the origination of its cycle as illustrated in Figure 8.9. Several forms of phase

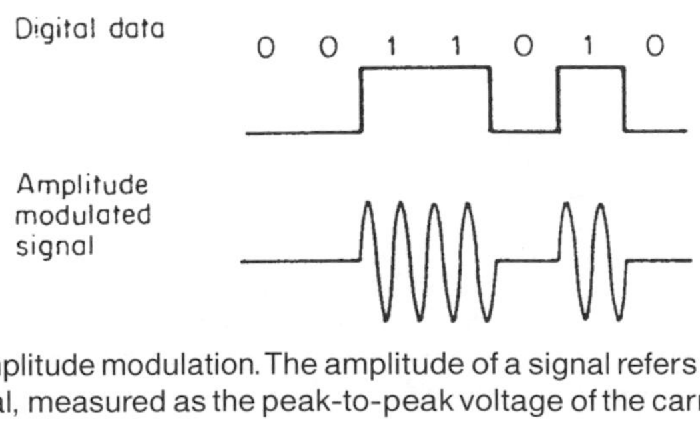

Figure 8.7 Amplitude modulation. The amplitude of a signal refers to the magnitude of the size of the signal, measured as the peak-to-peak voltage of the carrier signal

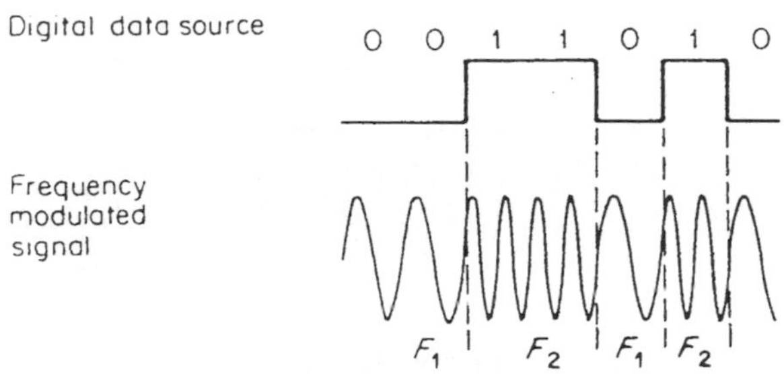

Figure 8.8 Frequency modulation. Frequency modulation refers to how frequently a signal repeats itself at a given amplitude. It is expressed in cycles per second (cps) or Hertz (Hz)

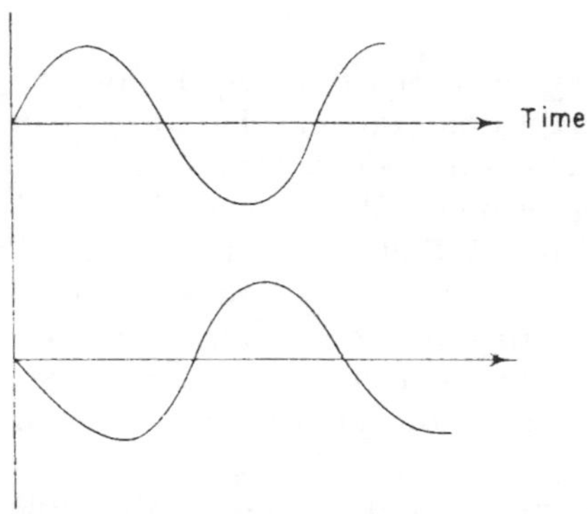

Figure 8.9 Phase modulation. Phase is the position of the wave form of a signal with respect to the origination of the carrier cycle. In this illustration, the bottom wave is 1808 out of phase with a normal sine wave illustrated at the top

modulation are used in modems to include single- and multiple-bit phase-shift keying (PSK) and the combination of amplitude and multiple-bit phase-shift keying.

In single-bit, phase-shift keying, the transmitter simply shifts the phase of the signal to represent each bit entering the modem. A binary one might thus be represented by a 90° phase change whereas a zero bit could be represented by a 270° phase change. Owing to the variance of phase between two-phase values to represent binary ones and zeros, this technique is known as two-phase modulation. Figure 8.10 illustrates the two-phase modulation where a 180° phase change is used to inform the receiver of a change in the value of the modulated bit. In two-phase modulation the transmitter shifts the phase of the carrier to correspond to a change in the value of the bit entering the modem.

Prior to discussing multiple-bit, phase-shift keying let us review the basic parameters of a voice circuit and the difference between the data rate and signaling speed. This will enable us to understand the rational for the utilization of multiple-bit, phase-shift keying, where two or more bits are grouped together and represented by one phase shift in a signal.

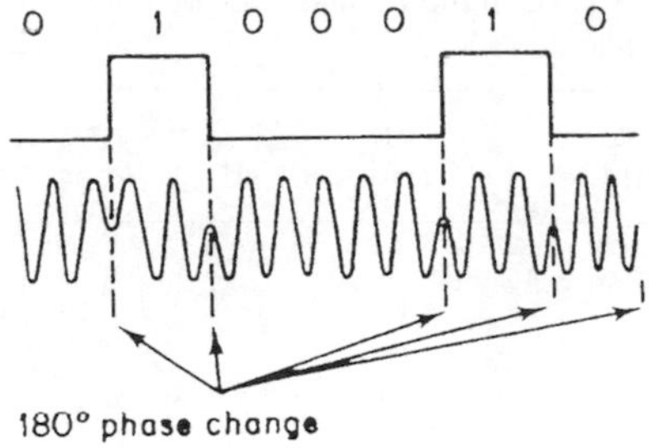

Figure 8.10 Two-phase modulation

8.2.2 BPS vs. baud

Bits per second is the number of binary digits transferred per second and represents the data transmission rate of a device. Baud is the signaling rate of a device such as a modem. If the signal of the modem changes with respect to each bit entering the device, then 1 bps = 1 baud. Suppose that a modem is constructed such that one signal change is used to represent two bits. Then the baud rate would be one-half of the bps rate.

When two bits are used to represent one baud, the encoding technique is known as dibit encoding. Similarly, the process of using three bits to represent one baud is known as tribit encoding and the bit rate is then one-third of the baud rate. Both dibit and tribit encoding are known as multilevel coding techniques and are commonly implemented using phase modulation.

8.2.3 Voice circuit parameters

Bandwidth is a measurement of the width of a range of frequencies. A voice-grade telephone channel has a passband, which defines its slot in the frequency spectrum, which ranges from 300 to 3300 Hz. The bandwidth of a voice-grade telephone channel is thus 3300 – 300 or 3000 Hz.

As data enter a modem it is converted into a series of analog signals, with the signal change rate of the modem known as its baud rate. In 1928, Nyquist developed the relationship between the bandwidth and the baud rate on a circuit as

$$B = 2W$$

where B = baud rate and W = bandwidth in Hz.

For a voice-grade circuit with a bandwidth of 3000 Hz, this relationship means that data transmission can only be supported at baud rates lower than 6000 symbols or signaling elements per second, prior to one signal interfering with another and causing intersymbol interference.

Since any oscillating modulation technique immediately halves the signaling rate, this means that most modems are limited to operating at one-half of the Nyquist limit. Thus, in a single-bit, phase-shift keying modulation

Table 8.3 Common, phase-angle values used in multilevel, phase-shift keying modulation

Coding technique	Bits transmitted	Possible phase-angle values (deg)		
Dibit				
	00	0	45	90
	01	90	135	0
	10	180	225	270
	11	270	315	180
Tribit				
	000	0	22.5	45
	001	45	67.5	0
	010	90	112.5	90
	011	135	157.5	135
	100	180	202.5	180
	101	225	247.5	225
	110	270	292.5	270
	111	315	337.5	315

technique, where each bit entering the modem results in a phase shift, the maximum data rate obtainable would be limited to approximately 3000 bps. In such a situation the bit rate would equal the baud rate, since there would be one signal change for each bit.

To overcome the Nyquist limit required engineers to design modems that first grouped a sequence of bits together, examined the composition of the bits, and then implemented a phase shift based upon the value of the grouped bits. This technique is known as multiple-bit, phase-shift keying, or multilevel, phase-shift keying. Two-bit codes called dibits and three-bit codes known as tribits are formed and transmitted by a single phase shift from a group of four or eight possible phase states.

It is important to note that the previous discussion of the relationship between bandwidth and baud rate is applicable to all types of communications to include cable television The reason modem designers initially developed techniques to pack two or more bits into one signal change was to overcome the limited 3000 Hz of bandwidth available on voice circuits. Even when a new generation of DSL and cable modems were developed to take advantage of significant additional bandwidth, modem designers continued to use techniques to encode multiple bits into each modem signal change, as this enables modems to achieve a higher data transfer capability.

Most modems operating at 600 to 4800 bps employ multilevel, phase-shift keying modulation. Some of the more commonly used phase patterns employed by modems using dibit and tribit encoding are listed in Table 8.3.

8.2.4 Combined modulation techniques

Since the most practical method to overcome the Nyquist limit is obtained by placing additional bits into each signal change, modem designers combined modulation techniques to obtain higher-speed data transmission over voice-grade circuits. One combined modulation technique commonly used involves both amplitude and phase modulation. This technique is known as quadrature amplitude modulation (QAM) and results in four bits being placed into each signal change, with the signal operating at 2400 (baud), causing the data rate to become 9600 bps.

The first implementation of QAM involved a combination of phase and amplitude modulation, in which 12 values of phase and 3 values of amplitude are employed to produce 16 possible signal states, as illustrated in Figure 8.11. One of the earliest modems to use QAM in the United States was the Bell System 209, which modulated a 1650 Hz carrier at a 2400 baud rate to effect data transmission at 9600 bps. Today, most 9600 bps modems manufactured adhere to the V.29 standard. The V.29 modem uses a carrier of 1700 Hz which is varied in both phase and amplitude, resulting in 16 combination of 8 phase angles and 4 amplitudes. Under the V.29 standard, fallback data rates of 7200 and 4800 bps are specified.

In addition to combining two modulation techniques, QAM also differs from the previously discussed modulation methods by its use of two carrier signals. Figure 8.12 illustrates a simplified block diagram of a modem's transmitter employing QAM. The encoder operates on four bits from the serial data stream and causes both an in-phase (IP) cosine carrier and a sine wave that serves as the quadrature component (QC) of the signal to be modulated. The IP and QC signals are then summed and they result in the transmitted signal being changed in both amplitude and phase, with each point placed at the x–y coordinates representing the modulation levels of the cosine carrier and the sine carrier.

If you plot the signal points previously illustrated in Figure 8.11 which represent all of the data samples possible in that particular method of QAM,

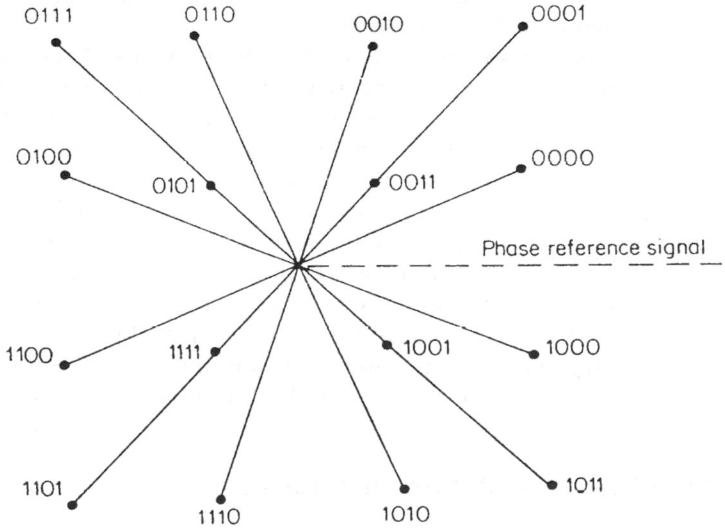

Figure 8.11 Quadrature amplitude modulation produces 16 signal states from a combination of 12 angles and 3 amplitude levels

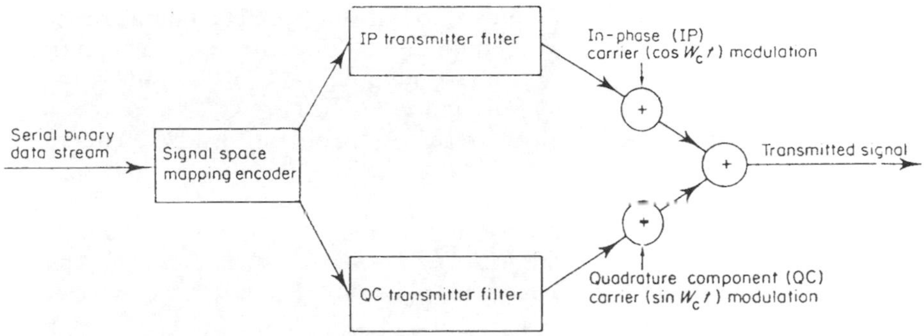

Figure 8.12 QAM modem transmitter

the series of points can be considered to be the signal structure of the modulation technique. Another popular term used to describe these points is the constellation pattern. By an examination of the constellation pattern of a modem, it becomes possible to pre-determine its susceptibility to certain transmission impairments. As an example, phase jitter which causes signal points to rotate about the origin can result in one signal being misinterpreted for another, which would cause four bits to be received in error. Since there are 12 angles in the QAM method illustrated in Figure 8.11, the minimum rotation angle is 30°, which provides a reasonable immunity to phase jitter.

Other modulation techniques

By the late 1980s several vendors began offering modems that operated at data rates up to 19 200 bps over leased voice-grade circuits. Originally, modems that operated at 14 400 bps employed a quadrature amplitude modulation technique, collecting data bits into a 6-bit symbol 2400 times per second, resulting in the transmission of a signal point selected from a 64-point signal constellation. The signal pattern of one vendor's 14 400 bps modem is illustrated in Figure 8.13. Note that this particular signal pattern appears to form a hexagon, and according to the vendor was used since it provides a better performance level with respect to signal-to-noise (S/N) ratio and phase jitter than conventional rectangular grid signal structures. However, in spite of hexagonal packed signal structures, it should be obvious that the distances between signal points for a 14 400 bps modem are less than for the resulting points from a 9600 bps modem. This means that a 14 400 bps conventional QAM modem is more susceptible to transmission impairments and the overall data throughput under certain situations can be less than that obtainable with 9600 bps modems. Figure 8.14 illustrates the typical throughput variance of 9600 and 14 400 bps modems with respect to the ratio of noise to the strength of the signal (N/S) on the circuit. From this illustration, it should be apparent that 14 400 bps modems using conventional quadrature amplitude modulation should only be used on high-quality circuits.

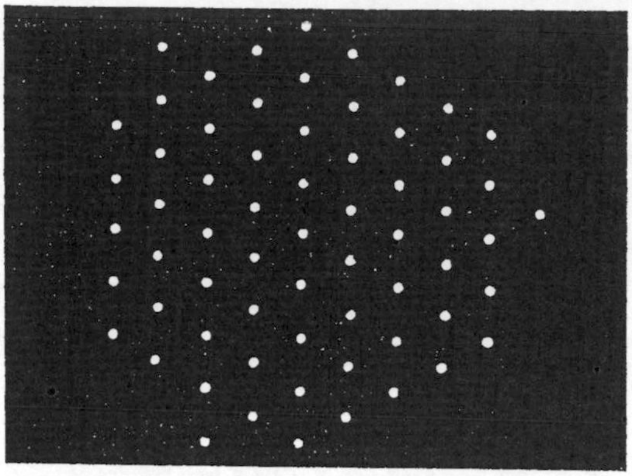

Figure 8.13 14 400 bps hexagonal signal constallation pattern

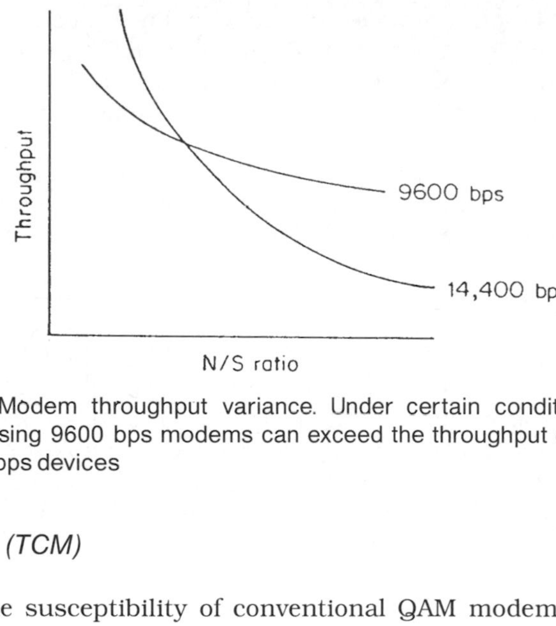

Figure 8.14 Modem throughput variance. Under certain conditions the throughput obtained by using 9600 bps modems can exceed the throughput obtained when using some 14 400 bps devices

Trellis coded modulation (TCM)

Owing to the susceptibility of conventional QAM modems to transmission impairments, a new generation of modems based upon Trellis coded modulation (TCM) was developed. In actuality, TCM is not a modulation technique. Instead, it represents a coding technique which adds one or more extra bits to a group of data bits that results in a mathematical relationship between preceding and succeeding data bits. This relationship not only enables transmission errors to be detected, but, in addition, many times permits such errors to be corrected. Modems employing TCM can tolerate more than twice as much noise power as conventional QAM modems, permitting operating rates up to 33 600 bps without compression over the switched telephone network and reliable data transmission at speeds ranging up to 33 600 bps over good-quality leased lines. The 33 600 bps operating rate limitation is applicable for modems when transmission results in two analog to digital conversions. As we will note later in this section, the most recent

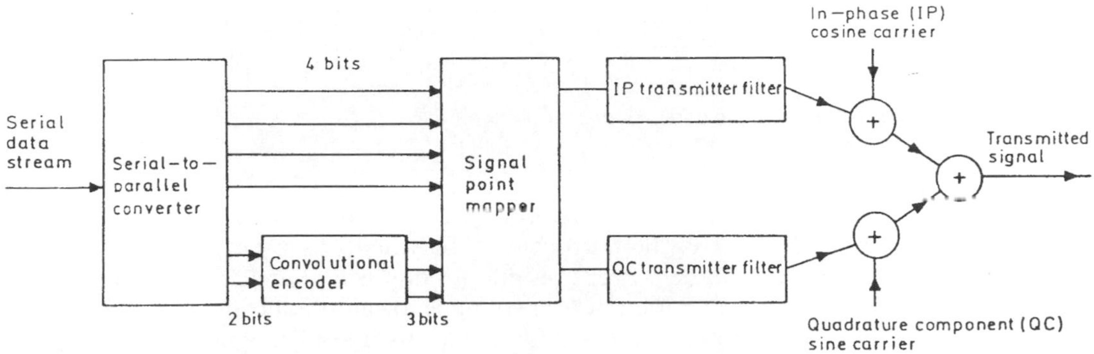

Figure 8.15 Trellis coded modulation

member of voice grade modems (the V.90 modem) theoretically provides a 56 kbps data transfer capability. However, that operating rate is only applicable if one end of the transmission line is directly connected to a digital infrastructure.

To understand how TCM provides a higher tolerance to noise and other line impairments, to include phase jitter and distortion, let us consider what happens when a line impairment occurs when conventional QAM modems are used. Here the impairment causes the received signal point to be displaced from its appropriate location in the signal constellation. The receiver then selects the signal point in the constellation that is closest to what it has received. Obviously, when line impairments are large enough to cause the received point to be closer to a signal point that is different from the one transmitted, an error occurs. To minimize the possibility of such errors, TCM employs an encoder that adds a redundant code bit to each symbol interval.

In actuality, at 14 400 bps the transmitter converts the serial data stream into 6-bit symbols and encodes two of the six bits employing a binary convolutional encoding scheme as illustrated in Figure 8.15. The encoder adds a code bit to the two input bits, forming three encoded bits in each symbol interval. As a result of this encoding operation, three encoded bits and four remaining data bits are then mapped into a signal point which is selected from a 128-point (2^7) signal constellation.

The key of the ability of TCM to minimize errors at high data rates is the employment of forward error correcting (FEC) in the form of convolutional coding. With convolutional coding, each bit in the data stream is compared with one or more bits transmitted prior to that bit. The value of each bit, which can be changed by the convolutional encoder, is therefore dependent on the value of other bits. In addition, a redundant bit is added for every group of bits compared in this manner. The following examination of the formation of a simple convolutional code clarifies how the convolutional encoder operates.

Convolutional encoder operation

Assuming that a simple convolutional code is formed by the modulo 2 sum of the two most recent data bits, then two output bits will be produced for each

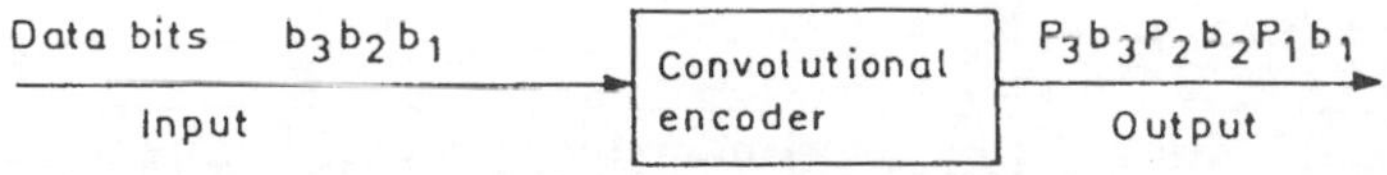

Figure 8.16 Simple convolutional code generation

data bit: a data bit and a parity bit. If we also assume that the first output bit from the encoder is the current data bit, then the second output bit is the modulo 2 sum of the current bit and its immediate predecessor. Figure 8.16 illustrates the generation of this simple convolutional code.

Because each parity bit is the modulo 2 sum of the two most recent data bits, the relationship between the parity bits and the data bits becomes

$$P_i = b_i + b_{i-1} \qquad i = 1, 2, 3 \ldots$$

If the composition of the first four data bits entering the encoder was 1101 ($b_4b_3b_2b_1$), the four parity bits are developed as follows:

$$\begin{aligned} P_1 &= b_1 + b_0 = 1 + 0 = 1 \\ P_2 &= b_2 + b_1 = 0 + 1 = 1 \\ P_3 &= b_3 + b_2 = 1 + 0 = 1 \\ P_4 &= b_4 + b_3 = 1 + 1 = 0 \end{aligned}$$

thus, the four-bit sequence 1101 is encoded as 01111011.

The preceding example also illustrates how dependencies can be constructed. In actuality, there are several trade-offs in developing a forward error correction scheme based upon convolutional coding. When a bit is only compared with a previously transmitted bit the number of redundant bits required for decoding at the receiver is very high. The complexity of the decoding process is, however, minimized. When the bit to be transmitted is compared with a large number of previously transmitted bits the number of redundant bits required is minimized. The processing required at both ends, however, increases in complexity.

In a 14 400 bps TCM modem, the signal point mapper uses the three encoded bits to select one of eight (2^3) subsets consisting of 16 points developed from the four data bits. This encoding process ensures that only certain points are valid. At the receiving modem, the decoder compares the observed sequence of signal points and selects the valid point closest to the observed sequence. The encoder makes this selection process possible by generating redundant information that establishes dependencies between successive points in the signal constellation. At the receiving modem, the decoder uses an algorithm that compares previously received data with currently received data. The convolutional decoding algorithm then enables the modem to select the optimum signal point. Because of this technique, a TCM modem is twice as immune to noise as a conventional QAM modem. In addition, the probability of an error occurring when a TCM modem is used is substantially lower than when an uncoded QAM modem is used.

TCM modem developments

The first TCM modems marketed in 1984 operated at 14.4 kbps. Since then, manufacturers introduced modems that use more complex TCM techniques. These modems operate at 16.8, 19.2 and 24.4 kbps. By 1997, the use of three-dimensional Trellis coding, in conjunction with a technology known as echo cancellation which is described later in this chapter, enabled modems to be developed that provide a full-duplex data transfer capability at 33.6 kbps on the switched telephone network. This operating rate more than doubled the transmission speed of the first generation of modems that used Trellis coding. In 1998 TCM was incorporated into several proprietary 56 kbps modem products and the resulting V.90 standard also includes this technology.

8.2.5 Mode of transmission

If a modem's transmitter or receiver sends or receives data in one direction only, the modem will function as a simplex modem. If the operations of the transmitter and receiver are combined so that the modem may transmit and receive data alternately, the modem will function as a half-duplex modem. In the half-duplex mode of operation, the transmitter must be turned off at one location, and the transmitter of the modem at the other end of the line must be turned on before each change in transmission direction. The process of turning the modem's transmitter ON and OFF results in the carrier signal being turned ON and OFF. Thus, another name used to refer to half-duplex modem operations is switched carrier.

The time interval required to turn the transmitter of one modem OFF and turn the transmitter of the other modem ON is referred to as line turnaround time. It is called this because the process of reversing the state of the transmitters of two half-duplex modems is necessary to change the direction of the flow of information.

On four-wire circuits, there are two distinct signaling paths that permit modems to be operated with continuous carrier. On a two-wire circuit, the path must be used for both sending and receiving. Although a two-wire circuit can be operated full-duplex at low data rates by splitting the telephone line's bandwidth into two channels by frequency, at higher data rates they must normally be operated in a half-duplex mode. One exception to this is obtained through the use of echo cancellation, which is described later in this chapter. At high data rates they must, thus, be operated with a controlled carrier. The delay resulting from the carrier being switched on and off results in a reduced level of transmission efficiency.

During the 1970s and 1980s prior to the development of echo cancellation technology, most high-speed modems used on the public switched two-wire network operated in a half-duplex transmission mode. This resulted in a turnaround delay that adversely affected transmission throughput.

To illustrate the potential effect of turnaround time, assume that a 2400 bps modem has a 120 ms turn-on delay. If you transmit a data block of 80 characters, the time to send the block would be (80 characters $\times$ 8 bits per

Table 8.4 Typical modem delay effect

Operation	Time (ms)
2400 bps carrier turn-on delay	120
Time to transmit 80-character block	267
Remote model turn-on delay	120
	507

Data transmitted 262/507 or 51.68% of the time.

character)/2400 bps or 267 ms. Based on the preceding, Table 8.4 summarizes the efficiency of transmission from a terminal if each 80-character block flowing in one direction results in line turnaround.

From an examination of the entries in Table 8.4, it is also apparent that transmission using a protocol which requires the acknowledgement of each data block via the switched telephone network can result in extended transmission times due to the effect of modem delays. Recognizing this problem, modem manufacturers developed several techniques to obtain a full-duplex transmission capability on the two-wire switched telephone network. The use of full-duplex transmission enables acknowledgements to flow over a separate return path while eliminating modem turnaround time, significantly improving data transfer performance.

8.2.6 Transmission techniques

Modems are designed for asynchronous or synchronous data transmission. Asynchronous transmission is also referred to as start–stop transmission, and is usually employed by unbuffered terminals where the time between character transmission occurs randomly.

Asynchronous

In asynchronous transmission, the character being transmitted is initialized by the character's start bit as a mark-to-space transition on the line and terminated by the character's stop bit which is converted to a 'space-marking' signal on the line. The digital pulses between the start and stop bits are the encoded bits which determine the type of character which was transmitted. Between the stop bit of one character and the start bit of the next character, the asynchronous modem places the line in the 'marking' condition. Upon receipt of the start bit of the next character the line is switched to a mark-to-space transition, and the modem at the other end of the line starts to sample the data.

Synchronous

Synchronous transmission permits more efficient line utilization since the bits of one character are immediately followed by bits of the next character,

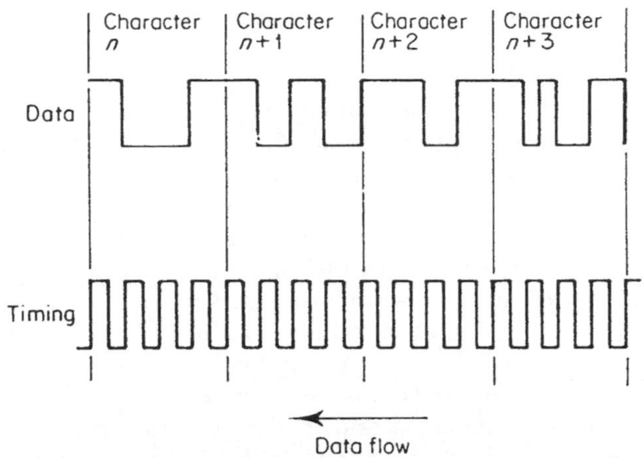

Figure 8.17 Synchronous timing signals. The timing signal is used to place the bits that form each character into a unique time period

with no start and stop bits required to delimit individual characters. In synchronous transmission, groups of characters are formed into data blocks, with the length of the block varying from a few characters to a thousand or more. Often, the block length is a function of the terminal's physical characteristics or its buffer size. As an example, for the transmission of data that represents punched card images, it may be convenient to transmit 80 characters of one card as a block, as there are that many characters if you construct the card image from an 80-column card deck. If punched cards are being read by a computer for transmission, and data is such that every two cards contain information about one employee, the block size could be increased to 160 characters.

In synchronous transmission, the individual bits of each of the characters within each block are identified based on a transmitted timing signal which is usually provided by the modem and which places each bit into a unique time period. This timing or clock signal is transmitted simultaneously with the serial bit stream, as shown in Figure 8.17.

8.2.7 Modem classification

Modems can be classified into many categories including the mode of transmission and transmission technique as well as by the application features they contain and the type of line they are built to service. Generally, modems can be classified into four line-servicing groups: subvoice or narrowband lines, voice-grade lines, wideband lines, and dedicated lines. Subvoice-band modems require only a portion of the voice-grade channel's available bandwidth and are commonly used with equipment operating at speeds up to 300 bps. On narrow-band facilities, modems can operate in the full-duplex mode by using one-half of the available bandwidth for transmission in each direction and use an asynchronous transmission technique.

Modems designed to operate on voice-grade facilities may be asynchronous or synchronous, half-duplex or full-duplex.

Voice-grade modems currently transfer data at rates up to 56 000 bps when only one analog to digital conversion occurs. However, when two analog to digital conversions occur, so-called 56K modems, which were standardized as V.90 devices, are limited to an operating rate of 33.6 kbps. Wideband modems, which are also referred to as group-band modems since a wideband circuit is a grouping of lower-speed lines, permit users to transmit synchronous data at speeds in excess of 56 000 bps. Although wideband modems are primarily used for computer-to-computer transmission applications, they are also used to service multiplexers which combine the transmission of many low- or medium-speed terminals to produce a composite higher transmission speed. Dedicated or limited-distance modems, which are also known by such names as shorthaul modems and modem bypass units, operate on dedicated solid conductor twisted pair or coaxial cables, permitting data transmission at distances ranging up to 15 to 20 miles, depending upon the modem's operating speed and the resistance of the conductor.

8.2.8 Limited-distance modems

Modems in this category can operate at speeds up to approximately 1.5 million bps and are particularly well suited for in-plant usage, where users desire to

Table 8.5 Common modem features

	Line type					
		Voice-grade				
Features	Subvoice (up to 300 bps)	Low, up to 1800 bps	Medium 2000–4800 bps	High 7200–56 000 bps	Wideband (above 56 000 bps)	Limited distance (up to 1.5 Mbps)
Asynchronous	✓	✓				✓
Synchronous			✓	✓	✓	✓
Switched network	✓	✓	✓	✓		
Leased only		✓	✓	✓	✓	✓
Half-duplex	✓	✓	✓	✓		
Full-duplex	✓	✓	✓	✓	✓	✓
Fast turnaround for dial-up use			✓	✓		
Reverse/ secondary channel	✓	✓	✓			✓
Manual equalization			✓			
Automatic equalization		✓	✓	✓		
Multiport capability			✓	✓		✓
Voice/data			✓	✓		✓

instal their own communications lines between terminals and a computer located in the same facility or complex. Also, in comparison with voice-band and wideband modems, these modems are relatively inexpensive since they are designed to operate only for limited distance and do not contain equalization circuitry which is described later in this section. In addition, by using this type of modem and stringing their own in-plant lines, users can eliminate a monthly telephone charge that would occur if the telephone company were to furnish the facilities. In Table 8.5 the common application features of modems are denoted by the types of lines to which they can be connected.

8.2.9 Line-type operations

Modems with a rated transmission speed of up to 33 600 bps that employ two analog to digital conversions can operate over the switched, dial-up telephone network. In addition, when only one analog/digital conversion occurs, so-called 56K modems may be able to provide a data transmission rate approaching 56 000 kbps in one direction. However, in the opposite direction, the transmission rate will be limited to 33.6 kbps. Later in this chapter we will review the operation of V.90 modems in detail.

8.2.10 Reverse and secondary channels

Modem manufacturers developed a reverse channel to eliminate turnaround time when transmission is over the two-wire switched network or to relieve the primary channel of the burden of carrying acknowledgement signals on four-wire dedicated lines. A reverse channel is used to provide a path for the acknowledgement of transmitted data at a slower speed than the primary channel. This reverse channel can be used to provide a simultaneous transmission path for the acknowledgement of data blocks transmitted over the higher speed primary channel at up to 150 bps.

A secondary channel is similar to a reverse channel. It can, however, be used in a variety of applications which include providing a path for a high-speed terminal and a low-speed terminal. When a secondary channel is used as a reverse channel, it is held at one state until an error is detected in the high-speed data transmission and is then shifted to the other state as a signal for retransmission. Another application where a secondary channel can be utilized is when a location contains a high-speed, synchronous terminal and a slow-speed, asynchronous terminal, such as a Teletype device. If both devices are required to communicate with a similar distant location, one way to alleviate dual line requirements as well as the cost of extra modems to service both devices is by using a pair of modems that have secondary channel capacity, as shown in Figure 8.18. Although a reverse channel is usable on both two-wire and four-wire telephone lines, the secondary channel technique is usable only on a four-wire circuit. A secondary channel modem derives two channels from the same line: a wide one to carry synchronous data, and a narrow channel to carry asynchronous teletype-like data. Some modems with the secondary channel option can actually provide two slow-speed

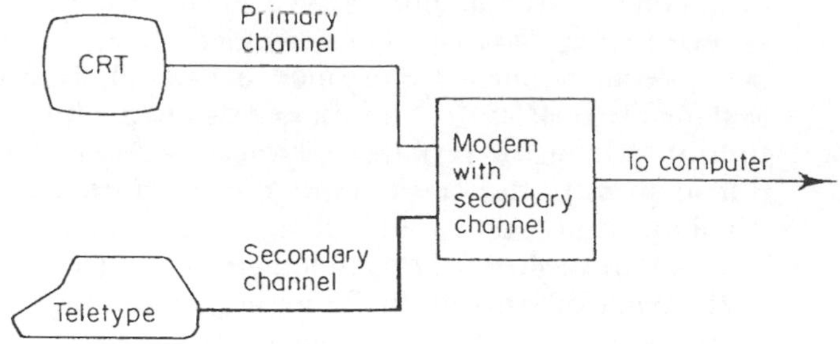

Figure 8.18 Secondary channel operation. Two terminals can communicate with a distant location by sharing a common line through the use of a modem with a secondary channel

channels as well as one high-speed channel, with the two slow-speed channels being capable of transmitting asynchronous data up to a composite speed of 150 bps.

Although the use of reverse and secondary channels in modems has greatly diminished over the past decade, their concept has been incorporated into other communications products. Thus, while it may be difficult today to purchase or lease a modem with a reverse or secondary channel, their concept lives on in other communications products.

8.2.11 Equalization

Modem manufacturers build equalizers into a modem to compensate for inconsistencies produced by the telephone circuit, amplifiers, switches and relays, as well as other equipment that data may be transmitted across in establishing a data link between two or more points. An equalizer is basically an inverse filter which is used to correct amplitude and delay distortion which, if uncorrected, could lead to intersymbol interference during transmission. The reader is referred to Chapter 3 in which the operational effect of delay and attenuation equalizers were both described and illustrated.

A well-designed equalizer matches line conditions by maintaining certain of the modem's electrical parameters at the widest range of marginal limits in order to take advantage of the data rate capability of the line while eliminating intersymbol interference. The design of the equalizer is critical, since if the modem operates too near or outside these marginal limits, the transmission error rate will increase. There are three basic methods for achieving equalization. The first method, the utilization of fixed equalizers, is typically accomplished by using marginally adjustable high-Q filter sections. Modems with transversal filters use a tapped delay line with manually adjustable variable tap gains, while automatic equalization is usually accomplished by a digital transversal filter with automatic tap gain adjustments. The faster the modem's speed, the greater the need for equalization and the more complex the equalizer. Until the mid-1980s, most modems with rated speeds up to 4800 bps incorporated non-adjustable, fixed equalizers which were designed to match the average line conditions that were found to occur on the dial-up network. Thus, most modems with fixed or non-adjustable equalizers were designed for a normal, randomly routed call between two locations over the dial-up network. If the modem is equipped with a signal-quality light which indicates an error rate that is unacceptable, or if the operator encounters difficulty with the connection, the problem can be alleviated by simply disconnecting the call and dialing a new call, which should reroute the connection through different points on the dial-up network.

The explosive growth in the manufacture of modems for use with personal computers resulted in a significant base over which modem manufacturers can amortize research and development and manufacturing costs. As a result of this growth vendors developed automatic equalization circuitry, which is described after we first examine manual equalization, and which is now a standard feature incorporated into most high speed modems.

Manual equalization

Manually adjusted equalization was originally employed on some 4800 bps modems used for transmission over leased lines, with the parameters being turned or preset at installation time, and re-equalization usually not required unless the lines are reconfigured.

Primarily designed to operate over unconditioned leased telephone lines, manually equalized modems were developed to allow users to eliminate the monthly expense associated with line conditioning. Owing to the incorporation of microprocessors into modems for signal processing they were soon employed to perform automatic equalization. This resulted in most modems incorporating automatic equalization, leaving manual equalization as an essentially obsolete technology.

Automatic equalization

Today, automatic equalization is used on just about all modems designed for operation over the switched telephone network. With automatic equalization, a certain initialization time is required to adapt the modem to existing line conditions. This initialization time becomes important during and after line outages, since line initial equalization times can extend otherwise short dropouts unnecessarily. Recent modem developments shortened the initial equalization time to between 15 and 25 ms, whereas only a few years ago a much longer time was commonly required. After the initial equalization, the modem continuously monitors and compensates for changing line conditions by an adaptive process. This process allows the equalizer to 'track' the frequently occurring line variations that occur during data transmission without interrupting the traffic flow. On one 9600 bps modem, this adaptive process occurs 2400 times a second, permitting the recognition of variations as they occur.

8.2.12 Synchronization

For synchronous communications, the start–stop bits characteristic of asynchronous communications can be eliminated. Bit synchronization is necessary so that the receiving modem samples the link at the exact moment that a bit occurs. The receiver clock is supplied by the modem in phase coherence with the incoming data bit stream, or more simply stated, tuned to the exact speed of the transmitting clock. The transmitting clock can be supplied by either the modem (internal) or the terminal device (external). A modem that is optioned for internal timing produces a train of timing signals derived from the received data. To ensure that the timing recovery circuit in the modem stays in synchronization, the actual data stream must contain a specific number of bit state changes, from 0 to 1 and vice versa. Although the bit stream provided by a terminal or computer port can consist of any arbitrary bit pattern, in certain cases a pattern will appear with a sufficient run of 0 or 1 bits that will not provide the receiver with a sufficient number of transitions for synchronization. To prevent this situation, synchronous modems include circuitry in their transmitter known as a scrambler, which will change the bit stream in a controller manner. The scrambler is designed to provide an equal probability of occurrence of each possible phase angle value during a set binary string length, providing the receiver demodulator with enough phase shifts to recover the clocking signal in the data. In the receiver a section of circuitry called a descrambler performs an operation which is the reverse of the scrambler, with the reconstructed bit stream then being transmitted on pin 3 of the DCE interface.

The transmission of synchronous data is generally under the control of a master clock which is the fastest clock in the system. Any slower data clock rates required are derived from the master clock by digital division logic, and those clocks are referred to as slave clocks. For instance, a master clock oscillating at a frequency of 96 kHz could be used to derive 9.6 kbps (1/10), 4.8 kbps (1/20), and 2.4 kbps (1/40) clock speeds.

Although the use of synchronous modems on leased lines has greatly diminished over the past decade due to the widespread availability of digital transmission facilities, the need for clocking and scramblers continues. When we examine service units that can be considered to represent digital modems later in this chapter, we will note that they also include a clocking facility, and transmission on digital media must include a sufficient number of 1 bits, referred to as 1s density.

8.2.13 Multiport capability

Modems with a multiport capability offer a function similar to that provided by a multiplexer. In fact, multiport modems contain a limited function time division multiplexer (TDM) which provides the user with the capability of transmitting more than one synchronous data stream over a single transmission line, as shown in Figure 8.19.

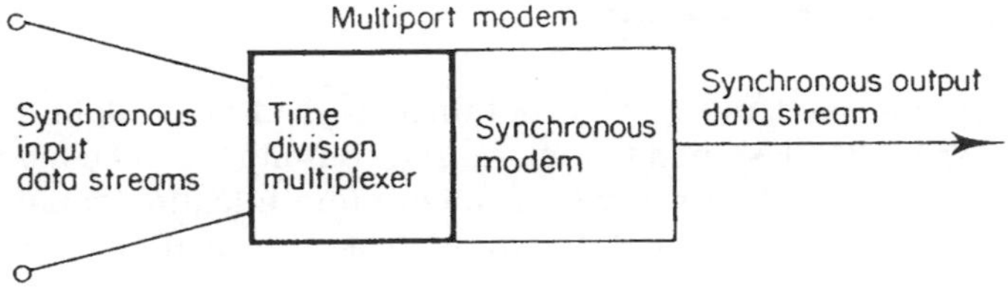

Figure 8.19 Multiport modem. Containing a limited function time division multiplexer, a multiport modem combines the input of a few synchronous input data streams for transmission

In contrast with typical multiplexers, the limited function multiplexer used in a multiport modem combines only a few high-speed synchronous data streams, whereas multiplexers can normally concentrate a mixture of asynchronous and synchronous, high- and low-speed data streams.

8.2.14 Security capability

To provide an additional level of network protection for calls originated over the PSTN several vendors market 'security' modems. In essence, these modems contain a buffer area into which a network administrator enters authorized passwords and associated telephone numbers. When a potential network user dials the telephone number assigned to the security modem, that device prompts the person to enter a password. If a valid match between the entered and previously stored password occurs, the security modem disconnects the line and then dials the telephone number associated with the password. Thus, a modem with a security capability provides a mechanism to verify the originator of calls over the PSTN by his telephone number.

Although dial-back verifies the originator's telephone number, it is difficult for people to use while travelling. Thus, other verification methods, to include the use of a credit card-sized device that generates a sequence of digits every minute, are more suitable for many applications.

8.2.15 Multiple speed selection capability

Dial backup supports data communication systems which require the full-time service of dedicated lines but need to access the switched network if the dedicated line should fail or degrade to the point where it cannot be used. Since transmission over dedicated lines usually occurs at a higher speed than is obtainable over the switched network, one method of facilitating dial backup is through switching down the speed of the modem. For example, a multiple speed modem which is designed to operate at 28 800 bps over dedicated lines may thus be switched down to 19 000 or 9600 bps for operation over the dial-up network until the dedicated lines are restored.

8.2.16 Voice/data capability

Some modems can be obtained with a voice/data option which permits a specially designed telephone set, commonly called a voice adapter, to be used to provide the user with a voice communication capability over the same line which is used for data transmission. Depending on the modem, this voice capability can be either alternate voice/data or simultaneous voice/data. Thus, the user may communicate with a distant location at the same time as data transmission is occurring, or the user may transmit data during certain times of the day and use the line for voice communications at other times. Voice/data capability can also be used to minimize normal telephone charges when data transmission sequences require voice coordination.

Modems that require the use of a voice adapter are primarily synchronous devices and should not be confused with intelligent modems which incorporate a 'Go to Voice' feature. This feature, when enabled by sending an appropriate code to an intelligent modem, tells the modem to cease the use of the connection, permitting voice communications to occur on the connection.

8.2.17 Modem handshaking

Modem handshaking is the exchange of control signals necessary to establish a connection between two data sets. These signals are required to set up and terminate calls, and the type of signaling used is predetermined according to one of three major standards, such as the Electronics Industry Association (EIA) RS-232 or RS-449 standard or the ITU-T V.24 recommendation. RS-232 and V.24 standards are practically identical and are used by over 95% of all modems currently manufactured. To better understand modem handshaking, let us examine the control signals used by 103-type modems. The handshaking signals of common asynchronous modems and their functions are listed in Table 8.6, and the handshaking sequence is illustrated in Figure 8.20.

The handshaking routine commences when an operator at a remote location dials the telephone number of a distant computer. At the computer site, a ring

Table 8.6 Modem handshaking signals and their functions

Control signal	Function
Transmit data	Serial data sent from device to modem
Receive data	Serial data received by device
Request to send	Set by device when user program wishes to transmit
Clear to send	Set by modem when transmission may commence
Data set ready	Set by modem when it is powered on and ready to transfer data; set in response to data terminal ready
Carrier detect	Set by modem when signal present
Data terminal ready	Set by device to enable modem to answer an incoming call on a switched line; reset by adaptor disconnect call
Ring indicator	Set by modem when telephone rings

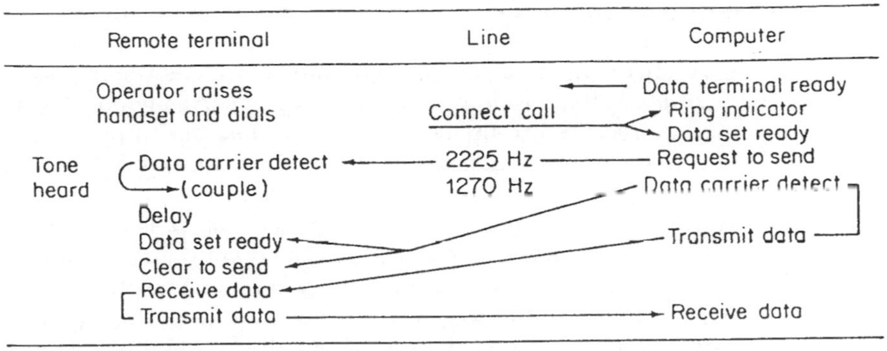

Figure 8.20 Modem handshake sequence

indicator (RI) signal at the answering modem is set and passed to the computer. The distant computer then sends a Data Terminal Ready (DTR) signal to its modem, which then transmits a tone signal to the modem connected to the originator. Upon hearing this tone, the terminal operator presses the data pushbutton on the modem, if it is so equipped for manual operation. Upon depression of the data button for manually operated modems, the originating modem sends a Data Set Ready (DSR) signal to the attached terminal device, and the answering modem sends the same signal to the computer. At this point in time both modems are placed in the data mode of operation.

The remote computer normally transmits a request for identification. To do this the computer sets Request to Send (RTS) which informs the terminal's modem that it wishes to transmit data. The terminal's modem will respond with the Clear to Send (CTS) signal and will transmit a carrier signal. The computer's modem detects the Clear to Send and Carrier ON signals and begins its data transmission to the terminal. When the computer completes its transmission it drops the RTS, and the terminal's modem then terminates its carrier signal. Depending on the type of circuit on which transmission occurs, some of these signals may not be required. For example, on a switched two-wire telephone line, the RTS signal determines whether a terminal is to send or receive data, whereas on a leased four-wire circuit RTS can be permanently raised. For further information, refer to specific vendor literature or appropriate technical reference publications.

8.2.18 Self-testing features

Many low-speed and most high-speed modems have a series of pushbutton test switches which may be used for local and remote testing of the data set and line facilities. Other modems that recognize codes can be placed in predefined test mode via the transmission of certain command codes from a DTE to the modem.

In the local or analog test mode, the transmitter output of the modem is connected to the receiver input, disconnecting the customer interface from the modem. A built-in word generator is used to produce a stream of bits which are checked for accuracy by a word comparator circuit, and errors are displayed on an error lamp as they occur. The local test is illustrated in Figure 8.21.

To check the data sets at both ends as well as the transmission medium, a digital loop-back, self-test may be employed. To conduct this test, personnel may be required to be located at each data set to push the appropriate test buttons, although a number of vendors introduced modems that can be automatically placed into the test mode at the distant end when the central site modem is switched into an appropriate test mode of operation. In the digital loop-back test, the modem at the distant end has its receiver connected to its transmitter, as shown in Figure 8.22. At the other end, the local modem transmits a test bit stream from its word generator, and this bit stream is looped back from the distant end to the receiver of the central site modem where it is checked by the comparator circuitry. Again, an error lamp indicates abnormal results and indicates that either the modems or the line may be at fault.

The analog loop-back self-test should normally be used to verify the internal operation of the modem, and the digital loop-back test will check both modems and the transmission medium. Although analog and digital tests are the main self-tests built into modems, several vendors offer additional diagnostic capabilities that may warrant attention. Two of these diagnostic tests that deserve mention include bit error rate testing and alarm threshold monitoring.

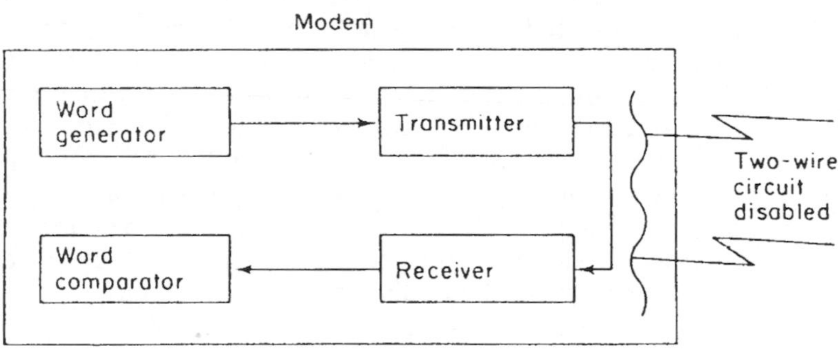

Figure 8.21 Local (analog) testing. In local testing the transmitter is connected to the receiver, and the bit stream produced by the word generator is checked by the word comparator

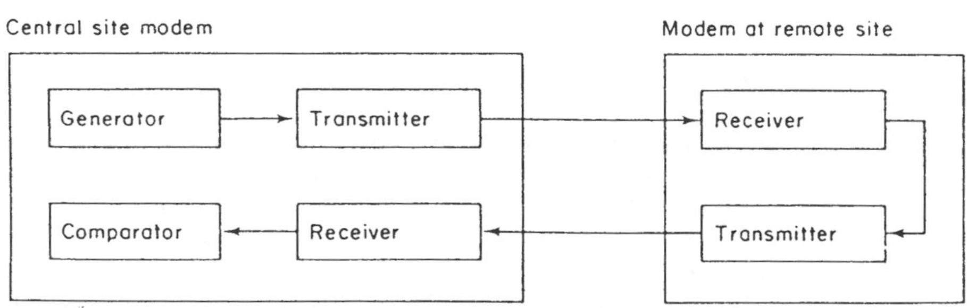

Figure 8.22 Digital loop-back self-test. In the digital loop-back test both the modems and the transmission facility are tested (— primary data, --- control channel)

8.2.19 Modem indicators

Most modems contain a series of light-emitting diode indicators on the front panel of the device that display the status of the modem's operation. Typical indicators include a power indicator which is illuminated when power to the modem is on, a terminal-ready indicator which is illuminated when an attached terminal is ready to send or receive data, a transmit-data indicator which illuminates when data are sent from the attached device to the modem, a receive-data indicator that illuminates when data are received from a distant computer or terminal, and a carrier-detect signal which illuminates when a carrier signal is received from a distant modem. Other indicators in some modems include an off-hook indicator that illuminates when the modem is using the telephone line and a high-speed indicator on dual speed modems that illuminates whenever the modem is operating at its high speed. Table 8.7 lists the most common modem indicator symbols, their meaning and the resulting status of the modem when the indicator is illuminated on the front panel of the modem.

8.2.20 Modem operations and compatibility

Many modem manufacturers describe their product offerings in terms of compatibility or equivalency with modems manufactured by Western Electric for the Bell System, prior to its breakup into independent telephone companies, or with International Telecommunications Union – Telecommunication Standardization Sector (ITU-T) recommendations. The International Telecommunications Union which is based in Geneva, developed a series of modem

Table 8.7 Common modem indicator symbols

Symbol	Meaning	Status
HS	High speed	ON when the modem is communicating with another modem at 2400 baud
AA	Auto answer/Answer	ON when the modem is in auto answer mode and when on-line in answer mode
CD	Carrier detect	ON when the modem receives a carrier signal from a remote modem. Indicates that data transmission is possible
OH	Off hook	ON when the modem takes control of the phone line to establish a data link
RD	Receive data	Flashes when a data bit is received by the modem from the phone line, or when the modem is sending result codes to the terminal device
SD	Send data	Flashes when a data bit is sent by the terminal device to the modem
TR	Terminal ready	ON when the modem receives a Data Terminal Ready signal
MR	Modem ready/Power	ON when the modem is powered on
AL	Analog loopback	ON when the modem is an analog loopback self-test mode

standards for recommended use. These recommendations are primarily adapted by the Post, Telephone and Telegraph (PTT) organizations that operate the telephone networks of many countries outside the United States; however, owing to the popularity of certain recommendations, they have also been followed in designing certain modems for operation on communications facilities within the USA. The following examination of the operation and compatibility of the major types of Bell System and ITU-T modems is based on their operating rate.

Although most modens that operate at data rates below 28 800 bps are considered to be obsolete, the description of low-speed modems will introduce you to several communications techniques applied to modern products. Such techniques as full-duplex transmission via separate channels and echo cancellation that were developed during the 1970s and 1980s are also used in broadband modems to provide the high-speed transmission many Internet users seek.

300 bps

Modems operating at 300 bps use a frequency shift keying (FSK) modulation technique. In this technique the frequency of the carrier is alternated to one of two frequencies, one frequency representing a space or zero bit while the other frequency represents a mark or a one bit. Table 8.8 lists the frequency assignments for Bell System 103/113 and ITU-T V.21 modems which represent the two major types of modems that operate at 300 bps.

Bell System 103 and 113 series modems are designed so that one channel is assigned to the 1070 to 1270 Hz frequency band, whereas the second channel is assigned to the 2025 to 2225 Hz frequency band. Modems that transmit in the 1070 to 1270 Hz band but receive in the 2025 to 2225 Hz band are designated as originate modems, whereas a modem which transmits in the 2025 to 2225 Hz band but receives in the 1070 to 1270 Hz band is designed as an answer modem. When using such modems, their correct pairing is important, since two originate modems cannot communicate with each other.

Bell System 113A modems were originate-only devices that were normally used when calls were placed in one direction. This type of modem was used to enable teletype-compatible terminals to communicate with time-sharing systems where such terminals only originate calls. Bell System 113B modems were answer-only and were primarily used at computer sites where users dialed in to establish communications. Since these modems were designed to

Table 8.8 Frequency assignments (Hz) for 300 bps modems

Major modem types	Originate	Answer
Bell System	Mark 1270	2225
(103/113 type)	Space 1070	2025
ITU-T V.21	Mark 980	1650
	Space 1180	1850

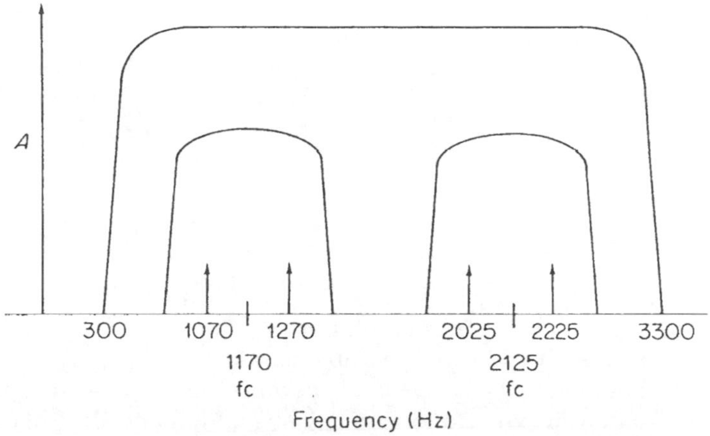

Figure 8.23 Bell System 103/113 frequency spectrum

transmit and receive on a single set of frequencies, their circuitry requirements were less than those of other modems and their costs were more economical.

Modems in the 103 series, could transmit and receive in either the low or the high band. This ability to switch modes is denoted as 'originate and answer', in comparison to the Bell 113A which operated only in the originate mode and the Bell 113B which operated only in the answer mode.

As indicated in Table 8.8, modems operating in accordance with the V.21 recommendation employ a different set of frequencies for the transmission and reception of marks and spaces. Thus, Bell System 103/113 type modems and V.21 devices could never communicate with one another. The two pairs of frequencies used by the modems listed in Table 8.8 permit the bandwidth of a communications channel to be split into two subchannels by frequency. This technique is illustrated in Figure 8.23 for Bell System 103/113 modems. Since each subchannel can permit data to be transmitted in a direction opposite that transmitted on the other subchannel, this technique permits full-duplex transmission to occur on the switched telephone network, which is a two-wire circuit that normally can only support half-duplex transmission.

300 to 1800 bps

There are several Bell System and ITU-T V series modems that operate in the range 300 to 1800 bps. Some of these modems such as the Bell System 212A and V.22 devices can operate at either of two speeds; and other modems such as the Bell System 202 and the V.23 only operate at one data rate. We will examine these modems in pairs, enabling their similarities and differences to be compared.

The Bell System 212A and V.22 modems

The Bell System 212A modem permits either asynchronous or synchronous transmission over the public switched telephone network. The 212A contains

Table 8.9 212A type modem phase shift encoding

Dibit	Phase shift (deg)
00	90
01	0
10	180
11	270

a 103-type modem for asynchronous transmission at speeds up to 300 bps. At this data rate FSK modulation is employed, using the frequency assignments previously indicated in Table 8.8. At 1200 bps, dibit phase shift keyed (DPSK) modulation is used which permits the modem to operate either asynchronously or synchronously. The phase shift encoding of the 212A type modem is illustrated in Table 8.9.

One advantage in the use of this modem is that it permits the reception of transmission from devices operating at two different transmission speeds. When the call is made, the answering 212A modem automatically switches to a 300 bps or 1200 bps operating speed. During data transmission, both modems remain in the same speed until the call is terminated, when the answering 212A can be set to the other speed by a new call. The dual-speed 212A permits both terminals connected to Bell System 100 series data sets operating at up to 300 bps or terminals connected to other 212A modems operating at 1200 bps to share the use of one modem at a computer site and thus can reduce central computer site equipment requirements.

Although the 212A modem has been obsolete for approximately 20 years, it represents the first dual-speed modem to be mass marketed. The technique used to lock on to the speed of the originating modem was based upon the use of destinct carrier tones (frequencies) at 300 and 1200 bps and is used today to enable modern modems to support over ten operating rates.

The V.22 standard is for modems that operate at 1200 bps on the PSTN or leased circuits and has a fallback data rate of 600 bps. The modulation technique employed is four-phase PSK at 1200 bps and two-phase PSK at 600 bps, with five possible operational modes specified for the modem at 1200 bps. Table 8.10 lists the V.22 modulation phase shifts with respect to the bit patterns entering the modem's transmitter. Modes 1 and 2 are for synchronous and asynchronous data transmission at 1200 bps, respectively, while mode 3 is for synchronous transmission at 600 bps. Mode 4 is for asynchronous

Table 8.10 V.22 modulation phase shift as opposed to bit patterns

Dibit values (1200 bps)	Bit values (600 bps)	Phase change modes 1, 2, 3, 4	Phase change mode V
00	0	90	270
01	—	0	180
11	1	270	90
10	—	180	0

transmission at 600 bps, while mode 5 represents an alternate phase change set for 1200 bps asynchronous transmission.

In comparing V.22 modems to the Bell System 212A devices it should be apparent that they are totally incompatible at the lower data rate, since both the operating speed and modulation techniques differ. At 1200 bps the modulation techniques used by a V.22 modem in modes one through four are exactly the same as that used by a Bell System 212A device. Unfortunately, a Bel 212A modem that answers a call sends a tone of 2225 Hz on the line that the originating modem is supposed to recognize. This frequency is used because of the construction of the switched telephone network in the United States and other parts of North America. Under V.22, the answering modem first sends a tone of 2100 Hz since this frequency is more compatible with the design of European switched telephone networks. Then, the V.22 modem sends a 2400 Hz tone that would not be any better except that the V.22 modem also sends a burst of data whose primary frequency is about 2250 Hz, which is close enough to the Bell standard of 2225 Hz that many Bell 212A-type modems will respond. Some Bell 212A modems can thus communicate with V.22 modems at 1200 bps, whereas other 212-type modems may not be able to communicate with V.22 devices, with the ability to successfully communicate being based upon the tolerance of the 212 type modem to recognize the V.22 modem's data burst at 2250 Hz.

Bell System 202 series modems

Bell System 202 series modems are designed for speeds up to 1200 or 1800 bps. The 202C modem can operate on either the switched network or on leased lines, in the half-duplex mode on the former and the full-duplex mode on the latter. The 202C modem can operate half-duplex or full-duplex on leased lines. This series of modems uses frequency shift keyed (FSK) modulation, and the frequency assignments are such that a mark is at 1200 Hz and a space at 2200 Hz. When either modem is used for transmission over a leased four-wire circuit in the full-duplex mode, modem control is identical to the 103 series modem in that both transmitters can be strapped on continuously which alleviates the necessity of line turnarounds.

Since the 202 series modems do not have separate bands, on switched network utilization half-duplex operation is required. This means that both transmitters (one in each modem) must be alternately turned on and off to provide two-way communication.

The Bell 202 series modems have a 5 bps reverse channel for switched network use, which employs amplitude modulation for the transmission of information. The channel assignments used by a Bell System 202 type modem are illustrated in Figure 8.24, where the 387 Hz signal represents the optional 5 bps AM reverse channel. Owing to the slowness of this reverse channel, its use is limited to status and control function transmission. Status information such as 'ready to receive data' or 'device out of paper' can be transmitted on this channel. Owing to the slow transmission rate, error detection of received messages and an associated NAK and request for retransmission is normally accomplished on the primary channel, since even

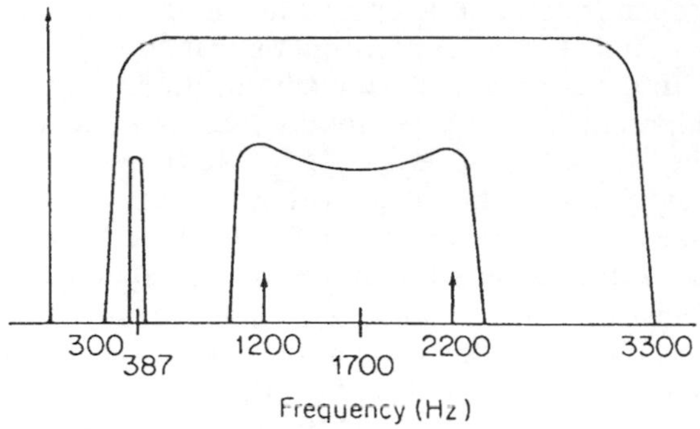

Figure 8.24 Bell System 202-type modem channel assignments

with the turnaround time, it can be completed at almost the same rate one obtains in using the reverse channel for that purpose. Non-Bell 202-equivalent modems produced by many manufacturers provide reverse channels of 75 to 150 bps which can be utilized to enhance overall system performance.

Advances in modem technology resulted in the 202 type modem becoming essentially obsolete, although this author is aware of several computer networks that still use this modem.

V.23 modems

The V.23 standard is for modems that transmit at 600 or 1200 bps over the PSTN. At 600 bps a mark is represented by a tone at 1300 Hz, and the space occurs at 1700 Hz. At 1200 bps the frequency used to represent a space is shifted to 2100 Hz.

Both asynchronous and synchronous transmission are supported by using FSK modulation; an optional 75 bps backwards or reverse channel can be used for error control. Figure 8.25 illustrates the channel assignments for a V.23 modem. In comparing Figure 8.25 with Figure 8.24, it is obvious that Bell System 202 and V.23 modems are incompatible with each other.

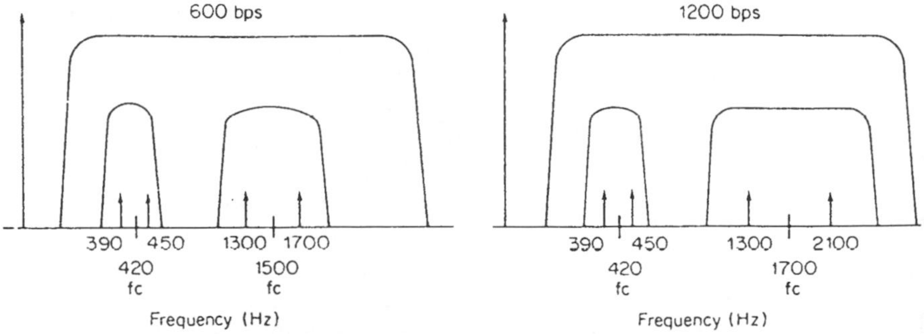

Figure 8.25 V.23 channel assignments

The V.23 modem is widely employed for use in a transmission application called by the generic term videotext in Europe. Communications for this application use the 75 bps reverse channel to transmit information from a terminal device, such as a specially equipped television. Since user selections from a menu painted on the screen or a selection of digits is small in comparison to the flow of data from a computer to the terminal, the secondary channel with respect to the terminal operates at a higher data rate and is the primary channel with respect to the computer's transmission.

2400 bps

Examples of modems that operate at 2400 bps include the ITU-T V.26 series and the V.22 bis modem. The V.26 series modems are designed for synchronous bit serial transmission at a data rate of 2400 bps, and the V.22 bis standard governs 2400 bps asynchronous transmission.

V.26 modem

The V.26 standard specifies the characteristics for a 2400 bps synchronous modem for use on a four-wire leased line. Modems operating according to the V.26 standard employ dibit phase shift keying, using one of two recommended coding schemes. The phase change based upon the dibit values for each of the V.26 coding schemes is listed in Table 8.11.

Two ITU-T recommendations similar to V.26 are V.26 bis and V.26 ter. The V.26 bis recommendation defines a dual speed 2400/1200 bps modem for use on the PSTN. At 2400 bps the modulation and coding method is the same as the V.26 recommendation for pattern B listed in Table 8.11. At the reduced data rate of 1200 bps a 2-phase shift modulation scheme is employed, with a binary zero represented by a 908 phase shift, and a binary one is represented by a 2708 phase shift. The V.26 bis recommendation also includes an optional reverse or backward channel that can be used for data transfer up to 75 bps. When employed, frequency shift keying is used to obtain this channel capacity, with a mark or one bit represented by a 390 Hz signal and a space or zero bit represented by a 450 Hz signal.

The V.26 ter recommendation uses the same phase shift scheme as the V.26 modem, but incorporates an echo-canceling technique that allows transmitted and received signals to occupy the same bandwidth. Thus, the V.26 ter

Table 8.11 V.26 modulation phase shift versus bit pattern

	Phase change	
Dibit values	Pattern A	Pattern B
00	0	45
01	90	135
11	180	225
10	270	315

modem is capable of operating in full duplex at 2400 bps on the PSTN. Echo canceling will be described later in this chapter when the V.32 modem is examined.

V.22 bis modem

The V.22 bis recommendation governs modems designed for asynchronous data transmission at 2400 bps over the PSTN, with a fallback rate of 1200 bps. Since V.22 bis defines operations at 1200 bps to follow the V.22 format, communications capability with Bell System 212A type modems at that data rate may not always be possible, owing to the answer tone incompatibility usually encountered between modems following Bell System specifications and ITU-T recommendations. In addition, V.22 bis modems manufactured in the USA may not be compatible with such modems manufactured in Europe, at fallback data rates. This is because V.22 bis modems manufactured in Europe follow the V.22 format, with fallback data rates of 1200 and 600 bps. At 1200 bps the incompatibility between most European telephone networks, which are designed to accept only 2100 Hz answer tones, whereas the US telephone network usually accepts an answer tone between 2100 and 2225 Hz, may preclude communications between a US and a European manufactured V.22 bis modem at 1200 bps. At a lower fallback speed the European modem will operate at 600 bps while the US V.22 bis modem operates at 300 bps, insuring incompatibility.

In spite of the previously mentioned problems, V.22 bis modems became a *de facto* standard for use with terminals and personal computers communicating over the PSTN. This is due to several factors, to include the manufacture in the United States of V.22 bis modems that are Bell System 212A compatible, permitting persons with such modems to be able to communicate with other personal computers and mainframe computers connected to either 212A or 103/113 type modems. In addition, at 2400 bps US V.22 bis modems can communicate with European V.22 bis, in effect providing worldwide communications capability over the PSTN.

4800 bps

The Bell System 208 series and ITU-T V.27 modems represent common types of modem designed for synchronous data transmission at 4800 bps. The Bell System 208 Series modems use a quadrature amplitude modulation technique. The 208A modem is designed for either half-duplex or full-duplex operation at 4800 bps over leased lines. The 208B modem is designed for half-duplex operation at 4800 bps on the switched network.

Newer versions of the 208A were offered by AT&T as the 2048A and 2048C models, which were also designed for four-wire leased line operation. The 2048C has a start-up time less than one half of the 2048A, which makes it more suitable for operations on multidrop lines.

Both Bell 208 type modems and ITU-T V.27 modems pack data three bits at a time, encoding them for transmission as one of eight phase angles.

Table 8.12 V.27 modulation phase shift versus bit pattern

Tribit values	Phase change (deg)
001	0
000	45
010	90
011	135
111	180
110	225
100	270
101	315

Unfortunately, since each type of modem uses different phase angles to represent a tribit value, they cannot talk to each other. Table 8.12 lists the V.27 modulation phase shifts with respect to each of the eight possible tribit values.

9600 bps

Three common modems that are representative of devices that operate at 9600 bps are the Bell System 209, and the ITU-T V.29 and V.32 modems.

Bell System 209 modem

Modems equivalent to the Bell System 209 and ITU-T V.29 devices are designed to operate in a full-duplex, synchronous mode at 9600 bps over private lines. The Bell System 209A modem operates by employing a quadrature amplitude modulation technique, as previously illustrated in Figure 8.11. Included in this modem is a built-in synchronous multiplexer which will combine up to four data rate combinations for transmission at 9600 bps. The multiplexer combinations are shown in Table 8.13. A newer version of the 209A is the 2096A. This modem is noteworthy because it has an EIA RS-449/423 interface with RS-232 compatibility.

The use of a built-in multiplexer permits several data sources to share the use of a leased line. Although the functionality of the built-in multiplexer is limited, as it supports only a few speed settings and is restricted to operating

Table 8.13 Bell 209A multiplexer combinations

2400–2400–2400–2400 bps
4800–2400–2400 bps
4800–4800 bps
7200–2400 bps
9600 bps

on synchronous data, the ability to share the housing and power supply of the modem made it extremely economical. While most organizations during the 1990s converted their analog leased lines to digital transmission facilities, you can still find some 209LA modems in operation. In addition, the concept of building a multiplexer into analog modems was carried over to digital modems, the latter being referred to as Data Service Units.

V.29 modem

With the exception of Bell System 209-type modems, a large majority of 9600 bps devices manufactured throughout the world adhere to the V.29 standard. The V.29 standard governs data transmission at 9600 bps for full- or half-duplex operation on leased lines, with fallback data rates of 7200 and 4800 bps allowed. At 9600 bps the serial data stream is divided into groups of four consecutive bits. The first bit in the group is used to determine the amplitude to be transmitted and the remaining three bits are encoded as a phase change, with the phase changes identical to those of the V.27 recommendation listed in Table 8.12.

Because the V.29 modem packs four bits into one signaling change, its baud rate is one-fourth of its transmission rate. This means that a V.29 modem operating at 9600 bps has a 2400 baud signaling rate.

Table 8.14 lists the relative signal element amplitude of V.29 modems, based on the value of the first bit in the quadbit and the absolute phase which is determined from bits two through four. Thus, a serial data stream composed of the bits 1 1 0 0 would have a phase change of 270° and its signal amplitude would be 5. The resulting signal constellation pattern of V.29 modems is illustrated in Figure 8.26.

V.32 modem

The V.32 modem represents the first of a series of communications devices developed to support relatively high-speed full-duplex transmission over the public switched telephone network. The V.32 modem, which has a maximum transmission rate of 9600 bps, was followed by the V.32 bis modem that has a maximum operating rate of 14 400 bps. Two additional modems that also support high speed full-duplex transmission on the PSTN include the V.34,

Table 8.14 V.29 signal amplitude construction

Absolute phase	1st bit	Relative signal element amplitude
0, 90, 180, 270	0	3
	1	5
45, 135, 225, 315	0	$\sqrt{2}$
	1	$3\sqrt{2}$

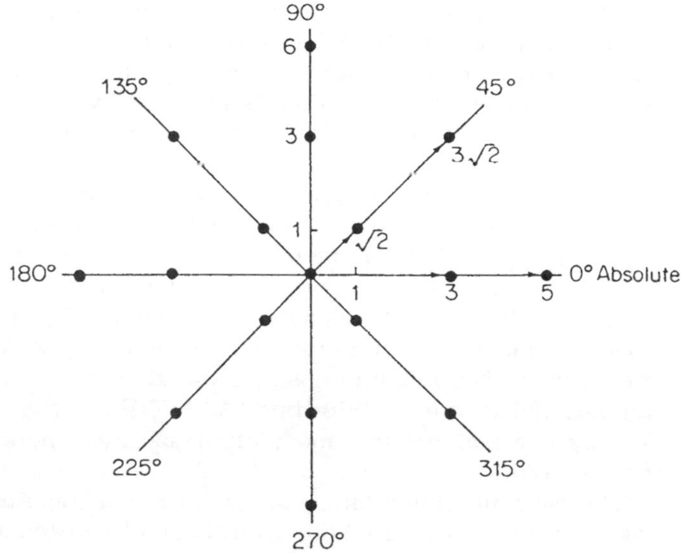

Figure 8.26 V.29 signal constellation pattern

which extended transmission to 33.6 kbps, and the V.90, which supports a theoretical data transfer of 56 kbps. Each of the previously mentioned modems uses echo-canceling technology to support full-duplex transmission over the two-wire PSTN. In addition, each of the previously mentioned modems incorporates an asynchronous to synchronous converter which increases transmission efficiency, since start and stop bits are eliminated for transmission. Although the V.32 modem is now obsolete, having been replaced by higher-speed operating devices, a review of its operations will provide us with an appreciation for coding techniques used in subsequent modems.

A V.32 modem establishes two high-speed channels in opposite directions, as illustrated in Figure 8.27. Each of these channels shares approximately the same bandwidth. V.32 modems employ an echo-canceling technique that permits transmitted and received signals to occupy the same bandwidth. This is made possible by intelligence designed into the modem's receiver that permits it to cancel out the effects of is own transmitted signal, enabling the

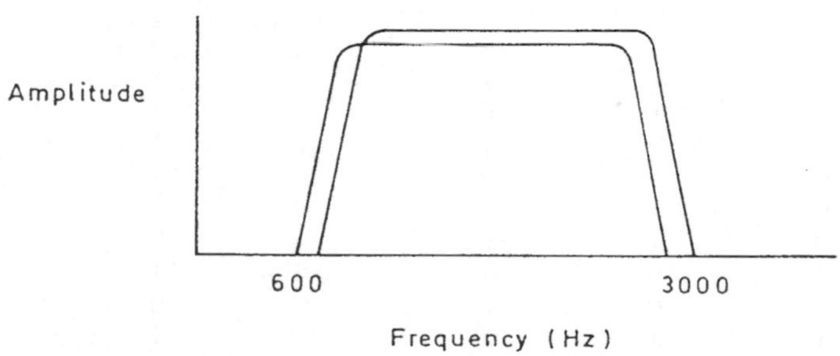

Figure 8.27 V.32 channel derivation

modem to distinguish its sending signal from the signal being received. To accomplish this task, the echo canceler in the modem must be adaptive, since the echo that must be modeled often changes during a communications session. The actual echo that is modeled by the modem's echo canceler includes the near-end and far-end echo paths. The resulting near-end and far-end echo replica is then added to a copy of the transmitted signal and subtracted from the modem's receiver before the signal is demodulated, resulting in the reconstruction of the far-end modem signal.

Under the V.32 recommendation, synchronous data signaling rates of 2400, 4800, and 9600 bps are supported for asynchronous data entering the modem at those rates. This support is accomplished using an asynchronous-to-synchronous converter built into the modem. A V.32 modem uses a carrier frequency of 1800 Hz and has a modulation rate of 2400 baud to support a data transfer rate of 9600 bps. At 9600 bps, the V.32 recommendation specifies two alternative modulation schemes: nonredundant coding and trellis coding.

The use of nonredundant coding results in a 16-point constellation pattern, whereas trellis coding results in a 32-point constellation pattern. Note that while all V.32 modems must be capable of interworking using the 16-point constellation pattern, not all V.32 modems include Trellis coding. Thus, users who require the better immunity to impairments afforded by trellis coding should ensure that the V.32 modem they are considering supports that alternative.

Under the nonredundant coding technique, the data to be transmitted at 9600 bps are divided into groups of four consecutive data bits. The value of the first two bits is used in conjunction with the value of the two bits last output to

Table 8.15 Differential quadrant coding for 4800 bps and nonredundant coding at 9600 bps

Inputs		Previous outputs		Phase quadrant change	Outputs		Signal state for 4800 bps
$Q1_n$	$Q2_n$	$Y1_{n-1}$	$Y2_{n-1}$		$Y1_n$	$Y2_n$	
0	0	0	0	+90	0	1	B
0	0	0	1		1	1	C
0	0	1	0		0	0	A
0	0	1	1		1	0	D
0	1	0	0	0	0	0	A
0	1	0	1		0	1	B
0	1	1	0		1	0	D
0	1	1	1		1	1	C
1	0	0	0	+180	1	1	C
1	0	0	1		1	0	D
1	0	1	0		0	1	B
1	0	1	1		0	0	A
1	1	0	0	+270	1	0	D
1	1	0	1		0	0	A
1	1	1	0		1	1	C
1	1	1	1		0	1	B

generate the value for the next two output bits. These two output bits are then used with the value of bits 3 and 4 of the quadbit to select an appropriate signal point. Table 8.15 indicates the dependencies of the output dibit selection on the value of the input dibit and the previous output dibits. Note that the values of the two input bits ($Q1_n$ and $Q2_n$) and the value of the previous dibit outputs ($Y1_{n-1}$ and $Y2_{n-1}$) are used to determine the phase quadrant change where the signal point will be located. However, the value of the two input bits does not actually locate the point in the quadrant. To perform the latter operation, the dibit outputs ($Y1_n$ and $Y2_n$) are used in conjunction with the values of the third and fourth bits in the quadbit input to select a position in the quadrant.

To select a quadrant position, the value of the dibit output ($Y1_n$ and $Y2_n$) is used in conjunction with the value of the third and fourth input (Q3 and Q4) bits. Table 8.16 indicates the selection of the nonredundant coding axis position based on the values of the dibit output and the second dibit input.

To understand the use of Tables 8.15 and 8.16, assume that a quadbit input has the value 0001 and the previous dibit output was 01. Because $Q1_n$ and $Q2_n$ have the value 00 while $Y1_{n-1}$ and $Y2_{n-1}$ have the value 01, from Table 8.15 the outputs ($Y1_n$ and $Y2_n$) become 11 and the phase quadrant change will be 90°. From Table 8.16, the axis position will be 3, 1 ($Y1 = Y2 = 1$, $Q3 = 0$, $Q4 = 1$), which will place the signal in the lower right position of the quadrant.

Figure 8.28 illustrates the 16-point constellation pattern generated by a V.32 modem using nonredundant coding. The four points that are circled represent valid signal points when the modem operates at 4800 bps. When operating at 4800 bps using nonredundant coding, the modem operates on two bits at a time, comparing the input bit values to the previous output dibit values to select one of four signal points.

Table 8.16 V.32 Nonredundant coding signal-state mappings for 9600 bps

Coded inputs				Nonredundant coding axis position	
Y1	Y2	Q3	Q4	X	Y
0	0	0	0	−1	−1
0	0	0	1	−3	−1
0	0	1	0	−1	−3
0	0	1	1	−3	−3
0	1	0	0	1	−1
0	1	0	1	1	−3
0	1	1	0	3	−1
0	1	1	1	3	−3
1	0	0	0	−1	1
1	0	0	1	−1	3
1	0	1	0	−3	1
1	0	1	1	−3	3
1	1	0	0	1	1
1	1	0	1	3	1
1	1	1	0	1	3
1	1	1	1	3	3

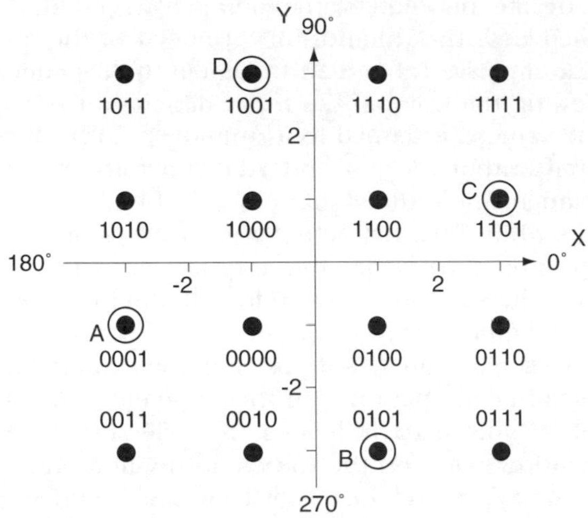

Figure 8.28 V.32 16-point signal constellation

When a V.32 modem employs Trellis coding, the input data stream is also divided into groups of four bits. As in the nonredundant coding method, the first two bits in each group ($Q1_n$ and $Q2_n$) are differentially encoded into $Y1_n$ and $Y2_n$, based on the previous dibit value output. Table 8.17 indicates the differential encoding used by a V.32 modem operating at 9600 bps and employing Trellis coding.

Table 8.17 V.32 differential encoding for use with trellis-coded alternative at 9600 bps

Inputs		Previous outputs		Outputs	
$Q1_n$	$Q2_n$	$Y1_{n-1}$	$Y2_{n-1}$	$Y1_n$	$Y2_n$
0	0	0	0	0	0
0	0	0	1	0	1
0	0	1	0	1	0
0	0	1	1	1	1
0	1	0	0	0	1
0	1	0	1	0	0
0	1	1	0	1	1
0	1	1	1	1	0
1	0	0	0	1	0
1	0	0	1	1	1
1	0	1	0	0	1
1	0	1	1	0	0
1	1	0	0	1	1
1	1	0	1	1	0
1	1	1	0	0	0
1	1	1	1	0	1

Unlike nonredundant coding, under trellis coding the two differentially encoded bits ($Y1_n$ and $Y2_n$) are used as the input to a convolutional encoder. The convolutional encoder results in the generation of three bits based on the two-bit input. Two of the three bits are the differentially encoded bits ($Y1_n$ and $Y2_n$) that are passed by the encoder. The third bit ($Y0_n$) is a redundant bit produced by the convolutional encoding process whose value is based on the value of $Y1_n$ and $Y2_n$.

At 9600 bps, the two passed through output bits ($Y1_n$ and $Y2_n$) and the redundant bit generated by the encoder ($Y0_n$) are used in conjunction with the value of the third and fourth bits in each quadbit to select one of 32 signal state mapping points. Table 8.18 indicates trellis coding signal points. Figure 8.29 illustrates the constellation pattern formed by a plot of all 32

Table 8.18 V.32 trellis coding at 9600 bps

	Coded inputs				Trellis coding	
(Y0)	Y1	Y2	Q3	Q4	Re	Im
0	0	0	0	0	−4	1
	0	0	0	1	0	−3
	0	0	1	0	0	1
	0	0	1	1	4	1
	0	1	0	0	4	−1
	0	1	0	1	0	3
	0	1	1	0	0	−1
	0	1	1	1	−4	−1
	1	0	0	0	−2	3
	1	0	0	1	−2	−1
	1	0	1	0	2	3
	1	0	1	0	2	−1
	1	1	0	0	2	−3
	1	1	0	1	2	1
	1	1	1	0	−2	−3
	1	1	1	1	−2	1
1	0	0	0	0	−3	−2
	0	0	0	1	1	−2
	0	0	1	0	−3	2
	0	0	1	1	1	2
	0	1	0	0	3	2
	0	1	0	1	−1	2
	0	1	1	0	3	−2
	0	1	1	1	−1	−2
	1	0	0	0	1	4
	1	0	0	1	−3	0
	1	0	1	0	1	0
	1	0	1	1	1	−4
	1	1	0	0	−1	−4
	1	1	0	1	3	0
	1	1	1	0	−1	0
	1	1	1	1	−1	4

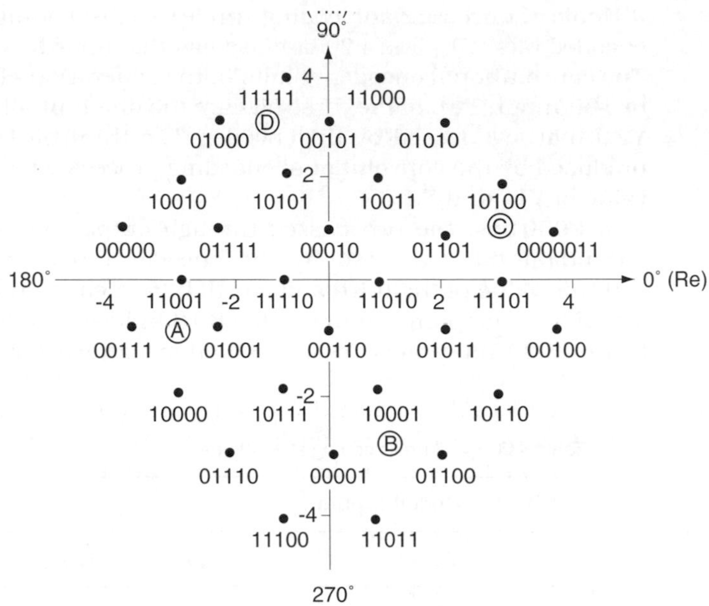

Figure 8.29 V.32 Trellis coding signal constellation pattern

signal points. When a V.32 modem operates at 4800 bps, the device operates on dibits instead of quadbits. When this occurs, the bits $Q1_n$ and $Q2_n$ are differentially encoded into $Y1_n$ and $Y2_n$ according to Table 8.15, which results in a constellation pattern of four signal points indicated in Figure 8.29 by the letters A, B, C and D.

Because the ITU-T V.32 recommendation was promulgated in 1984, numerous modem manufacturers used the V.32 modulation scheme as a platform for adding enhancements to this modem. These enhancements include error detection and correction capabilities and data compression. Refer to section 3 in this chapter for specific information concerning these features.

14 400 bps

There are two standardized modems whose primary operating rates are 14 400 bps. The V.32 bis modem was designed for use on the switched network, and the V.33 modem is designed for use on four-wire leased lines.

V.32 bis modem

The V.32 bis recommendation was promulgated in 1991, and it represents the ITU-T standard for modems operating on the switched telephone network at data rates up to 14 400 bps. Although this modem has been available for over a decade, until 1992 the average retail price of this modem exceeded $1000, and this limited its widespread use. Since then, manufacturers reduced the retail price of a V.32 bis modem to under $100, which significantly increased its acquisition by personal computer users in the home, business, government, and academia. Although millions of V.32 bis modems were manufactured by the year 2000, most vendors focused their attention upon manufacturing V.34 and V.90 modems. While the V.32 bis is now obsolete, its large, installed based insured that both V.34 and V.90 modems provide downward compatibility with this modem. In fact, V.34 and V.90 modems are usually downward compatible with V.32 bis, V.32, V.22 bis, and sometimes even with Bell 212A modems.

The V.32 bis recommendation is very similar to the V.32 recommendation in several key areas. Both recommendations specify the use of echo cancellation to obtain a full duplex transmission capability on the two-wire switched telephone network, and both specify the use of Trellis coding, which significantly reduces the probability of transmission errors by improving the signal-to-noise ratio of transmission. Major differences between the V.32 bis and V.32 recommendations are in the areas of operating rates supported, the number of encoded bits used per signal change and the resulting constellation pattern, the support of alternative operating rates during a transmission session, and the method and time required for retraining. Table 8.19 summarizes the major differences between the V.32 and V.32 bis recommendations.

As indicated in Table 8.19, the V.32 bis modem supports two operating rates above the maximum operating rate of the V.32 modem. Since those operating rates reflect the data transfer capability of each modem without the effect of data compression, the addition of that feature to each modem significantly increases the difference in throughput achievable through the use of each modem. For example, a V.32 modem that supports V.42 bis data compression, which provides an average compression ratio of 4 : 1, results in a throughput of 38.4 kbps when the modem operates at 9600 bps. In comparison, a V.32 bis modem operating at 14 400 bps employing V.42 bis compression provides a throughput of 57.6 kbps when the average compression ratio is 4 : 1. Thus, the operating rate difference of 4800 bps between a V.32 and a V.32 bis can expand to a throughput difference of almost 20 kbps when the effect of compression is considered.

Table 8.19 V.32 versus V.32 bis feature comparison

Feature	V.32	V.32 bis
Operating rates	9600 7200 (optional) 4800	14400 12000 9600 7200
Encoded bits/symbol and signal constellation points		6 (+TCM) 128
14400 bps		
1200 bps		5 (+TCM) 64
9600 bps	4 (+TCM) 32	
Fallback	yes	yes
Fall-forward	no	no
Retrain	15 seconds	10 seconds

Both V.32 and V.32 bis modems employ Trellis coding at operating rates of 9600 bps and above. Each modem operates at 2400 baud and packs either 4, 5, or 6 data bits plus the Trellis Coding Modulation (TCM) bit into each signal change. The constellation pattern thus increases from 32 signal points for a V.32 modem operating at 9600 bps to 64 and 128 signal points for a V.32 bis modem operating at 12 000 and 14 400 bps, respectively.

The fallback feature listed in Table 8.19 refers to the ability of a modem to change its operating rate downward automatically when it encounters a predefined signal-to-noise ratio that would result in an unacceptable error rate if transmission at the current operating rate were maintained. Unfortunately, under the V.32 recommendation there was no provision for restoring the original operating rate if line quality improved after a fallback. Under V.32 bis an automatic fall-forward capability is included in the recommendation. This feature enables a V.32 bis modem to return to a higher operating rate if line quality improves.

Some modem vendors have implemented an enhanced fall-forward capability that can improve the transmission capability of the modem. For example, US Robotics developed a feature known as Adaptive Speed Leveling for its V. 32 bis modems. This feature enables the data rate of each direction of transmission to vary. Thus, if the incoming data rate is lowered due to noise encountered in one direction or another impairment, the outgoing data can still be transmitted at the highest operating rate, and vice versa. In comparison, most V.32 bis modem manufacturers implement an auto-fall-forward capability symmetrically.

The last major difference between V.32 and V.32 bis modems concerns retraining. The V.32 modem has a 15 second retrain time. In comparison, a V.32 bis modem has a shorter retrain time, which minimizes the effect of a retrain upon modem throughput.

V.33 modem

The V.33 modem can be viewed as a less complex extension of V.32 technology. This is because although the operating rate of the V.33 modem increased to 14.4 kbps from the 9600 bps operating rate of the V.32 modem, it achieves full-duplex transmission without the use of echo cancellation. This is possible because the V.33 modem is designed to operate on four-wire leased lines which permit the use of two two-wire signal paths.

V.33-compatible modems can operate at data rates of 14.4, 12.0 and 9.6 kbps. When operating at 14.4 kbps, the V.33 modem uses QAM modulation, assigning 6 data bits to each signal change and a seventh bit for trellis coding. This results in a constellation pattern of 128 signal points. Although the V.33 standard is specific in its requirement for operation on four-wire leased lines, a few vendors developed proprietary half-duplex versions of the V.33 modem for use on the two-wire switched network. Those modems never achieved popularity, as most switched network users prefer using V.32 bis modems to obtain a full-duplex transmission capability at an operating rate of 14.4 kbps. In addition, the V.32 bis modem provides those modem users with the ability to communicate with a large potential audience, in comparison with the small base of installed modems using a proprietary half-duplex V.33 modulation scheme.

28 800/33 600 bps

When the ITU-T V.34 modem was standardized in 1994, its maximum operating rate was 28 800 bps. Until early 1996, many persons expected the ITU to promulgate a new standard for 33 600 bps operations. In fact, during 1996 several vendors referred to their non-standardized products that operated at 33 600 bps as V.34 bis compatible modems. However, instead of issuing a new standard, the ITU developed a technical specification that revised the V.34 standard and added two additional operating rates to a new maximum rate of 33 600 bps.

The V.34 standard is based upon the use of three-dimensional Trellis coding which results in a significantly higher level of performance than the single-dimension Trellis coding used in V.32 and V.32 bis modems. The increase in performance results from the higher data transfer rate of V.34 modems as well as their lower error rate resulting from the use of three-dimensional Trellis coding. Because the lower error rate reduces the need for retransmission, this improves modem performance.

The 1994 version of the V.34 standard can operate at six distinct signaling rates in its high-speed V.34 mode of operation. Table 8.20 indicates the modem signaling rate, carrier frequency, bandwidth requirements, and maximum bit rate for the V.34 modem's high speed mode of operation.

The initial top speed of the V.34 modem was 28 800 bps. To achieve this operating rate the V.34 modem operates at a signaling rate of 3200 baud, mapping nine bits into each signal change. The V.34 modem standard initially specified two additional mandatory baud rates of 2400 and 3000 and optional signaling rates of 2743, 2800, and 3429 baud.

The 2743 and 2800 baud signaling rates were defined to support a voice digitization technique referred to as Adaptive Differential Pulse Code Modulation (ADPCM). Unlike PCM which samples voice 8000 times per second and encodes the height of each sample into an 8-bit byte which results in a

Table 8.20 Initial V.34 carrier frequency, bandwidth, and maximum bit rate based on signaling rate

Signaling rate (Hz)	Carrier frequency (Hz)	Bandwidth requirements	Maximum bit rate (bps)
2400	1600	400–2800	21 600
2743	1646	274–3018	24 000
2800	1680	280–3080	24 000
3000	1800	300–3300	26 400
3200	1829	229–3429	28 800
3429	1959	244–3674	28 800

64 kbps digital data stream, ADPCM uses a predictor to encode each sample into a 4-bit word. While the resulting 32 kbps digital data stream is half that of PCM and provides very good voice reconstruction quality, it accomplishes this because voice does not rapidly vary. Unfortunately, the same is not true of modem constellation points that can rapidly vary in tandem with changes in a sequence of bits being operated on as an entity. Because ADPCM's predictor is not good enough to support signaling rates above 3000 baud, the optional 2743 Hz and 2800 Hz rates can be used to support high-speed modem communications over an ADPCM infrastructure. That infrastructure is sometimes found on international circuits as well as on private networks where network managers specified the use of ADPCM as a bandwidth conservation method.

In examining the entries in Table 8.20 you will note that the V,34 modem uses frequencies beyond the normal 300 to 3300 Hz voice passband to obtain higher signaling rates. The modem is able to accomplish this by transmitting at a low power level, since low and high pass filters used to form the passband result in a widening of the passband at lower power levels. Also note that, although several V.34 signaling rates result in a data rate nine times the baud rate, this is not always true. some V.34 signaling rates, such as 2743 baud, use a non-integral number of bits per symbol. When this occurs, a special technique referred to as a shell-mapping algorithm is used to generate the constellation pattern.

V.34 optional features

The V.34 standard includes several optional features whose use can significantly boost the capability of this modem over its V.32 bis predecessor. Those features include an asymmetrical transmission capability, an auxiliary channel, and a nonlinear encoding and precoding capability. Although these features are mandatory with respect to their incorporation into the modem's transmitter section, it is important to note they are optional with respect to their inclusion in the modem's receiver section. Thus, you should investigate the optional features supported in a V.34 modem's receiver if you are comparing products.

Asymmetrical transmission: The asymmetrical transmission capability of the V.34 modem enables it to transmit and receive data at different operating rates. This capability can be extremely useful if one modem user is located a relatively long distance away from their serving telephone company central office. In this situation the modem user might be limited to achieving an operating rate of 21 600 or 24 000 bps. However, if the other modem you wish to communicate with is located relatively close to the telephone office serving the subscriber using that device, that modem will probably be able to transmit at 28.8 kbps. Without an asymmetrical transmission capability, communications could be limited to either 21.6 or 24.0 kbps in both directions. However, through the support of an asymmetrical transmission capability in both modems, it becomes possible for each modem to support the highest operating rate obtainable in each direction, without having to have a common data rate in each direction. Thus, in this example communications could occur at 28.8 kbps in one direction and 21.6 kbps in the other direction.

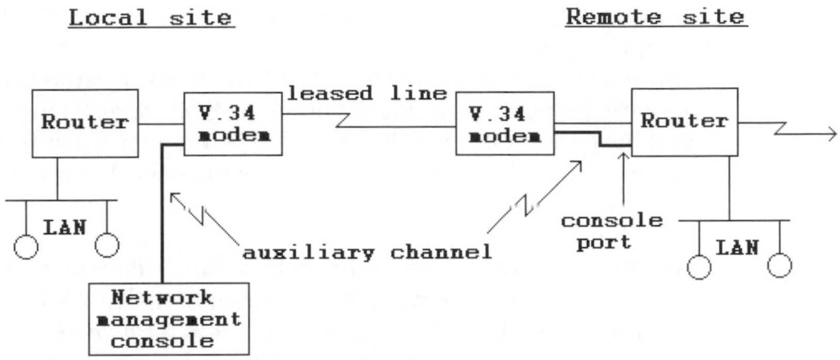

Figure 8.30 Using the V.34 modem auxiliary channel

Auxiliary channel: The addition of an auxiliary channel to a V.34 modem enables the support of management traffic without having to interrupt the primary data channel. The V.34 modem optionally supports a low-speed 200 bps management channel in addition to its primary high-speed data channel.

Figure 8.30 illustrates the use of a pair of V.34 modems that support the optional auxiliary channel to control a distant router at a remote location. In addition to alleviating the need for a separate modem to control the distant router, this configuration enhances security, as communications to and from the router console port flow over a dedicated connection. Thus, thc ability to configure the router via a LAN or the Internet could be disabled.

Nonlinear coding: To minimize the potential of bit errors due to signal distortion, nonlinear coding results in the constellation points most susceptible to noise being spaced further apart than those points less susceptible to noise. Thus, nonlinear coding is a technique that results in modem constellation points being spaced from one another at unequal distances. Through the use of nonlinear coding, the error rate of a V.34 modem operating at or near its highest data transfer rate be reduced by as much as 50%.

Precoding: Precoding represents another feature incorporated into V.34 modems. Through the use of precoding, which can be considered to represent a form of equalization, the amount of high-frequency noise on a telephone channel is reduced. This action decreases intersymbol interference, which in turn reduces potential transmission errors.

Compatibility issues

As previously discussed, until early 1996 it was expected that the ITU-T would issue a new standard for the support of data transmission rates beyond 28.8 kbps. Instead of issuing an expected V.34 bis standard the ITU-T modified the technical specifications of V.34 by adding support for two new operating rates – 31.2 kbps and 33.6 kbps. To accomplish this the newer V.34 modem uses either the 3200 Hz or 3429 Hz signaling rate previously listed in Table 8.20. By packing more bits per baud an operating rate of 31.2 kbps at a signaling rate of 3200 Hz and an operating rate of 33.6 kbps at a signaling rate of 3429 Hz become possible.

Although most of the original V.34 modems manufactured were capable of firmware updates, a significant percentage were not updated. This means that if you have the newer version of the V.34 modem, your ability to transmit and receive data at 33.6 kbps or 31.2 kbps will depend upon both the quality of the line connection and the version of the V.34 modem at the other end of the connection.

Another area of compatibility that requires consideration is the effect of a distant modem not being a V.34. Fortunately, a V.34 modem is downward compatible with V.32 bis and V.32 modems. Because both of those modems are compatible with the V.22 bis modem, this provides compatibility down to 2.4 kbps. In addition, since many modem vendors commonly add V.23, V.21, and Bell System 212A compatibility, a V.34 modem may be capable of supporting up to seven distinct modulation methods.

Prior to turning our attention to what will probably be the last switched network modem standard, a brief discussion of Universal Asynchronous Receiver Transmitter (UART) use is warranted. Thus, let's turn our attention to this important topic.

UART considerations

Because all modern modems now support V.42 bis data compression, the effective use of this capability requires the transfer rate from an attached PC or another terminal device connected to a modem to exceed the modem's operating rate. This means that when modem compression is enabled and the modem operates at 28.8 kbps, the connection from the DTE to the modem should be set to 115.2 kbps if the average compression ratio is 4:1. Unfortunately, the UART built into the serial port of almost all non-Pentium computers is either a National Semiconductor 8250 or 16450 chip or an equivalent produced by a different manufacturer.

The 8250 UART represents the first PC generation of this chip. This UART was used in all IBM PC, PC XT, and compatible computers during the early to mid-1980s and incorporates a one-character buffer, limiting its ability to transmit and receive data to 19.2 kbps prior to data loss occurring. Thus, older PCs cannot support the operating rate of modems beyond 19.2 Kbps unless their UART is replaced.

The 16450 UART can be considered to represent a second-generation PC UART. This UART was originally used in the IBM PC AT and compatible computers as well as most Intel 80386 and 80486 systems. This UART begins to lose data at rates above 57.6 kbps, limiting its use to V.32 bis modems when you want to use the compression capability of the modem.

The ability to support data transfer beyond 57.6 kbps requires the use of the more modern 16450 UART. This UART includes the ability to buffer a group of characters, allowing it to support a higher data transfer rate. Because just about all Pentium, Pentium II, and Pentium III-based computers use the 16450 UART, you should be able to successfully use any modern high-speed modem with those computers without having to replace their UART.

56 000 bps

The V.34 modem, when operating at 33.6 kbps, represents the maximum data transfe rate achievable on the PSTN based upon Shannon's Law. Thus, when the revision to the V.34 modem occurred which added a 33.6 kbps operating rate, it was expected that this would be the highest operating rate achievable on the PSTN.

Under Shannon's Law the signal-to-noise ratio governs the data transfer rate. By decreasing the level of noise it would then become possible to transmit at a higher data rate. While it is also possible to transmit at a higher data rate by using more bandwidth, the passband of a voice channel is fixed. Thus, engineers turned their attention towards reducing the level of noise.

When analog data is transmitted over the PSTN, there are normally two analog/digital conversions. One conversion occurs at the central office that serves one modem subscriber. A second conversion occurs at the distant central office serving the destination modem subscriber. The backbone of the PSTN, which is an all-digital transmission facility, requires the two conversions as the local loops are used to transmit analog data.

The difference between the height of the analog sample and its digital representation is referred to as quantization noise. When a digital signal is reconverted into an analog signal to be used by the distant modem quantization noise limits the ability of the modem to reconstruct the original modulated signal. Thus, quantization noise places a limit on the degree of signal changes that can flow end-to-end and limits the theoretical data transfer capability of a modem to 33.6 kbps.

If one analog to digital conversion can be eliminated the level of quantization noise would be reduced. This fact was recognized by modem designers as a mechanism to increase the data transfer rate of modems beyond 33.6 kbps and resulted in the development of the V.90 specification.

V.90 modem

The ITU-T V.90 modem represents a standardized approach to 56 kbps transmission pioneered by competing technologies from Rockwell International (K56flex) and U.S. Robotics (X2), the latter now part of 3Com.

Under the V.90 specification, developed during 1998, the modem operates as a V.34 when there are two analog/digital conversions. When one end of the connection terminates in a digital circuit, such as a channelized T1 or ISDN circuit, it becomes possible to support a 56 kbps operation. Developed during 1998, the modem operates as a V.34 when there are two analog/digital conversions. When one end of the connection terminates in a digital circuit, such as a channelized T1 or ISDN circuit, it becomes possible to support a theoretical 56 kbps data transfer rate towards the call originator. The term 'theoretical' is used because due to FCC regulations transmission from the digital side is limited with respect to signal power, which reduces transmission to approximately 53 kbps.

Table 8.21 Modem operational characteristics

Modem type	Maximum data rate	Transmission technique	Modulation technique	Transmission mode	Primary line use
Bell System					
103A, E	300	asynchronous	FSK	Half, Full	Switched
103F	300	asynchronous	FSK	Half, Full	Leased
201B	2400	synchronous	PSK	Half, Full	Leased
201C	2400	synchronous	PSK	Half, Full	Switched
202C	1200	asynchronous	FSK	Half	Switched
202S	1200	asynchronous	FSK	Half	Switched
202D/R	1800	asynchronous	FSK	Half	Leased
202T	1800	asynchronous	FSK	Half, Full	Leased
208A	4800	synchronous	PSK	Half, Full	Leased
208B	4800	synchronous	PSK	Half	Switched
209A	9600	synchronous	QAM	Full	Leased
212	0–300	asynchronous	FSK	Half, Full	Switched
	1200	asynchronous/ synchronous	PSK	Half, Full	Switched
CCITT					
V.21	300	asynchronous	FSK	Half, Full	Switched
V.22	600	asynchronous	PSK	Half, Full	Switched/ Leased
	1200	asynchronous/ synchronous	PSK	Half, Full	Switched/ Leased
V.22 bis	2400	asynchronous	QAM	Half, Full	Switched
V.23	600	asynchronous/ synchronous	FSK	Half, Full	Switched
	1200	asynchronous/ synchronous	FSK	Half, Full	Switched
V.26	2400	synchronous	PSK	Half, Full	Leased
	1200	synchronous	PSK	Half	Switched
V.26 bis	2400	synchronous	PSK	Half	Switched
V.26 ter	2400	synchronous	PSK	Half, Full	Switched
V.27	4800	synchronous	PSK	Half, Full	Switched/ Leased
V.29	9600	synchronous	QAM	Half, Full	Leased
V.32	9600	synchronous	TCM/QAM	Half, Full	Switched
V.32 bis	14400	synchronous	TCM	Half, Full	Switched
V.33	14400	synchronous	TCM	Half, Full	Leased
V.34	33600	synchronous	TCM	Half, Full	Switched
V.90	56000	synchronous	TCM	Half, Full	Switched

Because it is common for information utilities and Internet Service Providers to directly connect their network access controllers to a digital circuit, most uses for a V.90 modem will be for downloading information at a speed approaching 56 kbps. However, because transmission to the access controller requires two conversions, the data rate in the opposite direction will be limited to 33.6 kbps.

A summary of the operational characteristics of Bell System and ITU-T V series type modems is listed in Table 8.21.

8.3 INTELLIGENT MODEMS

Due to the popularity of the Hayes Microcomputer Products series of Smartmodems™, the command sets of those modems are the key to what the terms intelligent modems and 'Hayes compatibility' means. Since just about all modern personal computer communications software programs are written to operate with the Hayes command set, the degree of Hayes compatibility that a non-Hayes modem supports will affect the communications software that can be used with that modem. In some cases, non-Hayes modems will work as well as or even better than a Hayes modem if the software supports the non-Hayes features of that device. In other cases, the omission of one or more Hayes Smartmodem features may require the personal computer user to reconfigure his or her communications software to work with a non-Hayes modem, usually resulting in the loss of a degree of functionality.

In addition to the commands that an intelligent modem is manufactured to understand, there are several features and functions that are commonly associated with those modems. Those features and functions include error detection and correction, flow control and data compression, topics that will also be covered in this chapter.

8.3.1 Hayes command set modems

The Hayes command set actually consists of a basic set of commands and command extensions. The basic commands, such as placing the modem off-hook, dialing a number and performing similar operations are common to all Hayes modems. The command extensions, such as placing a modem into a specific operating speed, are only applicable to modems built to transmit and receive data at that speed.

The commands in the Hayes command set are initiated by transmitting an attention code to the modem, followed by the appropriate command or set of commands that one desires the modem to implement. The attention code is the character sequence AT, which must be specified as all uppercase or all lowercase letters. The requirement to prefix all command lines with the code AT has resulted in many modem manufacturers denoting their modems as Hayes AT compatible.

The command buffer in a Hayes Smartmodem holds 40 characters, permitting a sequence of commands to be transmitted to the modem on one command line. This 40-character limit does not include the attention code, nor does it include spaces included in a command line to make the line more readable. Table 8.22 lists the major commands included in the basic Hayes command set.

The basic format required to transmit commands to a Hayes compatible intelligent modem is

AT Command[Parameter(s)]Command[Parameter(s)]..Return

Each command line includes the prefix AT, followed by the appropriate command and the command's parameters. The command parameters are

Table 8.22 Hayes command set

Major commands	
Command	Description
A	Answer call
A/	Repeat last command
B	Select the method of modem modulation
C	Turn modem's carrier on or off
D	Dial a telephone number
E	Enable or inhibit echo of characters to the screen
F	Switch between half- and full-duplex modem opration
H	Hang up telephone (on-hook) or pick up telephone (off-hook)
I	Request identification code or request check sum
L	Select the speaker volume
M	Turn speaker off or on
N	Negotiate handshake options
O	Place modem on-line
P	Pulse dial
Q	Request modem to send or inhibit sending of result code
R	Change modem mode to 'originate-only'
S	Set modem register values
T	Touch-tone dial
V	Send result codes as digits or words
W	Negotiation progress message selection
X	Use basic or extended result code set
Z	Reset the modem
+++	Escape command

usually the digits 0 or 1, which serve to define a specific command state. As an example, H0 is the command that tells the modem to hang up or disconnect a call, and H1 is the command that results in the modem going off-hook, which is the term used to define the action that occurs when the telephone handset is lifted. Since many commands do not have parameters, those terms are enclosed in brackets to illustrate that they are optional. A number of commands can be included in one command line, so long as the number of characters does not exceed 40, which is the size of the modem's command buffer. Finally, each command line must be terminated by a carriage return character.

To illustrate the utilization of the Hayes command set let us assume that we desire to automatically dial New York City information. First, we must tell the modem to go off-hook, which is similar to one manually picking up the telephone handset. Then we must tell the modem the type of telephone system we are using, pulse or touch-tone, and the telephone number to dial. If we have a terminal or personal computer connected to a Hayes compatible modem, we would thus send the following commands to the modem:

```
AT H1
AT DT1,212-555-1212
```

In the first command, the 1 parameter used with the H command places the modem off-hook. In the second command, DT tells the modem to dial (D) a telephone number using touch-tone (T) dialing. The digit 1 was included in the telephone number because it was assumed that we have to dial long-distance, while the comma between the long-distance access number (1) and the area code (212) causes the modem to pause for 2 s prior to dialing the area code. This 2 s pause is usually of sufficient duration to permit the long-distance dial tone to be received prior to dialing the area code number.

Since a Smartmodem automatically goes off-hook when dialing a number, the first command line is not actually required and is normally used for receiving calls. In the second command line, the type of dialing does not have to be specified if a previous call was made, since the modem will then use the last type specified. Although users with only pulse dialing availability must specify P in the dialing command when using a Hayes Smartmodem, several vendors now offer modems that can automatically determine the type of dialing facility that the modem is connected to, and then use the appropriate dialing method without requiring the user to specify the type of dialing. For other non-Hayes modems, when the method of dialing is unspecified, such modems will automatically attempt to perform a touch-tone dial and, if unsuccessful, then redial using pulse dialing.

To obtain an appreciation of the versatility of operations that the Hayes command set provides, assume that two personal computers users are communicating with one another. If the users wish to switch from modem to voice operations without hanging up or redialing, one user would send a message via the communications program he or she is using to the other user indicating that voice communications is desired. Then, both users would lift their telephone handsets and type + + + (Return) ATH(Return) to switch from on-line operations to command mode (hand-up). This will cause the modems to hang-up, turning off the modem carrier signals and permitting the users to converse.

Result codes

The response of the Smartmodem to commands is known as result codes. The Q command with a parameter of 1 is used to enable result codes to be sent from the modem in response to the execution of command lines whereas a parameter of 0 inhibits the modem from responding to the execution of each command line.

If the result codes are enabled, the V command can be used to determine the format of the result codes. When the V command is used with a parameter of 0, the result codes will be transmitted as digits, while the use of a parameter of 1 will cause the modem to transmit the result codes as words. Table 8.23 lists the Basic Result Codes set of the Hayes Smartmodem 1200.

The introduction of higher speed modems that operate at different data rates while performing different methods of error detection and correction and data compression resulted in a significant increase in the number of result codes used to define the response of a modem to a command. Today many Hayes and Hayes-compatible modems support over 100 result codes

Table 8.23 Smartmodem 1200 basic result codes code set

Digit word	Word code	Meaning
0	OK	Command line executed without errors
1	CONNECT	Carrier detected
2	RING	Ring signal detected
3	NO CARRIER	Carrier signal lost or never heard
4	ERROR	Error detected in the command line

which provide informative information about the operation of a communications session or its reason for termination. As an example of the use of results codes, let us assume that the following commands were sent to a modem:

AT Q0
AT V1

The first command, ATQ0, would cause the modem to respond to commands by transmitting result codes after a command line is executed. The second command, ATV1, would cause the modem to transmit each result code as a word code. Returning to Table 8.23, this would cause the modem to generate the word code 'CONNECT' when a carrier signal is detected. If the command ATV0 were to be sent to the modem, a result code of 1 would be transmitted by the modem, since the 0 parameter would cause the modem to transmit result codes as digits.

By combining an examination of the result codes issued by a Smartmodem with the generation of appropriate commands, software can be developed to perform such operations as redialing a previously dialed telephone number to resume transmission in the event a communications session is interrupted, and automatically answering incoming calls when a ring signal is detected.

Modem registers

A third key to the degree of compatibility between non-Hayes and Hayes Smartmodems is the number, use and programmability of registers contained in the modem. Hayes Smartmodems contain a series of programmable registers that govern the function of the modem and the operation of some of the commands in the modem's command set. Table 8.24 lists the functions of the first 12 registers built into the Hayes Smartmodem 1200, to include the default value of each register and the range of settings permitted. These registers are known as S registers, since they are set with the S command in the Hayes command set. In addition, the current value of each register can be read under program control, permitting software developers to market communications programs that permit the user to easily modify the default values of the modem's S registers.

Although many Hayes and Hayes compatible modems can have in excess of 12 S registers whose functionalities significantly differ from one another, the

Table 8.24 S register control parameters

Register	Function	Default value	Range
S0	Ring to answer on		0–255
S1	Counts number of rings	0	0–255
S2	Escape code character	ASCII 43	ASCII 0–127
S3	Carriage return character	ASCII 13	ASCII 0–127
S4	Line feed character	ASCII 10	ASCII 0–127
S5	Backspace character	ASCII 8	ASCII 0–127
S6	Dial tone wait time (s)	2	2–255
S7	Carrier wait time (s)	30	1–255
S8	Pause time caused by comma (s)	2	0–255
S9	Carrier detect response time (1/10 s)	6	1–255
S10	Time delay between loss of carrier and hang-up (1/10 s)	7	1–255
S11	Touch-tone duration and spacing time (ms)	70	50–255

first 12 S registers normally provide an identical level of functionality. Thus, S register compatibility is normally limited to the initial group of registers listed in Table 8.24 regardless of the operating rate of the modem.

To understand the utility of the ability to read and reset the values of the modem's S registers, consider the time period that a Smartmodem waits for a dial tone prior to going off-hook and dialing a telephone number. Since the dial tone wait time is controlled by the S6 register, a program offering the user the ability to change this wait time might first read and display the setting of this register during the program's initialization. The reading of the S6 register would be accomplished by the program sending the following command to the modem:

AT S6?

The modem's response to this command would be a value between 2 and 255, indicating the time period in seconds that the modem will wait for a dial tone. Assuming that the user desires to change the waiting period, the communications program would then transmit the following command to the modem, where n would be a value between 2 and 255:

AT S6 = n

One of the key modem register settings that commonly causes problems in an era of high-speed modems commonly attempting to communicate with older modems is the value of the S7 register. That register defines the carrier wait time and has a default value of 30 seconds. When a high-speed modem attempts to communicate with a lower operating rate device, it first transmits a carrier tone at its current operating rate for 15 seconds, waiting for a response. If no response is received, the modem transmits a second carrier tone at a different frequency that represents a different modem operating rate. This second tone is also transmitted until recognized or for 15 seconds, whichever comes first. If the destination modem operates at a different data rate, by now 30 seconds have expired and the modem will terminate the connection. Thus, if you need more than two modem speeds to cycle through to reach the operating rate of a distant modem, both modems will need a higher value for their S7 register.

Extended AT Commands

Advances in the development of modem technology resulted in an increased level of functionality associated with this category of communications equipment. In an effort to control this increased functionality, Hayes Microcomputer Products extended the basic command set through the use of the ampersand (&) as a prefix to letters that might otherwise be considered to represent basic commands.

Table 8.25 lists the extended AT Command Set supported by the Hayes V-Series Smartmodem 2400. Many of those extended commands, such as the &S, &T, &W, &X and &Z commands are supported across all modern Hayes-manufactured modems, whereas other command, such as &M commands, may only be applicable to certain modems. Although many modem manufacturers incorporate extended AT commands into their product, other vendors support many extended modem features through the use of a large number of S registers. Due to this, the support of extended AT commands is normally different between vendor product lines.

Compatibility

For a non-Hayes modem to be considered to be compatible to a Hayes modem, command-set compatibility, result-codes compatibility and modem-register compatibility is required.

In considering command set compatibility, that compatibility is normally applicable to basic AT commands. Since many modem manufacturers use S registers to control advanced modem features in place of extended commands, S register compatibility is typically limited to the first 12 registers. Thus, an examination of modem compatibility is focused on a subset of modem operations, and you should check your communications program to ensure that the advanced features of a modem you wish to use are supported. Otherwise, you can consider using the manual setup feature of most communications programs in conjunction with your modem manual to configure software to initiate the modem features you desire.

Benefits of utilization

With appropriate software, intelligent modems can be employed in a variety of ways which may provide the potential to reduce the cost of communications as well as to increase the efficiency of the user's data-processing operations. To obtain an understanding of the benefits that may be derived from the use of intelligent modems, let us assume that your organization has a number of

Table 8.25 Extended AT command set (Hayes V-Series Smartmodem 2400)

Command	Description
&C0	Assume data carrier always present
&C1	Track presence of data carrier
&D0	Ignore DTR signal
&D1	Assume command state when an ON to OFF transition of DTR occurs
&D2	Hang up and assume command state when an ON-to-OFF transition of DTR occurs
&D3	Reset when an ON-to-OFF transition of DTR occurs
&D4	Reset and enter lower power mode when DTR is low
&G0	No guard tone
&G1	550-Hz guard tone
&G2	1800-Hz guard tone
&J0	RJ-11/RJ-41S/RJ-45S telco jack
&J1	RJ-12/RJ-13 telco jack
&M0	Asynchronous mode
&M1	Synchronous mode 1
&M2	Synchronous mode 2
&M3	Synchronous mode 3
&R0	Track CTD according to RTS
&R1	Ignore RTS; always assume presence of CTS
&S0	Assume presence of DSR signal
&S1	Track presence of DSR signal
&T0	Terminate test in progress
&T1	Initiate local analog loopback
&T3	Initiate local digital loopback
&T4	Grant request from remote modem for RDL
&T5	Deny request from remote modem for RDL
&T6	Initiate remote digital loopback
&T7	Initiate remote digital loopback with self-test
&T8	Initiate local analog loopback with self-test
&W0	Save storable parameters of active configuration as profile 0
&W1	Save storable parameters of active configuration as profile 1
&X0	Modem provides transmit clock signal
&X1	Data terminal provides transmit clock signal
&X2	Receive carrier provides transmit clock signal
&Z	Store telephone number

sales offices geographically dispersed throughout many States. Let us further assume that each sales office uses a personal computer to process orders, which are then mailed to company headquarters for fulfilment.

Due to postal delivery time or other factors, the delay between receiving an order at a sales office and its transmittal to company headquarters may be unacceptable. Since the order processing delay is making some customers unhappy, while other customers citing faster competitor delivery time have been canceling or reducing their orders, management is looking for a way to expedite orders at a minimum cost to the organization.

Although a person in each sales office could be delighted to call a computer system at corporate headquarters at the end of each day and use a mainframe program to enter orders, this activity would operate at the speed of the person entering the data, communications costs would be high since the session would occur during the day and last-minute orders might not get processed until the next day, unless the person performing the data entry activity agreed to stay late. Since each sales office is assumed to have a personal computer, another method you may wish to consider is the utilization of personal computers to expedite the transmittal of orders between the sales offices and the company headquarters.

Since the personal computer in each sales office is already used to process orders and prepare a report that is mailed to headquarters, one only has to arrange for the transmission of the order file each day, since the program that produces the report would only have to be sent to company headquarters once, unless the program was revised at a later date. Then, after a personal computer at company headquarters has received the order file, it would use that file as input to its copy of the order processing program, permitting the report to be produced at company headquarters.

Due to the desire to automate the ordering process at a minimal cost, it might be advisable to perform communications after 7 p.m. when dates for the use of the switched telephone network are usually at their lowest. Since it would defeat the purpose of communications economy to have an operator at each sales office late in the evening, a communications program that provides unattended operation capability would be probably be obtained. This type of communications program would require the use of an intelligent modem at the company headquarters location as well as at each sales office location. Then the communications software program operating on a personal computer at company headquarters could be programmed through the use of macrocommands or menu settings to automatically dial each sales office computer at a predefined time after 7 p.m., request the transmission of the order file and then disconnect after that file had been received. The communications program would then dial the next sales office, repeating the file transfer procedure. At each sales office, a similar unattended communications program would be operating in the personal computer at that office. On receiving a call, the intelligent modem connected to the personal computer would inform the computer that a call had been received and the program would then answer the call, receive the request to transfer the order file, transmit that file to the distant computer and then hang up the telephone. To prevent anyone from dialing each sales office, most unattended communications programs permit password access to be implemented, enabling the user to assign appropriate passwords to the call program that will enable access to the files on the called personal computer.

8.3.2 Key intelligent modem features

The inclusion of a microprocessor in a modem revolutionized modem technology, enabling many previously desired features to be added to a smart or intelligent modem. In addition to responding to commands, the use of a microprocessor enabled a modem to perform error detection and correction, use control characters for flow control, and even implement distinct modem protocols to support different communications functions. In this section we will turn our attention to obtaining an overview of error detection and correction and flow control. In subsequent sections in this chapter, we will examine a popular modem protocol as well as focus our attention upon one of the most popular functions incorporated into a modem, data compression.

Error detection and correction

Error detection and correction, also known as error control and error correction, has been implemented in a large number of switched network modems using a variety of techniques. Each technique follows a common methodology.

Data transmitted by an attached terminal device formally known as data terminal equipment (DTE) are first gathered into a block of characters.

An algorithm is applied to the block to generate one or more checksum characters that are appended to the block for transmission.

The receiving modem performs the same algorithm on the block it receives, less its checksum character or characters. The checksum computed by the receiving modem is referred to as the locally generated checksum, and it is compared to the transmitted checksum. If the locally generated and transmitted checksums are equal, the data block is assumed to have been received error-free. Otherwise, the data block is assumed to have one or more bits in error, and the receiving modem then requests the transmitting modem to retransmit the data block.

Error correction is accomplished by retransmission, so all modems that perform error detection and correction must have buffers in which they can store data temporarily until an acknowledgement from the receiving modem occurs. At that point, the transmitting modem can discard the stored data block that was received error-free by the modem at the opposite end of the transmission path. Because modem data buffers are finite, a mechanism is required to control the flow of data from the attached terminal device to the modem. This mechanism is known as flow control. Because all modems that perform error detection and correction must perform flow control, we will examine this subject area before investigating the key topic of this section.

Flow control

Flow control compensates for the difference between the rate at which data reaches a device and the rate at which the device processes and transmits data. For an illustration of the rationale behind flow control, see Figure 8.31. Assume that the modem in the figure has a modulation rate of 2400 baud and packs four bits into each signal change, in effect, transferring data at 9600 bps to the public switched telephone network (PSTN). If the compression ratio of the modem is 2 : 1, on average every two characters of data entering the modem are compressed into one character. Thus, the terminal device can be configured to transfer data to the modem at 19 200 bps, which is twice the data rate at which the modem transfers data onto the PSTN.

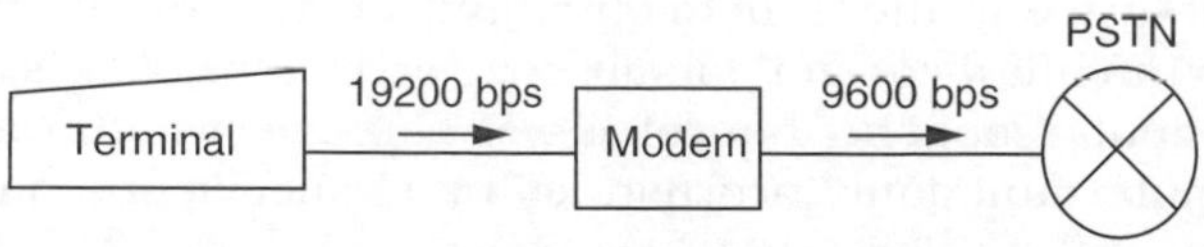

Figure 8.31 Rationale for flow control. When the modem achieves a 2 : 1 compression ratio, data entering the device at 19 200 bps are immediately placed on the line at 9600 bps. When the compression ratio is under 2 : 1 data cannot exit the modem as fast as they enter the modem, and are placed into a buffer area. To prevent the buffer from overflowing, the modem requires a mechanism to stop the terminal from sending additional data, a process known as flow control

So long as the modem can compress two characters into one, the terminal device can continue to transfer data to the modem at 19 200 bps. However, suppose that a portion of the data entering the modem cannot be compressed or is compressible at a ratio less than 2:1. Either situation would cause data flowing into the modem to be lost. To prevent a potential data loss, modems that compress data or perform other functions, including error detection and correction, will include data buffers. The buffers act as temporary storage locations to compensate for the difference between the data flow into the modem and the rate at which the mode can process and transmit data.

Because data buffers represent a finite amount of storage, a modem that accepts data at one rate and transfers them at a different rate must be able to control the flow of data into its buffers. Flow control, therefore, is the technique that prevents modem buffers from overflowing and losing data.

To illustrate how modem flow control operates, consider Figure 8.32, which illustrates a modem's buffer storage area. As data enter the modem from an attached terminal device at a faster rate than they can be placed on the line, the buffer begins to fill. When the occupancy of the buffer reaches a predefined

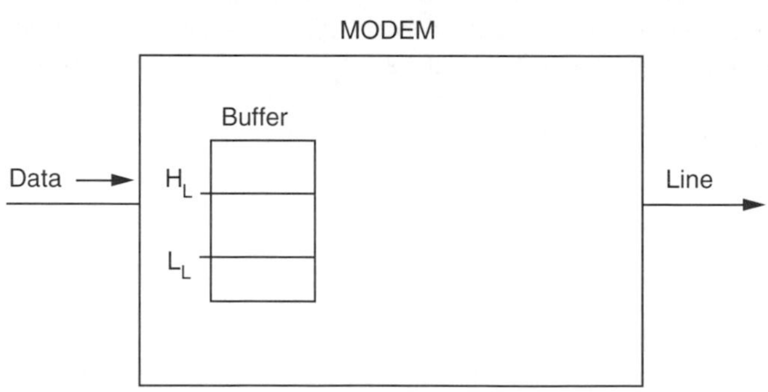

Legend
H_L = high level to disable data flow

L_L = low level to enable data flow

Figure 8.32 Flow control and buffer utilization

high level (H_L), the modem initiates flow control to inhibit additional data from flowing into its buffer. Otherwise, the buffer could continue to fill until data eventually could not be stored and would be lost. As data flow out of the buffer and are modulated and placed on the line, the point of low occupancy is reached (L_L). When this level of occupancy is reached, the modem uses flow control to enable transmission to resume.

Methods of flow control

There are three primary methods by which flow control can be implemented, including the use of RTS/CTS control signals and the transmission of the character pairs XON/XOFF and ENQ/ACK. In addition, some devices support the use of a mixture of two flow control methods.

RTS/CTS signaling

When a modem receives an RTS (Request to Send) signal from a terminal device and is ready to receive data, it responds by raising its CTS (Clear to Send) signal. Thus, one method to control the flow of data into the modem is for the modem to turn off its CTS signal whenever it wants to stop the flow of data into its buffers. Then, whenever the modem wants the flow of data to resume, it raises or turns on its CTS signal. This method is referred to as hardware flow control because the controlling signals at the interface govern the operation of the hardware.

XON/XOFF

The use for XON and XOFF characters for flow control is referred to as in-band signaling, because the characters flow over pin 3 (receive data) to the attached terminal device. Since the XON and XOFF characters are generated by software, this method of flow control is also commonly referred to as software handshaking. In comparison, the CTS control signal that is not a data signal is referred to as out-band signaling.

Many asynchronous terminal devices recognize an ASCII code of 19, which represents the CTRL-S character, as a signal to suspend transmission. This XOFF signal is also known as a DC3 (device control number 3) character and is issued by the modem when it wants an attached DTE to suspend the flow of data to the modem. Once the modem has emptied its buffer to a low level, the modem transmits an XON character to the DTE, which serves as a signal for the DTE to resume transmission. The XON character has an ASCII code of 17 and it represents the CTRL-Q character. This is also known as the DC1 (device control number 1) character.

ENQ/ACK

The enquire/acknowledge method of flow control is used with certain Hewlett-Packard computers and terminal devices. In this method of flow control, the

DTE sends an enquire (ENQ) message and then receives an acknowledgement (ACK) before actually transmitting data. Under the Hewlett Packard ENQ/ACK protocol, the DTE that receives an ACK in response to its ENQ transmits a block of data of variable size that can be up to approximately 2000 characters long.

When a modem is configured to use ENQ/ACK for control, it responds to the DTE's ENQ with an ACK when it can accept data. At that time it releases the ENQ to flow to the remote modem.

DTE flow control

The previously described methods of flow control are also applicable in the reverse direction, with DTEs controlling the flow of data.

DTEs are similar to modems that contain buffers, in that when they operate communications software, a finite area of storage is reserved for receiving data to be processed. If the received data are transferred to a peripheral device (such as a printer operating slower than the communications line) data could be lost. This is because the modem would be passing demodulated data to the DTE faster than it could empty its buffer area, eventually resulting in a loss of data when the buffer becomes full. Like modems, DTEs capable of performing flow control may be able to support three methods to accomplish this task: RTS/CTS signaling and the use of the character pairs XON/XOFF and ENQ/ACK.

Under RTS/CTS signaling, the DTE drops its RTS signal as an indication that it wants to suspend receiving data. In a full-duplex operation, the modem stops sending data to the DTE when RTS drops, and resumes transmission when the RTS signal is raised. Under half-duplex operations, the modem transmits the contents of its buffer to the DTE and drops its CTS signal in response to the DTE lowering its RTS signal.

When XON/XOFF flow control is used by the DTE, the local modem suspends sending data to that device when it receives an XOFF. On receipt of an XON, the modem resumes sending data to the DTE.

As previously mentioned, some Hewlett-Packard computers and terminals support the ENQ/ACK protocol. When both modems are configured to use the ENQ/ACK protocol, the receipt of an ENQ from a remote DTE is passed through both modems to the local DTE. Until the local DTE responds with an ACK, the local modem will not send data to that device.

Now that we have examined the methods by which modems perform flow control, let us focus on methods of detecting and correcting errors.

Methods of error detection and correction

Until 1989, there was a conspicuous absence of *de jure* standards for error detection and correction. Several error detection and correction techniques were developed based on the use of cyclic or polynomial code error detection schemes. Under each scheme a data block is treated as a data polynomial

D(X), which is divided by a predefined generating polynomial G(X), resulting in a quotient polynomial Q(X) and a remainder polynomial R(X), such that:

$$D(X)/G(X) = Q(X) + R(X)$$

The remainder of the division process is known as the cyclic redundancy check (CRC) and is normally 16 bits long, or two 8-bit bytes. The CRC is appended to the block of data to be transmitted. The receiving modem uses the same predefined generating polynomial to generate its own CRC based on the received block and then compares the locally generated CRC with the transmitted CRC. If the two match, the receiving modem transmits a positive acknowledgement to the transmitting modem, which not only informs the distant modem that the data was received correctly but can also inform the remote modem to send the next of any additional blocks of data that remain to be transmitted. If an error has occurred, the locally generated CRC will not match the transmitted CRC, and the receiving modem transmits a negative acknowledgement that informs the remote modem to retransmit the previously transmitted data block.

Although most modem manufacturers used the CRC-16 polynomial, $X^{16} + X^{15} + X^5 + 1$, which has the bit composition 11000000000010001, to operate against each data block, incompatibilities between the methods used to block data and transmit negative and positive acknowledgments made the error detection and correction method employed by one vendor incompatible with the method used by another. The one major exception to this incompatibility among vendor error detection and correction methods is the Microcom Networking Protocol (MNP), which has been licensed by Microcom to a large number of modem manufacturers. Until 1989, MNP was considered as a *de facto* standard due to a base of approximately 1 million modems supporting one or more MNP classes. In 1989 the MNP method of error detection and correction was recognized by the ITU-T as one of two methods for performing this function when the V.42 recommendation was promulgated.

Rationale

One of the issues that confuse many modem users is the rationale for using a modem's error detection and correction feature. After all, file transfer protocols, such as XMODEM, YMODEM, ZMODEM, and their derivatives, also provide an error detection and correction capability.

When a modem's error control feature is enabled and operates successfully in conjunction with a distant modem, error detection and correction is operating during the entire communications session. This means that regardless of the function you are performing, whether reading an electronic mail message, transferring a file, or sending a message to SYSOP, your transmission is protected. In comparison, the error detection and correction function embedded into a file transfer protocol operates only during the file transfer. Thus, a logical question you may have is why you should use a file transfer protocol with error detection and correction when your modem performs that function.

8.3.3 Microcom Networking Protocol (MNP)

The Microcom Networking Protocol (MNP) was developed by the modem manufacturer Microcom, Inc., to provide a sophisticated level of error detection and correction as well as to enhance the data file transfer of intelligent modems. Microcom has licensed their MNP for use by other modem vendors, resulting in a large number of manufacturers incorporating this protocol into their products.

The MNP protocol was designed in a layered fashion like the OSI Reference Model developed by the International Standards Organization. MNP contains three layers instead of the seven layers in the OSI Reference Model. Figure 8.33 illustrates the correspondence between the OSI Reference Model and the MNP protocol.

The MNP link layer is responsible for establishing a connection between two devices. Included in the link layer is a set of negotiations that are conducted between devices to enable them to agree upon such factors as the transmission mode (full- or half-duplex), how many data messages can be transmitted prior to requiring a confirmation and how much data can be contained in a single message. After these values have been established, the link layer initiates the data transfer process as well as performing error detection and correction through the use of a frame checking scheme.

Figure 8.34 illustrates the format of an MNP frame of information which has similarities to both bisynchronous and HDLC communications. Each frame contains three bytes which act as a 'start flag'. The SYN character tells the receiver that a message is about to arrive, the combination of data link escape (DLE) and start of text (STX) informs the receiver that everything following is part of the message. The first header describes the user data, such as the duplex setting, number of data messages before confirmation, etc. The session

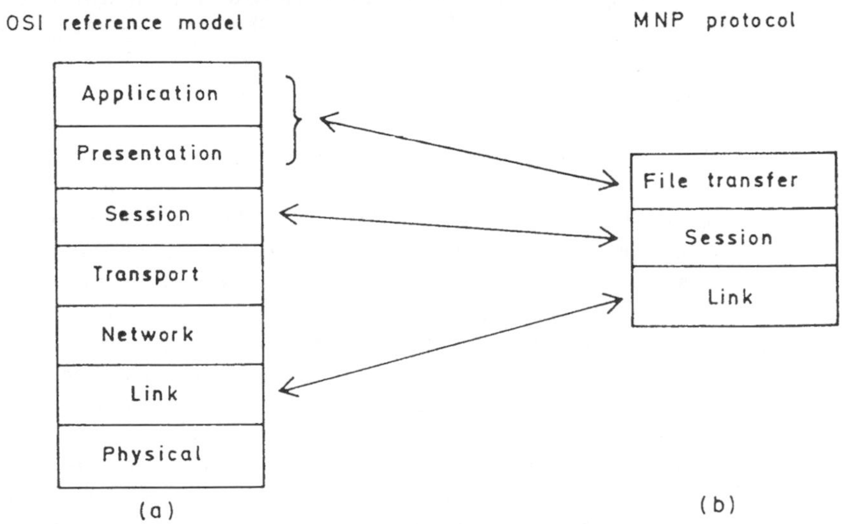

Figure 8.33 OSI reference model and MNP protocol

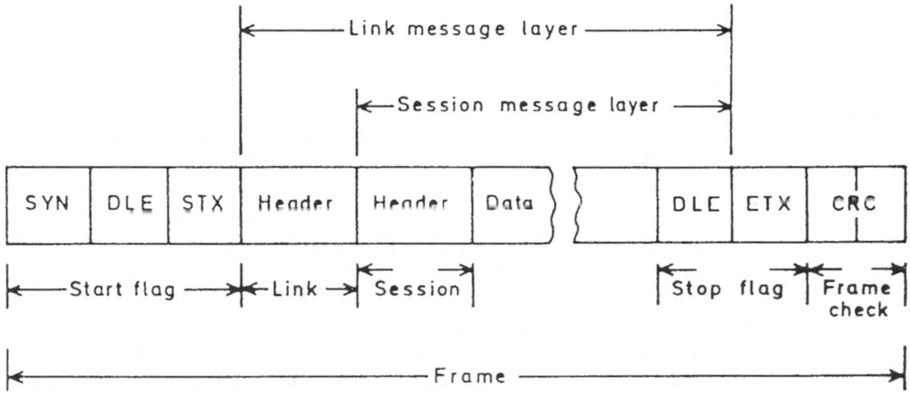

Figure 8.34 MNP block format

header defines additional information about the transmitted data which enables the automatic negotiation of the level of service that can be used between devices communicating with one another. Currently there are ten versions or classes of the MNP protocol, with each higher level adding more sophistication and efficiency. When an MNP link is established the protocol assumes that the devices on both sides can only operate at the lowest level. Then, the devices negotiate with each other to determine the highest mutually supported class of MNP services they can support. If a non-MNP device is encountered the MNP device reverts to a 'dumb' operating mode, providing an MNP modem with the ability to be used with non-MNP devices. Table 8.26 describes the ten MNP protocol classes.

The error correction capability of MNP actually occurs under is Class 4 operation. Thus, modems that are MNP error control-compatible will be advertised as MNP Class 4 compliant. Under Class 4 the actual framing of data depends on whether the data are asynchronous or synchronous. If the attached terminal device transmits asynchronous data, the MNP frame format is as indicated in Figure 8.35(a). If the attached terminal device transmits synchronous data, the frame format is as indicated in Figure 8.35(b). For both frame formats, the frame check sequence (FCS) characters are generated by the previously discussed CRC-16 polynomial.

LAP-M

The MNP error detection and correction method is one of two methods recognized under the ITU-T V.42 recommendation. In actuality, the MNP method was recognized as an alternative procedure under V.42. The primary method was known as Link Access Protocol-Modem or LAP-M.

Under V.42, the originating mode transmits an Originator Detection Pattern (ODP). The ODP is defined as the bit sequence

01000 10001 11...11 01000 10011 11...11

Table 8.26 MNP protocol classes

Protocol Class	Description
Class 1	The lowest performance level. Uses an asynchronous byte-oriented half-duplex method of exchanging data. The protocol efficiency of a Class 1 implementation is about 70% (a 2400 bps modem using MNP Class 1 will have a 1690 bps throughput).
Class 2	Uses asynchronous byte-oriented full-duplex data exchange. The protocol efficiency of a Class 2 modem is about 84% (a 2400 bps modem will realize a 2000 bps throughput).
Class 3	Uses synchronous bit-oriented full-duplex data exchange. This approach is more efficient than the asynchronous byte-oriented approach, which takes 10 bits to represent 8 data bits because of the 'start' and 'stop' framing bits. The synchronous data format eliminates the need for start and stop bits. Users still send data asynchronously to a Class 3 modem but the modems communicate with each other synchronously. The protocol efficiency of a Class 3 implementation is about 108% (a 2400 bps modem will actually run at a 2600 bps throughput).
Class 4	Adds two techniques: Adaptive Packet Assembly and Data Phase Optimization. In the former technique, if the data channel is relatively error-free, MNP assembles larger data packets to increase throughput. If the data channel is introducing many errors, then MNP assembles smaller data packets for transmission. Although smaller data packets increase protocol overhead, they concurrently decrease the throughput penalty of data retransmissions, so more data are successfully transmitted on the first try. Data Phase Optimization eliminates some of the administrative information in the data packets, which further reduces protocol overhead. The protocol efficiency of a Class 4 implementation is about 120% (a 2400 bps modem will effectively yield a throughput of 2900 bps).
Class 5	This class adds data compression, which uses a real-time adaptive algorithm to compress data. The real-time capabilities of the algorithm allow the data compression to operate on interactive terminal data as well as on file transfer data. The adaptive nature of the algorithm allows it to analyze user data continuously and adjust the compression parameters to maximize data throughput. The effectiveness of the data compression algorithm depends on the data pattern being processed. Most data patterns will benefit from data compression, with performance advantages typically ranging from 1.3 to 1.0 and 2.0 to 1.0, although some files may be compressed at an even higher ratio. Based on a 1.6 to 1 compression ratio, Microcom gives Class 5 MNP a 200% protocol efficiency, or 4800 bps throughput in a 2400 bps modem installation.
Class 6	This class adds 9600 bps V.29 modulation, universal line negotiation, and statistical duplexing to MNP Class 5 features. Universal link negotiation allows two unlike MNP Class 6 modems to find the highest operating speed (between 300 and 9600 bps) at which both can operate. The modems begin to talk at a common lower speed and automatically negotiate the use of progressively higher speeds. Statistical duplexing is a technique for simulating full-duplex service over half-duplex, high-speed carriers. Once the modem link has been established using full-duplex V.22 modulation, user data streams move via the carrier's faster half-duplex mode. However, the modems monitor the data streams and allocate each modem's use of the line to best approximate a full-duplex exchange. Microcom claims that a 9600 bps V.29 modem using MNP Class 6 (and Class 5 data compression) can achive 19.2 kbps throughput over dial circuits.
Class 7	Uses an advanced form of Huffman encoding called Enhanced Data Compression. Enhanced Data Compression has all the characteristics of Class 5 compression, but in addition predicts the probability of repetitive characters in the data stream. Class 7 compression, on the average, reduces data by 42%.
Class 8	Adds CCITT V.29 Fast-Train modem technology to Class 7 Enhanced Data Compression, enabling half-duplex devices to emulate full-duplex transmission.

Table 8.26 (*continued*)

Protocol Class	Description
Class 9	Combines CCITT V.32 modem modulation technology with Class 7 Enhanced Data Compression, resulting in a full-duplex throughput that can exceed that obtainable with a V.32 modem by 300%. Class 9 also employs selective retransmission, in which errors packets are retransmitted, and piggybacking, in which acknowledgment information is added to the data.
Class 10	Adds Adverse Channel Enhancement (ACE), which optimizes modem performance in environments with poor or varying line conditions, such as cellular communications, rural telephone service, and some international connections. Adverse Channel Enhancements fall into five categories: Negotiated Speed Upshift: modem handshake begins at the lowest possible modulation speed, and when line conditions permit, the modem upshifts to the highest possible speed. Robust Auto-Reliable Mode: enables MNP10 modems to establish a reliable connection during noisy call set-ups by making multiple attempts to overcome circuit interference. In comparison, other MNP classes make only one call set-up attempt. Dynamic Speed Shift: causes an MNP10 modem to adjust its operating rate continuously throughout a session in response to current line conditions. Aggressive Adaptive Packet Assembly: results in packet sizes varying from 8 to 256 bytes in length. Small data packets are used during the establishment of a link, and there is an aggressive increase in the size of packets as conditions permit. Dynamic Transmit Level Adjustment (DTLA): designed for cellular operations, DTLA results in the sampling of the modem's ransmit level and its automatic adjustment to optimize data throughput.

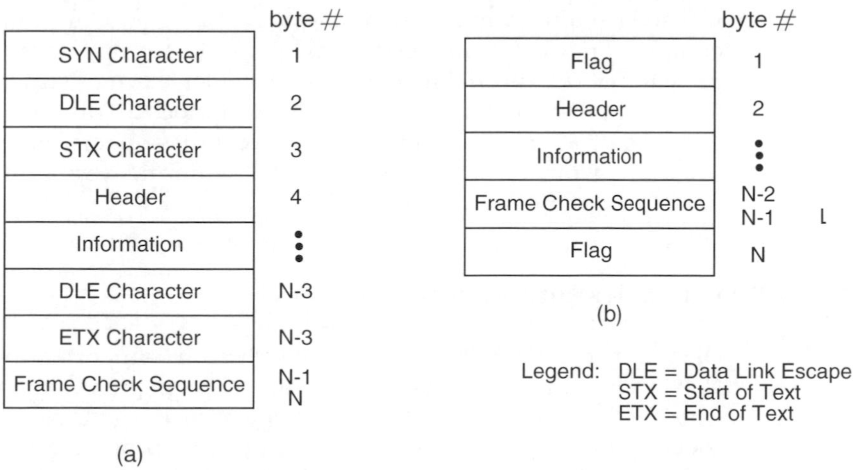

Figure 8.35 MNP frame format: (a) asynchronous data frame format; (b) synchronous data frame format

which represents the DC1 character with even parity, followed by 8 to 16 ones, followed by DC1 with odd parity, followed by 8 to 16 ones. A V.42-compatible modem responds to the ODP with an Answer Detection Pattern (ADP), whose bit format indicates that V.42 is supported or that no error-correcting protocol is desired. If the ADP is not observed within a predefined period of time (V.42 modems use a default value of 750 milliseconds), it is assumed that the distant modem does not possess V.42 error-correcting capability. In this situation, the originating modem may fall back to a non-error-correcting mode of operation, or if it incorporates MNP it can then attempt to negotiate an MNP error detection and correction mode of operation.

The LAP-M protocol uses a different frame structure from MNP. In addition, the frame check sequence characters are generated by a different polynomial. Thus, MNP and LAP-M are completely incompatible with each other.

8.3.4 Data compression

Similar to their steps to incorporate error detection and correction, modem manufacturers have included a variety of data compression algorithms in their products. Until 1989 most compression methods represented a proprietary scheme of one vendor, but were licensed by that vendor to other manufacturers. This resulted in several *de facto* data compression standards being used for modems manufactured by various vendors. Two popular *de facto* data compression methods are Microcom's MNP Class 5 and Class 7 compression procedures.

In 1990 the ITU-T promulgated the V.42 bis recommendation, which defines a new data compression method known as Lempel-Ziv as an international standard. Unlike V.42, which concerns error detection and correction and specifies MNP Class 4 as an alternative, V.42 bis does not specify an alternative method of data compression.

Today all high-speed modems support data compression, although the method used to compress data may vary. Since special coding is employed to compress data, a transmission error would result in the incorrect decompression of transferred information. A modem cannot support data compression without utilizing an error control protocol. This means that both the error control protocol and the data compression method supported by an originating modem must be supported by an answering modem to transfer data in a compressed mode.

8.3.5 MNP Class 5 compression

Of the two methods of data compression supported by the Microcom Networking Protocol, Class 5 is the more popular, although it predates Class 7. Because each MNP Class is downward negotiable for compatibility, a modem that supports MNP Class 7 can communicate with a modem that is MNP Class 5-compatible using MNP Class 5 data compression.

MNP Class 5 specifies that the sending modem apply two modifications to the transmitted data stream in an attempt to reduce the number of bits actually sent. The first manipulation or data compression method is run-length encoding and the second method of compression is adaptive frequency encoding.

MNP Class 5 uses run-length encoding to avoid sending long sequences of repeated data octets. Each octet represents 8 bits that can define a character in a particular character set or any binary value from 0 to 255. This value is represented by the individual bit settings within the octet. Under the MNP Class 5 version of run-length encoding, a repetition count is inserted into the data stream to represent the number of repeated data octets that follow the first three occurrences of a sequence. The first 3 repeated data octets that are actually sent signal the beginning of a run-length encoded sequence. The next octet is always a repetition count that has a maximum value of 250. If the repeated sequence is only 3 octets in length, a repetition count of 0 is used. Thus, 4 octets consisting of 3 repeated data octets and a count octet are used to compress any repeating sequence from 3 to 250 octets. Figure 8.36 illustrates the format of MNP Class 5 run-length encoding and gives a few examples of its operation.

MNP Class 5 run-length encoding can be considered as the first level of a two-level data compression scheme. The second level of compression, adaptive frequency encoding, is applied to the data stream after any repeated data octets are removed by the use of run-length encoding.

In adaptive frequency encoding, a compression token is substituted for the actually occurring data octet in an attempt to transmit fewer than 8 bits for each data octet. The token used changes with the frequency of occurrence of the actual data octet, so that shorter tokens are substituted for more frequently occurring data octets.

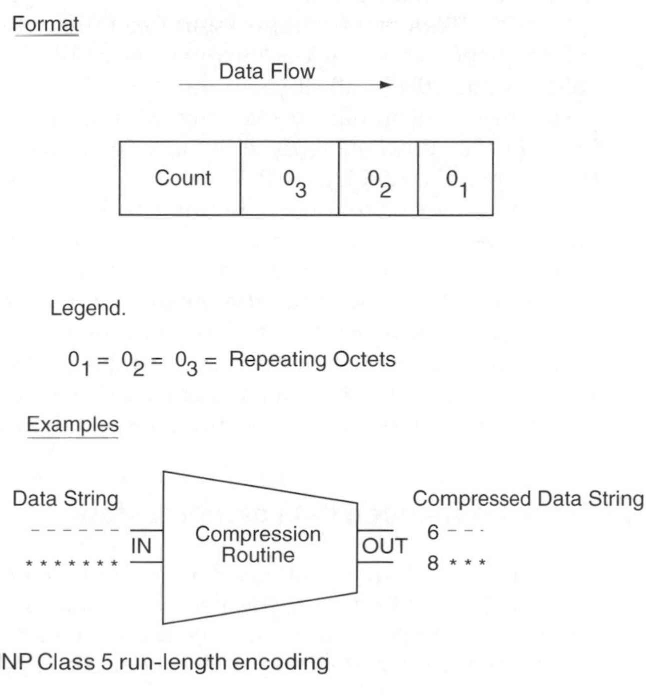

Figure 8.36 MNP Class 5 run-length encoding

The compression token used to represent a data octet is composed of two parts: a fixed-length header and a variable-length body. The header is 3 bits long and, in general, indicates the length of the body portion of the token. There are three special cases, however. There are two tokens with a header of 0; in these two cases, the true length of the body is 1. When the header indicates a length of 7 and the body is seven 1-bits, then the actual length is 8.

At the initiation of data compression, the relative frequency of occurrence of each data octet is 0. For purposes of octet/token mapping, however, data octet 0 (00000000 binary) is assumed to be the most frequently occurring octet and is represented by the first of the shortest tokens; data octet 1 (00000001 binary) is taken to be the next most frequently occurring and is represented by the next token. This continues to data octet 255 (11111111 binary), initially assumed to be the most infrequently occurring data octet. Data octet/token mapping at compression initialization is shown in Table 8.27.

For the encoding of a data octet, the token to which it is currently mapped is substituted for the actual data octet in the data stream. After this substitution, the frequency of occurrence of the current data octet is increased by one. If the frequency of this data octet is greater after incrementing than the frequency of the next most frequently occurring data octet, then the compression tokens of the current data octet and the next most frequently occurring data octet are exchanged. The frequency of the current data octet is then compared to the frequency of the data octet that is not the next most frequently occurring data octet. If the frequency of the current data octet is greater, then the compressed tokens are once again swapped. This cycle continues until no more swaps are needed, at which time the mapping of data octets and compression tokens is correctly adapted based on the relative frequency of the data octets.

Once the data octet/compression token mappings are sorted by frequency, the frequency count of the current character is compared to the fixed limit value of 255 (decimal). If this limit has been reached, then the frequency of occurrence of each data octet is scaled downward by dividing each frequency by 2. Integer division is used, so any remainder after division is discarded. (For example, 3 divided by $2 = 1$.)

As previously noted, the repetition count, the fourth octet in a run-length encoded sequence of repeated data octets is also mapped to a compression token. The token used is the one mapped to the count of the most frequently occurring data octet. Thus, the token used for a count of 5 would be 01001. Note also that this count octet does not increase the frequency of occurrence of the data octet to which the token is mapped. Further, the repetitions of the run-length encoded data octet represented by the count do not contribute to the frequency of occurrence of the repeated data octet.

8.3.6 MNP Class 7 enhanced data compression

MNP Class 7 enhanced data compression builds on the concept of combining run-length encoding with the use of an adaptive encoding table. The table contains a single column listing each character ordered by frequency of occurrence. Under MNP Class 7, run-length encoding is combined with the

Table 8.27 MNP Class octet/token mapping at compression initialization

Data octet (decimal value)	Header (MSB LSB)	Body (MSB LSB)
0	000	0
1	000	1
2	001	0
3	001	1
4	010	00
5	010	01
6	010	10
7	010	11
8	011	000
9	011	001
10	011	010
11	011	011
12	011	100
13	011	101
14	011	110
15	011	111
16	100	0000
17	100	0001
18	100	0010
19	100	0011
20	100	0100
21	100	0101
22	100	0110
23	100	0111
24	100	1000
25	100	1001
26	100	1010
27	100	1011
28	100	1100
29	100	1101
30	100	1110
31	100	1111
32	101	00000
33	101	00001
34	101	00010
[35–246 token header/body continues in same pattern]		
247	111	1110111
248	111	1111000
249	111	1111001
250	111	1111010
251	111	1111011
252	111	1111100
253	111	1111101
254	111	1111110
255	111	11111110

use of a first-order Markov model. This model is used to predict the probability of the occurrence of a character based on the value of the previous character. An adaptive table of 256-character columns represents the ordered frequency of occurrence of each succeeding character.

Markov model

To encode a character, the compressor selects a code that depends on the immediately preceding encoded character. The selected code is based on the frequency with which a character follows the previous character. For example, the probability of a U following a Q is very high; generally the U will be encoded as 1 bit. Likewise, an H following a C has a different probability than an R following a C, and will be coded according to its frequency of occurrence.

As noted in Table 8.28, the compressor keeps up to 256 coding tables, one for each possible 8-bit character (or pattern). To code a character for transmission, it uses the previous character to select the appropriate coding table. For example, when an A is transmitted, the model looks under the A pointer for the next character, which is ordered according to its frequency of occurrence. If a C is the next character, it is compressed based on its location in the table. The model looks next under the C pointer to find the following character, and so on. Each table contains the codes for characters following the previous character and is organized according to the rules of Huffman coding.

Huffman coding

In Table 8.28 each column of characters under the pointer character is compressed according to the rules of Huffman coding. Huffman coding changes the number of bits representing a character when the character's frequency of occurrence changes sufficiently. Huffman can adapt to various alphabets (for example, ASCII, EBCDIC, and all uppercase) and languages (natural language, compiler code, and spreadsheets) without being preinformed of the data used.

Unlike Class 5, Huffman coding can represent a character with only one bit, if it occurs often enough. In general, if one character occurs twice as often as another, its code is half as long.

MNP Class 7 is adaptive, meaning that it changes the coding of the data when the frequency of character occurrence changes. The compressor starts

Table 8.28 MNP Class 7 enhanced data compression

Pointers	A	B	C	D	E	
Characters coded according to their frequency of following the previous character, that is, the pointer	T H C M B	L E U . .	H O R . .	O A E . .	D R S N P	Up to maximum of 256 characters each

off with no assumptions about the data coding, and even learns that it is ASCII/English based on the data itself. The compression tables are empty at the start of each connection and are built as data are passed. When the data divide naturally into characters and the 'working set' of frequent characters is not too large (natural language is good here), then adaptation creates a coding structure that compresses the data well.

Run-length encoding

Multiple consecutive copies of the same character (or 8-bit pattern) are compressed for transmission using run-length encoding. If the encoder has sent the same character three consecutive times, the encoder sends the count of the remaining identical consecutive characters as a single 4-bit nibble. For example, a series of 'A's would be sent as AAA with the remaining number of 'A's sent as a 4-bit nibble. For example, a series of five 'A's would be sent as three Huffman encodings of 'A' and a nibble of binary 2.

Decoding

To decode an encoded data stream, the receiving modem assumes that it has an exact copy of the sending modem's compression data structures because: both modems reset their structures at the start of each connection; communication has been error-free (MNP ensures this); and decoding the data stream causes identical data structure updates to match the encoder's.

Because the receiving modem has the same compression tables as the sender, it knows which set of Huffman data decodes the incoming data stream. The receiving modem compares the incoming bits against the Huffman codes in the table until there is a complete match. At this time, the character is delivered to the DTE, and the table is updated.

8.3.7 V.42 bis

V.42 bis data compression is a modified version of the Lempel-Ziv method of compression that was developed approximately 20 years prior to the promulgation of the recommendation. The V.42 bis compression method uses an algorithm in which a string of information received from an attached DTE is encoded as a variable-length code word. To facilitate the development of code words, strings are stored in dictionaries at the encoding and decoding device and are dynamically updated to reflect changes in the composition of data.

The key to the V.42 bis compression process is its dictionary, which is dynamically built and modified. Think of the dictionary as representing a set of trees in which each root corresponds to a character in the alphabet, as illustrated in Figure 8.37. Each tree represents a set of known strings beginning with one specific character, and each node or point in the tree represents one set of strings. Thus, the trees in Figure 8.37 represents the strings A, B, BA, BAG, BAR, BAT, BI, BIN, C, D, DE, DO, and DOG.

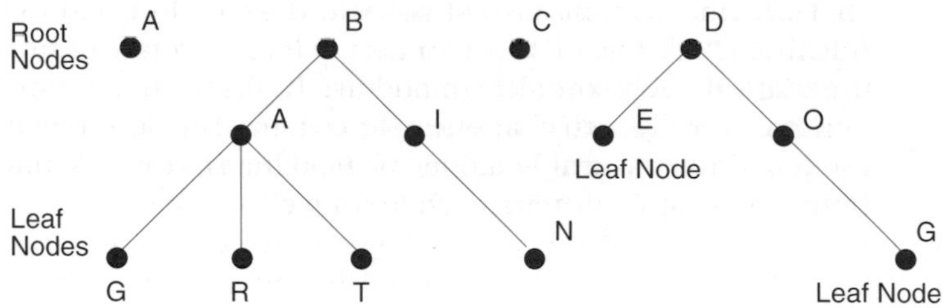

Figure 8.37 V.42 bis tree based dictionary representation

Each of the leaf nodes shown in Figure 8.37 represents a node that has no other dependent nodes, in effect representing the last character in a string. Conversely, a node that has no parent is known as a root node and it represents the first character in a string.

Initially, each tree in the dictionary consists of a root node, with a unique code word assigned to each node. As data are received from an attached DTE, a string-matching process occurs in which a sequence of DTE-originated characters is matched against the dictionary. This process begins with a single character. If the string matches a dictionary entry and the entry has not been created since the last invocation of the string-matching procedure, the next DTE-originated character is read and appended to the string. This process is repeated under the previously described conditions until the maximum string length is reached, the string does not match a dictionary entry, or the string matches an entry created since the last invocation of the string-matching procedure. At that time, the last character appended to the string is removed and it represents an unmatched character; the characters in the string are encoded as a code word.

Under V.42 bis the maximum string length can range from 6 to 250 and is negotiated between modems. The number of code words has a default value of 512, which is its minimum value. However, a maximum value is not specified, and any value above the default value can be negotiated between two modems.

During the compression process, the dictionary is dynamically modified by the addition of new strings based on the composition of the data. The new strings are formed by appending a single character that was not matched from a string-matching operation to an existing string, which results in the addition of a new node to a tree. As a result of the replacement of strings by code words, V.42 bis data compression is approximately 20 to 30% more efficient than MNP Class 5 compression and probably 5 to 10% more efficient than MNP Class 7 enhanced data compression. Although the increased efficiency of V.42 bis resulted in its widespread adoption by modem manufacturers, the large installed base of MNP modems resulted in most vendors incorporating both compression methods into their products.

8.4 BROADBAND MODEMS

In this section we turn our attention to two types of modems that provide a high data transmission operating rate approximately a thousand times or more that provided by the fastest modem designed for use on the public switched telephone network. Earlier in this section I referred to such devices as cable and digital subscriber line modems. That reference, which is commonly used, denotes the primary application or intended application of each type of modem. That is, a cable modem is designed to provide data transmission capability on cable TV systems, while a digital subscriber line modem is designed to support high-speed transmission via the subscriber line that connects homes and office to the public switched telephone network. Since the term 'broadband' is commonly used to represent a data transmission operating rate at or above 1 Mbps, which most cable and DSL modems support, I have taken author liberties and classified them collectively as broadband modems.

Both cable and DSL modems provide the potential to revolutionize communications technology as well as substantially alter the manner by which we commonly perform daily communications-related activities at work and at home. Through their use you can surf the Internet's World Wide Web via the twisted-pair telephone line or a coaxial TV cable at operating rates a thousand or more times greater than are obtainable through the use of analog voice-grade modems, order video-on-demand movies via either type of media, and use either transmission system to communicate by voice, picture phone or perhaps even a technology requiring megabyte operating rates that will be developed tomorrow.

Currently, the dropping of regulatory barriers as well as the availability and affordability of megabyte transmission technology is fueling a race between cable TV operators and local telephone companies as they develop and implement plans to enter each other's industry. Although a telecommunications reform bill was finally passed into law after years of debate, the actual implementation of various aspects of that law raises some questions concerning the degree of competition that will occur. However, it is with a degree of certainty that competition will occur through the use of cable and DSL modems that are the subject of this section. Thus, a good place to commence a discussion of broadband modems is by first reviewing the basic infrastructure of telephone and cable TV transmission, placing special emphasis on the methods used to connect subscribers to their backbone network via different local loop technologies.

8.4.1 Telephone and cable TV infrastructure

The cabling infrastructure developed for telephone and cable TV (CATV) operations represents two diverse but to a degree potentially merging technologies. Telephone company plant is based upon the routing of twisted-pair wire from a serving central office to individual subscribers to provide two-way communications. In comparison, cable TV was based upon the use of coaxial cable to provide a one-way transmission path for video distribution to subscribers.

Telephone

Although the routing of wire-pairs directly to individual subscribers was long ago replaced in many areas by the multiplexing of several subscriber wire-pairs from a common location near groups of homes or offices to a telephone company central office as an economy measure, multiplexing maintains a direct path from the subscriber to the central office by either time or frequency, depending upon the technique employed. In addition, that path is bidirectional, enabling two-way voice and data to be transported from one subscriber to another via the telephone company wiring infrastructure.

Figure 8.38 provides a general schematic that illustrates the present telephone company wiring infrastructure. Although almost all long-distance transmission is carried via fiber optic cable, almost all local loop wiring is currently carried via metallic twisted-pair. Since the present twisted-pair local loop is limited by filters to a bandwidth under 3300 Hz, this also limits the maximum bidirectional operating rate of voice-grade modems to approximately 33.6 kbps.

Based upon several field trials conducted during the mid-1990s telephone companies expand the use of fiber by installing so-called 'fiber to the

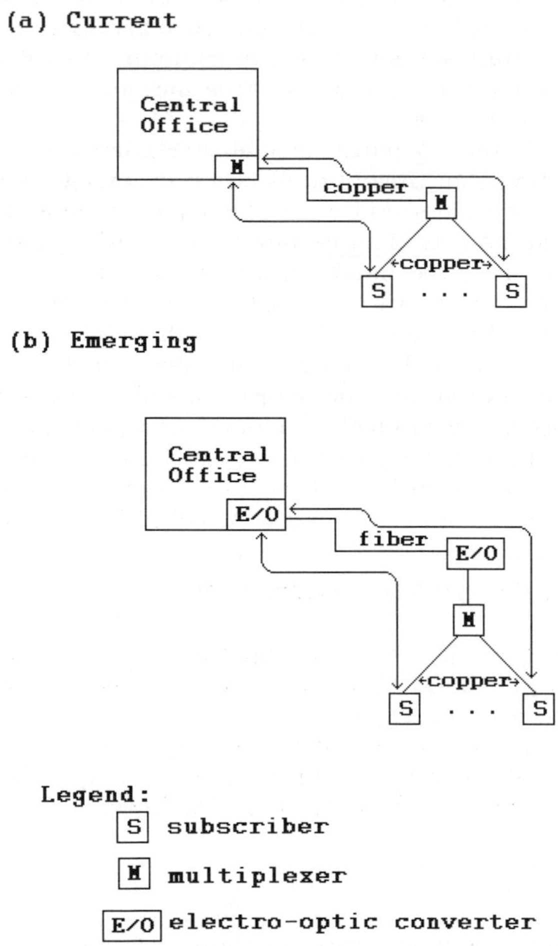

Figure 8.38 Current and emerging telephone company wiring infrastructure

neighborhood'. Under the 'fiber to the neighborhood' concept, fiber is routed from a central office to a common location in a housing subdivision, building complex, or similar site. Then, using electrical/optical converters the fiber trunk is connected to relatively short existing metallic twisted-pair wiring routed into homes and offices. Under this concept the high bandwidth of the fiber enables telephone companies to eliminate the filtering of twisted-pair local loops, which considerably expands the bandwidth available for use by modems. Thus, a new series of modems can be used on the local loop to provide a much higher operating rate than is obtainable by the use of conventional twisted-pair wiring. Figure 8.38 illustrates the emerging telephone company wiring infrastructure.

Due to the considerable time and expense associated with recabling the existing telephone local loop infrastructure, several communications carriers examined alternative methods to provide subscribers with high-speed transmission over existing local loop twisted-pair wire. One promising technique is known as asymmetric digital subscriber line (ADSL). This provides data rates up to 8 Mbps, which is sufficient for live television and video-on-demand applications. Although ADSL uses the existing twisted-pair local loop cable infrastructure, it requires special equipment, referred to as splitters, at the subscriber's premises and at the central office, whose operation will be described later in this section.

ADSL represents one of a family of DSL services and modems being offered to telephone company subscribers. A second DSL service oriented towards the mass market is referred to as G.lite, which can be considered to represent a lower operating rate splitterless version of ADSL. Later in this section we will examine both ADSL and G.lite.

Cable TV

The cable television infrastructure was originally developed to provide one-way video signals to subscribers. To do so, its wiring infrastructure was designed using a tree structure, with signals transmitted from a headend located at the beginning of the tree onto branches for distribution to subscribers. Signals on main branches were further split onto additional branches, with amplifiers installed to boost signal power since signal loss occurs due to cable distance as well as from the splitters encountered in the signal path.

The original bandwidth used on CATV systems ranged from 10 to 550 MHz, using 6 MHz per TV channel. Multiple signals are placed on a coaxial cable through the use of frequency division multiplexing equipment at the headend, with a tuner in subscriber set-top boxes used to extract the appropriate channel. This cabling and multiplexing scheme resulted in most CATV systems initially providing up to 83 channels to subscribers. The entire cable infrastructure of most pre-1995 CATV systems was based upon the use of coaxial cable as illustrated in Figure 8.39.

Similar to telephone field trials, the CATV industry has had its share of trials to test different types of cabling and the support of two-way transmission by several competing techniques. One technique involves the use of a second cable to provide a return path, while the second technique uses a different portion of the coaxial cable bandwidth to obtain a return path.

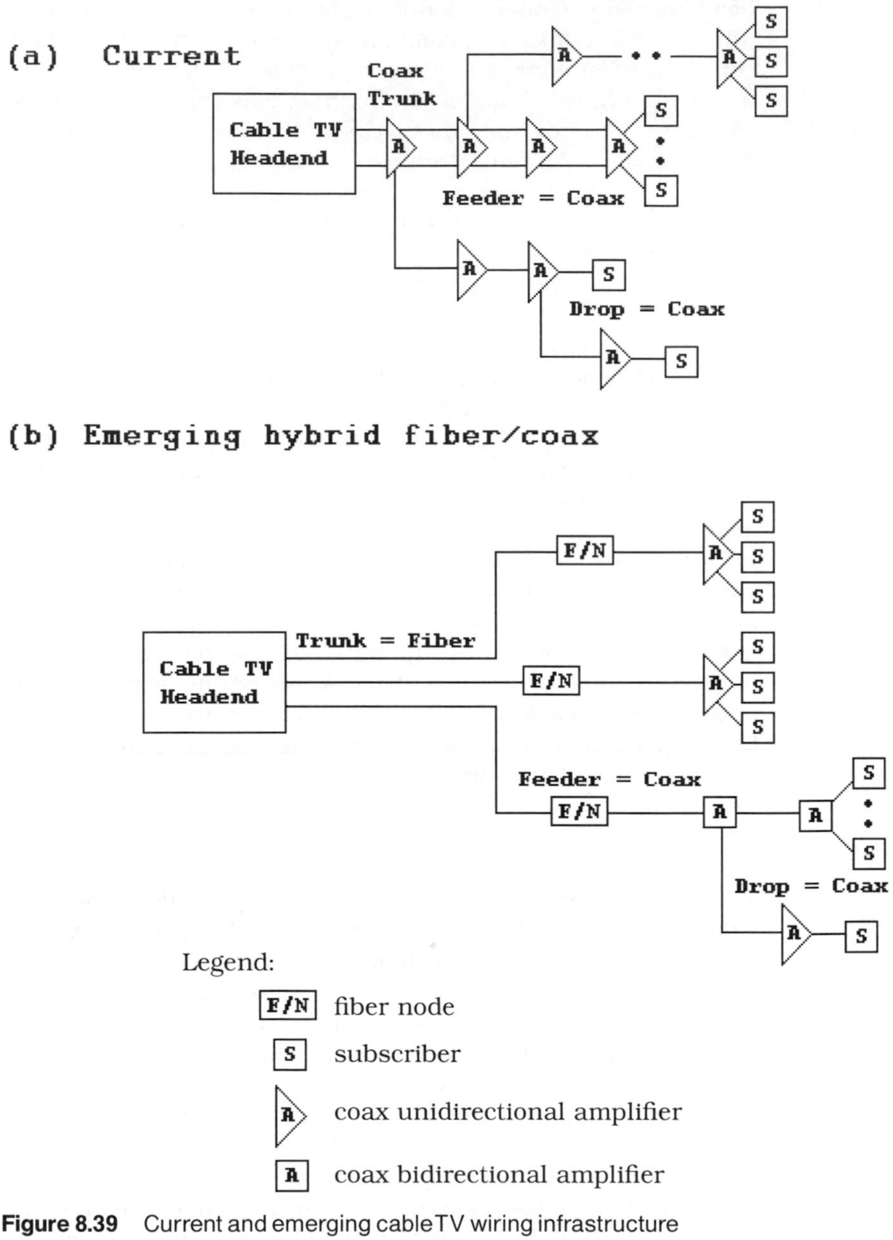

Figure 8.39 Current and emerging cable TV wiring infrastructure

Both techniques involve a considerable change to the existing CATV cabling infrastructure. For example, using a different portion of the frequency spectrum for the return path requires the installation of bidirectional amplifiers, while the use of a second cable requires both a new set of amplifiers and cabling.

Recognizing that the use of the CATV cabling infrastructure could not accommodate the requirements of a large subscriber base for video-on-

Table 8.29 CATV vs telephone operating characteristics

Operating characteristic	CATV	Telephone
Bandwidth use	6 MHz/TV channel	4 kHz/line
Current	Video	Voice and data
Evolving	Video, data, voice	Voice, data, video
Directionality		
Current	One-way	Twp-way
Evolving	Two-way	Two-way
Transmission method		
Current	Broadcast	Switched
Evolving	Broadcast/Switched	Switched
Media		
Current	Coaxial cable	Twisted-pair
Evolving	Fiber + coaxial cable	Fiber + twisted-pair

demand, high-speed data transmission, and even digitized telephone service, cable operators began to install hybrid fiber/coax (HFC) systems during the mid-1990s. Under HFC a star cabling infrastructure is used, with fiber cable routed from a cable TV switch to an optical distribution node commonly located in a subdivision or building complex. Then, coax is routed from the distribution node to individual subscribers. Under the HFC architecture, downstream frequencies of 50 MHz to 750 MHz can be subdivided to provide a variety of analog and video channels to include data services and telephony. Upstream operations are limited to the 5 MHz to 40 MHz frequency band for telephony, data services, and control channels used by interactive video set-top boxes.

Figure 8.39(b) provides a general schematic of the evolving CATV HFC cabling infrastructure. Note that the amplifiers shown in Figure 8.39(b) must be bidirectional to enable upstream transmission from subscribers to be supported. Table 8.29 provides a comparison of the current and evolving operating characteristics of CATV and telephone company transmission facilities. In examining the entries in Table 8.29 the evolving CATV transmission method requires a degree of elaboration. Currently one-way CATV systems broadcast frequency multiplexed signals. While the evolving HFC infrastructure will still use a broadcast transmission method for delivery of basic cable video services, it will also employ switching technology to provide routing and delivery of voice, data and video-on-demand in a manner similar to which the telephone company uses switching to establish a connection between a calling and called party for a voice or data communications session.

Now that we have an appreciation for the current and evolving cabling infrastructures used for CATV and telephone local loop operations, let's turn our attention to the modems that will provide high-speed transmission on each cabling system.

8.4.2 Cable modems

As an evolving technology there were several different engineering approaches used to develop cable modems. One approach is based upon the IEEE's

802.14 Working Group effort. This Working Group was formed in May 1994 under the name 'Standard Protocol for Cable-TV Based Broadband Communications' to develop standards for transmitting data over cable systems.

The original goal of the IEEE 802.14 Working Group was to submit a cable modem media access control (MAC) layer 2 and physical (PHY) layer 1 connection standard to the IEEE by December 1995. As the delivery of the standard slipped to late 1997, cable operators became impatient. In addition, while the IEEE 802.14 working group concentrated on engineering aspects of the standard, the cable operators were concerned about offering a viable, ready to use, economical product. Thus, the cable operators, including Comcast, Cox, TCI and Time Warner, formed a limited partnership known as Multimedia Cable Network System Partners Ltd (MCNS). MCNS released its Data Over Cable System Interface (DOCSI) specification for cable modem products to manufacturers in April 1997.

At the physical layer, both the IEEE 802.14 and DOCSI specifications define the same modulation methods. Both specify 64 QAM and 256 QAM to provide a degree of cable operator flexibility. A signaling rate of 5 MHz using a carrier frequency between 151 MHz and 749 MHz spaced 6 MHz apart is used for downstream transmission and corresponds to current TV channel assignments. At the MAC layer, the DOCSI specification uses the Internet Protocol (IP) via the assignment of time slots to enable subscribers within a neighborhood to contend for upstream and downstream access. In comparison, the IEEE 802.14 Working Group specified the use of Asynchronous Transfer Mode (ATM), a more expensive switching technology which is discussed later in this book. Over 20 cable modem manufacturers currently market products based upon the DOCSI specification. In comparison, I am not aware of any products currently being manufactured based upon the IEEE 802.14 standard.

To obtain an appreciation for the operation and utilization of cable modems and the cable system infrastructure, we will look at two products in the remainder of this section. First, we will examine the operation of the LANcity LCP, which was the world's fastest and least expensive cable TV modem when it was introduced in April 1995. We will then examine a more recent Scientific Atlanta product that was designed based upon the DOSCI specification which will also acquaint us with the modulation methods supported by that specification.

LANcity LC

One of the first modem manufacturers to provide a city-wide data communications capability via the development of a series of products to include a cable modem was LANcity of Andover, MA. Founded in 1990, by the summer of 1994 LANcity had introduced equipment to provide cable TV connectivity to the Internet. In April 1995, LANcity introduced the world's fastest and least expensive personal cable TV modem, named LCP (LANcity Personal Cable TV Modem). This modem can be set to operate in any 6 MHz TV channel in the transmit range from 5 to 42 MHz and the receive frequency range of 54 to 750 MHz and uses Quadrature Phase Shift Keying (QPSK) modulation to

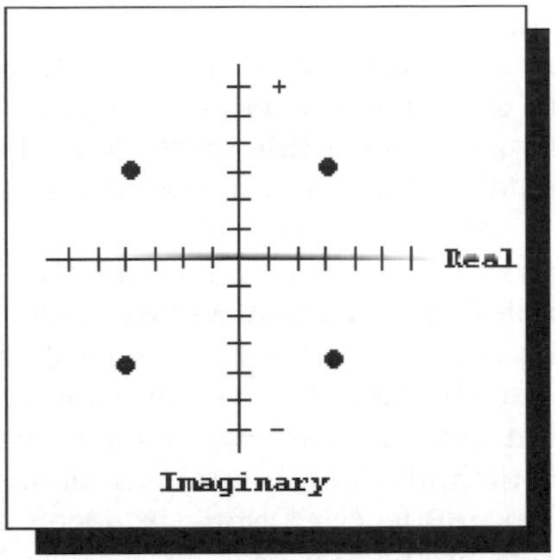

Figure 8.40 Four-point QPSK constellation

obtain a symmetrical 10 Mbps operating rate, requiring approximately 5 MHz of RF bandwidth in a 6 MHz channel.

Under QPSK the phase of the carrier signal is varied based upon the composition of the digital data to be transmitted. For example, a digital '1' could be transmitted by generating a 180° phase shift in the carrier, while a '0' might be represented by a 0° phase shift. The 'quadrature' aspect of the modulation scheme results in the carrier being capable of being shifted to one of four possible phases (0°, 90°, 180°, 270°) based upon the dibit value of the data to be transmitted. Figure 8.40 illustrates the four-point QPSK constellation pattern.

Although LANcity achieved a significant number of orders for its LCP modem, this modem predated the DOCSI specification, resulting in the vendor developing a new modem compliant with the specification. LANcity was acquired by Bay Networks, which in turn was later acquired by Nortel.

Access protocol

Based on the method by which CATV channels are routed on the main trunk and feeder cables through drop cables to the subscriber, bandwidth is shared among subscribers. In comparison, a local loop twisted-pair connection from the telephone company central office to a subscriber represents dedicated bandwidth. This means that access to the 6 MHz channel the cable modem uses must be shared among many subscribers, requiring an access protocol to govern the orderly flow of data onto the cable.

The primary method used to govern the flow of data from different subscribers onto shared upstream and downstream channels is via the use of time slots. Thus, a cable modem with data to transmit will wait for an appropriate time slot. Similarly, data received at the headend of a cable operator destined for a particular subscriber will be placed into an appropriate time slot on the downstream channel. Because the operating rate of a cable modem is beyond that supported by a PC's serial and parallel ports, a different interface is required to support a cable modem.

LANcity and most cable modem vendors support the Ethernet Carrier Sense Multiple Access/Collision Detection (CSMA/CD) protocol to provide a mechanism for cable subscribers to gain orderly access to the 6 MHz TV channel. To accomplish this the LANcity LCP cable modem is cabled to an Ethernet 10 Mbps adapter card installed in a personal computer. The PC uses software to establish a TCP/IP protocol stack, resulting in the computer becoming in effect a workstation on a LAN that can represent up to 200 miles of CATV cable infrastructure. While stand-alone cable modems primarily use an Ethernet 10 Mbps connection to a subscriber's PC, some vendors offer a Fast Ethernet 100 Mbps connection capability. Other cable modems that are fabricated on a PCI adapter card use the high-speed PCI bus for data transfer within the computer. In the future, as more PCs are manufactured with the Universal Serial Bus (USB), you can expect USB-compliant stand-alone cable modems to reach the market.

In addition to requiring an Ethernet adapter card in each PC that will be connected to the cable modem, the upgrade of CATV plant to support data transmission via cable modems requires several additional components. Those components, which are also applicable for DOSCI-compatible cable modems, include a frequency converter, configuration server and router. Figure 8.41 illustrates the use of those components on a CATV cabling infrastructure to provide subscribers with a bidirectional 10 Mbps operating capability using the LANcity LCP.

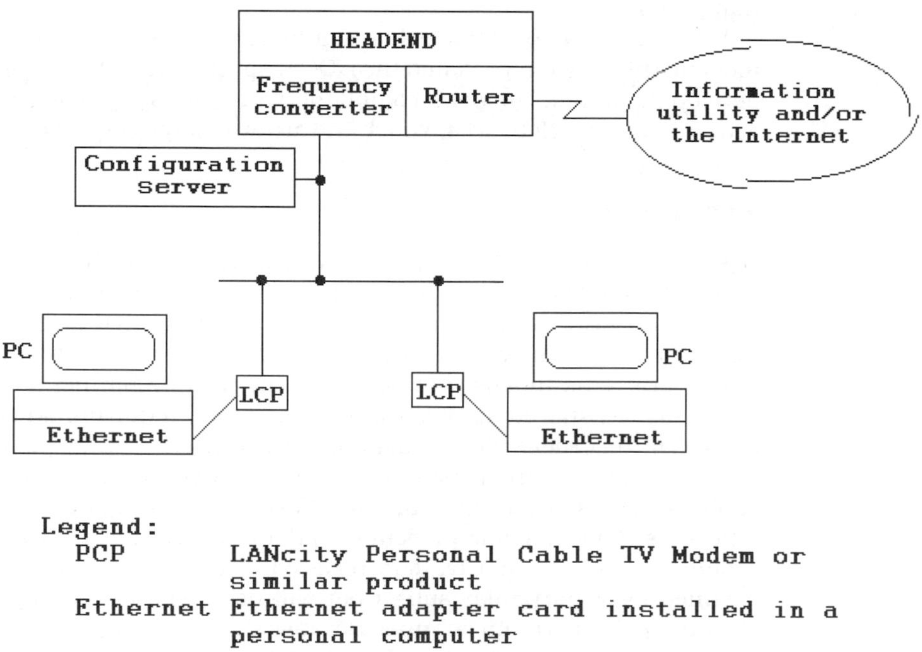

Figure 8.41 CATV plant components supporting data transmission

Frequency converter

The use of a frequency converter is required to change the reverse direction frequency to the forward direction frequency and vice versa. This conversion is required to enable one user to communicate with another as one channel provides a transmission path while a second channel is used for data reception. This conversion occurs at the headend and is limited to the 6 MHz channels used for data, ignoring other cable services. The LANcity product that performs frequency conversion is marketed as the LANcity TransMaster (LCT).

Configuration server

The configuration server supports the TCP/IP network formed on the CATV cable infrastructure. In doing so the server performs address resolution services as well as manages IP addresses for the client computers using cable modems.

Router

The function of the router is to encapsulate data into standardized frames for transport to external networks as well as to receive frames originating from outside the CATV LAN.

Scientific Atlanta

In concluding our examination of cable modems I will focus upon the asymmetric architecture of a Scientific Atlanta cable modem. The Scientific Atlanta cable modem we will examine is based upon an asymmetric design, using QAM in a 6 MHz downstream channel to obtain an operating rate of 27 MHz. In the opposite direction the modem uses QPSK modulation to provide an operating rate of 1.5 Mbps upstream. The modem supports downstream frequencies in the 54 to 750 MHz spectrum and frequencies in the 14 MHz to 26.5 MHz range for upstream communications.

The Scientific Atlanta cable modem's modulation method was proposed to the IEEE 802.14 Working Group and became the basis for use in both the IEEE standard and the DOCSI specification. Scientific Atlanta noted that QAM is non-proprietary and was previously selected as the European Telecommunications Standard. In the firm's proposal, two levels of modulation based upon 64 QAM and 256 QAM were defined to permit implementation flexibility. The standardization of QAM for downstream transmission results in a signaling rate of 5 MHz using a carrier frequency between 151 MHz and 749 MHz spaced 6 MHz apart to correspond to TV channel assignments.

The use of a 5 MHz signaling rate and 64 QAM which enables six bits to be encoded in one signal change permits a transmission rate of 6 bits/symbol × 5 MHz, or 30 Mbps. In comparison, the use of 256 QAM results in the packing of eight bits per signal change, resulting in a transmission rate of

8 bits/signal change × 5 MHz, or 40 Mbps. Through the use of forward error coding, the data rate throughput is slightly reduced from the modem's operating rate to 35.504 Mbps for 256 QAM and 27.37 Mbps for 64 QAM. This reduction results from extra parity bits becoming injected into the data stream to provide the forward error detection and correction capability.

Constellation patterns

Figure 8.42 illustrates the signal constellation pattern for 64 QAM modulation, while Figure 8.43 illustrates the pattern for one quadrant when 256 QAM modulation is used. Note that the signal element coding is differential quadrant coding using Gray coding within a quadrant. Under the Gray code the difference between two successive binary numbers is limited to one bit changing its state. Through the use of Gray Code encoding the most likely error during demodulation in which an incorrect adjacent code is selected will result in a one-bit error when decoded at the receiver. Table 8.30 contains the binary and Gray code equivalent for three-bit encoding.

For both 64 and 256 QAM element coding the outputs are encoded using a phase change based upon the inputs and previous inputs. Table 8.31 lists the proposed 64 and 256 QAM element coding.

Cable modem utilization

In comparing the use of cable modems to conventional analog modems designed for operation on the PSTN and to DSL devices, it is important to note

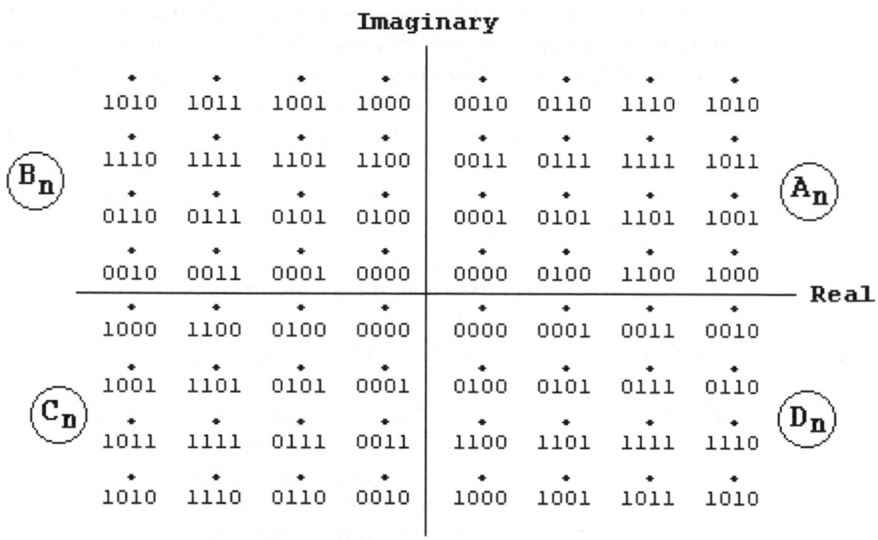

Figure 8.42 Signal constellation pattern for 64 QAM modulation. The binary numbers denote $b3_n$, $b2_n$, $b1_n$ and $b0_n$, and the letters A_n, B_n, C_n and D_n denote the four quadrants

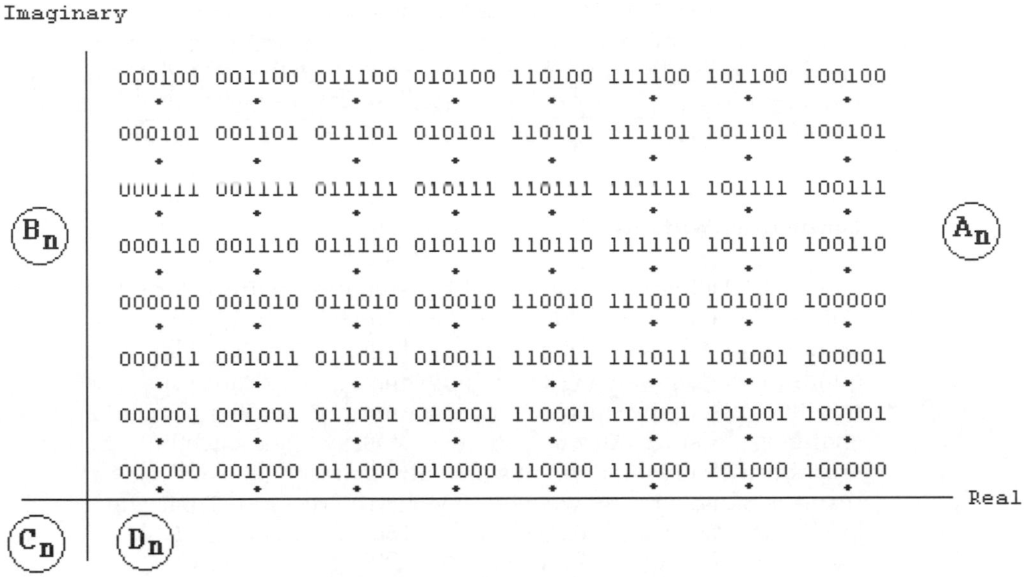

Figure 8.43 Constellation pattern for 256 QAM modulation (one quadrant illustrated)

that the former represents a shared media technology. While the DOCSI specification supports a 36 Mbps downstream channel operating rate, your actual data transfer depends both on the number of other cable subscribers downloading data, as well as the operating rate of the interface to your computer. Concerning the effect of other subscribers, assume the downstream channel is shared among 1000 homes and 10% of subscribers are downloading data. Then, the average downstream delivery rate you can expect would be 360 Kbps, which is not much more than from a V.90 modem operating with data compression enabled. Even for a best-case scenario in the unlikely event of no other subscribers downloading data, your data delivery rate will be limited by the connection from your cable modem to your PC. If you are using an Ethernet 10 Mbps connection, then the maximum data transfer rate will be limited to 10 Mbps. While the use of a cable modem can provide a significant

Table 8.30 Binary and Gray code equivalence

Decimal	Binary	Gray code
0	000	000
1	001	001
2	010	011
3	011	010
4	100	110
5	101	111
6	110	101
7	111	100

Table 8.31 64 and 256 QAM element coding

Inputs	Previous inputs	Phase change (degrees)	Outputs ($b5_n b4_n$)
A_n	A_{n-1}	0	00
A_n	B_{n-1}	270	10
A_n	C_{n-1}	180	11
A_n	D_{n-1}	90	01
B_n	A_{n-1}	90	01
B_n	B_{n-1}	0	00
B_n	C_{n-1}	270	10
B_n	D_{n-1}	180	11
C_n	A_{n-1}	180	11
C_n	B_{n-1}	90	01
C_n	C_{n-1}	0	00
C_n	D_{n-1}	270	10
D_n	A_{n-1}	270	10
D_n	B_{n-1}	180	11
D_n	C_{n-1}	90	01
D_n	D_{n-1}	0	00

data transfer rate beyond that obtainable via conventional analog modems, it is also important to note that you may only achieve a fraction of the stated transfer rate of the modem.

8.4.3 DSL modem

The selection of a 3 kHz passband on a voice-grade line was based upon economics. Although humans produce a range of frequencies below 300 Hz and above 3300 Hz, a vast majority of normal conversation is within the 300 Hz to 3300 Hz range. Thus, when the telephone company established their network they filtered frequencies below 300 Hz and above 3300 Hz, resulting in a 3000 Hz passband. This reduced passband allowed more conversations to be simultaneously multiplexed on a trunk interconnecting telephone company central offices, resulting in economic savings. Although this action made sense for the use of subscriber lines as a voice delivery system, it restricted the ability of modems to transfer data, since Nyquist showed that the maximum baud rate obtainable is twice the bandwidth prior to intersymbol interference adversely affecting transmission.

As we entered the era of the Internet, telephone companies looked for methods to provide higher speed transmission over their existing infrastructure to satisfy customer demand as well as to compete with cable modems. By removing the filters on the subscriber line, it became possible to use bandwidth up to and even beyond 1 MHz on the twisted-pair subscriber loop. This in turn enables the voice portion of the subscriber loop to be used to transport voice, while frequencies above the voice passband are modulated to carry data.

The DSL family

Over the past few years a series of DSL services and modems were developed to satisfy different transmission requirements. Two commonly available services and modems (HDSL and HDSL2) represent digital T1/E1 circuit replacement. HDSL modems are used to obtain a T1/E1 transmission capability without requiring the use of repeaters on a twisted-pair circuit. Thus, we will first focus our attention on these two members of the DSL family.

HDSL

High Bit Rate Digital Subscriber Line (HDSL) modems operate over two pairs of twisted-pair wire and allocate bandwidth symmetrically in both directions. Each pair of wires operates at one-half the T1 or E1 operating rate and uses the modulation technique employed by ISDN, referred to as 2B1Q coding. Both ISDN and 2B1Q coding are covered in detail later in this book.

HDSL represents the earliest member of the DSL family of products. Because HDSL permits subscribers to be located further from a central office without requiring digital repeaters, it provides an economical substitute for T1 and E1 local loops and is commonly used by businesses.

HDSL2

HDSL2, also referred to as Symmetric Digital Subscriber Line (SDSL), represents a single-pair version of HDSL. This means that HDSL2 can provide a T1 or E1 access line over a standard single-pair telephone subscriber line.

ADSL

Asymmetric Digital Subscriber Line (ADSL) allocates bandwidth asymmetrically in the frequency spectrum. ADSL is designed to provide a downstream data transfer rate up to approximately 8 Mbps and an upstream data transfer rate up to approximately 1.5 Mbps, at distances up to 18 000 feet from a central office. Later in this section we will examine the operation of ADSL in detail.

RDSL

Rate Adaptive Digital Subscriber Line (RDSL) represents a rate adaptive version of DSL in which the quality of the line is examined and different portions of the subscriber line bandwidth are used based upon varying line conditions. Although at one time RDSL was a separate category of DSL products, over the past few years rate adaption has been incorporated into other versions of DSL products.

G.lite

G.lite represents a speed-limited version of ADSL that does not require splitters to be installed at the subscriber residence. Although G.lite data transfer is limited to 1.5 Mbps in the downstream direction, its lower operating rate allows data transmission and telephone operations to occur without the necessity for the telephone company to install a filter, referred to as a splitter, at the subscriber's premises. This in turn enables G.lite modems to be purchased and installed in a manner similar to conventional analog modems. Later in this section we will examine G.lite modems in detail.

VDSL

Very-high-bit-rate Digital Subscriber Line (VDSL) represents a version of DSL developed to compete with cable TV's video-on-demand capability. VDSL provides a downstream transmission capability of up to 52 Mbps and an upstream transmission of 1.5 Mbps for limited transmission distances, typically up to a few thousand feet.

Of the six DSL technologies previously mentioned, two have a reasonable probability of being implemented as mass market products in the near future. Those DSL technologies are ADSL and G.lite.

ADSL modems

An ADSL circuit with ADSL modems connected to each end of a twisted-pair line has three channels: a high-speed downstream channel, a medium-speed upstream channel, and a standard voice telephone channel. The latter is split off from the ADSL modems by filters, which ensures that subscribers can continue to obtain the use of a voice telephone channel on the existing twisted-pair connection even if one or both ADSL modems fail.

Although data transmission is separated by frequency from the bandwidth of the subscriber line used to carry a voice conversation (crosstalk), the impulse noise generated by a telephone going off-hook, changing loop characteristics based upon temperature, and even signal intrusion from AM radio (since the subscriber line acts as an antenna) all affect transmission. Due to this, ADSL requires the use of splitters at both the central office and the subscriber's premises to separate low and high frequencies from one another. The low frequencies represent the voice conversation portion of the bandwidth, with the splitters ensuring that voice and data operations do not adversely interfere with one another. Because the installation of a splitter at a residence requires the services of a telephone company employee, the use of splitters adds to the cost of ADSL.

Figure 8.44 illustrates the basic operation of an ADSL circuit. The downstream operating rate depends upon several factors to include the length of the subscriber line, its wire gauge, the presence or absence of bridged taps and level of interference on the line. Based upon the fact that line attenuation

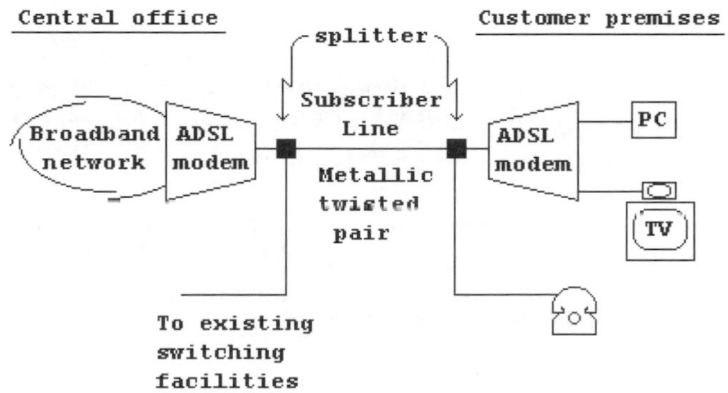

Figure 8.44 Asymmetric Digital Subscriber Line (ADSL). This supports three channels formed by frequency – a high-speed downstream channel, a medium-speed upstream channel, and a conventional telephone channel. The latter is formed through the use of filters

increases with line length and frequency, while it decreases as the wire diameter increases, we can note ADSL performance in terms of the wire gauge and subscriber line distance. Ignoring bridged taps that represent sections of unterminated twisted-pair cable connected in parallel across the cable under consideration, various tests of ADSL lines provided a general indication of their operating rate capability. That capability is summarized in Table 8.32.

Operation

ADSL operations are based upon advanced digital signal processing and the employment of specialized algorithms to obtain high data rates on twisted-pair telephone wire. Currently there are two competing technologies used to provide ADSL capabilities: Discrete Multitone (DMT) modulation and Carrierless Amplitude/Phase (CAP) modulation. The first technology, DMT, represents an American National Standards Institute (ANSI) standard. In comparison, CAP represents a proprietary technology originally developed by Paradyne, a former subsidiary of AT&T and now an independent company; however, at the time this book was prepared CAP had been licensed to a number of communications carriers throughout the world. Both DMT and CAP permit the transmission of high-speed data using Frequency Division Multiplexing (FDM) to create multiple channels on twisted-pair. Through the use of FDM the copper twisted-pair subscriber line is partitioned into three

Table 8.32 ADSL performance

Operating rate	Wire gauge	Subscriber line distance
1.5/2.0 Mbps	24 AWG	18 000 feet
1.5/2.0 Mbps	26 AWG	15 000 feet
6.1 Mbps	24 AWG	12 000 feet
6.1 Mbps	26 AWG	9 000 feet

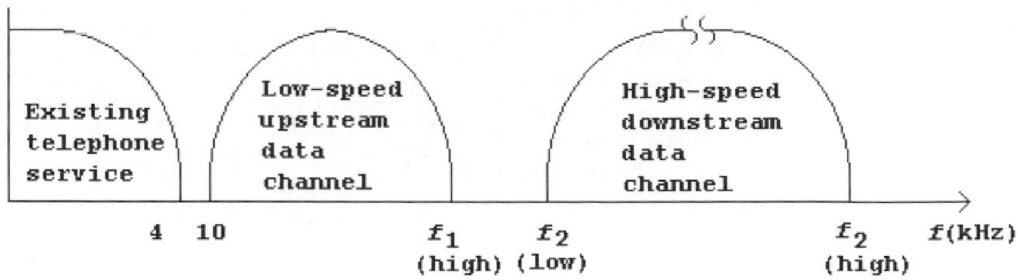

Figure 8.45 ADSL frequency spectrum

parts by frequency, as illustrated in Figure 8.45. FDM assigns one channel for downstream data and a second channel for upstream data, while the third channel from 0 to 4 kHz is used for normal telephone operations. The downstream path can be subdivided through time division multiplexing to derive several high and low speed subchannels by time. In a similar manner the upstream channel can be subdivided.

Discrete Multitone modulation: The concept behind Discrete Multitone (DMT) modulation is similar to that used in the Telebit Packetized Ensemble Protocol (PEP) modem. That is, under DMT modulation, available bandwidth is split or subdivided into a large number of independent subchannels. Since the amount of attenuation at high frequencies depends upon the length of the subscriber line and wire gauge, a DMT modem at the central office must determine which subchannels are usable. To do so that modem sends tones to the remote modem where they are analyzed. The remote modem responds to the central office modem subchannel scan at a relatively low speed, which significantly reduces the possibility of the signal analysis performed by the remote modem being misinterpreted. Based upon the returned signal analysis the central office modem will use up to 256 approximately 4 kHz wide subchannels for downstream transmission. Through a reverse measurement process the remote modem will use up to 32 4 kHz wide subchannels for upstream transmission.

Under the ANSI standard each of the 255 channels is 4.3125 kHz wide. Each channel has its own logical QAM modem that has a 4 kHz signaling rate. The bandwidths on the twisted-pair subscriber line in the form of 4.3125 kHz channels are referred to as bins, with bin 1 representing the frequency range from 0 to 4.3125 kHz. Thus, bin 1 is not actually used as it represents the bandwidth reserved for voice conversations.

To reduce potential interference between data transmission and voice conversations, upstream transmission is restricted to bins 6 through 38, representing the 25 kHz to approximately 163 kHz frequency spectrum. Downstream transmission can use bins 33 to 255, representing the 142 KHz to approximately 1.1 MHz frequency spectrum. The overlapping use of bins requires echo cancellation to be employed. Otherwise, conventional FDM is used and no overlapping of bins occurs.

Out of the 255 bins, bins 16 (69 kHz) and 64 (276 kHz) are used for pilot tones. The actual amount of data transferred per bin varies across bins. A maximum of 16 bits can be encoded per symbol via QAM.

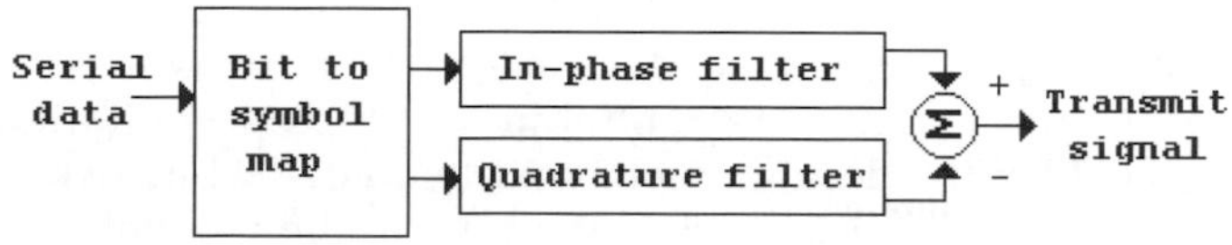

Figure 8.46 Carrierless Amplitude Phase modulation

One of the key advantages of DMT is its ability to take advantage of the characteristics of twisted-pair wire that can vary from one local loop to another. This makes DMT modulation well suited for obtaining a higher data throughput than is obtainable through the use of a single carrier transmission technique.

Carrierless Amplitude/Phase (CAP) modulation: Carrierless Amplitude/Phase (CAP) modulation is a derivative of QAM that was developed by AT&T Paradyne when Paradyne was a part of AT&T. Unlike DMT which subdivides the bandwidth of the wire into 4 kHz segments, CAP uses the entire bandwidth in the upstream and downstream channels. Under CAP serial data is encoded by mapping a group of bits into a signal constellation point using two-dimensional eight-state Trellis coding with Reed–Solomon forward error correction. The latter automatically protects transmitted data against impairments due to crosstalk, impulse noise and background noise.

Figure 8.46 illustrates the CAP modulation process. Once a group of bits are mapped to a predefined point in the signal constellation, the in-phase and quadrature filters are used to implement the positioning in the signal constellation. Since this technique simply adjusts the amplitude and phase without requiring a constant carrier, the technique is referred to as 'carrierless'.

The ADSL unit developed by Paradyne uses a CAP-256 line code (256-point signal constellation) for downstream operations, using bandwidth from 120 kHz to 1224 kHz. The composite signaling rate is 960 kbaud, and seven bits are packed into each signal change to provide a downstream operating rate of 6.72 Mbps. However, the use of Reed–Solomon forward error correction reduces the actual payload to 6.312 Mbps plus a 64 kbps control channel. In the upstream direction the Paradyne device uses a CAP-16 line code in the 35 kHz to 72 kHz frequency band to obtain a composite signaling rate of 24 kbaud across 16 subchannels. Packing three bits per signal change an upstream line rate of 72 kbps is obtained, of which 64 kbps is available for data.

Although several telephone companies offered ADSL service, the added cost associated with installing splitters resulted in the requirement for a more practical product for subscribers who require faster Internet access but do not need the capability to download information at 8 Mbps. Recognizing that more subscribers would be interested in high-speed Internet access than a video-on-demand capability resulted in the development of G.lite.

G.lite modems

The goal behind the development of G.lite was to provide a modem to be used for a DSL service that would be no more difficult to install than a conventional analog modem designed for use on the PSTN. G.lite, which was standardized by the ITU, represents a scaled-down splitterless version of ADSL.

G.lite modems use only bins through 127, limiting bandwidth usage to 546 kHz. In addition, each bin is limited to 8 bits per symbol instead of ADSL's 16, further limiting the complexity of the modem.

The reduction in the use of bins and the maximum of 8 bits per symbol reduces the operation of G.lite to providing up to 1.5 Mbps downstream and up to 512 kbps in the upstream direction. Although G.lite is referred to as a 'splitterless' version of ADSL, in actuality this is not true for two reasons. First, a splitter is still required at the central office to separate low and high frequencies. Secondly, due to noise from existing telephones in a home, several G.lite modem vendors include self-installable micro-filters with each G.lite modem. The micro filter is in a rectangular housing approximately the size of a quarter dollar coin. It is designed for insertion between a telephone outlet and a telephone, and filters out frequencies that could adversely affect the operation of the modem. During 1999 several vendors successfully conducted G.lite interoperability tests, and G.lite service was in the process of being offered by several telephone companies when this book revision was prepared.

8.5 SERVICE UNITS

Transmission on a digital network requires the conversion of unipolar signals generated by computers and terminal devices to a bipolar signal, as previously discussed in Chapter 4. There are two devices required to perform the signal conversion as well as amplify and filter the signal and perform clocking and pulse conversion. Those devices are collectively referred to as service units and are the focus of this section.

The Channel Service Unit (CSU) and Data Service Unit (DSU) were originally separate devices that were only obtainable from communications carriers for use on their digital networks. DSUs and CSUs became commercially available with the introduction of AT&T's Dataphone Digital Service during the early 1980s, and were bundled into the tariff of that service. Changes in telecommunications regulations first enabled third party vendors to market DSUs while allowing communications carriers to continue their monopoly on CSUs, as those devices include a protection feature which is designed to isolate malfunctioning equipment voltage or current levels from the digital circuit. Additional modifications to telecommunications tariffs in the era of deregulation resulted in the ability of third party vendors to market both DSUs and CSUs. When this occurred one of the first actions of those vendors was to combine the functionality of DSUs and CSUs into one device, a combined CSU/DSU. That combined device is currently marketed for use on digital circuits operating at or below 56 kbps. At data rates above 56 kbps the CSU is still separated from the DSU, and it performs a variety of framing functions that will be described later in this book when we examine the T-carrier digital facility. The DSU is normally built into equipment, such as T1 multiplexers or are obtainable as modular cards inserted into the chassis of communications equipment.

8.5.1 The DSU

The primary functions of a DSU include clocking and pulse conversion. The DSU is a synchronous device which derives timing from the received signal and passes it to the attached DTE. Since Dataphone Digital Service as well as competitive offerings from other communications carriers operate from one precise systemwide clock, data must enter the network at the precise circuit operating rate. Thus, the DSU's receive clock obtained from the received data is also used to generate the transmit clock for the attached DTE.

Since digital circuits use bipolar transmission the DSU also functions as a pulse converter. In doing so the DSU converts unipolar digital signals generated by the attached terminal equipment into the bipolar signal suitable for transmission on the digital circuit. Although the DSU does not modulate data in the manner by which a modem alters a carrier signal, its conversion of unipolar to bipolar is considered by many persons to represent a form of modulation. This explains why the combined DSU/CSU is commonly referred to as a digital modem.

Two additional functions performed by DSUs include maintaining network synchronizations and performing loopbacks.

To maintain network synchronization the DSU transmits pulse patterns known as control-idle pulses to the network. Those control-idle pulses are transmitted whenever the terminal device becomes inactive and functions as a mechanism to maintain network timing.

Unlike analog transmission facilities on which frequencies are used to initiate loopback, digital networks use sequences of 1s and 0s as well as coding violations to loop data for testing. The DSU includes circuitry to recognize and act on loopback requests which enables end-users or communications carrier personnel to test a circuit end-to-end from one end.

8.5.2 The CSU

Two key functions performed by CSUs include the protection of the digital network from malfunctioning customer equipment and the actual performing of loopback operations under the control of DSU circuitry.

The CSU is designed to accept nominal 50% duty cycle bipolar pulses from the DSU on the transmit and receive data leads. The pulses, synchronized with the digital circuit, are amplified, filtered, and passed on to the four-wire metallic telephone company cable. The signals on the receiver pair are amplified, equalized and sliced by the line receiver. The resultant bipolar pulses are then passed to the DSU or the DSU portion of a combined DSU/CSU which must recover the synchronous clock used for timing the transmitted data and sampling the received data. The customer's DSU must further detect DDS network codes, enter appropriate control states, and remove bipolar 'violations' from the data stream.

Figure 8.47 illustrates in a block diagram format the functions of the DSU and CSU as separate entities. As previously explained, the functionalities of DSUs and CSUs are now almost always combined within one device, which when used at data rates at or below 56 kbps for transmission on a digital

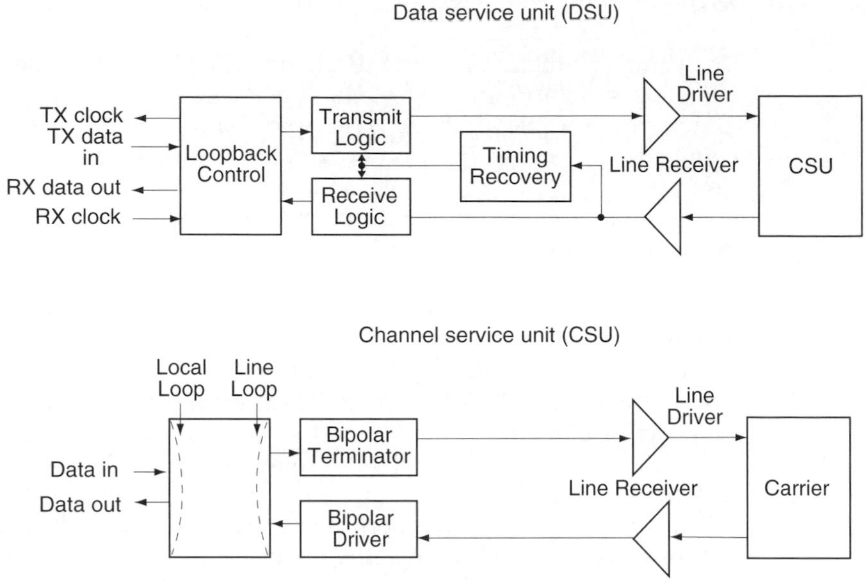

Figure 8.47 Service units for digital transmission. The DSU contains all of the circuitry necessary to make the device plug compatible with existing data terminal equipment and performs required clocking and initiates loopbacks. The CSU protects the digital circuit from malfunctioning end-user equipment and performs loopbacks under the control of the DSU

circuit is referred to as a DSU. Today almost all combined CSU/DSUs are connected to digital transmission facilities via a standard 8-pin RJ-45 plug and jack arrangement. Although in the past DSUs were sold to operate at a specific data rate such as 2400, 4800, 9600, 19 200 or 56 000 kbps, today most DSUs are manufactured as multirate devices. This enables end-users to simply change a knob or switch setting to use the device at a different operating rate; however, to do so the user must first obtain a new digital circuit. This is because digital circuits cannot be increased in operating speed by changing the operating rate of the DSU. Instead, a new digital circuit must be installed to support the desired operating rate.

8.6 REVIEW QUESTIONS

1. What is the general relationship between pulse width and data rate?

2. What is the general relationship between the transmission rate and cable length on an RS-232 serial connection?

3. Discuss an application for a line driver.

4. Describe the basic signal conversion performed by modems.

5. What is a carrier signal? By itself, does it convey any information? Why?

6. How can the characteristics of a carrier signal be altered?

7. What is the difference between a bit per second and a baud? When can they be equivalent? When are they not equivalent?

8. What is the Nyquist relationship and why does it require modem designers to develop multilevel phase-shift keying modulation schemes for modems to operate at high data rates?

9. Why does full-duplex transmission enhance transmission efficiency?

10. How could you use a reverse channel? What is the difference between a reverse channel and a secondary channel?

11. How does a modem obtain a multiport capability?

12. Describe the operation of a security modem.

13. What is the function of the modem handshaking process?

14. What is the difference between a local analog modem test and a digital loop-back modem test?

15. Assume that two modems connected to a leased line have both local and digital loop-back self-testing features. If communications were disabled, discuss the tests you would perform to determine if the line or one or both modems caused the communications failure.

16. What does the term 'Bell System' compatibility mean when discussing the operational characteristics of a modem?

17. If an originate mode modem transmits a mark at f_1 and a space at f_2 and receives a mark at f_3 and a space at f_4, what would be the corresponding frequencies of an answer mode modem to ensure communications compatibility?

18. Discuss the conventional utilization of originate and answer mode modems.

19. Why is it more likely than not that an American using his or her portable personal computer in Europe would not be able to communicate with a computer located in Europe?

20. What does the signal constellation pattern of a modem represent? What is the normal relationship between the signal constellation pattern and the susceptibility of a modem to transmission impairments?

21. Discuss the difference between Trellis coded modulation and conventional quadrature amplitude modulation with respect to the density of the signal constellation and the susceptibility of a modem employing each modulation technique to transmission impairments?

22. Why are Bell System 212-type modems operating at 1200 bps sometimes compatible with V.22 modems, whereas at other times they are incompatible?

23. Why are Bell System 202 type modems incompatible with ITU-T V.23 modems?

24. Discuss the compatibility of a ITU-T V.26 modem employing a pattern A phase change with a similar modem using the pattern B phase change.

25. Explain why the V.29 signal constellation pattern forms a mirror image.

26. What is echo cancellation and why is its use important for modem operations on the switched network?

27. Besides a faster operating rate, name another advantage associated with the use of a V.32 bis modem over a V.32 modem.

28. Compare the features of a V.32 bis modem to a V.33 modem.

29. What is the purpose of the V.34 modem supporting signaling methods below 300 Hz?

30. Discuss two potential problems when using a V.34 modem that could limit its ability to operate at 33.6 kbps.

31. How can the asymmetrical transmission capability of a V.34 modem improve its performance?

32. How does a reduction in quantization noise effect the transmission capability of a modem?

33. Why is a V.90 modem limited to a 56 kbps data rate in one direction?

34. Why is flow control important when using error detection and correction?

35. Describe two flow control methods.

36. What is the difference between V.42 and V.42 bis?

37. Discuss the reason why a cable modem can transmit data at a higher operating rate than a modem designed for use on the switched network when both modems use the same general modulation method.

38. How can an ADSL modem use the same portion of frequency for upstream and downstream transmission?

39. Describe two differences between an ADSL modem and a G.lite modem.

40. Describe two DSU functions in addition to network synchronization.

41. Describe two CSU functions.

42. How does a DSU perform network synchronization?

43. Discuss the relationship of the DSU to the CSU with respect to loopbacks.

9

REGULATORS AND CARRIERS

Prior to January 1, 1984, when a court-supervised agreement between AT&T and the United States Justice Department came into effect, the selection of a communications service and determining the price paid for the service was a relatively simple process. Although there were and still are over 2000 independent telephone companies operating in the United States, by 1984 the AT&T Bell System accounted for over 80% of all local and 95% of long-distance calls. The costs of such calls were highly regulated by local, state and federal tariffs. While the divestiture of AT&T of its 22 local telephone operating companies resulted in an increase in the competitiveness of the communications industry, it was not until the 1996 Telecommunications Act and the expenditure of tens to hundreds of billions of dollars that local and long-distance competition became a true reality. Today many residential and business customers can select local service from an incumbent local exchange carrier (ILEC), competitive local exchange carrier (CLEC), or bypass carrier. In addition, through the use of a '1010' prefix you can literally dynamically select a long-distance carrier from an ever-expanding base of hundreds of carriers. Although there is a high degree of competition for your communications dollar, that level of competition may further increase in the future as a result of a series of actual and pending mergers whose effect may revolutionize the manner by which we communicate.

In the international area a process similar to divestiture occurred in many countries. For example, in the United Kingdom, British Telecom, once owned by the UK government, is now a private company owned by its shareholders. Similarly, control of Nippon Telephone & Telegraph passed from the Japanese government to shareholders when the company was privatized. Like the United States, many mergers have occurred in the international telecommunications area. In 1999 Olivetti acquired Telecom Italia while Global Crossing, a firm headquartered in Bermuda, was in the process of competing with Quest Communications to acquire Frontier Corporation and US West. Thus, the communications carrier market can be easily categorized as a dynamic environment.

The previously mentioned dynamic environment does not mean that communications carriers are free to offer any service at any cost. At the national level, many services are regulated, especially those that require a portion of the frequency spectrum or cross national boundaries. At the state and local level, various regulatory commissions in the United States maintain a considerable influence over the cost and type of certain offerings. In addition, as we will note later in this chapter, the Telecommunications Act of 1996 provides certain criteria that must be met to enable long distance and local competition to occur.

In this chapter, we will first focus our attention upon the regulatory process, examining a few of the general functions performed by a regulatory body. In the second section of this chapter, we will review the operational characteristics and utilization of several popular carrier offerings. Since AT&T and the Regional Bell Operating Companies (RBOCs) have a considerable influence on communications in the United States, we will examine them in detail in this section.

9.1 REGULATORS

9.1.1 US regulatory evolution

Until 1984, the AT&T Bell System had a *de facto* monopoly on telephone service in the United States. Although telephone service started in 1876, the current era has its beginnings in the Communications Act of 1934, by which Congress sought to provide a telephone service to all people at reasonable prices, the so-called universal service concept. It reasoned that elimination of competition and regulation by the government would bring about what private industry could not. Thus, with its blessing, AT&T grew to control more than 80% of local calls and more than 95% of long distance calls through its 22 operating companies and Long Lines Division. Superiority was assured by the research and development efforts of Bell Laboratories and the manufacturing facilities of Western Electric.

Communications Act of 1934

The Communications Act of 1934 established a rate and operations regulation context, through which the Bell System could introduce different services at predefined rates established by the approval of a tariff. As a result of the Communications Act the government, in effect, legally sanctioned the *de facto* monopoly of AT&T on telephone services. At the same time, the Communications Act of 1934 was structured to encourage growth in basic telecommunications services by approving tariffs which subsidized home service at the expense of long distance and business services.

AT&T enjoyed free rein in the telephone industry until about 1949, embracing not only telephone operations but such businesses as data processing. The US Justice Department then filed an antitrust suit that was eventually settled when AT&T signed the 1956 Consent Decree, which mandated that AT&T stay out of unregulated businesses. The late 1960s brought increased competition, aided by a series of court decisions to include the 1968 Carterfone case and Federal Communications Commission (FCC) rulings, as well as the Telecommunications Act of 1996.

The Carterfone Decision

Until 1968, AT&T had a monopoly on providing equipment for use on the public switched telephone network (PSTN). In that year, a small Texas company called Carterfone petitioned the Federal Communications Commission to direct telephone companies to permit the interconnection of non-telephone company supplied equipment, referred to as 'foreign' equipment, to the PSTN. At that time, Carterfone marketed a product line of oil field communications equipment which, in some instances, required interconnection to the PSTN for its use.

After a series of hearings, the FCC ruled that telephone companies could not unreasonably and arbitrarily forbid the connection of foreign equipment. In the same ruling, the FCC recognized the right of telephone companies to protect the PSTN from harmful equipment. This ruling, known as the Carterfone Decision, resulted in the manufacture and use of Data Access Arrangements (DAAs).

The DAA is a small device that limits the power level and frequency of a signal produced by non telephone company equipment prior to its entering the PSTN. Thus, under the Carterfone Decision, non-telephone company equipment could be connected to the PSTN by the use of a DAA located between the 'foreign' equipment and the telephone network.

FCC equipment registration program

Two key problems associated with the use of DAAs was their cost and inconvenience. Initially, DAAs were primarily obtainable from telephone companies on a monthly rental basis. In addition to the problem of having to order equipment, such as a modem from one vendor and a DAA from the telephone company, the cost of both often exceeded the cost of a modem from the telephone company. As an example of the economics associated with the use of DAA's, consider the cost comparison presented in Table 9.1. Although the monthly cost of a telephone company supplied modem might be 10% above the cost of a foreign device, when the cost of the DAA was included, telephone company supplied equipment became more economical.

Table 9.1 Equipment cost comparison

	Monthly rental by equipment source	
Type of equipment	Foreign	Telephone company
Modem	$50	$55
DAA	$10	not required

As a result of the problems of cost and inconvenience associated with DAAs, numerous third-party manufacturers registered complaints with the FCC. The FCC eventually adopted an equipment registration program. Under this program, a manufacturer of interconnect equipment, such as a telephone or modem, certifies to the FCC that the design of the equipment incorporates a built-in DAA to limit the equipment's operational frequency and power levels. To certify the equipment's frequency and power levels, vendors can submit their equipment to independent testing laboratories for compliance testing. This testing should not be confused with operational testing which is used to verify the functional characteristics of equipment and which is of no concern to the FCC.

Under the current FCC equipment registration program, approved equipment is assigned to a unique 14-digit number as well as a 3-digit ringer equivalence number. The latter designates the power and impedance of the equipment, and both numbers 'theoretically' must be provided to the telephone company when ordering a circuit to which 'foreign' equipment is to be attached.

Divestiture

A second significant force responsible for shaping the regulatory environment in the United States was the US Justice Department. In 1974 the Justice Department again filed suit against AT&T, charging restraint of trade. Another seven years passed before the lawsuit was resolved with AT&T's 1981 agreement to divest itself of 22 local telephone operating companies. AT&T's divestiture plan was approved by the courts in 1983 and was implemented on January 1, 1984.

Under the settlement between AT&T and the Justice Department, AT&T divested its ownership of the 22 Bell Operating Companies responsible for providing local service. AT&T retained ownership of the installed base of customer-premises equipment, such as telephone sets and modems, and was required to provide research, manufacturing, development and other services to the RBHCs until September 1, 1987. Of key importance to the development of competitive communications, RBHCs were required to provide equal access to local facilities for both AT&T and its competitors, resulting in the growth of such alternative long distance carriers as MCI Communications and Sprint during the 1980s.

Tariffs

Until AT&T's divestiture took effect, there were only two types of tariff that governed the cost of communications: interstate and intrastate. Interstate tariffs were filed by AT&T and other common carriers (OCCs), such as MCI and Sprint, with the Federal Communications Commission for approval. Intrastate tariffs which govern communications within a state were filed by AT&T's local operating companies as well as by approximately 2000 small independent telephone companies with the appropriate state public utility commission.

Although the distinction between interstate and intrastate communications with respect to federal or state regulatory authority survived divestiture, a new criterion was added governing tariffs. This criterion is based upon the establishment of Local Access and Transport Areas (LATAs) that roughly correspond to the Standard Metropolitan Statistical Areas defined by the US Commerce Department.

If communication occurs-within a LATA, the provided service is intra-LATA. If communication requires the routing of a path linking two or more LATAs, the provided service is inter-LATA. Since some LATAs cross state boundaries and other LATAs and groups of LATAs are within the geographical boundary of a state, there are federal and state tariffs governing inter- and intra-LATA service. In addition, to facilitate equal access to long distance carriers, there are now access tariffs filed by the local operating companies that govern the use of their facilities to access inter-LATA carrier networks, such as MCI Communications, Sprint and other carriers. These tariffs, known as LATA access, also exist for both interstate and intrastate communications. The two types of tariffs prior to divestiture have thus been replaced by six: interstate and intrastate tariffs for inter-LATA, intra-LATA and LATA access.

The location within a LATA where the local operating company interfaces a intra- LATA carrier is known as a point-of-presence (POP), a location AT&T calls a 'serving office'. Regardless of the name used, these are the only locations within a LATA, where an intra-LATA carrier can transmit and receive traffic.

To avoid the cost associated with using a local operating company to access an intra-LATA carrier, many large organizations use a variety of methods to bypass the local operating company. These bypass techniques include using cable TV transmission facilities and the construction of microwave towers, and the use of radio relay systems. Due to the economic advantages associated with bypass technology, it has achieved a significant rate of growth which, if unchecked by regulatory authorities, may result in an increase in local operating company rates to home and retail subscribers that cannot take advantage of this technology.

The communications carrier that originally was provided with a monopoly for intra-LATA communications is referred to as an incumbent local exchange carrier (ILEC). The ILEC not only provides intra-LATA communications but typically provides a degree of inter-LATA communications as long as the connected LATAs do not cross a state boundary. From the point of presence in a LATA, calls that cross state boundaries are handed off from the local exchange to a long distance communications carrier. Because this carrier provides the facilities to connect exchanges it is also referred to as an interexchange carrier (IXC). In general, you can consider AT&T, MCI Worldcom and Spring to represent IXCs, while BellSouth, US West and other RBOCs can be considered to represent ILECs.

Telecommunications Act of 1996

The Telecommunications Act of 1996 resulted from a recognition that the telecommunications industry has significantly changed since the last major legislation affecting its operations. That legislation did not take note of the fact that technology now enables cable TV to transport voice and data communications, whereas the local twisted-pair wire or the replacement of that wire by an optical fiber cable by the telephone company would enable voice, data and images to be carried by the local telephone company. Thus, regulatory bills in the United States were addressing the ability of telephone and cable television to compete with one another, because until 1996 most areas of the country restricted competition, either by Federal or by state or local law. For example, competition for local telephone service was primarily restricted based on Federal rulings, whereas competition for cable television service was primarily restricted by local municipalities which normally granted licenses to a vendor to serve a community for a period of time based on the payment of a fee to the municipality. Other regulations at the Federal level restricted the rates the cable TV operator could charge.

During 1995 several bills were introduced at the Federal level to reform the telecommunications industry. In addition to enabling local telephone companies and cable television operators to compete with one another there were several interesting additional 'reforms' associated with those bills. Those reforms were developed to enable local telephone companies after a period of time to compete with long distance communications carriers, and allow electric utilities to compete with local telephone companies and cable TV operators.

The series of bills introduced in the US Congress during 1995 resulted in the passage of Senate bill S.652 on February 1, 1996. That bill, known as the Telecommunications Act of 1996, represents the first comprehensive rewrite of US communications laws. The stated purpose of the Telecommunications Act of 1996 was to promote competition and reduce regulation as a mechanism to obtain lower prices and higher quality service for American consumers.

The provisions of the Act fall into five major areas – telephone service, telecommunications equipment manufacturing, cable television, radio and television broadcasting, and online computer services to include the Internet. In this section we will focus our attention primarily on the first and last areas and briefly discuss the other areas of the Act.

Telephone service

Under the Telecommunications Act of 1996 pricing authority in the local access market is explicitly assigned to the states, where it has always been. This was not changed, since the US Constitution provides states with the right to regulate commerce within their own borders. What did change was existing and potential state restrictions on competition in local and long-distance services. Under the Telecommunications Act of 1996 the ability of states to restrict competition was eliminated, enabling the former Regional Bell Operating Companies (RBOCs) to immediately provide long-distance telephone service outside their serving area and within their service area, with the latter occurring once they open their service area to competition. Under the 1996 Act a series of 14 steps must be taken by each RBOC in order to obtain the ability to offer long-distance service within their region. In addition to enabling the incumbent local exchange carrier (ICLEC) to offer long-distance service within their serving area, the Telecommunications Act of 1996 permits long distance interexchange carriers (IXCs) to offer local service. However, because it would be too expensive for an IXC to rewire a city or rural area, most IXCs look for alternative mechanisms, such as reselling ICLEC services, using wireless transmission, or 'cherry picking', the latter being a term used to describe wiring directly into an office or apartment building that bypasses the local loop to provide service to a group of users, which is also referred to as bypass routing. Although approximately four years have passed since the 1996 Act, at the time this book revision occurred only one RBOC had obtained the right to offer long-distance service within their service area. Because the ability of long-distance companies to offer local services using ILEC facilities was severely restricted by the prices ILECs charged for resale, the goal of the Telecommunications Act of 1996 concerning competition was not fully met, and this explains why AT&T was in the process of spending approximately $100 billion to acquire cable television systems as a mechanism to bypass the local loop controlled by ILECs.

Telecommunications equipment manufacturing

Under the 1996 Act RBOCs can manufacture telephone equipment. This capability was not permitted when divestiture occurred.

Cable television

Under the Telecommunications Act of 1996 regulations governing cable television were removed over a three-year period which ended on March 31, 1999. Although most consumers noted significant cable system price increases, the Act also permitted telephone companies to either offer cable television services or carry video programming. The intent of the Act concerning cable television is to foster competition as a mechanism to enhance service offerings and limit rate hikes.

Radio and television broadcasting

The primary effect of the Telecommunications Act of 1996 concerning radio and television broadcasting was to relax FCC rules concerning the ability of a company to own multiple radio and TV stations.

Internet and online computer services

Under the Telecommunications Act of 1996 criminal penalties were established for persons who knowingly transmit obscene material on an interactive computer service. Another provision makes it a crime to repeatedly harass a person electronically.

The CLEC

In addition to the previously mentioned areas, the Telecommunications Act of 1996 defined a new class of communications carrier called a competitive LEC, or CLEC. Under the Act CLECs are entitled to resell any or all services offered by the ILEC. By the turn of the millennium over 150 CLECs were in operation throughout the United States, offering 5 million access lines. Many Internet Service Providers (ISPs) expanded their dial-up customer base by becoming CLECs, allowing them to offer leased lines to include Digital Subscriber Line (DSL) services.

9.1.2 International regulatory authorities

In a majority of countries telephone networks are owned and administered by a government department responsible for Posts, Telegraphs and Telephones (PTT). Although the actual title of the department varies from country to country, they are known collectively as PTTs.

Each PTT normally determines what equipment can be permitted to be connected to its telephone system. Most PTTs publish their requirements which, if met, permit equipment to be connected to the telephone network they own and administer. Normally, PTT requirements focus on the safety of their plant and personnel as well as the potential interference with other equipment and facilities.

To ensure that 'foreign' equipment has a correct level of performance, a PTT will establish an approval process similar to the previously discussed FCC Equipment Registration program. Under the PTT approval process the manufacturer or supplier of equipment prepares a document that describes the technical specifications of the equipment for which approval is requested. The PTT then reviews the technical specifications and requests the loan of equipment for evaluation. If no problems occur during the evaluation process, the PTT will issue a letter or other document stating that the equipment is approved and denote in which conditions such equipment can be used.

In some countries the equipment approval process has been removed from the PTT and placed under the auspices of a third party. As an example, in the United Kingdom the British Approval Board for Telecommunications (BABT) was responsible for certification of equipment for connection on the PSTN. Under the UK Telecommunications Act of 1984, BABT is still the principal evaluation authority; however, the power to grant approval of an apparatus for use on the UK PSTN now rests with the Director General of the Office of Telecommunications (Oftel). Another variance between PTTs is in the area of communications equipment acquisition. Some PTTs require users to obtain communications equipment directly from the PTT. In comparison, other PTTs require users only to obtain approved equipment.

Communications policies

In comparing communications policies in the United States to those of other countries, a trend toward substantial liberalization and government deregulation is common. Where the United States differs from other countries is in the method used to foster competition.

In the United Kingdom, Japan and other countries, competition was introduced without breaking up the existing national carrier. In both the United Kingdom and Japan, government ownership of the national carrier was transferred to shareholders and at the same time competition was allowed by the licensing of other carriers to provide communications services.

In most European countries national telephone companies were privatized during the late 1990s. In fact, the European Union, a cooperative of 15 European countries, stipulated that telephone services were to be deregulated by 1998 in most countries, with Spain, Greece, Ireland and Portugal having to 2003 to do so.

Similar to the United States, equipment to be utilized on most European telephone networks must be certified. Certification can occur by the telephone network operator or an independent testing laboratory. For most countries the certification process permits any vendor to design and manufacture equipment for use with the telephone network once the equipment's operational parameters have been certified to be within acceptable limits with respect to frequency and power levels.

In the area of basic services, such as access to the PSTN and long distance communications, most countries limit local access to one communications carrier, in effect, promoting a monopoly for local access. Most countries by regulation, however, permit a local telephone subscriber to select a primary long distance carrier, resulting in competition among long distance telephone service.

Enhanced services, such as the utilization of value-added networks, are limited in most countries with respect to regulation and can be considered as regulated competition. This is because most countries regulate the interconnection of national packet networks to other packet networks as well as monopolizing access to the packet network since a local loop connection is regulated to service provided by only one vendor.

9.2 CARRIER OFFERINGS

Today, many large business organizations maintain a staff of communications analysts to keep abreast of current carrier offerings, tariffs and the effect of changes in each on corporate operations. To provide you with an appreciation for the diversity of carrier offerings, we will examine the effect of divestiture on AT&T in the United States. This examination will include an overview of Regional Bell Operating Company service areas as well as the types of services currently offered by AT&T and RBOCs.

9.2.1 AT&T system evolution

We can obtain an appreciation for the general evolution of the communications industry in the United States by examining the effect of divestiture upon AT&T, and the manner by which several RBOCs and non-RBOCs initiated mergers and acquisitions with both the divestitured AT&T and the resulting independent Regional Bell Operating Companies.

Figure 9.1 provides via a horizontal tree structure the divestiture and acquisition trail of companies that at one time were part of the Bell System. As mentioned earlier in this chapter, at one time the Bell System consisted of 22 operating companies. The 22 operating companies were grouped into seven regional companies, resulting in the term RBOC being used to describe a divested company that contained two or more former operating companies. In examining Figure 9.1 the seven companies listed in the column of blocks on the left, commencing with Ameritech and ending with US West, represent the seven original divestitured RBOCs.

In 1991 AT&T acquired NCR Corporation. In 1997 AT&T voluntarily split itself into three, in effect privatizing its equipment manufacturing arm which was then renamed Lucent Technologies as well as spinning off its previously acquired NCR Corporation. At this point in time AT&T eliminated its research capability as Bell Laboratories, where the transistor was invented, became the research and development arm of Lucent Technologies.

In examining Figure 9.1 note that at the time this book revision occurred, AT&T was in the process of spending over $100 billion to acquire both Tele-Communications (TCI) and MediaOne, two very large cable television

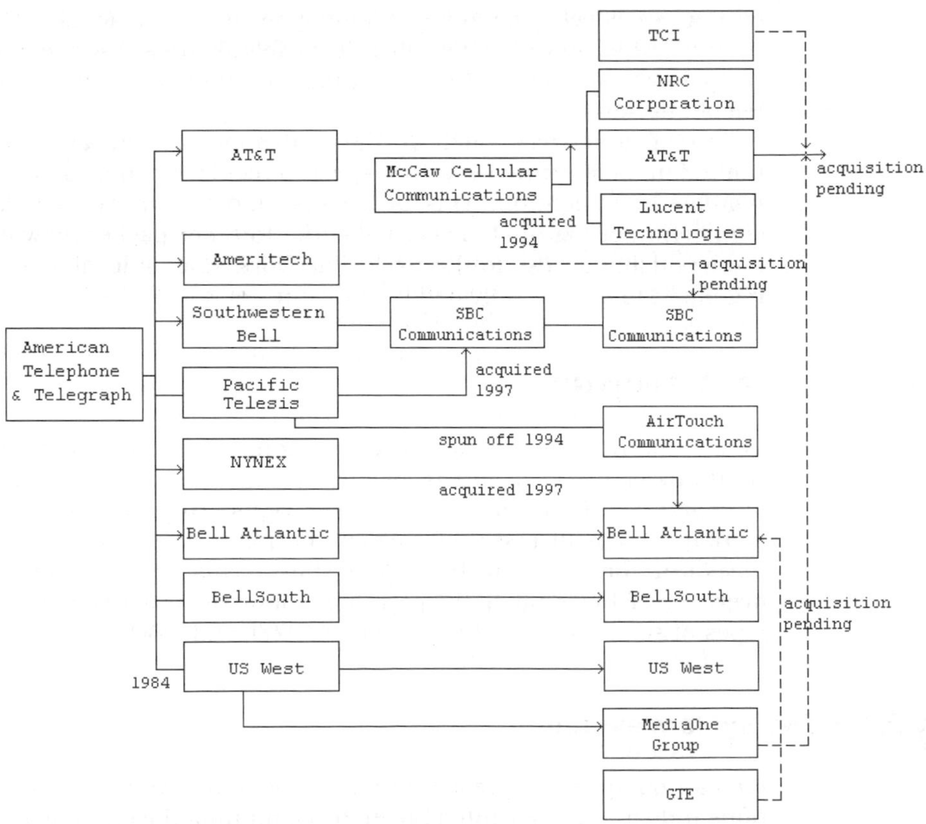

Figure 9.1 The evolution of AT&T

operators that could provide AT&T access to approximately 60% of US households. These acquisitions will provide AT&T with a mechanism to bypass RBOCs and offer voice via cable in a package that could include cable television and high-speed Internet access.

Although almost all of the companies shown in Figure 9.1 resulted from the breakup of the Bell System, there are three exceptions. The first exception is McCaw Cellular Communications which was acquired by AT&T in 1994. The second is TCI, shown in the top right portion of Figure 9.1. TCI is an independent cable television operator that was one of two cable systems being acquired by AT&T. The third exception is GTE which is shown in the lower right portion of Figure 9.1. GTE is the largest independent telephone company in the United States, providing local access in approximately 25 states and serving several million customers. Because GTE was never part of the Bell System, it can and does offer long distance service in its service area and does not have to comply with the 14 provisions of the Telecommunications Act of 1996.

Prior to its split-up in 1997, AT&T was organized into four broad groups: Communications Services, Multimedia Products, Global information Solutions, and Network Systems.

The Communications Services Group had the responsibility for US and international long-distance operations to include WATS, telephone conference services, electronic mail, and telecommunications relay services, the latter a service which assists individuals with hearing loss to use the phone to communicate with persons that are not hearing impaired. When AT&T reorganized itself in 1997, the Communications Services group formed the core of the new AT&T.

The Multimedia Products Group manufactured, sold and leased such communications products as PBXs, modems, multiplexers, CSUs and DSUs. One member of that group, AT&T Paradyne, was sold by AT&T and is currently one of the leading developers of Digital Subscriber Line products.

AT&T Global Information Solutions primarily consisted of the formerly independent NCR Corporation that in 1997 again became an independent company that reused the NCR name.

The fourth AT&T group, Network Systems, manufactured network telecommunications equipment for telephone, cable television, and wireless service providers located throughout the world. This group also designed and manufactured large central office switches, optical and fiber cable, and structured cabling systems.

As part of AT&T's reorganization its Network Systems group was spun off as a private company called Lucent Technologies. This reorganization positions AT&T as telecommunications provider and enables Lucent Technologies to better compete for equipment sales to other carriers that in the past might be adverse to purchasing products from the same company they compete with for providing communications services.

9.2.2 The Bell system

The 22 local Bell Operating Companies (BOCs) provide local residential and business services. Although the BOCs remained virtually intact and kept their original names (e.g. New Jersey Bell, New York Telephone), they were initially controlled by seven Regional Bell Operating Companies (RBOCs).

While continuing to provide basic phone service, the RBOCs acquired some additional privileges. Each RBOC can enter the unregulated, competitive side of the telecommunications market. Through separate subsidiaries, they are free to sell equipment manufactured by any company, not just AT&T. Equipment sales pit the individual RBOC against AT&T. Every RBOC has taken advantage of this opportunity by announcing new product lines.

In addition to these ventures, the BOCs assumed control of the Yellow Pages. Publishing and selling of advertisements for the Yellow Pages generated more than $4 billion a year for the BOCs, capital that helps subsidize the less profitable regulated side of the business.

Although the regulated business results in each RBOC having in effect a monopoly concerning local telephone services, that monopoly was on the verge of extinction when this book revision was prepared. Several alternative communications carriers, such as MFS (formerly Metropolitan Fiber Systems and now a subsidiary of MCI Worldcom), which for years were wiring buildings within cities for local bypass, reached agreement with some telephone companies to become an alternative local carrier. This enables the customer to retain their current telephone number when switching to a different local access provider and the inability to do so was a significant restriction on the ability of businesses to easily change their local service provider. In addition, under recently passed legislation RBOCs obtained permission to reenter the long distance market while long distance companies obtained the ability to enter the local telephone market.

9.2.3 The regional Bell operating companies

The RBOCs cover regions that vary in size from 2 to 14 states. Some of the RHCs retained the Bell name and its conservative business approach, whereas others hired advertising agencies to devise catchy new names and consulting firms to devise aggressive competitive strategies.

The geographic distribution of each of the original seven regional companies is illustrated in Figure 9.2. (Alaska and Hawaii are not included because they are served by independent telephone companies.) One measure of financial strength is the percentage of local exchange revenues, of which the seven orignal RBOCs generate more than 80% in the US. The service areas of the RBOCs are briefly described in the following seven sections, with a brief description of unregulated service provided for several RBOCs to provide an indication of how their business evolved beyond regulated communications.

NYNEX

NYNEX combined New York Telephone and New England Telephone. It provides regulated service to Connecticut, Maine, Massachusetts, New Hampshire, New York, Rhode Island and Vermont. NYNEX was acquired by Bell Atlantic in 1997.

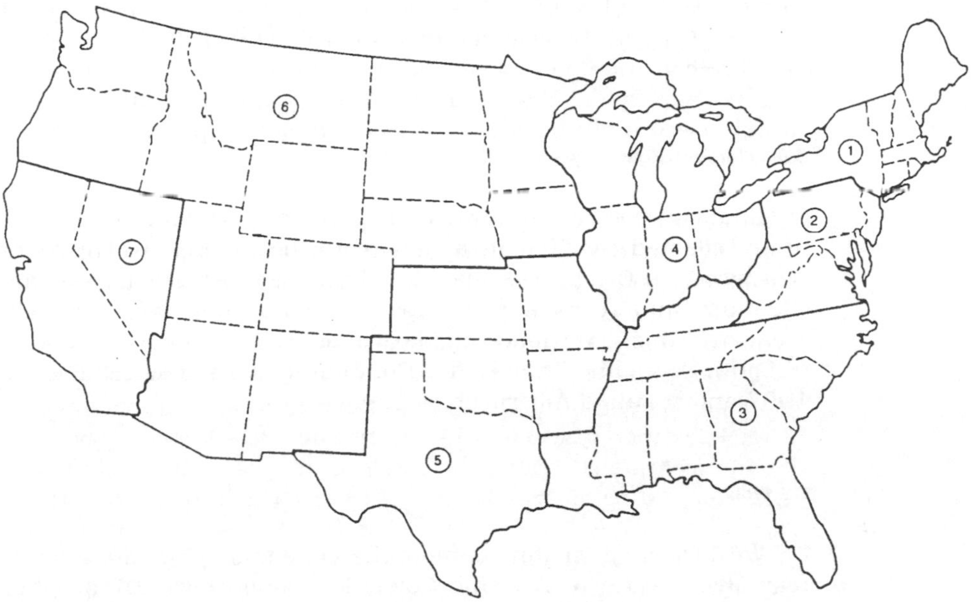

Figure 9.2 Regional Bell operating company distribution. 1. NYNEX, 2. Bell Atlantic, 3. BellSouth, 4. Ameritech, 5. Southwestern Bell (renamed SBC Communications), 6. US West, 7. Pacific Telesis

Bell Atlantic Corporation

A highly regimented marketing approach, not unlike that of the old Bell System, is the trademark of Bell Atlantic. It originally provided regulated service to Delaware, Maryland, New Jersey, Pennsylvania, Virginia and West Virginia and acquired NYNEX in 1997, which considerably expanded its service area.

BellSouth

BellSouth provides regulated service to Alabama, Florida, Georgia, Kentucky, Louisiana, Mississippi, North Carolina, South Carolina and Tennessee. In addition, BellSouth markets a premises-based PBX, frame relay packet service, ISDN service, and a high speed connectionless packet-switched data transmission service known as SMDS which is described later in this chapter. BellSouth International, a subsidiary of BellSouth, teamed with the UK's Cable & Wireless and Australian investors to operate Optus, Australia's second largest communications carrier. The firm's international subsidiary also operates cellular networks in several areas of South America.

American Information Technologies (Ameritech)

This RHC provided regulated service to Illinois, Indiana, Michigan, Ohio and Wisconsin. Ameritech also offers cellular, paging, interactive video and wireless communications for much of the United States and many parts of Europe. In 1999 the acquisition of Ameritech by SBC Communications was consumated.

SBC Communications

Originally known as Southwestern Bell, this company originally relied on unprecedented growth in its region to provide the impetus for revenue growth. Southwestern Bell provides regulated service to Arkansas, Kansas, Missouri, Oklahoma and Texas and has expanded to the south with a $1 billion investment in Telmex (Telefonos de Mexico) as well as initiating cellular ventures in South Korea and France. SBC Communications acquired Pacific Telesis in 1997 and acquired Ameritech in 1999.

US West

US West covers a territory of more than one million square miles. It provides regulated service to Arizona, Colorado, Idaho, Iowa, Minnesota, Montana, Nebraska, New Mexico, North Dakota, Oregon, South Dakota, Utah, Washington and Wyoming. At the time this book revision was prepared, both Global Crossings and Quest Communications were competing to acquire US West.

Pacific Telesis

The smallest RBOC, Pacific Telesis provides regulated service to California and Nevada. Other services offered by Pacific Telesis beyond local access include voice mail, directory publishing, wireless voice and data transmission and interactive video. Pacific Telesis was acquired by SBC Communications in 1997.

Although the primary business emphasis of each RBOC was to provide a regulated telephone service to their customers, they have also branched out into numerous business ventures. This diversification includes such communications-related activities as mobile telephone service and PBX distribution as well as such non-related activities as marketing EDP and financial services.

The Bell operating companies

The 22 operating companies are the heart of the regulated telephone service in the 48 contiguous states of the US. As subsidiaries of the seven RBOCs, they provide exchange service between telephone subscribers and interchange carriers (e.g. AT&T, GTE Sprint, MCI Worldcom). Intra-LATA service falls under their control, and toll service can be provided by a BOC as long as the call remains within the confines of a single LATA or connects LATAs within a state. The RBOCs and their respective territories are listed in Table 9.2.

Table 9.2 The original structure of the RBOCs

RBOC	BOC	States
NYNEX	New England Telephone	Connecticut, Maine, Massachusetts, New Hampshire, Rhode Island, Vermont
	New York Telephone	New York
Bell Atlantic	Bell of Pennsylvania	Pennsylvania
	New Jersey Bell	New Jersey
	Diamond State Telephone	Delaware
	Chesapeake and Potomac Telephones (three companies)	District of Columbia, Maryland, Virginia, West Virginia
BellSouth	Southern Bell	Alabama, Florida, Georgia, North Carolina, South Carolina
	South Central Bell	Kentucky, Louisiana, Mississippi, Tennessee
Ameritech	Illinois Bell	Illinois
	Indiana Bell	Indiana
	Michigan Bell	Michigan
	Ohio Bell	Ohio
	Wisconsin Telephone	Wisconsin
SBC Communications	Southwestern Bell	Arkansas, Kansas, Missouri, Oklahoma, Texas
US West	Mountain States Bell	Arizona, Colorado, Idaho, Montana, New Mexico, Utah, Wyoming
	Northwestern Bell	Iowa, Minnesota, Nebraska, North Dakota, South Dakota, Oregon, Washington
Pacific Telesis	Pacific Bell	California
	Nevada Bell	Nevada

9.2.4 AT&T service offerings

As previously indicated in this chapter, AT&T provides a large number of communications facilities that can be used for data communications. Since data can be transmitted over communications facilities originally designed for voice, and videoconferencing and emerging multimedia can be considered to represent data communications applications due to the digitization of voice, image and video, it is often difficult to distinguish between voice and data communications. In this section we will obtain an overview of a variety of AT&T facilities and services which can be used to transport data. It should be noted that most long distance communications carriers as well as the previously described RBOCs offer competitive facilities and services.

In this section we will focus our attention upon the following facilities and services marketed by AT&T for data communications users.

Switched analog
Dedicated analog
Terrestrial digital
Public services network
Satellite

Switched analog facilities

The AT&T system's extensive voice-oriented telephone network is also used to provide switched line facilities for data transmission. The four types of switched services are:

Long distance service (LDS)
Wide area telecommunications service (WATS)
800 service to include 888 and 887 toll-free dialing
900 service

These switched facilities consist of unconditioned voice-grade lines. Because these lines are subject to distortions caused by environmental conditions and network traffic and because several generations of switching equipment are used at various parts of the network, the quality of the lines must be set at the lowest common denominator. Modems and communications processing equipment can be used to effectively enhance voice-grade line performance characteristics. Less than a decade ago dial-up switched lines could be used to send data at a maximum speed of 2.4 kbps. Recent equipment improvement has pushed this ceiling to 33.6 kbps.

Long distance service

Long distance (LDS) service provides public dial-up facilities for inter-LATA point-to-point data transmission over conventional (switched) lines. It can also be used for infrequent or low-volume data transmission to diverse points by low- to medium-speed terminal equipment. Intra-LATA LDS is also available; however, data transfer rates are controlled by the RBOCs and BOCs and vary from state to state.

LDS is identical to the voice-oriented telephone network; using it is the same as making a local or toll call. Service charges are based on an initial one minute minimum with one minute increments. The user has the flexibility of calling any point in the network and is charged only for the total connect time. Dial-up lines are often used as backup facilities for private dedicated lines. In many data communications systems, dial-up lines are also used to reach a private communications system's multiplexing or message-switching equipment.

The best feature of LDS is its broad coverage. The network reaches approximately 200 million telephone stations in the US; it is connected to the telephone network in Canada and, through international direct-distance dialing (IDDD), to approximately 140 other countries.

Wide area telecommunications service

WATS is a bulk rate service for high volume users of the voice and data communications switched network. Several packages are available based on the geographic area covered, the amount of time per day allowed and the calling direction.

Geographically, WATS coverage can be intrastate or interstate. Intrastate WATS covers a specific state and is available in all 48 contiguous states from the local operating company. The rate for each state varies, and each state can have slightly different regulations regarding WATS use. Interstate WATS is also available to and from Puerto Rico and US Virgin Islands, while it is also possible to provide certain foreign locations with toll-free access to the United States.

The cost of WATS is based on several factors, including the service area selected, the number of access lines used, when calls occur and the total number of hours of calls during the month.

800 service

800 is an inward service that covers calls from outlying points to the 800 subscriber. The service is basically the same as WATS but in the opposite direction. As 800 numbers became depleted, inward WATS service was extended to use the 888 and 887 area code prefixes.

900 service

AT&T's DIAL-IT 900 Service can be used as a polling service to sample public opinion or to deliver recorded or live information to distant audiences by way of the telephone. For example, when used to measure public opinion, 7000 calls can be simultaneously serviced, usually with one number assigned to persons responding positively to a question and a second number for persons responding negatively to a question. Unlike 800, 888 and 887 Services, the call originator pays for the cost of 900 Service.

Dedicated analog facilities

Dedicated lines, also known as private or leased lines, account for most line facilities used for data communications. Although the basic transmission media used to carry private line traffic from one point to another may be the same as those being used to carry switched line traffic, with a dedicated line, the routing and transmission qualities are engineered to meet specific parameters. A voice-grade private line that is conditioned can be used to transmit data at up to 33.6 kbps and should yield a lower error rather than a switched voice-grade line. The carrier's tariff filed for each specific dedicated line service specifies the maximum distortions allowed on that type of line.

Basically, all dedicated lines are leased by the subscriber for full-time use; the choice for the user is really one of bandwidth, which influences the maximum speed possible. AT&T offers dedicated lines for such services as telegraph, voice, data, audio, television and wideband.

Telegraph service

These channels, formerly called the Series 1000, provide low-speed data and voice transmission services at speeds of 30 to 150 bps. The channels are designated for remote metering and signaling, Morse code teletypewriter, radio-telegraph traffic, foreign exchange, teletypewriter and data transmissions.

Voice and data service

Formerly the Series 2000, 3000 and 4000, these channels handle voice, radiotelephone, radiotelegraph, foreign exchange, remote metering, supervisory control, signaling and data transmissions. Data transfer rates range from 300 bps to 33.6 kbps based on the type of channel selected and the modem used on the channel.

Alliance 1300 and 2000 teleconferencing service

This service handles audio and analog graphics for teleconferencing with up to 59 conferees.

Wideband service

This service was formerly the Series 8000. Its wideband channels can transmit high-speed data at up to 230.4 kbps and facsimile traffic at up to 56 kbps.

Terrestrial digital facilities

Digital transmission facilities are not new to AT&T. The Bell System first introduced and put into service its T1 carrier system in the early 1960s. The use of cables for short-haul digital transmission is also common. What have become known as digital line facilities are carrier-provided, full digital long distance transmission services that are available on an end-to-end basis. Based on this definition, the AT&T system currently offers a family of digital services called ACCUNET.

DATAPHONE digital service

DATAPHONE Digital Service (DDS) is a fully digital transmission service for medium- and high-speed data and facsimile transmissions, available at speeds of 2.4, 4.8, 9.6, 19.2 and 56 kbps.

The two major advantages of this service are reliability and low cost. This service is designed to provide 99.5% error-free seconds at a 56 kbps speed and should perform even better at slower speeds. The phrase '99.5% error-free seconds' means that data transmitted in blocks of approximately 1 s duration, of which no more than 1 out of 200 should contain an error.

ACCUNET T1.5 service

AT&T's high capacity terrestrial digital service (HCTDS) provides T1 lines operating at 1.544 Mbps to customers with large digital traffic requirements.

Each T1 carrier can carry 24 circuits that individually have a speed of 56 or 64 kbps. Customers may lease as many of these terrestrial circuits as needed to create their own private data and voice communications systems.

ACCUNET Reserved 1.5/3.0 service

This high-speed switched digital service (HSSDS) is for the occasional user who needs to reserve sufficient capacity for expected peak periods. Speeds of either 1.566 or 3.0 Mbps are available on satellite and terrestrial circuits.

ACCUNET Switched Digital Services

This member of the ACCUNET digital services family is available at 56, 384 and 1536 kbps for transmitting synchronous data. This service is typically used for videoconferencing as well as to supplement or complement the use of leased digital lines.

ACCUNET Spectrum of Digital Services

Under this trademark AT&T provides fractional T1 service based on user selected 56/64 kbps increments from 56/64 kbps to 768 kbps. This end-to-end digital service is available within the United States as well as to Hawaii, Puerto Rico and the US Virgin Islands.

ACCUNET Fractional T45 Service

This digital service permits users to obtain the use of a fraction of a T3 transmission facility operating at 44.736 Mbps. Currently, AT&T offers fractional T3 service under the trademark T45 at data rates of 4.6, 6.2, 7.7 and 10.8 Mbps. In addition to Fractional T45 Service AT&T also offers full T3 transmission marketed as ACCUNET T45 Service.

Public network services

AT&T currently provides several complementary types of public network services. Two existing public network services are ACCUNET packet service and ACCUNET frame relay service. A new service offered by AT&T is a cell-based asynchronous transfer mode (ATM) service which supports a mixture of voice, data and video and is described later in this book.

ACCUNET packet service

This AT&T digital service is the backbone of several other services, including local area data transport (LADT) and circuit-switching digital capability

(CSDC). Formerly the basic packet switching service (BPSS), ACCUNET Packet Service is available for data communications at speeds of 4.8, 9.6 and 56 kbps. This service can be used to develop private packet switched networks by individual companies. Customers must lease the access and trunk circuits as well as provide their own packet assembly and disassembly equipment.

ACCUNET frame relay service

The key difference between packet and frame relay services involves the flow of data through the public network and the maximum operating rate supported.

ACCUNET packet service is based on ITU-T X.25 packet technology in which data are checked at each node as they flow through the network. Since each data packet must be received in its entirety to determine whether or not it was received correctly and, if not, to request a retransmission, the cumulative delays adversely effect throughput.

Although error checking at each node was a necessity during the 1970s and 1980s when analog transmission facilities provided the majority of the infrastructure for packet networks, the rapid installation of optical fiber transmission facilities lowered transmission error rates by several orders of magnitude. This essentially made error checking at each node in a packet network unnecessary and resulted in the introduction of frame relay services by AT&T and other communications carriers.

Under frame relay data are only checked for routing at each node, with the upper layer application responsible for error detection and correction at end nodes. This relatively simple change results in a significant increase in throughput since delays associated with an X.25 packet network due to error checking of data at each node are eliminated.

A second major change between X.25 packet networks and frame relay concerns the operating rate of information entering and exiting each network. Currently X.25 packet networks are limited to supporting a maximum operating rate of 56 kbps. In comparison, frame relay orignally supported a maximum data rate of 1.544 Mbps in North America which is the T1 operating rate, and 2.048 Mbps in Europe which is the E1 operating rate. In 1999 the operating rate of frame relay was extended to the T3 rate of 44.736 Mbps.

Satellite facilities

AT&T has grouped its family of services provided by satellite under the heading SKYNET. The satellite digital circuits (SDCs) are similar to the digital services provided by terrestrial digital circuits (TDCs). Because of limited availability of the SDCs, SKYNET 1.5 service is a high-volume, private-line communications service offering a transmission speed of 1.544 Mbps. Other services offered by SKYNET include a Television Service for monochrome or color television transmission and an Audio Service for stereo transmissions using two 7.5 or 15 kHz signals.

IP services

AT&T offers a range of Internet Protocol (IP) services for commercial and residential users. For residential users, AT&T provides dial-up Internet access. For business, the company can provide dedicated connections to the Internet via different types of digital leased lines, establish a virtual private network (VPN) through the Internet to minimize the cost of customer communications, perform Web site hosting and provide a variety of electronic commerce operations that results in AT&T functioning as an Internet Service Provider (ISP).

9.2.5 Regional Bell operating company offerings

Each regional Bell operating company (RBOC) is structured to provide more than a simple telephone service. Each RBOC has product lines that include telephone and data communications equipment Many of them also established unregulated subsidiaries or divisions to offer customer premise equipment (CPE) from multiple vendors.

Most RBOC offerings are similar and complement the services previously described for AT&T. This includes the public and private switched analog and digital services and they parallel the Private Line and ACCUNET families from AT&T.

The one major exception to RBOC offerings that are similar to those of AT&T concerns switched multimegabit data service (SMDS). Although SMDS is offered by many US carriers, to include Ameritech, Bell Atlantic, BellSouth, GTE, MCI Worldcom and Sprint, it was not provided by AT&T when this book was revised.

SMDS

Switched multimegabit data service was developed to facilitate the transmission of connectionless data packets throughout a metropolitan area within a radius of approximately 50 miles, typically using T1, T3 and, when available, Synchronous Optical (SONET) transmission facilities. Here the term connectionless is used to note that a transmitting device does not first have to establish a session connection with a receiver prior to transmitting data.

Connectionless transmission is primarily used on local area networks. Thus, SMDS can be viewed as providing an extension of LAN transmission across a metropolitan area. Since SMDS is a public network service, it provides a communications capability for both inter- and intra-company communications.

Since its original offering requiring a minimum access rate of 1.544 Mbps SMDS service availability has been lowered to 56/64 kbps access by some carriers. Figure 9.3 illustrates the difference in operating rate support between X.25 packet switching, frame relay and SMDS. The shaded portion of SMDS indicates that access below an operating rate of 1.544 Mbps is not universally available. Similar to frame relay, a connection to an SMDS packet

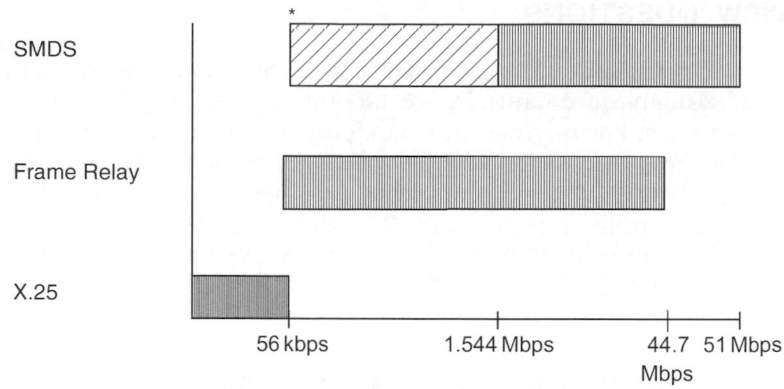

* optionally supported by some communications carriers

Figure 9.3 Comparing packet service operating rates

network is obtained via the installation of an appropriate digital leased line to the communications carrier, with a CSU/DSU required to be installed at the customer premises to terminate the digital circuit.

9.3 ATM OVERVIEW

Asynchronous Transfer Mode (ATM) represents an evolving networking technology designed to facilitate the integration of voice and data on both LANs and WANs. Since voice communication is time-dependent and cannot tolerate delays, a mechanism was required to transport time-dependent voice as well as data, so one would not interfere with the other. The mechanism developed was the use of very short, fixed length packets known as cells. By multiplexing 53 byte fixed length cells (5 byte header and 48 byte information field) ATM can transport both packetized voice and data without delaying voice so that it becomes distorted at the receiver due to transmission delays.

ATM has been developing at a rapid pace since 1993. Although ATM was originally designed to operate at the desktop at 155 Mbps its technology was modified to be used on a 25 Mbps local area network.

For wide area networks ATM cells are designed to be transported via T1 or T3 to a central office and to be switched onto a SONET transport system at operating rates up to 2.4 Gbps and higher between central offices. Thus, ATM represents a scalable technology by which voice and data is transported end-to-end in fixed size cells that can be carried at different data rates as cells flow onto and off different transport facilities.

As a relatively new technology, the cost associated with installing ATM represents a considerable expenditure for many organizations. Although communications carriers are implementing ATM as a transport mechanism, its widespread adoption as a LAN technology is far from certain due to competition from other technologies, such as 100 Mbps Ethernet and Gigabit Ethernet, both of which are described later in this book.

9.4 REVIEW QUESTIONS

1. What was the primary effect of the Communications Act of 1934 on the Bell System and its subscribers?

2. What was the impact of the Carterfone Decision upon the attachment of third-party products to the telephone network in the United States?

3. What is the purpose of a data access arrangement (DAA)? Where was the DAA originally located? Where is the DAA primarily located today?

4. What is the purpose of a point-of-presence?

5. If the use of bypass technology continues to increase, what effect do you think it will have upon existing local operating company subscribers? Why?

6. How does the Telecommunications Act of 1996 promote competition?

7. Discuss the relationships between an ILEC, CLEC, and IXC.

8. Discuss the competition that you can expect for both residential and business local telephone service as a result of the revision of the Communications Act of 1934 by the Telecommunications Act of 1996.

9. Describe the primary difference between the policy of the US government and other countries with respect to telecommunications deregulation.

10. Discuss the divestiture, merger, and acquisition trend in the telecommunications industry from 1984 to the present. Do you see a trend occurring?

11. What is the major difference between X.25 and frame relay packet services with respect to the operating rates they support and data flow through each network?

10

TRANSMISSION ERRORS: CAUSES, MEASUREMENTS AND CORRECTION METHODS

Transmission errors are similar to taxes in that they are basically unavoidable when a communications circuit is used. In this chapter we will first review the primary causes of transmission errors and commonly employed methods to determine line error rates. This information will provide you with the ability to determine the quality of a circuit with respect to an international standard which is discussed in this chapter as well as a basis for comparing circuit quality. The information previously presented will then be used as a foundation for the discussion of error detection and correction methods presented later in this chapter.

10.1 CAUSES OF TRANSMISSION ERRORS

As a signal propagates down a transmission medium, several factors can cause it to be received in error: the transmission medium employed and impairments caused by nature and machinery.

The transmission medium will have a certain level of resistance to current flow that will cause signals to attenuate. In addition, inductance and capacitance will distort the transmitted signals and there will be a degree of leakage which is the loss in a transmission line due to current flowing across, through insulators, or changes in the magnetic field.

Transmission impairments result from numerous sources. First, Gaussian or white noise is always present as it is the noise level that exists due to the thermal motions of electrons in a circuit. Next, impulse noise can occur from line hits due to atmospheric static or poor contacts in a telephone system.

Common causes of transmission errors in addition to noise include attenuation distortion, delay distortion jitter, microwave fading and harmonic distortion.

As discussed in Chapter 3, attenuation distortion results from a deterioration of the strength of a signal as it propagates over a transmission line. Although amplifiers can be used to rebuild the signal, since higher frequencies attenuate more rapidly than lower frequencies, the spacing of telephone company equipment, as well as the setting of the equipment, may not have the desired effect in rebuilding the signal. A similar problem occurs with respect to delay distortion, since frequencies propagate at different rates through a transmission medium. Due to this, the use of delay equalizers and their spacing in telephone company facilities may still result in some frequencies being received at the receiver at slightly different times although they were originally transmitted at the same time.

Although not a common occurrence due to shielding, power lines routed near transmission lines can cause a shift in the phase of an analog signal. This phase shift can result in the demodulated analog signal being received either too early or too late, a condition known as phase jitter. Two additional common causes of errors include microwave fading and harmonic distortion.

Microwave fading occurs due to such atmospheric conditions as heavy rain, thunderstorms and sunspots. Each of these disturbances can result in either the loss of a signal or a reduction in the quality of a signal. Because a large portion of communications between cities is via microwave transmission, it is not uncommon for many circuit quality tests to exhibit high error rates when performed during a period of atmospheric disturbances.

The harmonics of a signal are integral multiples of the signal, such as $3f$, $5f$, $7f$ and so on. Harmonic distortion results from the creation of unwanted harmonic frequencies which, when combined, produce a distorted signal.

With the exception of microwave fading, the previously mentioned transmission impairments are applicable to both LANs and WANs; however, their causes can considerably differ. For example, impulse noise on a WAN is primarily attributable to lightning and poor contacts in the relatively obsolete mechanical switches used on some telephone systems. In comparison, impulse noise on a LAN primarily results from electromagnetic interference, such as a LAN cable being routed to close to a florescent ballast, One impairment common to LANs is near end crosstalk (NEXT), which for obvious reasons does not occur on WANs,

10.2 PERFORMANCE MEASUREMENTS

Two common analog and digital circuit performance measurements are the bit error rate and block error rate.

10.2.1 Bit error rate

The bit error rate (BER) provides a measure of circuit quality. The BER is derived by dividing the number of bits received in error by the total number of bits transmitted during a predefined period of time. Thus

$$\text{BER} = \frac{\text{bits received inerror}}{\text{number of bits transmitted}}$$

The BER provides an indication of end-to-end channel performance. As an example, during the 1970s AT&T did extensive testing of the PSTN and determined that the typical data call conducted at 1200 bps over that network could be expected to have an error rate of 1 bit per 10^5 bits. Thus, without using a method of error detection and correction, communications over the PSTN could expect to result in 1 erroneous bit in every 100 000 bits transmitted.

Currently WAN digital transmission facilities usually provide a BER between 10^{-7} and 10^{-8}. In comparison, WAN analog leased lines provide a BER between 10^{-6} and 10^{-7}, whereas the error rate on most LANs falls below one bit in 10^8.

10.2.2 Bit error rate tester

To determine the bit error rate, a device called a bit error rate tester (BERT) is used. Bit error rate testing (BERT) involves generating a known data sequence into a transmission device and examining the received sequence at the same device or at a remote device for errors.

Normally, BERT testing capability is built into another device, such as a 'sophisticated' break-out box or a protocol analyzer; however, several vendors manufacture hand-held BERT equipment.

Since a BERT generates a sequence of known bits and compares a received bit stream to the transmitted bit stream, it can be used to test both communications equipment and line facilities. Figure 10.1 illustrates the typical employment of a bit error rate tester on an analog leased line. You would employ a BERT in the same manner to determine the bit error rate on a digital circuit, with the BERT used with CSU/DSUs instead of modems.

At the top of Figure 10.1, the placement of the modem closest to the terminal into a loop can be used to test the modem. Since a modem should always correctly modulate and demodulate data, if the result of the BERT shows even one bit in error, the modem is considered to be defective. If the distant modem is placed into a digital loop-back mode of operation where its transmitter is tied to its receiver to avoid demodulation and remodulation of data the line quality in terms of its BER can be determined. This is because the data stream from the BERT is looped back by the distant modem without that modem functioning as a modem.

In the lower portion of Figure 10.1, the use of two BERTs on a full-duplex transmission system is shown. Since a leased line is a pair of two wires, this type of test could be used to determine if the line quality of one pair was better than the other pair. On occasion, due to the engineering of leased lines through telephone company offices and their routing into microwave facilities, it is quite possible that each pair is separated by a considerable bandwidth. Since some frequencies are more susceptible to atmospheric disturbances

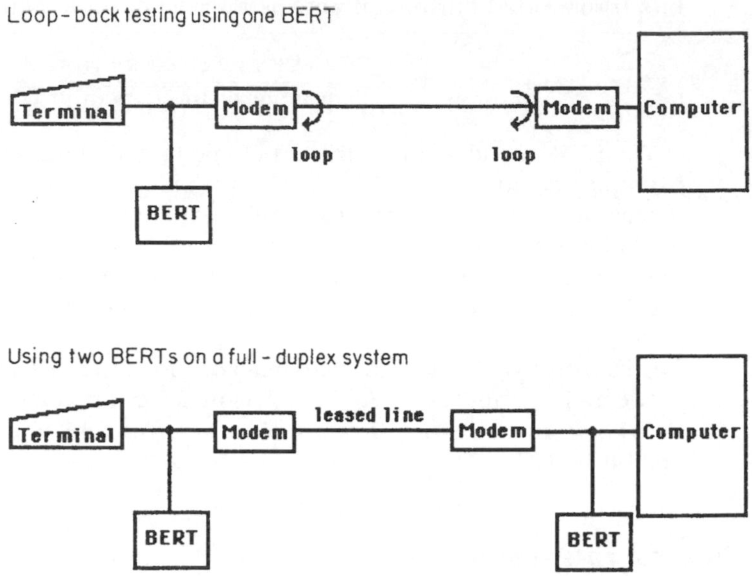

Figure 10.1 Using a bit error rate tester (BERT). If a single BERT is used you cannot isolate a line problem to a specific wire pair

than other frequencies, it becomes quite possible to determine that the quality of one pair is better than the other pair. In one situation the author is aware of, an organization that originally could not transmit data on a leased line determined that one wire pair had a low BER while the other pair had a very high BER. Rather than turn the line over to the communications carrier for repair during the workday, this organization switched their modems from full-duplex to half-duplex mode of operation and continued to use the circuits. Then after business hours they turned the circuit over to the communications carrier for analysis and repair.

10.2.3 BERT time

One common problem associated with bit error rate testing is in determining the amount of time to conduct a test. Table 10.1 lists the time required for three common bit error rate tests based on ten distinct data rates.

As an example of the use of Table 10.1, consider the 300 bps data rate for testing purposes. If during a test time of 5 minutes and 33 seconds exactly 7 bit errors occurred, then the bit error rate is 7×10^{-5}.

10.2.4 Performance classifications

ITU-T Recommendation G.821 defines four error rate performance categories. 'Available and Acceptable' involves intervals of test time of at least 1 minute,

Table 10.1 Bit error rate as opposed to test times

Data rate (bps)	Bit error rate 1×10^{-5}	1×10^{-6}	1×10^{-9}
300	5 min 33 s	55 min 33 s	55 555 min
600	2 min 47 s	27 min 47 s	27 777 min
1200	1 min 23 s	13 min 53 s	13 888 min
2400	42 s	6 min 57 s	6 944 min
4800	21 s	3 min 28 s	3 472 min
9600	11 s	1 min 44 s	1 736 min
19.2 K	6 s	52 s	868 min
56 K	–	17.8 s	297 min
1.544 M	–	15.6 s	260 min
		–	10.8 min
2.048 M	–	–	8.14 min

during which the error rate is under 10^{-6}. 'Available but Degraded' involves intervals of test time of at least 1 minute, during which the error rate is between 10^{-3} and 10^{-6}. 'Available but Unacceptable' involves intervals of test time of at least 1 s but less than 10 consecutive seconds, during which the error rate is greater than 10^{-3}, while the last performance category, 'Unavailable', involves intervals of test time of at least 10 consecutive seconds, during which the error rate is greater than 10^{-3}. Figure 10.2 illustrates in graphic form the four ITU G.821 performance classifications.

It is important to note that the performance classifications associated with G.821 were developed when the primary use of communications transmission facilities was to transport digitized voice. Because the original method used to digitize voice, known as Pulse Code Modulation (PCM), resulted in a 64 kbps data stream, an error rate between 10^{-3} and 10^{-4} which G.821 defines as available but degraded provides a level of voice communications that is still very understandable at the receiver. In comparison, an error rate between 10^{-3} and 10^{-4} would seriously impair the transmission of data. This is

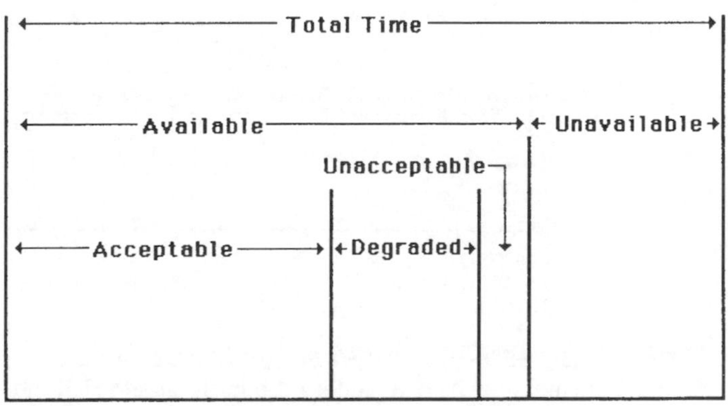

Figure 10.2 ITU G.821 performance classifications

because data are normally transmitted in blocks with an error checking mechanism operating on each block. Many modem data communications systems use block lengths from 512 bytes upward. At 8 bits per byte this involves the transfer of more than 4000 bits as an entity. Thus, an error rate of one bit in 1000 (10^{-3}) would result in each block having more than one bit in error. Because the receiver on detection of a transmission error requests the retransmission of the data block an error rate of 10^{-3} would in effect shut down the transfer of data.

10.2.5 Block error rate testing

Block error rate testing (BLERT) is used to analyze transmission in which data are grouped into blocks and the detection of a block error results in the retransmission of the block. For this type of transmission, BLERT provides a more realistic performance indicator than is obtainable by an examination of a circuit's BER. To understand this, consider the example of bit errors occurring on two circuits shown in Figure 10.3. Although 8 bit errors occurred on line A and only 4 on line B, line B probably had a higher block error rate since the errors are spaced further apart in time, thus affecting more blocks.

Similarly to the computation of a bit error rate, the block error rate (BLER) is obtained by dividing the number of blocks in which one or more bits are in error by the total number of blocks transmitted. Thus

$$\text{BLER} = \frac{\text{blocks with 1 or more bits in error}}{\text{total blocks transmitted}}$$

In general a high block error rate requires a lower block size to obtain a reasonable level of throughput. A BER with errors grouped in bursts affects fewer blocks and results in a lower BLERT than an equivalent BER with errors evenly distributed over time. As a result of BERT and BLERT testing, it becomes possible to improve network performance by adjusting the block size of a transmission system to the characteristics of the transmission facility used.

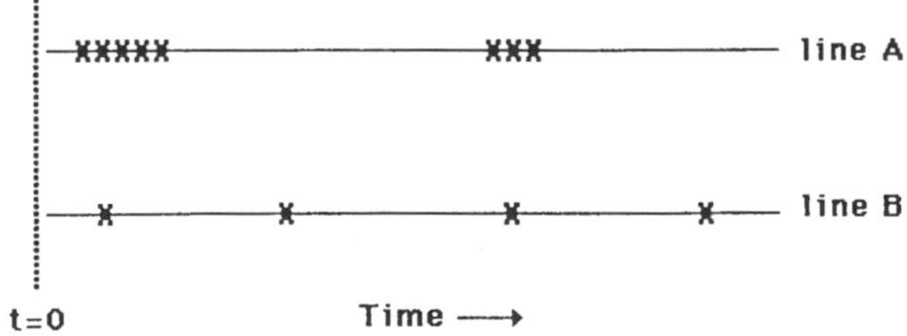

Figure 10.3 BER versus BLERT. Since the bit errors occurring on line B are spaced further apart in time, they have a higher probability of affecting more blocks than groups of bit errors occurring on line A. The BLERT of line B is thus probably higher than line A's, even though the BER on line A is higher than line B's (X = line impairment resulting in a bit error)

Table 10.2 Digital circuit error performance

	Error free seconds (%)	Error seconds in 8-hour day
ITU Recommendation G.821 for 64 kbps service	98.8	346
British Telecom 64 kbps KilostreamlMegastream Goal	99.5	144

10.2.6 Error-free second testing

The introduction of high-speed digital communications facilities has resulted in the use of a third error rate parameter to define network performance: errored seconds or error free seconds expressed as a percentage.

In an error-free second (EFS) test, received data are analyzed on a per second basis. If one or more bit errors occurs during a 1 s interval, the interval is recorded as an errored second. Thus

$$\text{EFS}(\%) = 100\% - \frac{\text{error seconds}}{\text{total seconds}} \times 100$$

Many communications carriers specify the availability of their high-speed digital facilities in terms of errored seconds or error-free seconds over a specified time period. Table 10.2 compares the ITU-T G.821 error performance recommendation with British Telecom's 64 kbps digital service goal.

In addition to error-free second testing there are several related tests used to define the characteristics of a digital transmission facility. Those tests determine the number of errored seconds (ES), severely errored seconds (SES), and failed seconds (FS). An errored second is a second that contains one or more bit errors, while a severely errored second is considered to be a second with 320 or more bits in error. If ten consecutive severely errored seconds occur, this condition is considered to represent a failed signal state. Then each signal in a failed state is considered to be a failed second.

10.3 ERROR DETECTION AND CORRECTION TECHNIQUES

To improve transmission, a variety of error detection and correction schemes have been developed. Since these schemes are typically based upon the type of transmission (asynchronous or synchronous) we will examine them with respect to each transmission category.

10.3.1 Asynchronous transmission

In asynchronous transmission the most common form of error control is the use of a single bit, known as a parity bit, for the detection of errors. Owing to the proliferation of personal computer communications, more sophisticated error detection methods have been developed which resemble the methods employed with synchronous transmission.

Parity checking

Character parity checking, which is also known as vertical redundancy checking (VRC), requires an extra bit to be added to each character in order to make the total quantity of 1s in the character either odd or even, depending upon whether you are employing odd parity checking or even parity checking. When odd parity checking is employed, the parity bit is set to 1 if the number of 1s in the character's data bits is even; or it is set at 0 if the number of 1s in the character's data bits is odd. When even parity checking is used, the parity bit is set to 0 if the number of Is in the character's data bits is even; or it is set to 1 if the number of 1s in the character's data bits is odd.

Two additional terms used to reference parity setting are 'mark' and 'space'. When the parity bit is set to a mark condition the parity bit is always 1, whereas space parity results in the parity bit always set to 0. Although not actually a parity setting, parity can be set to none, in which case no parity checking will occur. When transmitting binary data asynchronously, such as between personal computers, parity checking must be set to none or off. This enables all 8 bits to be used to represent a character. Table 10.3 summarizes the effect of five types of parity checking on the eighth data bit in asynchronous transmission.

For an example of parity checking, let us examine the ASCII character R whose bit composition is 1 0 1 0 0 1 0. Since there are three 1 bits in the character R, a 0 bit would be added if odd parity checking is used or a 1 bit would be added as the parity bit if even parity cheking is employed. Thus, the ASCII character R would appear as follows.

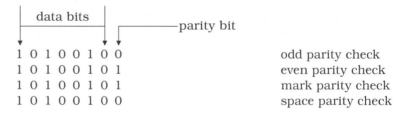

data bits	parity bit	
1 0 1 0 0 1 0	0	odd parity check
1 0 1 0 0 1 0	1	even parity check
1 0 1 0 0 1 0	1	mark parity check
1 0 1 0 0 1 0	0	space parity check

Table 10.3 Parity effect upon eighth data bit

Parity type	Parity effect
Odd	Eighth data bit is logical zero if the total number of logical 1s in the first seven bits is odd.
Even	Eighth data bit is logical zero if the total number of logical 1s in the first seven data bits is even.
Mark	Eighth data bit is always logical 1.
Space	Eighth data bit is always logical zero.
None/Off	Eighth data bit is ignored.

Since there are three bits set in the character R, a 0 bit is added if odd parity checking is employed while a 1 bit is added if even parity checking is used. Similarly, mark parity results in the parity bit being set to 1 regardless of the composition of the data bits in the character, while space parity results in the parity bit always being set to 0.

Undetected errors

Although parity checking is a simple mechanism to investigate if a single-bit error has occurred, it can fail when multiple-bit errors occur. This can be visualized by returning to the ASCII R character example and examining the effect of two bits erroneously being transformed as indicated in Table 10.4. Here the ASCII R character has three set bits and a one-bit error could transform the number of set bits to four. If parity checking is employed, the received set parity bit would result in the character containing five set bits, which is obviously an error since even parity checking is employed. Now suppose that two bits are transformed in error as indicated in the lower portion of Table 10.4. This would result in the reception of a character containing six set bits, which would appear to be correct under even parity checking. Thus, two bit errors in this situation would not be detected by a parity error detection technique.

In addition to the potential of undetected errors, parity checking has several additional limitations. First, the response to parity errors will vary depending on the type of computer with which you are communicating. Certain mainframes will issue a 'Retransmit' message on detection of a parity error. Some mainframes will transmit a character that will appear as a 'fuzzy box' on your computer screen in response to detecting a parity error, whereas the other mainframes will completely ignore parity errors.

When transmitting data asynchronously on a personal computer, most communications programs permit the user to set parity to odd, even, off, space or mark. Off or no parity would be used if the system with which you are communicating does not check the parity bit for transmission errors. No parity would be used when you are transmitting 8-bit EBCDIC or an extended 8-bit ASCII coded data, such as that available on the IBM PC and similar personal computers. Mark parity means that the parity bit is set to 1, and space parity means that the parity bit is set to 0.

In the asynchronous communications world, two common sets of parameters are used by most bulletin boards and information utilities, and

Table 10.4 Character parity cannot detect an even number of bit errors

ASCII character R	1	0	1	0	0	1	0	
Adding an even parity bit	1	0	1	0	0	1	0	1
		1						
1 bit in error	1	Ø	1	0	0	1	0	1
		1		1				
2 bits in error	1	Ø	1	Ø	0	1	0	1

are supported by mainframe computers. The first set consists of 7 data bits and I stop bit with even parity checking employed, and the second set consists of 8 data bits and 1 stop bit using no parity checking.

File transfer problems

Although visual identification of parity errors in an interactive environment is possible, what happens when you transfer a large file over the switched telephone network? During the 1970s a typical call over the switchboard telephone network resulted in the probability of a random bit error occurring approximately 1 in 100 000 bits at a data transmission rate of 1200 bps. To upload or download a 1000-line program containing an average of 40 characters per line a total of 320 000 data bits would have to be transmitted. During the 4.4 minutes required to transfer this file you could expect 3.2 bit errors to occur, probably resulting in several program lines being received incorrectly if the errors occur randomly. In such situations you would prefer an alternative to visual inspection. Thus, a more efficient error detection and correction method was needed for large data transfers.

Block checking

In this method, data are grouped into blocks for transmission. A checksum character is generated and appended to the transmitted block and the checksum is also calculated at the receiver, using the same algorithm. If the checksums match, the data block is considered to be received correctly. If the checksums do not match, the block is considered to be in error and the receiving station will request the transmitting station to retransmit the block.

One of the most popular asynchronous block checking methods is included in the XMODEM protocol, which was the first method developed to facilitate file transfers and is still extensively used in personal computer communications. This protocol blocks groups of asynchronous characters together for transmission and computes a checksum which is appended to the end of the block.. The checksum is obtained by first summing the ASCII value of each data character in the block and dividing that sum by 255. Then, the quotient is discarded and the remainder is appended to the block as the checksum. Thus, mathematically the XMODEM checksum can be represented as

$$\text{Checksum} = R\left(\frac{\sum_{i=1}^{128} \text{ASCII value of characters}}{255}\right)$$

where R is the remainder of the division process.

When data are transmitted using XMODEM protocol, the receiving device at the other end of the link performs the same operation upon the block being received. This 'internally' generated checksum is compared to the transmitted checksum. If the two checksums match, the block is considered to have been received error-free. If the two checksums do not match, the block is considered

Start of header	Block number	One's complement block number	128 data characters	Checksum

Figure 10.4 XMODEM protocol block format. The start of header is the ASCII SOH character whose bit composition is 00000001, while the one's complement of the block number is obtained by subtracting the block number from 255. The checksum is formed by first adding the ASCII values of each of the characters in the 128 character block, dividing the sum by 255 and using the remainder

to be in error and the receiving device will then request the transmitting device to resend the block.

Figure 10.4 illustrates the XMODEM protocol block format. The Start of Header is the ASCII SOH character whose bit composition is 00000001, and the one's complement of the block number is obtained by subtracting the block number from 255. The block number and its complement are contained at the beginning of each block to reduce the possibility of a line hit at the beginning of the transmission of a block causing the block to be retransmitted.

The construction of the XMODEM protocol format permits errors to be detected in one of three ways. First, if the Start of Header is damaged, it will be detected by the receiver and the data block will be negatively acknowledged. Next, if either the block number of the one's complement field is damaged, they will not be the one's complement of each other, resulting in the receiver negatively acknowledging the data block. Finally, if the checksum generated by the receiver does not match the transmitted checksum, the receiver will transmit a negative acknowledgement. For all three situations, the negative acknowledgement will serve as a request to the transmitting station to retransmit the previously transmitted block.

Data transparency

Since the XMODEM protocol supports an 8-bit, no parity data format it is transparent to the data content of each byte. This enables the protocol to support ASCII, binary and extended ASCII data transmission, where extended ASCII is the additional 128 graphic characters used by the IBM PC and compatible computers through the employment of an 8-bit ASCII code.

Error detection efficiency

Although the employment of a checksum reduces the probability of undetected errors in comparison to parity checking, it is still possible for undetected errors to occur under the XMODEM protocol. This can be visualized by examining the construction of the checksum character and the occurrence of multiple errors when a data block is transmitted.

0 10

..10011000~~1~~|00110001|001100~~01~~|...

Figure 10.5 Multiple errors on an XMODEM data block may not be detected

Assuming that a 128-character data block of all 1s is to be transmitted, each data character has the format 01100010, which is an ASCII 49. When the checksum is computed the ASCII value of each data character is first added, resulting in a sum of 6272 (128 × 49). Next, the sum is divided by 255, with the remainder used as the checksum, which in this example is 152.

Suppose that two transmission impairments occur during the transmission of a data block under the XMODEM protocol affecting two data characters as illustrated in Figure 10.5. Here the first transmission impairment converted the ASCII value of the character from 49 to 48, and the second impairment converted the ASCII value of the character from 49 to 50. Assuming that no other errors occurred, the receiving device would add the ASCII value of each of the 128 data characters and obtain a sum of 6272. When the receiver divides the sum of 255, it obtains a checksum of 152, which matches the transmitted checksum and the errors remain undetected. Although the preceding illustration was contrived, it illustrates the potential for undetected errors to occur under the XMODEM protocol. To make the protocol more efficient with respect to undetected errors, some bulletin boards have implemented a cyclic redundancy checking (CRC) method into the protocol. The use of CRC error detection reduces the probability of undetected errors to less than one in a million blocks and is the preferred method for insuring data integrity. The concept of CRC error detection is explained later in this section under synchronous transmission, as it was first employed with this type of transmission.

The XMODEM protocol will be explained in more detail in Chapter 11.

10.3.2 Synchronous transmission

The majority of error detection schemes employed in synchronous transmission involve geometric codes or cyclic code. However, several modifications to the original XMODEM protocol, such as XMODEM-CRC, use a cyclic code to protect asynchronously transmitted data.

Geometric codes

Geometric codes attack the deficiency of parity by extending it to two dimension. This involves forming a parity bit on each individual character as well as on all the characters in the block. Figure 10.6 illustrates the use of block parity checking for a block of 10 data characters. As indicated, this block parity character is also known as the 'longitudinal redundancy check' (LRC) character.

			Character parity bit
Character	1	1 0 1 1 0 1 1	0
Character	2	0 1 0 0 1 0 1	0
	3	0 1 1 0 1 0 0	0
	4	1 0 0 1 0 0 1	0
	5	0 1 1 1 1 0 1	0
	6	1 0 1 0 0 0 0	1
	7	0 1 0 1 1 1 0	1
	8	0 1 1 1 0 0 1	1
	9	1 0 0 0 1 1 0	0
	10	0 1 1 0 1 0 1	1
Block parity character (LRC)		1 1 1 0 1 0 1	1

Figure 10.6 VRC/LRC geometric code (odd parity checking)

Geometric codes are similar to the XMODEM error detection technique in that they are also far from foolproof. As an example of this, suppose that a 2-bit duration transmission impairment occurred at bit positions 3 and 4 when characters 7 and 9 in Figure 10.6 were transmitted. Here the two 1s in those bit positions might be replaced by two 0s. In this situation each character parity bit as well as the block parity character would fail to detect the errors.

A transmission system using a geometric code for error detection has a slightly better capability to detect errors than the method used in the XMODEM protocol and is hundreds of times better than simple parity checking. Although block parity checking substantially reduces the probability of an undetected error in comparison to simple parity checking on a character by character basis, other techniques can be used to further decrease the possibility of undetected errors. Among these techniques is the use of cyclic or polynomial code.

Cyclic codes

When a cyclic of polynomial code error detection scheme is employed, the message block is treated as a data polynomial $D(x)$, which is divided by a predefined generating polynomial $G(x)$, resulting in a quotient polynomial $Q(x)$ and a remainder polynomial $R(x)$, such that

$$D(x)/G(x) = Q(x) + R(x)$$

The remainder of the division process is known as the cyclic redundancy check (CRC) and is normally 16 bits in length or two 8-bit bytes. The CRC checking method is used in synchronous transmission similar to the manner in which the checksum is employed in the XMODEM protocol previously discussed, that is, the CRC is appended to the block of data to be transmitted. The receiving device uses the same predefined generating polynomial to generate its own CRC based on the received message block and then compares the 'internally' generated CRC with the transmitted CRC. If the two

Table 10.5 Common generating polynomials

Standard	Polynomial
CRC-16 (ANSI)	$X^{16}+X^{15}+X^5+1$
CRC (ITU-T)	$X^{16}+X^{12}+X^5+1$
CRC-12	$X^{12}+X^{11}+X^3+1$
CRC-32	$X^{32}+X^{26}+X^{22}+X^{16}+X^{12}+X^{11}+X^{10}+$ $X^8+X^7+X^5+X^4+X^2+X+1$

match, the receiver transmits a positive acknowledgement (ACK) communications control character to the transmitting device which not only informs the distant device that the data was received correctly but also serves to inform the device that if additional blocks of data remain to be transmitted the next block can be sent. If an error has occurred, the internally generated CRC will not match the transmitted CRC and the receiver will transmit a negative acknowledgement (NAK) communications control character which informs the transmitting device to retransmit the block previously sent.

Table 10.5 lists four generating polynomials in common use today. The CRC-16 is based upon the American National Standards Institute and is commonly used in the United States. The ITU-T CRC is commonly used in transmissions in Europe, whereas the CRC-12 is used with six-level transmission codes and has been basically superseded by the 16-bit polynomials. The 32-bit CRC is defined for use in local networks by the Institute of Electrical and Electronic Engineers (IEEE) and the American National Standards Institute (ANSI). For further information concerning the use of the CRC-32 polynomial you are is referred to the IEEE/ANSI 802 standards publications.

The column labeled polynomial in Table 10.5 actually indicates the set bits of the 16-bit or 12-bit polynomial. Thus, the CRC-16 polynomial has a bit composition of 11000000000010001.

International transmission

Due to the growth in international communications, one frequently encountered transmission problem is the employment of dissimilar CRC generating polynomials. This typically occurs when an organization in the United States attempts to communicate with a computer system in Europe or a European organization attempts to transmit to a computer system located in the United States. When dissimilar CRC generating polynomials are employed, the 2-byte block check character appended to the transmitted data block will never equal the block check character computed at the receiver. This will result in each transmitted data block being negatively acknowledged, eventually resulting in a threshold of negative acknowledgements being reached. When this threshold is reached the protocol aborts the transmission session, causing the terminal operator to reinitiate the communications procedure required to access the computer system they wish to connect to. Although the solution to this problem requires either the terminal or a port on the computer system to be

Table 10.6 Hamming code error correction example

Message length	= 7 bits
Information bits	= 4 bits = 1100
Parity	= 3 bits = P_1, P_2, P_3

changed to use the appropriate generating polynomial, the lack of publicity of the fact that there are different generating polynomials has caused many organizations to expend a considerable amount of needless effort. One bank, which the author is familiar with monitored transmission attempts for almost 3 weeks. During this period, they observed each block being negatively acknowledged and blamed the communications carrier, insisting that the quality of the circuit was the culprit. Only after a consultant was called and spent approximately a week examining the situation was the problem traced to the utilization of dissimilar generating polynomials.

Forward error correcting

During the 1950s and 1960s when mainframe computers used core memory circuits, designers spent a considerable amount of effort developing codes that carried information which enabled errors to be detected and corrected. Such codes are collectively called forward error correction (FEC) and have been employed in Trellis Coded Modulation modems under the ITU-T V.32, V.32bis, V.33, V.34 and V.90 recommendations. The FEC scheme used in those modems is called convolutional coding and was previously described in Chapter 8.

Another example of a forward error correcting code is the Hamming code. This code can be used to detect one or more bits in error at a receiver as well as determine which bits are in error. Since a bit can only have one of two values, knowledge that a bit is in error allows the receiver to reverse or reset the bit, correcting its erroneous condition.

The Hamming code uses m parity bits with a message length of n bits, where $n = 2^m - 1$. This permits k information bits where $k = n - m$. The parity bits are then inserted into the message at bit positions 2^{j-1} where $j = 1,2,\ldots,m$. Table 10.6 illustrates the use of a Hamming code error correction for $m = 3$, $k = 4$ and $n = 7$.

$$[P_1 P_2 1 P_3 100] \begin{bmatrix} 0 & 0 & 1 \\ 0 & 1 & 0 \\ 0 & 1 & 1 \\ 1 & 0 & 0 \\ 1 & 0 & 1 \\ 1 & 1 & 0 \\ 1 & 1 & 1 \end{bmatrix} = \begin{cases} P_1 + 0 + 1 + 0 + 1 + 0 + 0 = 0 \\ 0 + P_2 + 1 + 0 + 0 + 0 + 0 = 0 \\ 0 + 0 + 0 + P_3 + 1 + 0 + 0 = 0 \end{cases}$$

Figure 10.7 The Hamming code encoding process. Using exclusive OR arithmetic the three equations yield $P_1 = 0$, $P_2 = 1$, $P_3 = 1$

$$[0\ 1\ 1\ 1\ 0\ 0\ 0]\begin{bmatrix}0&0&1\\0&1&0\\0&1&1\\1&0&0\\1&0&1\\1&1&0\\1&1&1\end{bmatrix}=\begin{cases}0\ 0\ 1\ 0\ 0\ 0\ 0=1\\0\ 1\ 1\ 0\ 0\ 0\ 0=0\\0\ 0\ 0\ 1\ 0\ 0\ 0=1\end{cases}$$

Figure 10.8 The Hamming code decoding process

Encoding and decoding

In the Hamming encoding process the data and parity bits are exclusive ORed with all possible data values to determine the value of each parity bit. This is illustrated in Figure 10.7 which results in P_1, P_2 and P_3 having values of 0, 1 and 1, respectively.

Using the values obtained for each parity bit results in the transmitted message becoming 0111100. Now assume that an error occurs in bit 5, resulting-in the received message becoming 0111000. Figure 10.8 illustrates the Hamming code decoding process in which the received message is exclusive ORed against a matrix of all possible values of the three parity bits, yielding the sum 101_2, which is the binary value for 5. Thus, bit 5 is in error and is corrected by reversing or resetting its value.

In addition to being used in conventional high-speed modems, FEC is employed in digital subscriber line modems and cable modems. The extra parity bits required to detect errors are not used to convey data, explaining why the actual throughput obtainable through the use of DSL modems and cable modems is less than their transmission rate.

10.4 REVIEW QUESTIONS

1. Name five causes of transmission errors. Discuss the effect of each upon a signal.

2. Explain how atmospheric disturbances can affect transmission.

3. Describe and discuss a transmission impairment unique to WANs and one unique to LANs.

4. Compare the typical bit error rates on WAN analog and digital transmission lines to the error rates on a local area network. Why would you expect the LAN bit error rate to be lower?

5. What is meant by bit error rate testing? How is it performed?

6. Explain how a full-duplex four-wire circuit could be used for communications even if one wire pair had a very high bit error rate.

7. Assume that the average bit error rate on a circuit is 10^{-4} and data are to be transmitted in blocks of 1024 eight-bit characters. What would you expect to happen to the throughput of data being transmitted? Why

8. Assume that your transmission line operates at 2400 bps. What BERT test times should be used to determine the bit error rate in terms of bit errors per 100 000 bits?

9. Under what conditions does a block error rate provide a more realistic performance indicator than obtainable by an examination of a circuit's BER?

10. What is an errored second?

11. What is a severely errored second?

12. What is a failed second?

13. What is vertical redundancy checking? Discuss five types of vertical redundancy checking.

14. What would the parity bit setting be for the character whose bit composition is 1010100 if odd parity checking is employed? What would it be if even parity checking is employed?

15. What is the purpose of the XMODEM protocol checksum? How is that checksum formed?

11

THE WAN DATA LINK LAYER

In the ISO model, the data link layer is responsible for the establishment, control and termination of connections among network devices. To accomplish these tasks the data link layer assumes responsibility for the flow of user data as well as for detecting and providing a mechanism for recovery from errors and other abnormal conditions, such as a station failing to receive a response during a predefined time interval.

In this chapter we will first examine the key element that defines the data link layer: its protocol. In this examination we will differentiate between terminal protocols and data link protocols to eliminate this terminology as a potential area of confusion. Next, we will focus our attention on several specific protocols, starting with simple asynchronous line-by-line protocols. Protocols covered in the second section of this chapter include: an asynchronous teletypewriter; IBM's character-oriented binary synchronous communication, commonly referred to as BSC or bisync; Digital Equipment Corporation's digital data communications message protocol (DDCMP); and bit-oriented higher level data link control (HDLC) as well as its IBM near-equivalent, synchronous data link control (SDLC).

As described in Chapter 6, the IEEE 802 committee responsible for LAN standardization subdivided the data link layer into two or more entities, with the number of sublayers varying between two and four depending on the technology to include the signaling method associated with a specific local area network. All LANs have at least two data link sublayers, known as logical link control (LLC) and media access control (MAC). Although there are certain similarities between the WAN HDLC data link layer protocol discussed in this chapter and a LAN logical link control protocol, we will defer a discussion of the latter until we examine basic LAN operations as a separate entity in a later chapter in this book.

Because the focus of this chapter is upon the data link layer, we will not cover popular Internet protocols in this chapter. We will defer a discussion of Internet protocols until later in this book when we examine the TCP/IP protocol suite in detail to include its use for transporting such popular applications as electronic mail, file transfers and, of course, Web access.

11.1 TERMINAL AND DATA LINK PROTOCOLS: CHARACTERISTICS AND FUNCTIONS

Two types of protocol should be considered in a data communications environment: terminal protocols and data link protocols.

The data link protocol defines the control characteristics of the network, and it is a set of conventions that are followed which govern the transmission of data and control information. A terminal or a personal computer can have a predefined control character or set of control characters which are unique to the terminal and are not interpreted by the line protocol. This internal protocol can include such control characters as the bell, line feed and carriage return for conventional teletype terminals, blink and cursor positioning characters for a display terminal and form control characters for a line printer.

To experiment with members of the IBM PC series and compatible computers readers can execute the one-line BASIC program 10 PRINT CHR$(X)"4DEMO", substituting different ASCII values for the value of X to see the effect of different PC terminal control characters. As an example, using the value 7 for X, the IBM PC will beep prior to displaying the message DEMO, since ASCII 7 is interpreted by the PC as a request to beep the speaker. Using the value 9 for X will cause the message DEMO to be printed commencing in position 9, since ASCII 9 is a tab character which causes the cursor to move on the screen 8 character positions to the right. Another example of a terminal control character is ASCII 11, which is the home character. Using the value 11 for X will cause the message DEMO to be printed in the upper left-hand corner of the screen since the cursor is first placed at that location by the home character.

Although poll and select is normally thought of as a type of line discipline or control, it is also a data link protocol. In general, the data link protocol enables the exchange of information according to an order or sequence by establishing a series of rules for the interpretation of control signals which will govern the exchange of information. The control signals govern the execution of a number of tasks which are essential in controlling the exchange of information via a communications facility. Some of these information control tasks are listed in Table 11.1.

Although all of the tasks listed in Table 11.1 are important, not all are required for the transmission of data, since the series of tasks required is a function of the total data communications environment. As an example, a single terminal or personal computer connected directly to a mainframe or another terminal device by a leased line may not require the establishment and verification of the connection. Several devices connected to a mainframe computer on a multidrop or multipoint line would, however, require the verification of the identification of each terminal device on the line to ensure that data transmitted from the computer would be received by the proper

Table 11.1 Information control tasks

Connection establishment	Transmission sequence
Connection verification	Data sequence
Connection disengagement	Error control procedures

device. Similarly, when a device's session is completed, this fact must be recognized so that the mainframe computer's resources can be made available to other users. Connection disengagement on devices other than those connected on a point-to-point leased line thus permits a port on the front-end processor to become available to service other users.

11.1.1 Transmission sequence

Another important task is the transmission sequence which is used to establish the precedence and order of transmission, including both data and control information. As an example, this task defines the rules for when devices on a multipoint circuit may transmit and receive information. In addition to the transmission of information following a sequence, the data may be sequenced. Data sequencing themselves occurs when a long block is broken into smaller blocks for transmission, with the size of the blocks being a function of the personal computer's or terminal's buffer area and the error control procedure employed. By dividing a block into smaller blocks for transmission, the amount of data that must be retransmitted, in the event that an error in transmission is detected, is reduced.

Although the error checking techniques currently employed are more efficient when short blocks of information are transmitted, the efficiency of transmission correspondingly decreases since an acknowledgement (negative or positive) is returned to the device transmitting after each block is received and checked. For communications between remote job entry terminals and computers, blocks of up to several thousand characters are typically used. Block lengths from 80 to 1024 characters are, however, the most common sizes. Although some protocols specify block length, most protocols permit the user to set the size of the block and in the absence of a setting use a predefined default value.

11.1.2 Error control

Pertaining to error control procedures, the most commonly employed method to correct transmitted errors is to inform the transmitting device simply to retransmit a block. This procedure requires coordination between the sending and receiving devices, with the receiving device either continuously informing the sending device of the status of each previously transmitted block or transmitting a negative acknowledgement only when a block is received in error.

If the protocol used requires a response to each block and the block previously transmitted contained no detected errors, the receiver will transmit a positive acknowledgement and the sender will transmit the next block. If the receiver detects an error, it will transmit a negative acknowledgement and discard the block containing an error. The transmitting station will then retransmit the previously sent block. Depending on the protocol employed, a number of retransmissions may be attempted. If, however, a default limit is reached owing to a bad circuit or other problems, then the computer or terminal device acting as the master station may terminate the session, and the operator will have to reestablish the connection.

If the protocol supports transmission of a negative acknowledgement only when a block is received in error, additional rules are required to govern transmission. As an example, the sending device could transmit several blocks and, in fact, could be transmitting block $n + 4$ prior to receiving a negative acknowledgement concerning block n. Depending on the protocol's rules, the transmitting device could retransmit block n and all blocks after that block or finish transmitting block $n + 4$, then transmit block n and resume transmission with block $n + 5$.

11.2 TYPES OF PROTOCOL

In this section we will examine the characteristics, operation and utilization of several types of protocol that provide a predefined agreement for the orderly exchange of information. To facilitate this examination we will start with an overview of one of the simplest protocols in use and structure our overview of protocols with respect to their complexity.

11.2.1 Teletypewriter protocols

Teletypewriter terminals support relatively simple protocols that are used for conveying information. In general, a teletypewriter protocol is a line by line protocol that requires no acknowledgement of line receipt. Thus, the key elements in this protocol define how characters are displayed, when a line is terminated and when the next line is to be displayed. Some additional elements included in line by line teletypewriter protocols are actually part of the terminal protocol, since they define how the terminal should respond to specific control characters.

Teletype Model 33

One commonly used teletypewriter protocol is the Teletype™ Model 33 data terminal. Although this terminal has been obsolete for several decades, its large base of installed devices resulted in modern PCs providing compatibility with its protocol. In fact, many minicomputers and mainframes were designed to support the teletypewriter protocol and support for the so-called 'TTY' protocol is included in most communications software programs.

The Teletype Model 33 terminal transmits and receives data asynchronously on a line-by-line basis using a modified ASCII code in which lower-case characters received by a Model 33 are actually printed as their upper-case equivalent, a term known as 'fold-over-printing'. Although the ASCII code defines the operation of 32 control characters (refer to Table 4.12), only 11 control characters can be used for communications control purposes. Prior to examining the use of communications control characters in the teletypewriter protocol, let us first review the operational function and typical use of each control character. These characters were previously listed in Table 4.13 with the two-character designator CC following their meaning, and will be reviewed in the order of their appearance in the referenced table.

Communications control characters

As its name implies, the Null (NUL) character is a non-printable time delay or filler character. This character is primarily used for communicating with printing devices that require a defined period of time after each carriage return in which to reposition the printhead to the beginning of the next line. Many mainframe computers and bulletin boards operating on personal computers will prompt users to 'Enter the number of nulls'; this is a mechanism to permit conventional terminals, personal computers and personal computers with a variety of printers to use the system without obtaining garbled output.

The Start of Heading (SOH) is a communications control character used in several character-oriented protocols to define the beginning of a message heading data block. In synchronous transmission on a multipoint or multidrop line structure, the SOH is followed by an address which is checked by all devices on the common line to ascertain if they are the recipient of the data. In asynchronous transmission, the SOH character can be used to signal the beginning of a filename during multiple file transfers, permitting the transfer to occur without treating each file transfer as a separate communications session. Since asynchronous communications typically involve point-to-point communications, no address is required after the SOH character; however, both devices must have the same communications software program that permits multiple file transfers in this manner.

The Start of Text (STX) character signifies the end of heading data and the beginning of the actual information contained within the block. This communications control character is used in the bisynchronous protocol that will be examined later in this chapter.

The End of Text (ETX) character is used to inform the receiver that all the information within the block has been transmitted. This character is also used to denote the beginning of the block check characters appended to a transmission block as an error detection mechanism. This communications control character is primarily used in the bisynchronous protocol.

The End of Transmission (EOT) character defines the end of transmission of all data associated with a message transmitted to a device. If transmission occurs on a multidrop circuit the EOT also informs other devices on the line to check later transmissions for the occurrence of messages that could be addressed to them. In the XMODEM protocol the EOT is used to indicate the end of a file transfer operation.

The Enquiry (ENQ) communications control character is used in the bisynchronous protocol to request a response or status from the other station on a point to point line or to a specifically addressed station on a multidrop line. In response to the ENQ character, the receiving station may respond with the number of the last block of data that it successfully received. In a multidrop environment, the mainframe computer would poll each device on the line by addressing the ENQ to one particular station at a time. Each station would respond to the poll positively or negatively, depending on whether or not they had information to send to the mainframe computer at that point in time.

The Acknowledgement (ACK) character is used to verify that a block of data was received correctly. After the receiver computes its own 'internal' checksum or cyclic code and compares it to the one appended to the transmitted block, it will transmit the ACK character if the two checksums match. In the XMODEM protocol the ACK character is used to inform the transmitter that the next block of data can be transmitted. In the bisynchronous protocol the Data Link Escape (DLE) character is normally used in conjunction with the 0 and 1 characters in place of ACK character. Alternating DLE0 and DLE1 as positive acknowledgement to each correctly received block of data eliminates the potential of a lost or garbled acknowledgement resulting in the loss of data.

The Negative Acknowledgement (NAK) communications control character is transmitted by a receiving device to request the transmitting device to retransmit the previously sent data block. This character is transmitted when the receiver's internally generated checksum or cyclic code does not match the one transmitted, indicating that a transmission error has occurred. In the XMODEM protocol this character is used to inform the transmitting device that the receiver is ready to commence a file transfer operation as well as to inform the transmitter of any blocks of data received in error.

The Synchronous Idle (SYN) character is employed in the bisynchronous protocol to maintain line synchronization between the transmitter and receiver during periods when no data is transmitted on the line. When a series of SYN characters is interrupted, this indicates to the receiver that a block of data is being transmitted.

The End of Transmission Block (ETB) character is used in the bisynchronous protocol in place of an ETX character when data is transmitted in multiple blocks. This character then indicates the end of a particular block of transmitted data.

Information flow

Figure 11.1 illustrates one possible flow of information between a teletype compatible terminal and a computer system employing a basic teletypewriter protocol. In this protocol the terminal operator might first transmit the ENQ character, which is formed by pressing the Shift and E keys simultaneously. If the call originated over the PSTN, the ENQ character, in effect, tells the computer to respond with its status. Since the computer is beginning its servicing of a new connection request, it normally responds with a log-on message. This log-on message can contain one or more lines of data.

The first line of the log-on message shown in Figure 11.1 is prefixed with a carriage return (CR) line feed (LF) sequence, which positions the printhead (or cursor on a PC's monitor when the device operates a teletype emulation program) to the first column on a new line prior to printing the data in the received log-on message line. The log-on message line, as well as all following lines transmitted by the computer, will have a CR LF suffix, in effect,

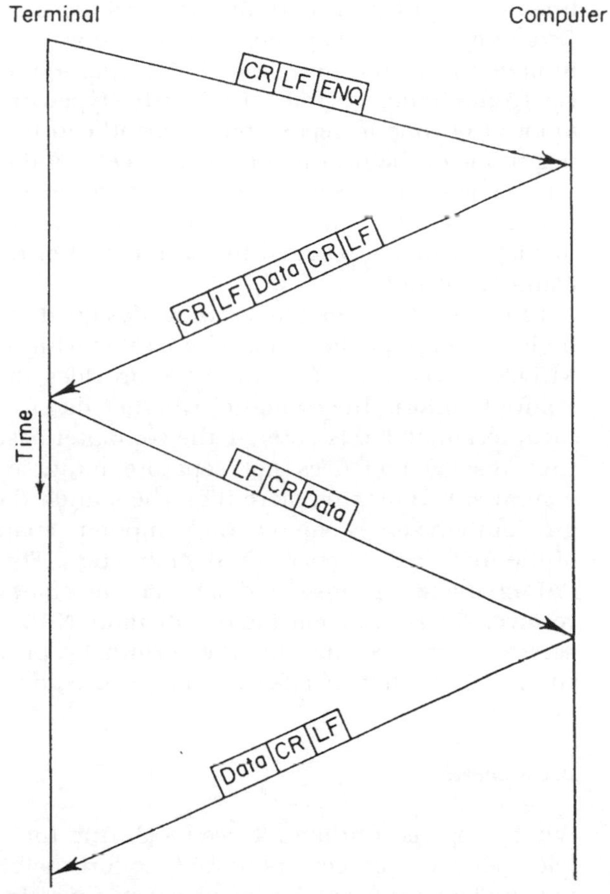

Figure 11.1 Basic teletypewriter protocol

preparing the terminal for the next line of data. On receipt of the log-on message the terminal operator keys in his or her log-on code, which is transmitted to the computer as data followed by the CR LF suffix which terminates the line entry.

Variations

There are numerous variations to the previously discussed teletypewriter protocol, of which space permits mentioning only two.

Some computers will not recognize an ENQ character on an asynchronous ASCII port. Those computers are normally programmed to respond to a sequence of two or more carriage returns. Thus, the sequence ENQ LF CR would be replaced by CR CR or CR CR CR.

With the growth in popularity and use of personal computers as terminals, it was found that the time delay built into computers to separate multiple lines of output from partially overstriking one another was not necessary. Originally, the transmission of one line was separated by several character intervals from the next line. This separation was required to provide the electromechanical printer used on teletypewriter terminals with a sufficient amount of time to reposition its printhead from the end of one line to the beginning of the next line prior to receiving the first character to be printed on the next line. Since a cursor on a video display can be repositioned almost instantly, the growth in the use of personal computers and video display terminals resulted in the removal of time delays between computer transmitted lines.

Today, some computer software designed to service asynchronous terminals will prompt the terminal operator with a message similar to "ENTER NULLS (0 TO 5)=". This message provides the terminal operator with the ability to inform the computer whether he or she is using an electromechanical terminal. If 0 is entered, the computer assumes that the terminal has a CRT display and does not separate multiple lines transmitted from the computer by anything more than the standard CR LF sequence. If a number greater than zero is entered, the computer separates multiple lines by the use of the indicated number of null characters. The NUL character, also called a PAD character, is considered to be a blank character which is discarded by the receiver. Thus, transmitting one or more NUL characters between lines only serves to provide time for the terminal's printhead to be repositioned to column 1 and has no effect on the received data.

Error control

What happens if a line hit occurs during the transmission of data when a teletypewriter protocol is used? Unfortunately, the only error detection mechanism employed by teletypewriter terminals and computer ports that supports this protocol is parity checking.

If parity checking is supported by the terminal, it may simply substitute and display a special error character for the character received with a parity error. This places the responsibility for error detection and correction on the terminal operator, who must first virtually observe the error and then request the computer to retransmit the line containing the parity error.

As previously discussed in Chapter 10, the response of a computer to a parity error can range from no action to the generation of a special symbol to denote the occurrence of a parity error. In fact, most asynchronous line by line protocols do not check for parity errors. These protocols use echoplex, a technique which involves the retransmission of each character received by the computer. Under echoplexing, the terminal operator is responsible for examining locally printed characters, each of which was first transmitted to the computer and retransmitted from the computer back to the terminal. Although echoplex does not isolate where in the round trip an error occurred, it at least provides an error indication to an operator that has the fortitude to examine the locally printed copy of transmitted data. If an error is then visually observed, the operator can correct a previously transmitted line by retransmitting the entire line of data.

11.2.2 PC file transfer protocols

Recognizing the limitations associated with teletypewriter protocols, a better mechanism than echoplexing was required for the transfer of files between bulletin boards and PCs. This mechanism was accomplished through the segmentation of portions of a file into a series of data blocks and the addition of a check sequence to each block that represents the operation of an algorithm applied to the block. This allowed a receiver to apply the same algorithm to each data block and compare the results to the transmitted check sequence appended to the block. In this section we will examine several popular file transfer protocols developed during the 1970s and 1980s and which remain in use today.

XMODEM protocol

The XMODEM protocol originally developed by Ward Christensen has been implemented in many asynchronous personal computer communications software programs and a large number of bulletin boards. Figure 11.2 illustrates the use of the XMODEM protocol for a file transfer consisting of two blocks of data. As illustrated, under the XMODEM protocol the receiving device transmits a Negative Acknowledgement (NAK) character to signal the transmitter that it is ready to receive data. In response to the NAK the transmitter sends a Start of Header (SOH) communications control character followed by two characters that represent the block number and the one's complement of the block number. Here the one's complement is obtained by subtracting the block number from 255. Next a 128-character data block is transmitted which in turn is followed by the checksum character. As previously discussed in Chapter 10, the checksum is computed by first adding the ASCII values of each of the characters in the 128-character block and dividing the sum by 255. Next, the quotient is discarded and the remainder is retained as the checksum.

If the data blocks are damaged during transmission, the receiver can detect the occurrence of an error in one of three ways. If the Start of Header is damaged, it will be detected by the receiver and the data block will be negatively acknowledged. If either the block count or the one's complement field are damaged, they will not be the one's complement of each other. Finally, the receiver will compute its own checksum and compare it to the transmitted checksum. If the checksums do not match, this is also an indicator that the transmitted block was received in error.

If the two checksums do not match or the SOH is missing, or the block count and its complement field are not the one's complement of each other, the block is considered to have been received in error. The receiving station will then transmit a NAK character which serves as a request to the transmitting station to retransmit the previously transmitted block. As illustrated in Figure 11.2, a line hit occurring during the transmission of the second block resulted in the receiver transmitting a NAK and the transmitting device resending the second block. Suppose more line hits occur, which affect the

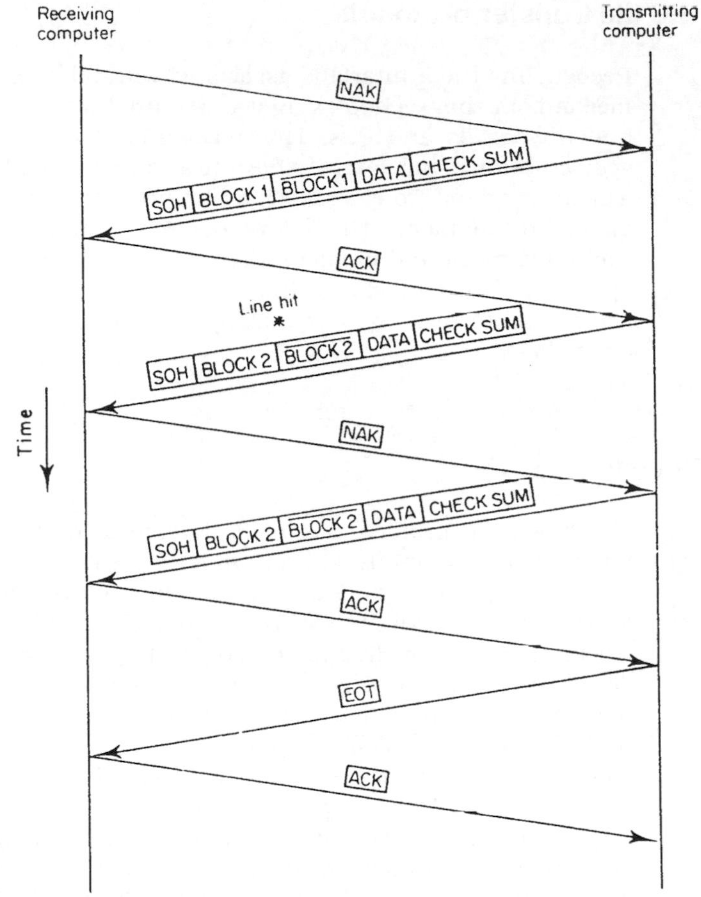

Figure 11.2 XMODEM protocol file transfer operation

retransmission of the second block. Under the XMODEM protocol the retransmission process will be repeated until the block is correctly received or until nine retransmission attempts occur. If, due to a thunderstorm or other disturbance, line noise is a problem, after ten attempts to retransmit a block the file transfer process will be aborted. This will require a manual operator intervention to restart the file transfer at the beginning and is one of the major deficiencies of the XMODEM protocol.

Three of the key limitations associated with the XMODEM protocol include its error detection scheme, its half-duplex transmission method and its block length. Its use of a checksum instead of a CRC makes the protocol more susceptible to undetected errors. Its half-duplex transmission scheme, in which each transmitted block must be acknowledged, significantly reduced throughput, whereas its 128 character block length is a holdover from an era when long distance transmission occurred primarily via analog transmission facilities and had a relatively high bit error rate. By recognizing the lower bit error rate associated with modern long distance transmission facilities other protocols substituted either a variable block sizing or extended the XMODEM data block to 1000 or more characters, resulting in a significant improvement in throughput during file transfer operations.

In spite of the limitations of the XMODEM protocol, it is still one of the most popular protocols employed by personal computer users for asynchronous data transfer because of several factors. First, the XMODEM protocol is in the public domain which means it is readily available at no cost for software developers to incorporate into their communications programs. Secondly, the algorithm employed to generate the checksum is easy to implement using a higher level language such as BASIC or Pascal. In comparison, a CRC-16 block-check character is normally generated using assembly language. In addition, the simplistic nature of the protocol is also easy to implement in BASIC or Pascal which enables many personal computer users to write their own routines to transfer files to and from bulletin boards using this protocol. Since the XMODEM protocol only requires a 256-character communications receiver buffer, it can be easily incorporated into communications software that will operate on personal computer systems with limited memory, such as the early systems that were produced with 64K or less RAM.

Several variations of the original XMODEM protocol have been introduced into the public domain. Some of these XMODEM protocols incorporate a true CRC block-check character error detection scheme in place of the checksum character, resulting in a much higher level of error detection capability.

Other variations of the XMODEM protocol support the transfer of multiple files, initiate file transfers without error detection and correction, or support full-duplex transmission using extended error detection and correction. Prior to examining synchronous protocols, we will examine some XMODEM derivatives as well as a file transfer protocol originally developed to support file transfer from mainframes to microcomputers.

XMODEM/CRC

An early modification to the XMODEM protocol was the replacement of the one byte checksum used with that protocol with a two byte Cyclic Redundancy Check (CRC-16). This replacement resulted in the name XMODEM/CRC being used to reference this protocol.

Figure 11.3 illustrates the block format of the XMODEM/CRC protocol. In comparing that block format to the XMODEM block format previously illustrated in Figure 10.4, you will note the similarity between the two formats, because only the error detection mechanism has changed.

SOH	Block Number	1's Complement of Block Number	128 Data Characters	CRC High Byte	CRC Low Byte

Figure 11.3 XMODEM/CRC block format

Through the use of a CRC, the probability of an undetected error is significantly reduced in comparison to the use of the XMODEM checksum. The CRC will detect all single- and double-bit errors, all errors with an odd number of bits, all bursts of errors up to 16 bits in length, 99.997% of 17 bit error bursts, and 99.998% of 18 bit and longer bursts.

To differentiate between the XMODEM and XMODEM/CRC protocols during the start-up or synchronization process, the XMODEM/CRC receiver transmits an ASCII C (43 h) instead of a NAK when requesting the sender to transmit the first block. Although the XMODEM/CRC protocol significantly reduces the probability of an undetected error, it is a half-duplex protocol similar to XMODEM and uses the same size data block. Thus, it removes only one of the three constraints associated with the XMODEM protocol.

YMODEM and YMODEM BATCH

Under the YMODEM protocol a header block is used to relay the filename and other information, and multiple files can be transmitted in a batch mode. In addition, data are normally transferred in 1024 byte blocks, which results in more time being spent actually transferring data and less time spent computing checksums or CRCs and sending acknowledgements.

The original development of the YMODEM protocol only included support for the transfer of one file at a time using 1024 byte (1 K) blocks. Although many communications software programs implement YMODEM correctly as it was designed (as a single file protocol) most modern communications programs implement it as a multiple-file protocol. In actuality the multiple-file protocol version of YMODEM is normally and correctly referred to as YMODEM BATCH. Since YMODEM BATCH is the same as YMODEM, except that the former allows multiple-file (batch) transfers, we will examine both protocols in this section and collectively refer to them as YMODEM, although this is not technically correct.

The format of the YMODEM protocol is illustrated in Figure 11.4. Under this protocol the Start of Text (STX) character whose ASCII value is 02 h replaces the SOH character used by the XMODEM and XMODEM/CRC protocols. The use of the STX character informs the receiver that the block contains 1024 data characters; however, the receiver can also accept 128 data character blocks. When 128 data character blocks are sent, the SOH character replaces the STX character.

When a YMODEM data transfer is initiated, the receiver transmits the ASCII C to the sender to synchronize transmission startup as well as indicate that CRC

STX	Block Number	1's Complement of Block Number	1024 Data Characters	CRC High Byte	CRC Low Byte

Figure 11.4 YMODEM block format

checking is to be employed. The sender then opens the first file and transmits block number 0 instead of block number 1, used with the XMODEM and XMODEM/CRC protocols. Block number 0 will contain the filename of the file being transmitted and may optionally contain the file length, and creation date.

Based upon the manner in which most personal computer operating systems work, the creation or modification date of a file being downloaded is modified to the current date when the file is received. For example, if the file being downloaded was created on JAN 5 1999 and today's date is JLY 19 2000, the file date would be changed to JLY 19 2000 when the file is downloaded. The remaining data characters in the block are set to null. Once block 0 containing the filename and any optional information has been correctly received, it will be ACKed if the receiver can perform a 'write open' operation. If the receiver cannot perform a 'write open' operation it will transmit a CAN character to cancel the file transfer operation. Once an ACK has been received by the sender, it will commence transferring the contents of the file similarly to the manner in which data are transmitted using the XMODEM/CRC protocol.

During the data transfer the sender can switch between 128 and 1024 data character blocks by prefixing 128 data character blocks with the SOH character and 1024 data character block with the STX character. After the contents of a file have been successfully transmitted, the receiver will transmit an ASCII C, which serves as a request for the next file. If no additional files are to be transmitted the sender will transmit a data block of 128 data characters, with the value of each character set to an ASCII 00 h or NUL character.

Figure 11.5 illustrates the transmission of the file named INVOICE.DAT which was last modified on JLY 19 2000 at 18:45 hours and which contains 2274 characters of information. Figure 11.5 uses a time chart to illustrate the file transfer.

To initiate the file transfer, the receiver transmits an ASCII C to the sender. On its receipt the sender transmits a 128 data block numbered as block 0. This block is prefixed with the SOH character to differentiate it from a 1024-data character block. The 'file info' field in block number 0 contains the filename (INVOICE.DAT) followed by the time the file was last modified (18:45), the date the file was last modified (JLY 19 2000), and the file size in bytes (2274). A single space is used to separate the modification date from the file size, resulting in a total of 30 characters used to convey file information. Since the smallest data block supported contains 128 characters, 98 NULs are added to complete the block. When this block is acknowledged by the receiver, the sender transmits the first 1024 data characters of the file through the use of a 1024-data block. This block is prefixed with the STX character to distinguish it from a 128-data character block.

Since the file size was 2274 characters, the YMODEM protocol attempts to use as many 1024-data character blocks as possible. Thus, blocks 01 and 02 are transmitted as 1024-data character size blocks, which leaves 226 bytes remaining to be transmitted. Block 3 is then transmitted as a 128-data character block, since the number of bytes remaining to be transmitted would result in the waste of space if transmitted in a 1024-data character block. Thus, the transmission of block 3 leaves 98 bytes remaining in the file to be transmitted. The remaining bytes in the file are then transmitted using another 128-data character block in which 30 characters are set to NULs.

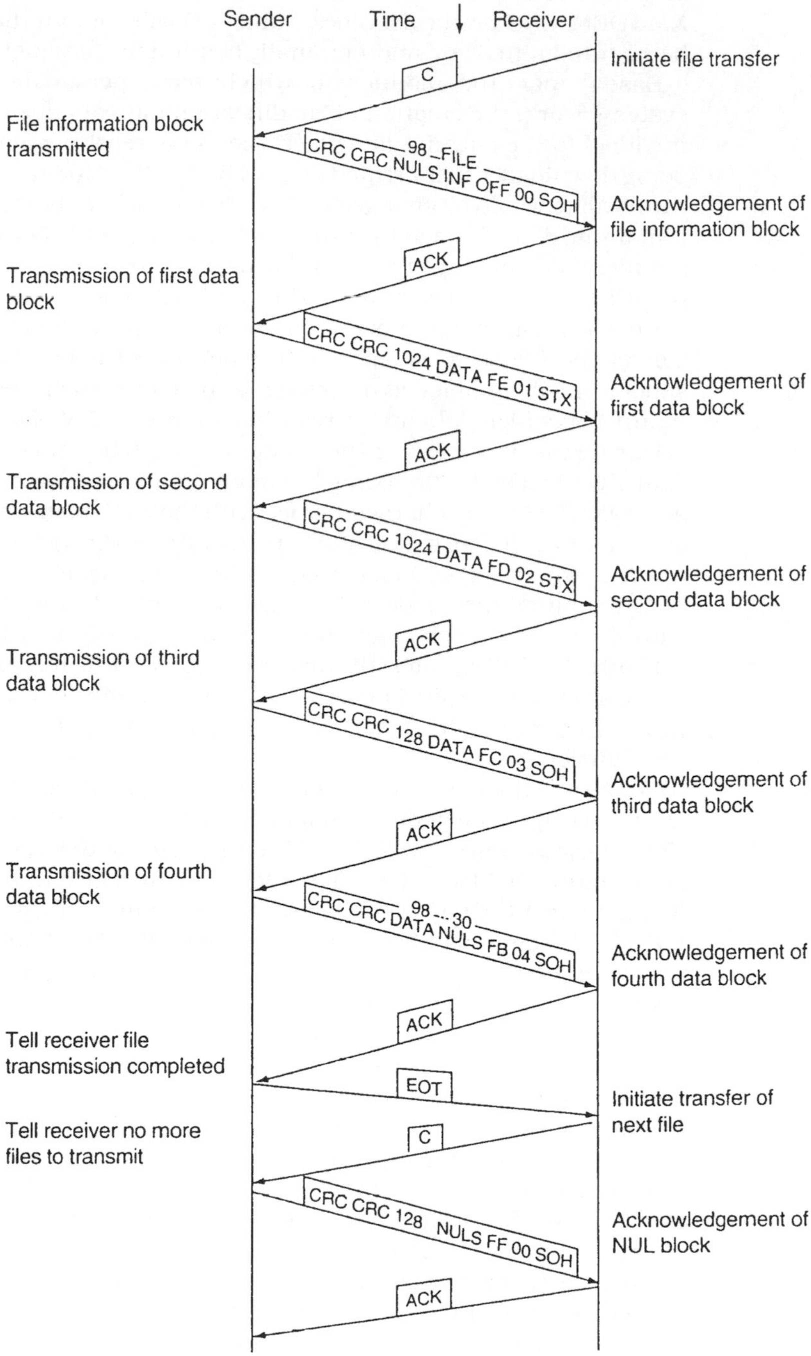

Figure 11.5 YMODEM protocol file transfer example

Once the last block has been successfully transferred, the sender transmits the EOT character to denote the completion of the transfer of the file. The receiver then transmits an ASCII C as an indicator to the sender to initiate the transfer of the next file. Since only one file was to be transmitted, the sender transmits a new block number 00 h that contains 128 NUL characters, signifying that no more files remain to be transmitted. Once this block has been acknowledged, the transmission session is complete.

The header information carried by YMODEM enables communications programs to compute the expected duration of the file transfer operation. This explains why most programs will visually display the file transfer time and why some programs will provide an updated bar chart of the progress of a YMODEM file transfer, although they cannot do the same when the XMODEM or XMODEM/CRC protocols are used.

XMODEM-1K

The XMODEM-1K protocol is a derivative of the XMODEM standard. The XMODEM-1K protocol follows the previously described XMODEM protocol, substituting 1024-byte blocks in place of byte data blocks. The XMODEM-1K is not compatible with the YMODEM nor the YMODEM BATCH protocols, as the former does not send or accept a block 0, which contains file information. Since the block size of this protocol is significantly longer than that of the XMODEM protocol, you can expect a higher level of throughput when transmitting on good quality circuits using XMODEM-1K.

YMODEM-G and YMODEM-G BATCH

The development of error correction and detection modems essentially made the use of checksum and CRC checking within a protocol redundant. In recognition of this, a G option was originally added to the YMODEM protocol, which changed it into a 'streaming' protocol in which all data blocks are transmitted one after another, with the receiver acknowledging the entire transmission. This acknowledgement simply acknowledges the entire transmission without the use of error detection and correction. In fact, the 2-byte CRC field is set to zero during a YMODEM-G transmission. Thus, this protocol should only be used with error-correcting modems that provide data integrity.

Like the YMODEM BATCH protocol, the YMODEM-G BATCH protocol permits multiple files to be transmitted, and it transmits the first 128-data character block with file information in the same manner as carried by the YMODEM BATCH protocol. To differentiate YMODEM-G BATCH from YMODEM-G, the receiver will initiate a batch transfer by sending the ASCII G instead of the ASCII C. When the sender recognizes the ASCII G, it bypasses the wait for an ACK to each transmitted block and sends succeeding blocks one after another, subject to any flow control signals issued by an attached modem or by a packet network if that network is used to obtain a transmission path. When the transmission has been completed, the sender transmits an EOT character and the receiver returns an ACK, which serves

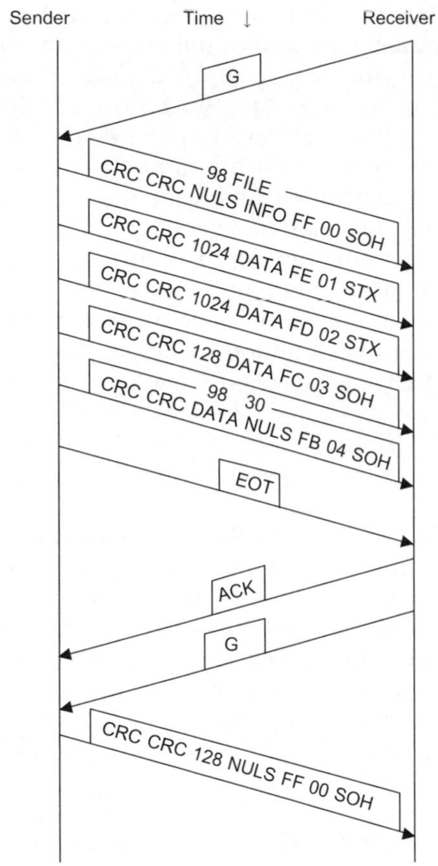

Figure 11.6 YMODEM-G protocol file transfer example

to acknowledge the entire file transmission. The ACK is then followed by the receiver transmitting another ASCII G to initiate the transmission of the next file. If no additional files are to be transmitted, the sender then transmits a block of 128 characters with each character set to an ASCII 00 h or NUL character.

Figure 11.6 illustrates the transmission of the previously described INVOICE.DAT file using the YMODEM-G protocol. In comparing Figure 11.6 to Figure 11.5 it becomes obvious that the streaming nature of YMODEM-G and YMODEM-G BATCH increases transmission throughput, resulting in a decrease in the time required to transmit a file or group of files.

ZMODEM

The development of the ZMODEM protocol was funded by Telenet which is now operated by Sprint as SprintNet. This packet switching vendor turned to Mr. Chuck Forsberg, the author of the original YMODEM protocol to develop a file transfer protocol that would provide a more suitable mechanism for transferring information via packet networks. The resulting file transfer protocol, which was called ZMODEM, corrected many of the previously described constraints associated with the use of the XMODEM and YMODEM protocols. Significant features of the ZMODEM protocol include its streaming file transfer operation, an extended error detection capability, automatic file transfer capability, the use of data compression and downward compatibility with the XMODEM-1K and YMODEM protocols.

The streaming file transfer capability of ZMODEM is similar to that incorporated into YMODEM-G; that is, the sender will not receive an acknowledgement until the file transfer operation has been completed. In addition to the streaming capability, ZMODEM supports the transmission of conventional 128- and 1024-byte block lengths of XMODEM-based protocols. In fact, ZMODEM is backward compatible with XMODEM-1K and YMODEM.

The extended error detection capability of ZMODEM is based on the ability of the protocol to support both 16- and 32-bit CRCs. According to the protocol developer, the use of a 32-bit CRC reduces the probability of an undetected error by at least five orders of magnitude below that obtainable from the use of a 16-bit CRC. In fact, the use of 32-bit CRCs are commonly used with local area network protocols to reduce the probability of undetected errors occurring on LANs.

The automatic file transfer capability of ZMODEM enables a sending or receiving computer to trigger file transfer operations. In comparison, XMODEM and YMODEM protocols and their derivatives are receiver-driven. Concerning the file transfer start-up process, a file transfer begins immediately under ZMODEM, whereas XMODEM and YMODEM protocols and their derivatives have a 10 second delay as the receiver transmits NAKs or another character during protocol start-up operations.

An additional significant feature associated with the ZMODEM protocol is its support of data compression. When transmitting data between Unix systems, ZMODEM compresses data using a 12-bit modified Lempel-Ziv compression technique, similar to the modified Lempel-Ziv technique incorporated into the ITU-T V.42 bis modem recommendation. When ZMODEM is used between non-Unix systems, compression occurs through the use of Run-Length Encoding similar to MNP Class 5.

Kermit

Kermit was developed at Columbia University in New York City primarily as a mechanism for downloading files from mainframes to microcomputers. Since its original development this protocol has evolved into a comprehensive communications system which can be employed to transfer data between most types of intelligent devices. Although the name might imply some type of acronym, in actuality this protocol was named after Kermit the Frog, the star of the well known Muppet television show.

Kermit is a half-duplex communications protocol which transfers data in variable sized packets, with a maximum packet size of 96 characters. Packets are transmitted in alternate directions, because each packet must be acknowledged in a manner similar to the XMODEM protocol.

In comparison with the XMODEM protocol which permits seven- and eight-level ASCII as well as binary data transfers in their original data composition, all Kermit transmissions occur in seven-level ASCII. The reason for this restriction is the fact that Kermit was originally designed to support file transfers to seven-level ASCII mainframes. Binary file transfers are supported by the protocol prefixing each byte whose eighth bit is set by the ampersand ($) character. In addition, all characters transmitted to include seven-level ASCII must be printable, resulting in Kermit transforming each ASCII control character with the pound (£) character. This transformation is accomplished through the complementation of the seventh bit of the control character 64 modulo 64 is thus added or subtracted from each control character encountered in the input data stream. When an 8-bit byte is encountered whose low order 7 bits represent a control character, Kermit appends a double prefix to the character. Thus, the byte 100000001 would be transmitted as &£A.

Although character prefixing adds a considerable amount of overhead to the protocol, Kermit includes a run length compression facility which may partially reduce the extra overhead associated with the control character and binary data transmission. Here, the tilde (˜) character is used as a prefix character to indicate run length compression. The character following the tilde is a repeat count, while the third character in the sequence is the character to be repeated. Thus, the sequence XA is used to indicate a series of 88 As, since the value of X is 1011000 binary or decimal 88. Through the use of run length compression the requirement to transmit printable characters results in an approximate 25% overhead increase in comparison to the XMODEM protocol for users transmitting binary files. If ASCII data are transmitted, Kermit's efficiency can range from more efficient to less efficient in comparison to the XMODEM protocol, with the number of control characters in the file to be transferred and the susceptibility of the data to run length compression the governing factors in comparing the two protocols.

Figure 11.7 illustrates the format of a Kermit packet. The Header field is the ASCII Start of Header (SOH) character. The Length field is a single character whose value ranges between 0 and 94. This one-character field defines the packet length in characters less two, since it indicates the number of characters to include the checksum that follow this field.

The Sequence field is another one-character field whose value varies between 0 and 63. The value of this field wraps around to 0 after each group of 64 packets is transmitted.

The Type field is a single printable character which defines the activity the packet initiates. Packet types include D (data), Y (acknowledgement), N (negative acknowledgement), B (end of transmission or break), F (file header), Z (end of file) and E (error).

Header	Length	Sequence	Type	Data	Check

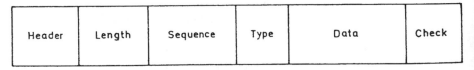

Figure 11.7 The Kermit packet format. The first three fields in the Kermit packet are one character in length and the maximum total packet length is 96 or less characters

The information contents of the packet are included in the Data field. As previously mentioned, control characters and binary data are prefixed prior to their placement in this field.

The Check field can be one, two or three characters in length, depending on which error detection method is used since the protocol supports three options. A single character is used when a checksum method is used for error detection. When this occurs, the checksum is formed by the addition of the ASCII values of all characters after the Header character through the last data character, and the low order 7 bits are then used as the checksum. The other two error detection methods supported by Kermit include a two-character checksum and a three-character 16-bit CRC. The two- character checksum is formed similar to the one-character checksum; however, the low order 12 bits of the arithmetic sum are used and broken into two 7-bit printable characters. The 16-bit CRC is formed using the CCITT standard polynomial, with the high order 4 bits going into the first character while the middle 6 and low order 6 bits are placed into the second and third char-acters, respectively.

By providing the capability to transfer both the filename and contents of files, Kermit provides a more comprehensive capability for file transfers than XMODEM. In addition, Kermit permits multiple files to be transferred in comparison to XMODEM, which requires the user to initiate file transfers on an individual basis.

11.2.3 Bisynchronous protocols

During the 1970s IBM's BISYNC (binary synchronous communications) protocol was one of the most frequently used for synchronous transmission. This particular protocol is actually a set of very similar protocols that provides a set of rules which effect the synchronous transmission of binary-coded data.

Although there are numerous versions of the bisynchronous protocol in existence, three versions account for the vast majority of devices operating in a bisynchronous environment. These three versions of the bisynchronous protocol are known as 2780, 3780 and 3270. The 2780 and 3780 bisynchronous protocols are used for remote job entry communications into a mainframe computer, with the major difference between these versions the fact that the 3780 version performs space compression, whereas the 2780 version does not incorporate this feature. In contrast to the 2780 and 3780 protocols that are designed for point to point communications, the 3270 protocol is designed for operation with devices connected to a mainframe on a multidrop circuit or device connected to a cluster controller which, in turn, is connected to the mainframe. Thus, 3270 is a poll and select software protocol.

Originally, 2780 and 3780 workstations were large devices that controlled such peripherals as card readers and line printers. Today, an IBM PC or compatible computer can obtain a bisynchronous communications capability through the installation of a bisynchronous communications adapter card into the PC's system unit. This card is designed to operate in conjunction with a bisynchronous communications software program which with the adapter card enables the PC to operate as an IBM 2780 or 3780 workstation or as an IBM 3270 type of interactive terminal.

The bisynchronous transmission protocol can be used in a variety of transmission codes on a large number of medium- to high-speed equipment. Some of the constraints of this protocol are that it is limited to half-duplex transmission and that it requires the acknowledgement of the receipt of every block of data transmitted. A large number of protocols have been developed owing to the success of the BISYNC protocol. Some of these protocols are bit-oriented, whereas BISYNC is a character-oriented protocol; and some permit full-duplex transmission, whereas BISYNC is limited to half-duplex transmission.

Data code use

Most bisynchronous protocols support several data codes, including ASCII and EBCDIC. Error control is obtained by using two-dimensional parity check (LRC/VRC) when transmission is in ASCII. When transmission is in EBCDIC the CRC-16 polynomial is used to generate a block-check character.

Figure 11.8 illustrates the generalized bisynchronous block structure. For synchronization, most BISYNC protocols require the transmission and detection of two successive synchronization (SYN) characters. The start of message control code is normally the STX communications control character. The end of message control code can be either the End of Text (ETX), End of Transmission Block (ETB), or the End of Transmission (EOT) character; the actual character, however, depends upon whether the block is one of many blocks, the end of the transmission block, or the end of the transmission session.

The ETX character is used to terminate a block of data started with a SOH or STX character which was transmitted as an entity. SOH identifies the beginning of a block of control information, such as a destination address, priority and message sequence number. The STX character denotes both the end of the message header and the beginning of the actual content of the message. A BCC character always follows an ETX character. Since the ETX only signifies the end of a message, it requires a status reply from the receiving station prior to subsequent communications occurring. A status reply can be a DLE0, DLEI, NAK, WACK or RVI character, with the meaning of the last three characters discussed later in this section.

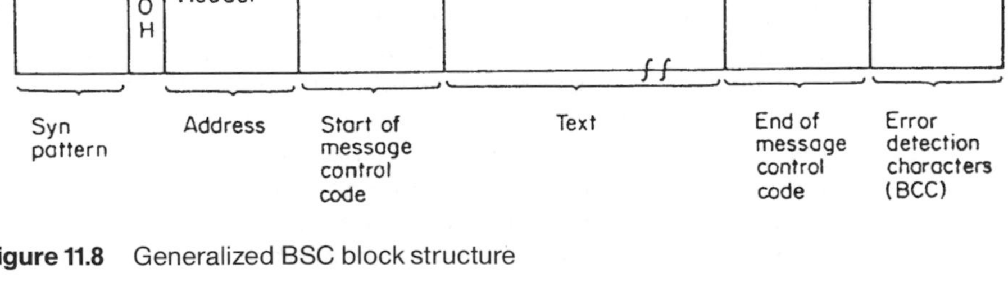

Figure 11.8 Generalized BSC block structure

The ETB character identifies the end of a block that was started with a SOH or STX. Similar to ETX, a BCC is sent immediately after the FTB and the receiving station is required to furnish a status reply.

The EOT code defines the end of message transmission for a single or multiple block message. The effect of the EOT is to reset all receiving stations. In a multidrop environment, the EOT is used as a response to a poll when an addressed station has no data to transmit.

Figure 11.9 illustrates the error control mechanism employed in a bisynchronous protocol to handle the situation where a line hit occurs during transmission or if an acknowledgement to a previously transmitted data block becomes lost or garbled.

In the example on the left-hand portion of Figure 11.9, a line hit occurs during the transmission of the second block of data from the mainframe computer to a terminal or a personal computer. Note that although Figure 11.9 is an abbreviated illustration of the actual bisynchronous block structure and

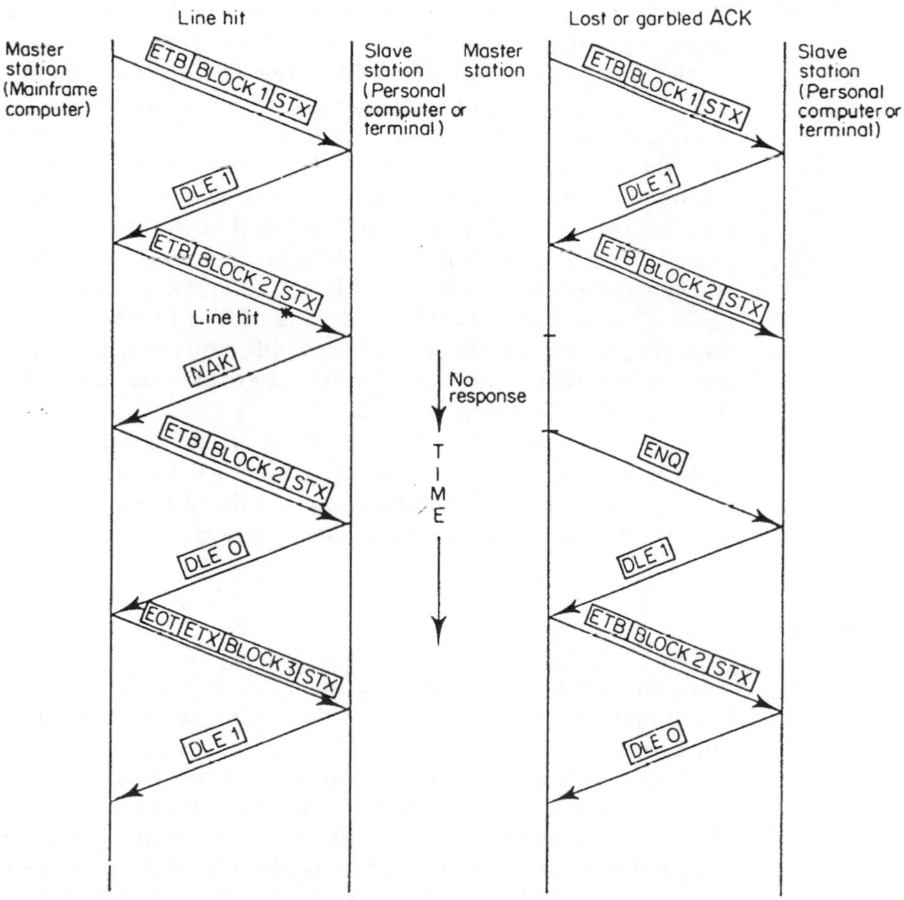

Figure 11.9 BSC error control methods

does not show the actual block-check characters in each block, in actuality they are contained in each block. The line hit which thus occurs during the transmission of the second block results in the 'internally' generated BCC being different from the BCC that was transmitted with the second block. This causes the terminal device to transmit a NAK to the mainframe, which results in the retransmission of the second block.

In the example of the right-hand part of Figure 11.9, let us assume that the terminal received block 2 and sent an acknowledgement which was lost or garbled. After a predefined timeout period, the master station transmits an ENQ communications control character to check the status of the terminal. On receipt of the ENQ, the terminal will transmit the alternating acknowledgement, currently DLE1; however, the mainframe was expecting DLE0. Thus, the mainframe is informed by this that block 2 was never acknowledged and as a result retransmits that block.

Other control codes

Three additional transmission codes commonly used in a bisynchronous protocol are the WACK, RVI and TTD characters.

The Wait-Before-Transmit Affirmative Acknowledgement (WACK) code is used by a receiving station to inform a transmitting station that the former is a temporary not-ready-to-receive condition. In addition to denoting the previously described condition, the WACK also functions as an affirmative acknowledgement (ACK) of the previously received data block. Once a WACK has been received by the sending station, that station will normally transmit ENQs at periodic intervals to the receiving station. The receiving station will continue to respond to each ENQ with a WACK until it is ready to receive data.

The Reverse Interrupt (RVI) code is used by a receiving station to request the termination of a current session to enable a higher priority message to be sent. Similar to a WACK, the RVI also functions as a positive acknowledgement to the most recently received block.

The Temporary Text Delay (TTD) is used by a sending station to keep control of a line. TTD is normally transmitted within two seconds of a previously transmitted block and indicates that the sender cannot transmit the next block within a predefined timeout period.

Timeouts

Timeouts are incorporated into most communications protocols to preclude the infinite seizure of a facility due to an undetected or detected but not corrected error condition. The bisynchronous protocol defines four types of timeout: transmit, receive, disconnect and continue.

The transmit timeout defines the rate of insertion of synchronous idle character sequences used to maintain synchronization between a transmitting and receiving station. Normally, the transmitting station will insert SYN SYN or DLE SYN sequences between blocks to maintain synchronization. Transmit timeout is normally set for 1 s.

The receive timeout can be used to limit the time a transmitting station will wait for a reply, signal a receiving station to check the line for synchronous idle characters or to set a limit on the time a station on a multidrop line can control the line. The typical default setting of the receive timeout is three seconds.

The disconnect timeout causes a station communicating on the switched network to disconnect from the circuit after a predefined period of inactivity. The default setting for a disconnect timeout is normally 20 s of inactivity.

The fourth timeout supported by bisynchronous protocols is the continue timeout. This timeout causes a sending station transmitting a TTD to send another TTD character if it is unable to send text. A receiving station must transmit a WACK within 2 s of receiving the TTD if it is unable to receive.

Although the default timeout values are sufficient for most applications, there is one area where they almost always result in unnecessary problems: the situation where satellite communications facilities are used. Satellite communications add at least a 52 000 mile round trip delay to signal propagation, resulting in a built-in round trip delay of approximately 0.5 s for bisynchronous transmission; thus transmit and continue timeouts may always be experienced and even many receive timeouts that are unwarranted if default timeout values are used. To eliminate the occurrence of unwarranted timeouts 1 s should be added to the default timeout values for each satellite 'hop' in a communications path, where a 'hop' can be defined as the transmission from one earth station to another earth station via the use of a satellite.

To illustrate the deterioration in a bisynchronous protocol when transmission occurs on a satellite circuit, assume that you wish to transmit 80-character data blocks at 9.6 kbps and use modems whose internal delay time is 5 ms. Let us further assume that there is a single satellite hop that transmission will flow over, resulting in a one-way propagation delay of 250 ms. Since each message block must be acknowledged prior to the transmission of the next block, let us assume there are eight characters in each acknowledgement message. Based on those assumptions, Table 11.2

Table 11.2 Bisynchronous protocol efficiency example

Message transmission time $\frac{80 \text{ characters} \times 8 \text{ bit/character}}{9600 \text{ bps}}$ (ms)	67
Propagation delay (ms)	250
Modem delay time (ms)	10
Acknowledgement delay time $\frac{8 \text{ characters} \times 8 \text{ bit/character}}{9600 \text{ bps}}$ (ms)	7
Propagation delay (ms)	250
Modem delay (ms)	10
	594
Efficiency $= \frac{\text{Time spent transmitting}}{\text{Total time to transmit and acknowledge}}$	$= \frac{67}{594} = 11.3\%$

lists each of the delay times associated with the transmission of one message block until an acknowledgement is received as well as the computation of the protocol efficiency.

There are three methods that you can consider to improve throughput efficiency. You can increase the size of the message block, use high-speed modems or employ a full-duplex protocol. The first two methods have distinct limitations. As the size of the data block increases, a point will be reached where the error rate on the data link results in the retransmission of the larger size message every so often, negating the efficiency increase from an increased block size. Since the data rate obtainable is a function of the bandwidth of a channel, it may not be practical to increase the data transmission rate, resulting in a switch to a full-duplex protocol being the method used by most organizations to increase efficiency when transmitting via a satellite link.

Data transparency

In transmitting data between two devices there is always a probability that the composition of an 8-bit byte will have the same bit pattern as a bisynchronous control character. This probability significantly increases if, as an example, you are transmitting the binary representation of a compiled computer program.

Since 8-bit groupings are examined to determine if a specific control character has occurred, a bisynchronous protocol would normally be excluded from use when transmitting binary data. To overcome this limitation, protocols have what is known as a transparent mode of operation.

The control character pair DLE STX is employed to initiate transparent mode operations while the control character pairs DLE ETB or DLE ETX are used to terminate this mode of operation. Any control characters formed by data when the transparent mode is in operation are ignored. In fact, if a DLE character should occur in the data during transparent mode operations, a second DLE character will be inserted into the data by the transmitter. Similarly, if a receiver recognizes two DLE characters in sequence, it will delete one and treat the second one as data, eliminating the potential of the composition of the bit patterns of the data causing a false ending to the transparent mode of operation.

11.2.4 Digital Data Communications Message Protocol (DDCMP)

Digital Equipment Corporation's Digital Data Communications Message Protocol (DDCMP) is a character-oriented data link protocol similar to IBM's bisynchronous protocol. Unlike IBM's protocol, that is restricted to synchronous transmission, DDCMP operates asynchronously or synchronously, over switched or non-switched facilities in full- or half-duplex.

Figure 11.10 illustrates the DDCMP protocol format, in which the header contains 56 bits partitioned into six distinct fields.

SYN	SYN	Type	Header						Information (any number of 8-bit characters)	CRC-16 (16 bits)
			Count (14 bits)	Flags (2 bits)	Response (8 bits)	Sequence (8 bits)	Address (8 bits)	CRC-16 (16 bits)		

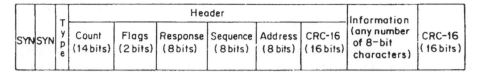

Figure 11.10 Digital data communications message protocol (DDCMP) format

Structure

Like IBM's bisynchronous protocol, DDCMP uses two SYNC characters for synchronization. The type field is a one-character field which defines the type of message being transmitted. Data messages are indicated by a SOH character, and control messages which in DDCMP are either an ACK or NAK, are indicated by the ENQ character. A third type of message, maintenance, is denoted by the use of the DLE character in the type field.

When data are transferred the count field defines the number of bytes in the information field, including CRC bytes. One advantage of this structure is its inherent transparency, since the count field defines the number of bytes in the information field, the composition of the bytes will not be misinterpreted as they are not examined as part of the protocol. If the message is a control message, the count field is used to clarify the type of NAK indicated in the type field. Although there is only one type of ACK, DDCMP supports several types of NAK, such as buffer overrun or the occurrence of a block-check error on a preceding message.

Table 11.3 lists the composition of the rightmost 6 bits of the count field which are used to define the reason for a NAK. Both ACK and NAK are denoted by an ENQ in the type field (00000101), so the count field is also used to distinguish between an a ACK and a NAK. If the first 8 bits in the count field are binary 00000001, an ACK is defined, whereas, if the first 8 bits in the count field are binary 00000010, this bit composition defines a NAK. Thus, the bit compositions listed in Table 11.3 are always prefixed by binary 00000010 which defines a NAK.

The high order bit of the flag field denotes the occurrence of a SYNC character at the end of the current message. This allows the receiver to reinitialize its synchronization detection logic. The low order bit of the FLAG field indicates the current message to be the last of a series the transmitter

Table 11.3 Count field NAK definitions

Count field value (rightmost 6 bits)	NAK definition
000001	CRC header error
000010	CRC data error
000011	Reply response
001000	Buffer unavailable
001001	Receiver overrun
010000	Message too long
010001	Header format error

intends to send. This allows the addressed station to begin transmission at the end of the current message.

Both the response and sequence fields are used to transmit message numbers. DDCMP stations assign a sequence number to each message they transmit, placing the number in the sequence field. If message sequencing is lost, the control station can request the number of the last message previously transmitted by another station. When this request is received, the answering station will place the last accepted sequence number in the response field of the message that it transmits back to the control station.

The address field is used in a multipoint line configuration to denote stations destined to receive a specific message. The following CRC1 field provides a mechanism for the detection of errors in the header portion of the message. This CRC field is required since error-free transmission depends on the count field being detected correctly. The actual data is placed in the information field and, as previously mentioned, can include special control characters. Finally, the CRC2 field provides an error detection and correction mechanism for the data in the information field.

Operation

Unlike IBM's bisynchronous protocol, DDCMP does not require the transmission of an acknowledgement to each received message. Only when a transmission occurs or if traffic is light in the opposite direction, a condition in which no data messages are to be sent, is it necessary to transmit a special NAK or ACK.

The number in the response field of a normal header or in either a special NAK or ACK message is used to specify the sequence number of the last good message received. To illustrate this, assume that messages 3, 4, 5 and 6 were received since the last time an acknowledgement was sent and message 7 contains an error. The header in the NAK message would then have a response field value of 6, indicating that messages 3, 4, 5 and 6 were received correctly and message 7 was received incorrectly. Under the DDCMP protocol up to 255 messages can be outstanding due to the use of an 8-bit response field.

Another advantage of DDCMP over IBM's bisynchronous protocols is the ability of DDCMP to operate in a full-duplex mode. This eliminates the necessity of line turnarounds and results in an improved level of throughput. Another function of the response field is to inform a transmitting station of the occurrence of a sequence error. This is accomplished by the transmitting station examining the contents of the response field. For example, if the next message the receiver expects is 4 and it receives 5, it will not change the response field of its data messages which contains a 3. In effect, this tells the transmitting station that the receiving station has accepted all messages up through message 3 and is still awaiting message 4.

11.2.5 Bit-oriented line control procedures

A number of bit-oriented line control procedures were implemented by computer vendors that were based upon the International Standards Organization

(ISO) procedure known as High-level Data Link Control (HDLC). Various names for line control procedures similar to HDLC include IBM's Synchronous Data Link Control (SDLC) and Burrough's Data Link Control (BDLC) (the latter firm is now known as Unisys).

The advantages of bit-oriented protocols are three-fold. First, their full-duplex capability supports the simultaneous transmission of data in two directions, resulting in a higher throughput than is obtainable in BISYNC. Secondly, bit-oriented protocols are naturally transparent to data, enabling the transmission of pure binary data without requiring special sequences of control characters to enable and disable a transparency transmission mode of operation as required with BISYNC. Lastly, most bit-oriented protocols permit multiple blocks of data to be transmitted one after another prior to requiring an acknowledgement. Then, if an error affects a particular block, only that block need be retransmitted.

High-level Data Link Control (HDLC) structure

Under the HDLC transmission protocol one station on the line is given the primary status to control the data link and supervise the flow of data on the link. All other stations on the link are secondary stations and respond to commands issued by the primary station.

The vehicle for transporting messages on an HDLC link is called a frame and is illustrated in Figure 11.11.

The HDLC frame contains six fields, wherein two fields serve as frame delimiters and are known as the HDLC flag. The HDLC flag has the unique bit combination of 01111110, which defines the beginning and end of the frame. To protect the flag and assure transparency the transmission device will always insert a zero bit after a sequence of five one bits occurs to prevent data from being mistaken as a flag. This technique is known as zero insertion. The receiver will always delete a zero after receiving five ones to insure data integrity.

The address field is an 8-bit pattern that identifies the secondary station involved in the data transfer while the control field can be either 8 or 16 bits

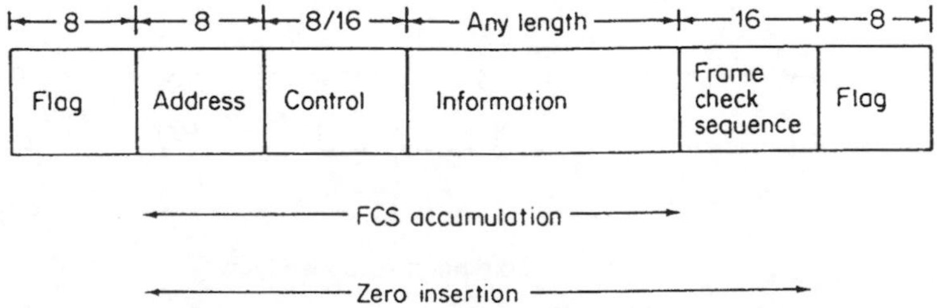

Figure 11.11 HDLC frame format. HDLC flag is 01111110 which is used to delimit an HDLC frame. To protect the flag and assure transparency the transmitter will insert a zero bit after a fifth one bit to prevent data from being mistaken as a flag. The receiver always deletes a zero after receiving five ones

in length. This field identifies the type of frame transmitted as either an information frame or a command–response frame. The information field can be any length and is treated as pure binary information, whereas the frame check sequence (FCS) contains a 16-bit value generated using a cyclic redundancy check (CRC) algorithm.

Control field formats

The 8-bit control field formats are illustrated in Figure 11.12. *N*(S) and *N*(R) are the send and receive sequence counts. They are maintained by each station for Information (I-frames) sent and received by that station. Each station increments its *N*(S) count by one each time it sends a new frame. The *N*(R) count indicates the expected number of the next frame to be received.

Using an 8-bit control field, the *N*(S)/*N*(R) count ranges from 0 to 7. Using a 16-bit control field the count can range from 0 to 127. The P/F bit is a poll/final bit. It is used as a poll by the primary (set to 1) to obtain a response from a secondary station. It is set to 1 as a final bit by a secondary station to indicate the last frame of a sequence of frames.

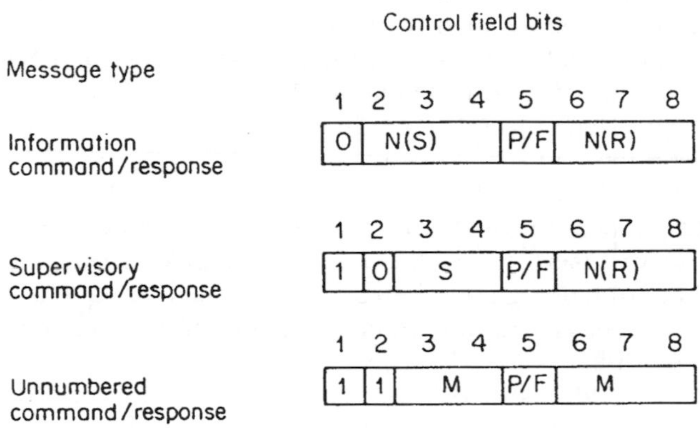

Figure 11.12 HDLC control field formats. *N*(S) = send sequence count; *N*(R) = receive sequence count; S = supervisory function bits *M* = modifier function bits; *P/F* = poll/final bit

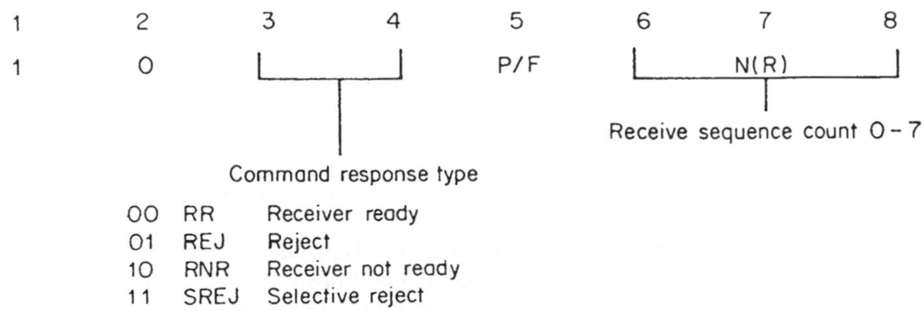

Figure 11.13 Supervisory control field

The supervisory command/response frame is used in HDLC to control the flow of data on the line. Figure 11.13 illustrates the composition of the supervisory control field: supervisory frames (S-frames) contain an *N*(R) count and are used to acknowledge I-frames, request retransmission of I-frames, request temporary suspension of I-frames, and perform similar functions.

To illustrate the advantages of HDLC over BISYNC transmission, consider the full-duplex data transfer illustrated in Figure 11.14. For each frame transmitted, this figure shows the type of frame, *N*(S), *N*(R) and poll/final (P/F) bit status.

In the transmission sequence illustrated in the left-hand portion of Figure 11.14 the primary station has transmitted five frames, numbered zero through four, when its poll bit is set in frame four. The poll bit is interpreted by the secondary station as a request for it to transmit its status and it responds by transmitting a Receiver Ready (RR) response, indicating that it expects to receive frame five next. This serves as an indicator to the primary

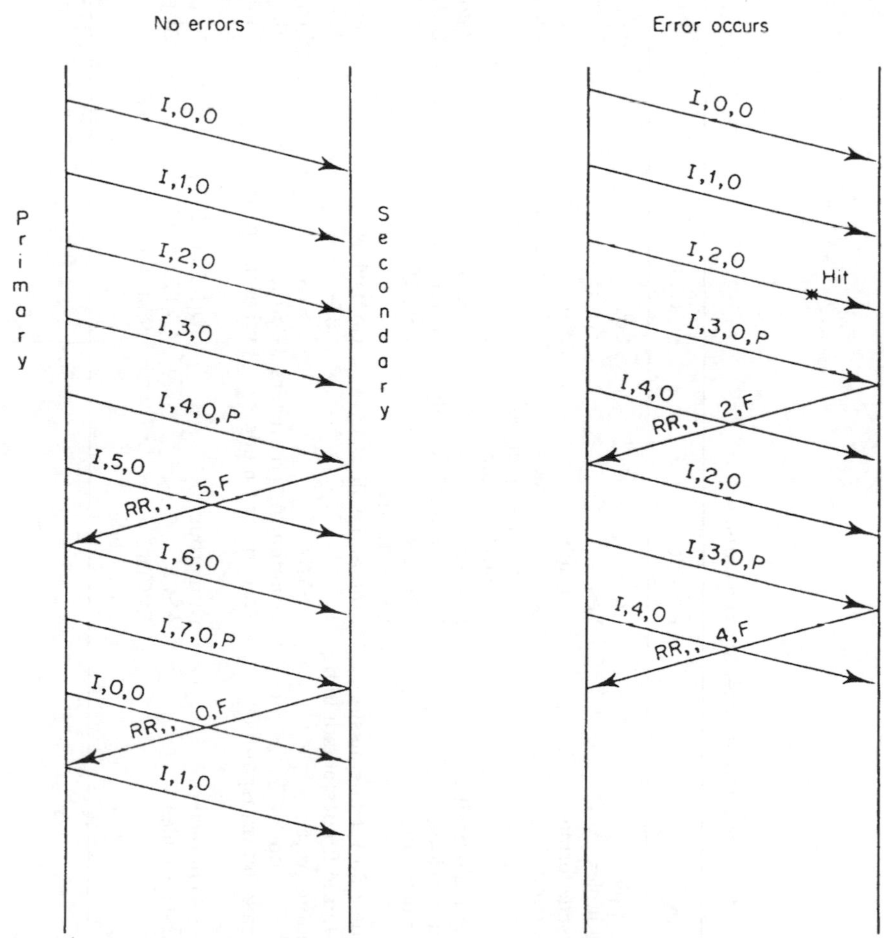

Figure 11.14 HDLC full-duplex data transfer. Format: type, N(S), *N*(R), *P/F*

Table 11.4 Protocol characteristic comparison

Feature	BISYNC	DDCMP	SDLC	HDLC
Full-duplex	No	Yes	Yes	Yes
Half-duplex	Yes	Yes	Yes	Yes
Message format	Variable	Fixed	Fixed	Fixed
Link control	Control character, character sequences, optional header	Header (fixed)	Control field (8 bits)	Control field (8/16 bits)
Station addressing	Header	Header	Address field	Address field
Error checking	Information field only	Header, information field	Entire frame	Entire frame
Error detection	VRC/LRC-8 VRC/CRC-16	CRC-16	CRC-CCITT	CRC-CCITT
Request for retransmission	Stop and wait	Go back N	Go back N	Go back N, select reject
Maximum frames outstanding	1	255	7	127
Framing start	2 SYNs	2 SYNs	Flag	Flag
end	Terminating characters	Count	Flag	Flag
Information transparency	Transparent mode	Inherent (count)	Inherent (zero insertion/deletion)	Inherent (zero insertion/deletion)
Control characters	Numerous	SOH, DLE, ENQ	None	None
Character codes	ASCII EBCDIC Transcode	ASCII (control character only)	Any	Any

station that frames zero through four were received correctly. The secondary station sets its poll/final bit as a final bit to indicate to the primary station that its transmission has been completed.

Note that since full-duplex transmission is permissible under HDLC, the primary station continues to transmit information (I) frames and the secondary station is responding to the primary's polls. If an 8-bit control field is used, the maximum frame number that can be outstanding is limited to seven, because 3 bit positions are used for *N*(S) frame numbering. Thus, after frame number seven is transmitted, the primary station then begins frame numbering again at N(S) equal to zero. Notice that when the primary station sets its poll bit when transmitting frame seven the secondary station responds, indicating that it expects to receive frame zero. This indicates to the primary station that frames five through seven were received correctly, since the previous secondary response acknowledge frames zero through four.

In the transmission sequence indicated on the right-hand side of Figure 11.14, assume that a line hit occurs during the transmission of frame two. Note that in comparison to BISYNC, under HDLC the transmitting station does not have to wait for an acknowledgement of the previously transmitted data block; and it can continue to transmit frames until the maximum number of frames outstanding is reached; or, it can issue a poll to the secondary station to query the status of its previously transmitted frames while it continues to transmit frames up until the maximum number of outstanding frames is reached.

The primary station thus polled the secondary in frame three and then sent frame four while it waited for the secondary's response. When the secondary's response was received, it indicated that the next frame the secondary expected to receive *N*(R) was two. This informed the primary station that all frames after frame one would have to be retransmitted. Thus, after transmitting frame four the primary station then retransmitted frames two and three prior to retransmitting frame four.

It should be noted that if selective rejection is implemented, the secondary could have issued a Selective Reject (SREJ) of frame two. Then, on its receipt, the primary station would retransmit frame two and would have then continued its transmission with frame five. Although selective rejection can considerably increase the throughput of HDLC, even without its use this protocol will provide the user with a considerable throughput increase in comparison to BISYNC.

For comparison purposes Table 11.4 compares the major features of BISYNC, DDCMP, IBM's SDLC and the CCITT HDLC protocols.

11.3 REVIEW QUESTIONS

1. What is the difference between a terminal protocol and a data link protocol?

2. What is the purpose of data sequencing in which a large block is broken into smaller blocks for transmission?

3. Define the characteristics of a teletypewriter protocol.

4. What is the purpose of using one or more null characters after a carriage return, line feed sequence?

5. What error detection method is used in the teletypewriter protocol? How are errors corrected when they are detected?

6. What is echoplex? When echoplex is used who is responsible for examining locally printed characters?

7. How can a receiver detect the occurrence of an error when the XMODEM protocol is used?

8. Discuss three limitations of the XMODEM protocol.

9. What is the difference between the XMODEM and XMODEM/CRC file transfer protocols? Which provides a more reliable method of error detection and correction?

10. Describe why a communications program can provide a display of the status of a file transfer operation when the YMODEM protocol is used.

11. Under what conditions should you consider the use of the YMODEM-G or YMODEM-G batch protocols?

12. How does the ZMODEM protocol provide an extended error detection capability?

13. What are the major differences between the 2780, 3780 and 3270 protocols?

14. What is the purpose of a bisynchronous protocol transmitting alternating acknowledgements (DLE1 and DLE0)?

15. How does the DDCMP protocol provide data transparency?

16. Assume that the response field of a NAK message in a DDCMP protocol has a value of 14 and messages 12, 13, 14, 15 and 16 are outstanding. What does this indicate?

17. What is the purpose of the HDLC flag? How is its integrity protected?

18. What is the advantage of a selective reject command?

12

INCREASING WAN LINE UTILIZATION

In this chapter we will examine several categories of communications equipment whose acquisition is primarily justified by the economic savings their utilization promotes. Each of the devices we will examine provides users with potential economic savings by permitting many data sources to share the use of a common wide area network (WAN) transmission facility. Due to this, these devices can also be classified as line sharing equipment.

Devices discussed in this chapter, such as different types of multiplexers, and control units, were originally developed to support the communications requirements of organizations that established mainframe-based networks. Although the conversion of many mainframe-based networks to LAN-based client–server computing resulted in the development of bridges, routers and gateways that replaced many communications devices discussed in this chapter, to paraphrase Mark Twain the death of the mainframe has been prematurely reported. Thus, for some organizations the devices covered in this chapter will continue to be relevant. For other organizations that have either migrated to, or are in the process of migrating to, LAN based client–server computing, the next two chapters in this book will be more relevant.

Although remote bridges and routers can be classified as communications devices that increase WAN line utilization, they are not covered in this chapter. Because they are more commonly associated with the transmission of data between LANs, we will focus our attention on those devices in the next two chapters of this book. Another reason for deferring coverage of bridges and routers to later chapters is the fact that their operation is based upon the interleaving of packets onto a serial transmission facility with no guarantee concerning the positioning of packets in the data stream. This means random delays occur between packets, which makes it difficult for routers to transport digitized voice, video and other multimedia applications that require a uniform data flow in order to correctly reconstruct the digitized data. In comparison, multiplexers use frequency or time slots that allow data to flow in a predictable manner and enable voice and video to be easily reconstructed.

In the first section of this chapter we will investigate the operation and utilization of several types of multiplexer. This will be followed by focusing attention upon a second category of line sharing equipment known by such terms as modem, line and' port sharing units. Although Tl multiplexers logically fall into the category of multiplexing equipment, their operation and utilization is deferred to Chapter 15 when digital transmission is examined.

12.1 MULTIPLEXERS

With the establishment of distributed computing, the cost of providing the required communications facilities became a major focus of concern to users. Numerous network structures were examined to determine the possibilities of using specialized equipment to reduce these costs. For many networks where geographically distributed users accessed a common computational facility, a central location could be found which would serve as a hub to link those users to the computer. Even when terminal traffic was low and the cost of leased lines could not be justified on an individual basis, quite often the cumulative cost of providing communications to a group of users could be reduced if a mechanism was available to enable many terminals to share common communications facilities. This mechanism was provided by the utilization of multiplexers whose primary function is to provide the user with a reduction of communications costs. This device enables one high-speed line to be used to carry the formerly separate transmissions of a group of lower speed lines. The use of multiplexers should be considered when a number of data terminals communicate from within a similar geographical area or when a number of leased lines run in parallel for any distance.

12.1.1 Evolution

From the historical perspective, multiplexing technology can trace its origin to the early development of telephone networks. Then, as today, multiplexing was the employment of appropriate technology to permit a communications circuit to carry more than one signal at a time.

In 1902, 26 years after the world's first successful telephone conversation, an attempt to overcome the existing ratio of one channel to one circuit occurred. Using specifically developed electrical network terminations, three channels were derived from two circuits by telephone companies.

The third channel was denoted as the phantom channel, hence the name 'phantom' was applied to this early version of multiplexing. Although this technology permitted two pairs of wires to effectively carry the load of three, the requirement to keep the electrical network finely balanced to prevent crosstalk limited its practicality.

12.1.2 Device support

In general, any device that transmits or receives a serial data stream can be considered a candidate for multiplexing. Data streams produced by the devices listed in Table 12.1 are among those that can be multiplexed. The

Table 12.1 Candidates for data stream multiplexing

Analog network private line modems
Analog switched network modems
Digital network data service units
Digital network channel service units
Data terminals
Data terminal controllers
Minicomputers
Concentrators
Computer ports
Computer–computer links
Other multiplexers

intermix of devices as well as the number of any one device whose data stream is considered for multiplexing is a function of the multiplexer's capacity and capabilities, the economics of the application, and cost of other devices which could be employed in that role, as well as the types and costs of high-speed lines being considered. As we will note later in this chapter and in succeeding chapters, modern high-capacity multiplexers can also multiplex the serial output of routers and video conference equipment.

12.1.3 Multiplexing techniques

Today, two basic techniques are commonly-used for multiplexing: frequency division multiplexing (FDM) and time division multiplexing (TDM). Within the time division technique, two versions are available: fixed time slots which are employed by traditional TDMs and variable use of time slots which are used by statistical and intelligent TDMs.

Frequency division multiplexing (FDM)

In the FDM technique, the available bandwidth of the line is split into smaller segments called data bands or derived channels. Each data band in turn is separated from another data band by a guard band which is used to prevent signal interference between channels, as shown in Figure 12.1. Typically, frequency drift is the main cause of signal interference and the sizes of the guard

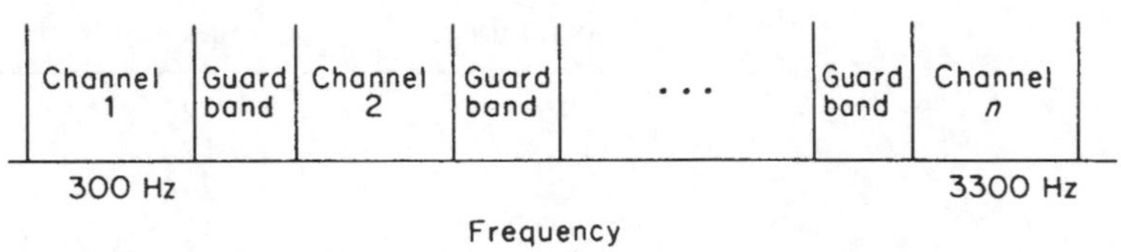

Figure 12.1 FDM channel separations. In frequency division multiplexing the 3 kHz bandwidth of a voice-grade line is split into channels or data bands separated from each other by guard bands

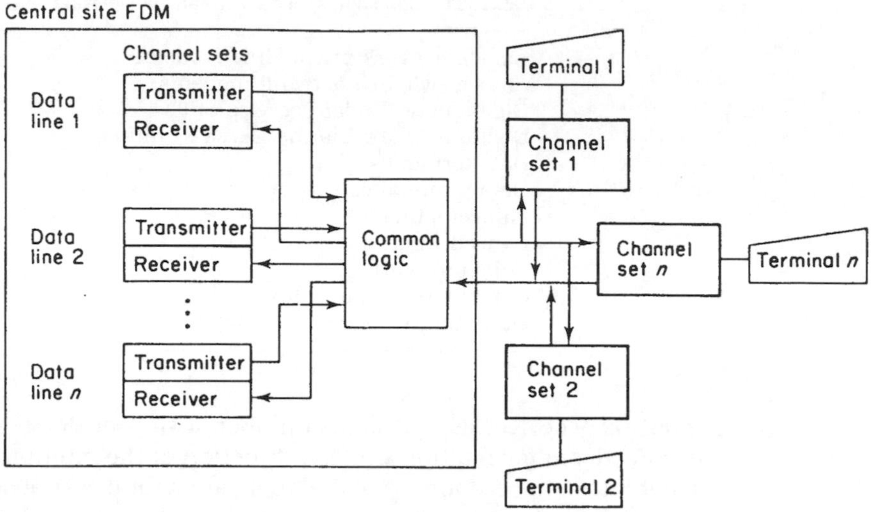

Figure 12.2 Frequency division multiplexing. Since the channel sets modulate the line at specified frequencies, no modems are required at remote locations

bands are structured to prevent data in one channel from drifting into another channel.

Physically, an FDM contains a channel set for each data channel as well as common logic, as shown in Figure 12.2. Each channel set contains a transmitter and receiver tuned to a specific frequency, with bits being indicated by the presence or absence of signals at each of the channel's assigned frequencies. In FDM, the width of each frequency band determines the transmission rate capacity of the channel, and the total bandwidth of the line is a limiting factor in determining the total number or mix of channels that can be serviced. Although a multipoint operation is illustrated in Figure 12.2, FDM equipment can also be utilized for the multiplexing of data between two locations on a point-to-point circuit. Data rates up to 1200 bps can be multiplexed by FDM onto an analog voice-grade line. Typical FDM channel spacings required at different data rates are listed in Table 12.2.

The overall FDM's aggregate data handling capacity depends upon the mixture of subchannels as well as the type of line conditioning added to the circuit. A chart of FDM subchannel allocations is given in Figure 12.3. This

Table 12.2 FDM channel spacings

Speed (bps)	Spacing (Hz)
75	120
110	170
150	240
300	480
450	720
600	960
1200	1800

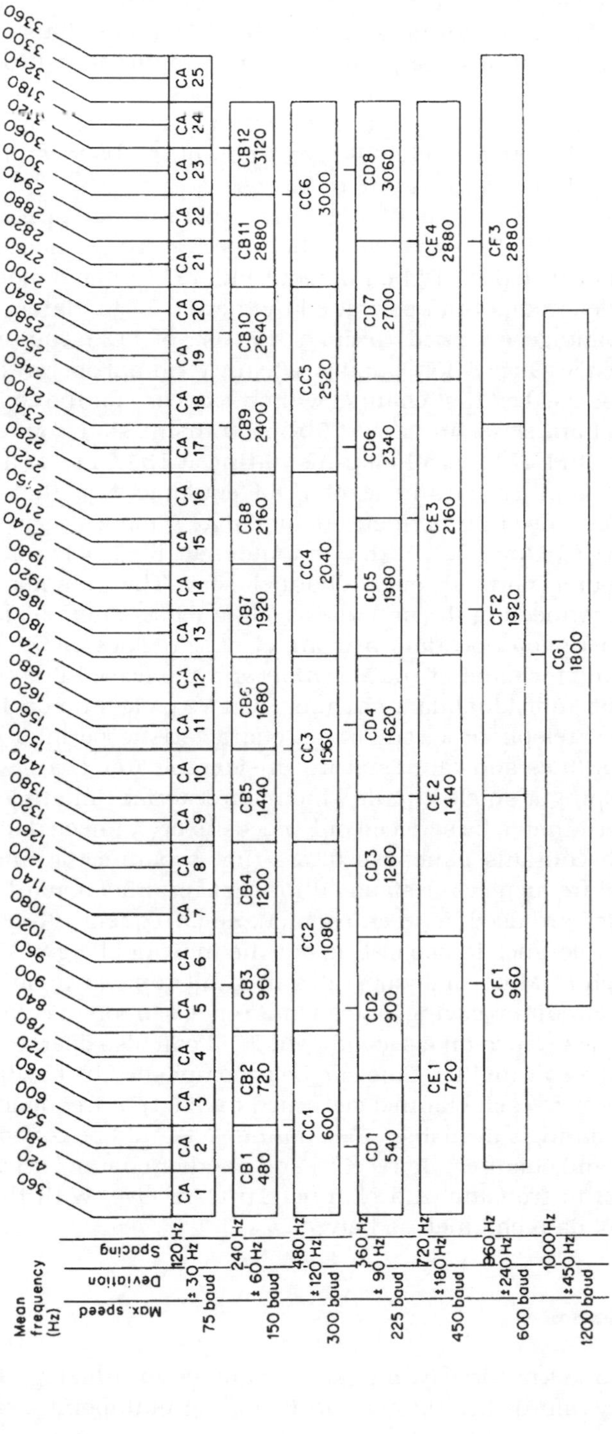

Figure 12.3 ITU-T FDM subchannel allocations

chart can be used to compute the mixture of data subchannels that can be transmitted via a single voice-grade channel when frequency division multiplexing is employed. The referenced chart is based on data subchannel spacing standards formulated by the ITU-T. As illustrated, 17 75 bps subchannels can be multiplexed on an unconditioned (C0) circuit, 19 subchannels on a Cl conditional circuit, 22 subchannels on a C2 circuit and 24 channels on a C4 conditioned circuit. For higher data rates the ITU-T standards allocate a fractional proportion of the bandwidth allocated to the previously discussed data rate of 75 bps.

With the development of terminals operating at speeds that were not multiples of ITU-T frequencies, such as 134.5 bps teleprinters, a number of vendors developed FDM equipment tailored to make more efficient use of voice-grade circuits than permitted by the ITU-T standards.

An advantage obtained through the use of FDM equipment is its code transparency. Once a data band is set, any terminal operating at that speed or less can be used on that channel without concern for the code of the terminal. Thus, a channel set to carry 300 bps transmission could also be used to service an IBM 2741 terminal transmitting at 134.5 bps or a Teletype 110 bps terminal. Another advantage of FDM equipment is that no modems are required because the channel sets modulate the line at specified frequencies, as shown in Figure 12.2. At the computer site, the FDM multiplexer interfaces the computer ports through channel sets. The common logic acts as a summer, connecting the multiplexer channel sets to the leased line.

At each remote location, a channel set provides the necessary interface between the terminal at that location and the leased line. When using FDM equipment, individual data channels can be picked up or dropped off at any point on a telephone circuit. This characteristic permits the utilization of multipoint lines and can result in considerable line charge reductions based on developing a single circuit which can interface multiple terminals. Each remote terminal to be serviced only needs to be connected to an FDM channel set which contains bandpass filters that separate the line signal into the individual frequencies designated for that terminal. Guard bands of unused frequencies are used between each channel frequency to permit the filters a degree of tolerance in separating out the individual signals.

Although FDM normally operates in a full-duplex transmission mode on a four-wire circuit by having all transmit tones sent on one pair of wires and all receive tones return on a second pair, FDM can also operate in the full-duplex mode on a two-wire line. This can be accomplished by having the transmitter and receiver of each channel set tuned to different frequencies. For example, with 16 channels available, one channel set could be tuned for channel 1 to transmit and channel 9 to receive, and another channel set would be tuned to channel 2 to transmit and channel 10 to receive. With this technique, the number of data channels is halved.

FDM utilization

As mentioned previously, one key advantage in utilizing FDM equipment is the ability afforded to the user in installing multipoint circuits for use in a

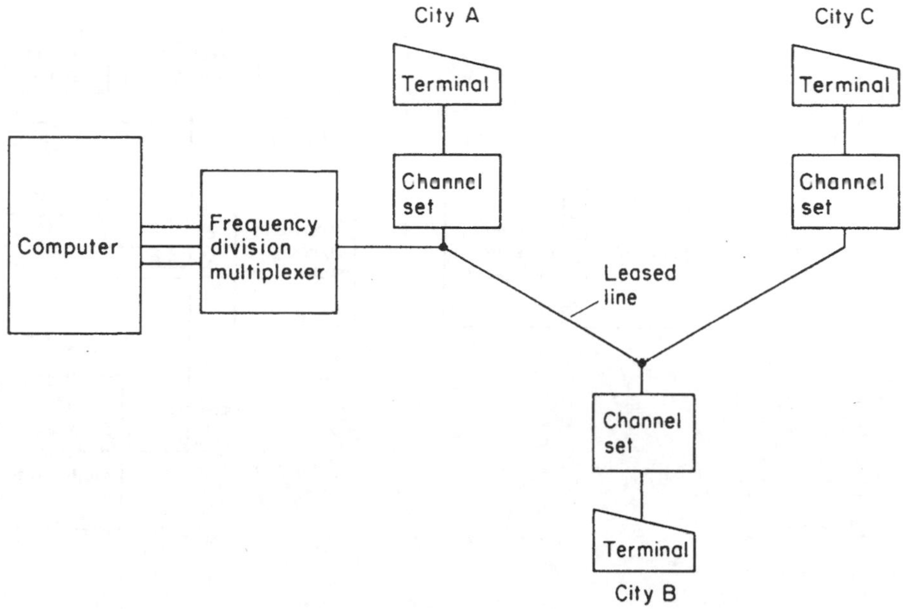

Figure 12.4 Frequency division multiplexing permits multipoint circuit operations. Each terminal on an FDM multipoint circuit is interfaced through the multiplexer to an individual computer port

communications network. This can minimize line costs because a common line, optimized in routing, can now be used to service multiple terminal locations. An example of FDM equipment used on a multipoint circuit is shown in Figure 12.4, where a four-channel FDM is used to multiplex traffic from terminals located in four different cities. Although the entire frequency spectrum is transmitted on the circuit, the channel set at each terminal location filters out the preassigned bandwidth for that location, in effect producing a unique individual channel that is dedicated for utilization by the terminal at each location. This operation is analogous to a group of radio stations transmitting at different frequencies and setting a radio to one frequency, so as always to be able to receive the transmission from a particular station.

In contrast to poll and select, multipoint line operations where one computer port is used to transmit and receive data from many buffered terminals connected to a common line, FDM used for multipoint operations as shown requires one computer port for each terminal. Such terminals do not, however, require a buffer area to recognize their addresses, nor is poll and select software required to operate in the computer. When buffered terminals and poll and select software are available, polling by channel can take place, as illustrated in Figure 12.5. In this example, channels 1 and 2 are each connected to a number of relatively low-traffic terminals which are polled through the multiplexer system. Terminals 3 and 6 are presumed to be higher traffic stations and are thus connected to individual channels of the FDM or to individual channel sets.

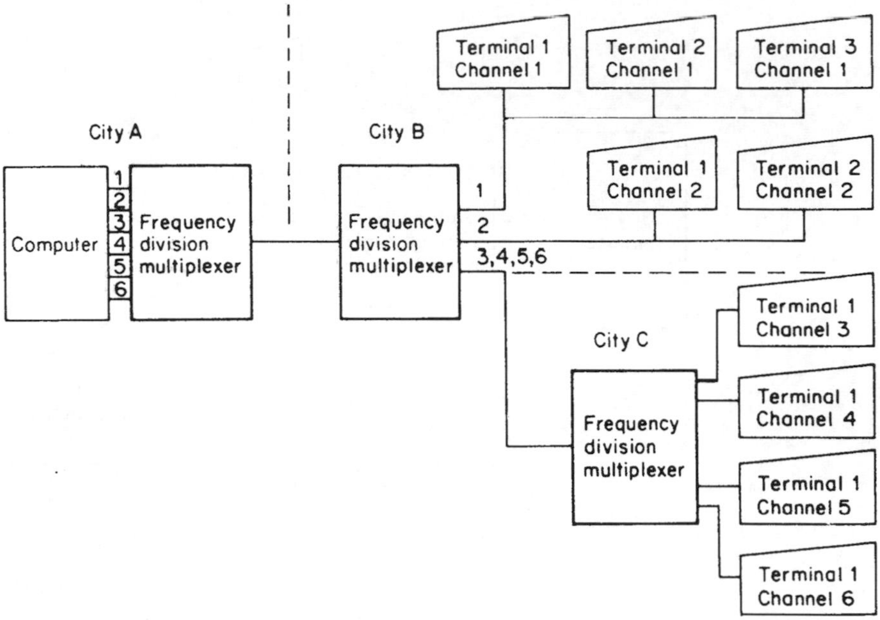

Figure 12.5 FDM can intermix polled and dedicated terminals in a network. Of the six channels used in this network, channels 1 and 2 service a number of polled terminals while channels 3 through 6 are dedicated to service individual terminals

Time division multiplexing (TDM)

In the FDM technique, the bandwidth of the communications line serves as the frame of reference. The total bandwidth is divided into subchannels consisting of smaller segments of the available bandwidth, each of which is used to form an independent data channel. In the TDM technique, the aggregate capacity of the line is the frame of reference, since the multiplexer provides a method of transmitting data from many terminals over a common circuit by interleaving them in time. The TDM divides the aggregate transmission on the line for use by the slower-speed devices connected to the multiplexer. Each device is given a time slot for its exclusive use so that at any one point in time the signal from one terminal is on the line. In the FDM technique, in which each signal occupies a different frequency band, all signals are being transmitted simultaneously.

The fundamental operating characteristics of a TDM are shown in Figure 12.6. Here, each low- to medium-speed terminal device is connected to the multiplexer through an input/output (I/O) channel adapter. The I/O adapter provides the buffering and control functions necessary to interface the transmission and reception of data to and from the multiplexer. Within each adapter, a buffer or memory area exists which is used to compensate for the speed differential between the terminals and the multiplexer's internal operating speed. Data are shifted from the terminal to the I/0 adapter at different rates, depending on the speed of the input data source; but when

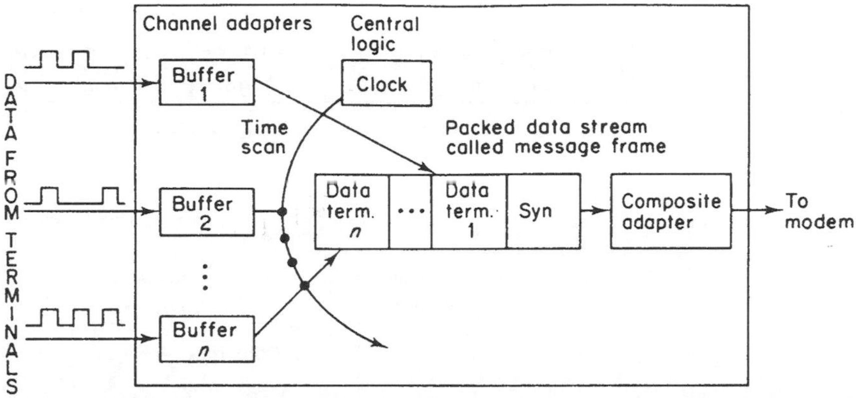

Figure 12.6 Time division multiplexing. In time division multiplexing, data is first entered into each channel adapter buffer area at a transfer rate equal to the device to which the adapter is connected. Next, data from the various buffers are transferred to the multiplexer's central logic at the higher rate of the device for packing into a message frame for transmission

data are shifted from the I/O adapter to the central logic of the multiplexer, or from central logic to the composite adapter, it is at the much higher fixed rate of the TDM. On output from the multiplexer the reverse is true, since data are first transferred at a fixed rate from central logic to each adapter and then from the adapter to the terminal device at the data rate acceptable to the terminal. Depending on the type of TDM system, the buffer area in each adapter will accommodate either bits or characters.

The central logic of the TDM contains controlling, monitoring and timing circuitry which facilitates the passage of individual terminal data to and from the high-speed transmission medium. The central logic will generate a synchronizing pattern which is used by a scanner circuit to interrogate each of the channel adapter buffer areas in a predetermined sequence, blocking the bits of characters from each buffer into a continuous, synchronous data stream which is then passed to a composite adapter. The composite adapter contains a buffer and functions similar to the I/O channel adapters. However, it now compensates for the difference in speed between the high-speed transmission medium and the internal speed of the multiplexer.

TDM techniques

There are two TDM techniques – bit interleaving and character interleaving. Bit interleaving is generally used in systems which service synchronous terminals, whereas character interleaving is generally used to service asynchronous terminals. When interleaving is accomplished on a bit-by-bit basis, the multiplexer takes 1 bit from each channel adapter and then combines them as a word or frame for transmission. As shown in Figure 12.7 (top), this technique produces a frame containing one data element from each channel adapter. When interleaving is accomplished on a character-by-character

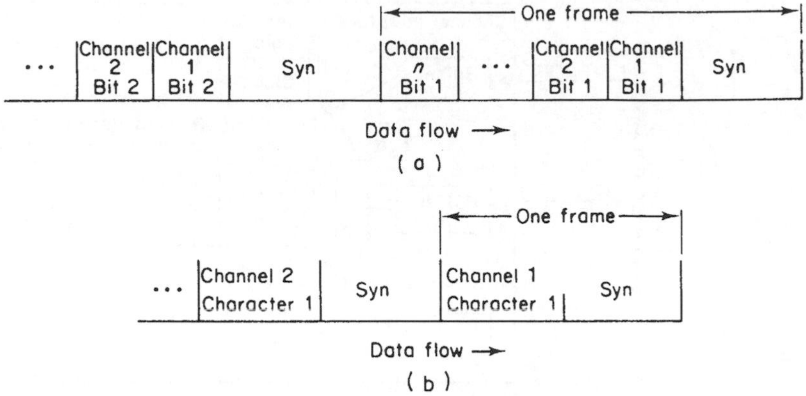

Figure 12.7 Time division interleaving bit-by-bit and character-by-character. When interleaving is accomplished bit-by-bit (top), the first bit from each channel is packed into a frame for transmission. Bottom: Time division multiplexing character-by-character. When interleaving is conducted on a character-by-character basis, one or more complete characters are grouped with a synchronization character into a frame for transmission

basis, the multiplexer assembles a full character into one frame and then transmits the entire character, as shown in Figure 12.7 (bottom). Although a frame containing only one character of information is illustrated in Figure 12.7, to increase transmission efficiency most multiplexers transmit long frames containing a large number of data characters to reduce the synchronization overhead associated with each frame. Thus, while a frame containing one character of information has a synchronization overhead of 50%, a frame containing four data characters has its overhead reduced to 20%, and a frame containing nine data characters has a synchronization overhead of only 10%, assuming constant slot sizes for all characters.

Since the character-by-character interleaved method preserves all bits of a character in sequence, the TDM equipment can be used to strip the character of any recognition information that may be sent as part of that character. Examples of this would be the servicing of such terminals as a Teletype Model 33 or an IBM PC transmitting asynchronous data, where a transmitted character contains 10 or 11 bits which include a start bit, 7 data bits, a parity bit, and 1 or 2 stop bits. When the bit-interleaved method is used, all 10 or 11 bits would be transmitted to preserve character integrity, whereas in a character-interleaved system, the start and stop bits can be stripped from the character, with only the 7 data bits and on some systems the parity bit warranting transmission.

To service terminals with character codes containing different numbers of bits per character, two techniques are commonly employed in character interleaving. In the first technique, the time slot for each character is of constant size, designed to accommodate the maximum bit width or highest level code. Making all slots large enough to carry American Standard Code for Information Interchange (ASCII) characters makes the multiplexer an inefficient carrier of a lower level code such as five-level Baudot. The electronics required in the device and its costs are, however, reduced. The second technique used is to proportion the slot size to the width of each character according to its bit size. This technique maximizes the efficiency of the multiplexer, although the complexity of the logic and the cost of the multiplexer increases. Due to the reduction in the cost of semiconductors, most character interleaved multiplexers marketed are designed to operate on the proportional assignment method.

Although bit interleaving equipment is less expensive, it is also less efficient when used to service asynchronous terminals. On the positive side, bit-interleaved multiplexers offer the advantage of faster resynchronization and shorter transmission delay, since character-interleaved multiplexers must wait to assemble the bits into characters; whereas a bit-interleaved multiplexer can transmit each bit as soon as it is received from the terminal. Multiplexers, which interleave by character use a number of different techniques to build the character, with the techniques varying between manufacturers and by models produced by manufacturers.

A commonly utilized technique is the placement of a buffer area for each channel adapter which permits the character to be assembled within the channel adapter and then scanned and packed into a data stream. Another technique which can be used is the placement of programmed read only memory within the multiplexer so that it can be used to assemble characters for all the input channels. The second technique permits many additional functions to be performed in addition to the assembly and disassembly of characters. Such multiplexers with programmed memory are referred to as intelligent multiplexers, and are discussed later in this chapter.

TDM applications

The most commonly used TDM configuration is the point-to-point system, which is shown in Figure 12.8. This type of system, which is also called a two-point multiplex system, links a mixture of terminals to a centrally located multiplexer. As shown, the terminals can be connected to the multiplexer in a variety of ways. Terminals can be connected by a leased line running from the terminal's location to the multiplexer, by a direct connection if the user's terminal is within the same building as the multiplexer and a cable can be laid to connect the two, or terminals can use the switched network to call the multiplexer over the dial network. For the latter method, since the connection is not permanent, several terminals can share access to one or more multiplexer channels on a contention basis.

As shown in Figure 12.8, terminals in cities B and C use the dial network to contend for one multiplexer channel which is interfaced to an automatic answer unit on the dial network. Whichever terminal accesses that channel maintains use of it and thus excludes other terminals from access, to that particular connection to the system. As an example, one might have a network which contains 50 terminals within a geographical area wherein between 10 to 12 are active at any time; and one method to deal with this environment would be through the installation of a 12-number rotary interfaced to a 12-channel multiplexer. If all of the terminals were located within one city, the only telephone charges that the user would incur in addition to

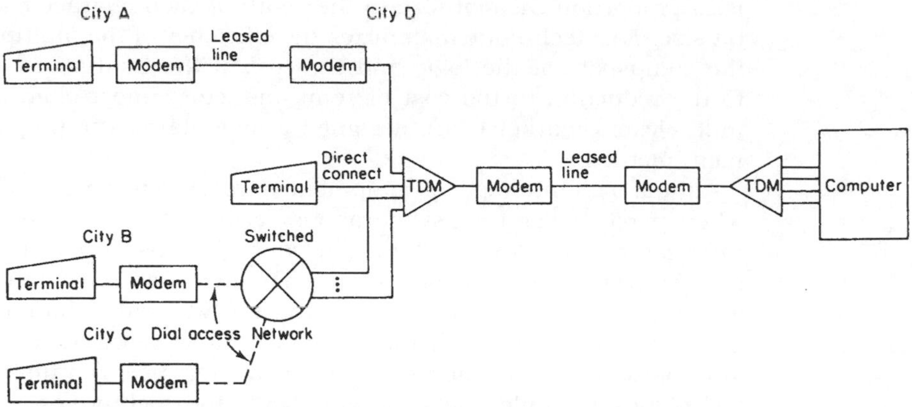

Figure 12.8 Time division multiplexing point-to-point. A point-to-point or two-point multiplexing system links a variety of data users at one or more remote locations to a central computer facility

those of the leased line between multiplexers would be local call charges each time a terminal user were to dial the local multiplexer number.

Traffic engineering

The procedure used to determine the number of dial-in lines required to service a population of terminal devices is referred to as traffic engineering. Traffic engineering was first used by telephone companies to determine the number of trunks to install between switches based upon the probability of different numbers of telephone describers requiring the ability to make long-distance calls.

Under traffic engineering the number of trunks is determined based upon anticipated traffic and a desired grade of service, where the grade of service represents the probability that a caller receives a busy signal. For example, under one anticipated traffic load a grade of service of 0.05, equivalent to 1 in 20 calls being dropped, might require 10 trunks. In comparison, a grade of service of 0.01, equivalent to 1 in 100 calls being dropped, might require 50 trunks based upon the same level of anticipated traffic.

Traffic engineering techniques developed to facilitate the construction of telephone networks during the early 1900s were 'borrowed' and applied to similar problems. Such problems include determining the number of ports and dial-in lines for multiplexers serving a population of terminal users. While this type of traffic engineering was very popular during the 1970s through 1980s, the growth in LANs resulted in a new application for traffic engineering. This new application involves determining the number of ports to install on access concentrators which receive dial-in calls from persons wishing to access a LAN, to obtain access to either a corporate network or the Internet. Because most Internet Service Providers (ISP) use access controllers to service dial-in modem users, your ability to connect to the Internet without encountering a busy signal depends upon the manner by which the ISP sized their access controller for a particular grade of service.

Multiplexing schemes

Over the past 30 years a series of multiplexing schemes were developed as a mechanism both to satisfy different communications requirements as well as to reduce the cost of transmission. In this section we will examine several schemes developed to use multiplexers to satisfy different types of communications requirements.

Series multipoint multiplexing

One popular form of multiplexing results from linking the output of one multiplexer into a second multiplexer. Referred to as series multipoint multiplexing, this technique is most effective when terminals are distributed at two or more locations and the user desires to alleviate the necessity of obtaining two long-distance leased lines from the closer location to the computer. As shown in Figure 12.9, four low-speed terminals are multiplexed at city A onto one high-speed channel which is transmitted to city B, where this line is in turn multiplexed along with the data from a number of other terminals at city B. Although the user requires a leased line between city A and city B, only one line is now required to be installed for the remainder of the distance from city B to the computer at city C. If city A is located 50 miles from city B, and city B is 2000 miles from city C, 2000 miles of duplicate leased lines are avoided by using this multiplexing technique.

Hub-bypass multiplexing

A variation of series multipoint multiplexing is hub-bypass multiplexing. To be effectively used, hub-bypass multiplexing can occur when a number of remote locations have the requirement to transmit to two or more locations. To satisfy this requirement, the remote terminal traffic is multiplexed to a central location which is the hub, and the terminals which must communicate

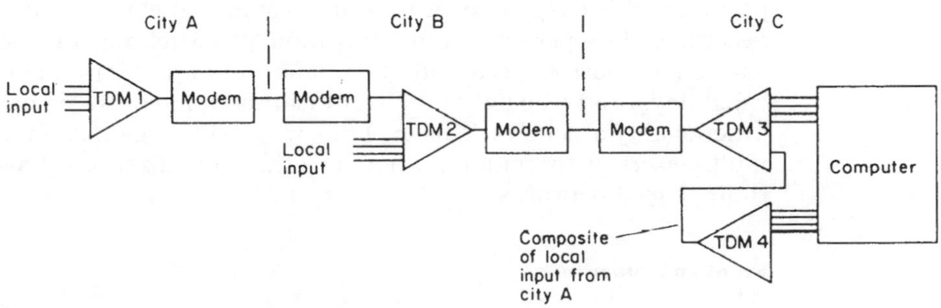

Figure 12.9 Series multipoint multiplexing. Series multipoint multiplexing is accomplished by connecting the output of one multiplexer as input to a second device

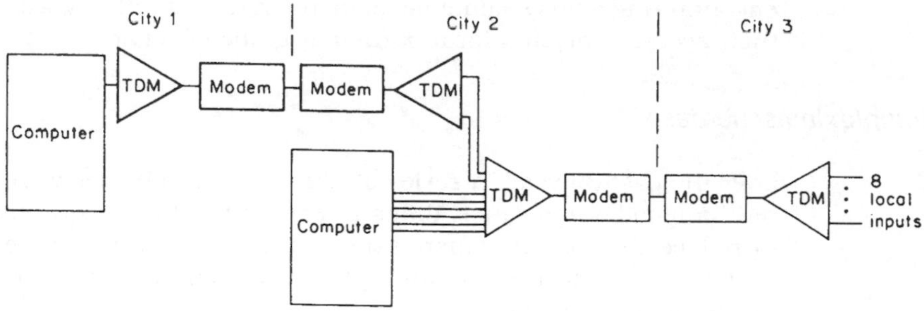

Figure 12.10 Hub-bypass multiplexing. When a number of terminals have the requirement to communicate with more than one location, hub-bypass multiplexing should be considered

with the second location are cabled into another multiplexer which transmits this traffic, bypassing the hub.

Figure 12.10 illustrates one application where hub bypassing might be utilized. In this example, terminals at city 3 require a communications link with one of two computers; six terminals always communicate with the computer at city 2, and two terminals use the facilities of the computer at city 1. Data from all eight terminals are multiplexed over a common line to city 2, where the two channels that correspond to the terminals which must access the computer at city 1 are cabled to a new multiplexer, which then remultiplexes the data from those terminals to city 1. When many terminal locations have dual location destinations, hub-bypassing can become very economical. Since the data flows in series, an equipment failure will, however, terminate access to one or more computational facilities, depending on the location of the break in service.

Although hub-bypass multiplexing can be effectively used to connect collocated terminals to different destinations, if more than two destinations exist a more efficient switching arrangement can be obtained by the employment of a port selector or a multiplexer that has port selection capability.

A port selector or multiplexer with port selection capability functions as a dynamic data switch, establishing a temporary connection between a port on the input side of the device or, in the case of a multiplexer, a channel on the multiplexing frame and its output destination. The reader is referred to the portion of this chapter covering statistical and intelligent multiplexers which describes the operation and utilization of switching and port contention features of those devices. Although modern multiplexers include a switching capability, that capability is significantly limited in comparison to the functionality of routers. This explains why many organizations have replaced multiplexers with routers, whereas other organizations have constructed router-based networks.

Front end substitute

Although not commonly utilized, a TDM may be installed as an inexpensive front end for a computer, as shown in Figure 12.11. When used as a front end,

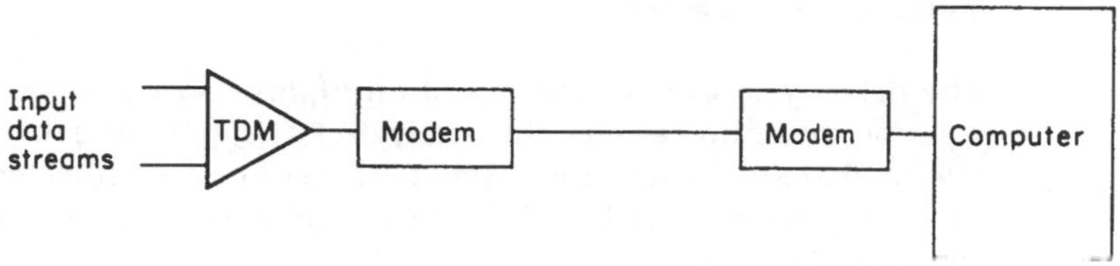

Figure 12.11 TDM system used as a front end substitute. When a TDM is used as a front-end processor, the computer must be programmed to perform demultiplexing

only one computer port is then required to service the terminals which are connected to the computer through the TDM. The TDM can be connected at the computer center, or it can be located at a remote site and connected over a leased line and a pair of modems or DSUs. Since demultiplexing is conducted by the computer's software, only one multiplexer is necessary.

Due to the wide variations in multiplexing techniques of each manufacturer, no standard software has been written for demultiplexing; and, unless multiple locations can use this technique, the software development costs may exceed the hardware savings associated with this technique. In addition, the software overhead associated with the computer performing the demultiplexing may degrade its performance to an appreciable degree and must be considered.

Inverse multiplexing

A multiplexing system which is coming into widespread usage is the inverse multiplexing system. As shown in Figure 12.12, inverse multiplexing permits a high-speed data stream to be split into two or more slower data streams for transmission over lower-cost transmission facilities.

Originally developed to support high speed transmission between computers via analog transmission facilities during the 1970s, a new series of similar products were introduced during the 1990s to operate on digital transmission facilities. Referred to as bandwidth on demand multiplexers, these inverse multiplexers are used to support videoconferencing via the switched digital network by aggregating two or more 56 kbps or 64 kbps calls to obtain an operating rate required by videoconferencing equipment.

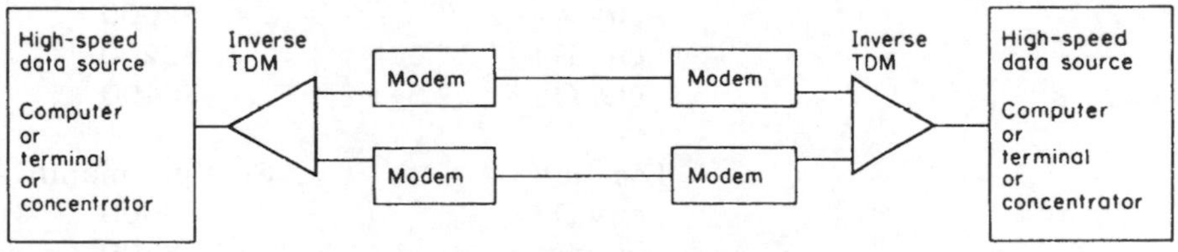

Figure 12.12 Inverse multiplexing. An inverse multiplexer splits a serial data stream into two or more individual data streams for transmission at lower data rates

Multiplexing economies

The primary motive for the use of multiplexers in a network is to reduce the cost of communications. In analyzing the potential of multiplexers, one should first survey terminal users to determine the projected monthly connect time of each terminal device. Then the most economical method of data transmission from each individual terminal to the computer facility can be computed. To do this, direct dial costs should be compared with the cost of a leased line from each terminal device to the computer site.

Once the most economical method of transmission for each individual terminal device to the computer is determined, this cost should be considered the 'cost to reduce'. The telephone mileage costs from each terminal city location to each other terminal city location should be determined in order to compute and compare the cost of utilizing various techniques, such as multiplexing of data by combining several low- to medium-speed terminals' data streams into one high-speed line for transmission to the central site.

In evaluating multiplexing costs, the cost of telephone lines from each terminal location to the 'multiplexer center' must be computed and added to the cost of the multiplexer equipment. Then, the cost of the high-speed line from the multiplexer center to the computer site must be added to produce the total multiplexing cost. If this cost exceeds the cumulative most economical method of transmission for individual terminals to the central site, then multiplexing is not cost-justified. This process should be reiterated by considering each city as a possible multiplexer center to optimize all possible network configurations. In repeating this process, terminals located in certain cities will not justify any calculations to prove or disprove their economic feasibility as multiplexer centers, because of their isolation from other cities in a network.

An example of the economics involved in multiplexing is illustrated in Figure 12.13. In this example, assume that the volume of terminal traffic from the devices located in cities A and B would result in a dial-up charge of $3000 per month if access to the computer in city G was over the switched network. The installation of leased lines from those cities to the computer at city G would cost $2000 and $2200 per month, respectively. Furthermore, let us assume that the terminals at cities C, D, and E only periodically communicate with the computer, and their dial-up costs of $400, $600, and $500 per month, respectively, are much less than the cost of leased lines between those cities and the computer. Then without multiplexing, the network's most economical communications cost would be:

Location	Cost per month
city A	$2000
city B	$2200
city C	$400
Location	Cost per month
city D	$600
city E	$500
Total cost	$5700

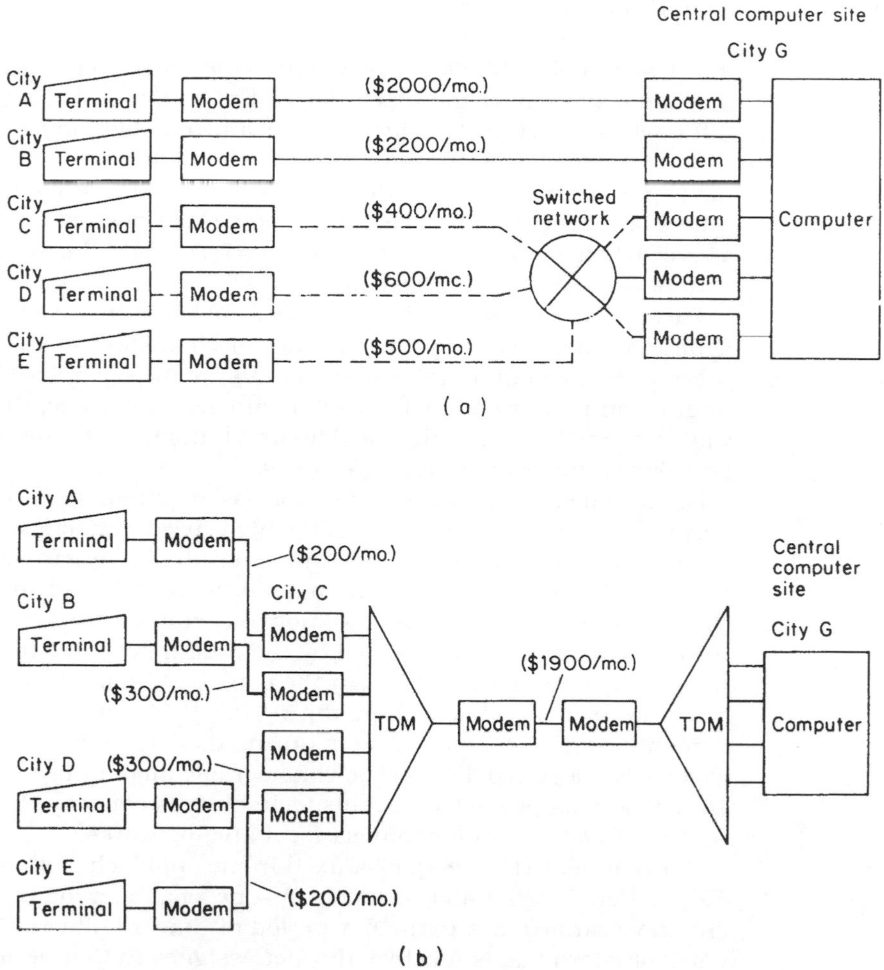

Figure 12.13 Multiplexing economics. On an individual basis, the cost of five terminals accessing a computer system (top) can be much more expensive than when a time division multiplexer is installed (bottom)

Let us further assume that city C is centrally located with respect to the other cities so we could use it as homing point or multiplexer center. In this manner, a multiplexer could be installed in city C, and the terminal traffic from the other cities could be routed to that city, as shown in the bottom portion of Figure 12.13. Employing multiplexers would reduce the network communications cost to $2900 per month which produces a potential savings of $2800 per month, which should now be reduced by the multiplexer costs to determine net savings. If each multiplexer costs $500 per month, then the network using multiplexers will save the user $1800 each month. Exactly how much saving can be realized, if any, through the use of multiplexers depends not only on the types, quantities, and distributions of terminals to be serviced but also on the leased line tariff structure and the type of multiplexer employed.

Statistical and intelligent multiplexers

In a traditional TDM, data streams are combined from a number of devices into a single path so that each device has a time slot assigned for its use. Although such TDMs are inexpensive and reliable, and can be effectively employed to reduce communications costs, they make inefficient use of the high-speed transmission medium. This inefficiency is due to the fact that a time slot is reserved for each connected device, whether or not the device is active. When the device is inactive, the TDM pads the slot with nulls and cannot use the slot for other purposes.

These pad characters are inserted into the message frame because demultiplexing occurs by the position of characters in the frame. Thus, if these pads are eliminated, a scheme must then be employed to indicate the origination port or channel of each character. Otherwise, there would be no way to correctly reconstruct the data and route them to the correct computer port during the demultiplexing process.

By dynamically allocating time slots as required, statistical multiplexers permit more efficient utilization of the high-speed transmission medium. This permits the multiplexer to service more terminals without an increase in the high-speed link as would a traditional multiplexer. The technique of allocating time slots on a demand basis is known as statistical multiplexing, and this means that data are transmitted by the multiplexer only from the terminals that are actually active. Although the concept of statistical multiplexing was first used by data concentrators during the 1960s, it was not until the development of statistical multiplexers during the late 1970s that the technology was widely deployed. Due to the considerable amount of network bandwidth statistical multiplexing saves, this technology is employed in modern routers that are designed to interconnect local area networks.

Conventional TDMs employ a fixed frame approach as illustrated in Figure 12.14. Here, each frame consists of one character or bit for each input channel scanned at a particular period of time. As illustrated, even when a particular terminal is inactive, the slot assigned to that device is included in the message frame transmitted because the presence of a pad or null character in the time slot is required to correctly demultiplex the data. In the lower portion of Figure 12.14 the demultiplexing process which is accomplished by time slot position is illustrated. Because a typical terminal may be idle 90% of the time, this technique contains obvious inefficiencies.

Statistical frame construction

A statistical multiplexer employs a variable frame building technique which takes advantage of terminal idle times to enable more terminals to share access to a common circuit. The use of variable frame technology permits previously wasted time slots to be eliminated, because control information is transmitted with each frame to indicate which terminals are active and have data contained in the message frame.

One of many techniques that can be used to denote the presence or absence of data traffic is the activity map which is illustrated in Figure 12.15. When

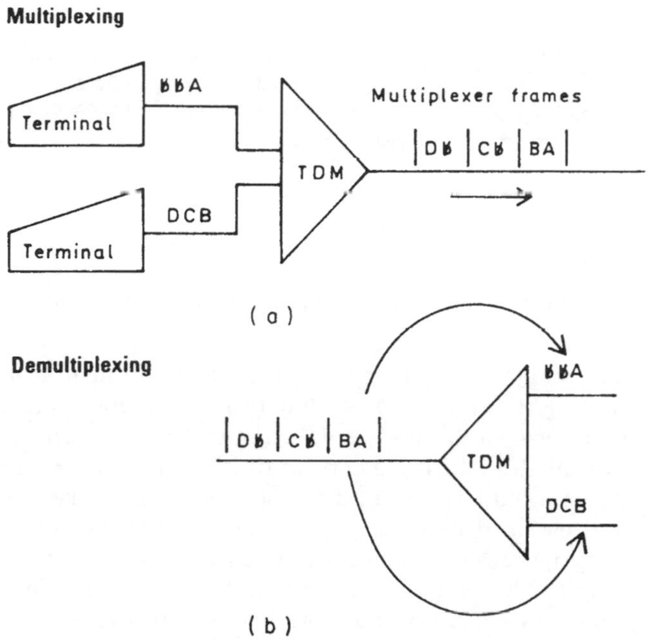

Figure 12.14. Conventional TDMs use fixed time slots. (a) Multiplexing and demultiplexing by TDMs. (b) Absence of line activity during multiplexer scan, null character inserted into message frame. b̸ indicates blank or null characters

an activity map is employed, the map itself is transmitted before the actual data. Each bit position in the map is used to indicate the presence or absence of data from a particular multiplexer time slot scan. The two activity maps and data characters illustrated in Figure 12.15 represent a total of 10 characters which would be transmitted in place of the 16 characters that would result from two scans of an 8 port conventional TDM.

Another statistical multiplexing technique involves buffering data from each data source and then transmitting the data with an address and byte count. The address is used by the demultiplexer to route the data to the

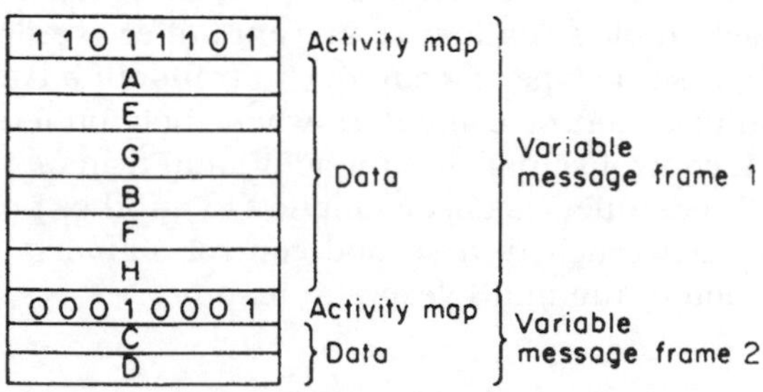

Figure 12.15 Activity mapping to produce variable frames. Using an activity map where each bit position indicates the presence or absence of data for a particular data source permits variable message frames to be generated

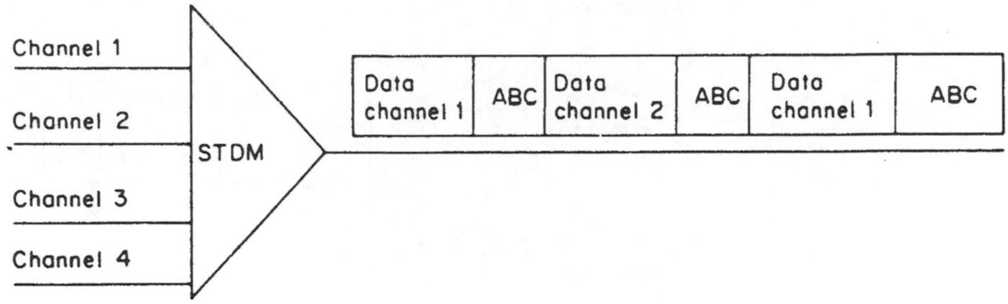

Figure 12.16 Address and byte count (ABC) frame composition

correct port, and the byte count indicates the quantity of data to be routed to that port. Figure 12.16 illustrates the message frame of a four-channel statistical multiplexer employing the address and byte count frame composition method during a certain time interval. Note that since channels 3 and 4 had no data traffic during the two particular time intervals, there was no address and byte count nor data from those channels transmitted on the common circuit. Also note that the data from each channel are of variable length. Typically, statistical multiplexers employing an address and byte count frame composition method wait until either 32 characters or a carriage return are encountered prior to forming the address and byte count and forwarding the buffered data. The reason that 32 characters are selected as the decision criterion is that this represents the average line length of an interactive transmission session.

A few potential technical drawbacks of statistical multiplexers exist which users should note. These problems include the delays associated with data blocking and queuing when a large number of connected terminals become active or when a few terminals transmit large bursts of data. For either situation, the aggregate throughput of the multiplexer's input active data exceeds the capacity of the common high-speed line, causing data to be placed into buffer storage.

Another reason for delays is that a circuit error causes one or more retransmissions of message frame data to occur. Since the connected terminals may continue to send data during the multiplexer-to-multiplexer retransmission cycle, this can also fill up the multiplexer's buffer areas and cause time delays.

If the buffer area should overflow, data would be lost which would create an unacceptable situation. To prevent buffer overflow, all statistical multiplexers employ some type of technique to transmit a traffic control signal to attached terminals and/or computers when their buffers are filled to a certain level. Such control signals inhibit additional transmission through the multiplexer until the buffer has been emptied to another predefined level. Once this level has been reached, a second control signal is issued which permits transmission to the multiplexers to resume.

Buffer control

The three major buffer or flow control techniques employed by statistical multiplexers include inband signaling, outband signaling, and clock reduction.

Inband signaling involves transmitting XOFF and XON characters to inhibit and enable the transmission of data from terminals and computer ports that recognize these flow control characters. Since many terminals and computer ports do not recognize these control characters, a second common flow control method involves raising and lowering the Clear to Send (CTS) control signal on the RS-232 or V.24 interface. Since this method of buffer control is outside the data path where data are transmitted in pin 2, it is known as outband signaling.

Both inband and outband signaling are used to control the data flow of asynchronous devices. Since synchronous devices transmit data formed into blocks or frames, you would most likely break a block or frame by using either inband or outband signaling. This would cause a portion of a block or frame to be received, which would result in a negative acknowledgement when the receiver performs its cyclic redundancy computation. Similarly, when the remainder of the block or frame is allowed to resume its flow to the receiver, a second negative acknowledgement would result.

To alleviate these potential causes of decrease of throughput, multiplexer vendors typically reduce the clocking speed furnished to synchronous devices. Thus, a synchronous terminal operating at 4800 bps might first be reduced to 2400 bps by the multiplexer halving the clock. Then, if the buffer in the multiplexer continues to fill, the clock might be further reduced to 1200 bps.

Service ratio

The measurement used to denote the capability of a statistical multiplexer is called its service ratio, which compares its overall level of performance to that of a conventional TDM. Since synchronous transmission by definition denotes blocks of data with characters placed in sequence in each block, there are no gaps in this mode of transmission. In comparison, a terminal operator transmitting data asynchronously may pause between characters to think prior to pressing each key on the terminal. Thus, the service ratio of STDMs for asynchronous data is higher than the service ratio for synchronous data. Typically, STDM asynchronous service ratios range between 2:1 and 3.5:1, whereas synchronous service ratios range between 1.25:1 and 2:1, with the service ratio dependent upon the efficiency of the STDM as well as its built-in features to include the stripping of start and stop bits from asynchronous data sources. In Figure 12.17, the operational efficiency of both a statistical and a conventional TDM are compared. Here we have assumed that the STDM has an efficiency of twice that of the TDM.

Assuming four 9600 bps data sources are to be multiplexed, the conventional TDM illustrated in the top part of Figure 12.17 would be required to operate at a data rate of at least 28 400 bps. This would then require the use of a 56 kbps DSU connected to a 56 kbps digital transmission facility. For the STDM shown in the lower portion of this illustration, assuming a two-fold increase in efficiency over the conventional TDM, the composite data rate required will be19 200 bps. This permits the employment of a lower operating rate modem and enables an analog transmission facility to be used, an important consideration if digital service is not available at the locations to be serviced. Alternatively, you could connect eight 9600 bps data sources or two

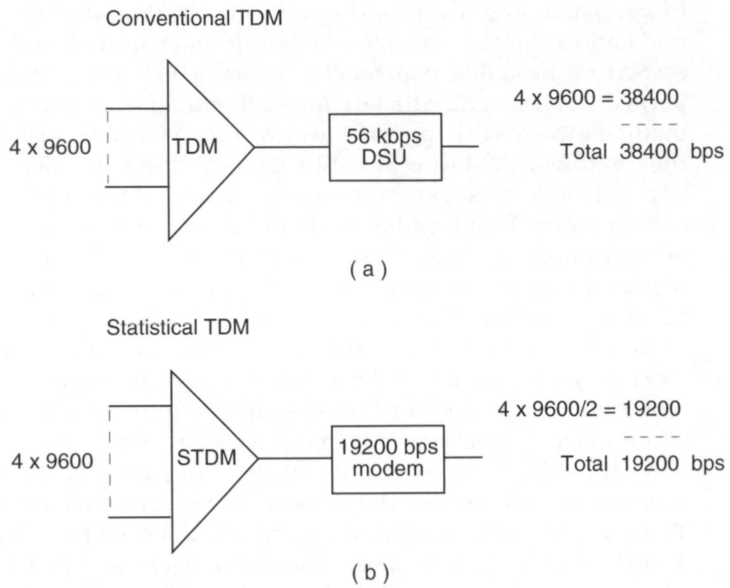

Figure 12.17 Comparing conventional and statistical TDMs. An STDM typically has an efficiency of two to four times a conventional TDM. Using an efficiency level twice the conventional TDM results in a composite operating data rate requirement of 19 200 bps which is serviced by the use of 19 200 bps modem

19 200 bps data sources or a more flexible combination of data sources to the STDM and use 56 kbps DSU. Thus, the use of an STDM provides you with the ability to support both additional data sources as well as a more flexible selection of data sources than a traditional TDM.

Data source support

Some statistical multiplexers only support asynchronous data whereas other multiplexers support both asynchronous and synchronous data sources. When a statistical multiplexer supports synchronous data sources it is extremely important to determine the method used by the STDM vendor to implement this support.

Some statistical multiplexer vendors employ a band pass channel to support synchronous data sources. When this occurs, not only is the synchronous data not multiplexed statistically, but the data rate of the synchronous input limits they capability of the device to support asynchronous transmission. Figure 12.18 illustrates the effect of multiplexing synchronous data via the use of a band pass channel. When a band pass channel is employed, a fixed portion of each message frame is reserved for the exclusive multiplexing of synchronous data, with the portion of the frame reserved proportional to the data rate of the synchronous input to the STDM. This means that only the remainder of the message frame is then available for the multiplexing of all other data sources.

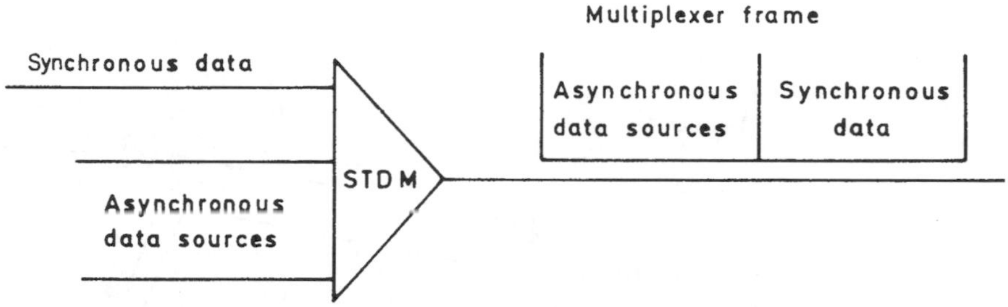

Figure 12.18 The use of a band pass channel to multiple synchronous data

As an example of the limitations of band pass multiplexing, consider an STDM that is connected to a 19 200 bps modem and supports a synchronous terminal operating at 9600 bps. If band pass multiplexing is employed, only 9600 bps is then available in the multiplexer for the multiplexing of other data sources. In comparison, assume that another STDM statistically multiplexes synchronous data. If this STDM has a service ratio of 2:1, then a 9600 bps synchronous input to the STDM would on the average take up 4800 bps of the 19 200 bps operating line. Since the synchronous data are statistically multiplexed, when that data source is not active other data sources serviced by the STDM will flow through the system more efficiently. In comparison, the band pass channel always requires a predefined portion of the high-speed line to be reserved for synchronous data, regardless of the activity of the data source.

Switching and port contention

Two features normally available with more sophisticated statistical multiplexers are switching and port contention. Switching capability is also referred to as alternate routing and it requires the multiplexer to support multiple high-speed lines whose connection to the multiplexer is known as a node. Thus, switching capability normally refers to the ability of the multiplexer to support multiple nodes. Figure 12.19 illustrates how alternate routing can be used to compensate for a circuit outage. In the example shown if the line connecting locations 1 and 3 should become inoperative an alternate route through location 2 could be established if the STDMs support data switching.

Port contention is normally incorporated into large capacity multinodal statistical multiplexers that are designed for installation at a central computer facility. This type of STDM may demultiplex data from hundreds of data channels; however, since many data channels are usually inactive at a given point in time, it is a waste of resources to provide a port at the central site for each data channel on the remote multiplexers. Port contention thus results in the STDM at the central site containing a smaller number of ports than the number of channels of the distant multiplexers connected to that device. Then, the STDM at the central site directs the data sources entered through remote multiplexer channels to the available ports on a demand

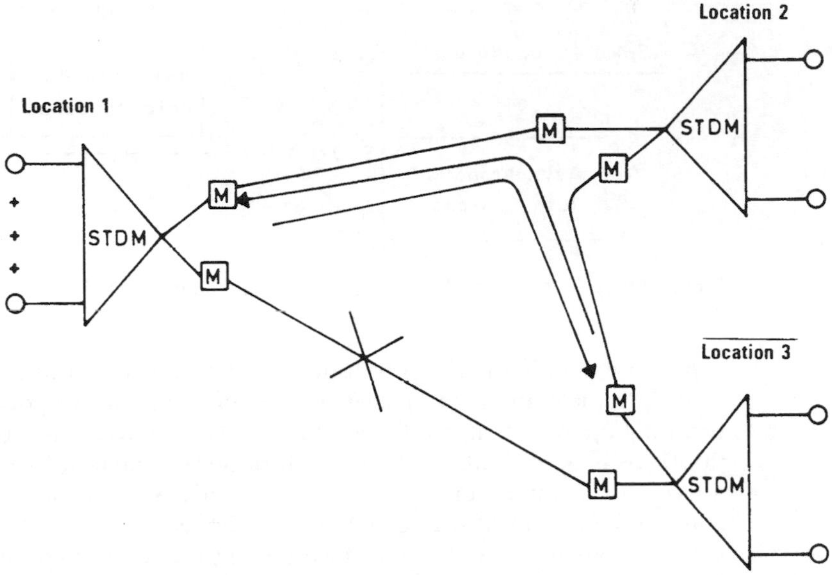

Figure 12.19 Switching permits load balancing and alternate routing if a high-speed line should become inoperative (M = Modem)

basis. If no ports are available, the STDM may issue a 'NO PORTS AVAILABLE' message and disconnect the user or put the user into a queue until a port becomes available.

Intelligent time division multiplexers (ITDMs)

One advancement in statistical multiplexer technology resulted in the introduction of data compression into STDMs. Such devices intelligently examine data for certain characteristics and are known as intelligent time division multiplexers (ITDM). These devices take advantage of the fact that different characters occur with different frequencies and use this quality to reduce the average number of bits per character by assigning short codes to frequently occurring characters and long codes to seldom-encountered characters.

The primary advantage of the intelligent multiplexer lies in its ability to make the most efficient use of a high-speed data circuit in comparison with the other classes of TDMs. Through compression, synchronous data traffic which normally contains minimal idle times during active transmission periods can be boosted in efficiency. Intelligent multiplexers typically permit an efficiency four times that of conventional TDMs for asynchronous data traffic and twice that of conventional TDMs for synchronous terminals.

Statistical (STDM) and intelligent (ITDM) division multiplexer statistics

Although the use of statistical and intelligent multiplexers can be considered on a purely economic basis to determine if the cost of such devices is offset by

Table 12.3 Intelligent multiplexer statistics

Multiplexer loading: % of time device not idle
Buffer utilization: % of buffer storage in use
Number of frames transmitted
Number of bits of idle code transmitted
Number of negative acknowledgements received

$$\text{Traffic density} = \frac{\text{non-idle bits}}{\text{total bits}}$$

$$\text{Error density} = \frac{\text{NAKs received}}{\text{frames transmitted}}$$

$$\text{Compression efficiency} = \frac{\text{total bits received}}{\text{total bits compressed}}$$

$$\text{Statistical loading} = \frac{\text{number of actual characters received}}{\text{maximum number which could be received}}$$

$$\text{Character error rate} = \frac{\text{characters with bad parity}}{\text{total characters received}}$$

the reduction in line and modem or DSU/CSU costs, the statistics that are computed and made available to the user of such devices should also be considered. Although many times intangible, these statistics may warrant consideration even though an economic benefit may at first be hard to visualize. Some of the statistics normally available on statistical and intelligent multiplexers are listed in Table 12.3. Through a careful monitoring of these statistics, network expansion can be preplanned to cause a minimum amount of potential busy conditions to users. In addition, frequent error conditions can be noted prior to user complaints and remedial action taken earlier than normal when conventional devices are used.

Features to consider

In Table 12.4, you will find a list of the primary selection features to consider when evaluating statistical multiplexers. Although many of these features were previously discussed, a few features were purposely omitted from consideration until now. These features include auto baud detect, flyback delay and echoplex, and they primarily govern the type of terminal devices that can be efficiently supported by the statistical multiplexer.

Auto baud detect is the ability of a multiplexer to measure the pulse width of a data source. Since the data rate is proportional to the pulse width, this feature enables the multiplexer to recognize and adjust to different speed terminals accessing the device over the switched telephone network.

On electromechanical printers, a delay time is required between sending a carriage return to the terminal and then sending the first character of the next line to be printed. This delay time enables the print head of the terminal to be

Table 12.4 Statistical multiplexer selection features

Feature	Parameters to consider
Auto baud detect	Data rates detected
Flyback delay	Settings available
Echoplex	Selectable by channel or device
Protocols supported	2780/3780, 3270, HDLC/SDLC, other
Data type supported	Asynchronous, synchronous
Service ratios	Asynchronous, synchronous
Flow control	XON-XOFF, CTS, clocking
Multinodal capability	Number of nodes
Switching	Automatic or manual
Port contention	Disconnect or queued when all ports in use
Data compression	Stripping bits or employs compression algorithm

repositioned prior to the first character of the next line being printed. Many statistical multiplexers can be set to generate a series of fill characters after detecting a carriage return, enabling the print head of an electromechanical terminal to return to its proper position prior to receiving a character to be printed. This feature is called flyback delay and can be enabled or disabled by channel on many multiplexers.

Since some networks contain full-duplex computer systems that echo each character back to the originating terminal, the delay from twice traversing through statistical multiplexers may result in the terminal operator obtaining the feeling that his or her terminal is non-responsive. When echoplexing is supported by an STDM, the multiplexer connected to the terminal immediately echoes each character to the terminal, whereas the multiplexer connected to the computer discards characters echoed by the computer. This enables data flow through the multiplexer system to be more responsive to the terminal operator. Since error detection and correction is built into all statistical multiplexers, a character echo from the computer is not necessary to provide visual transmission validation and is safely eliminated by echoplexing.

The other options listed in Table 12.4 should be self-explanatory, and the user should check vendor literature for specific options available for use on different devices.

Utilization considerations

Although your precise network requirements govern the type of multiplexer that will result in the best price-performance, several general comparisons can be made between devices. In comparison to FDM, the principal advantages of TDM include the ability to service high input data sources, the capacity for a greater number of individual inputs, the performance of data compression (intelligent multiplexers), the detection of errors, and the request for retransmission of data (statistical and intelligent multiplexers). The key differences between multiplexers are tabulated in Table 12.5.

Table 12.5 Multiplexer comparisons

	FDM	TDM	STDM	ITDM
Efficiency	poor	good	better	best
Channel capacity	poor	good	better	best
High-speed data	very poor	poor	better	best
Configuration change				
Data rate	good	fair	good	good
Number of channels	poor	good	better	better
Installation ease	poor	poor	good	good
Problem isolation	poor	poor	good	good
Error detection/retransmission	n/a	n/a–good	automatic	automatic
Multidrop capability	good	n/a	possible	possible

Inverse multiplexers

The inverse multiplexer was introduced during the 1970s to provide transmission at wideband rates through the utilization of two to six voice-grade analog lines. These devices also provide network configuration flexibility and provide reliable back-up facilities during leased line outage situations. Although the widespread availability of digital transmission facilities substantially reduced the market for inverse multiplexers during the 1980s, the demand for economical videoconferencing resulted in the development and use of a new type of inverse multiplexer during the 1990s. That inverse multiplexer is commonly known as a bandwidth on demand multiplexer, and it is used to dial multiple 56 or 64 kbps digital connections to construct an aggregate transmission rate of $n \times 56/64$ kbps. In addition to supporting videoconferencing, some bandwidth on demand multiplexers function as a mechanism to supplement leased digital transmission facilities with dial-up digital transmission. That is, when the occupancy of an existing digital transmission facility exceeds a predefined level the bandwidth on demand multiplexer will establish one or more new connections via the use of the switched digital network to provide additional bandwidth for transmission between two locations.

Operation

The original inverse multiplexer operates by splitting a data stream at the transmitting station, and two to six substreams travel down different paths to a receiving station. Such a data communications technique has several distinct advantages over single-channel wideband communications lines. Using multiple leased lines increases network routing flexibility and permits the use of the direct distance dialing (DDD) network as a back-up in the event of the failure of one or more leased lines.

Similar in design and operation, the original series of inverse multiplexers developed by several manufacturers to aggregate transmission over multilpe analog leased lines permit data transmission at speeds up to 38 400 bps by combining the transmission capacity of two voice-grade circuits. Their

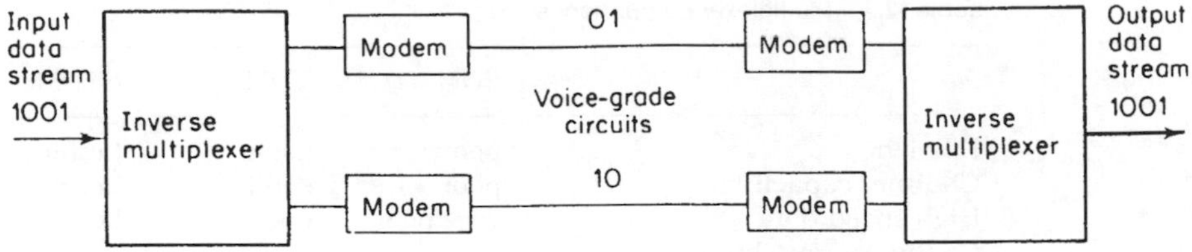

Figure 12.20 An inverse multiplexer splits a transmitted data stream into odd and even bits which are transmitted over two voice-grade circuits and recombined at the distant end

operation can be viewed as reverse time division multiplexing. Input data streams are split into two paths by the unit's transmitting section. In its simplified form of operation, all odd bits are transmitted down one path and all even bits down the other. At the other end, the receiver section continuously and adaptively adjusts for differential delays caused by two-path transmission and recombines the dual bit streams into one output stream, as illustrated in Figure 12.20.

Each inverse multiplexer contains a circulating memory that permits an automatic training sequence, triggered by modem equalization, to align the memory to the differential delay between the two channels. This differential delay compensation allows, for example, the establishment of a 38 400 bps circuit consisting of a 19 200 bps satellite link, and a 19 200 bps ground or undersea cable, as shown in Figure 12.21.

Since the propagation delay on the satellite circuit would be 250 ms or more, whereas the delay on the cable link would be less than 30 ms, without adjusting for the difference in transmission delays the bit stream could not be reassembled correctly. In this type of application, the failure of one circuit can be compensated for by transmitting the entire data load over the remaining channel at one-half the normal rate.

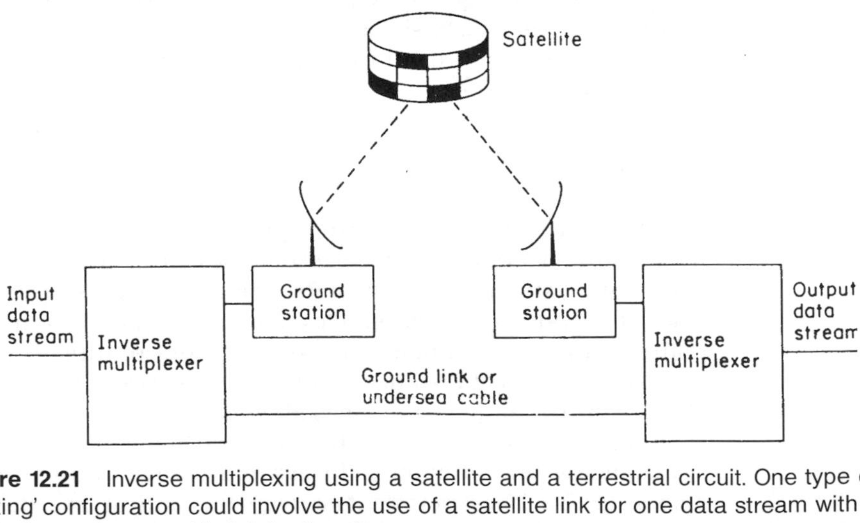

Figure 12.21 Inverse multiplexing using a satellite and a terrestrial circuit. One type of 'plexing' configuration could involve the use of a satellite link for one data stream with a ground or undersea cable link for the other

Table 12.6 Wideband transmission combinations using 19 200 bps modems

Subchannel A	Subchannel B	Transmission speed
19 200	19 200	38 400
14 400	19 200	33 600
14 400	14 400	28 800
14 400	12 000	26 400
12 000	12 000	24 000
12 000	9600	21 600
9600	9600	19 200
9600	7200	16 800
7200	7200	14 400
7200	4800	12 000

Typical applications

As stated previously, one key advantage obtained by using inverse multiplexers is the cost savings associated with two analog voice-grade lines in place of the use of analog wideband facilities. Another advantage afforded by these devices gives the user the ability to configure and reconfigure a network based upon the range of distinct speeds available to meet changing requirements. For example, if synchronous 19 200 bps speed-selectable modems are installed with the inverse multiplexers, up to ten possible throughput bit rates may be transmitted, as shown in Table 12.6. Using more modern 33.6 kbps modems allows even more combinations

Economics of use

One advantage of the use of an inverse multiplexer is the cost savings associated with using analog voice-grade lines in comparison to wideband facilities. Although both voice-grade and wideband circuit tariffs follow a sliding scale of monthly per mile fees based upon distance, the cost of wideband facility is often up to 16 times more expensive than a similar-distance voice-grade line. This means that the cost of four modems, two inverse multiplexers, and two voice-grade lines will normally be less expensive than one wideband circuit and two wideband modems.

Since tariffs follow a sliding scale, with the monthly cost per mile of a short circuit much more expensive than a long circuit, there are certain situations where the preceding does not hold true. This economic exception is usually encountered when connecting locations 80 to 100 miles or less distant from one another. Normally, if the two locations to be connected are over 80 to 100 miles apart, inverse multiplexing is a more economic method of performing communications, whereas wideband is more economically advantageous at distances under that mileage range.

The economics of inverse multiplexing are illustrated in Figure 12.22. Note that the 80 to 100 mile economic decision point range results from the fact that the monthly lease cost of inverse multiplexers varies between vendors.

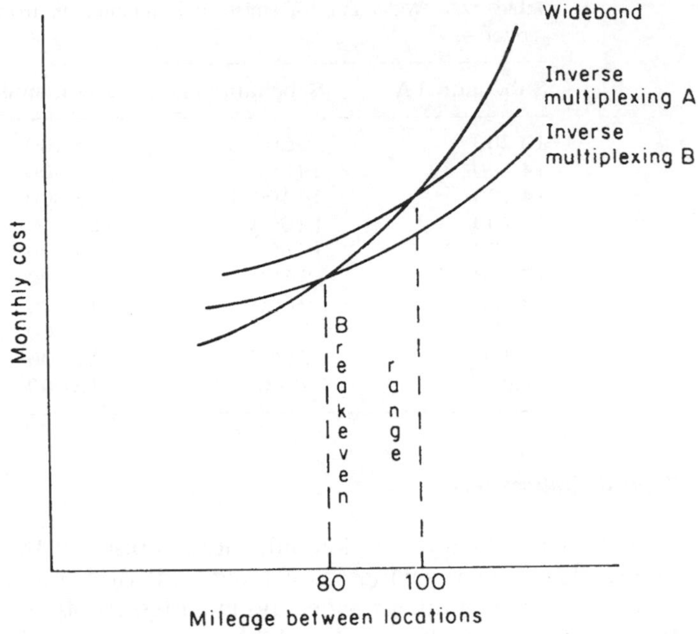

Figure 12.22 Economics of inverse multiplexing. In comparing the monthly cost of one wideband circuit and two wideband modems to the cost of two voice-grade lines, two inverse multiplexers, and four voice-grade modems, the breakeven range will be between 80 and 100 miles

Bandwidth on demand multiplexers

As previously noted, a bandwidth on demand multiplexer can be considered to represent a modern type of inverse multiplexer developed to aggregate two or more dialed digital circuits to support a higher operating rate than that obtainable from conventional 56/64 kbps dial digital transmission facilities. Figure 12.23 illustrates an example of the use of a pair of bandwidth on demand multiplexers to support dial-up videoconferencing. In this example each multiplexer is shown connected to multiple digital circuits via individual DSUs. In actuality, most bandwidth on demand multiplexers are connected to

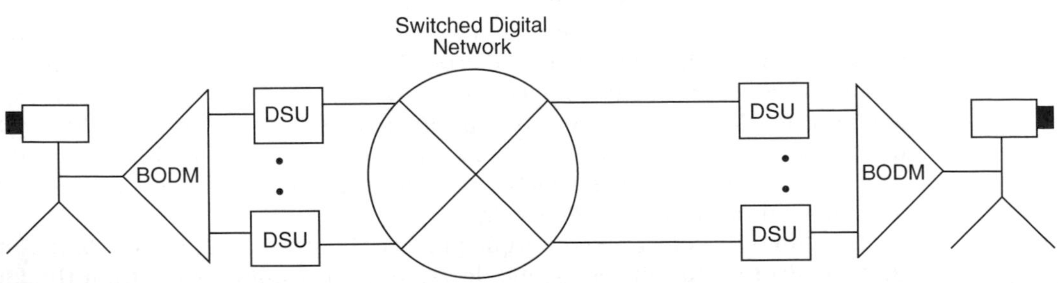

Figure 12.23 Using bandwidth on demand multiplexers to establish a videoconference session

only one digital circuit; however, that circuit is either an ISDN 23B+D line or a conventional T1 circuit with 24 subchannels that can individually operate at 56 or 64 kbps. Those circuits are routed from the customer's premises to a local telephone company office, where a predefined number of channels on the circuit are connected to the switched digital network. Then, the bandwidth on demand multiplexer 'dials' calls on individual channels on the ISDN or T1 line which become calls on the switched digital network.

Economics

Similar to the original inverse multiplexers developed for use on analog transmission facilities, bandwidth on demand multiplexers provide a series of economic advantages obtained through their use. Those economic advantages result from the ability of the multiplexer to allow users to replace costly digital leased lines with a grouping of aggregated digital dial-up transmissions. In addition, because dial-up allows one location to communicate with other locations connected to the digital dial network, the use of bandwidth on demand multiplexers allows support of videoconferencing from one location to multiple locations without having to install an expensive mesh-structured network of digital leased lines.

To illustrate the economic advantages associated with the use of bandwidth on demand multiplexers, let's assume your organization has offices in New York, Chicago and Miami. Let's further assume you would like to videoconference New York and Miami, New York and Chicago, and Chicago and Miami offices for 2 hours per week at a data rate of 384 kbps. If you are using ISDN your telephone company may bill you 10 cents/minute for a 64 kbps long-distance call. Because a 384 kbps data rate requires six 64 kbps calls to be aggregated, your organization's digital dial cost becomes 60 cents/minute, or $36.00 per hour. With three locations each using dial transmission for 2 hours per week, the total weekly transmission cost becomes $216.00. Based upon a four-week month, the monthly cost of dial-up service becomes $864.00. If you priced digital leased lines required to interconnect the previously mentioned locations, you would more than likely have to spend well over $3000 per month to obtain a 384 kbps fractional T1 line between each location. Thus the use of bandwidth on demand multiplexers can save organizations a considerable amount of money when videoconferencing or similar applications requiring high-speed transmission applications occur on only a periodic basis.

12.2 CONTROL UNITS

The function of a control unit is to economize upon computer ports and transmission facilities. To do so the control unit operates in a poll-and-select environment, allowing the computer to poll terminals connected to the control unit, functioning as a traffic cop since the computer tells each terminal when it can transmit data.

12.2.1 Control unit concept

The concept of the control unit originated in the 1960s with the introduction of the IBM.3270 Information Display System which consisted of three basic components: a control unit, display station and printer. The control unit enabled many display stations and printers to share access to a common port on IBM's version of a front-end processor, marketed as a communication controller, or via a common channel on a host computer, when the control unit was directly attached to a host computer channel. Here the term 'display station' represents a CRT display with a connected keyboard.

Since the introduction of the 3270 Information Display Unit, most computer vendors have manufactured similar products, designed to connect their terminals and printers to their computer systems. Control units vary considerably in configuration, ranging from a single control unit built into a display station to a configuration in which the control unit directs the operation of up to 64 attached display stations and printers. These attached display stations and printers are usually referred to by the term cluster; hence, another popular name for a control unit is cluster controller.

The major differences between IBM and other vendor control units is in the communications protocol employed for data transmission to and from the control unit, the types of terminals and printers that can be attached to the control unit, and the physical number of devices the control unit supports. In this section we will focus our attention on IBM control units; however, when applicable, differences in the operation and utilization of IBM and other vendor devices will be discussed.

12.2.2 Attachment methods

IBM control units can be connected to a variety of IBM computer systems, ranging in scope from System/360 through System/380 and 390 processors so 43XX computers as well as the modem Enterprise series of mainframes. The control unit can be connected either directly to the host computer via a channel attachment to a selector, multiplexer, or byte multiplexer channel on the host; or it can be link-attached to IBM's version of the front-end processor known as a communication controller as illustrated in Figure 12.24. Once connected, the control unit directs the operation of attached display stations and printers, with such devices connected via a coaxial cable to the control unit. Note that a generic type of control unit is indicated in Figure 12.24 by the use of the term 327*x*. Several control unit models were marketed by IBM over the past two decades, with numerous types of each model offered. The actual control unit installed depends upon the communications protocol (bisync or synchronous data-link control) used for communications between the control unit and the host computer, the type of host system to which the control unit is to be connected, and the method of attachment (channel- or link-attached), and the number of devices that the control unit is to support.

Two popularly used IBM control units which are now essentially obsolete are the 3274 and 3276 devices. The 3274 enables up to 32 devices to be

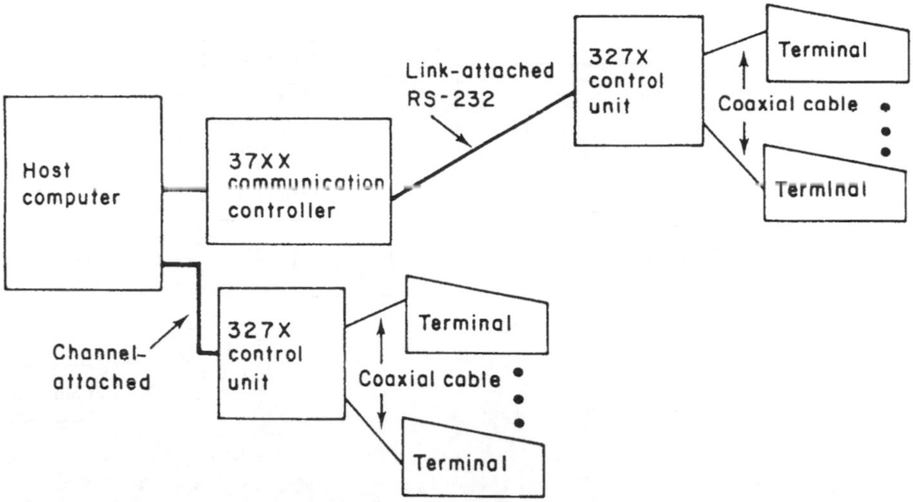

Figure 12.24 IBM 3270 Information Display System. Control units used in the IBM 3270 Information Display System can be either link-attached to a communication controller or channel-attached to a host computer system

connected, and the 3276 is a table-top control unit with an integrated display station that permits up to seven additional devices to be connected to it. In 1987 IBM introduced its 3174 control unit which can be obtained in several different configurations supports a maximum of 64 devices. The major difference between the 3174 and 3274 is the availability of both Ethernet and Token-Ring LAN connection options and a protocol conversion option for the newer 3174 control unit. The Ethernet and Token-Ring option permits the control unit to be connected to either an Ethernet local area network or a Token-Ring network, whereas the protocol conversion option permits asynchronous ASCII terminals to be connected to the control unit in place of normally more expensive EBCDIC devices.

Since the attachment of terminal devices to an IBM control unit requires constant polling from the control unit to the terminal, either dedicated or leased lines are normally used to connect the two devices together. Due to this, the 3270 Information Display System is considered to be a 'closed system'. In addition, many IBM control units require terminals to be connected by the use of coaxial cable. Thus, this normally precludes the use of asynchronous terminals using the PSTN to access a 3270 Information Display System.

The devices marketed by other computer vendors vary considerably in comparison to the method of attachment employed by IBM. Most non-IBM control units permit terminal devices to be connected via an RS-232/V.24 interface while this capability is only available on certain 3174 and 3274 models. Thus, terminal devices remotely located from the main cluster of terminals can be connected to such control units via tail circuits as illustrated in Figure 12.25. Another difference between IBM and other computer vendor control units concerns the method of attachment of the control unit to the host computer system. IBM control units can be channel-attached or

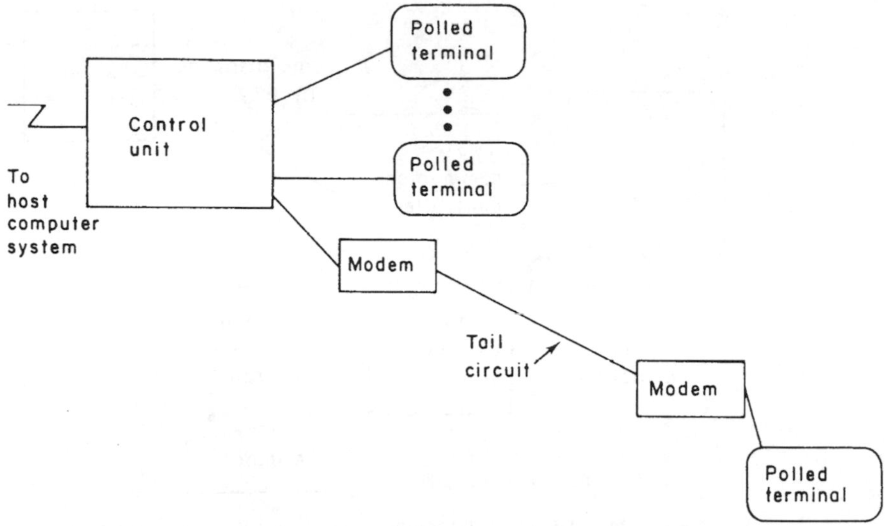

Figure 12.25 Non-IBM control units can connect terminals via tail circuits. Control units produced by many computer manufacturers other than IBM have RS-232 V.24 interfaces for attaching terminal devices. This type of interface permits distant terminals to be connected to the controller via tail circuits

link-attached, whereas most non-IBM control units are link-attached to the vendor's front-end processor, even when the control unit is located in close proximity to the computer room.

12.2.3 Unit operation

The 3270 data stream that flows between the control unit and the host computer system contains orders and control information in addition to the actual data. Both buffer control and printer format orders are sent to the control unit, where they are interpreted and acted upon. Buffer control orders are used by the control unit to perform such functions as positioning, defining, modifying and assigning attributes, and formatting data that is then written into a display character buffer that controls the display of information on the attached terminal. Thus, the terminal display information flows from the host computer to the control unit in an encoded, compressed format, where it is interpreted by the control unit and acted upon. Similarly, printer format orders are stored in the printer character buffer in the control unit as data and are interpreted and executed by the printer's logic when encountered in the print operation. Since a network can consist of many control units, with each control unit having one or more attached terminal devices, both a control unit address and terminal address are required to address each terminal device.

Protocol support

Both binary synchronous communications and synchronous data link control (SDLC) communications are supported by most members of the IBM family of 327*x* control units. Some control units are soft-switch selectable, incorporating a built-in diskette drive that permits a communications module incorporating either communication protocol to be loaded into the control unit; other models may require a field modification to change the communications discipline. By employing a communications discipline, end-to-end error detection and correction is accomplished between the control unit and the host computer system.

12.2.4 Breaking the closed system

One of the key constraints of the IBM family of control units is the fact that it is a closed system, requiring terminal devices to be directly connected to the control unit. Without the introduction of a variety of third party products, this cabling restriction would preclude the utilization of personal computers and asynchronous terminals in a 3270 network. Fortunately, a variety of third party products have reached the communications market that enable users operating 3270 networks to overcome the cable restriction of control unit connections. Foremost among such products are protocol converters and terminal interface units.

Protocol converters

When used in a 3270 network, a protocol converter will function as either a terminal emulator or a control unit/terminal emulator, depending on the method employed to attach a non-3270 terminal into a 3270 network. When functioning as a terminal emulator, the protocol converter is, in essence, an asynchronous to synchronous converter that converts the line by line transmission display of an asynchronous terminal or personal computer into the screen-oriented display on which the control unit operates. Similarly, the protocol converter changes the full screen-oriented display image transmitted by the control unit into the line-by-line transmission which the terminal or personal computer was built to recognize.

During this translation process cursor positioning, character attributes, and other control codes are mapped from the 3270 format into the format which the terminal or personal computer was built to recognize, and vice versa. A protocol converter functioning as a terminal emulator is illustrated at the top of Figure 12.26 It should be noted that some protocol converters are stand-alone devices, whereas other protocol converters are manufactured as adapter boards that can be inserted into a system expansion slot within a personal computer's system unit.

The lower portion of Figure 12.26, a protocol converter functioning as a control unit/terminal emulator is illustrated. Since all terminals access a 3270

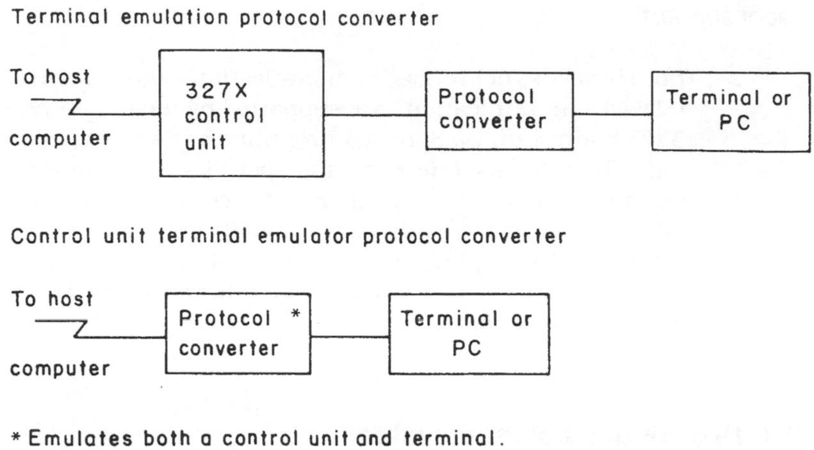

Figure 12.26 Protocol converters open the closed 3270 system. Through the use of protocol converters a variety of asynchronous terminals and personal computers can be connected to a 3270 network

network through control units, stand-alone terminals or personal computers must also emulate the control unit function if they are to be directly attached to the host computer system.

Terminal interface unit

Another interesting device that warrants attention prior to concluding our discussion of control units is the terminal interface unit. This device is basically a coaxial cable to RS-232 converter, which enables coaxial cabling terminals to access asynchronous resources to include public databases and computer systems that support asynchronous transmission.

In an IBM 3270 network, terminal devices were originally connected directly to a control unit, precluding the use of the terminal from accessing other communications facilities. The terminal interface unit breaks this restriction, since it converts a coaxial interface used to connect many 3270 terminals to control units into an RS-232 interface at the flick of the switch on the unit. The use of this device is illustrated in Figure 12.27. When in a power-off state, the terminal interface unit is transparent to the data flow and the terminal operates directly connected to the control unit. When in an operational mode, the terminal interface unit converts the keyboard-entered data into ASCII characters for transmission and similarly converts received characters for display on the IBM terminal's screen. Thus, the terminal interface unit can be employed to make a coaxial cabled terminal multifunctional, permitting it to access the coaxial cable port on a control unit as well as an RS-232/V.24 device, such as an asynchronous modem. During the 1990s IBM introduced control units that added support for twisted-pair wiring. The introduction of such control units greatly diminished the market for terminal interface units.

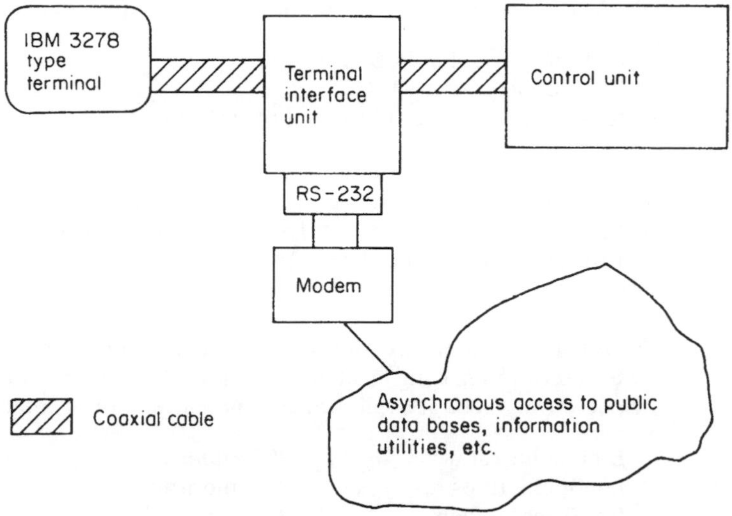

Figure 12.27 The terminal interface unit makes a coaxial cabled terminal multifunctional

12.3 REVIEW QUESTIONS

1. What are the major differences between time division multiplexing and frequency division multiplexing?

2. How many 300 bps data sources could be multiplexed by a frequency division multiplexer and by a traditional time division multiplexer?

3. Why does frequency division multiplexing inherently provide multidrop transmission capability without requiring addressable terminals and poll and select software?

4. Assume that your organization has two computer systems, one located in city A and another located in city B. Suppose that your organization has eight terminals in city C and 10 terminals in city B, with the following computer access requirements:

Terminal location	Terminal destination	
	City A computer	City B computer
City B	6	4
City A	4	4

Draw a network schematic diagram illustrating the use of time division hub-bypass multiplexing to connect the terminals to the appropriate computers.

5. Why is a front-end processor substitution for a multiplexer at a computer site the exception rather than the rule?

6. Assume that the cost of communications equipment and facilities is as follows:

8-channel TDM at $500
28 800 modem at $400
leased line city A to city B $1000/month

If twelve 2400 bps terminals are located in city A and the computer is located in city B, draw a schematic of the multiplexing network required to permit the terminals to access the computer, assuming each terminal can be directly connected to the multiplexer. What is the cost of this network for one year of operation?

7. Assume that the requirements of Question 6 are modified, resulting in all twelve terminals being located beyond cabling distance of the multiplexer. Assume the following cost of communications equipment and facilities:

8 position rotary	$50/month
telephone lines at	$30/month
2400 bps modems at	$30

Draw a revised schematic of the multiplexing network.What is the cost of this network for one year of operation?

8. Why must a statistical multiplexer add addressing information to the data it multiplexes?

9. Why are most statistical multiplexers ill-suited for multiplexing multidrop circuits?

10. Assume that the statistical muiltiplexer you are considering using has an efficiency 2.5 times that of a conventional TDM. If you anticipate connecting the multiplexer to a 28 800 bps modem, how many 2400 bps data sources should the multiplexer multiplex.

11. Assume that the multiplexer discussed in Question 10 services synchronous data by the use of a bandpass channel. What would be the effect upon the number of 2400 bps asynchronous data sources supported in Question 10 if you must service a 9600 bps synchronous data source by the use of a bandpass channel?

12. Assume that you are considering the use of an intelligent multiplexer that has an efficiency three times that of a conventional multiplexer for asynchronous data traffic and 1.5 that of a conventional multiplexer for synchronous data traffic. If you have to multiplex 18 1200 bps asynchronous terminals and four 2400 bps synchronous terminals on a line operating at 9600 bps, could the multiplexer support you requirement? Why?

13. How could you use the statistical loading data available from an intelligent multiplexer to determine if the vendor's literature concerning its efficiency in comparison to a conventional TDM is reasonable?

14. Draw a network schematic showing how the port contention option of a multiplexer could be used in a hub-bypass network.

15. You are comparing the cost of inverse multiplexing to using wideband facilities to transmit data at 38 400 bps, and you determine the monthly cost of the following facilities and devices.

Facility/device	Monthly cost
Wideband line	\$2820
Voice-grade line	\$750
Inverse multiplexer at	\$225
19 200 bps modem at	\$35

Would it be economical with these figures to use inverse multiplexers?

16. What are the primary advantages of inverse multiplexing in comparison to wideband transmission?

17. Describe two applications suitable for a bandwidth on demand multiplexes.

18. What is the major difference between a channel-attached and a link-attached communications controller?

19. Assume the cost of a fractional T1 (FT1) line operating at 384 kbps between New York and Chicago is \$2100 per month. If the use of 64 kbps ISDN costs 10 cents/minute, how many minutes per month can you operate a 384 kbps videoconference between locations prior to exceeding the cost of the FT1 line? (For the problem, do not consider the cost of inverse multiplexers.)

20. Why is it difficult to attach RS-232/V.24 devices to an IBM 327Q Information Display System network? Discuss some of the products one can obtain to connect RS-232/V.24 devices into a 327Q network to include their operational features.

13

LOCAL AREA NETWORKS

One of the limiting factors of personal computers is the primary use of such devices as isolated workstations. This means that in most organizations it is difficult for personal computer users to share data and the use of peripheral devices, since such sharing normally requires physical activity to occur. Both the direct economical benefits of sharing data and the use of peripheral devices as well as productivity gains resulting from the integration of personal computers into a common network have contributed to an increasing demand for local area network products which is the focus of this chapter.

In this chapter we will first examine the origins and major benefits derived from the utilization of local area networks and their relationship to typical network applications. In doing so we will compare and contrast LANs and WANs to obtain a better understanding of the similarities and differences between the two types of networks. Next we will look at the major areas of local area network technology and the effect these areas have upon the efficiency and operational capability of such networks. Here our examination will focus upon network topology, transmission media and the major access methods employed in LANs. Using the previous material as a base, we will then focus our attention on the operation of several types of local area networks.

13.1 ORIGIN

The origin of local area networks can be traced, in part, to IBM terminal equipment introduced in 1974. At that time, IBM introduced a series of terminal devices designed for use in transaction-processing applications for banking and retailing. What was unique about those terminals was their method of connection; a common cable that formed a loop provided a communications path within a localized geographical area. Unfortunately, limitations in the data transfer rate, incompatibility between individual IBM loop systems, and other problems precluded the widespread adoption of this method of networking. The economics of media-sharing and the ability to provide common access to a centralized resource were, however key advantages, and they resulted in IBM and other vendors investigating the use of different techniques to provide a localized communications capability between different devices. In 1977, Datapoint Corporation began selling its

Attached Resource Computer Network (Arcnet), considered by most people to be the first commercial local area networking product. Since then, several types of LANs to include different versions of Ethernet, Token-Ring, and Fiber Data Distributed Interface (FDDI) local area network technologies were standardized and hundreds of companies have developed local area networking products, and the installed base of terminal devices connected to such networks has increased exponentially. They now number in the hundreds of millions.

13.2 COMPARISON WITH WANs

Local area networks can be distinguished from wide area networks by geographical area of coverage, data transmission and error rates, ownership, government regulation, data routing and, in many instances, by the type of information transmitted over the network.

13.2.1 Geographical area

The name of each network provides a general indication of the scope of the geographical area in which it can support the interconnection of devices. As its name implies, a LAN is a communications network that covers a relatively small local area. This area can range in scope from a department located on a portion of a floor in an office building, to the corporate staff located on several floors in the building, to several buildings on the campus of a university.

Regardless of the LAN's area of coverage, its geographical boundary will be restricted by the physical transmission limitations of the local area network. These limitations include the cable distance between devices connected to the LAN and the total length of the LAN cable. In comparison, a wide area network can provide communications support to an area ranging in size from a town or city to a state, country, or even a good portion of the entire world. Here, the major factor governing transmission is the availability of communications facilities at different geographical areas that can be interconnected to route data from one location to another.

While the geographical area of a LAN typically is limited to a building or floor within a building, the interconnection of local area networks via wide area networks can span cities, countries and continents. In fact, the global Internet can be considered to represent a connection of hundreds of thousands of local area networks.

13.2.2 Data transmission and error rates

Two additional areas that differentiate LANs from WANs and explain the physical limitation of the LAN geographical area of coverage are the data transmission rate and error rate for each type of network. LANs normally operate at a low megabit-per-second rate, typically ranging from 4 Mbps to 16 Mbps, with several recently standardized LANs operating at 100 Mbps, and one type of LAN referred to as Gigabit Ethernet operating at 1 Gbps. In comparison, the communications facilities used to construct a major portion of most WANs provide a data transmission rate at or under the T1 and E1 data rates of 1.544 Mbps and 2.048 Mbps.

Since LAN cabling is primarily within a building or over a small geographical area, it is relatively safe from natural phenomena, such as thunderstorms and lightning. This safety enables transmission at a relatively high data rate, resulting in a relatively low error rate. In comparison, since wide area networks are based on the use of communications facilities that are much farther apart and always exposed to the elements, they have a much higher probability of being disturbed by changes in the weather, electronic emissions generated by equipment, or such unforeseen problems as construction workers accidentally causing harm to a communications cable. Because of these factors, the error rate on WANs is considerably higher than the rate experienced on LANs. On most WANs you can expect to experience an error rate between 1 in a million and 1 in 10 million (1×10^6 to 1×10^7) bits. In comparison, the error rate on a typical LAN may be less than that range by one or more orders of magnitude, resulting in an error rate from 1 in 10 million to 1 in 100 million bits.

13.2.3 Ownership

The construction of a wide area network requires the leasing of transmission facilities from one or more communications carriers. Although your organization can elect to purchase or lease communications equipment, the transmission facilities used to connect diverse geographical locations are owned by the communications carrier. In comparison, an organization that installs a local area network normally owns all of the components used to form the network, including the cabling used to form the transmission path between devices.

13.2.4 Regulation

Since wide area networks require transmission facilities that may cross local, state and national boundaries, they may be subject to a number of governmental regulations at the local, state and national levels. In comparison, regulations affecting local area networks are primarily in the areas of building codes. Such codes regulate the type of wiring that can be installed in a building and whether the wiring must run in a conduit.

13.2.5 Data routing and topology

In a local area network, data are routed along a path that defines the network. That path is normally a bus, ring, tree or star structure, and data always flow on that structure. The topology of a wide area network can be much more complex. In fact, many wide area networks resemble a mesh structure,

Table 13.1 Comparing LANs and WANs

Characteristics	Local area network	Wide area network
Geographic area of coverage	Localized to a building, group of buildings, or campus	Can span an area ranging in size from a city to the globe
Data transmission rate	Typically 4 Mbps to 16 Mbps, with some copper and fibre optic based networks operating at 100 Mbps	Normally operate at or below T1 and E1 translation rates of 1.544 Mbps and 2.048 Mbps
Error rate	1 in 10^7 to 1 in 10^8	1 in 10^6 to 1 in 10^7
Ownership	Usually with the implementor	Communications carrier retains ownership of line facilities
Data routing	Normally follows fixed route	Switching capability of network allows dynamic alteration of data flow
Topology	Usually limited to bus, ring, tree and star	Virtually unlimited design capability
Type of information carried	Primarily data	Voice, data and video commonly integrated

including equipment used to reroute data in the event of communications circuit failure or excessive traffic between two locations. Thus, the data flow on a wide area network can change, whereas the data flow on a local area network primarily follows a single basic route.

13.2.6 Type of information carried

The last major difference between local and wide area networks is the type of information carried by each network. Many wide area networks support the simultaneous transmission of voice, data and video information. In comparison, most local area networks are currently limited to carrying data. In addition, although all wide area networks can be expanded to transport voice, data and video, many local area networks are restricted by design to the transportation of data. Table 13.1 summarizes the general similarities and differences between local and wide area networks.

13.3 UTILIZATION BENEFITS

In its simplest form, a local area network is a cable that provides an electronic highway for the transportation of information to and from different devices connected to the network. Because a LAN provides the capability to route data between devices connected to a common network within a relatively limited distance, numerous benefits can accrue to users of the network. These can include the ability to share the use of peripheral devices, thus obtaining common access to data files and programs, the ability to communicate with other people on the LAN by electronic mail, and the ability to access the larger processing capability of mainframes or minicomputers through the common gateways that link a local area network to larger computer systems. Here the gateway can be directly cabled to the mainframe or minicomputer if they reside at the same location, or it may be connected remotely via a corporate wide area network.

Another key benefit of LANs in our modern information age is the ability to provide all workstations on the network with communications to information utilities, other corporate locations, or the Internet via one shared transmission facility. For example, connecting a LAN to the Internet via a high-speed T1 line would provide all users on the LAN with the ability to exchange email, surf the Web, and perform other communications related activities.

13.3.1 Peripheral sharing

Peripheral sharing allows network users to access color laser printers, CD-ROM jukebox systems, and other devices that may be needed for only a small portion of the time that a workstation is in operation. Thus, users of a LAN can obtain access to resources that would probably be too expensive to justify for each individual workstation user.

13.3.2 Common software access

The ability to access data files and programs from multiple workstations can substantially reduce the cost of software. In addition, shared access to database information allows network users to obtain access to updated files on a real-time basis.

13.3.3 Electronic mail

One popular type of application program used on LANs enables users to transfer messages electronically. Commonly referred to as electronic mail or e-mail, this type of application program can be used to supplement and, in many cases, eliminate the need for paper memoranda.

13.3.4 Gateway access to mainframes

For organizations with mainframe or minicomputers, a local area network gateway can provide a common method of access to those computers. Without the use of a LAN gateway, each personal computer requiring access to a mainframe or minicomputer would require a separate method of access. This might increase both the complexity and the cost of providing access.

13.3.5 Internet access

If your organization has 20, 30, or 50 or more employees at one location, the cost of providing each person with individual Internet access on a cumulative basis could be prohibitive. In addition to providing each employee with an Internet account that could cost $20 per month, you might have to install a separate telephone line and modem. Instead of supporting individual dial access to the Internet you could install a router and leased line to an Internet Service Provider (ISP) for a monthly charge that is a fraction of the cost associated with a large number of individual Internet dial network accounts. In addition to being more economical, the use of a T1 or fractional T1 (FT1) access line would result in much faster access than is possible when users access the Internet via a modem connection.

13.3.6 Virtual private network operations

A virtual private network (VPN) represents a network created via the routing of data between two or more locations over a public packet network, like the Internet. If your organization has two or more geographically separated LANs it may be less expensive to connect them to the Internet and use the Internet as a VPN than to interconnect the LANs via a leased line. However, since the Internet is a public network you would also need to consider the cost of security equipment, such as a firewall, authentication server, and encryption equipment to secure communications between organizational locations via the Internet.

13.4 TECHNOLOGICAL CHARACTERISTICS

Although a local area network is a limited distance transmission system the variety of options available for constructing such networks is anything but limited. Many of the options available for the construction of local area networks are based on the technological characteristics that govern their operation. These characteristics include different topologies, signaling methods, transmission media, access methods used to transmit data on the network, and the hardware and software required to make the network operate.

13.4.1 Topology

The topology of a local area network is the structure or geometric layout of the cable used to connect stations on the network. Unlike conventional data communications networks, which can be configured in a variety of ways with the addition of hardware and software, most local area networks are designed to operate based upon the interconnection of stations that follow a specific topology. The most common topologies used in LANs include the loop, bus, ring, star and tree, as illustrated in Figure 13.1.

Loop

As previously mentioned in this chapter, IBM introduced a series of transaction-processing terminals in 1974 that communicated through the

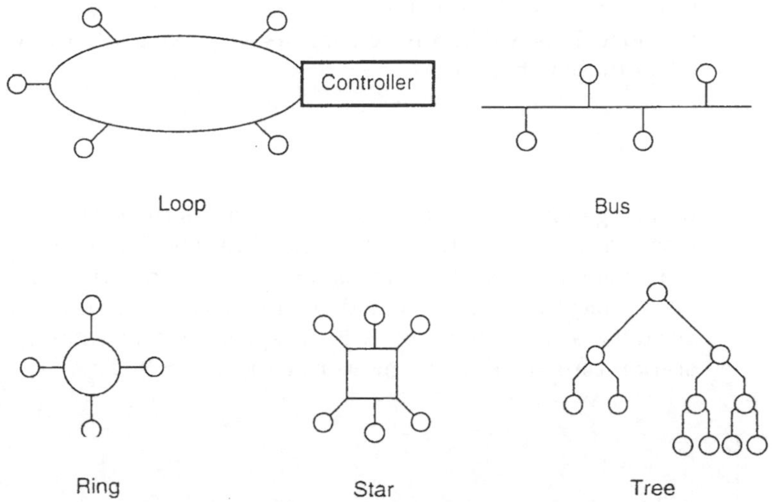

Figure 13.1 LAN topology. The five most common geometric layouts of local area network cabling form a loop, bus, ring, star, or tree structure

use of a common controller on a cable formed into a loop. This type of topology is illustrated at the top of Figure 13.1.

Since the controller employed a poll-and-select access method, terminal devices connected to the loop required a minimum of intelligence. Although this reduced the cost of terminals connected to the loop, the controller lacked the intelligence to distribute the data flow evenly among terminals. A lengthy exchange between two terminal devices or between the controller and a terminal would thus tend to bog down this type of network structure. A second problem associated with this network structure was the centralized placement of network control in the controller. If the controller failed, the entire network would become inoperative. Due to these problems, the use of loop systems is restricted to several niche areas, and they are essentially considered a derivative of a local area network.

Bus

In a bus topology structure, a cable is usually laid out as one long branch, onto which branches are used to connect each station on the network to the main data highway. Although this type of structure permits any station on the network to talk to any other station, rules are required to recover from such situations as when two stations attempt to communicate at the same time. Later in this chapter we will examine the relationships among the network topology, the method employed to access the network, and the transmission medium employed in building the network.

Readers familiar with LANs might be wondering why a hub is omitted from Figure 13.1. In actuality, a single hub functions either as a data regenerator, in effect representing a bus, or as a loop. Thus, depending upon the type of

LAN a hub is designed to support, it can represent either a bus or ring topology. This will become clearer as we move further into our examination of LANs in this chapter.

Ring

In a ring topology, a single cable that forms the main data highway is shaped into a ring. As with the bus topology, branches are used to connect stations to one another via the ring. A ring topology can thus be considered to be a looped bus. Typically, the access method employed in a ring topology requires data to circulate around the ring, with a special set of rules governing when each station connected to the network can transmit data.

Star

The fourth major local area network topology is the star structure, illustrated in the lower portion of Figure 13.1. In a star network, each station on the network is connected to a network controller. Then access from any one station on the network to any other station can be accomplished through the network controller. Here the network controller can be viewed as functioning similarly to a telephone switchboard, because access from one station to another station on the network can occur only through the central device.

Tree

A tree network structure represents a complex bus. In this topology, the common point of communications at the top of the structure is known as the headend. From the headend, feeder cables radiate outward to nodes, which in turn provide workstations with access to the network. There may also be a feeder cable route to additional nodes, from which workstations gain access to the network.

Mixed topologies

Some networks are a mixture of topologies. For example, as previously discussed, a tree structure can be viewed as a series of interconnected buses. Another example of the mixture of topologies is a type of Ethernet known as 10BASE-T. That network can actually be considered a star-bus topology; workstations are first connected to a common device known as a hub, which in turn can be connected to other hubs to expand the network,

13.4.2 Comparison of topologies

Although there are close relationships among the topology of the network, its transmission media and the method used to access the network, we can examine topology as a separate entity and make several generalized observations. First, in a star network, the failure of the network controller will render the entire network inoperative. This is because all data flow on the network must pass through the network controller. On the positive side, the star topology normally consists of telephone wires routed to a switchboard. A local area network that can use in-place twisted-pair telephone wires in this is simple to implement and usually very economical.

In a ring network, the failure of any node connected to the ring normally inhibits data flow around the ring. Due to the fact that data travel in a circular path on a ring network, any cable break has the same effect as the failure of the network controller in a star-structured network. Since each network station is connected to the next network station, it is usually easy to install the cable for a ring network. In comparison, a star network may require cabling each section to the network controller if existing telephone wires are not available, and this can result in the installation of very long cable runs.

In a bus-structured network, data are normally transmitted from a single station to all other stations located on the network, with a destination address appended to each transmitted data block. As part of the access protocol, only the station with the destination address in the transmitted data block will respond to the data. This transmission concept means that a break in the bus affects only network stations on one side of the break that wish to communicate with stations on the other side of the break. Thus, unless a network station functioning as the primary network storage device becomes inoperative, a failure in a bus-structured network is usually less serious than a failure in a ring network. However, some local area networks, such as Token-Ring and FDDI, were designed to overcome the effect of certain types of cable failures. Token-Ring networks include a backup path which, when manually placed into operation, may be able to overcome the effect of a cable failure between hubs (referred to as multistation access units or MAUs). In an FDDI network, a second ring can be activated automatically as part of a self-healing process to overcome the effect of a cable break.

A tree-structured network is similar to a star-structured network in that all signals flow through a common point. In the tree-structured network the common signal point is the headend. Failure of the headend renders the network inoperative. This network structure requires the transmission of information over relatively long distances. For example, communications between two stations located at opposite ends of the network would require a signal to propagate twice the length of the longest network segment. Due to the propagation delay associated with the transmission of any signal, the use of a tree structure may result in a response time delay for transmissions between the nodes that are most distant from the headend.

13.4.3 Signaling methods

The signaling method used by a local area network refers to both the way data are encoded for transmission and the frequency spectrum of the media. To a large degree, the signaling method is related to the use of the frequency spectrum of the media.

Broadband versus baseband

Two signaling methods used by LANs are broadband and baseband. In broadband signaling, the bandwidth of the transmission medium is subdivided by frequency to form two or more subchannels, with each subchannel permitting data transfer to occur independently of data transfer on another subchannel. In baseband signaling only one signal is transmitted on the medium at any point in time.

Broadband is more complex than baseband, because it requires information to be transmitted via the modulation of a carrier signal, thus requiring the use of special types of modems.

Figure 13.2 illustrates the difference between baseband and broadband signaling with respect to channel capacity. It should be noted that although a twisted-pair wire system can be used to transmit both voice and data, the data transmission is baseband, since only one channel is normally used for data. In comparison, a broadband system on coaxial cable can be designed to carry voice and several subchannels of data, as well as fax and video transmission.

Broadband signaling

A broadband local area network uses analog technology, in which high frequency (HF) modems operating at or above 4 kHz place carrier signals onto the transmission medium. The carrier signals are then modified: a process known as modulation, which impresses information onto the carrier. Other modems connected to a broadband LAN reconvert the analog signal block into its original digital format: a process known as demodulation.

The most common modulation method used on broadband LANs is frequency shift keying (FSK), in which two different frequencies are used, one to represent a binary 1 and another frequency to represent a binary 0. Another popular modulation method uses a combination of amplitude and

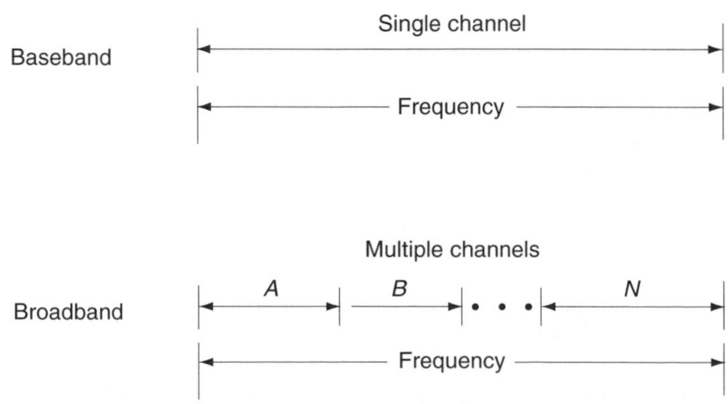

Figure 13.2 Baseband versus broadband signaling. In baseband signaling the entire frequency bandwidth is used for one channel. In comparison, in broadband signaling the channel is subdivided by frequency into many subchannels

phase shift changes to represent pairs of bits. Referred to as amplitude modulation phase shift keying (AM PSK), this method of analog signaling is also known as duobinary signaling because each analog signal represents a pair of digital bits.

When LANs were initially developed it was not economically feasible to design amplifiers that boost signal strength to operate in both directions. Thus, broadband LANs are unidirectional. To provide a bidirectional information transfer capability, a broadband LAN uses one channel for inbound traffic and another channel for outbound traffic. These channels can be defined by differing frequencies or obtained by the use of a dual cable.

Baseband signaling

In comparison to broadband local area networks, which use analog signaling, baseband LANs use digital signaling to convey information.

To understand the digital signaling methods used by many baseband LANs, let us first review the method of digital signaling used by computers and terminal devices. In that signaling method a positive voltage is used to represent a binary 1, whereas the absence of voltage (0 volts) is used to represent a binary 0. If two successive 1 bits occur, two successive bit positions then have a similar positive voltage level or a similar zero voltage level. Since the signal goes from 0 to some positive voltage and does not return to 0 between successive binary 1s, it is referred to as a unipolar non-return to zero signal (NRZ). This signaling technique is illustrated at the top of Figure 13.3.

Although unipolar non-return to zero signaling is easy to implement, its use for transmission has several disadvantages. One of the major disadvantages associated with this signaling method involves determining where one bit ends and another begins. Overcoming this problem requires synchronization between a transmitter and receiver by the use of clocking circuitry, which can be relatively expensive.

To overcome the need for clocking, baseband LANs use Manchester or Differential Manchester encoding. In Manchester encoding a timing transition always occurs in the middle of each bit, and an equal amount of positive and negative voltage is used to represent each bit. This coding technique provides a good timing signal for clock recovery from received data, due to its timing transitions. In addition, since the Manchester code always maintains an equal amount of positive and negative voltage, it prevents direct current (DC) voltage buildup, enabling repeaters to be spaced further apart from one another.

The middle portion of Figure 13.3 illustrates an example of Manchester coding. Note that a low to high voltage transition represents a binary 1, and a high to low voltage transition represents a binary 0. Differential Manchester encoding is illustrated in Figure 13.3(c). The difference between Manchester encoding and Differential Manchester encoding occurs in the method by which binary 1s are encoded. In Differential Manchester encoding, the direction of the signal's voltage transition changes whenever a binary 1 is transmitted, but remains the same for a binary 0. The IEEE 802.3 standard specifies the use of Manchester coding for baseband Ethernet operating at data rates up to 10 Mbps. The IEEE 802.5 standard specifies the use of

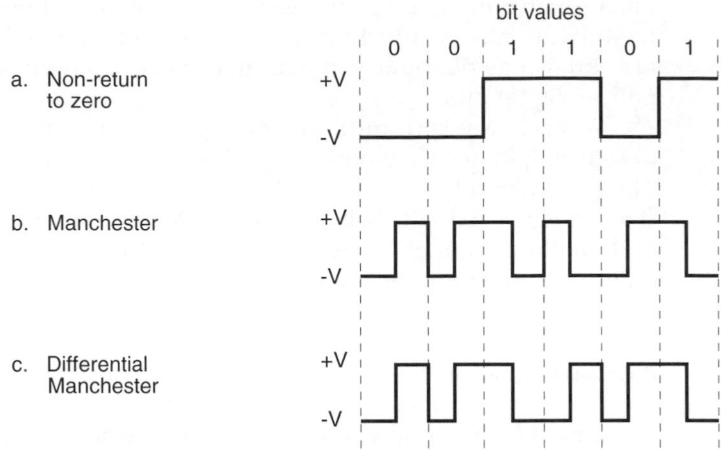

Figure 13.3 Common baseband signaling techniques

Differential Manchester encoding for Token-Ring networks at the physical layer to transmit and detect four distinct symbols: a binary 0, a binary 1 and two non-data symbols.

13.4.4 Transmission medium

The transmission medium used in a local area network can range in scope from twisted-pair wire, such as is used in conventional telephone lines, to coaxial cable, fiber optic cable, and electromagnetic waves such as those used by FM radio and infrared. Each transmission medium has a number of advantages and disadvantages. The primary differences between media are their cost and ease of installation; the bandwidth of the cable, which may or may not permit several transmission sessions to occur simultaneously; the maximum speed of communications permitted; and the geographical scope of the network that the medium supports. Readers are referred to Chapter 3, which discussed the characteristics of different types of transmission media to include structured wiring.

13.4.5 Access methods

If the topology of a local area network can be compared to a data highway, then the access method might be viewed as the set of rules that enable data from one workstation to successfully reach the destination via the data highway. Without such rules, it is quite possible for two messages sent by two different workstations to collide, with the result that neither message reaches its destination. Two common access methods primarily employed in local area networks are carrier-sense multiple access/collision detection (CSMA/CD) and token passing. Each of these access methods is uniquely structured to address the previously mentioned collision and data destination problems.

Prior to discussing how access methods work, let us first examine the two basic types of devices that can be attached to a local area network to gain an appreciation for the work that the access method must accomplish.

Listeners and talkers

We can categorize each device by its operating mode as being a listener or a talker. Some devices, like printers, only receive data (modern printers can return status messages; however, for the sake of this discussion, we can consider a printer to be a listener), and thus operate only as listeners. Other devices, such as personal computers, can either transmit or receive data and are capable of operating in both modes. In a baseband signaling environment where only one channel exists, or on an individual channel on a broadband system, if several-talkers wish to communicate at the same time a collision will occur. Therefore, a scheme must be employed to define when each device can talk and, in the event of a collision, what must be done to keep it from happening again.

For data to reach the destination correctly, each listener must have a unique address, and its network equipment must be designed to respond to a message on the network only when it recognizes its address. The primary goals in the design of an access method are to minimize the potential for data collision, to provide a mechanism for corrective action when data collide, and to ensure that an addressing scheme is employed to enable messages to reach their destination.

Carrier-Sense Multiple Access with Collision Detection (CSMA/CD)

Carrier-Sense Multiple Access with Collision Detection (CSMA/CD) can be categorized as a listen then send access method. CSMA/CD is one of the earliest developed access techniques, and it is the technique used in Ethernet.

Under the CSMA/CD concept, when a station has data to send, it first listens to determine if any other station on the network is talking. The fact that the channel is idle is determined in one of two ways, based on whether the network is broadband or baseband.

In a broadband network, the fact that a channel is idle is determined by carrier-sensing, or noting the absence of a carrier tone on the cable.

Most versions of Ethernet, like other baseband systems, use one channel for data transmission and do not employ a carrier. Instead, most versions of Ethernet encode data using a Manchester code, in which a timing transition always occurs in the middle of each bit as previously illustrated in Figure 13.3. Although Ethernet does not transmit data via a carrier, the continuous transitions of the Manchester code can be considered as equivalent to a carrier signal. Carrier-sensing on a baseband network is thus performed by monitoring the line for activity.

In a CSMA/CD network, if the channel is busy, the station will wait until it becomes idle before transmitting data. Since it is possible for two stations to listen at the same time and discover an idle channel, it is also possible that the two stations could then transmit at the same time. When this situation arises, a collision will occur. On sensing that a collision has occurred, a delay scheme will be employed to prevent a repetition of the collision. Typically, each station will use either a randomly generated or predefined time-out period before attempting to retransmit the message that collided. Since this access method requires hardware capable of detecting the occurrence of a collision, additional circuitry required to perform collision detection adds to the cost of such hardware.

Figure 13.4 illustrates a CSMA/CD bus-based local area network. Each workstation is attached to the transmission medium, such as coaxial cable, by a device known as a bus interface unit (BIU). To obtain an overview of the operation of a CSMA/CD network, assume that station A is currently using the channel and that stations C and D wish to transmit. The BIUs connecting stations C and D to the network would listen to the channel and note that it is busy. Once station A has completed its transmission, stations C and D attempt to gain access to the channel. Because station A's signal takes longer to propagate down the cable to station D than to station C, C's BIU notices that the channel is free slightly before station D's BIU. However, as station C gets ready to transmit, station D now assumes that the channel is free. Within an infinitesimal period of time, C starts transmission, followed by D, resulting in a collision. Here, the collision is a function of the propagation delay of the signal and the distance between two competing stations. CSMA/CD networks therefore work better as the main cable length decreases.

The CSMA/CD access technique is best suited for networks with intermittent transmission, since an increase in traffic volume causes a corresponding increase in the probability of the cable being occupied when a station wishes to talk. In addition, as traffic volume builds under CSMA/CD, throughput may decline, because there will be longer waits to gain access to the network, as well as additional time-outs required to resolve collisions that occur. In spite

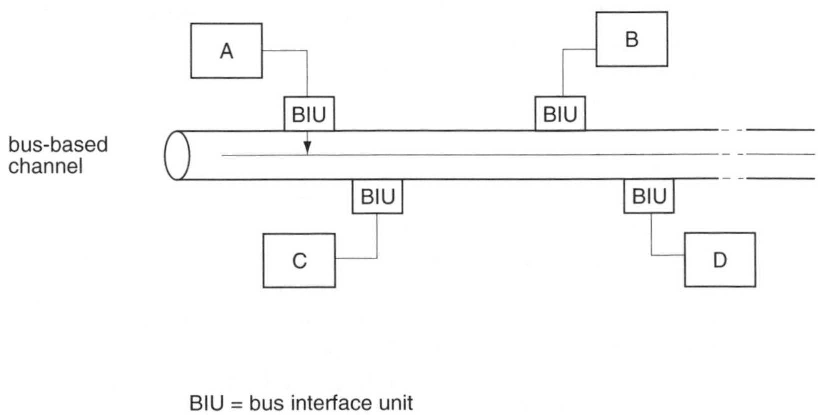

Figure 13.4 CSMA/CD network operation. In a CSMA/CD network, as the distance between workstations increases, the resulting increase in propagation delay time increases the probability of collisions

of those deficiencies it is relatively easy to implement the CSMA/CD access protocol, resulting in Ethernet networks currently accounting for well over 85% of all LANs.

Token passing

In a token passing access method, each time the network is turned on, a token is generated. The token, consisting of a unique bit pattern, travels the length of the network, either around a ring or along the length of a bus. When a station on the network has data to transmit, it must first seize a free token. On a Token- Ring network, the token is then transformed to indicate that it is in

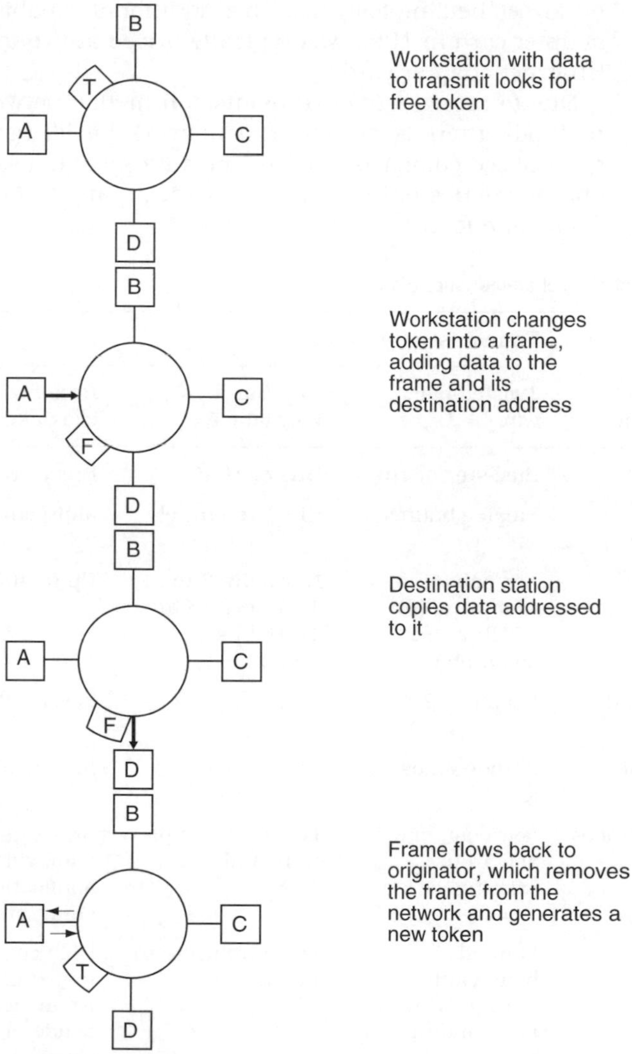

Figure 13.5 Token-Ring operation

use. Information is added to produce a frame, which represents data being transmitted from one station to another. During the time that the token is in use, other stations on the network remain idle, eliminating the possibility of collisions. Once the transmission is completed, the token is converted back into its original form by the station that transmitted the frame, and becomes available for use by the next station on the network.

Figure 13.5 illustrates the general operation of a token-passing Token-Ring network using a ring topology. Since a station on the network can only transmit when it has a free token, token passing eliminates the requirement for collision detection hardware. Due to the dependence of the network on the token, the loss of a station can bring the entire network down. To avoid this, the design characteristics of Token-Ring networks include circuitry that automatically removes a failed or failing station from the network as well as other self-healing features. This additional capability is costly; a Token-Ring adapter card in 1999 was typically priced at two to three times the cost of an Ethernet adapter card.

Due to the variety of transmission media, network structures, and access methods, there is no one best network for all users. Table 13.2 provides a generalized comparison of the advantages and disadvantages of the technical characteristics of local area networks, using the transmission medium as a frame of reference.

Table 13.2 Technical characteristics of LANs

	Transmission medium			
Characteristic	Twisted-pair wire	Baseband coaxial cable	Broadband coaxial cable	Fibre optic cable
Topology	Bus, star or ring	Bus or ring	Bus or ring	Bus, ring or star
Channels	Single channel	Single channel	Multi-channel	Single, multi-channel
Data rate	Normally 2 to 4 Mbps, up to 16 Mbps obtainable	Normally 2 to 10 Mbps, up to 100 Mbps obtainable	Up to 400 Mbps	Up to Gbps
Maximum nodes on net	Usually <255	Usually <1024	Several thousand	Several thousand
Geographical coverage	In thousands of feet	In miles	In tens of miles	In tens of miles
Major advantages	Low cost, may be able to use existing wiring	Low cost, simple to install	Supports voice, data, video applications simultaneously	Supports voice, data, video applications simultaneously
Major disadvantages	Limited bandwidth, requires conduits, low immunity to noise	Low immunity to noise	High cost, difficult to install, requires RF modems and headend	Cable cost, difficult to splice

13.5 ETHERNET NETWORKS

From the title of this section it is apparent that there is more than one type of Ethernet network. From a network access perspective, there is actually only one Ethernet network. However, the CSMA/CD access protocol used by Ethernet, as well as its general frame format and most of its operating characteristics, were used by the Institute of Electrical and Electronic Engineers (IEEE) to develop a series of Ethernet-type networks under the IEEE 802.3 umbrella. Thus, this section will focus on the different types of Ethernet networks by closely examining the components and operating characteristics of Ethernet and then comparing its major features to the different networks defined by the IEEE 802.3 standard. Once this has been accomplished, we will focus our attention on the wiring, topology, and hardware components associated with each type of IEEE 802.3 Ethernet network, as well as the Ethernet frame used to transport data

13.5.1 Original network components

The 10 Mbps bus-based Ethernet network standard originally developed by Xerox, Digital Equipment Corporation and Intel was based on the use of five hardware components. Those components include a coaxial cable, a cable tap, a transceiver, a transceiver cable, and an interface board (also known as an Ethernet controller). Figure 13.6 illustrates the relationship among the original 10 Mbps bus-based Ethernet network components.

Coaxial cable

One of the problems faced by the designers of Ethernet was the selection of an appropriate medium. Although twisted-pair wire is relatively inexpensive and

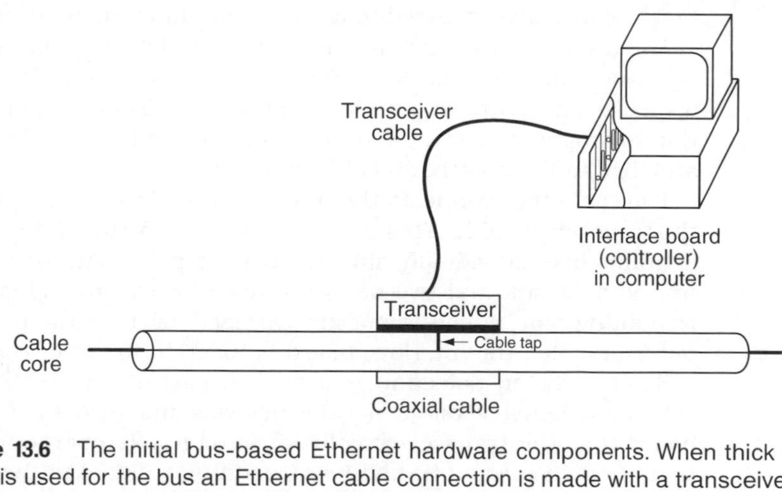

Figure 13.6 The initial bus-based Ethernet hardware components. When thick coaxial cable is used for the bus an Ethernet cable connection is made with a transceiver cable and a transceiver tap tapped into the cable

easy to use, the short distances between twists serve as an antennae for receiving electromagnetic and radio frequency interference in the form of noise. Thus, the use of twisted-pair cable restricts the network to relatively short distances. Coaxial cable, however, has a dielectric shielding the conductor. As long as the ends of the cable are terminated, coaxial cable can transmit over greater distances than twisted-pair cable. Thus, the initial selection for Ethernet transmission medium was coaxial cable.

There are two types of coaxial cable that can be used to form the main Ethernet bus. The first type of coaxial cable specified for Ethernet was a relatively thick 50Ω cable, which is normally colored yellow and is commonly referred to as thick Ethernet. This cable has a marking every 2.5 m to indicate where a tap should occur, if one is required to connect a station to the main cable at a particular location. These markings represent the minimum distance by which one tap must be separated from another on an Ethernet network.

A second type of coaxial cable used with Ethernet is smaller and more flexible; however, it is capable of providing a transmission distance only one-third of that obtainable on thick cable. This lighter and more flexible cable is referred to as thin Ethernet and also has an impedance of 50Ω.

Two of the major advantages of thin Ethernet over thick cable are its cost and its use of BNC connectors. Thin Ethernet is significantly less expensive than thick Ethernet. Thick Ethernet requires connections via taps, whereas the use of thin Ethernet permits connections to the bus via industry standard BNC connectors that form T-junctions.

Transceiver and transceiver cable

Transceiver is a shortened form of transmitter-receiver. This device contains electronics to both transmit onto and receive signals carried by the coaxial cable. The transceiver contains a tap that, when pushed against the coaxial cable, penetrates the cable and makes contact with the core of the cable. In books and technical literature the transceiver, its tap, and its housing are often referred to as the medium attachment unit (MAU).

The transceiver is responsible for carrier detection and collision detection. When a collision is detected during a transmission, the transceiver places a special signal, known as a jam on the cable. This signal is of sufficient duration to propagate down the network bus and inform all the other transceivers attached to the bus that a collision occurred.

The cable that connects the interface board to the transceiver is known as the transceiver cable. This cable can be up to 50 m (165 feet) in length, and it contains five individually shielded twisted pairs. Two pairs are used for data in and data out, and two pairs are used for control signals in and out. The remaining pair, which is not always used, permits the power from the computer in which the interface board is inserted to power the transceiver.

Since collision detection is a critical part of the CSMA/CD access protocol, the original version of Ethernet was modified to inform the interface board that the transceiver collision circuitry is operational. This modification resulted in each transceiver's sending a signal to the attached interface board after every transmission, informing the board that the transceiver's collision circuitry is operational. This signal is sent by the transceiver over the collision pair of the transceiver cable, and must start within 0.6 ms after each frame is transmitted. The duration of the signal can vary between 0.5 and 1.5 ms. Known as the Signal Quality Error and also referred to as the SQE or heartbeat, this signal is supported by Ethernet Version 2.0, published as a standard in 1982, and by the IEEE 802.3 standard.

Interface board

The interface board, or network interface card (NIC), is inserted into an expansion slot within a computer, and is responsible for transmitting frames to and receiving frames from the transceiver. This board contains several special chips, including a controller chip that assembles data into an Ethernet frame and computes the cyclic redundancy check used for error detection. Thus, this board is also referred to as an Ethernet controller.

Repeaters

A repeater is a device that receives, amplifies and retransmits signals. Since a repeater operates at the physical layer, it is transparent to data and simply regenerates signals. Figure 13.7 illustrates the use of a repeater to connect two Ethernet cable segments. As indicated, a transceiver is taped to each cable segment to be connected, and the repeater is cabled to the transceiver. When used to connect cable segments, a repeater counts as one station on

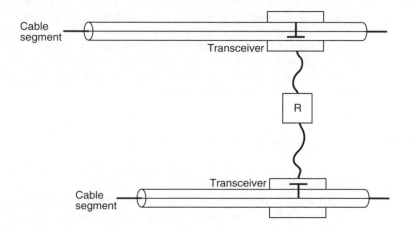

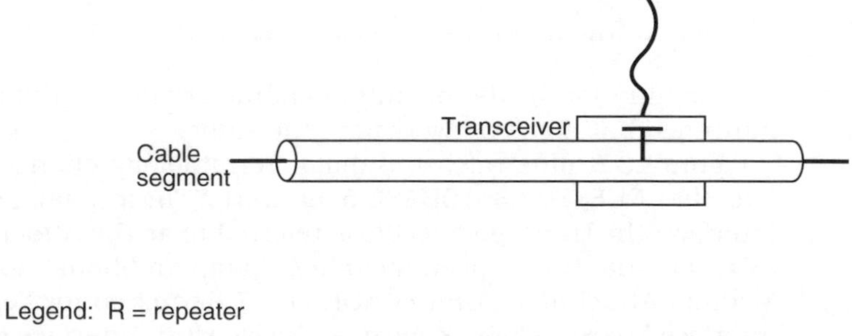

Figure 13.7 Using a repeater. Cable segments can be joined together by a repeater to expand the network. The repeater counts as a stations on each cable segment

each connected segment. Thus, a segment capable of supporting up to 100 stations can support only 99 additional stations when a repeater is used to connect cable segments.

13.5.2 IEEE 802.3 networks

The IEEE 802.3 standard is based on Ethernet. However, it has several significant differences, particularly its support of multiple Physical layer options, which include 50 and 75Ω coaxial cable, unshielded twisted-pair wire, an operating rate of 100 Mbps for two standardized versions of Ethernet, and support for Gigabit Ethernet over several types of fiber optic and copper media. Other differences between various types of IEEE 802.3 networks and Ethernet include the data rates supported by some 802.3 networks, their method of signaling, the maximum cable segment lengths permitted prior to the use of repeaters, and their network topologies.

Network names

The standards that define IEEE 802.3 networks have been given names that generally follow the form 's type 1'. Here, s refers to the speed of the network in Mbps, type is BASE for baseband and BROAD for broadband, and 1 refers to the maximum segment length in 100 m multiples. Thus, 10BASE-5 refers to an IEEE 802.3 baseband network that operates at 10 Mbps and has a maximum segment length of 500 m. One exception to this general form is 10BASE-T, which is the name for an IEEE 802.3 network that operates at 10 Mbps using unshielded twisted-pair (UTP) wire.

Table 13.3 compares the operating characteristics of six currently defined IEEE 802.3 networks to Ethernet. Because there are several versions of Gigabit Ethernet that have different operating characteristics based upon the type of media they support, the last column in Table 13.3 is limited as far as being specific. Later in this chapter when we examine Gigabit Ethernet in detail, we will also examine the different versions of Gigabit Ethernet based upon the media they operate over.

10BASE-5

An examination of the operating characteristics of Ethernet and 10BASE-5 indicates that these networks are the same.

Figure 13.8 illustrates the major terminology changes between Ethernet and the IEEE 802.3 10BASE-5 network. These changes are in the media interface: the transceiver cable is referred to as the Attachment Unit Interface (AUI), and the transceiver, including its tap and housing, is referred to as the Medium Attachment Unit (MAU). The Ethernet controller, also known as an interface board, is now known as the Network Interface Card (NIC).

Both Ethernet and the IEEE 802.3 10BASE-5 standards support a data rate of 10 Mbps and a maximum cable segment length of 500 m. 10BASE-5, like

Table 13.3 Ethernet and IEEE 802.3 network characteristics

Operational characteristics	Ethernet	10BASE-5	10BASE-2	1BASE-T	10BASE-T	10BROAD-36	100BASE-T	1000BASE-T
Operating rate Mbps	10	10	10	1	10	10	100	1000
Access protocol	CSMA/CD	CSMA/CD	CSMA/CD	CSMA/CD	CSMA/CD	CSMA/CD	CSMA/CD	CSMA/CD
Type of signaling	basebase	baseband	baseband	baseband	baseband	broadband	baseband	baseband
Data encoding	Manchester	Manchester	Manchester	Manchester	Manchester	Manchester	8B6T or 4B5B coding	varies based upon media
Maximum segment length (meters)	500	500	185	250	100	1800	100	varies based upon media
Stations/segment	100	100	30	12/hub	12/hub	100	12/hub	varies based upon media
Medium	50Ω coaxial (thick)	50Ω coaxial (thick)	50Ω coaxial (thin)	unshielded twisted pair	unshielded twisted pair	75Ω coaxial	unshielded twisted pair (Category 5)	several types of optical fiber and copper
Topology	bus	bus	bus	star	star	bus	star	star

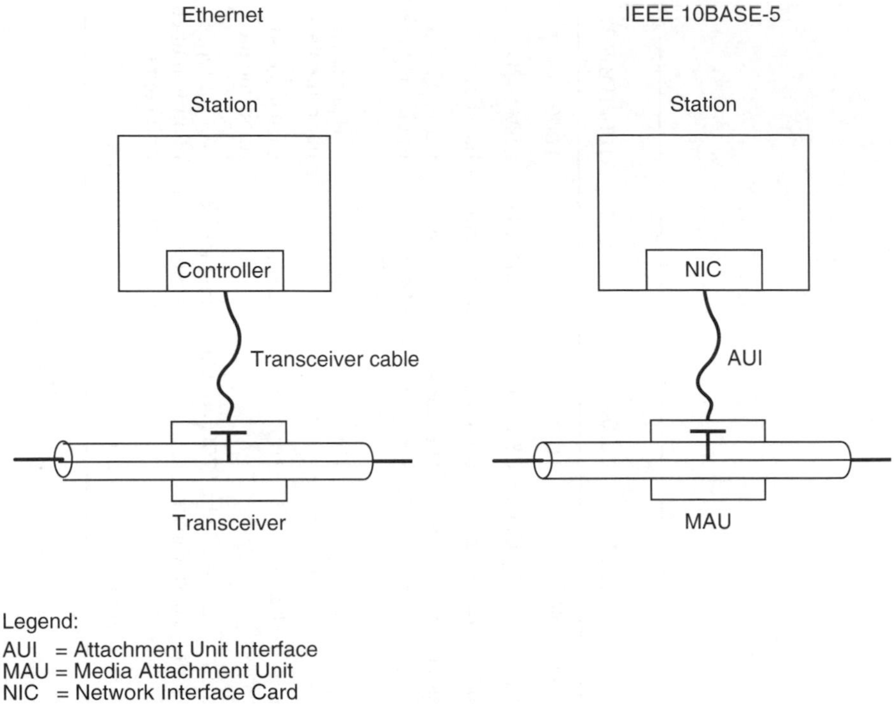

Figure 13.8 Ethernet and 10BASE-5 media interface terminology differences. Terminology changes under the IEEE 10BASE-5 standard resulted in the transceiver being called the media attachment unit, while the transceiver cable is known as the attachment unit interface

Ethernet, requires a minimum spacing of 2.5 m between MAUs and supports a maximum of five segments in any end-to-end path through the traversal of up to four repeaters in any path. Within any path, no more than three cable segments can be populated (have stations attached to the cable) and the maximum number of attachments per segment is limited to 100.

10BASE-2

10BASE-2 is a smaller and less expensive version of 10BASE-5. This standard uses a thinner RG-58 coaxial cable, thus earning the names 'cheapnet' and 'thinnet', as well as 'thin Ethernet'. Although 10BASE-2 cable is both less expensive and easier to use than 10BASE-5 cable, it cannot carry signals as far as 10BASE-5 cable.

Under the 10BASE-2 standard, the maximum cable segment length is reduced to 185 m (607 feet), with a maximum of 30 stations per segment. Another difference between 10BASE-5 and 10BASE-2 concerns the integration of transceiver electronics into the network interface card under the 10BASE-2 standard. This permits the NIC to be directly cabled to the main

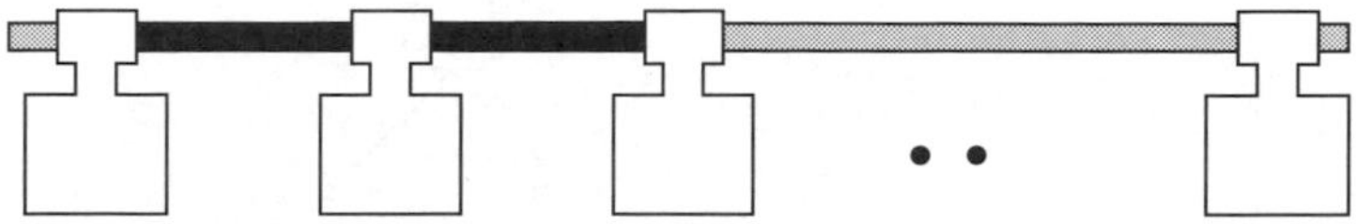

Figure 13.9 Cabling a 10BASE-2 network. A 10BASE-2 cable segment cannot exceed 185 meters and is limited to supporting up to 30 nodes or stations

trunk cable. In fact, under 10BASE-2 the thin Ethernet cable is routed directly to each workstation location and routed through a BNC T-connector, one end of which is pressed into the BNC connector built into the rear of the network interface card.

Figure 13.9 illustrates the cabling of a one-segment 10BASE-2 network, which can support a maximum of 30 nodes or stations. BNC barrel connectors can be used to join two lengths of thin 10BASE-2 cable to form a cable segment, as long as the joined cable does not exceed 185 m in length. A BNC terminator must be attached to each end of each 10BASE-2 cable segment. One of the two terminators on each segment contains a ground wire that should be connected to a ground source, such as the screw on an electrical outlet.

10BROAD-36

10BROAD-36 is the only broadband network based on the CSMA/CD access protocol standardized by the IEEE. Unlike a baseband network, in which Manchester encoded signals are placed directly onto the cable, the 10BROAD-36 standard requires the use of radio frequency (RF) modems. Those modems modulate non-return to zero (NRZ) encoded signals for transmission on one channel at a specified frequency, and demodulate received signals by listening for tones on another channel at a different frequency.

A 10BROAD-36 network is constructed with a 75Ω coaxial cable, similar to the cable used in modern cable television (CATV) systems. Under the IEEE 802.3 broadband standard, either single or dual cables can be used to construct a network. If a single cable is used, the end of the cable (referred to as the headend) must be terminated with a frequency translator. That translator converts the signals received on one channel to the frequency assigned to the other channel, retransmitting the signal at the new frequency. Since the frequency band for a transmitted signal is below the frequency band 10BROAD-36 receivers scan, we say the frequency translator upconverts a transmitted signal and retransmits it for reception by other stations on the network. If two cables are used, the headend simply functions as a relay point, transferring the signal received on one cable onto the second cable.

A broadband transmission system has several advantages over a baseband system. Two of the primary advantages of broadband are its ability to support multiple transmissions occurring on independent frequency bands simultaneously, and its ability to support a tree structure topology carrying multiple simultaneous transmissions. Using independent frequency bands, you can

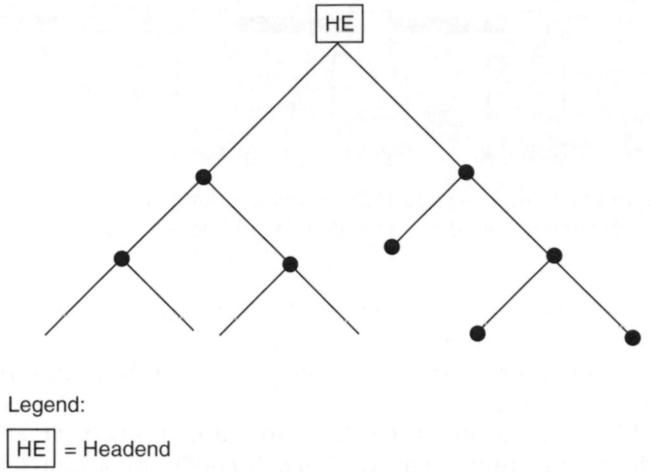

Figure 13.10 Broadband tree topology support. The headend receives transmissions at one frequency and regenerates the received signals back onto the network at another predefined frequency

establish several independent networks. In fact, each network can be used to carry voice, data and video over a common cable. As for topology, broadband permits the use of a tree structure network, such as the structure shown in Figure 13.10. In this example, the top of the tree would be the headend; it would contain a frequency translator, which would regenerate signals received at one frequency back onto the cable at another predefined frequency.

The higher noise immunity of 75Ω coaxial cable permits a 10BROAD-36 network to span 3600 m, making this medium ideal for linking buildings on a campus. In addition the ability of 10BROAD-36 to share channel space on a 75Ω coaxial cable permits organizations that have an existing CATV system, such as one used for video security, to use that cable for part or all of their broadband CSMA/CD network. Although these advantages can be significant, the cost associated with RF modems has limited the use of broadband primarily to campus environments that require an expanded network span. In addition, the rapid development and acceptance of 10BASE-T and the decline in the cost of fiber cable to extend the span of 10BASE-T networks have severely limited what once many persons anticipated as a promising future for 10BROAD-36 networks.

1BASE-5

The 1BASE-5 standard was based on AT&T's low-cost CSMA/CD network known as StarLan. Thus, 1BASE-5 is commonly referred to as StarLan, although AT&T uses that term to refer to CSMA/CD networks operating at both 1 and 10 Mbps using unshielded twisted-pair cable. The latter is considered the predecessor to 10BASE-T.

The 1BASE-5 standard differs significantly from Ethernet and 10BASE-5 standards in its use of media and topology, and in its operating rate. The 1BASE-5 standard operates at 1 Mbps and uses unshielded twisted pair (UTP) wiring in a star topology; all stations are wired to a hub, which is known as a multiple access unit (MAU). To avoid confusion with the term media access unit, we will refer to this wiring concentrator as a hub.

Each station in a 1BASE-5 network contains a network interface card (NIC), cabled via UTP on a point-to-point basis to a hub port. The hub is responsible for repeating signals and detecting collisions.

The maximum cabling distance from a station to a hub is 250 m; up to five hubs can be cascaded together to produce a maximum network span of 2500 m. The highest level hub is known as the header hub, and it is responsible for broadcasting news of collisions to all other hubs in the network. These hubs, which are known as intermediate hubs, are responsible for reporting all collisions to the header hub.

AT&T's 1 Mbps StarLan network, along with other 1BASE-5 systems, initially received a degree of acceptance for use in small organizations. However, the introduction of 10BASE-T, which provided an operating rate 10 times that obtainable under 1BASE-5, severely limited the further acceptance of 1BASE-5 networks.

10BASE-T

In the late 1980s a committee of the IEEE recognized the requirement of organizations for transmitting Ethernet at a 10 Mbps operating rate over low-cost and readily available unshielded twisted-pair cable. Although several vendors had already introduced equipment that permitted Ethernet signaling via UTP cabling, such equipment was based on proprietary designs and was not interoperable. Thus, a new task of the IEEE was to develop a standard for 802.3 networks operating at 10 Mbps using UTP cable. The resulting standard was approved by the IEEE as 802.3i in September 1990, and is more commonly known as 10BASE-T.

The 10BASE-T standard supports an operating rate of 10 Mbps at a distance of up to 100 m (328 feet) over UTP Category 3 cable without the use of a repeater. The UTP cable requires two pairs of twisted wire. One pair is used for transmitting, and the other pair is used for receiving. Each pair of wires is twisted together, and each twist is 180°. Any electromagnetic interference (EMI) or radio frequency interference (RFI) is therefore received 180° out of phase; this theoretically cancels out EMI and RFI noise while leaving the network signal. In reality, the wire between twists acts as an antenna and receives noise. This noise reception resulted in a 100 m cable limit, until repeaters were used to regenerate the signal.

Network components

A 10BASE-T network can be constructed with network interface cards, UTP cable, and one or more hubs. Each NIC is installed in the expansion slot of a computer and wired on a point-to-point basis to a hub port. When all of the ports on a hub are used, one hub can be connected to another to expand the network, resulting in a physical star, logical bus network structure.

Most 10BASE-T network interface cards contain multiple connectors, which enable the card to be used with different types of 802.3 networks. For example, most modern NICs include a RJ-45 jack as well as BNC and DB-15 connectors. The RJ-45 jack supports the direct attachment of the NIC to a 10BASE-T network, and the BNC connector permits the NIC to be mated to a 10BASE-2 T-connector. The DB-15 connector enables the NIC to be cabled to a transceiver, and is more commonly referred to as the NIC's attachment unit interface (AUI) port.

The wiring hub in a 10BASE-T network functions as a multiport repeater: it receives, retimes and regenerates signals received from any attached station. The hub also functions as a filter: it discards severely distorted frames.

A 10BASE-T hub tests the integrity of the link from each hub port to a connected station by transmitting a special signal to the station. If the device does not respond, the hub will automatically shut down the port, and may illuminate a status light-emitting diode (LED) to indicate the status of each port.

Hubs monitor, record and count consecutive collisions that occur on each individual station link. Since an excessive number of consecutive collisions will prevent data transfer on all of the attached links, hubs are required to cut off or partition any link on which too many collisions occurred. This partitioning enables the remainder of the network to operate in situations where a faulty NIC transmits continuously. Although the IEEE 802.3 standard does not specify a maximum number of consecutive collisions, the standard does specify that partitioning can be initiated after 30 or more consecutive collisions occur. Thus, some hub vendors initiate partitioning when 31 consecutive collisions have occurred, whereas other manufacturers use a higher value.

Although a wiring hub is commonly referred to as a concentrator, this term is not technically correct. A 10BASE-T wiring hub is a self-contained unit that typically includes 8, 10 or 12 RJ-45 ports for direct connection to stations, and a BNC and/or DB-15 AUI port to expand the hub to other network equipment. The BNC and AUI ports enable the 10BASE-T hub to be connected to 10BASE-2 and 10BASE-5 networks, respectively. For the latter, the AUI port is cabled to a 10BASE-5 MAU (transceiver), which is tapped into thick 10BASE-5 coaxial cable. One 10BASE-T hub can be connected to another with a UTP link between RJ-45 ports on each hub.

Figure 13.11 illustrates the connectors on a typical 10BASE-T hub. On some hubs, one RJ-45 jack is labeled uplink/downlink for use in cascading hubs, whereas other vendors permits any RJ-45 port to be used for connecting hubs.

Unlike a hub, a concentrator consists of a main housing into which modular cards are inserted. Although some modular cards may appear to represent hubs, and do indeed function as 10BASE-T hubs, the addition of other modules permit the network to be easily expanded from one location and this allows additional features to be supported. For example, the insertion of a fiber optic inter-repeater module permits concentrators to be interconnected over relatively long distances of approximately 3 km.

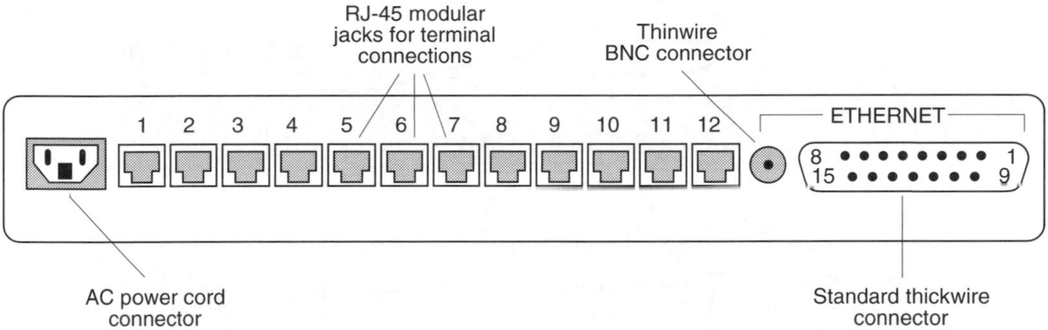

Figure 13.11 10BASE-T hub connectors. In addition to 8, 10, or 12 RJ-45 modular jacks for terminal connections, most 10BASE-T hubs contain a BNC and DB-15 port to permit attachment to thin and thick backbone networks

Expanding a 10BASE-T network

A 10BASE-T network can be expanded with additional hubs once the number of stations serviced has used up the hub's available terminal ports. In expanding a 10BASE-T network, the wiring that joins each hub together is considered to represent a cable segment, and each hub is considered as a repeater. Under the 802.3 specification, no two stations can be separated by more than four hubs connected together by five cable segments.

The 5-4-3 rule

When creating an Ethernet network that requires more than two repeaters there are certain signal limitations that must be considered. Those signal limitations are also applicable to interconnected hubs, since a hub can be considered to represent a repeater. To facilitate persons remembering the cabling limitation associated with Ethernet signals flowing across interconnected local networks, those limitations are collectively referred to as the 5-4-3 rule.

Under the 5-4-3 rule an Ethernet signal flowing from source to destination station can traverse a maximum of 3 populated segments, 4 repeaters or hubs, and 5 segments. If we examine Figure 13.12 that illustrates the expansion of a 10BASE-T network, we will note that a signal flowing from station A to station B passes through only four hubs. If one of those hubs is unpopulated then the signal will comply with the 5-4-3 rule. If the rule is violated there is no guarantee that the network will be reliable; hence, it is best to attempt to comply with this rule.

100BASE-T

The standardization of 100BASE-T, commonly known as Fast Ethernet, required an extension of previously developed IEEE 802.3 standards. In the

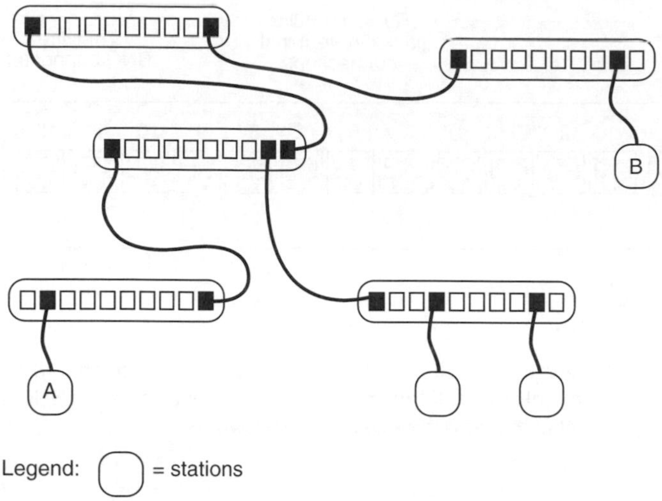

Figure 13.12 Expanding a 10BASE-T network. No two stations can be separated by more then four hubs in a 10BASE-T network

definition process of standardization development, both the Ethernet Media Access Control (MAC) and physical layer required adjustments to permit 100 Mbps operational support. For the MAC layer, scaling its speed to 100 Mbps from the 10BASE-T 10 Mbps operational rate required a minimal adjustment, since in theory the 10BASE-T MAC layer was developed independently of the data rate. For the physical layer, more than a minor adjustment was required since Fast Ethernet was designed to support three types of media, resulting in three new names which fall under the 100BASE-T umbrella. 100BASE-T4 uses three wire pairs for data transmission and a fourth for collision detection resulting in T4 being appended to the 100BASE mnemonic. 100BASE-TX uses two pairs of category 5 UTP with one pair employed for transmission, and the second is used for both collision detection and reception of data. The third 100BASE-T standard is 100BASE-FX which represents the use of fiber optic media.

Using work developed in the standardization process of FDDI in defining 125 Mbps full-duplex signaling to accommodate optical fiber, UTP and STP through Physical Media Dependent (PMD) sublayers, Fast Ethernet borrowed this strategy. Because a mechanism was required to map the PMD's continuous signaling system to the start-stop 'half-duplex' system used at the Ethernet MAC layer, the physical layer was subdivided. This subdivision is illustrated in Figure 13.13. The PMD sublayer supports the appropriate media to be used, whereas the convergence sublayer (CS), which was later renamed the physical coding sublayer, performs the mapping between the PMD and the Ethernet MAC layer.

Although Fast Ethernet represents a tenfold increase in the LAN operating rate from 10BASE-T, to ensure proper collision detections the 100BASE-T network span was reduced to 250 m, with a maximum of 100 m permitted between a network node and a hub. The smaller network diameter reduces

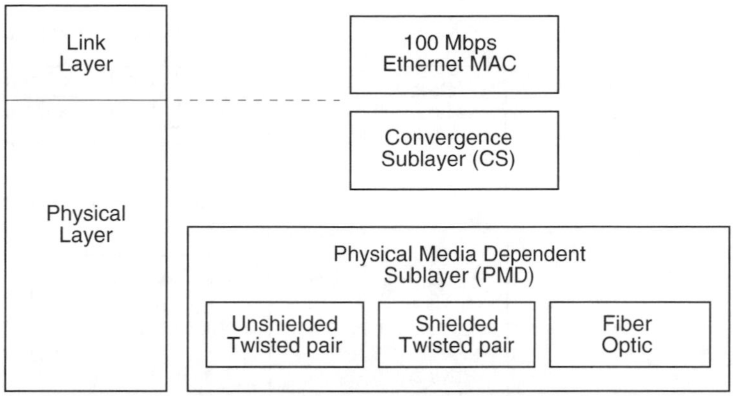

Figure 13.13 Overview of the Fast Ethernet physical layer subdivision

potential propagation delay. When coupled with a tenfold operating rate increase and no change in network frame size, the ratio of frame duration to network propagation delay for a 100BASE-T network is the same as for a 10BASE-T network.

Physical layer

The physical layer subdivision previously illustrated in Figure 13.13, as indicated in the title of the figure, presents an overview of the true layer subdivision. In actuality, a number of changes were required at the physical layer to obtain a 10 Mbps operating rate. Those changes include the use of three wire pairs for data (the fourth is used for collision detection), 8B6T ternary coding (for 100BASE-T4) instead of Manchester coding, and an increase in the clock signaling speed from 20 MHz to 25 MHz.

When the specifications for Fast Ethernet were being developed it was recognized that the physical signaling layer would incorporate medium-dependent functions if support was extended to two pair cable (100BASE-TX) operations. To separate medium-dependent interfaces to accommodate multiple physical layers, a common interface referred to as the Medium Independent Interface (MII) was inserted between the MAC layer and the physical encoding sublayer. The MII represents a common point of interoperability between the medium and the MAC layer. The MII can support two specific data rates, 10 Mbps and 100 Mbps, permitting older 10BASE-T nodes to be supported at Fast Ethernet hubs. To reconcile the MII signal with the MAC signal, a reconciliation sublayer was added under the MAC layer, resulting in the subdivision of the link layer into three parts: a logical link control layer, a media access control layer and a reconciliation layer. The top portion of Figure 13.14 illustrates this subdivision.

That portion of Fast Ethernet below the MII, which is the new physical layer, is now subdivided into three sublayers. The lower portion of Figure 13.14 illustrates the physical sublayers for 100BASE-T4 and 100BASE-TX.

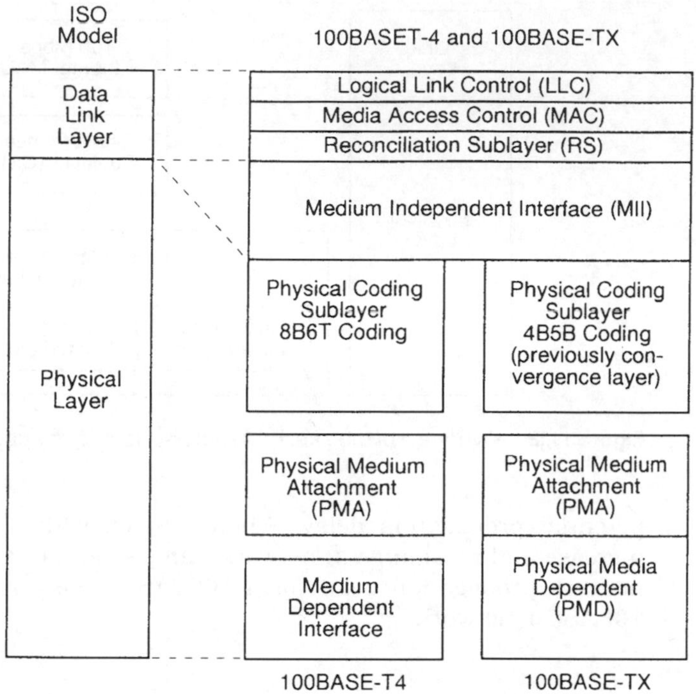

Figure 13.14 100BASE-T4 versus 100BASE-TX physical and data link layers

The physical coding sublayer performs the data encoding, transmit, receive and carrier sense functions. Since the data coding method differs between 100BASE-T4 and 100BASE-TX, this difference requires distinct physical coding sublayers for each version of Fast Ethernet.

The Physical Medium Attachment (PMA) sublayer maps messages from the physical coding sublayer (PCS) onto the twisted-pair transmission media and vice versa.

The Medium Dependent Interface (MDI) sublayer specifies the use of a standard RJ-45 connector. Although the same connector is used for 100BASE-TX, the use of two pairs of cable instead of four results in different pin assignments.

100BASE-T4

Figure 13.15 illustrates the RJ45 pin assignments of wire pairs used by 100BASE-T4. Note that wire pairs D1 and D2 are unidirectional. As indicated in Figure 13.15, three wire pairs are available for data transmission and reception in each direction whereas the fourth pair is used for collision detection.

The 100BASE-T4 physical coding sublayer implements 8B6T block coding. Under this coding technique, each block of eight-input bits is transformed into a unique code group of six ternary symbols. Figure 13.16 provides an overview of the 8B6T coding process used by 100BASE-T4.

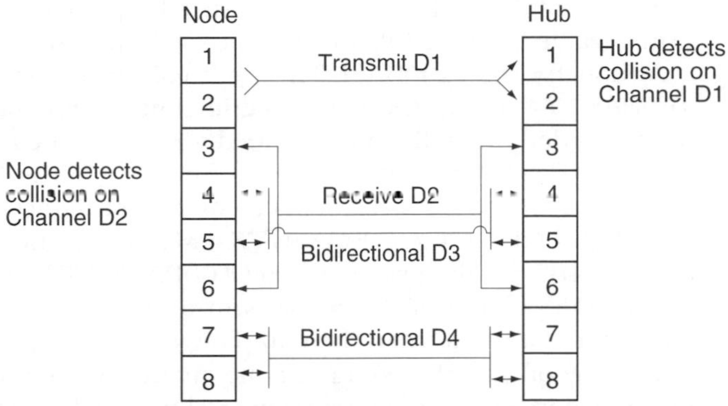

Figure 13.15 100BASE-T4 Pin assignments

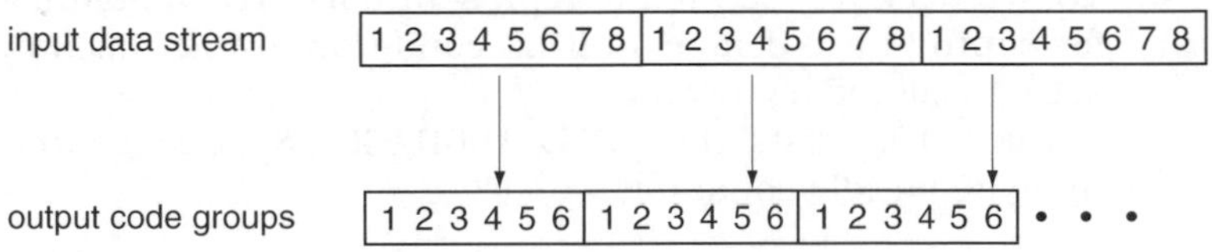

Figure 13.16 8B6T coding process

The output code groups resulting from 8B6T coding flow out to three parallel channels that are placed on three twisted pairs. Thus, the effective data rate on each pair is 100 Mbps/3 or 33.33 Mbps. Since 6 bits are represented by 8 bit positions, the signaling rate or baud rate on each cable pair becomes 33 Mbps $\times$ 6/8 or 25 MHz, which is the clock rate used at the MII sublayer.

100BASE-TX

100BASE-TX represents 100BASE-T which supports the use of two pairs of category 5 UTP cabling with RJ-45 connectors. A 100BASE-TX network requires a hub, and the maximum cable run is 100 m from hub port to node, with a maximum network diameter of 250 m.

Figure 13.17 illustrates the cabling of two pairs of UTP wires between a hub and node to support 100BASE-TX transmission. One pair of wires is used for transmission, and the second pair is used for collision detection and

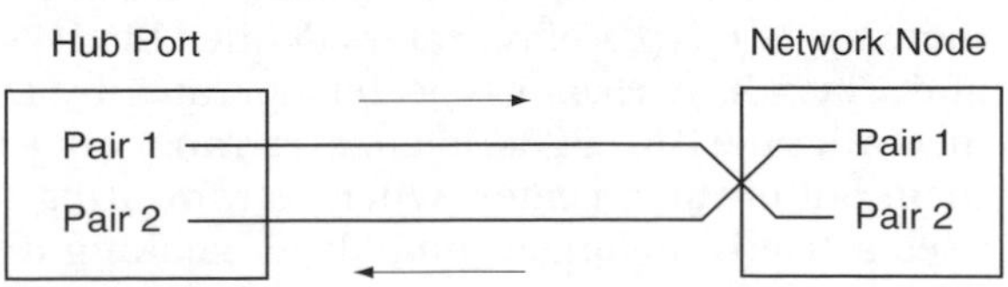

Figure 13.17 100BASE-TX cabling

reception of data. The use of a 125 MHz frequency requires the use of a 'data grade' cable. Thus, 100BASE-TX is based upon the use of category 5 UTP.

Although the 100BASE-TX physical layer structure resembles the 100BASE-T4 layer, there are significant differences between the two to accommodate the differences in media used. At the physical coding sublayer, the 100 Mbps start-stop bit stream from the MII is first converted to a full-duplex 125 Mbps bit stream. This conversion is accomplished by the use of the FDDI PMD as the 100BASE-TX PMD. Next, the data stream is encoded using a 4B5B coding scheme. The 100BASE-TX PMD decodes symbols from the 125 Mbps continuous bit stream and converts the stream to 100 Mbps start-stop data bits when the data flow is reversed.

The use of a 4B5B coding scheme enables data and control information to be carried in each symbol represented by a five bit code group. In addition, an inter-stream fill code (IDLE) is defined, as well as a symbol used to force signaling errors. Since four data bits are mapped into a five bit code, only 16 symbols are required to represent data. The remaining symbols not used for control or to denote an IDLE condition are not used by 100BASE-TX and are considered as invalid.

Table 13.4 lists the 4B5B 100BASE-TX code groups. Note that all 1s indicate an idle condition.

100BASE-FX

10BASE-FX represents the third 100BASE-T wiring scheme, defining Fast Ethernet transmission over fiber-optic media. 100BASE-FX requires the use of two-strand 62.5/125 micron multimode fiber media and supports the 4B5B coding scheme, identical to the one used by 100BASE-TX. Each 100BASE-FX adapter, repeater, and optical port on a hub has dual connectors labeled TX for transmit and RX for receive. When cabling two 100BASE-FX devices you would connect the TX connector on one device to the RX connector on the other device. Similarly, you would use the second optical cable to connect the RX connector of the local device to the TX connector on the distant device, providing an optical path from transmitter to receiver in each direction.

A 100BASE-FX media system is designed to support transmission between network segments at distances up to 412 meters. This distance limitation ensures that the round trip delay does not adversely affect Fast Ethernet operations and has no bearing on the ability of the media, which can transport optical signals considerably further than the 100BASE-FX Fast Ethernet limitation.

When constructing a 100BASE-FX network, it is important to note that there are two types of repeaters defined by the standard, referred to as Class I and Class II. A class I repeater operates by translating light signals on an incoming port to digital signal, retranslating the signal when transmitting them out of the repeater. When retranslating the signal the Class I repeater uses a relatively large timing delay, making it possible to connect segments that use different signaling techniques, such as a 100BASE-T4 segment and a 100BASE-TX segment. Because the delay occurs from a Class I repeater that

Table 13.4 4B/5B code groups

Code group 4320	Name	TXD/RD 3210	Interpretation
DATA			
11110	0	0000	Data 0
01001	1	0001	Data 1
10100	2	0010	Data 2
10101	3	0011	Data 3
01010	4	0100	Data 4
01011	5	0101	Data 5
01110	6	0110	Data 6
01111	7	0111	Data 7
10010	8	1000	Data 8
10011	9	1001	Data 9
10110	A	1010	Data A
10111	B	1011	Data B
11010	C	1100	Data C
11011	D	1101	Data D
11100	E	1110	Data E
11101	F	1111	Data F
IDLE			
11111	I		IDLE: Used as inter-stream fill code
CONTROL			
11000	J		Start-of-Stream Delimiter, Part 1 of 2; always used in pairs with K
10001	K		Start-of-Stream Delimiter, Part 2 of 2; always used in pairs with J
01101	T		End-of-Stream Delimiter, Part 1 of 2; always used in pairs with R
00111	R		End-of-Stream Delimiter, Part 2 of 2; always used in pairs with T
INVALID			
00100	V		Transmit Error; used to force signaling errors
00000	V		Invalid code
00001	V		Invalid code
00010	V		Invalid code
00011	V		Invalid code
00101	V		Invalid code
00110	V		Invalid code
01000	V		Invalid code
01100	V		Invalid code
10000	V		Invalid code
11001	V		Invalid code

uses 92 bit times, you can use only one such repeater in a given collision domain when maximum cable lengths are used.

The second type of 100BASE-FX repeater is a Class I repeater. This type of repeater immediately repeats an incoming signal onto all other ports without a translation process, resulting in a smaller timing delay. Due to their inability to perform a translation, a Class II repeater can only connect segments that employ the same signaling technique, such as a 100BASE-FX and a 100BASE-TX segment. A maximum of two Class II repeaters are supported within a given collision domain when maximum cable lengths are used.

Although the 100BASE-FX standard specifies a maximum point-to-point cabling limitation of 412 meters, several vendors market 100BASE-TX to 100BASE-FX media converters that considerably extend transmission distance. To accomplish this, such media converters have round-trip delays of only 33 bit times. In comparison, a Class I repeater's delay is 92 bit times. When used in pairs, the media converter can extend distances between devices up to 2000 meters.

Network utilization

Since 100BASE-T4 and 100BASE-TX preserve the 10BASE-T MAC layer, both standards are capable of interoperability with existing 10BASE-T networks, as well as with other low speed Ethernet technology. Through the use of NWay auto-sensing logic, Fast Ethernet adapters, hub and switch ports can determine if attached equipment can transmit at 10 or 100 Mbps and adjust to the operating rate of the distant device.

NWay is a cable and transmission auto-sensing scheme proposed by National Semiconductor to the IEEE 802.3 standards group in May 1994. The NWay auto-sensing scheme permits Ethernet circuits to detect both the cable type and speed of incoming Ethernet data, as well as enabling Ethernet repeaters to configure themselves for correct network operations. Since NWay can detect 10 Mbps versus 100 Mbps operations, as well as half- and full-duplex transmission, it permits Ethernet circuits to be developed to adjust automatically to the operating rate and cabling scheme used. This in turn can be expected to simplify the efforts of network managers and administrators, because products incorporating NWay are self-configurable and do not require the setting of DIP switches or software parameters.

There are several network configurations that you can consider for Fast Ethernet operations. First, you can construct a 100BASE-TX network similar to a 10BASE-T network, using an appropriate hub and workstations that support the specific network type. This type of network, commonly referred to as a shared-media hub-based network, should be considered when most, if not all, network users access servers to perform graphic intensive or similar bandwidth intensive operations. Figure 13.18(a) illustrates a Fast Ethernet shared-media hub network configuration.

A more common use for Fast Ethernet is its incorporation into one or more ports in an Ethernet switch. An Ethernet switch can be viewed as a sophisticated hub which can be programmed to transmit packets arriving on one input port to a predefined output port. This is accomplished by the switch reading

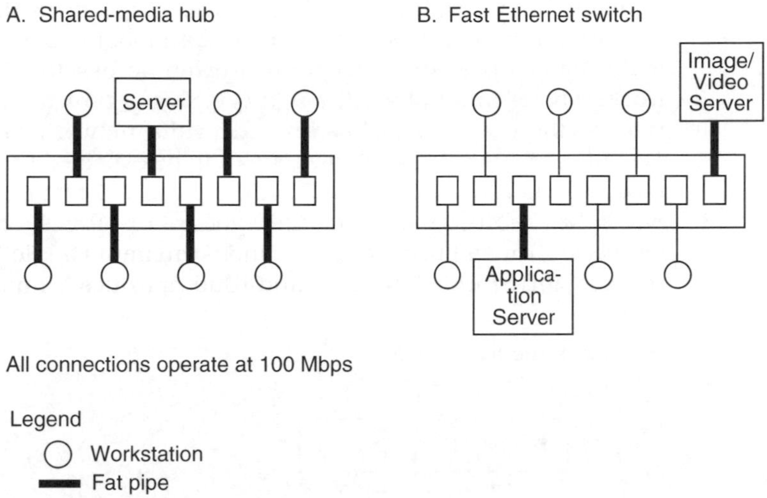

Figure 13.18 Fast Ethernet network applications

the destination address in the frame and comparing that address to a table of pre-configured address–port relationships. Most Ethernet switches contain one or two 100 Mbps operating ports, and the other ports are conventional 10BASE-T ports. Typically, you would connect servers that have a heavy workload to 100 Mbps Fast Ethernet ports, such as an image/video server and a database server. Figure 13.18(b) illustrates the use of a Fast Ethernet switch. In examining Figures 13.18(a) and (b), note that the heavily shaded connections to all workstations in Figure 13.18(a) and to the servers in Figure 13.18(b) represent 100 Mbps Fast Ethernet ports that require Fast Ethernet adapter cards to be installed in each server or workstation connected to a 100 Mbps port. A common term used to refer to the 100 Mbps connection is 'fat pipe'.

Since a Fast Ethernet port provides downward compatibility with 10BASE-T, you can interconnect conventional 10BASE-T hubs to a Fast Ethernet shared-media hub or Fast Ethernet switch,

Detailed information concerning the operation and utilization of LAN switches is presented in Chapter 14.

3.5 Gigabit Ethernet

Gigabit Ethernet represents an extension to the 10 Mbps and 100 Mbps IEEE 802.3 Ethernet standards. Providing a data transmission capability of 1000 Mbps, Gigabit Ethernet supports the CMSA/CD access protocol, which makes various types of Ethernet networks scalable from 10 Mbps to 1 Gbps.

Components

Similar to 10BASE-T and Fast Ethernet, Gigabit Ethernet can be used as a shared network through the attachment of network devices to a 1 Gbps repeater hub providing shared use of the 1 Gbps operating rate or as a switch, the latter providing 1 Gbps ports to accommodate high-speed access to servers while lower operating rate ports provide access to 10 Mbps and 100 Mbps workstations and hubs. Although very few organizations can be expected to require the use of a 1 Gbps shared media network as illustrated in Figure 13.19(a), the use of Gigabit switches can be expected to play an important role in providing a high-speed backbone linking 100 Mbps network users to large databases, mainframes, and other types of resources that can tax lower speed networks. In addition to hubs and switches, Gigabit Ethernet operations require workstations, bridges and routers to use a network interface card to

(a) Shared media hub use

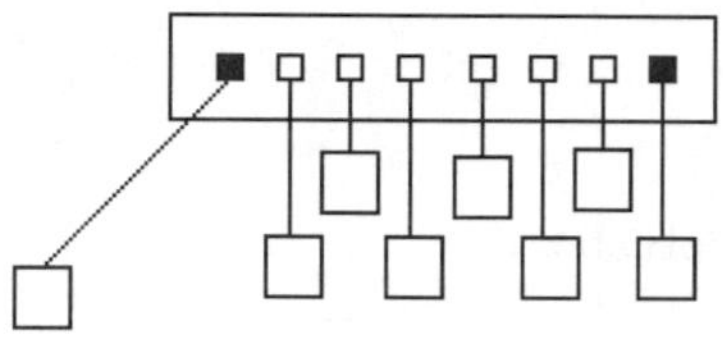

In a shared media environment the 1 Gbps bandwidth provided by Gigabit Ethernet is shared among all users

(b) Switching hub use

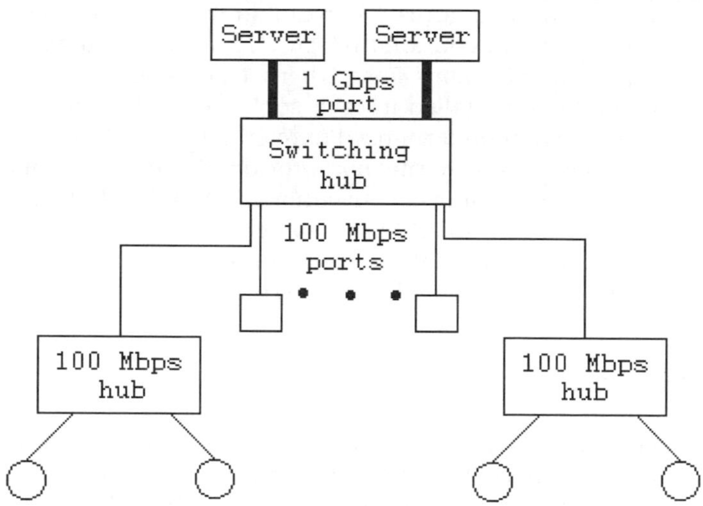

In a switch environment each 1 Gbps port can provide a full-duplex 2 Gbps data transfer capability

Legend: ○ workstations

Figure 13.19 Using Gigabit Ethernet

connect to a 1 Gbps network. The Gigabit Ethernet NICs introduced when this book was written were designed for PCI bus operations and use an SC fiber connector to support 62.5/125 and 50/125 micron (μm) fiber. Such adapters provide 1 Gbps of bandwidth for shared media operations and 2 Gbps aggregate bandwidth when used for a full-duplex connection to a switch port. In Chapter 14 when we review switching we will examine in more detail how Gigabit switches can be used to provide a backbone network capability.

Media support

Similar to the recognition that Fast Ethernet would be required to operate over different types of media, the IEEE 802.3z committee recognized that Gigabit Ethernet would also be required to operate over multiple types of media. This

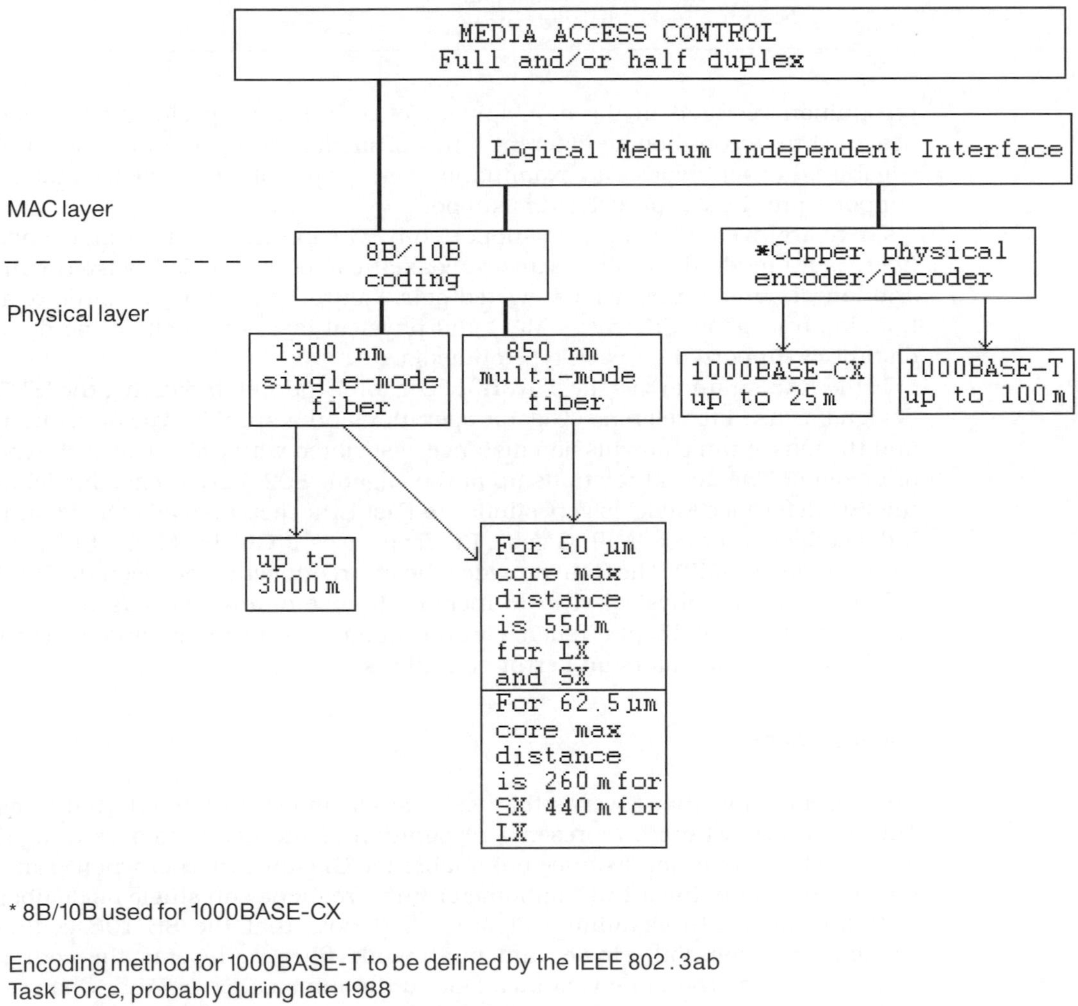

Figure 13.20 Gigabit Ethernet media support

Table 13.5 The flavors of Gigabit Ethernet

Media designator	Media type	Transmission distance
1000BASE-LX	SMF	3 km
1000BASE-LX	MMF, 50μ	550 m
1000BASE-LX	MMF, 62.5μ	440 m
1000BASE-SX	MMF, 50μ	550 m
1000BASE-SX	MMF, 62.5μ	260 m
1000BASE-CX	shielded balanced copper (coax or STP)	25 m
1000BASE-T	UTP, Category 5	100 m

Key: SMF single-mode fiber
MMF multi-mode fiber
UTP unshielded twisted pair
STP shielded twisted pair

recognition resulted in the development of a series of specifications, each designed to accommodate different types of media. Thus, any discussion of Gigabit Ethernet involves an examination of the types of media the technology supports and how it provides this support.

There are five types of media supported by Gigabit Ethernet – single-mode fiber, multi-mode fiber, short runs of coaxial cable or shielded twisted pair, and longer runs of unshielded twisted pair. Figure 13.20 illustrates the relationship of Gigabit Ethernet's MAC and physical layers to include the drive distances supported for each type of media.

Table 13.5 summarizes the 'flavors' of Gigabit Ethernet, indicating the IEEE designator used to reference Gigabit operations on a specific type of medium and the maximum transmission distance associated with the use of each type of medium. The actual relationship of the Gigabit 802.3z reference model to the ISO Reference Model is very similar to Fast Ethernet. Instead of a Medium Independent Interface (MII), Gigabit Ethernet uses a Gigabit Media Independent Interface (GMII). The GMII provides the interconnection between the MAC sublayer and the physical layer to include the use of an 8-bit data bus that operates at 125 MHZ plus such control signals as transmit and receiver clocks, carrier indicators and error conditions.

Single-mode fiber

The specification that governs the use of single-mode fiber is referred to as 1000BASE-LX, where L represents recognition of the use of long-wave light pulses. The maximum distance obtainable for Gigabit Ethernet when transmission occurs using a 1330 nanometer (nm) frequency on single-mode fiber is 3000 meters. In examining Figure 13.20 note that the 8B/10B coding scheme used for both single- and multi-mode fiber represents the coding scheme used by the Fibre Channel. Due to the importance of Fibre Channel technology incorporated into Gigabit Ethernet, a short digression to discuss that technology is warranted.

The Fibre Channel

The Fibre Channel actually represents a rather old technology, dating back to 1988 when the American National Standards Institute charted a working group to develop a high-speed data transfer capability for supporting data transfers between computers and peripheral devices. Initially, the Fibre Channel was developed to support fiber optic cabling. When support for copper media was added, an International Standards Organization (ISO) task force revised the spelling of fiber to reduce the association of fiber optics while maintaining the name recognition of the technology. Today both optical and electrical copper-based media are supported by the Fibre Channel, with operating rates ranging from 133 Mbps to 1.062 Gbps provided by most equipment. In 1995 ANSI approved 2.134 and 4.25 Gbps speed rates as enhancements to the previously developed Fibre Channel specifications.

The Fibre Channel supports point-to-point, shared media and switch network topologies, with Fibre Channel hubs, switches, loop access controllers and NICs installed on computers used to provide different network structures. Fibre Channel supports a five-layer protocol stack from FC-0 through FC-4. The three lower layers form the Fibre Channel physical standard while the two upper layers provide the interface to network protocols and applications. The FC-0 layer defines the physical characteristics of the media, transmitters, receivers and connectors available for use, transmission rates supported, electrical and optical characteristics, and other physical layer standards. The second layer, FC-1, defines the 8B/10B encoding and decoding method used, a serial physical transport, timing recovery and serial line balance. The 8B/10B encoding and decoding scheme was patented by IBM and is the same technique used in that vendor's 200 Mbps ESCON channel technology. Under 8B/10B coding eight data bits are transmitted as a 10-bit group. The two extra bits are used for error detection and correction.

To enable 1 Gbps operations the transmission rate of the Fibre Channel was raised to 1.25 Gbps. Then, the use of an 8B/10B coding technique permits data transfer at 80% of the operating rate, or 1.256×80, resulting in the modified Fibre Channel technology being capable of supporting the 1 Gbps data transfer of Gigabit Ethernet.

The third layer, FC-2, functions as the transport mechanism. This layer defines the framing rules of data transmitted between devices and how data flow is regulated. Although Ethernet frames are encapsulated within Fibre Channel frames for transmission between devices, the encapsulation is transparent to the Ethernet frame. Thus, the use of a modified Fibre Channel technology enabled the IEEE to use a proven technology as a transport mechanism for connecting Gigabit Ethernet devices while enabling the CSMA/CD protocol and Ethernet frame to be retained.

Multimode fiber

The support of multimode fiber is applicable to both long-wave (LX) and short-wave (SX) versions of Gigabit Ethernet. When transmission occurs on multimode fiber the short-wave version of Gigabit Ethernet, which is referred to as 1000BASE-SX, uses an 850 nm frequency, providing a 260 meter maximum distance when transmission occurs over a 62.5 micron (μm) core. Note that in some trade publications and books the maximum transmission distance may be indicated as 300 m; however, in September 1997 the IEEE task force working on the development of Gigabit Ethernet standards lowered the maximum distance from 300 meters to 260 meters. When a 1300 nm frequency is used on a 62.5 μm core which represents 1000BASE-LX, a maximum transmission distance of 440 m becomes possible. This distance was formerly 500 meters and was also reduced by the IEEE Gigabit Ethernet task force. When a 50 micron core fiber is used, the maximum distance is 550 meters for both LX and SX specifications.

Copper media

Gigabit Ethernet supports three types of copper media – Category 5 shielded twisted pair (STP), coaxial cable, and unshielded twisted pair (UTP). The use of the first two types of copper media is defined by the 1000BASE-CX standard which governs the use of patch cords and jumpers that provide a maximum cabling distance of 25 m. This standard also uses the Fibre Channel based 8B/10B coding method at a serial line operating rate of 125 Gbps, and runs over a 150-ohm balanced, shielded cable assembly referred to as twinax cable as well as on STP. The 25 meter maximum cabling distance makes 1000BASE-CX suitable for use in a wiring closet or a computer room as a short jumper interconnection cable. In comparison, the 1000BASE-T standard defines the use of four-pair UTP cable, with each wire operating at a modulation rate of 125 MHz in each direction simultaneously. The use of a 125 MHz signal rate permits Gigabit to operate over Category 5 cable at a distance up to 100 meters. To obtain a 1 Gbps transmission rate, symbols to be transmitted are selected from a four-dimensional (4D) code group of five level symbols, resulting in an operating rate of 250 Mbps per wire or an aggregate rate of 1 Gbps over the four Category 5 copper wires.

Duplex capability

Gigabit Ethernet supports a full-duplex operating mode for switch-to-switch and switch-to-end station connections, while a half-duplex operating mode is used for shared connections using repeaters. Although not supported in the current draft version of the 802.3z standards, several vendors announced full-duplex, multi-port, hub-like devices to be used to interconnect two or more 802.3 links operating at 1 Gbps. Referred to as a buffered distributor, this device functions as a repeater, forwarding all incoming data received on one port onto all ports other than the original port. However, unlike conventional repeaters, the buffered distributor has the ability to buffer one or more inbound frames on each port prior to forwarding them.

In addition to having memory, the buffered distributor supports the IEEE 802.3z flow control standard. These two features enable each port on a buffered distributor to provide transmission support at a rate that matches the maximum rate of the 1 Gbps shared bus used to broadcast frames received on one port to all other ports. Since the buffered distributor supports

802.3z full-duplex flow control, the situation where an offered load exceeds the bandwidth of the shared bus is easily handled. This is accomplished by a port transmitting a flow control frame to the transmitting system indicating that the port cannot accommodate additional data. The 802.3z compliant transmitting device will then cease transmission until the port forwards a frame indicating it can accept additional data.

To illustrate the operation of a buffered distributor, consider Figure 13.21 which illustrates the use of a four-port buffered distributor. Let's assume that at a certain point in time devices connected to ports 1 and 3 are transmitting data to the buffered distributor at 700 Mbps and 400 Mbps, respectively. It should be noted that those data rates represent traffic carried and not the operating rate of those devices which would be 1 Gbps. Since the aggregate rate exceeds 1 Gbps, a normal repeater would not be able to support this traffic load. However, a buffered repeater includes a memory buffer area into which the excessive 100 Mbps transmission (1100 – 1000) rapidly fills. To ensure that its buffers do not overflow, the buffered repeater will issue a flow control signal telling the transmitters to reduce their transmission rate. This flow control signal will occur once the occupancy of the buffer reaches a predefined level of occupancy as indicated in the right portion of Figure 13.21. Once data from the buffer is serviced to the point where occupancy is at a predefined low level, the buffered distributor will use flow control to enable transmission to the distributor by disabling flow control.

To provide an equitable method for sharing the 1 Gbps bus, the buffered distributor uses a round-robin algorithm to determine which port can transmit onto the bus. This round robin method occurs one frame at a time, with port 1 serviced, followed by port 2, and so on. Through the use of a buffered distributor it becomes possible to support applications that require a 'many-to-one' transmission capability, such as a LAN with a local server used for graphic database queries, email, and similar applications. Instead of migrating the LAN to a switch environment, a buffered distributor may represent another possible network solution to network bottlenecks for power users.

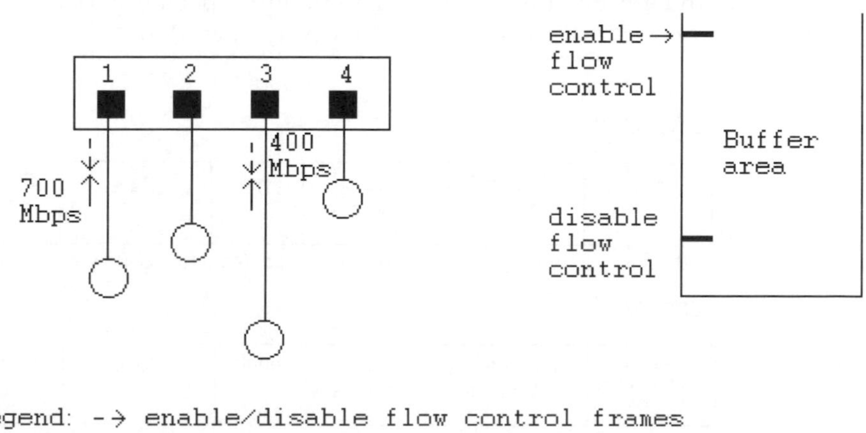

Figure 13.21 A buffered distributor uses IEEE 802.3z flow of control whenever the offered traffic exceeds the bandwidth of its shared bus

13.5.3 Frame composition

Figure 13.22 illustrates the general frame composition of Ethernet and IEEE 802.3 frames. You will note that they differ slightly. An Ethernet frame contains an eight-byte preamble, whereas the IEEE 802.3 frame contains a seven-byte preamble followed by a one-byte start-of-frame delimiter field. A second difference between the composition of Ethernet and IEEE 802.3 frames concerns the two-byte Ethernet type field. That field is used by Ethernet to specify the protocol carried in the frame, enabling several protocols to be carried independently of one another. Under the IEEE 802.3 frame format, the type field was replaced by a two-byte length field, which specifies the number of bytes that follow that field as data.

Not shown in Figure 13.22 is the format of the Gigabit Ethernet frame. As we will note later in this section, the Gigabit Ethernet frame format is the same as the IEEE 802.3 frame format; however, if the frame length is less than 512 bytes carrier extension symbols are added to ensure that the minimum length of the frame is 512 bytes in an adapter or 520 bytes when the preamble and start of frame delimiter fields are added when a frame is transmitted.

Now that we have an overview of the structure of Ethernet and 802.3 frames, let us probe deeper and examine the composition of each frame field. We will take advantage of the similarity between Ethernet and IEEE 802.3 frames to examine the fields of each frame on a composite basis, noting the differences between the two when appropriate.

Preamble field

The preamble field consists of eight (Ethernet) or seven (IEEE 802.3) bytes of alternating 1 and 0 bits. The purpose of this field is to announce the frame and to enable all receivers on the network to synchronize themselves to the incoming frame. In addition, this field by itself (under Ethernet) or in conjunction with the start-of-frame delimiter field (under the IEEE 802.3 standard) ensures there is a minimum spacing period of 9.6 ms between frames for error detection and recovery operations.

Ethernet

Preamble	Destination Address	Source Address	Type	Data	Frame Check Sequence
8 bytes	6 bytes	6 bytes	2 bytes	46–1500 bytes	4 bytes

IEEE 802.3

Preamble/ Start of Frame Delimiter	Destination Address	Source Address	Length	Data	Frame Check Sequence
8 bytes	6 bytes	6 bytes	2 bytes	46–1500 bytes	4 bytes

Figure 13.22 Ethernet and IEEE 802.3 frame formats

Start of frame delimiter field

This field is applicable only to the IEEE 802.3 standard, and can be viewed as a continuation of the preamble. In fact, the composition of this field continues in the same manner as the format of the preamble, with alternating 1 and 0 bits used for the first six bit positions of this one-byte field. The last two bit positions of this field are 11; this breaks the synchronization pattern and alerts the receiver that frame data follows.

It is important to note that the preamble and start of frame delimiter fields are only applicable for frames flowing on a network and are not included when a frame is formed in a computer, bridge or router. Thus, the addition of those fields occurs automatically by the network adapter when a frame is transmitted. This explains an area of confusion concerning minimum and maximum Ethernet frame lengths in many trade publications that deserves an elaboration.

The minimum length Ethernet frame is 64 bytes when in an adapter card and 72 bytes when placed on a LAN. Similarly, the maximum length Ethernet frame is 1518 bytes when in an adapter and 1526 bytes when placed on a LAN. The preceding is applicable for both Ethernet and IEEE 802.3 versions of Ethernet, with the exception of Gigabit. As previously mentioned in this section, a Gigabit Ethernet frame must be a minimum of 512 bytes in length when in an adapter and 520 bytes when placed on a LAN.

Destination address field

The destination address identifies the recipient of the frame. Although this may appear to be a simple field, in reality its length can vary between IEEE 802.3 and Ethernet frames. In addition, each field can consist of two or more subfields, whose settings govern such network operations as the type of addressing used on the LAN, and whether or not the frame is addressed to a specific station or more than one station. To obtain an appreciation for the use of this field, let us examine how this field is used under the IEEE 802.3 standard as one of the two field formats applicable to Ethernet.

Figure 13.23 illustrates the composition of the source and destination address fields. As indicated, the two-byte source and destination address fields are applicable only to IEEE 802.3 networks, whereas the six-byte source and destination address fields are applicable to both Ethernet and IEEE 802.3 networks. A user can select either a two- or six-byte destination address field; however, with IEEE 802.3 equipment, all stations on the LAN must use the same addressing structure. Today, almost all 802.3 networks use six-byte addressing.

I/G subfield

The one-bit I/G subfield is set to a 0 to indicate that the frame is destined to an individual station, or 1 to indicate that the frame is addressed to more than one station: a group address. One special example of a group address is the

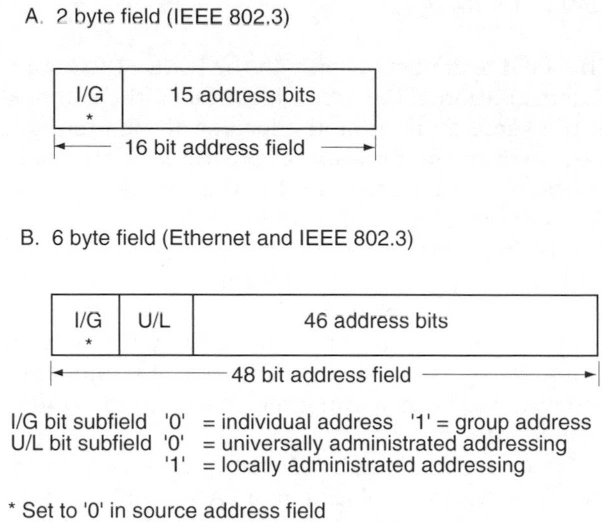

Figure 13.23 Source and destination address field formats

assignment of all 1s to the address field. Hex FFFFFFFFFFFF is recognized as a broadcast address, and each station on the network will receive and accept frames with that destination address.

When a destination address specifies a single station, the address is referred to as a unicast address. A group address that defines multiple stations is known as a multicast address, whereas a group address that specifies all stations on the network is, as previously mentioned, referred to as a broadcast address. As an example of addressing, assume a frame has a destination address of all 1s, or hex FFFFFF. This denotes a broadcast address, resulting in each adapter card in a station on a LAN copying the frame from the network. In comparison, if the address was a unicast address only the adapter in the station that has the destination address of the frame would copy the frame off the network.

U/L subfield

The U/L subfield is applicable only to the six-byte destination address field. The setting of this field's bit position indicates whether the destination address is an address that was assigned by the IEEE (universally administered) or is assigned by the organization via software (locally administered).

Universal versus locally administered addressing

Each Ethernet Network Interface Card (NIC) contains a unique address burned into its read-only memory (ROM) at the time of manufacture. To ensure that this universally administered address is not duplicated, the IEEE assigns blocks of addresses to each manufacturer. These addresses normally include a three-byte prefix, which identifies the manufacturer and is assigned by the IEEE, and a three-byte suffix, which is assigned by the adapter manufacturer to its NIC. For example, the prefix 02608C identifies an NIC manufactured by 3Com, and a prefix of hex 08002 identifies an NIC manufactured by Digital Equipment Company, the latter now owned by Compaq.

Although the use of universally administered addressing eliminates the potential for duplicate network addresses, it does not provide the flexibility obtainable from locally administered addressing. For example, under locally administered addressing, you can configure mainframe software to work with a predefined group of addresses via a gateway PC. Then, as you add new stations to your LAN, you simply use your installation program to assign a locally administered address to the NIC instead of using its universally administered address. As long as your mainframe computer has a pool of locally administered addresses that includes your recent assignment, you do not have to modify your mainframe communications software configuration. Since the modification of mainframe communications software typically requires recompiling and reloading, the attached network must become inoperative for a short period of time. Because a large mainframe may service hundreds or thousands of users, such changes are normally performed late in the evening or on a weekend. Thus, the changes required for locally administered addressing are more responsive to users than those required for universally administered addressing.

Source address field

The source address field identifies the station that transmitted the frame. Like the destination address field, the source address can be either two or six bytes in length.

The two-byte source address is supported only under the IEEE 802.3 standard and requires the use of a two-byte destination address; all stations on the network must use two-byte addressing fields. The six-byte source address field is supported by both Ethernet and the IEEE 802.3 standard. When a six-byte address is used, the first three bytes represent the address assigned by the IEEE to the manufacturer for incorporation into each NIC's ROM. The vendor then normally assigns the last three bytes for each of its NICs.

Type field

The two-byte type field is applicable only to the Ethernet frame. This field identifies the higher-level protocol contained in the data field. Thus, this field tells the receiving device how to interpret the data field.

Under Ethernet, multiple protocols can exist on the LAN at the same time. Xerox served as the custodian of Ethernet address ranges licensed to NIC manufacturers and defined the protocols supported by the assignment of type field values. Under the IEEE 802.3 standard, the type field was replaced by a length field, which precludes compatibility between pure Ethernet and 802.3

frames. While a pure IEEE 802.3 frame is limited to transmitting only one protocol, the IEEE recognized the necessity for its version of Ethernet to support multiple protocols. This was accomplished by the IEEE subdividing the data field into a series of fields to form what is referred to as an Ethernet-SNAP frame whose operation and utilization are described later in this section.

Length field

The two-byte length field, applicable to the IEEE 802.3 standard, defines the number of bytes contained in the data field. Under both Ethernet and IEEE 802.3 standards, the minimum size frame must be 64 bytes in length from preamble through FCS fields. This minimum size frame ensures that there was sufficient transmission time to enable Ethernet NICs to detect collisions accurately based on the maximum Ethernet cable length specified for a network and the time required for a frame to propagate the length of the cable. Based on the minimum frame length of 64 bytes and the possibility of using two-byte addressing fields, this means that each data field must be a minimum of 46 bytes in length.

Data field

As previously discussed, the data field must be a minimum of 46 bytes in length to ensure that the frame is at least 64 bytes in length. This means that the transmission of 1 byte of information must be carried within a 46-byte data field; if the information to be placed in the field is less than 46 bytes, the remainder of the field must be padded. Although some publications subdivide the data field to include a PAD subfield, the latter actually represents optional fill characters that are added to the information in the data field to ensure a length of 46 bytes. The maximum length of the data field is 1500 bytes.

Frame check sequence field

The frame check sequence field, applicable to both Ethernet and the IEEE 802.3 standard, provides a mechanism for error detection. Each transmitter computes a cyclic redundancy check (CRC) that covers both address fields, the type/length field and the data field. The transmitter then places the computed CRC in the four-byte FCS field.

The CRC treats the previously mentioned fields as one long binary number. The n bits to be covered by the CRC are considered to represent the coefficients of a polynomial $M(X)$ of degree $n-1$. Here, the first bit in the destination address field corresponds to the X^{n-1} term, whereas the last bit in the data field corresponds to the X^0 term. Next, $M(X)$ is multiplied by X^{32} and the result of that multiplication process is divided by the following polynomial:

$$G(X) = X^{32} + X^{26} + X^{23} + X^{22} + X^{16} + X^{12} + X^{11} + X^{10}$$
$$+X^8 + X^7 + X^5 + X^4 + X^2 + X + 1$$

Note that the term X^n represents the setting of a bit to a 1 in position n. Thus, part of the generating polynomial $X^5 + X^4 + X^2 + X^1$ represents the binary value 11011.

This division produces a quotient and remainder. The quotient is discarded, and the remainder becomes the CRC value placed in the four-byte FCS field. This 32-bit CRC reduces the probability of an undetected error to 1 bit in every 4.3 billion, or approximately 1 bit in $2^{32} - 1$ bits.

Once a frame reaches its destination, the receiver uses the same polynomial to perform the same operation upon the received data. If the CRC computed by the receiver matches the CRC in the FCS field, the frame is accepted. Otherwise, the receiver discards the received frame, as it is considered to have one or more bits in error. The receiver will also consider a received frame to be invalid and discard it under two additional conditions. Those conditions occur when the frame does not contain an integral number of bytes, or when the length of the data field does not match the value contained in the length field. The latter condition is obviously only applicable to the 802.3 standard, since an Ethernet frame uses a type field instead of a length field.

13.5.4 Media access control overview

Under the IEEE 802 series of standards, the data link layer of the OSI Reference Model was originally subdivided into two sublayers: logical link control (LLC) and medium access control (MAC). The frame formats examined in Figure 13.22 represent the manner in which LLC information is transported. Directly under the LLC sublayer is the MAC sublayer. The MAC sublayer is responsible for checking the channel and transmitting data if the channel is idle, checking for the occurrence of a collision, and taking a series of predefined steps if a collision is detected. Thus, this layer provides the required logic to control the network.

13.5.5 Logical link control overview

The logical link control (LLC) sublayer was defined under the IEEE 802.2 standard to make the method of link control independent of a specific access method. Thus, the 802.2 method of link control spans Ethernet (IEEE 802.3), Token Bus (IEEE 802.4), and Token-Ring (IEEE 802.5) local area networks. Functions performed by the LLC include generating and interpreting commands to control the flow of data, including recovery operations for when a transmission error is detected.

Link control information is carried within the data field of an IEEE 802.3 frame as an LLC Protocol Data Unit. When a frame is constructed using LLC it is referred to as an Ethernet Subnetwork Access Protocol (SNAP) frame. Figure 13.24 illustrates the relationship between the IEEE 802.3 Ethernet-SNAP frame and the LLC Protocol Data Unit carried in the frame.

Service Access Points (SAPs) function much like a mailbox. Since the LLC layer is bounded below the MAC sublayer and bounded above by the network layer, SAPs provide a mechanism for exchanging information between the

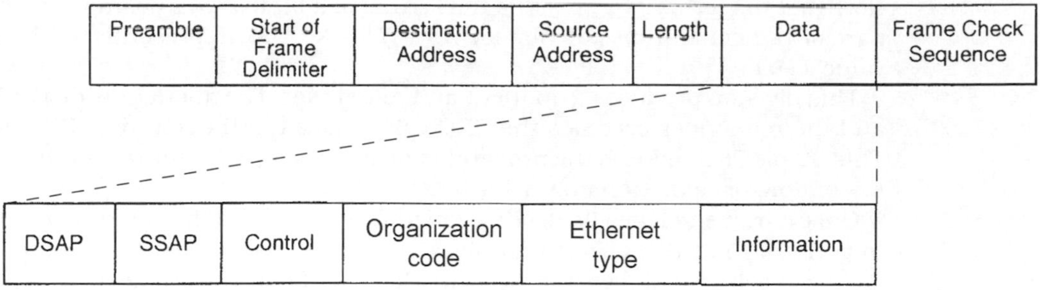

Figure 13.24 Formation of LLC protocol data unit in an Ethernet-SNAP frame. Control information is carried within a MAC frame

LLC layer and the MAC and network layers. For example, from the network layer perspective, a SAP represents the place to leave messages about the services requested by an application. The Destination Services Access Point (DSAP) is one byte in length, and it is used to specify the receiving network layer process. The Source Service Access Point (SSAP) is also one byte in length. The SSAP specifies the sending network layer process. Both DSAP and SSAP addresses are assigned by the IEEE. For example, hex address FF represents a DSAP broadcast address.

In an Ethernet-SNAP frame the value hex AA is placed into the DSAP and SSAP fields, while hex 03 is placed into the control field to indicate that a SNAP frame is being transported. The hex 03 value in the control field defines the use of an unnumbered format, which is the only format supported by a SNAP frame.

The organization code field references the organizational body that assigned the value placed in the following field, the Ethernet type field. A hex value of 00-00-00 in the organization code field indicates that Xerox assigned the value in the Ethernet type field. Through the use of the Ethernet-SNAP frame, you obtain the ability to transport multiple protocols in a manner similar to the original Ethernet frame that used the type field for this purpose.

Types and classes of service

Under the 802.2 standard, there are three types of service available for sending and receiving LLC data. These types are discussed in the next three paragraphs. Figure 13.25 provides a visual summary of the operation of each LLC service type.

Type 1

Type 1 is an unacknowledged connectionless service. The term connectionless refers to the fact that transmission does not occur between two devices as if a logical connection were established. Instead, transmission flows on the channel to all stations; however, only the destination address acts on the data.

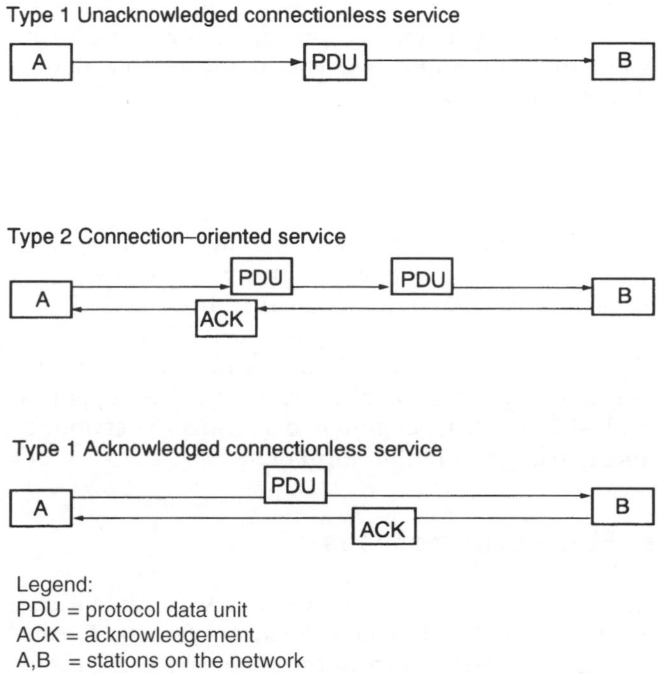

Figure 13.25 Local link control service types

As the name of this service implies, there is no provision for the acknowledgement of frames. Neither are there provisions for flow control or for error recovery. Therefore, this is an unreliable service.

Despite those shortcomings, Type 1 is the most commonly used service, since most protocol suites use a reliable transport mechanism at the transport layer, thus eliminating the need for reliability at the link layer. In addition, by eliminating the time needed to establish a virtual link and the overhead of acknowledgements, a Type 1 service can provide a greater throughput than other LLC types of services.

Type 2

The Type 2 connection-oriented service requires that a logical link be established between the sender and the receiver prior to information transfer. Once the logical connection is established, data will flow between the sender and receiver until either party terminates the connection. During data transfer, a Type 2 LLC service provides all of the functions lacking by in a Type 1 service, with a sliding window used for flow control.

Type 3

The Type 3 acknowledged connectionless service contains provision for the setup and disconnection of transmission; it acknowledges individual frames using the stop-and-wait flow control method. Type 3 service is primarily used in an automated factory process-control environment, where one central computer communicates with many remote devices that typically have a limited storage capacity.

Classes of service

All logical link control stations support Type 1 operations. This level of support is known as Class I service. The classes of service supported by LLC indicate the combinations of the three LLC service types supported by a station. Class I supports Type 1 service, Class II supports both Type 1 and Type 2, Class III supports Type 1 and Type 3 service, and Class IV supports all three service types. Since service Type 1 is supported by all classes, it can be considered a least-common denominator, enabling all stations to communicate using a common form of service.

13.5.6 Other Ethernet frame types

Three additional frame types that warrant discussion are Ethernet-802.3, Fast Ethernet, and Gigabit Ethernet frames. Because the latter two involve changes to the structure of the delimiters of the frame and not its contents, we will first turn our attention to the Ethernet-802.3 frame format. Since the composition of a frame can result in it being classified as an Ethernet frame, Ethernet-802.3 frame, or Ethernet-SNAP frame, we will discuss the manner by which a frame can be identified prior to examining the composition of Fast Ethernet and Gigabit Ethernet frames.

Ethernet-802.3

The Ethernet-802.3 frame represents a proprietary subdivision of the IEEE 802.3 data field to transport NetWare. Ethernet-802.3 is one of several types of frames that can be used to transport NetWare. The actual frame type used is defined at system setup by designating NetWare to a specific type of frame.

Figure 13.26 illustrates the format of the Ethernet-802.3 frame. Due to the absence of LLC fields, this frame is often referred to as *raw 802.3*.

For those using or thinking of using NetWare, a word of caution is in order concerning frame types. Novell uses the term Ethernet-802.2 to refer to the IEEE 802.3 frame. Thus, if you set up NetWare for Ethernet-802.2 frames, in effect your network is IEEE 802.3-compliant.

Frame determination

Through software, a receiving station can determine the type of frame and correctly interpret the data carried in the frame. To accomplish this, the value of the two bytes that follow the source address is first examined. If the value is greater than 1500, this indicates the occurrence of an Ethernet frame. If the

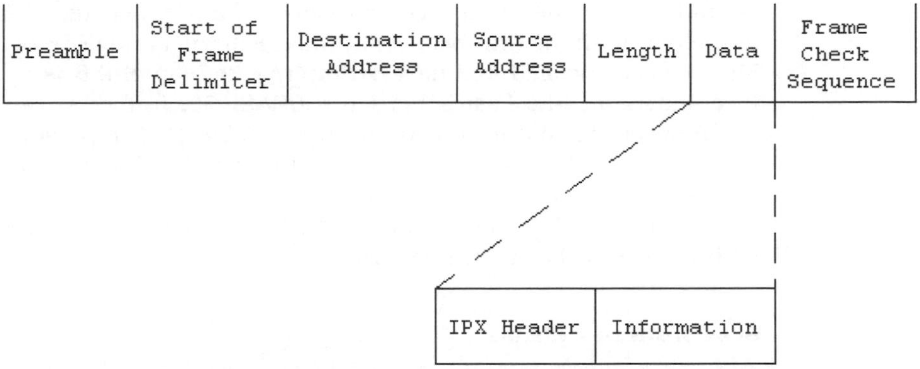

Figure 13.26 Novell's NetWare Ethernet-802.3 frame. An Ethernet-802.3 frame subdivides the data field into an IPX header field and an information field

value is less than or equal to 1500, the frame can be either a pure IEEE 802.3 frame or a variation of that frame. Thus, more bytes must be examined.

If the next two bytes have the hex value FF:FF, the frame is a NetWare Ethernet-802.3 frame. This is because the IPX header has hex FF:FF in the checksum field contained in the first two bytes in the IPX header. If the two bytes contain the hex value AA:AA, this indicates that it is an Ethernet-SNAP frame. Any other value determined to reside in those two bytes then indicates that the frame must be an Ethernet-802.3 frame.

Fast Ethernet

The frame composition associated with each of the three Fast Ethernet standards is illustrated in Figure 13.27. In comparing the composition of the Fast Ethernet frame with Ethernet and IEEE 802.3 frame formats previously illustrated in Figure 13.22, you will note that other than the addition of starting and ending stream delimiters, the Fast Ethernet frame duplicates the older frames. A third difference between the two is not shown, as it is not

SSD 1 byte	Preamble 7 bytes	SFD 1 byte	Destination Address 6 bytes	Source Address 6 bytes	L/T 2 bytes	Data 46 to 1500 bytes	FCS 1 byte	ESD 1 byte

Legend:
SSD = Start of Stream Delimiter
SFD = Start of Frame Delimiter
L/T = Length (IEEE 802.3)/Type (Ethernet)
ESD = End of Stream Delimiter

Figure 13.27 Fast Ethernet frame. The 100BASE-TX frame differs from the IEEE 802.3 MAC frame through the addition of a byte at each end to mark the beginning and end of the stream delimiter

actually observable from a comparison of frames, because this difference is associated with the time between frames. Ethernet and IEEE 802.3 frames are Manchester encoded and have an interpacket gap of 9.6 μs between frames. In comparison, the Fast Ethernet 100BASE-TX frame is transmitted using 4B5B encoding, and IDLE codes (refer to Table 13.4) representing sequences of 1 (binary 11111) symbols are used to mark a 0.96-μs interpacket gap.

Now that we have an overview of the differences between Ethernet/IEEE 802.3 and Fast Ethernet frames, let's focus upon the new fields associated with the Fast Ethernet frame format.

Start-of-stream delimiter

The start-of-stream delimiter (SSD) is used to align a received frame for subsequent decoding. The SSD field consists of a sequence of J and K symbols, which defines the unique code 11000 10001. This field replaces the first octet of the preamble in Ethernet and IEEE 802.3 frames whose composition is 10101010.

End-of-stream delimiter

The end-of-stream delimiter (ESD) is used as an indicator that data transmission terminated normally, and a properly formed stream was transmitted. This one-byte field is created by the use of T and R codes (see Table 13.4), whose bit composition is 01101 00111. The ESD field lies outside the Ethernet/IEEE 802.3 frame and for comparison purposes can be considered to fall within the interframe gap of those frames.

Gigabit Ethernet

Earlier in this chapter it was briefly mentioned that the Ethernet frame was extended for operations at 1 Gbps. In actuality the Gigabit Ethernet standard resulted in two modifications to conventional CSMA/CD operations. The first modification, which is referred to as carrier extension, is only applicable for half-duplex links and was required to maintain an approximate 200 meter topology at gigabit speeds. Instead of actually extending the frame, as we will shortly note, the time the frame is on the wire is extended. A second modification, referred to as packet burst, enables gigabit-compatible network devices to transmit bursts of relatively short packets without having to relinquish control of the network. Both carrier extension and packet bursting represent modifications to the CSMA/CD protocol to extend the collision domain and enhance the efficiency of Gigabit Ethernet, respectively. Both topics are covered in detail in this section.

Carrier extension

In an Ethernet network the attachment of workstations to a hub creates a segment. That segment or multiple segments interconnected via the use of one or more repeaters forms a collision domain. The latter term is formally defined as a single CSMA/CD network in which a collision will occur if two devices attached to the network transmit at or approximately at the same time. We can say approximately because there is a propagation delay time associated with the transmission of signals on a conductor. Thus, if one station is relatively close to another the propagation delay time is relatively short, requiring both stations to transmit data at nearly the same time for a collision to occur. If two stations are at opposite ends of the network the propagation delay for a signal placed on the network by one station to reach the other station is much greater. This means that one station could initiate transmission and actually transmit a portion of a frame while the second station might 'listen' to the network, hear no activity, and begin to transmit, resulting in a collision.

Figure 13.28 illustrates the relationship between a single collision domain and two collision windows. Note that as stations are closer to one another the collision window which represents the propagation delay time during which one station could transmit and another would assume there is no network activity decreases.

Ethernet requires that a station should be able to hear any resulting collision for the frame it is transmitting before it completes the transmission of the entire frame. This means that the transmission of the next to last bit of a frame that results in a collision should allow the transmitting station to 'hear' the collision voltage increase before it transmits the last bit. Thus, the maximum allowable cabling distance is limited by the bit duration associated with the network operating rate and the speed of electrons on the wire.

When Ethernet operates at 1 Gbps the allowable cabling distance would be reduced to approximately 10 meters or 33 feet. Clearly this would be a major restriction on the ability of Gigabit Ethernet to be effectively used in a shared media half-duplex environment. To overcome this transmission distance limitation Sun Microsystems, Inc. suggested the carrier extension scheme which became part of the Gigabit Ethernet standard for half-duplex operations.

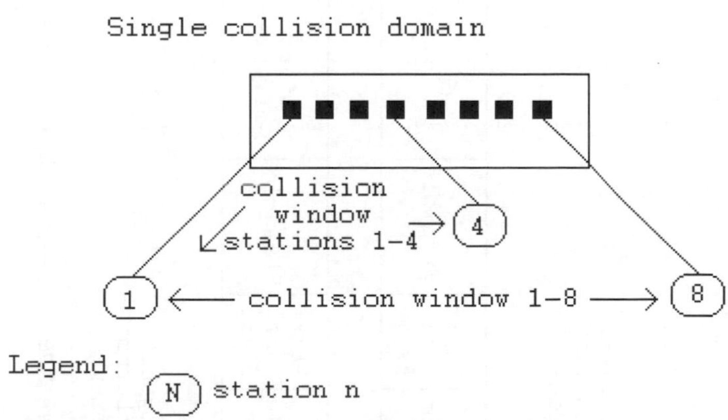

Figure 13.28 Relationship between a collision domain and collision windows

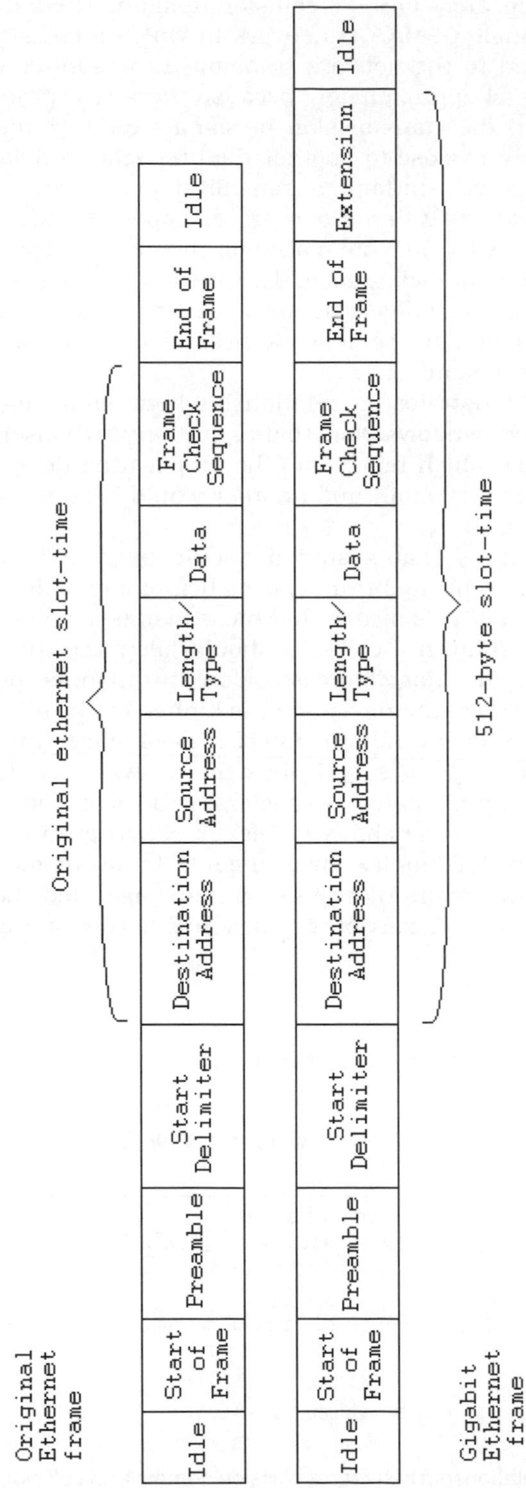

Figure 13.29 Half-duplex Gigabit Ethernet uses a carrier extension scheme to extend timing so that the slot time consists of at least 512 bytes

Under the carrier extension scheme the original Ethernet frame is extended by increasing the time the frame is on the 'wire'. The timing extension occurs after the end of the standard CSMA/CD frame, as illustrated in Figure 13.29. The carrier extension extends the frame timing to guarantee at least a 512-byte slot time for half-duplex Ethernet. Note that Ethernet's slot time is considered as the time from the first bit of the destination address field reaching the wire through the last bit of the Frame Check Sequence field. The increase in the minimum length frame does not change the frame size and only alters the time the frame is on the wire. Due to this, compatibility is maintained between the original Ethernet frame and the Gigabit Ethernet frame.

Although the carrier extension scheme enables the cable length of a half-duplex Gigabit network to be extended to a 200 m diameter, that extension is not without a price. That price is one of overhead, since extension symbols attached to a short frame waste bandwidth. For example, a frame with a 64-byte data field would have 448 bytes of wasted carrier extension symbols attached to it. To further complicate bandwidth utilization, when the data field is less than 46 bytes in length nulls are added to produce a 64-byte minimum length data field. Thus, a simple query to be transported by Ethernet, such as 'Enter your age', consisting of 14 data characters, would be padded with 32 null characters when transported by Ethernet to ensure a minimum 72-byte length frame. Under Gigabit Ethernet the minimum 512-byte time slot would require the use of 448 carrier extension symbols to ensure that the time slot from destination address through any required extension is at least 512 bytes in length.

In examining Figure 13.29 it is important to note that the carrier extension scheme does not extend the Ethernet frame beyond a 512-byte time slot. Thus, Ethernet frames with a time slot equal to or exceeding 512 bytes have no carrier extension. Another important item to note concerning the carrier extension scheme is that it has no relationship to a 'Jumbo Frames' feature that is proprietary to a specific vendor. That feature is supported by a switch manufactured by Alteon Networks and is used to enhance data transfers between servers, permitting a maximum frame size of up to 9 kbytes to be supported. Since 'Jumbo Frames' are not part of the Gigabit Ethernet standard, you must disable that feature to obtain interoperability between that vendor's 1 Gbps switch and other vendors' Gigabit Ethernet products. However, it should be noted that at the time this book revision was prepared the Internet Engineering Task Force (IETF) had published a proposal for extending the Ethernet frame as well as a mechanism to designate the fact that the frame is an extended frame in its header. Because a 9-kbyte frame requires one-sixth of the processing of a sequence of six 1500-byte frames, the use of jumbo frames makes it easier for adapter cards and computers to operate at or closer to the 1 Gbps operating rate of Gigabit Ethernet.

Packet bursting

Packet bursting represents a scheme added to Gigabit Ethernet to counteract the overhead associated with transmitting relatively short frames. This scheme was proposed by NBase Communications and is included in the Gigabit Ethernet standard as an addition to carrier extension.

Under packet bursting each time the first frame in a sequence of short frames successfully passes the 512-byte collision window using the carrier extension scheme previously described, subsequent frames are transmitted without including the carrier extension. The effect of packet bursting is to average the wasted time represented by the use of carrier extension symbols over a series of short frames. The limit on the number of frames that can be burst is a total of 1500 bytes for the series of frames, which represents the longest data field supported by Ethernet. To inhibit other stations from initiating transmission during a burst carrier extension, signals are inserted between frames in the burst.

In addition to enhancing network utilization and minimizing bandwidth overhead, packet bursting also reduces the probability of collisions occurring. This is because the burst of frames are only susceptible to a collision during the first frame in the sequence. Thereafter, carrier extension symbols between frames followed by additional short frames are recognized by all other stations on the segment, and inhibit those stations from initiating a transmission that would result in the occurrence of a collision.

13.6 TOKEN-RING

There are five major types of Token-Rings. The first three types of Token-Ring networks are the focus of this section: 4, 16 and 100 Mbps Token-Ring networks that operate according to the IEEE 802.5 standard. Two additional Token-Ring networks are the Fiber Distributed Data Interface (FDDI) that operates at 100 Mbps and FDDI transmission over copper wiring, an evolving standard commonly referred to as Copper Distributed Data Interface (CDDI).

13.6.1 Topology

Although the term Token-Ring implies a ring structure, in actuality this type of LAN is either a star or star-ring structure, with the actual topology based upon the number of stations to be connected. The term star is derived from the fact that a grouping of stations and other devices, including printers, plotters, repeaters, bridges, routers and gateways, are connected in groups to a common device called a Multistation Access Unit (MAU).

Figure 13.30 illustrates a single ring formed through the use of one MAU in which up to eight devices are interconnected. Thus, for a very small Token-Ring LAN consisting of a mixture of eight or less devices, the structure actually resembles a star.

When IBM introduced its 4 Mbps Token-Ring network, its first MAU, known as the 8228, was a 10-port device, of which two ports were used for Ring-In (RI) and Ring-Out (RO) connectors which enable multiple MAUs to be connected to one another to expand the network. The remaining eight ports on the 8228 are designed to connect devices to the ring. Since IBM's eight-port 8228 reached the market, other vendors have introduced similar products with different device support capacities. You can now commonly obtain MAUs that support 4, 8, and 16 devices.

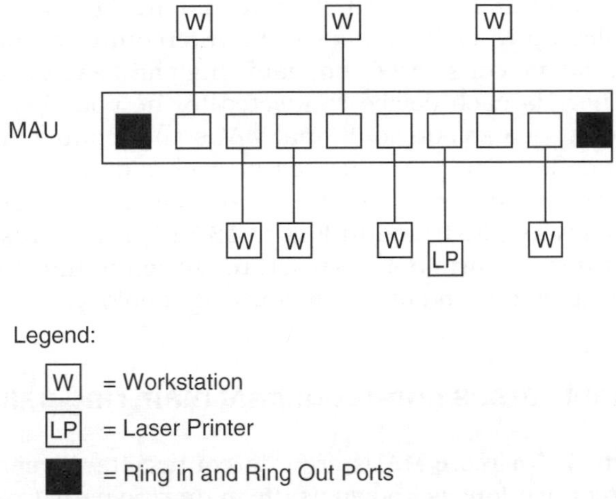

Figure 13.30 Single-ring LAN. A single-ring LAN can support up to eight devices through their attachment to a common MAU

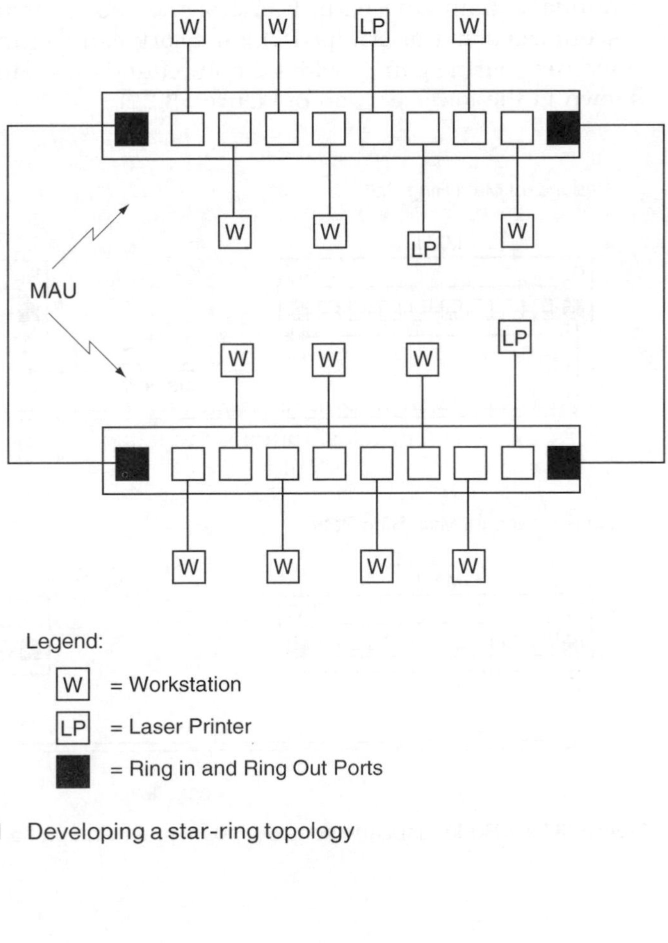

Figure 13.31 Developing a star-ring topology

In examining Figure 13.30, note that an eight-port MAU is illustrated, which enables up to eight devices to be interconnected to a Token-Ring LAN. The MAU can be considered the main ring path, as data will flow from one port to another via each device connected to the port. If you have more than eight devices, you can add additional MAUs, interconnecting the MAUs via the Ring-In and Ring-Out ports located at each side of each MAU. When this interconnection occurs, by linking two or more MAUs together you form a star-ring topology, as illustrated in Figure 13.31. In this illustration, the stations and other devices form a star structure, while the interconnection of MAUs forms the ring; hence you obtain a star-ring topology.

13.6.2 Redundant versus non-redundant main ring paths

When two or more MAUs are interconnected, the serial path formed by those interconnections is known as the main ring path. Connections between MAUs can be accomplished through the use of one or two pairs of wiring. One pair will be used as the primary data path, and the other pair functions as a backup data path.

The top of Figure 13.32 illustrates the formation of a ring consisting of two MAUs in which both primary and backup paths are established to provide a redundant main ring path. If one of the cables linking the MAUs becomes disconnected, cut or crimped, the network can continue to operate since the remaining wiring pair provides a non-redundant main ring path capability as shown in the lower portion of Figure 13.32.

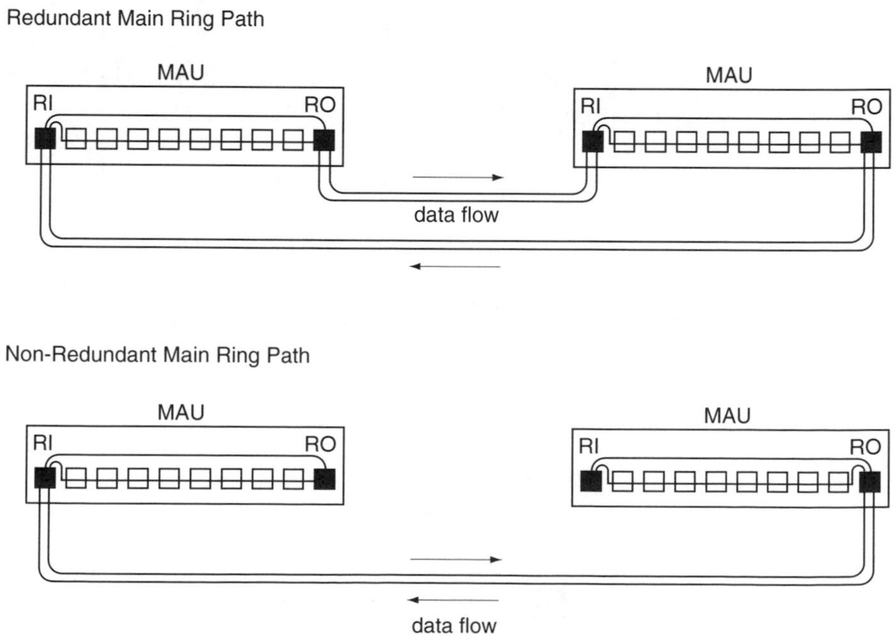

Figure 13.32 Redundant and non-redundant main-ring paths

The backup capability provided by redundant main ring paths is established through the use of loopback plugs or a built-in MAU self-shorting feature, both of which are discussed in additional detail later in this chapter. Since it is both difficult and tedious to draw wire pairs, in the remainder of this section we will use a single line to indicate wiring pairs between MAUs and MAUs and stations connected to MAUs.

13.6.3 Cabling and device restrictions

The type of cable or wiring used to connect devices to MAUs and interconnect MAUs is a major constraint that governs the size of a Token-Ring network. Since the IBM cabling system provides a large number of common types of wiring, let us first examine the type of cable defined by that cabling system.

IBM cabling system

The IBM cabling system was introduced in 1984 as a mechanism to support the networking requirements of office environments. By defining standards for cables, connectors, faceplates, distribution panels, and other facilities, IBM's cabling system is designed to support the interconnection of personal computers, conventional terminals, mainframe computers, and office systems. In addition, this system permits devices to be moved from one location to another, or added to a network through a simple connection to the cabling system's wall plates or surface mounts.

The IBM cabling system specifies seven different cabling categories. Depending on the type of cable selected, you can install the selected wiring indoors, outdoors, under a carpet, or in ducts and other air spaces.

The IBM cabling system uses wire which conforms to the American Wire Gauge or AWG. The IBM cabling system uses wire between 22 AWG (0.644 mm) and 26 AWG (0.405 mm) Since a large-diameter wire has less resistance to current flow than a small one, a smaller AWG permits cabling distances to be extended in comparison with a higher AWG cable.

It is important to note that the IBM cabling system was introduced prior to the EIA/TIA-568 specification that defined five categories of unshielded twisted-pair cabling as well as fiber for use in buildings. Although the IBM cabling system is no longer marketed by IBM, many independent vendors continue to offer cables according to the various categories supported by that system. More modern Token-Ring products are designed for UTP connectors, with Category 3 UTP able to support 4 and 16 Mbps Token-Ring communications, while Category 5 cable is required to support communications to devices that operate according to the recently standardized 100 Mbps High Speed Token-Ring specification. In most organizations it is quite common for all copper cabling to involve the use of Category 5 cable, as this enables an organization to retain its cabling infrastructure even if it replaces one type of network with another.

Type 1

The IBM cabling system Type 1 cable contains two twisted pairs of 22 AWG conductors. Each pair is shielded with a foil wrapping, and both pairs are surrounded by an outer braided shield or with a corrugated metallic shield. One pair of wires uses shield colors of red and green, and the second pair of wires uses shield colors of orange and black. The braided shield is used for indoor wiring, whereas the corrugated metallic shield is used for outdoor wiring. Type 1 cable is available in two different designs: plenum and non-plenum. Plenum cable can be installed without the use of a conduit, and non-plenum cable requires a conduit. Type 1 cable is typically used to connect a distribution panel or multistation access unit and the faceplate or surface mount at a workstation.

Type 2

Type 2 cable is actually a Type 1 indoor cable with the addition of four pairs of 22 AWG conductors for telephone usage. Due to this, Type 1 cable is also referred to as data-grade twisted-pair cable, while Type 2 cable is known as two data-grade and four-grade twisted pair. Due to its voice capability, Type 2 cable can support PBX interconnections. Like Type 1 cable, Type 2 cable supports plenum and non-plenum designs. Type 2 cable is not available in an outdoor version.

Type 3

Type 3 cable is conventional twisted pair telephone wire, with a minimum of two twists per foot. Both 22 AWG and 24 AWG conductors are supported by this cable type. One common use of Type 3 cable is to connect PCs to MAUs in a Token-Ring network.

Type 5

Type 5 cable is fiber optic cable. Two 100/140 μm optical fibers are contained in a Type 5 cable. This cable is suitable for indoor non-plenum installation or outdoor aerial installation. Due to the extended transmission distance obtainable with fiber-optic cable, Type 5 cable is used in conjunction with the IBM 8219 Token-Ring Network Optical Fiber Repeater to interconnect two MAUs up to 6600 feet (2 km) from one another.

Type 6

Type 6 cable contains two twisted pairs of 26 AWG conductors for data communications. It is available for non-plenum applications only and its smaller diameter than Type 1 cable makes it slightly more flexible. The primary use of Type 6 cable is for short runs as a flexible path cord. This type of cable is often used to connect an adapter card in a personal computer to a faceplate which, in turn, is connected to a Type 1 or Type 2 cable which forms the backbone of a network.

Type 8

Type 8 cable is designed for installation under a carpet. This cable contains two individually shielded, parallel pairs of 26 AWG conductors with a plastic ramp designed to make under-carpet installation as unobtrusive as possible. Although Type 8 cable can be used in a manner similar to Type 1, it only provides half of the maximum transmission distance obtainable through the use of Type 1 cable.

Type 9

Type 9 cable is essentially a low-cost version of Type 1 cable. Like Type 1, Type 9 cable consists of two twisted pairs of data cable; however, 26 AWG conductors are used in place of the 22 AWG wire used in Type 1 cable. As a result of the use of a smaller diameter cable, transmission distances on Type 9 cable are approximately two-thirds of those obtainable through the use of Type 1 cable. The color coding on the shield of Type 9 cable is the same as that used for Type 1 cable.

All seven types of cables defined by the IBM cabling system can be used to construct Token-Ring networks. However, the use of each type of cable has a different effect on the ability to connect devices to the network, the number of devices that can be connected to a common network, the number of wiring closets in which MAUs can be installed to form a ring, and the ability of the cable to carry separate voice conversations. The latter capability enables a common cable to be routed to a user's desk where a portion of the cable is connected to their telephone, and another portion of the cable is connected to their computer's Token-Ring adapter card.

Table 13.6 summarizes the performance characteristics of the cables defined by the IBM cabling system. The drive distance entry indicates the rela-

Table 13.6 IBM cable system cable performance characteristics

	Cable type						
Performance characteristics	1	2	3	5	6	8	9
Drive distance (relative to type 1)	1.0	1.0	0.45	3.0	0.75	0.5	0.66
Data rate (Mbps)	16	16	4*	250	16	16	16
Maximum devices per ring	260	260	72	260	96	260	260
Maximum closets per ring	12	12	2	12	12	12	12
Voice support	no	yes	yes	no	no	no	no

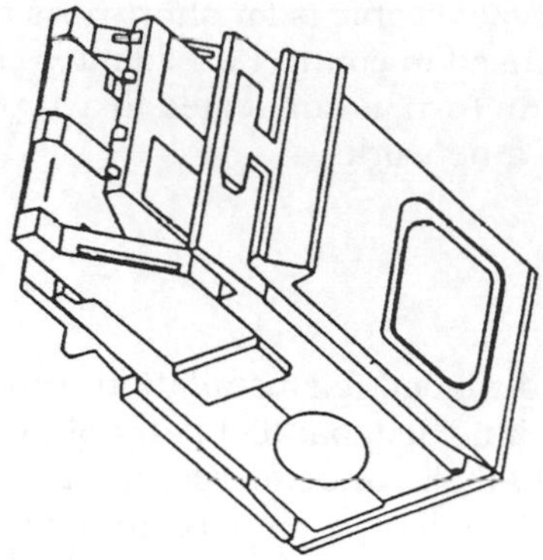

Figure 13.33 IBM cabling system data connector

tive relationship between different types of cable with respect to the maximum cabling distance between a workstation and an MAU as well as between MAUs. Type 1 cable provides a maximum drive distance of 100 m between a workstation and an MAU and 300 m between MAUs for a network operating at 4 Mbps. Other drive distance entries in Table 13.5 are relative to the drive distance obtainable when Type 1 cable is used.

Connectors

The IBM cabling system includes connectors to terminate both data and voice conductors. The data connector has a unique design based on the development of a latching mechanism which permits it to mate with another identical connector.

Figure 13.33 illustrates the IBM cabling system data connector. Its design makes it self-shorting when disconnected from another connector. This provides a Token-Ring network with electrical continuity when a station is disconnected. Unfortunately, the data connector is very expensive in comparison to RS-232 and RJ telephone connectors with the typical retail price of the data connector between $4 and $5, whereas RS-232 connectors cost approximately $1 and an RJ telephone connector can be purchased for a dime or so.

Due to the high cost of data connectors and cable, the acceptance of the IBM cabling system by end-users never reached its potential. Instead, Category 3 and Category 5 structured wiring with RJ45 connectors is commonly used to construct Token-Ring networks.

13.6.4 Constraints

There are several cabling and device constraints you must consider when designing a Token-Ring network to ensure that the network will work correctly. First, you must consider the maximum cabling distance between each device and an MAU that will service the device. The cable between the

MAU and the device is referred to as a lobe, with the maximum lobe distance being 100 m (330 feet) at both 4 and 16 Mbps. This means that you must consider the lobe distance in conjunction with the cabling distance restrictions between MAUs if you have more than eight devices to be connected to a Token-Ring LAN. In addition, for larger networks, you must also consider restrictions on the number of MAUs in the network and their placement in wiring closets, as well as a parameter known as the adjusted ring length, because they collectively govern the maximum number of devices that can be supported.

As we will note when we discuss 100 Mbps Token-Ring operations later in this section, its use is also restricted to a 100 m span. However, instead of supporting shared media operations, high-speed Token-Ring is limited in use to a switch environment, providing a 100 Mbps connection from servers to a switch or for interconnecting two switches.

Intra-MAU cabling distances

Table 13.7 lists the maximum intra-MAU cabling distances permitted on a Token-Ring network for the two most commonly used types of IBM cables. Those distances can be extended through the use of repeaters; however, their use adds to both the complexity and cost of the network. Because 100 Mbps Token-Ring does not presently support shared media operations, it is not used to interconnect MAUs and was omitted from the table.

As indicated in Table 13.7, the cabling distance between MAUs depends on both the operating rate of the LAN (4 Mbps or 16 Mbps) and the type of cable used. Type 1 is a double-shielded pair cable, and Type 3 is non-shielded twisted-pair telephone wire.

In examining the entries in Table 13.7, it may appear odd that the maximum intra-MAU cable distance is the same for both 4 and 16 Mbps networks when Type 1 cable is used. This situation occurred because, at the time IBM set a 100 m recommended limit for a 4 Mbps network, the company took into consideration the need to reuse the same cabling when customers upgraded to a 16 Mbps operating rate. When using Type 3 cable, distances are shorter because signal line noise increases in proportion to the square root of frequency. Thus, upgrading the operating rate from 4 to 16 Mbps with Type 3 cable decreases the maximum permissible distance.

As you plan to extend your network to interconnect additional devices, you must also consider the maximum number of MAUs and devices supported by

Table 13.7 Intra-MAU cabling distance (feet)

	Type of cable	
Operating rate	Type 1	Type 3
4 Mbps	330	1000
16 Mbps	330	250

a Token-Ring network. If you use Type 1 cable, you are limited to a maximum of 33 MAUs and 260 devices. If you use Type 3 cable, you are limited to a maximum of 9 MAUs and 72 devices. These limitations are applicable to both 4 and 16 Mbps networks; however, they represent a maximum number of MAUs and devices and do not indicate reality in which a lesser number of MAUs may be required due to the use of multiple wiring closets or a long adjusted ring length. Due to the role played by the adjusted ring length in governing the number of MAUs, let us examine what an adjusted ring length is and how it functions as a constraint.

Adjusted ring length

To fully understand the reason why we must consider the adjusted ring length (ARL) of a Token-Ring network requires a discussion on the network's ability to operate with a faulty cabling section. To illustrate this capability, consider the

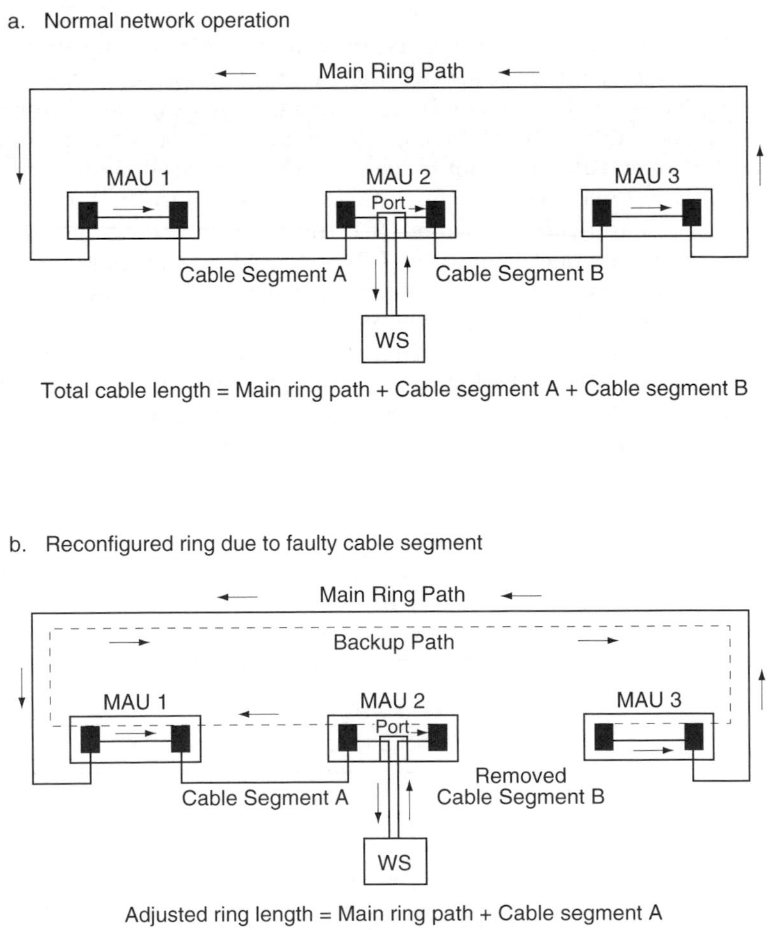

Figure 13.34 Computing the adjusted ring length

three-MAU network illustrated in Figure 13.34(a). Under normal network operation, data are transmitted from RO to RI between MAUs and over the main ring path which connects the last MAU's RO port to the first MAU's RI port.

If an attached device or a lobe cable fails, the lack of voltage on the MAU's port causes the port to be bypassed, permitting information from other stations to flow on the main ring path. If an MAU or a cable interconnecting two MAUs fails, the previously described built-in backup capability of MAUs permits the network to continue operating. This backup capability permits the Token-Ring to be re-established in one of two ways, dependent on the capability of the MAUs used in the network. Some MAUs have a self-shorting capability, which means that RI and RO connectors are joined together without requiring the use of a cable or plug to complete a ring. Other MAUs require the use of a cable or plug between the RI and RO connectors. By using the self-shorting capability, or using a cable or plug in the input and output connectors of the MAUs located at both ends of a failed cable, the ring can be reconfigured for operation Figure 13.34(b) illustrates the reconfigured ring. Note that the total cable length of the main ring path and available cable segments represents the adjusted ring length.

In actuality, the adjusted ring length is the total ring length (main ring path plus all cable segments) less the shortest cable segment between MAUs. The reason why we subtract the shortest cable segment is that doing so provides the longest total cable distance for a reconfigured ring. It is that distance that a signal must be capable of flowing around a ring without excessive distortion adversely affecting the signal. For example, suppose that the main ring path is 300 feet and cable segments A and B are 200 and 150 feet, respectively. Then, the total drive distance is $300+200+150$, or 650 feet, and the adjusted ring length is $650-150$, or 500 feet.

Other ring size considerations

The adjusted ring length and the length of the main ring path are two constraints that govern the size of a Token-Ring network. Other constraints include the operating rate of the ring, the length of the longest lobe, the type and number of MAUs in the network, and the number of distinct locations where the MAUs are installed, the latter referred to by IBM as wiring closets. The type of MAU is equivalent to the type of cabling used. A Type 1 MAU is cabled using Type 1 (double-shielded pair cable), while a Type 3 MAU uses Type 3 (non-shielded twisted-pair telephone wire) cable.

When Type 1 MAUs are used for 4 or 16 Mbps operation and all lobe cables terminate in a single wiring closet, you can interconnect up to 33 IBM 8228 MAUs, which can serve 260 devices, with each device cabled using up to a 100 m lobe cable. Here, the number of 8228s is

$$8228\text{s} = \text{Int}\frac{(\text{device} + 0.5)}{8} \leq 33$$

For 2 to 12 wiring closets, IBM provides a series of charts in the firm's *Token-Ring Network, Introduction and Planning Guide*, relationship between the lobe length, adjusted ring length, number of MAUs, and wiring closets when

Type 1 cable is used. Those charts are two-dimensional arrays with the number of wiring closets listed across the horizontal heading and number of MAUs listed down the vertical heading. Each entry in the array provides, in feet, the sum of the longest lobe cable and adjusted ring length permitted. For example, a two-wiring-closet network with two MAUs can have a total of 1192 feet of cable, made up of the sum of the longest lobe and the adjusted ring length. For a network with two wiring closets and three MAUs, the maximum cable distance decreased to 1163 feet, whereas a three-wiring-closet three-MAU network has a maximum cable distance of 1148 feet.

13.6.5 High speed Token-Ring

Based upon the need for additional bandwidth by users operating Token-Ring networks, the IEEE 802.5 working group developed a standard for 100 Mbps operations on copper cable that is commonly referred to as high-speed Token-Ring (HSTR). HSTR represents the latest revision to Token-Ring since its original specification was developed in 1985.

The first version of Token-Ring specified data rates of 1 and 4 Mbps; however, the lower rate was never actually developed by vendors into products. In 1989 the 16 Mbps version of Token-Ring was introduced and that remained the maximum operating rate through 1999. As an interim measure to provide additional capability to Token-Ring users, the IEEE defined switching and a full-duplex transmission capability at 16 Mbps between stations directly connected to a switch and between switches. This capability was added to Token-Ring during the mid-1990s, with the resulting array of interconnection options called dedicated Token-Ring (DTR) since the technology was limited to connections to dedicated Token-Ring switch ports.

When developing the HSTR standard, the IEEE noted that organizations operating Token-Ring networks commonly had capacity on existing 16 Mbps rings. However, bottlenecks were appearing on backbone rings used to interconnect rings or in switches that were developed to provide an interconnection capability between rings. Thus, the HSTR standard was developed as a switch-only standard for interconnecting switches or connecting servers to a switch.

Characteristics

High-speed Token-Ring adapted the 100BASE-TX physical layer to support both 100-ohm Category 5 UTP and 150-ohm shielded Category 5 twisted pair (STP) cabling. By supporting STP, HSTR is compatible with IBM Type 1 cabling, which allows the new technology to be used in many locations without having to replace the existing cable infrastructure. Although the HSTR standard does not include support for auto-negotiation in the manner that Fast Ethernet does through the Nways specification, equipment vendors can implement an auto-negotiation feature while adhering to the HSTR standard. This allows vendors to develop products that can automatically determine if they should insert at 4, 16 or 100 Mbps. Thus, you can purchase

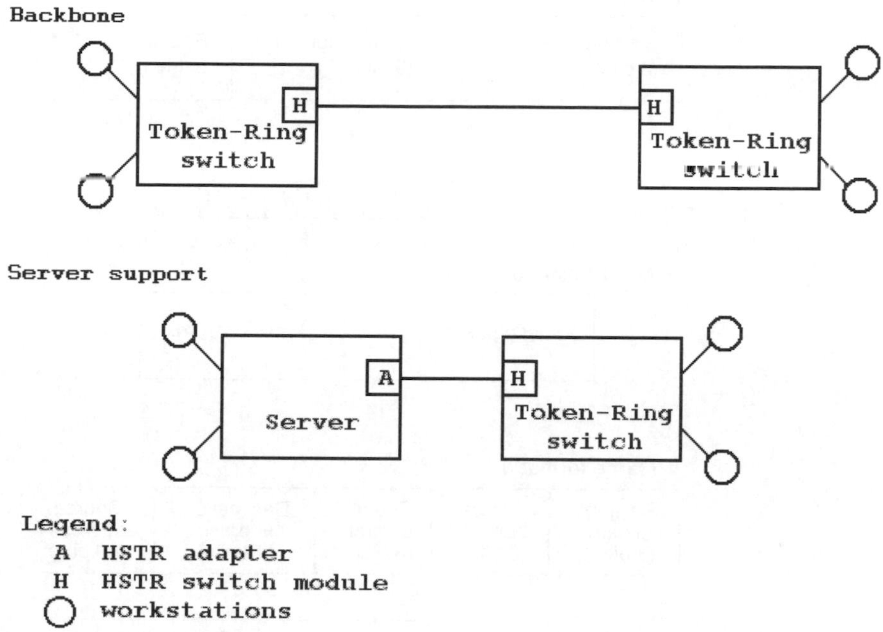

Figure 13.35 High-speed Token-Ring applications

an adapter card and use it at 16 Mbps today and have the option to use it at 100 Mbps at a later date if you upgrade a switch to support 100 Mbps HSTR.

Figure 13.35 illustrates the two connectivity methods supported by the HSTR standard. The upper portion of Figure 13.35 illustrates the use of HSTR modules in Token-Ring switches as a mechanism to interconnect switches at 100 Mbps. The lower portion of Figure 13.35 illustrates the use of a HSTR adapter card in a server and a HSTR switch module to allow the server to respond to queries at 100 Mbps.

Although it was necessary to indicate the placement of Token-Ring switches to understand the manner by which HSTR applications are oriented, we will defer a discussion of LAN switches until Chapter 14.

13.6.6 Transmission formats

Three types of transmission formats are supported on a Token-Ring network: token, abort and frame.

Token

The token format, as illustrated in the top of Figure 13.36, is the mechanism by which access to the ring is passed from one computer attached to the network to another device connected to the network. Here the token format consists of three bytes, of which the starting and ending delimiters are used to

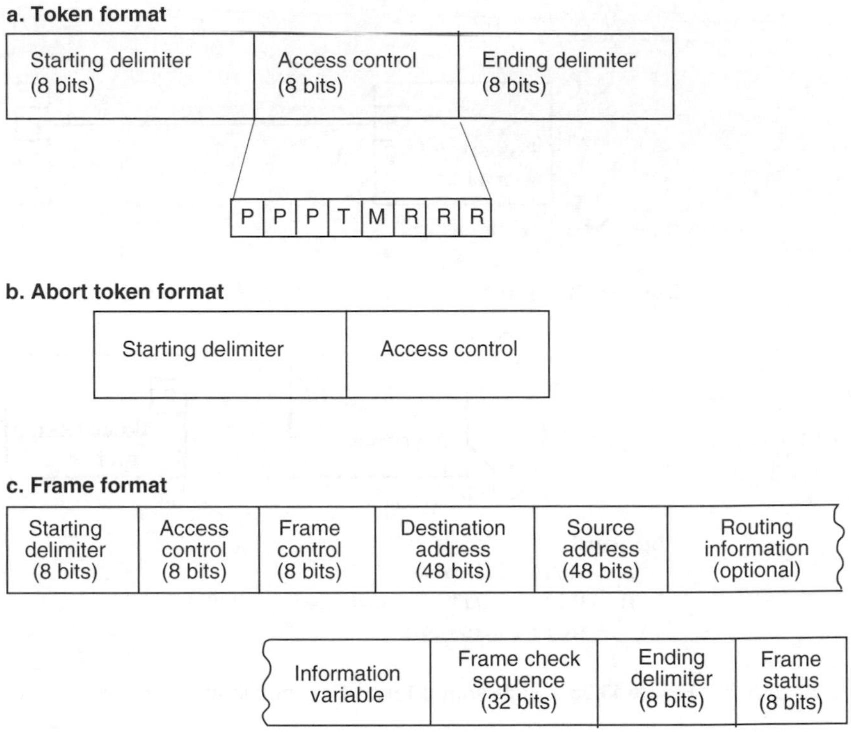

Figure 13.36 Token, abort, and frame formats (P: priority bits, T: token bit, M: monitor bit, R; reservation bits)

indicate the beginning and end of a token frame. The middle byte of a token frame is an access control byte. Three bits are used as a priority indicator, three bits are used as a reservation indicator, and one bit is used for the token bit, and another bit position functions as the monitor bit.

When the token bit is set to a binary 0, it indicates that the transmission is a token. When it is set to a binary 1, it indicates that data in the form of a frame is being transmitted.

Abort

The second Token-Ring frame format signifies an abort token. In actuality, there is no token, since this format is indicated by a starting delimiter followed by an ending delimiter. The transmission of an abort token is used to abort a previous transmission. The format of an abort token is illustrated in Figure 13.36(b).

Frame

The third type of Token-Ring frame format occurs when a station seizes a free token. At that time the token format is converted into a frame which includes the addition of frame control, addressing data, an error detection field and a frame status field. The format of a Token-Ring frame is illustrated in Figure 13.36(c). By examining each of the fields in the frame, we will also examine the token and token abort frames, due to the commonality of fields between each frame.

Starting/ending delimiters

The starting and ending delimiters mark the beginning and ending of a token or frame. Each delimiter consists of a unique code pattern which identifies it to the network.

Non-data symbols

Under Manchester and Differential Manchester encoding there are two possible code violations. Each code violation produces what is known as a non-data symbol, and it is used in the Token-Ring frame to denote starting and ending delimiters similar to the use of the flag in an HDLC frame. However, unlike the flag whose bit composition 01111110 is uniquely maintained by inserting a 0 bit after every sequence of five set bits and removing a 0 following every sequence of five set bits, Differential Manchester encoding maintains the uniqueness of frames by the use of non-data J and non-data K symbols. This eliminates the bit stuffing operations required by HDLC.

The two non-data symbols each consist of two half-bit times without a voltage change. The J symbol occurs when the voltage is the same as that of the last signal, and the K symbol occurs when the voltage becomes opposite of that of the last signal. Figure 13.37 illustrates the occurrence of the J and K non-data symbols based upon different last bit voltages.

The start delimiter field marks the beginning of a frame. The composition of this field is the bits and non-data symbols JK0JK000. The end delimiter field marks the end of a frame, as well as denoting whether or not the frame is the last frame of a multiple frame sequence using a single token or if there are

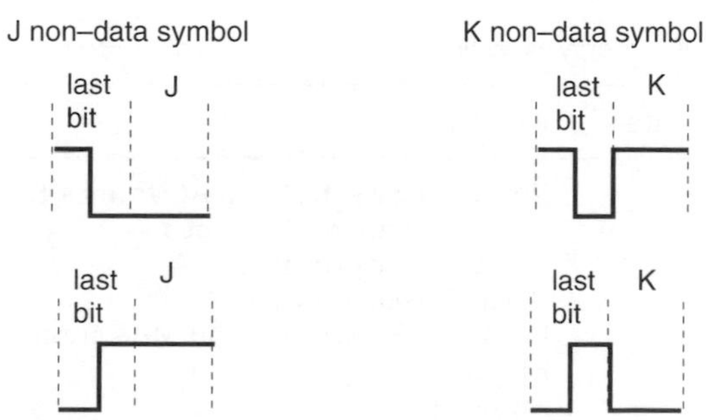

Figure 13.37 J and K non-data symbol composition. J and K non-data symbols are distinct code violations that can be mistaken for data

additional frames following this frame. The format of the end delimiter field is JK1JK1IE, where I is the intermediate frame bit. If I is set to 0, this indicates that it is the last frame transmitted by a station. If I is set to 1, this indicates that additional frames follow this frame. E is an Error-Detected bit. The E bit is initially set to 0 by the station transmitting a frame, token, or abort sequence. As the frame circulates the ring, each station checks the transmission for errors. On detection of a Frame Check Sequence (FCS) error, inappropriate non-data symbol, illegal framing, or another type of error, the first station detecting the error will set the E bit to a value of 1. Since stations keep track of the number of times they set the E bit to a value of 1, it becomes possible to use this information as a guide to locating possible cable errors. For example, if one workstation accounted for a very large percentage of E bit settings in a 72-station network, there is a high degree of probability that there is a problem with the lobe cable to that workstation. The problem could be a crimped cable or a loose connector and represents a logical place to commence an investigation in an attempt to reduce E bit errors.

Access control

The second field in both token and frame formats is the access control byte. As illustrated at the top of Figure 13.36, this byte consists of four subfields, and it serves as the controlling mechanism for gaining access to the network. When a free token circulates the network, the access control field represents one-third of the length of the frame since it is prefixed by the start delimiter and suffixed by the end delimiter.

The lowest priority that can be specified by the priority bits in the access control byte is 0 (000), whereas the highest is seven (111), providing eight levels of priority. Table 13.8 lists the normal use of the priority bits in the access control field. Workstations have a default priority of three, whereas bridges have a default priority of four.

To reserve a token, a workstation inserts its priority level in the priority reservation subfield. Unless another station with a higher priority bumps the requesting station, the reservation will be honored and the requesting station will obtain the token. If the token bit is set to 1, this serves as an indication that a frame follows instead of the ending delimiter.

Table 13.8 Priority bit settings

Priority bits	Priority
000	Normal user priority, MAC frames that do not require a token and response type MAC frames
001	Normal user priority
010	Normal user prioity
011	Normal user priority and MAC frames that require tokens
100	Bridge
101	Reserved
110	Reserved
111	Specialized station management

A station that needs to transmit a frame at a given priority can use any available token that has a priority level equal to or less than the priority level of the frame to be transmitted. When a token of equal or lower priority is not available, the ring station can reserve a token of the required priority through the use of the reservation bits. In doing so, the station must follow two rules. First, if a passing token has a higher priority reservation than the reservation level desired by the workstation, the station will not alter the reservation field contents. Secondly, if the reservation bits have not been set or if they indicate a lower priority than that desired by the station, the station can now set the reservation bits to the required priority level.

Once a frame is removed by its originating station, the reservation bits in the header will be checked. If those bits have a non-zero value, the station must release a non-zero priority token, with the actual priority assigned based on the priority used by the station for the recently transmitted frame, the reservation bit settings received on the return of the frame, and any stored priority.

On occasion, the Token-Ring protocol will result in the transmission of a new token by a station prior to that station having the ability to verify the settings of the access control field in a returned frame. When this situation arises, the token will be issued according to the priority and reservation bit settings in the access control field of the transmitted frame.

The monitor bit

The monitor bit is used to prevent a token with a priority exceeding zero or a frame from continuously circulating on the Token-Ring. This bit is transmitted as a 0 in all tokens and frames, except for a device on the network which functions as an active monitor, and it thus obtains the capability to inspect and modify that bit.

When a token or frame is examined by the active monitor, it will set the monitor bit to a 1 if it was previously found to be set to 0. If a token or frame is found to have the monitor bit already set to 1, this indicates that the token or frame has already made at least one revolution around the ring and an error condition has occurred, usually caused by the failure of a station to remove its transmission from the ring, or the failure of a high-priority station to seize a token. When the active monitor finds a monitor bit set to 1, it assumes that an error condition has occurred. The active monitor then purges the token or frame and releases a new token onto the ring. Now that we have an understanding of the role of the monitor bit in the access control field and the operation of the active monitor on that bit, let us focus our attention upon the active monitor.

Active monitor

The active monitor is the device that has the highest address on the network. All other stations on the network are considered as standby monitors and watch the active monitor.

As previously explained, the function of the active monitor is to determine if a token or frame is continuously circulating the ring in error. To accomplish this, the active monitor sets the monitor count bit as a token or frame goes by. If a destination workstation fails or has its power turned off, the frame will circulate back to the active monitor, where it is then removed from the network. In the event that the active monitor should fail or be turned off, the standby monitors watch the active monitor by looking for an active monitor frame. If one does not appear within seven seconds, the standby monitor that has the highest network address then takes over as the active monitor.

Frame control

The frame control field informs a receiving device on the network of the type of frame that was transmitted and how it should be interpreted. Frames can be either logical link control (LLC) or reference physical link functions according to the IEEE 802.5 media access control (MAC) standard. A media access control frame carries network control information and responses, and a logical link control frame carries data.

The eight-bit frame control field has the format FFZZZZZZ, where FF are frame definition bits. The top part of Table 13.9 indicates the possible settings of the frame bits and the assignment of those settings. The ZZZZZZ bits convey media access control (MAC) buffering information when the FF bits are set to 00. When the FF bits are set to 01 to indicate an LLC frame, the ZZZZZZ bits are split into two fields, designated rrrYYY. Currently, the rrr bits are reserved for future use and are set to 000. The YYY bits indicate the priority of the logical link control (LLC) data. The lower portion of Table 13.9 indicates

Table 13.9 Frame control field subfields

F bit settings	Assignment
00	MAC frame
01	LLC frame
10	Undefined (reserved for future use)
11	Undefined (reserved for future use)

Z bit settings	Assignment*
000	Normal buffering
001	Remove ring station
010	Beacon
011	Claim token
100	Ring purge
101	Active monitor present
110	Standby monitor present

* When F bits set to 00, Z bits are used to notify an adapter that the frame is to be express buffered.

48 bits

I/G	U/L	Reserved	FAI	31 bits

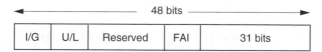

Figure 13.38 Destination address subfields (I/G: individual or group bit address identifier, U/L: universally or locally administered bit identifier, FAI: functional address indicator). The reserved field contains the manufacturer's identification in 22 bits represented by 6 hex digits

the value of the Z bits when used in MAC frames to notify a Token-Ring adapter that the frame is to be expressed buffered.

Destination address

Although the IEEE 802.5 standard is similar to the 802.3 standard in its support of 16- and 48-bit address fields, almost all implementations of Token-Ring use 48-bit addresses. The destination address field is made up of five subfields as illustrated in Figure 13.38. The first bit in the destination address identifies the destination as an individual station (bit set to 0) or as a group (bit set to 1) of one or more stations. The latter provides the capability for a message to be broadcast to a group of stations.

Universally and locally administered addresses

Similar to the IEEE 802.3 standard, universally administered addresses are assigned in blocks of numbers by the IEEE to each manufacturer of Token-Ring equipment, with the manufacturer encoding a unique address into each adapter card. Locally administered addressing permits users to temporarily override universally administered addressing and can be used to obtain addressing flexibility.

Functional address indicator

The functional address indicator subfield in the destination address identifies the function associated with the destination address, such as a bridge, active monitor, or configuration report server.

The functional address indicator indicates a functional address when set to 0 and the I/G bit position is set to a 1, the latter indicating a group address. This condition can only occur when the U/L bit position is also set to a 1, and it results in the ability to generate locally administered group addresses that are called functional addresses.

Source address

The source address field always represents an individual address which specifies the adapter card responsible for the transmission. The source address

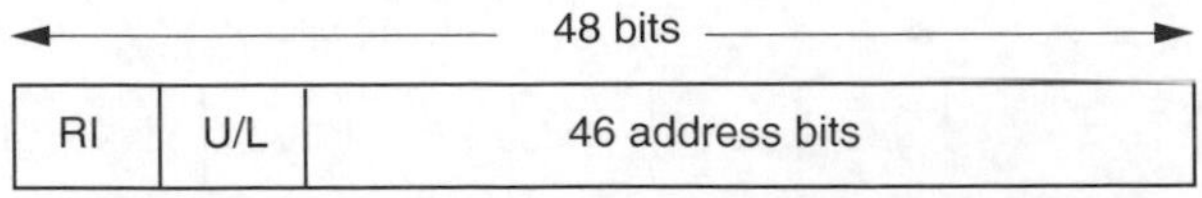

Figure 13.39 Source address field (RI: routing information bit identifier, U/L: universally or locally administered bit identifier). The 46 address bits consist of 22 manufacturer identification bits and 24 universally administered bits when the U/L bit is set to 0. If set to 1, a 31-bit locally administered address is used with the manufacturer's identification bits set to 0

field consists of three major subfields as illustrated in Figure 13.39. When locally administered addressing occurs, only 24 bits in the address field are used, because the 22 manufacturer identification bit positions are not used.

The routing information bit identifier identifies the fact that routing information is contained in an optional routing information field. This bit is set when a frame is routed across a bridge using IBM's source routing technique, which is described in detail in Chapter 14.

Routing information

The routing information field is optional and is included in a frame when the RI bit of the source address field is set. Figure 13.40 illustrates the format of the optional routing information field. If this field is omitted, the frame cannot leave the ring where it originated under IBM's source routing bridging method. Under transparent bridging, the frame can be transmitted onto another ring. Both source routing bridging and transparent bridging are covered in detail in Chapter 14. The routing information field is of variable length, and contains a control subfield and one or more two-byte route designator fields when included in a frame, as the latter are required to control the flow of frames across one or more bridges.

The maximum length of the routing information field (RIF) supported by IBM is 18 bytes. Since each RIF field must contain a two-byte routing control field, this leaves a maximum of 16 bytes available for use by up to eight route designators. As illustrated in Figure 13.40, each two-byte route designator consists of a 12-bit ring number and a four-bit bridge number. Thus, a maximum total of 16 bridges can be used to join any two rings in an Enterprise Token-Ring network.

Information field

The information field is used to contain Token-Ring commands and responses, as well as to carry user data. The type of data carried by the information field depends on the F bit settings in the frame type field. If the F bits are set to 00, the information field carries media access control (MAC) commands and responses that are used for network management operations. If the F bits are set to 01, the information field carries logical link control (LLC) or user data. Such data can be in the form of portions of a file being transferred on the network or an electronic mail message being routed to another workstation on the network. The information field is of variable

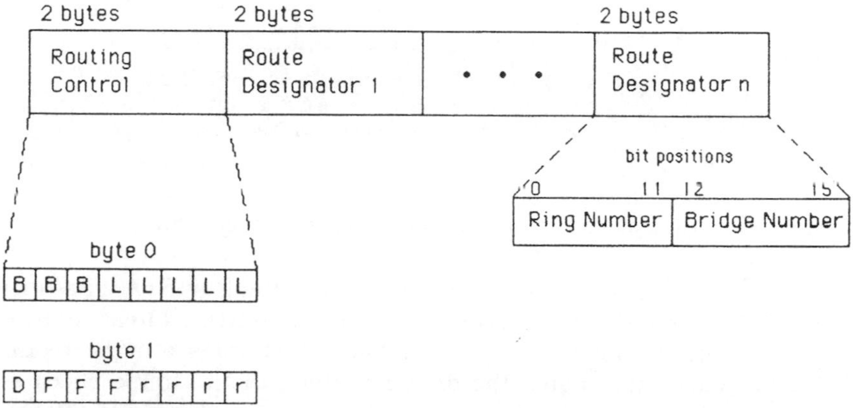

B = Broadcast Bits

- 0xx non-broadcast
- 10x all-routes broadcast
- 11x single route broadcast

L = Length Bits

Indicates the size in bytes of the routing information field

D = Direction Bit

- 0 = interpret route designators left to right
- 1 = interpret route designators right to left

FFF = Largest Frame Size of information field

- 000 up to 516 bytes
- 001 up to 1470 bytes
- 010 up to 2052 bytes
- 011 up to 4472 bytes
- 100 up to 8144 bytes
- 101 up to 11,407 bytes
- 110 up to 17,800 bytes
- 111 used in all-routes broadcast frames

r = Reserved Bits

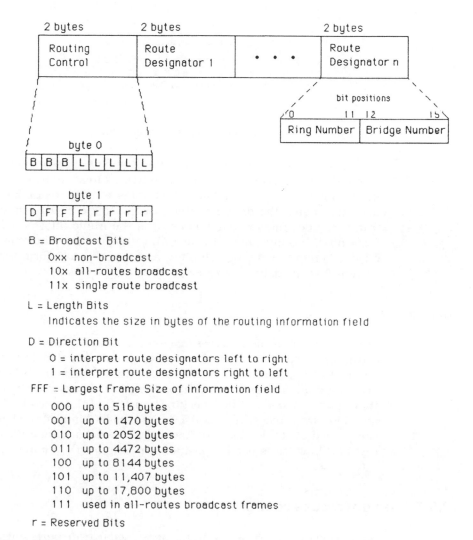

Figure 13.40 Routing information field

length and can be considered to represent the higher level protocol enveloped in a Token-Ring frame.

In the IBM implementation of the IEEE 802.5 Token-Ring standard, the maximum length of the information field depends on the Token-Ring adapter used and the operating rate of the network. Token-Ring adapters with 64 kbytes of memory can handle up to 4.5 kbytes on a 4 Mbps network, and up to 18 kbytes on a 16 Mbps network or at 100 Mbps on a HSTR connection.

Frame check sequence

The frame check sequence field contains four bytes which provide the mechanism for checking the accuracy of frames flowing on the network. The cyclic

A	C	r	r	A	C	r	r

A = Address–Recognized Bits
B = Frame–Copied Bits
r = Reserved Bits

Figure 13.41 Frame status field. The frame status field denotes whether the destination address was recognized and whether the frame was copied. Since this field is outside of CRC checking its subfields are duplicated for accuracy

redundancy check data included in the frame check sequence field cover the frame control, destination address, source address, routing information and information fields. If an adapter computes a cyclic redundancy check that does not match the data contained in the frame check sequence field of a frame, the destination adapter discards the frame information and sets an error bit (E bit) indicator. This error bit indicator actually represents a ninth bit position of the ending delimiter, and it serves to inform the transmitting station that the data were received in error.

Frame status

The frame status field serves as a mechanism to indicate the results of a frame's circulation around a ring to the station that initiated the frame, Figure 13.41 indicates the format of the frame status field. The frame status field contains three subfields that are duplicated for accuracy purposes, since they reside outside of CRC checking. One field (A) is used to denote whether an address was recognized, and a second field (C) indicates whether the frame was copied at its destination. Each of these fields is one bit in length. The third field, which is two bit positions in length (rr), is currently reserved for future use.

13.6.7 Medium access control

As previously discussed, a MAC frame is used to transport network commands and responses. As such, the MAC layer controls the routing of information between the LLC and the physical network. Examples of MAC protocol functions include the recognition of adapter addresses, physical medium access management, and message verification and status generation.

A MAC frame is indicated by the setting of the first two bits in the frame control field to 00. When this situation occurs, the contents of the information field which carries MAC data are known as a vector. Table 13.10 lists currently defined vector identifier codes for six MAC control frames defined under the IEEE 802.5 standard.

MAC control

As discussed earlier in this section, each ring has a station known as the active monitor which is responsible for monitoring tokens and taking action

Table 13.10 Vector identifier codes

Code value	MAC frame meaning
010	Beacon (BCN)
011	Claim token (CL TK)
100	Purge MAC frame (PRG)
101	Active monitor present (AMP)
110	Standby monitor present (SMP)
111	Duplicate address test (DAT)

to prevent the endless circulation of a token on a ring. Other stations function as standby monitors and one such station will assume the functions of the active monitor if that device should fail or is removed from the ring. For the standby monitor with the highest network address to take over the functions of the active monitor, the standby monitor needs to know there is a problem with the active monitor. If no frames are circulating on the ring but the active monitor is operating, the standby monitor might falsely presume the active monitor has failed. Thus, the active monitor will periodically issue an active monitor present (AMP) MAC frame. This frame must be issued every 7 seconds to inform the standby monitors that the active monitor is operational. Similarly, standby monitors periodically issue a standby monitor present (SMP) MAC frame to denote that they are operational.

If an active monitor fails to send an AMP frame within the required time interval, the standby monitor with the highest network address will continuously transmit claim token (CL TK) MAC frames in an attempt to become the active monitor. The standby monitor will continue to transmit CL TK MAC frames until one of three conditions occurs: a MAC CL TK frame is received and the sender's address exceeds the standby monitor's station address; a MAC beacon (BCN) frame is received; a MAC purge (PRG) frame is received.

If one of the preceding conditions occurs, the standby monitor will cease its transmission of CL TK frames and resume its standby function.

Purge frame

If a CL TK frame issued by a standby monitor is received back without modification, and neither a beacon nor purge frame is received in response to the CL TK frame, the standby monitor becomes the active monitor and transmits a purge MAC frame. The purge frame is also transmitted by the active monitor each time a ring is initialized or if a token is lost. Once a purge frame is transmitted, the transmitting device will place a token back on the ring.

Beacon frame

In the event of a major ring failure, such as a cable break or the continuous transmission by one station (known as jabbering), a beacon frame will be transmitted. The transmission of BCN frames can be used to isolate ring faults.

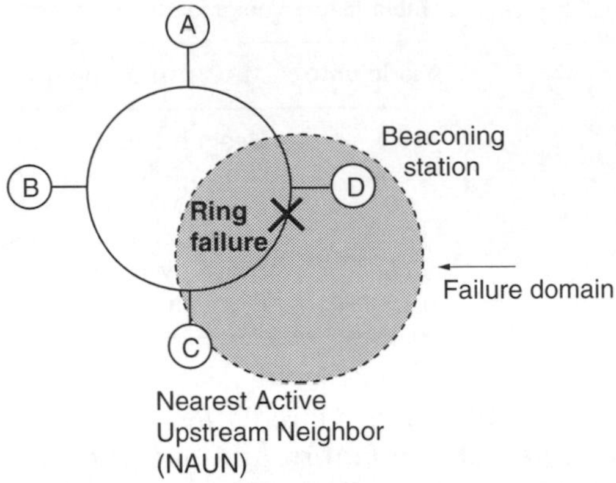

Figure 13.42 Beaconing. A beaconing frame indicates a failure occurring between the beaconing station and its nearest active upstream neighbor, an area referred to as a failure domain

For an example of the use of a beacon frame, consider Figure 13.42 in which a cable fault results in a ring break. When a station detects a serious problem with the ring, such as the failure to receive a frame or token, it transmits a beacon frame. That frame defines a failure domain which consists of the station reporting the failure via the transmission of a beacon and its nearest active upstream neighbor (NAUN), as well as everything between the two.

If a beacon frame makes its way back to the issuing station, that station will remove itself from the ring and perform a series of diagnostic tests to determine if it should attempt to reinsert itself into the ring. This procedure ensures that a ring error caused by a beaconing station can be compensated for by having that station remove itself from the ring. Since beacon frames indicate a general area where a failure occurred, they also initiate a process known as auto-reconfiguration. The first step in the auto-reconfiguration process is the diagnostic testing of the beaconing station's adapter. Other steps in the auto-reconfiguration process include diagnostic tests performed by other nodes located in the failure domain in an attempt to reconfigure a ring around a failed area.

Duplicate address test frame

The last type of MAC command frame is the duplicate address test (DAT) frame. This frame is transmitted during a station initialization process when a station joins a ring. The station joining the ring transmits a MAC DAT frame with its own address in the frame's destination address field. If the frame returns to the originating station with its address-recognized (A) bit in the frame control field set to 1, this means that another station on the ring is assigned that address. The station attempting to join the ring will send a message to the ring network manager concerning this situation and will not join the network.

13.6.8 Logical link control

In concluding this section, we will examine the flow of information within a Token-Ring network at the logical link control (LLC) sublayer. Similar to its use on Ethernet, the LLC sublayer is responsible for performing routing, error control and flow control. In addition, this sublayer is responsible for providing a consistent view of a LAN to upper OSI layers, regardless of the type of media and protocols used on the network.

Figure 13.43 illustrates the format of an LLC frame which is carried within the information field of the Token-Ring frame. As previously discussed in this section, the setting of the first two bits in the frame control field of a Token-Ring frame to 01 indicates that the information field should be interpreted as an LLC frame. The portion of the Token-Ring frame which carries LLC information is known as a protocol data unit, and it consists of either three or four fields, depending on the inclusion or omission of an optional information field. The control field is similar to the control field used in the HDLC protocol, and it defines three types of frames: information (I-frames) are used for

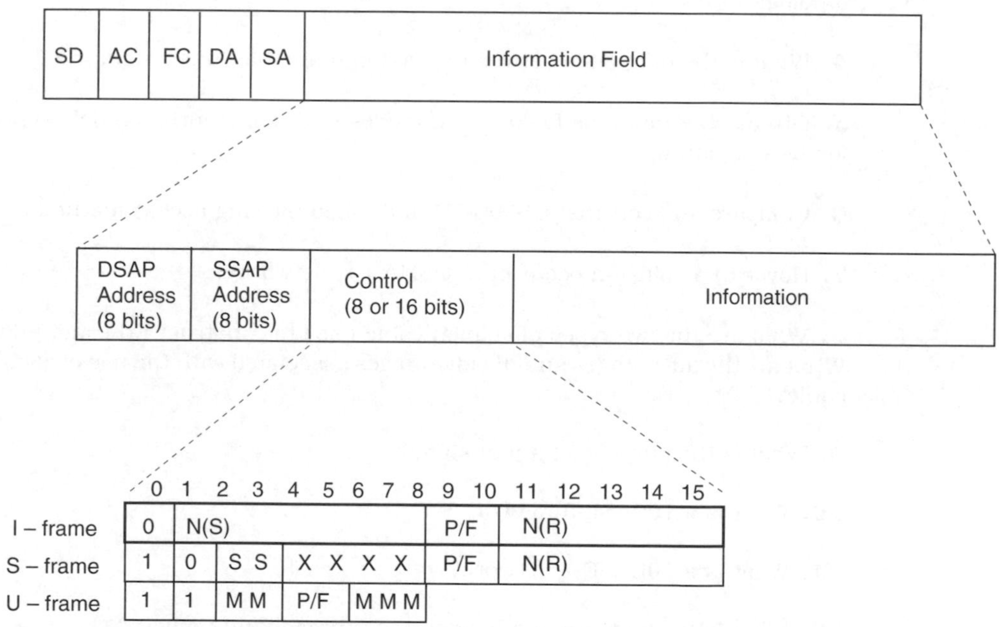

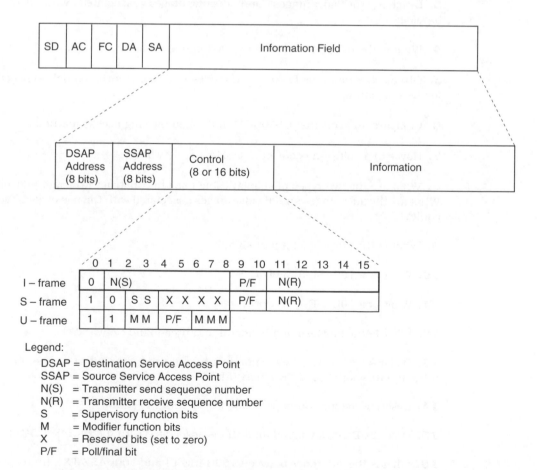

Legend:
DSAP = Destination Service Access Point
SSAP = Source Service Access Point
N(S) = Transmitter send sequence number
N(R) = Transmitter receive sequence number
S = Supervisory function bits
M = Modifier function bits
X = Reserved bits (set to zero)
P/F = Poll/final bit

Figure 13.43 Logical link control frame format

sequenced messages, supervisory (S-frames) are used for status and flow control, and unnumbered (U-frames) are used for unsequenced, unacknowledged messages. Both Service Access Points (SAPs) and Destination Service Access Points (DSAPs) function in the same manner as previously covered in our examination of Ethernet. In addition, Token-Ring, like Ethernet, supports the same types of connectionless and connection-oriented services at the data link layer.

13.7 REVIEW QUESTIONS

1. Discuss the differences between LANs and WANs with respect to their geographical area of coverage, data transmission and error rates, ownership, government regulation and data routing.

2. Describe and discuss five possible LAN applications.

3. Describe the advantages and disadvantages associated with five LAN topologies.

4. What is the difference between broadband and baseband signaling?

5. Discuss the rationale for using Manchester or Differential Manchester coding for LAN signaling.

6. Compare and contrast CSMA/CD and token passing access methods.

7. How can a collision occur on a CSMA/CD network?

8. What are the two types of coaxial cable used to construct Ethernet networks? What are the advantages and disadvantages associated with the use of each type of cable?

9. What is the purpose of a jam signal?

10. What is a 10BASE-5 network?

11. What is a 10BASE-2 network?

12. What is the purpose of a headend in a broadband network?

13. Discuss the three key components required to construct the physical infrastructure for a 10BASE-T network.

14. Describe the functions performed by a 10BASE-T hub.

15. Describe Ethernet signal limitations associated with the so-called '5-4-3' rule.

16. What is the difference between 100BASE-T4 and 100BASE-TX with respect to cable use and signaling method employed?

17. Why is NWay important?

18. Why are there dual connectors labeled Tx and Rx on optical ports built into adapter cards and other devices that support transmission over optical fiber?

19. Describe two 100BASE-T applications.

20. What type of media does Gigabit Ethernet operate over?

21. What is a buffered distributor?

22. What is the difference between universally administered addressing and locally administered addressing?

23. Why must an Ethernet frame have a minimum length of 64 bytes in an adapter and 72 bytes when flowing on a LAN?

24. What is the purpose of an Ethernet-SNAP frame?

25. What is meant by the term connectionless transmission?

26. How can software determine a particular type of Ethernet frame as the frame flows on a network?

27. Why are carrier extensions required on certain types of Gigabit Ethernet frames when half-duplex shared media transmission is supported?

28. What is the purpose of a Jumbo frame?

29. Draw a two MAU Token-Ring network illustrating ring-in and ring-out connections.

30. How does a Token-Ring MAU provide a redundant ring path?

31. Why is the use of RJ-45 connectors more popular than the use of IBM cabling system data connectors for cabling Token-Ring network components?

32. What type of application is 100 Mbps Token-Ring designed to be used for?

33. What is the purpose of an abort token?

34. What are J and K symbols?

35. What is the purpose of the monitor bit? Who is responsible for monitoring the monitor bit?

36. How is a MAC frame distinguished from an LLC frame on a Token-Ring network?

37. What is the purpose of the E bit in a Token-Ring frame?

38. What is the purpose of a beacon frame?

39. What is a failure domain?

40. What is the purpose of a duplicate address test frame?

14

BASIC LAN INTERNETWORKING

In Chapter 5 a brief overview of bridge and router operations was presented, along with information concerning the functionality of other local area network hardware and software components. That chapter deferred until now a detailed examination of bridge and router operating methods and their network utilization. In this chapter we will focus our attention on the operation of bridges and routers. Since the operation of gateways is highly related to the different types of network architecture that they are designed to interface, we will defer a discussion of this communications device until we examine different network architectures in Chapter 16.

14.1 BRIDGE OPERATIONS

Bridges operate by examining MAC layer addresses, using the destination and source addresses within a frame as a decision criteria to make their forwarding decisions. Operating at the MAC layer, bridges are not addressed and must therefore examine all frames that flow on a network.

14.1.1 Types of bridge

There are two primary methods used by bridges to connect local area networks: transparent or self-learning, and source routing. Transparent bridges were originally developed to support the connection of Ethernet networks, whereas source routing bridges support Token-Ring network operations.

Transparent bridges

The first type of bridge we will examine is referred to as a transparent bridge. This is the most basic type of bridge and, as its name implies, its use is transparent to the networks it connects. That is, a transparent bridge is a 'plug and play' device that does not require the setup or configuration of any hardware or software. You just plug cables from each transparent bridge port into an appropriate network connector and the bridge operates or 'plays'.

To illustrate the operation of a transparent bridge requires at least two networks. Thus, let's create several simple LANs so we can examine how a transparent bridge operates. The top portion of Figure 14.1 illustrates the use of a transparent bridge to connect workstations on two LANs. For simplicity we will assume that source addresses of stations on each LAN are represented by the letter contained in the squares attached to each network. In actuality source addresses are contained in a 48-bit field represented by 12 hex characters, of which the first six represent the manufacturer of the network adapter card, while the last six represent a unique number associated with the manufacturer of the card.

A transparent bridge operates in promiscuous mode, which means that it reads every frame transmitted on each LAN it is connected to. In the network example illustrated in Figure 14.1, a two-port bridge is shown connected to two LANs, with three stations shown connected to each network. Once the bridge is connected to each network and powered on, it examines each frame and either discards or forwards it. This decision is performed by the bridge using the destination address of the frame and its comparison to entries in an address-port table it maintains. That table is constructed by the bridge noting the source address contained in each frame received on a port and using those addresses to construct the table. For example, assume a frame with source address A and destination address B is received on port 1. If this is the first frame received on port 1, the bridge enters address A into its address-port

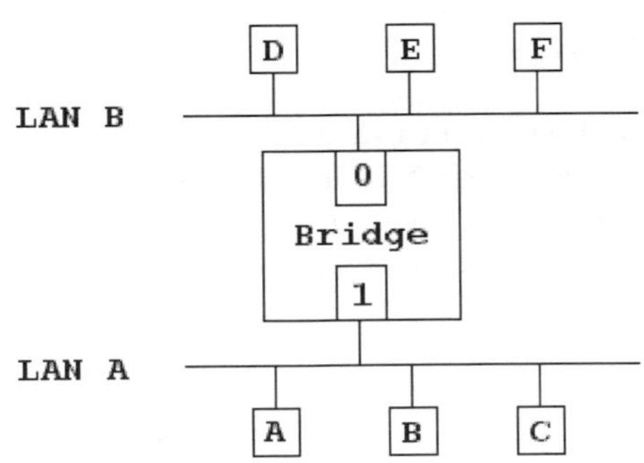

Steady state port address table

Port	Address
0	D
0	E
0	F
1	A
1	B
1	C

Figure 14.1 Using a transparent bridge to connect two networks

table. Since the frame is destined to address B, which is not in the table, the bridge at this point in time does not know where the destination resides. Thus, it 'floods' the frame, a term used to reference its transmission onto all ports other than the port it was received on. Since the bridge in Figure 14.1 is a two-port bridge, this means that the frame is forwarded onto port 0 even though the destination resides on LAN A. However, the frame will not be received by any station on LAN B and its forwarding does not affect its transmission on LAN A. Thus, the forwarding operation has no adverse effect on the two networks other than using some bandwidth on LAN B that might otherwise be used more productively. Next, let's assume station B transmits a frame to station A. The bridge reads the frame on port 1 and stores the source address in the address-port table. Since address A is already in the address-port table, the bridge knows the frame is destined to a station on the LAN connected to port 1 and discards it.

Although the previously described sequence of operations can be used as a decision criterion for forwarding or discarding frames, address tables are constructed more quickly using a reverse learning process. Due to this, transparent bridges use a backward learning algorithm.

The backward learning algorithm

Under the backward learning algorithm, a bridge examines the source address of frames flowing on each port to learn the addresses of destinations reachable on that port. Returning to Figure 14.1, eventually the bridge would associate source addresses D, E and F to port 0, and A, B and C to port 1. Then, by examining the destination address of each frame against the addresses stored in its port-address table, the bridge can determine whether to discard or forward the frame.

Frame forwarding

There are two types of forwarding operations a bridge can perform – specific and flooding. A specific forwarding operation occurs when the bridge can match a destination against an entry in its port-address table. Then, the bridge can determine the specific port onto which it should forward a frame. In comparison, flooding represents the forwarding of a frame onto all ports other than the port the frame was received on. This action is performed by a bridge when it does not have an entry in its port-address table that matches the destination address contained in a frame. Since the destination address may not have been learned at this point in time, the bridge sends the frame to all possible locations via flooding.

To illustrate how a flooding operation can occur, consider Figure 14.2 in which a second bridge was added to the network previously illustrated in Figure 14.1. In this network expansion the second bridge is used to connect LANs C and D to LAN A, which in effect results in LANs C and D being linked to LAN B.

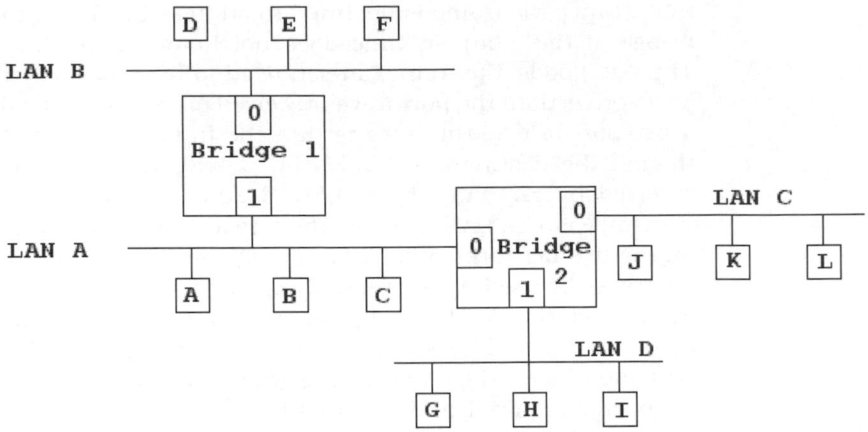

Figure 14.2 An expanded network using two bridges to interconnect four LANs

Suppose station D on LAN B transmits a frame to station G on LAN D. If station G did not previously transmit any frames to LAN A or LAN B, its address is only known to bridge 2. Thus, the port-address table in bridge 1 does not have an entry for address G. This means that bridge 1 forwards the frame onto port 1 where it is received by port 0 of bridge 2. If station G was active on port 1, bridge 2 would know to forward the frame onto port 1. However, if station G had just powered up, bridge 2 would not have an entry in its port-address table. Thus, bridge 2 would then flood the frame, forwarding it onto both ports 1 and 2 which represent all connections other than the connection the frame was received on.

Table entry considerations

As networks connected via bridges add workstations, this results in an expansion of the number of entries in their port-address tables. Although a variety of routines, to include starting entries in memory and using a hashing algorithm to perform matching or attempted matching of destination addresses against table entries, has significantly improved search time, as the number of entries increases so does the search time. This in turn can degrade bridge performance and even result in the dropping of frames if buffer areas in the bridge fill as it temporarily stores frames while performing a searching algorithm against its tables. Recognizing this problem, as well as the fact that a topology is dynamic and stations can be moved or powered on and off throughout the day, a bridge will include the current time with the entries in its port-address table. When a frame with a previously known destination is read, the bridge updates only the time associated with the address. Thus, the bridge maintains a last used time that is associated with each address stored in memory. Then, it periodically uses a timer to initiate a process that purges all entries older than a predefined time. This process enables port-address tables to shrink when users power off their stations or are not performing LAN-related activities. In addition, it allows the bridge to automatically adjust itself to adds, moves and changes, which is why another name for this type of bridge is the self-adjusting bridge. However, this process also results in a bridge periodically having to flood frames when a station is simply quiet for a period of time and another station needs to send data to that station.

We can summarize the operation of a bridge as follows:

1. If the destination and source addresses are on the same LAN, discard the frame.
2. If the destination and source addresses are on different LANs, forward the frame.
3. If the destination location is not known, flood the frame.

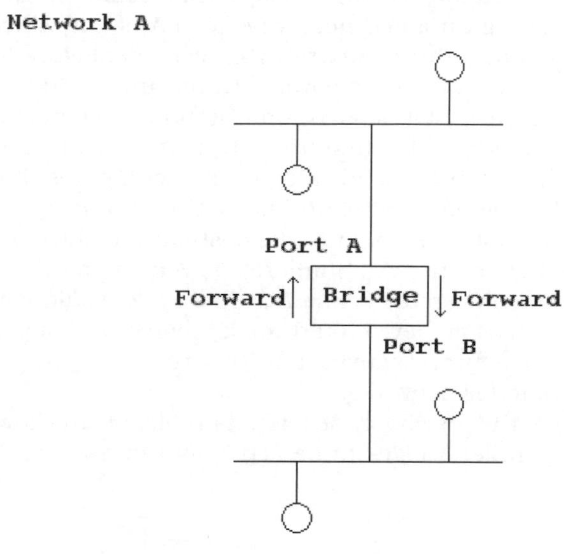

Port A Operation	Port B operation
1. Read source address of frames on LAN B to construct a table of destination addresses.	1. Read source address of frames on LAN A to construct a table of destination address.
2. Read destination address in frames and compare to addresses in port address table.	2. Read destination address in frames and compare to addresses in port address table.
3. If destination address on same LAN do nothing. If destination address in table forward on to port B. If destination address unknown flood the frame.	3. If destination address on same LAN do nothing. If destination address in table forward on to port A. If destination address unknown flood the frame.

Figure 14.3 Bridge switching operations

Figure 14.3 summarizes the operation of a two-port bridge. Although a destination address that is unknown results in the bridge flooding the frame, in actuality a two-port bridge would simply forward the frame onto the port other than the port it was received on.

Loops

Although the use of bridges plays an important role in extending networks and enabling two or more segments to operate as an entity, they also represent a critical point of failure. Thus, to increase reliability, you would probably expect network managers and administrators to use two or more bridges in parallel to interconnect pairs of networks. Depending upon the type of networks being interconnected, the use of parallel bridges may or may not be possible. If transparent bridges are used, only one bridge out of two or more used in parallel can be operational at any point in time. The reason for this is to preclude the creation of closed loops that could result in frames continuously circulating through interconnected LANs and which would literally consume their bandwidth. To illustrate this, consider Figure 14.4 which shows two LANs connected by a pair of parallel transparent bridges.

In examining Figure 14.4, let's assume station A generates an initial frame (A_1) at time $t = 0$. If this is the first frame transmitted and there are no entries in the port-address table, each bridge uses flooding, resulting in two initial frames, A_2 and A_3, being received on LAN B. Once frame A_3 is copied onto LAN B, bridge 1 sees it but does not know its destination. Thus, it copies the frame back onto LAN 1 generating A_4. Similarly, A_2 generated by bridge 1 is received by bridge 2 which generates frame A_5. This cycle would continue, which, if frames represented money, would make bankers happy. Unfortunately, frames represent network activity and this repeated cycle of activity would not be appropriate for networks.

The solution to the previously described problem is to allow only one bridge from multiple parallel bridges to be active and in a forwarding state at any

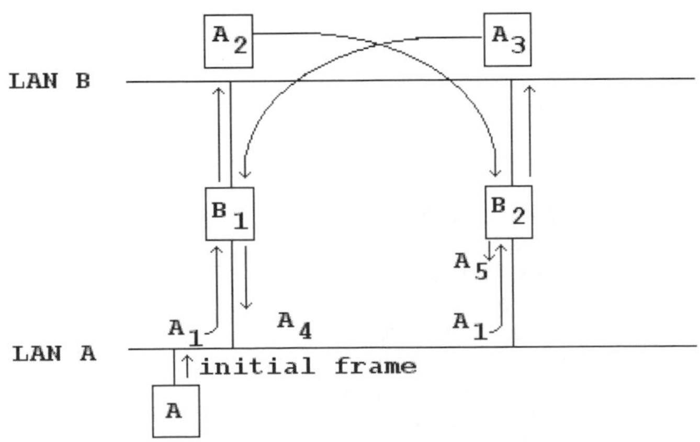

Figure 14.4 Active parallel transparent bridges would continuously copy frames between networks

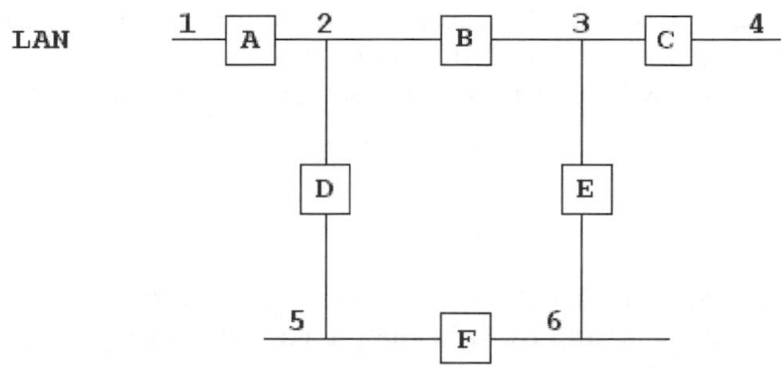

Four spanning trees are possible for this network configuration by placing either bridge B, D, E or F into a standby mode of operation.

Figure 14.5 Interconnecting six LANs with six bridges

point in time. To accomplish this, transparent bridges use a spanning tree algorithm and communicate information about their position in the tree using Bridge Protocol Data Units (BPDUs).

Through the use of the spanning tree algorithm, a unique path is logically defined between each interconnected network. Those paths preclude the use of closed loops, and any bridges in a network that would form a physically closed loop are then placed into a standby state. To illustrate the formation of a logical spanning tree upon a physical network topology, consider Figure 14.5 which illustrates six LANs interconnected via the use of six bridges. In examining Figure 14.5 note that bridges B, D, E and F result in the formation of a closed loop. By breaking the loop by placing either bridge B, D, E or F in a standby mode of operation, we can ensure there is a unique path between each LAN without the formation of a loop. Since either bridge B, D, E or F could be made inactive to satisfy providing one path from every LAN to every other LAN, we need a methodology by which bridges can communicate with each other to enable them to decide which bridge or bridges must go into a standby mode of operation. Thus, let's turn our attention to the method used by the spanning tree algorithm to construct a spanning tree without loops.

Spanning tree protocol

The problem of active loops was addressed by IEEE Committee 802 in the 802.1D standard with an intelligent algorithm known as the Spanning Tree Protocol (STP). The STP is based on graph theory and converts a loop into a tree topology by disabling a link. This action ensures there is a unique path from any node in a network to every other node. Disabled nodes are then kept in a standby mode of operation until a network failure occurs. At that time, the spanning tree protocol will attempt to construct a new tree using any of the previously disabled links.

To illustrate the operation of the spanning tree protocol, we must first become familiar with the difference between the physical and active topology of bridged networks. In addition, we should become familiar with a number of terms defined by the protocol and associated with the spanning tree algorithm. Thus, we will also review those terms prior to discussing the operation of the algorithm.

Physical versus active topology

In transparent bridging, a distinction is made between the physical and active topology resulting from bridged local area networks. This distinction enables the construction of a network topology in which inactive but physically constructed routes can be put into operation if a primary route should fail, and in which the inactive and active routes would form an illegal circular path violating the spanning tree algorithm if both routes were active at the same time.

The top of Figure 14.6 illustrates one possible physical topology of bridged networks. In this example it is assumed that the bridges between Token-Ring and Ethernet LAN while translating frames are transparent bridges. The cost *C* assigned to each bridge will be discussed later in this section. The lower portion of Figure 14.6 illustrates a possible active topology for the physical configuration shown at the top of that illustration.

When a bridge is used to construct an active path, it will forward frames through those ports used to form active paths. The ports through which frames are forwarded are said to be in a forwarding state of operation. Ports that cannot forward frames due to their operation forming a loop are said to be in a blocking state of operation.

Under the spanning tree algorithm, a port in a blocking state can be placed in a forwarding state, and it provides a path that becomes part of the active network topology. This new path must not form a closed loop, and it usually occurs due to the failure of another path, bridge component, or the reconfiguration of interconnected networks.

Spanning tree algorithm

The basis for the spanning tree algorithm is a tree structure because a tree forms a pattern of connections that has no loops. The term spanning is used because the branches of a tree structure span or connect subnetworks.

Root bridge and bridge identifiers

Similar to the root of a tree, one bridge in a spanning tree network will be assigned to a unique position in the network. Known as the root bridge, this bridge is assigned as the top of the spanning tree and it has the potential to carry the largest amount of network traffic due to its position.

Since bridges and bridge ports can be active or inactive, a mechanism is required to identify bridges and bridge ports. Each bridge in a spanning tree network is assigned a unique bridge identifier. This identifier is the MAC address on the bridge's lowest port number and a two-byte bridge priority level. The priority level is defined when a bridge is installed and functions as a bridge number. Similar to the bridge priority level, each adapter on a bridge

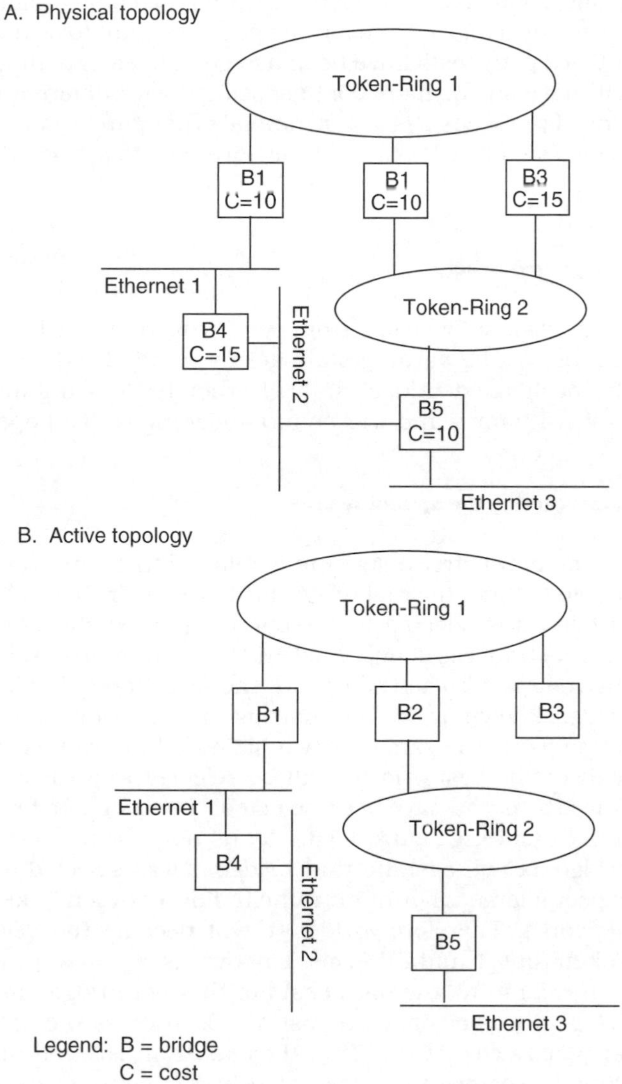

Figure 14.6 Physical versus active topology. When transparent bridges are used, the active topology cannot form a closed loop in the network

which functions as a port has a two-byte port identifier. Thus, the unique bridge identifier and port identifier enables each port on a bridge to be uniquely identified.

Path cost

Under the spanning tree algorithm, the difference in physical routes between bridges is recognized, and a mechanism is provided to indicate the preference for one route over another. That mechanism is accomplished by the ability to assign a path cost to each path. Thus, you could assign a low cost to a preferred route and a high cost to a route you only want to be used in a backup situation.

Once path costs have been assigned to each path in a network, each bridge will have one or more cost associated with different paths to the root bridge. One of those costs is lower than all other path costs. That cost is known as the bridge's root path cost and the port used to provide the least path cost toward the root bridge is known as the root port.

Designated bridge

To prevent active loops from occurring, only one bridge linking two networks can be in a forwarding state at any particular time. That bridge is known as the designated bridge, and all other bridges linking two networks will not forward frames and will be in a blocking state of operation.

Constructing the spanning tree

The spanning tree algorithm employs a three-step process to develop an active topology. First, the root bridge is identified. In Figure 14.6B we will assume that bridge 1 was selected as the root bridge. Next, the path cost from each bridge to the root bridge is determined and the minimum cost from each bridge becomes the root path cost. The port in the direction of the least path cost to the root bridge, known as the root port, is then determined for each bridge. If the root path cost is the same for two or more bridges linking LANs, then the bridge with the highest priority will be selected to furnish the minimum path cost. Once the paths have been selected, the designated ports are activated.

In examining Figure 14.6A, let us now use the cost entries assigned to each bridge. Let us assume that bridge 1 was selected as the root bridge, as we expect a large amount of traffic to flow between Token-Ring 1 and Ethernet 1 networks. Therefore, bridge 1 will become the designated bridge between Token-Ring 1 and Ethernet 1 networks.

In examining the path costs to the root bridge, note that the path through bridge 2 was assigned a cost of 10, whereas the path through bridge 3 was assigned a cost of 15. Thus, the path from Token-Ring 2 via bridge 2 to Token-Ring 1 becomes the designated bridge between those two networks. Hence, Figure 14.6B shows bridge 3 inactive by the omission of a connection to the Token-Ring 2 network. Similarly, the path cost for connecting the Ethernet 3 network to the root bridge is lower by routing through the Token-Ring 2 and Token-Ring 1 networks. Thus, bridge 5 becomes the designated bridge for the Ethernet 3 and Token-Ring 2 networks.

Bridge protocol data unit

One question that is probably in reader's minds by now is: how does each bridge know whether or not to participate in a spanned tree topology? Bridges obtain topology information by the use of Bridge Protocol Data Unit (BPDU) frames.

The root bridge is responsible for periodically transmitting a HELLO BPDU frame to all networks to which it is connected. According to the spanning tree protocol, HELLO frames must be transmitted every 1 to 10 seconds. A designated bridge will then update the path cost and timing information and forward the frame. A standby bridge will monitor the BPDUs but does not update nor forward them.

When a standby bridge is required to assume the role of the root or designated bridge, as the operational states of other bridges change, the HELLO BPDU will indicate that a standby bridge should become a designated bridge. The process by which bridges determine their role in a spanning tree network is an iterative process. As new bridges enter a network, they assume a listening state to determine their role in the network. Similarly, when a bridge is removed, another iterative process occurs to reconfigure the remaining bridges.

Source routing

Source routing is a bridging technique developed by IBM for connecting Token-Ring networks. The key to the implementation of source routing is the use of a portion of the information field in the Token-Ring frame to carry routing information and the transmission of 'discovery' packets to determine the best route between two networks.

The presence of source routing is indicated by the setting of the first bit position in the source address field of a Token-Ring frame to a binary one. When set, this indicates that the information field is preceded by a route information field (RIF) which contains both control and routing information.

The RIF field

Figure 13.40 previously illustrated the composition of a Token-Ring (RIF). This field is variable in length and is developed during a discovery process which is described later in this section.

The control field contains information which defines how information will be transferred and interpreted, as well as the size of the remainder of the RIF. The three broadcast bit positions indicate a non-broadcast, all-routes broadcast or single-route broadcast situation. A non-broadcast designator indicates a local or specific route frame. An all-routes broadcast designator indicates that a frame will be transmitted along every route to the destination station. A single-route broadcast designator is used only by designated bridges to relay a frame from one network to another. In examining the broadcast bit settings shown in Figure 13.40, readers should note that the letter X indicates a 'don't care bit' setting that can be either 1 or 0.

The length bits identify the length of the RIF in bytes, and the D bit indicates how the field is scanned, left to right or right to left. Since vendors have incorporated different memory in bridges which may limit frame sizes, the LF bits enable different devices to negotiate the size of the frame. Normally a default setting indicates a frame size of 512 bytes. Each bridge can select a number and, if supported by other bridges, that number is then used to represent the negotiated frame size. Otherwise, a smaller number used to represent a smaller frame size is selected and the negotiation process is repeated. Note that a 1500-byte frame is the largest frame size supported by Ethernet IEEE 802.3 networks. Thus, a bridge used to connect Ethernet and Token-Ring networks cannot support the use of Token-Ring frames exceeding 1500 bytes.

Up to eight route number subfields, each consisting of a 12-bit ring number and a 4-bit bridge number, can be contained in the routing information field. This permits two to eight route designators, enabling frames to traverse up to eight rings across seven bridges in a given direction. Both ring numbers and bridge numbers are expressed as hexadecimal characters, with three hex characters used to denote the ring number and one hex character used to identify the bridge number.

Operation example

To illustrate the concept behind source routing, consider the network illustrated in Figure 14.7. In this example, let us assume two Token-Ring networks are located in Atlanta and one network is located in New York. Each Token-Ring and each bridge are assigned ring and bridge numbers. For simplicity, ring numbers R1, R2 and R3 were used, although those numbers are actually represented in hexadecimal. Similarly, for simplicity bridge numbers are shown as B1, B2, B3, B4 and B5, instead of a hexadecimal character.

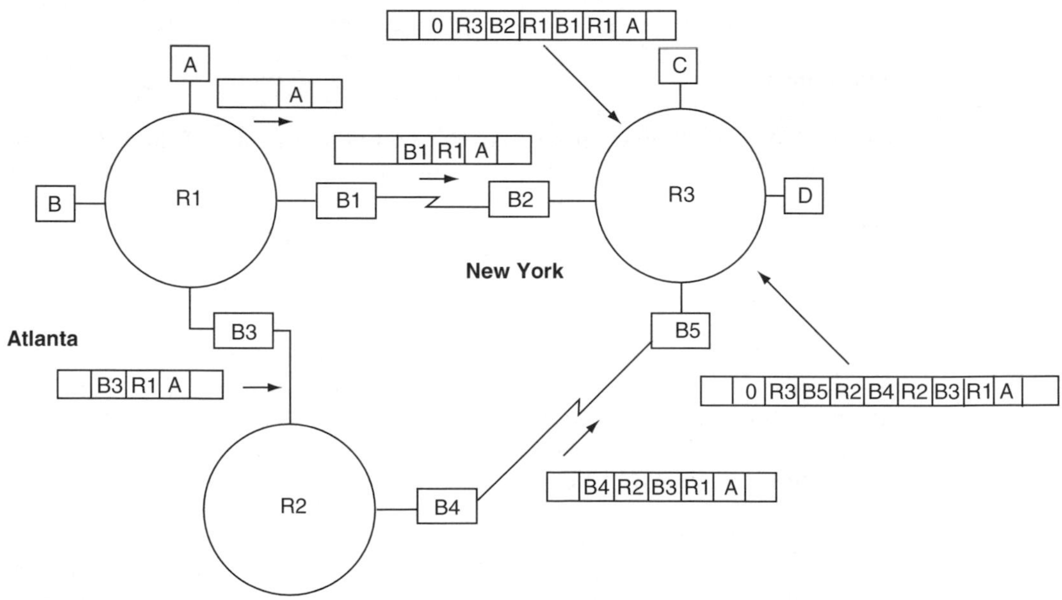

Figure 14.7 Source routing discovery operation. The route discovery process results in each bridge entering the originating ring number and its bridge number into the route information field

When a station wants to originate communications, it is responsible for finding the destination by transmitting a discovery packet to network bridges and other network stations whenever it has a message to transmit to a new destination address. Assuming that station A wishes to transmit to station C, it sends a route discovery packet which contains an empty route information field and its source address as indicated in the upper left portion of Figure 14.7. This packet is recognized by each source routing bridge in the network. When received by a source routing bridge, the bridge enters the ring number from which the packet was received and its own bridge identifier in the packet's routing information field. The bridge then transmits the packet to all its connections with the exception of the connection on which the packet was received, a process known as flooding. Depending on the topology of the interconnected networks, multiple copies of the discovery packet will more than likely reach the recipient. This is illustrated in the upper right corner of Figure 14.7, in which two discovery packets reach station C. Here one packet contains the sequence R1B1R1B2R30, where the zero indicates there is no bridging in the last ring. The second packet contains the route sequence R1B3R2B4R2B5R30. Station C then picks the best route based upon either the most direct path or the earliest arriving packet, and transmits a response to the discovery packet originator. The response indicates the specific route to use and station A then enters that route into memory for the duration of the transmission session.

Under source routing, bridges do not keep routing tables like transparent bridges. Instead, tables are maintained at each station throughout the network. Thus, each station must check its routing table to determine the route frames must traverse to reach their destination station. This routing method results in source routing using distributed routing tables in comparison to the centralized routing tables used by transparent bridges. This routing method also allows support for logical and physical loop topologies since the path to the destination is transported in a frame's RIF field. In addition, Token-Ring LANs employ an active monitor that will automatically purge a frame that attempts to circulate a ring more than once. Thus, between explicit paths carried in a frame's RIF field and the ability to prevent frames from continuously looping within a ring, source routing does not worry about loops nor does it need a spanning tree algorithm to ensure a logical closed loop is not formed.

Source routing transparent bridges

A source routing transparent bridge supports both IBM's source routing and the IEEE transparent spanning tree protocol operations. This type of bridge can be regarded as two bridges in one, and it has been standardized by the IEEE 802.1 committee as the IEEE 802.1D standard.

Operation

Under source routing, the media access control packets contain a status bit in the source field which identifies whether or not source routing is to be used for a message. If source routing is indicated, the bridge forwards the frame as a source routing frame. If source routing is not indicated, the bridge determines the destination address and processes the packet using a transparent mode of operation, using routing tables generated by a spanning tree algorithm.

Advantages

There are several advantages associated with the use of source routing transparent bridges. First and perhaps foremost, their use enables different networks to use different local area network operating systems and protocols. This capability enables you to interconnect networks developed independently of one another, and allows organization departments and branches to use LAN operating systems without restriction. Secondly, and also very importantly, source routing transparent bridges can connect Ethernet and Token-Ring networks while preserving the ability to mesh or loop Token-Ring networks. Thus, their use provides an additional level of flexibility for network construction.

14.1.2 Network utilization

Bridges can be used to form two basic network structures based on connecting LANs in series or in parallel.

Serial and sequential bridging

The top of Figure 14.8 illustrates the basic use of a bridge to interconnect two networks serially. Suppose that monitoring of each network indicates a high level of intra-network utilization. One possible configuration to reduce intra-LAN traffic on each network can be obtained by moving some stations off each of the two existing networks to form a third network. The three networks would then be interconnected through the use of an additional bridge as illustrated in the middle portion of Figure 14.8. This extension results in sequential bridging and is appropriate when intra-LAN traffic is necessary but minimal. Both serial and sequential bridging are applicable to transparent, source routing, and source routing transparent bridges which do not provide redundancy nor the ability to balance traffic flowing between networks. Each of these deficiencies can be alleviated through the use of parallel bridging. However, this bridging technique creates a loop and is only applicable to source routing and source routing transparent bridges.

Parallel bridging

The lower portion of Figure 14.8 illustrates the use of parallel bridges to interconnect two Token-Ring networks. This bridging configuration permits

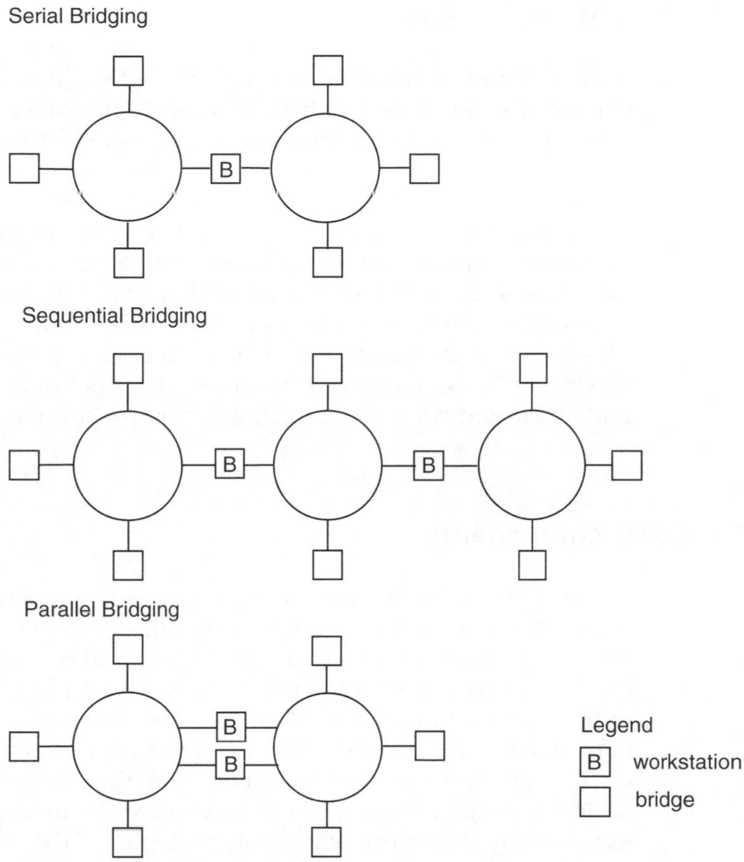

Figure 14.8 Serial, sequential, and parallel bridging

one bridge to back up the other, providing a level of redundancy to link the two networks, as well as a significant increase in the availability of one network to communicate with another. For example, assume the availability of each bridge used at the top of Figure 14.8 (serial bridging) and bottom of Figure 14.8 (parallel bridging) is 90%. The availability through two serially connected bridges would be 0.9×0.9, or 81%. In comparison, the availability through parallel bridges would be

$$1 - (\text{unavailability of bridge } 1 \times \text{unavailability of bridge } 2)$$

or $1 - 0.1 \times 0.1$, which is 99%.

The dual paths between networks also improve inter-LAN communications performance as communications between stations on each network can be load balanced. Thus, the use of parallel bridges can be expected to provide a higher level of inter-LAN communications than the use of serial or sequential bridges.

14.2 THE SWITCHING HUB

Bridges were developed for use with shared media LANs. Thus, the bandwidth constraints associated with shared media networks are also associated with bridges. That is, a conventional bridge can only route one frame at a time received on one port onto one or more ports, with multiple port frame broadcasting occurring during flooding or when a broadcast frame is received.

Recognizing the limitations associated with the operation of bridges, vendors incorporated parallel switching technology into devices known as switching hubs. This device was developed based upon technology used in matrix switches, which for decades have been successfully employed in telecommunications operations. By adding buffer memory to store address tables, frames flowing on LANs connected to different ports could be simultaneously read and forwarded via the switch fabric to ports connected to other networks.

14.2.1 Basic components

Figure 14.9 illustrates the basic components of a four-port intelligent switch. Although some switches function similar to bridges that read frames flowing on a network to construct a table of source addresses, other switches require their tables to be preconfigured. Either method allows the destination address to be compared to a table of destination addresses and associated port numbers. When a match occurs between the destination address of a frame flowing on a network connected to a port and the address in the port's address table, the frame is copied into the switch and routed through the switch fabric to the destination port, where it is placed onto the network connected to that port. If the destination port is in use due to a previously established cross-connection between ports, the frame is maintained in the buffer until it can be switched to its destination.

To illustrate the construction and utilization of a switch's port-address table, let's assume we are using a four-port switch, with one station connected to each port. Let's further assume that the MAC addresses of each station are A, B, C and D as indicated in Figure 14.10. If ports 0, 1, 2 and 3 are associated with addresses A, B, C and D, then the table in the lower portion of Figure 14.10 indicates how the port-address table associated with each port learns the addresses to place in its port-address table. For example, when address A needs to send a frame to address C after the switch is initialized, the port-address table for each port is filled with nulls. Since port 0 cannot find a destination address, the switch floods the frame onto all other ports than the port it was received on, just like a bridge. This action, coupled with internal switch logic, enables ports 1, 2 and 3 to note that address A is associated with port 0. Similarly, when station C next transmits a frame destined to station B, port 0 cannot find station B in its port-address table. Therefore, it floods the frame onto ports 0, 1 and 3, and those ports now associate station C with port 2. Thus, when the third function listed in the table in Figure 14.10 occurs (C sends frame to A), port 2 has the entry 0-A in its port-address table and can tell the switch to establish a cross-connection between port 2 and port 0.

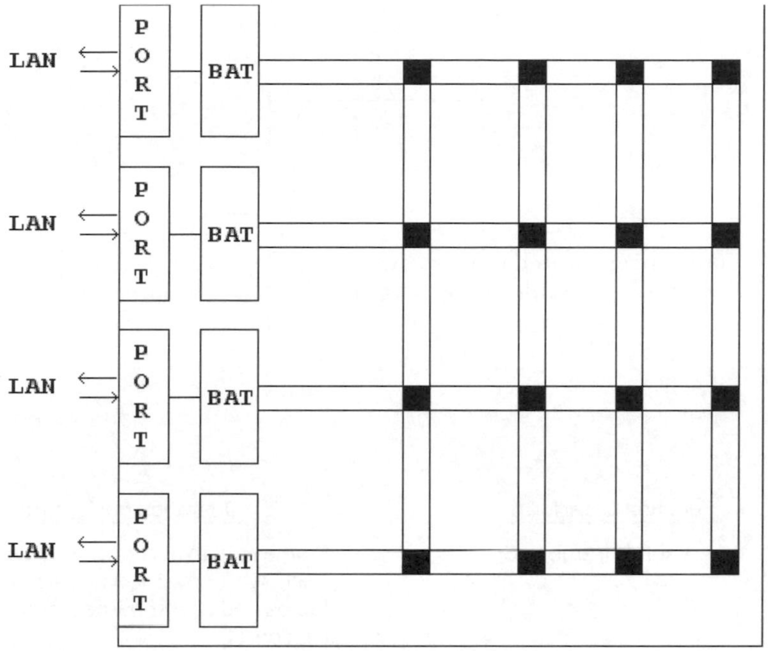

A network switch consists of buffers and address tables (BAT), logic and a switching fabric which permits frames entering one port to be routed to any port in the switch. The destination address in a frame is used to determine the associated port with that address via a search of the address table, with the port address used by the switching fabric for establishing the cross-connection.

Figure 14.9 Basic components of a network switch

14.2.2 Delay times

Switching occurs on a frame-by-frame basis, with the cross-connection torn down after being established for routing one frame. Thus, frames can be interleaved from two or more ports to a common destination port with a minimum of delay. For example, consider a maximum length Ethernet frame of 1526 bytes to include a 1500 byte data field and 26 overhead bytes. At a 10 Mbps operating rate, each bit time is $1/10^7$ seconds or 100 ns. For a 1526 byte frame the minimum delay time if one frame precedes it in attempting to be routed to a common destination becomes:

$$1526\,\text{bytes} \times \frac{8\,\text{bits}}{\text{byte}} \times \frac{100\,\text{ns}}{\text{bit}} = 1.22\,\text{ms}$$

This delay time represents blocking resulting from frames on two service ports having a common destination and should not be confused with another delay time referred to as latency. Latency represents the delay associated with

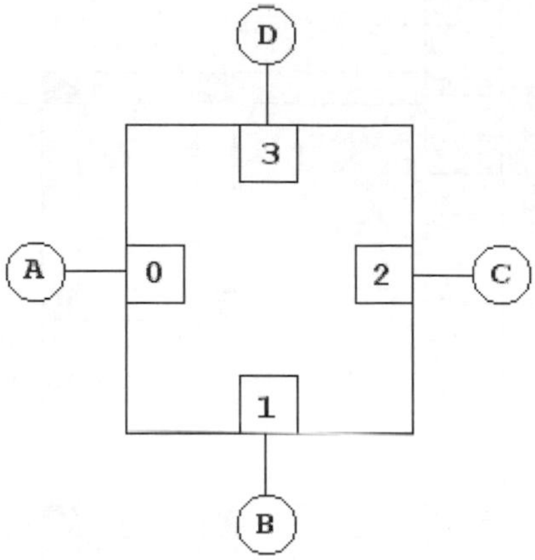

Function to perform	Operation performed
1. A sends frame to C	Frame sent to ports 1, 2, 3 as destination unknown. Ports 1, 2, 3 update port-address tables and now know address A associated with port 0.
2. C sends frame to B	Frame sent to ports 0, 1, 3 as destination unkown. Ports 0, 1, 3 update their port-address tables and now know address C associated with port 2.
3. C sends frame to A	Port-address table for port 2 has entry for destination A (see function 1) and causes cross-connection to port 0.

Figure 14.10 Switch functions vs. switch operations

the physical transfer of a frame from one port via the switch to another port, and is based upon the architecture of the switch which adds additional delay above and beyond the delay associated with the physical length of the frame being transported through the switch. In comparison, blocking delay depends upon the number of frames from different ports attempting to access a common destination port and the method by which the switch is designed to respond to blocking. Some switches simply have large buffers for each port and service ports in a round-robin fashion when frames on two or more ports attempt to access a common destination port. This method of service differs from politics as it does not show favoritism; however, it also does not consider the fact that some attached networks may have operating rates different from other attached networks. Other switch designs recognize that port buffers are filled based upon both the number of frames having a destination address of a different network and the operating rate of the network. Such switch designs use a priority service scheme based upon the occupancy of the port buffers in the switch.

14.2.3 Key advantages of use

A key advantage associated with the use of switching hubs results from their ability to support parallel switching, permitting multiple cross-connections between source and destination to occur simultaneously. For example, if four 10BASE-T networks were connected to the four-port switch shown in Figure 14.9, two simultaneous cross-connections, each at 10 Mbps, could occur, resulting in an increase in bandwidth to 20 Mbps. Here each cross-connection represents a dedicated 10 Mbps bandwidth for the duration of a frame. Thus, from a theoretical perspective, an N-port switching hub supporting a 10 Mbps operating rate on each port provides a throughput up to $(N/2) \times 10$ Mbps. For example, a 128-port switching hub would support a throughput up to $(128/2) \times 10$ Mbps or 640 Mbps, while a network constructed using a series of conventional shared media hubs connected to one another would be limited to an operating rate of 10 Mbps, with each workstation on that network having an average bandwidth of 10 Mbps/128 or 78 kbps.

Through the use of switching hubs you can overcome the operating rate limitation of a local area network. In an Ethernet environment, the cross-connection through a switching hub represents a dedicated connection so there will never be a collision. This fact enabled many switching hub vendors to use the collision wire-pair from conventional Ethernet to support simultaneous transmission in both directions between a connected node and hub port, resulting in a full-duplex transmission capability. In fact, a similar development permits Token-Ring switching hubs to provide full-duplex transmission, since if there is only one station on a port there is no need to pass tokens and repeat frames, raising the maximum bidirectional throughput between a Token-Ring device and a switching hub port to 32 Mbps for a 16 Mbps connection and to 200 Mbps when a 100 Mbps High Speed Token Ring connection is used. Thus, the ability to support parallel switching as well as initiate dedicated cross-connections on a frame-by-frame basis can be considered the key advantages associated with the use of switching hubs. Both parallel switching and dedicated cross-connections permit higher bandwidth operations.

Now that we have an appreciation for the general operation of switching hubs, let's focus our attention upon the different switching techniques that can be incorporated into this category of communications equipment.

14.2.4 Switching techniques

There are three switching techniques used by switching hubs: cross-point, also referred to as cut-through or 'on the fly', store-and-forward, and a hybrid method which alternates between the first two methods based upon the frame error rate. As we will soon note, each technique has one or more advantages and disadvantages associated with its operation.

Cross-point switching

The operation of a cross-point switch is based upon an examination of the destination of frames as they enter a port on the switching hub. The switch uses the destination address as a decision criterion to obtain a port destination from a look-up table. Once a port destination is obtained a cross-connection through the switch is initiated, resulting in the frame being routed to a destination port where it is placed onto a network where its frame destination address resides.

As previously noted earlier in this chapter, a switch can be considered to represent a more sophisticated type of bridge. Thus, it should come as no surprise that a cross-point switch which is limited to providing cross-connections between LAN stations of a similar type commonly uses a backward learning algorithm to construct a port-destination address table. That is, the switch monitors the MAC source addresses encountered on each port to construct a port-destination address table. If the destination address resides on the same port the frame was received from, this indicates that the frame's destination is on the current network and no switching operation is required. Thus, the switch discards the frame. If the destination address resides on a different port, the switch obtains the port destination and initiates a cross-connection through the switch, routing the frame to the appropriate destination port where it is placed onto a network where a node with the indicated destination address resides. If the destination address is not found in the table, the switch floods the frame onto all ports other than the port it was received on. Although flooding adversely affects the capability of a switch to perform multiple simultaneous cross-connections, the majority of this activity

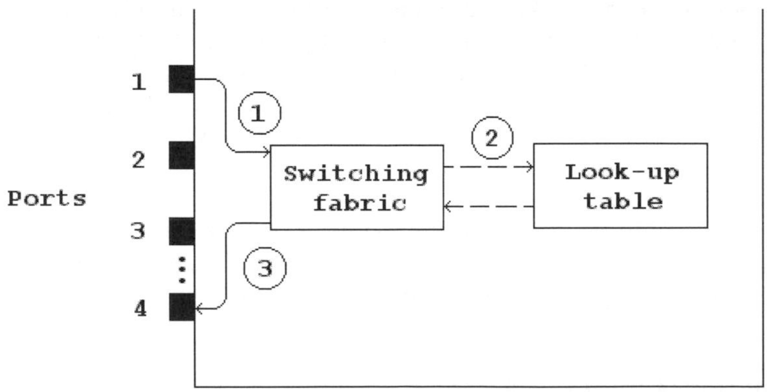

A cross-point or cut-through operating switch reads the destination address in a frame prior to storing the entire frame (1). It forwards that address to a look-up table (2) to determine the port destination address which is used by the switching fabric to provide a cross-connection to the destination port (3).

Figure 14.11 Cross-point/cut-through switching

occurs when a switch is powered on and its port-address table is empty. Thereafter, flooding occurs periodically after an entry is purged from the table due to aging and a new request to a purged destination occurs or when a broadcast address is encountered.

Figure 14.11 illustrates the basic operation of cross-point or cut-through switching. Under this technique the destination address in a frame is read prior to the frame being stored (1). That address is forwarded to a look-up table (2) to determine the port destination address which is used by the switching fabric to initiate a cross-connection to the destination port (3). Since this switching method only requires the storage of a small portion of a frame until it is able to read the destination address and perform its table look-up operation to initiate switching to an appropriate output port, latency through the switch is minimized.

Latency functions as a brake on two-way frame exchanges. For example, in a client–server environment the transmission of a frame by a workstation results in a server response. Thus, the minimum wait time is 2 $\times$ latency for each client–server exchange, lowering the effective throughput of the switch. Since a cross-point switching technique results in a minimal amount of latency, the effect upon throughput of the delay attributable to a switching hub using this switching technique is minimal.

Store-and-forward

In comparison to a cut-through switching hub, a store-and-forward switching hub first stores an entire frame in memory prior to operating on the data fields within the frame. Once the frame is stored, the switching hub checks the frame's integrity by performing a cyclic redundancy check (CRC) upon the contents of the frame, comparing its computed CRC against the CRC contained in the frame's Frame Check Sequence (FCS) field. If the two match, the frame is considered to be error-free and additional processing and switching will occur. Otherwise, the frame is considered to have one or more bits in error and will be discarded.

In addition to CRC checking, the storage of a frame permits filtering against various frame fields to occur. Although a few manufacturers of store-and-forward switching hubs support different types of filtering, the primary advantage advertised by such manufacturers is data integrity and the ability to perform translation switching, such as switching a frame between an Ethernet network and a Token-Ring network. Since the translation process is extremely difficult to accomplish on the fly due to the number of conversions of frame data, most switch vendors first store the frame, resulting in store-and-forward switches supporting translation between different types of connected networks. Concerning the data integrity capability of store-and-forward switches, whether or not this is actually an advantage depends upon how you view the additional latency introduced by the storage of a full frame in memory as well as the necessity for error checking. Concerning the latter, switches should operate error-free, so a store-and-forward switch only removes network errors, which should be negligible to start with.

When a switch removes an errored frame, the originator will retransmit the frame after a period of time. Since an errored frame arriving at its destination network address is also discarded, many persons question the necessity of error checking by a store-and-forward switching hub. However, filtering capability, if offered, may be far more useful as you could use this capability, for example, to route protocols carried in frames to destination ports far more easily than by frame destination address. This is especially true if you have hundreds or thousands of devices connected to a large switching hub. You might set up two or three filters instead of entering a large number of destination addresses into the switch. When a switch performs filtering of protocols, it really becomes a router. This is because it is now operating at Layer 3 of the OSI Reference Model.

Figure 14.12 illustrates the operation of a store-and-forward switching hub. Note that a common switch design is to use shared buffer memory to store entire frames, which increases the latency associated with this type of switching hub. Since the minimum length of an Ethernet frame is 72 bytes, then the minimum one-way delay or latency at 10 Mbps, not counting the switch overhead associated with the look-up table and switching fabric operation, becomes:

$$9.6\,\mu\text{s} + (72\,\text{bytes} \times 8\,\text{bits/byte} \times 100\,\text{ns/bit})$$

or

$$9.6 \times 10^{-6} + 576 \times 100 \times 10^{-9}\ \text{seconds}$$

or

$$67.2 \times 10^{-6}\ \text{seconds}$$

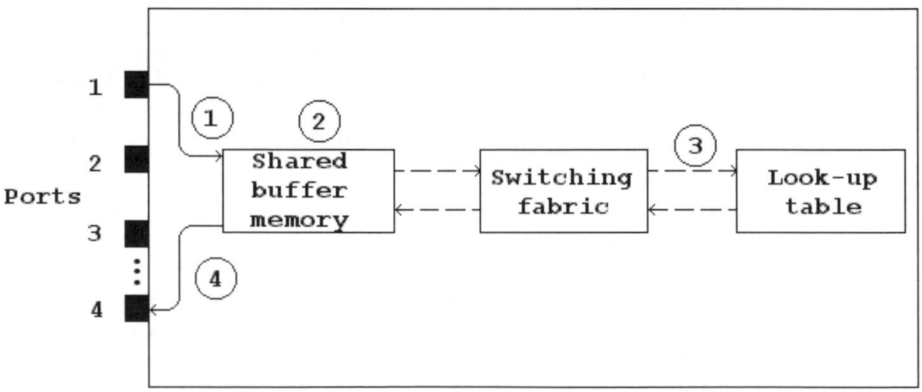

A store-and-forward switching hub reads the frame destination address (1) as it is placed in buffer memory (2). As the entire frame is being read into memory, a look-up operation (3) is performed to obtain a destination port address. Once the entire frame is in memory, a CRC check is performed and one or more filtering operations may be performed. If the CRC check indicates the frame is error-free, it is fowarded from memory to its destination address (4), otherwise it is disregarded.

Figure 14.12 Store-and-forward switching

Here 9.6 μs represents the Ethernet interframe gap, while 100 ns/bit is the bit duration of a 10 Mbps Ethernet LAN. Thus, the minimum one-way latency of a store-and-forward Ethernet switching hub is 0.000 067 2 seconds, while a round-trip minimum latency is twice that duration. For a maximum length Ethernet frame with a data field of 1500 bytes, the frame length becomes 1526 bytes. Thus, the one-way maximum latency at 10 Mbps becomes:

$$9.6\,\mu\text{s} + (1526\,\text{bytes} \times 8\,\text{bits/byte} \times 100\,\text{ns/bit})$$

or

$$9.6 \times 10^{-6} + 12\,208 \times 100 \times 10^{-9}\,\text{seconds}$$

or

$$0.001\,230\,4\,\text{seconds}$$

Hybrid

A hybrid switch supports both cut-through and store-and-forward switching, selecting the switching method based upon monitoring the error rate encountered by reading the CRC at the end of each frame and comparing its value to a computed CRC performed 'on the fly' on the fields protected by the CRC. Initially the switch might set each port to a cut-through mode of operation. If too many bad frames are noted occurring on the port, the switch will automatically set the frame processing mode to store-and-forward, permitting the CRC comparison to be performed prior to the frame being forwarded. This permits frames in error to be discarded without having them pass through the switch. Since the 'switch' (no pun intended) between cut-through and store-and-forward modes of operation occurs adaptively, another term used to reference the operation of this type of switch is adaptive.

The major advantages of a hybrid switch are that it provides minimal latency when error rates are low and discards frames by adapting to a store-and-forward switching method, so it can discard errored frames when the frame error rate rises. From an economics perspective, the hybrid switch can logically be expected to cost more than a cut-through or store-and-forward switch as its software development effort is a bit more comprehensive. However, due to the competitive market for communications products, upon occasion its price may be reduced below competitive switch technologies.

14.2.5 Port address support

In addition to being categorized by their switching technique, switching hubs can be classified by their support of single or multiple addresses per port. The former method is referred to as port-based switching, while the latter switching method is referred to as segment-based switching.

Port-based switching

A switching hub, which performs port-based switching only, supports a single address per port. This restricts switching to one device per port; however, it results in a minimum amount of memory in the switch as well as providing for a relatively fast table look-up when the switch uses a destination address in a frame to obtain the port for initiating a cross-connect.

Figure 14.13 illustrates an example of the use of a port-based switching hub. In this example M user workstations use the switch to contend for the resources of N servers. If $M > N$, then a switching hub connected to Ethernet 10 Mbps LANs can support a maximum throughput of $(N/2) \times 10$ Mbps, since up to $N/2$ simultaneous client–server frame flows can occur through the switch.

It is important to compare the maximum potential throughput through a switch to its rated backplane speed. If the maximum potential throughput is less than the rated backplane speed, the switch will not cause delays based upon the traffic being routed through the device. For example, consider a 64-port switch that has a backplane speed of 400 Mbps. If the maximum port rate is 10 Mbps, then the maximum throughput, assuming 32 active cross-connections were simultaneously established, becomes 320 Mbps. In this example the switch has a backplane transfer capability sufficient to handle the worst-case data transfer scenario. Now let's assume that the maximum backplane data transfer capability was 200 Mbps. This would reduce the maximum number of simultaneous cross-connections capable of being serviced to 20 instead of 32 and adversely affect switch performance under certain operational conditions.

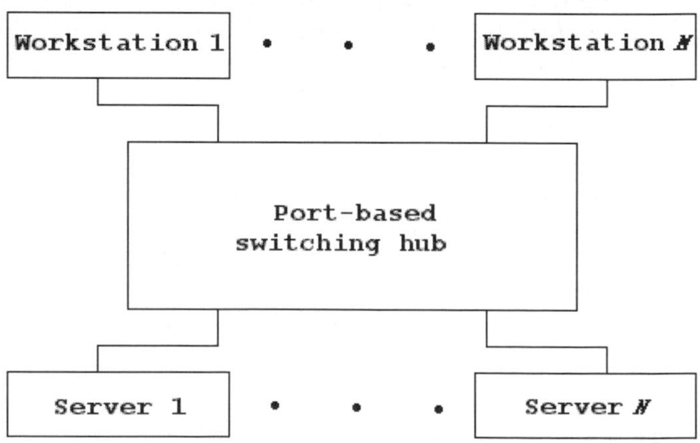

A port-based switching hub associates one address with each port, minimizing the time required to match the destination address of a frame against a table of destination addresses and associated port numbers.

Figure 14.13 Port-based switching

Since a port-based switching hub has to store only one address per port, search times are minimized. When combined with a pass-through or cut-through switching technique, this type of switch results in a minimal latency to include the overhead of the switch in determining the destination port of a frame.

Segment-based switching

A segment-based switching technique requires a switching hub to support multiple addresses per port. Through the use of this type of switch, you achieve additional networking flexibility, since you can connect other hubs to a single segment-based switching hub port.

Figure 14.14 illustrates an example of the use of a segment-based switching hub in an Ethernet environment. Although two segments in the form of conventional hubs with multiple devices connected to each hub are shown in

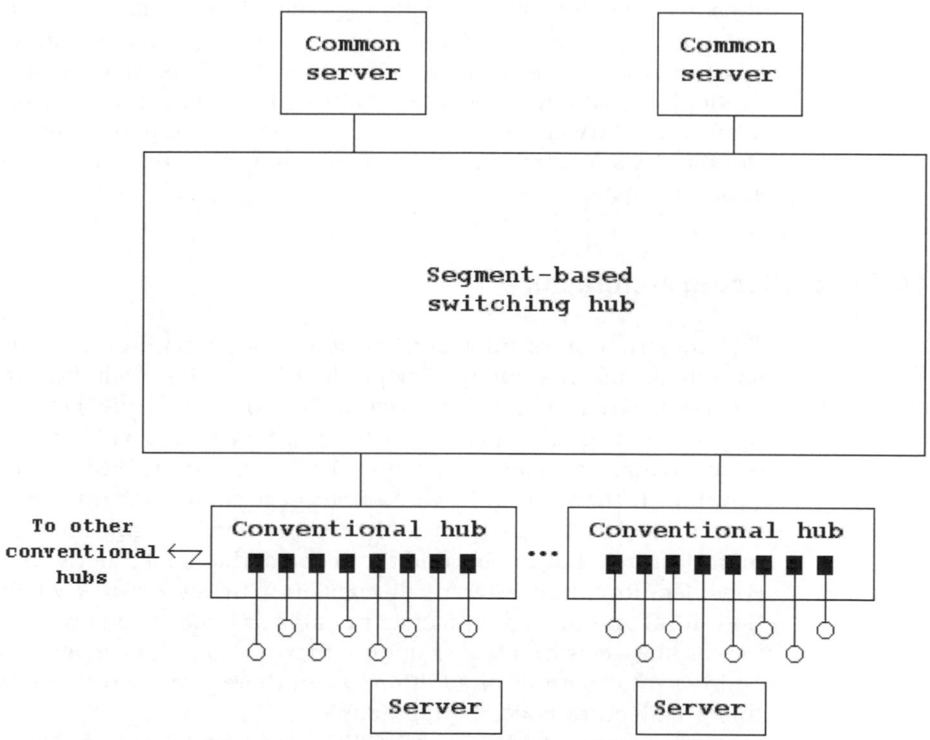

Through the use of a segment-based switching hub, you can maintain servers for use by workstations on a common network segment as well as provide access by all workstations to common servers.

Figure 14.14 Segment-based switching

the lower portion of Figure 14.14, note that a segment can consist of a single device, resulting in the connection of one device to a port on a segment-switching hub being similar to a connection on a port-switching hub. However, unlike a port-switching hub that is limited to supporting one address per port, the segment-switching hub can, if necessary, support multiple devices connected to a port. Thus, the two servers connected to the switch at the top of Figure 14.14 could, if desired, be placed on a conventional hub or a high-speed hub, such as a 100BASE-T hub, which in turn would be connected to a single port on a segment-switching hub.

In Figure 14.14 each conventional hub acts as a repeater and forwards every frame transmitted on that hub to the switching hub, regardless of whether or not the frame requires the resources of the switching hub. The segment-switching hub examines the destination address of each frame against addresses in its look-up table, forwarding only those frames that warrant being forwarded. Otherwise, frames are discarded as they are local to the conventional hub. Through the use of a segment-based switching hub, you can maintain the use of local servers with respect to existing LAN segments as well as install servers whose access is common to all network segments. The latter is illustrated in Figure 14.14 by the connection of two common servers shown at the top of the switching hub. If you obtain a store-and-forward segment-switching hub that supports filtering, you could control access to common servers from individual workstations or by workstations on a particular segment. In addition, you can also use the filtering capability of a store-and-forward segment-based switching hub to control access from workstations located on one segment to workstations or servers located on another segment.

14.2.6 Switching architecture

The construction of intelligent switches varies both between manufacturers as well as within some vendor product lines. Most switches are based upon the use of either Reduced Instruction Set Computer (RISC) microprocessors or Application Specific Integrated Circuit (ASIC) chips, while a few products use conventional Complex Instruction Sets Computer (CISC) microprocessors.

Although there are a large number of arguable advantages and disadvantages associated with each architecture from the standpoint of the switch manufacturer that are beyond the scope of this book, there are also some key considerations that warrant discussion with respect to evolving technology. Both RISC and CISC architectures enable switches to be programmed to make forwarding decisions based upon either the data link layer or network layer address information. In addition, when there is a need to modify the switch, this architecture is easily upgradable.

In comparison to RISC and CISC based switches, an ASIC based device represents the use of custom designed chips to perform specific switch functions in hardware. Although ASIC based switches are faster than RISC and CISC based switches, there is no easy way to upgrade this type of switch. Instead, the vendor will have to design and manufacture new chips and install the hardware upgrade in the switch.

Today most switches use an ASIC architecture as its speed enables the support of cut-through switching. While ASIC based switches provide the speed necessary to minimize latency, readers should carefully check vendor upgrade support as modifications to an ASIC based switch may be difficult and relatively expensive.

Now that we have an appreciation for the general operation and utilization of switching hubs, let's obtain an appreciation for the high-speed operation of switch ports which enables dissimilar types of networks to be connected and which can result in data flow compatibility problems, along with methods used to alleviate such problems.

14.2.7 High-speed port operations

There are several types of high-speed port connections intelligent switches may support. Those high-speed connections include 1 Gbps Gigabit Ethernet, 100 Mbps Fast Ethernet, 100 Mbps FDDI, 155 Mbps ATM, full-duplex Ethernet and Token-Ring, and fat pipes, with the latter referencing a grouping of ports treated as a transmission entity. The most common use of one or more high-speed connections on an intelligent switching hub is to support highly used

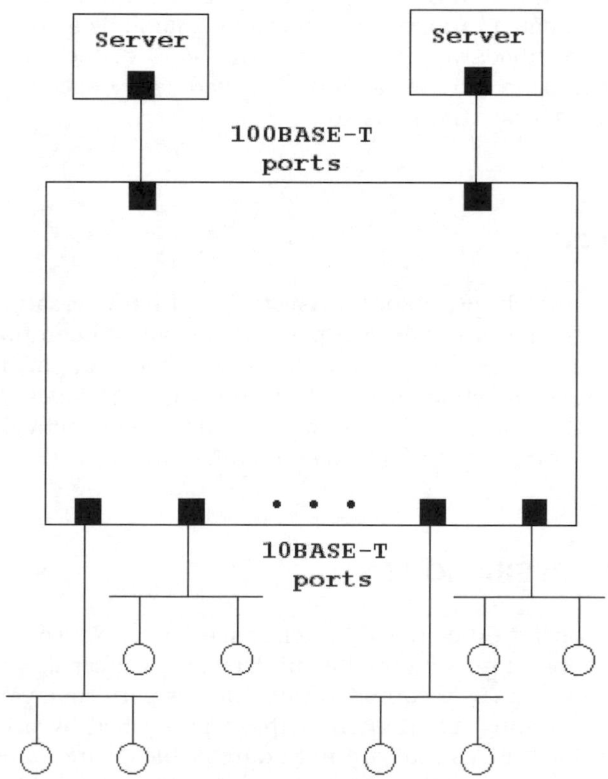

Figure 14.15 Using high-speed connections to servers

devices, such as network servers and color printers. Figure 14.15 illustrates the use of an Ethernet switch, with two 100BASE-T Fast Ethernet adapters built into the switch to provide a high-speed connection from the switch to each server. Through the use of high-speed connections the cross-connection time from server to client when the server responds to a client query is minimized. Since most client queries result in server responses containing many multiples of characters in the client query, this allows the server to respond to more queries per unit of time. Thus, the high-speed connection can enhance client–server response times through a switch.

In examining Figure 14.15, let's assume a small query results in the server responding by transmitting the contents of a large file back to the client. If data flows into the switch at 100 Mbps and flows from the switch to the client at 10 Mbps, any buffer area in the switch used to provide temporary storage for speed incompatibilities between ports will rapidly be filled and eventually overflow, resulting in the loss of frames which, when compensated for by retransmission, compounds the problem. Thus, a mechanism is required to regulate the flow of data into and out of switch ports. That mechanism is known as flow control, and specific methods used to implement flow control can include backpressure, inband signaling or the use of a special frame. Backpressure represents a term used to reference a false collision signal that results in the opposite end of the connection initiating a random delay prior to retransmitting. Inband signaling references the use of software that interprets data in frames for a special signal to delay transmission. While the first two methods are proprietary, the IEEE defined a flow control standard for Ethernet referred to as 802.3z which uses a special frame to inform the recipient to delay transmission.

14.2.8 Summary

The switching hub or LAN switch is a highly versatile device that may support single or multiple devices per port and whose operation can vary based upon its architecture. By providing the capability for supporting multiple simultaneous cross-connections, the LAN switch can significantly increase network bandwidth, and its ability to support high-speed network connections enhances its versatility as a networking tool.

14.3 ROUTER OPERATIONS

By operating at the ISO Reference Model Network Layer, a router becomes capable of making intelligent decisions concerning the flow of information in a network. To accomplish this, routers perform a variety of functions that are significantly different from those performed by bridges and most switches. Unlike bridges that are not addressable, routers are. Thus, routers examine frames that are directly addressed to them by looking at the network address within each frame to make their forwarding decision.

14.3.1 Basic operation and use of routing tables

To illustrate the basic operation of routers, consider the simple mesh structure formed by the use of three routers labeled R1, R2, and R3 in Figure 14.16(a). In this illustration, three Token-Ring networks are interconnected through the use of three routers. The initial construction of three routing tables is shown in Figure 14.16(b). Unlike bridges which learn MAC addresses, most routers are initially configured, with routing tables established at the time of equipment installation. Thereafter, periodic communications between routers dynamically updates routing tables to take into consideration changes in internet topology and traffic.

In examining Figure 14.16(b), note that the routing table for router R1 indicates the routers it must communicate with to access each interconnected Token-Ring network. Hence, router R1 would communicate with router R2 to

A. Simple mesh structure

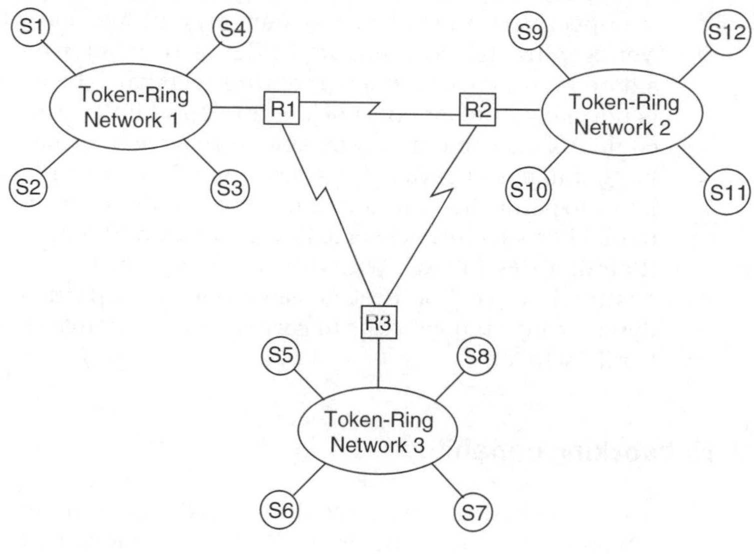

B. Routing tables

R1

1	*
2	R2
3	R3

R2

1	R1
2	*
3	R2

R3

1	R1
2	R2
3	*

C. Packet composition

MAC	LLC	Destination 2.S12	Source 1.S2	DATA

Legend: R router S network station

Figure 14.16 Basic router operation

reach Token-Ring network 2 and communicate with the router R3 to reach Token-Ring network 3.

Figure 14.16(c) illustrates the composition of a packet originated by station S2 on Token-Ring 1 that is to be transmitted to station S12 on Token-Ring 2. Router R1 first examines the destination network address and notes that it is on another network. Thus, the router searches its routing table and finds that the frame should be transmitted to router R2 to reach Token-Ring network 2. Hence, router R1 forwards the frame to router R2. Router R2 then places the frame onto Token-Ring network 2 for delivery to station S12 on that network.

Because routers use the network addresses instead of MAC addresses for making their forwarding decisions, it is possible to have duplicate locally administered MAC addresses on each network interconnected by a router. In comparison, the use of bridges would require you to first review and then eliminate any duplicate locally administered addresses common to networks to be interconnected, a process that can be time consuming when large networks are connected.

Another difference between bridges and routers is the ability of a router to support the transmission of data on multiple paths between local area networks. Although a multiport bridge with a filtering capability can be considered to perform intelligent routing decisions, the result of a bridge operation is normally valid for only one point-to-point link within a wide area network. In comparison, a router may be able to acquire information about the status of a large number of paths and select an end-to-end path consisting of a series of point-to-point links. In addition, most routers can fragment and reassemble data. This permits packets to flow over different paths and to be reassembled at their final destination. With this capability a router can route each packet to its destination over the best possible path at a particular instant in time and dynamically change paths to correspond to changes in network link status on traffic activity.

14.3.2 Networking capability

To better illustrate the networking capability of routers, consider Figure 14.17 which shows three geographically dispersed locations that have a total of four Ethernet and three Token-Ring networks interconnected through the use of four routers and four wide area network transmission circuits or links. For simplicity, the use of modems or DSUs on the wide area network is not shown. This illustration will be referred to several times in this section to denote different types of router operation.

In addition to supporting a mesh structure that is not obtainable from the use of transparent bridges, the use of routers offers other advantages in the form of addressing, message processing, link utilization and priority of service. Routers are known to stations that use its service. Hence, packets can be directly addressed to a router. This eliminates the necessity for the device to examine in detail every packet flowing on a network, and it results in the router only having to process messages that are addressed to it by other devices. Concerning link utilization, assume that a station on E1 transmits to a station on TR3. Depending on the status and traffic on network links,

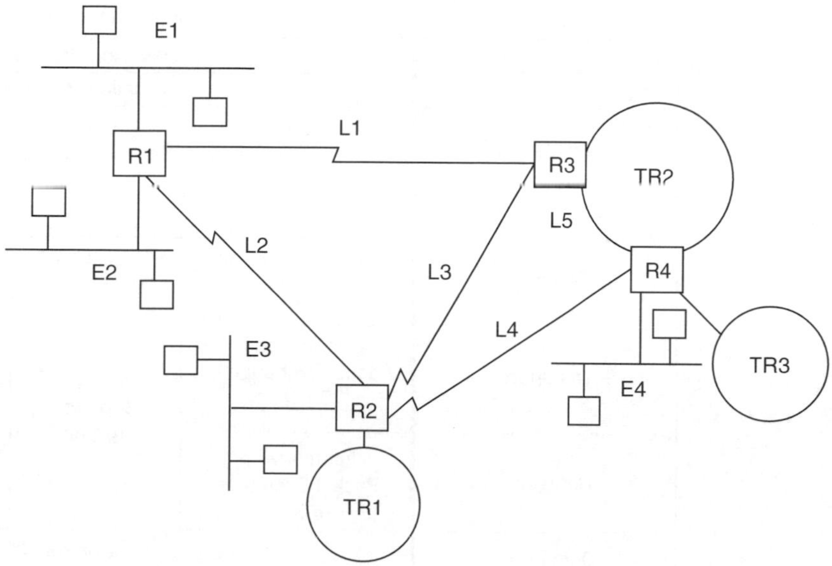

Figure 14.17 Router operation. The use of routers enables the transmission of data over multiple paths, alternate path routing, and the use of a mesh topology which transparent bridges cannot support

packets could be routed via L1 and use TR2 to provide a transport mechanism to R4, from which the packets are delivered to TR3. Alternatively, links L2 and L4 could be used to provide a path from R1 to R4. Although link availability and link traffic usually determines routing, routers can support prioritized traffic and may store low priority traffic for a small period of time to allow higher priority traffic to gain access to the wide area transmission facility. Due to these features which are essentially unavailable from the use of bridges, the router is a more complex and costly device.

14.3.3 Communication, transport and routing protocols

For routers to be able to operate in a network, they must normally be able to speak the same language at both the data link and network layers. The key to accomplishing this is the ability of routers to support common communication, transport and routing protocols.

Communication protocol

Communication protocols support the transfer of data from a station on one network to a station on another network, and they occur at the OSI network layer. In examining Figure 14.18, which illustrates several common protocol implementations with respect to the OSI Reference Model, note that Novell's NetWare originally used IPX as their network communications protocol,

<table>
<tr><th>OSI Layer</th><th colspan="4">Common Protocol Implementation</th></tr>
<tr><td rowspan="2">Application</td><td colspan="4">Application Programs</td></tr>
<tr><td colspan="4">Application Protocols</td></tr>
<tr><td>Presentation</td><td colspan="2">Novell Network File Server Protocol (NFSP)</td><td>IBM Server Message Block (SMB)</td><td>Microsoft LAN Manager</td></tr>
<tr><td>Session</td><td>Xerox Networking System (XNS)</td><td rowspan="2">NetBIOS</td><td>NetBIOS</td><td>NetBIOS Advanced Peer-to Peer Communications</td></tr>
<tr><td>Transport</td><td>Sequenced Packet Exchange (SPX)</td><td rowspan="2">PC LAN Support Program</td><td>Transmission Control Protocol (TCP) User Datagram Protocol (UDP)</td></tr>
<tr><td>Network</td><td colspan="2">Internet work Packet Exchange (IPX)</td><td>Internet Protocol (IP)</td></tr>
<tr><td rowspan="2">Data Link</td><td colspan="4">Logical Link Control 802.2</td></tr>
<tr><td colspan="4">Media Access Control</td></tr>
<tr><td>Physical</td><td colspan="4">Transmission Media
Twisted pair, coax, fiber optic</td></tr>
</table>

Figure 14.18 Common protocol implementations. Although Novell, IBM and Microsoft LAN operating system software support standardized physical and data link operations, they differ considerably in their use of communication and routing protocols

whereas IBM LANs use the PC LAN Support Program and Microsoft's LAN Manager uses the Internet Protocol (IP). Even though Novell now supports the TCP/IP suite, this means that a router linking networks based on different operating systems will more than likely have to support multiple communication protocols. Thus, router communication protocol support is a most important criteria in determining whether a particular product is capable of supporting your internetworking requirements.

Routing protocol

The routing protocol references the method used by routers to exchange routing information and forms the basis for providing a connection across interconnected networks. In evaluating routers, it is important to determine the efficiency of the routing protocol, its effect upon the transmission of information, the method used and memory required to construct routing tables, and the time required to dynamically adjust those tables. Examples of router-to-router protocols include Xerox Network Systems' (XNS) Routing Information Protocol (RIP) and the Transmission Control Protocol/Internet Protocol's (TCP/IP) RIP, Open Shortest Path First (OSPF), and Hello routing protocols.

Table 14.1 Representative communications protocols

AppleTalk	ISO CLNS
Applo Domain VINES	HDLC
Banyan	NOVELL IPX
CHAOSnet	SDLC
DECnet Phase IV	TCP/IP
DECnet Phase V	Xerox XNS
DDN X.25	X.25
Frame Relay	Ungermann-Bass Net/One

Transport protocol

The transport protocol represents the format by which information is physically transported between two points. Do not confuse the transport protocol with the Transport Layer that is used to distinguish one application from another as well as to support error detection and correction through a network and other functions. Here, the transport protocol represents the data link layer illustrated in Figure 14.18. Examples of transport protocols include Token-Ring and Ethernet MAC for LANs as well as such WAN protocols as X.25 and Frame Relay,

There is a wide variety of communication and transport protocols in use today. Some of these protocols were designed specifically to operate on local area networks, such as Apple Computer's AppleTalk. Other protocols, such as X.25 and Frame Relay, were developed as wide area network protocols.

Table 14.1 lists examples of 16 communication and transport protocols. Readers are cautioned that many routers support only a subset of the protocols listed in Table 14.1.

14.3.4 Router classifications

Depending on their support of communication and transport protocols, routers can be classified into two classes: protocol-dependent and protocol-independent.

Protocol-dependent routers

To understand the characteristics of a protocol-dependent router, consider the network previously illustrated in Figure 14.17. If a station on network E1 wishes to transmit data to a second station on network E3, router R1 must know that the second station resides on network E3 and the best path to use to reach that network. The method used to determine the network where the destination station resides determines the protocol dependency of the router.

If the station on network E1 tells router R1 the destination location, it must supply a network address in every LAN packet it transmits. This means that all routers in the network must support the protocol used on network E1. Otherwise, stations on network E1 could not communicate with stations residing on other networks and vice versa.

NetWare IPX example

To illustrate the operation of a protocol-dependent router, let us assume that networks E1 and E3 use Novell's NetWare as their LAN operating system. The routing protocol used at the network layer between a station and server on a Novell network is known as IPX. This protocol can also be used between servers as well as other protocols.

Under NetWare's IPX, a packet addressed to a router will contain the destination address in the form of network and host addresses as well as the origination address in the form of the source network and source host addresses. Here the IPX term host is actually the physical address of a network adapter card.

Figure 14.19(a) illustrates in simplified format the IPX packet composition for workstation A on network E1 transmitting data to workstation B on network E3 under Novell's NetWare IPX protocol. After router R1 has received and examined the packet, it notes that the destination address E3 requires the routing of the packet to router R2. Thus, it converts the first packet into a router (R1) to router (R2) packet, as illustrated in Figure 14.19(b). At router R2 the packet is again examined. Router R2 notes that the destination network address (E3) is connected to that router. Thus, router R2 reconverts the packet for delivery onto network E3 by converting the destination router address to a source router address and transmitting the packet onto network E3. This is illustrated in Figure 14.19(c).

A. Packet from workstation A, network E1 to router R1

B	E3	* * *	A	E1	R1	* * *	Data

B. Router (R1) to router (R2) packet

B	E3	R2	* * *	A	E1	R1	* * *	Data

C. Router R2 converts packet for placement on network E3

B	E3	* * *	A	E1	R2	* * *	Data

Figure 14.19 NetWare IPX routing

Addressing differences

In the preceding example note that each router uses the destination workstation and network addresses to transfer packets. If all protocols used the same format and addressing structure, routers would be protocol-insensitive at the network layer. Unfortunately this is not true. For example, under TCP/IP, addressing conventions are very different from that used by NetWare. This means that networks using different operating systems require the use of multiprotocol routers that are configured to perform address translation. To accomplish this, each multiprotocol router must maintain separate routing tables for each supported protocol, requiring additional memory and time to perform this task.

Other problems

Two additional problems associated with protocol-dependent routers are the time required for packet examination and the fact that not all LAN protocols are routable. Concerning packet examination, if a packet must traverse a large network the time required by a series of routers to both modify the packet and assure its delivery to the next router can significantly degrade router performance. To overcome this problem, organizations should consider the use of a frame relay service.

In addition to providing an enhanced data delivery service by eliminating error detection and correction occurring within the network, the use of a frame relay service can significantly reduce the cost of routers. To illustrate this, consider the network previously illustrated in Figure 14.17 in which four routers are interconnected through the use of five links. To support transmission on five links, the routers require 10 ports. Normally, each router port is obtained as an adapter card installed in a high-performance computer. If a frame relay service is used, the packet network providing that service also provides the routing paths to interconnect routers as illustrated in Figure 14.20. This reduces the number of required router ports to four, which can result in a considerable hardware savings.

A second problem associated with protocol-dependent routers is the fact that some LAN protocols cannot be routed using that type of device. This is because some LAN protocols, such as NetBIOS and IBM's LAN Server, unlike NetWare, DECnet and TCP/IP, do not include routing information within a packet. Instead, those protocols employ a user-friendly device-naming convention instead of using network and device addresses, permitting such names as Gil's PC, Accounting Printer, and other descriptors to be used. For example, IBM's Network Basic Input/Output System (NetBIOS) was designed to facilitate program-to-program communication by hiding the complexities of network addressing from the user. Thus, NetBIOS uses names that can be up to 16 alphanumeric characters in length to define clients and servers on a network. Unfortunately, NetBIOS does not include a facility to distinguish one network from another since it lacks a network addressing capability. Such protocols are restricted to using the physical addresses of adapter cards, such as Token-Ring source and destination addresses. Since a protocol-dependent

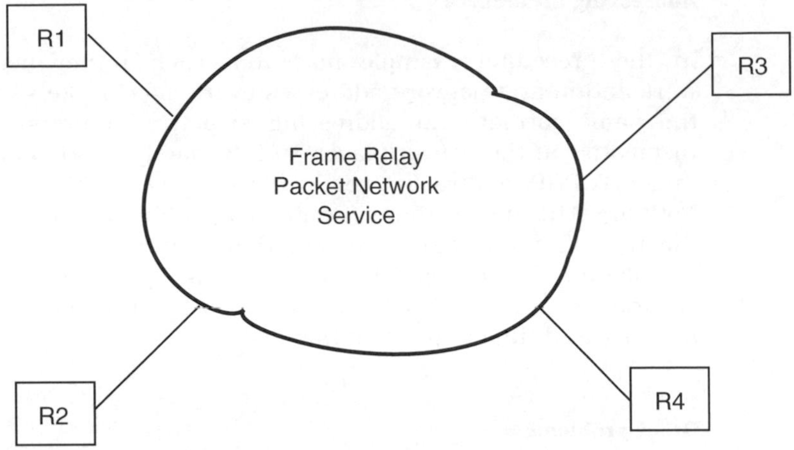

Figure 14.20 Using a frame relay service. If a frame relay service is used, the packet network provides the capability for interconnecting each network access port to other network access ports.Thus, only one router port is required to obtain an interconnection capability to numerous routers connected to the network

router must know the network on which a destination address is located, it cannot route such protocols. Thus, a logical question is: how does a router interconnect networks using an IBM LAN protocol? The answer to this question is bridging. That is, a protocol-dependent router must operate as a bridge in the previously described situation.

Protocol-independent routers

A protocol-independent router can be considered to function as a sophisticated transparent bridge. That is, it addresses the problem of network protocols that do not have network addresses, by examining the source addresses on connected networks to automatically learn what devices are on each network. The protocol-independent router assigns network identifiers to each network whose operating system does not include network addresses in its network protocol. This activity enables both routable and non-routable protocols to be serviced by a protocol-dependent router.

In addition to automatically building address tables like a transparent bridge, a protocol-independent router exchanges information concerning its routing directions with other internet routers. This enables each router to build a map of the internet. The method used to build the network map falls into the category of a link state routing protocol, which is described later in this chapter.

14.3.5 Routing protocols

The routing protocol is the key element which enables the transfer of information across interconnected networks in an orderly manner. The protocol is responsible for developing paths between routers, which is accomplished by a predefined mechanism by which routers exchange routing information.

Types of routing protocol

There are two general types of routing protocol: interior and exterior domain. Here, we use the term domain to refer to the connection of a group of networks to form a common entity, such as a corporate or university enterprise network.

An interior domain routing protocol is used to control the flow of information within a series of separate networks interconnected to form an internet. Thus, interior domain routing protocols provide a mechanism for the flow of information within a domain and are known as intra-domain routing protocols. Such protocols create routing tables for each autonomous system within the domain, and use such metrics as the hop count or time delay to develop routes from one network to another within the domain. Examples of interior domain routing protocols include RIP, OSPF, and Hello.

Exterior domain routing protocols

Exterior domain routing protocols are used to connect separate domains together. Thus, they are also referred to as inter-domain routing protocols. Currently, inter-domain routing protocols are only defined for OSI and TCP/IP. Example of inter-domain routing protocols include the Exterior Gateway Protocol (EGP), the Border Gateway Protocol (BGP) and the Inter-Domain Routing Protocol (IDRP). In comparison to interior domain routing protocols which are focused on the construction of routing tables for data flow within a domain, inter-domain routing protocols specify the method by which routers exchange information concerning which networks they can reach on each domain.

Figure 14.21 illustrates the use of interior and exterior domain routing protocols. In this example, OSPF is the intra-domain protocol used in Domain A, whereas RIF is the intra-domain protocol used in Domain B. Routers in Domains A and B use the inter-domain routing protocols EGP and/or BGP to determine the networks on other domains they can reach.

Exterior Gateway Protocol

There are four basic functions performed by the Exterior Gateway Protocol. First, the EGP performs an acquisition function which enables a router in one domain to request a router on another domain to exchange information. Since each router serves as a gateway to the domain, they are also referred to as gateways. A second function performed by the router gateway is to periodically test whether or not its EGP neighbors are responding. The third and most important function performed by the EGP is to enable router gateways to exchange information concerning the networks in each domain by

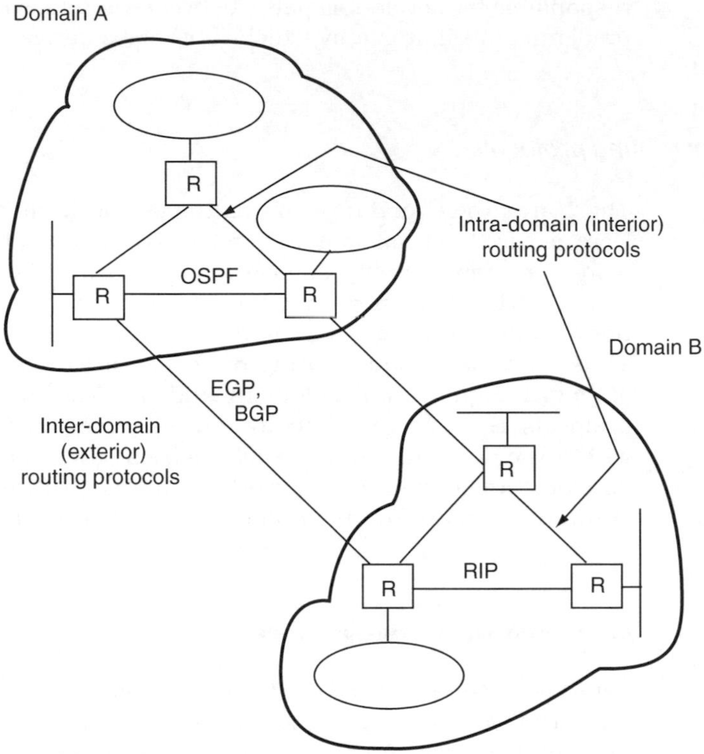

Figure 14.21 Interior and exterior routing protocols. An interior routing protocol controls the flow of information within a collection of interconnected networks known as a domain. An exterior routing protocol provides routers with the ability to determine which networks on other domains they can reach

transmitting routing update messages. The fourth function involves terminating an established neighbor relationship between gateways on two domains.

To accomplish its basic functions, EGP defines nine message types. Figure 14.22 illustrates EGP message types associated with each of the three basic features performed by the protocol.

Under the EGP, once a neighbor has been acquired Hello messages must be transmitted at a minimum of 30 s intervals. In addition, routing updates must be exchanged at intervals of at least two minutes. This exchange of information at two-minute intervals can result in the use of a considerable amount of the bandwidth linking domains when the number of networks on each domain is large or the circuits linking domains consist of low-speed lines. To alleviate those potential problems, the Border Gateway Protocol was developed.

Border Gateway Protocol

The Border Gateway Protocol represents a follow-on to the EGP. Unlike the EGP, in which all network information is exchanged at two-minute or less intervals, the BGP results in incremental updates being transmitted as

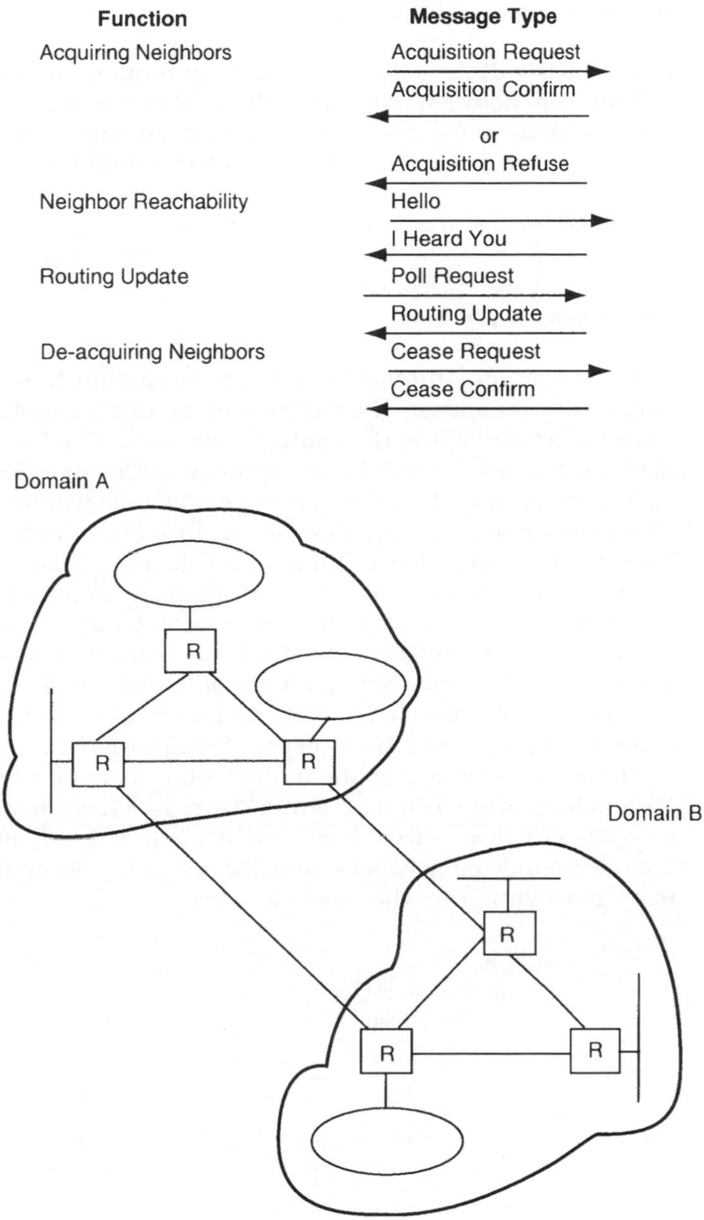

Figure 14.22 Exterior gateway protocol message types

changes occur. This can significantly reduce the quantity of data exchanged between router gateways and frees up a considerable amount of circuit bandwidth for the transmission of data. Both the EGP and the BGP run over TCP/IP and are standardized by Internet documents RFC904 and RFC1105, respectively.

Interior domain routing protocols

As previously discussed, interior domain routing protocols govern the flow of information between networks. Thus, this represents the type of routing protocol that is of primary interest to most organizations. Interior domain routing protocols can be further subdivided into two broad categories based on the method they use to build and update the contents of their routing tables: vector distance and link state.

Vector distance protocol

A vector distance protocol constructs a routing table in each router and periodically broadcasts the contents of the routing table across the networks linked together. When the routing table is received at another router, that device examines the set of reported network destinations and the distance to each destination. The receiving router then determines whether it knows a shorter route to a network destination, or finds a destination that it does not have in its routing table, or finds a route to a destination through the sending router where the distance to the destination changed. If any one of these situations occurs, the receiving router will change its routing tables.

The term 'vector distance' relates to the information transmitted by routers. Each router message contains a list of pairs known as vector and distance. The vector identifies a network destination, whereas the distance is the distance in hops from the router to that destination.

Figure 14.23 illustrates the initial vector distance routing table for routers R1 and R2 previously illustrated in Figure 14.17. Each table contains an entry for each directly connected network and is broadcast periodically throughout the interconnected networks. Here the distance column indicates the distance to each network from the router in hops.

a. Router R1

Destination	Distance
E1	0
E2	0

b. Router R2

Destination	Distance
E3	0
TR1	0

Figure 14.23 Initial vector-routing distance tables

a. Router R1

Destination	Distance	Route
E1	0	Direct
E2	0	Direct
E3	1	R2
TR1	1	R2

b. Router R2

Destination	Distance	Route
E1	1	R1
E2	1	R1
E3	0	Direct
TR1	0	Direct

Figure 14.24 Initial routing table update

At the same time that router R1 is constructing its initial vector distance table, other routers are performing a similar operation. The lower portion of Figure 14.23 illustrates the composition of the initial vector distance table for router R2.

As previously mentioned, under a vector distance protocol the contents of each router's routing table are periodically broadcast. Assuming that routers R1 and R2 broadcast their initial vector distance routing tables, each router uses the received routing table to update its initial routing table. Figure 14.24 illustrates the result of this initial routing table update process for routers R1 and R2.

As additional routing tables are exchanged, the routing table in each router will converge with respect to the internetwork topology. However, to ensure that each router knows the state of all links, routing tables are periodically broadcast by each router. Although this process has a minimal effect on small networks, its use with large networks can significantly reduce available bandwidth for actual data transfer. This is because the transmission of lengthy router tables will require additional transmission time in which data cannot flow between routers.

Popular vector distance routing protocols include the TCP/IP Routing Information Protocol (RIP), the AppleTalk Routing Table Management Protocol (RTMP), and Cisco's Interior Gateway Routing Protocol (IGRP).

Routing Information Protocol

Under RIP participants are either active or passive. Active participants are normally routers that transmit their routing tables, while passive machines listen and update their routing tables based upon information supplied by other devices. Normally host computers operate as passive participants, while routers operate as active participants.

Under RIP an active router broadcasts its routing table every 30 seconds. Each routing table entry contains a network address and the hop count to the network. To illustrate an example of the operation of RIP, let us redraw the network previously shown in Figure 14.17 in terms of its links and nodes, replacing the four routers by the letters A, B, C and D for simplicity of illustration. Figure 14.25 contains the revised network consisting of four nodes and five links.

When the routers are powered up they only have knowledge of their local conditions. Thus, each routing table would contain a single entry. For example, the table of router n would have the following value:

From n to	Link	Hop count
n	local	0

For the router represented by node A, its table would then become:

From A to	Link	Hop count
n	local	0

Thirty seconds after being turned on, node A will broadcast its distance vector (A = 0) to all its neighbors, which in Figure 14.25 are nodes B and C. Node B receives on link 1 the distance vector A = 0. Upon receipt of this message, it updates its routing table as follows, adding 1 to the hop count associated with the distance vector supplied by node 1:

From B to	Link	Hop count
B	local	0
A	1	1

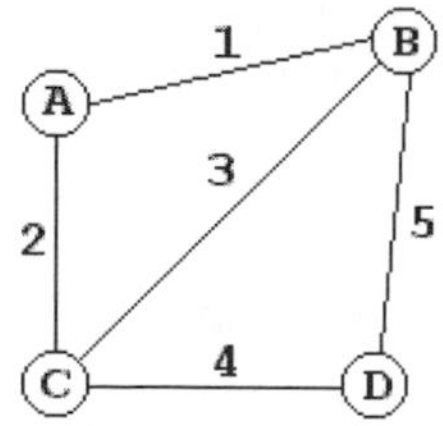

Figure 14.25 Redrawing the network in Figure 14.17 in terms of its links and nodes

Node B can now prepare its own distance vector (B=0, A=1) and transmit that information on its connections (links 1, 3 and 5).

During the preceding period node C would have received the initial distance vector transmission from node A. Thus, node C would have updated its routing table as follows:

From C to	Link	Hop count
C	local	0
A	2	1

Once it updates its routing table, node C will then transmit its distance vector (C=0, A=1) on links 2, 3 and 4.

Assuming the distance vector from node B is now received at nodes A and C, each will update their routing tables. Thus, their routing tables would appear as follows:

From A to	Link	Hop count
A	local	0
B	1	1

From C to	Link	Hop count
C	local	0
A	2	1
B	3	1

At node D, its initial state is first modified when it receives the distance vector (B=O, A=1) from node B. Since D received that information on link 5, it updates its routing table as follows, adding 1 to each received hop count:

From D to	Link	Hop count
D	local	0
B	5	1
A	5	2

Now, let's assume node D receives the update of node C's recent update (C=0, A=1, B=1) on link 4. As it does not have an entry for node C, it will add it to its routing table by entering C=1 for link 4. When it adds 1 to the hop count for A received on link 4, it notes that the value is equal to the current hop count for A in its routing table. Thus, it discards the information about node A received from node C. The exception to this would be if the router maintained alternate routing entries to use in the event of a link failure. Next, node D would operate upon the vector B=1 received on link 4, adding 1 to the hop count to obtain B=2. Since that is more hops than the current entry, it would discard the received distance vector. Thus, D's routing table would appear as follows:

From D to	Link	Hop count
D	local	0
C	4	1
B	5	1
A	5	2

The preceding example provides a general indication of how RIP enables nodes to learn the topology of a network. In addition, if a link should fail, the condition can be easily compensated for as, like bridge table entries, those of routers are also time stamped and the periodic transmission of distance vector information would result in a new route replacing the previously computed one.

Link state protocols

A link state routing protocol addresses the traffic problem associated with large networks that use a vector distance routing protocol. It does this by transmitting routing information only when there is a change in one of its links. A second difference between vector distance and link state protocols concerns the manner in which a route is selected when multiple routes are

Table 14.2 Common routing protocols

AURP	AppleTalk Update Routing Protocol. This routing protocol is implemented in Apple networks and sends changes to routing tables via updates.
BGP	Border Gateway Protocol. This is a TCP/IP interdomain routing protocol.
CNLP	Connectionless Network Protocol. This is the OSI version of the IP routing protocol.
DDP	Datagram Delivery Protocol. This routing protocol is used in Apple's AppleTalk network.
EGP	Exterior Gateway Protocol. This TCP/IP protocol is used to locate networks on another domain.
IDRP	Inter-Domain Routing Protocol. This is the OSI interdomain routing protocol.
IGP	Interior Gateway Protocol. This TCP/IP protocol is used by routers to move information within a domain.
IGRP	Interior Gateway Routing Protocol. This is a proprietary routing protocol developed by Cisco Systems,
IP	Internet Protocol. The network layer protocol of the TCP/IP (Transmission Control Protocol/ Internet Protocol) suite of protocols.
IPX	Internet Packet Exchange. This routing protocol is based on Xerox's XNS, was developed by Novell, and is implemented in Novell's NetWare.
IS-IS	Intermediate System to Intermediate System. This is an OSI link-state routing protocol which routes both OSI and RIP traffic.
NCP	NetWare Core Protocol. This is Novell's NetWare specific routing protocol.
OSPF	Open Shortest Path First. This is a TCP/IP link state routing protocol which can be viewed as an alternative to RIP.
RIP	Routing Information Protocol. This routing protocol is used in TCP/IP, XNS, and IPX. Under RIP, a message is broadcast to find the shortest route to a destination based on a hop count.
SPF	Shortest Path First. This link state routing protocol uses a set of user-defined parameters to find the best route between two points.

available between destinations. For a vector distance protocol, the best path is the one that has the fewest number of intermediate routers on hops between destinations. In comparison, a link state protocol can use multiple paths to provide traffic balancing between locations. In addition, a link state protocol permits routing to occur based on link delay, capacity and reliability. This provides the network manager with the ability to specify a variety of route development situations.

Over the past decade over 50 routing protocols have been developed, ranging from 'standardized' protocols that are defined in RFCs to vendor proprietary protocols, Table 14.2 provides a summary of 16 common routing protocol mnemonics, what the mnemonics represent, and a short description of each protocol.

14.4 REVIEW QUESTIONS

1. How does a transparent bridge construct its address table?

2. Why do transparent bridges time-stamp entries in their port-address tables?

3. What is the major weakness of a transparent bridge?

4. What is the difference between the physical and logical structure of a network?

5. What is the major function performed by the spanning tree protocol?

6. What is the function of a HELLO Bridge Protocol Data Unit?

7. Why is it important to ensure that a bridge used to connect Ethernet and Token-Ring LANs does not support the use of Token-Ring frames exceeding 1500 bytes?

8. What is the purpose of discovery packets?

9. Discuss three operations routers can perform that bridges cannot.

10. Where are tables maintained under transparent bridging and source routing bridging?

11. What are the advantages associated with using source routing transparent bridges?

12. What is the key advantage associated with the use of a switching hub?

13. Assume you purchased a 64-port 10BASE-T switching hub. What is the minimum backplane operating rate required to ensure no blocking occurs?

14. Describe three switching techniques.

15. Describe the difference between port and segment switching with respect to the number of addresses required to be supported by a switch.

16. Compare and contrast the basic operations of bridges and routers.

17. What does the term routing protocol refer to?

18. Discuss a hardware advantage obtained from the use of a frame relay service.

19. Discuss the reason why some LAN protocols are not routable.

20. How does a router interconnect LANs using an IBM LAN protocol?

21. What is the difference between an interior domain and exterior domain routing protocol?

22. What do the terms vector and distance reference in the vector distance router protocol?

23. Assume the following network:

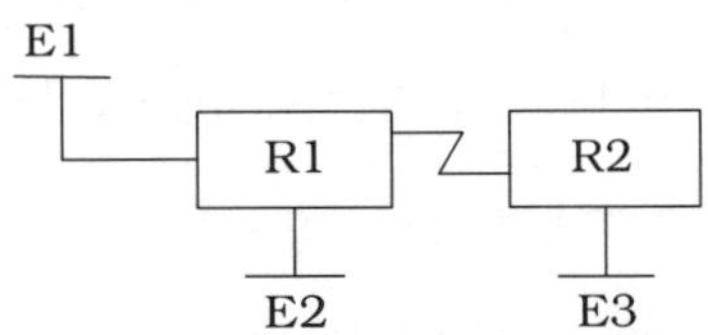

Define the initial routing tables for routers R1 and R2.

24. What is the advantage of a link state protocol in comparison to a vector distance protocol?

15

DIGITAL TRANSMISSION SYSTEMS AND EQUIPMENT

In this chapter we will focus our attention on digital transmission systems and equipment. Since digital transmission facilities were originally restricted to use by communications carriers, we will first examine the evolution of the primary digital communications facility which is also popularly referred to as the T1 carrier in North America and the E1 carrier in Europe. This examination will include the method by which telephone companies originally digitized voice conversations for transmissions on a T1 or E1 carrier and the framing used to provide synchronization between equipment operating on those carriers.

In the second section of this chapter we will examine the operation and utilization of the T1/E1 multiplexer. Although similar in operation to a time division multiplexer, the T1/E1 multiplexer can support voice, data and video, while a conventional TDM is normally restricted to supporting data.

Based on our discussion of the T1 and E1 carrier and multiplexer we will examine transmission services whose capacity is derived from the use of those carriers. For our discussion of T1 and E1 carriers and derived services we will focus attention on North America and European facilities, providing us with the ability to compare and contrast service offerings.

The explosive growth in the use of the Internet for electronic commerce resulted in many organizations requiring multiple T1 connections to support Web server connections when their server was accessed by tens of thousands of persons each day. Although a T3 transmission facility operating at approximately 45 Mbps was very costly only a few years ago, advances in the installation of fiber connections and the demand for higher Internet access data rates resulted in its increased use by business, government agencies and academic institutions. Due to its declining cost T3 transmission facilities, while still relatively expensive, are increasingly being used and are also covered in this chapter.

Since no discussion of digital transmission would be complete without mentioning the Integrated Services Digital Network (ISDN), we will conclude this chapter on this subject. Although ISDN field trials were completed at many locations in the early 1990s, its adoption until recently was slow to materialise. The growth in the use of video conferencing, LAN interconnectivity and similar applications can be expected to increase the use of this technology.

15.1 THE T AND E CARRIERS

The T1 and E1 carriers were originally developed in the early 1960s as facilities to relieve cable congestion problems in urban areas. The use of T1 and E1 carriers was originally restricted to communications carriers and not offered on a subscriber basis to commercial customers. During the 1970s T1 and E1 circuits became commercially available; however, their relatively high monthly lease restricted their use primarily to Fortune 500 corporations and large government agencies. During the 1980s communications carriers installed tens of thousands of miles of fiber-optic cable which greatly increased the number of T1 and E1 circuits available for customers. Since the use of fiber-optics resulted in a significant increase in the availability of bandwidth, communications carriers significantly lowered monthly cost of T1 and E1 circuit miles. For some users and potential users of T1 circuits the monthly cost per mile was reduced from $14 to below $3, resulting in many organizations replacing multiple low speed analog or digital circuits by a single T1 or E1 carrier. Recognizing that even though the reduction in the cost of T1 and E1 circuits made them economically viable for an expanded base of users, and that their cost was still prohibitive for other users, communications carriers introduced fractional T1 (FT1) and fractional E1 (FE1) services, leasing increments of T1 and E1 circuits based on a fraction of the cost of full capacity T1 and E1 circuits. This resulted in an expanded market for digital transmission facilities and was followed by communications carriers introducing a similar marketing strategy concerning the T3 carrier, with fractional T3 becoming available for customers during the mid-1990s.

15.1.1 Channel banks

When used by a communications carrier, a T1 or E1 system is formed by the use of two channel banks connected to one another by a span line as illustrated in Figure 15.1. The channel bank contains a group of coder–decoder equipment whose acronym is codec, as well as a time division multiplexer and line driver. Each codec performs analog-to-digital (A/D) and digital-to-analog (D/A) signal conversion. Channel banks used in North America contain 24 codecs, whereas those used in Europe normally contain 30.

Digital signaling

As previously discussed in Chapter 4 a new method of digital signaling was developed for use on digital transmission facilities resulting from the desire of communications carriers to use a common cable for both powering repeaters as well as for transferring information. That signaling is bipolar non-return to zero with a 50% duty cycle and is commonly referred to as Alternate Mark

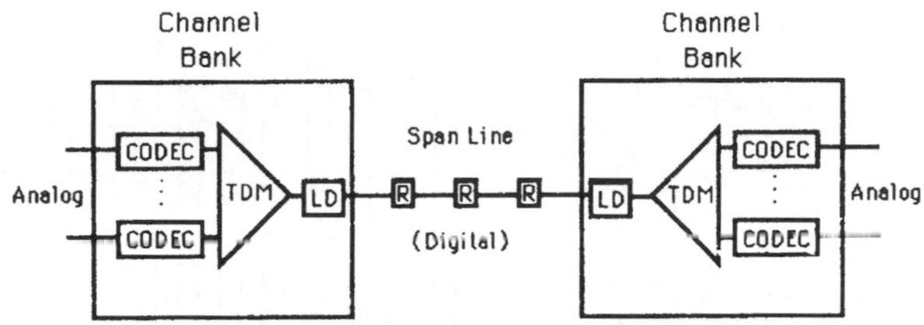

Figure 15.1 Communications carrier T1 system. A communications carrier T1 system includes two channel banks connected to one another by a span line operating at a 1.544 Mbps data rate (LD line driver, R repeater, TDM time division multiplexer)

Inversion (AMI). You should review the digital signaling section in Chapter 4 to obtain an appreciation of the signaling used on span lines.

Line driver operation

The line driver converts the unipolar signaling output of the multiplexer in the channel bank into a bipolar signal which propagates down the span line. Other functions performed by the line drivers in each channel bank relate to the electrical characteristics of the span line; that is, the line driver ensures that pulses are of uniform height and width and of correct voltage to match the specifications of the electrical characteristics of the span line. Thus, a line driver in a channel bank essentially functions as a DSU/CSU.

Pulse code modulation

Over the last 30 years, a variety of techniques were developed to digitize voice signals. Among the earliest techniques, and the one most often employed by telephone companies to convert analog voice conversations into a digital data stream, is pulse code modulation, commonly referred to as PCM.

The key to the PCM process is the use of the codec which performs four functions: sampling, quantizing, coding and decoding. The first three functions performed by the codec result in the conversion of an analog signal into a digital signal, whereas the fourth function results in the reconstruction of the original analog signal from the received digital signal.

Sampling

Nyquist's theorem states that all of the information of a sampled message can be obtained if the samples are taken at a rate which is at least twice that of the value of the highest significant signaling frequency in the bandwidth. Since a voice-grade telephone call has its highest frequency at 4 kHz, even though actual speech is transmitted from 300 to 3300 Hz, then 8000 samples must be taken each second.

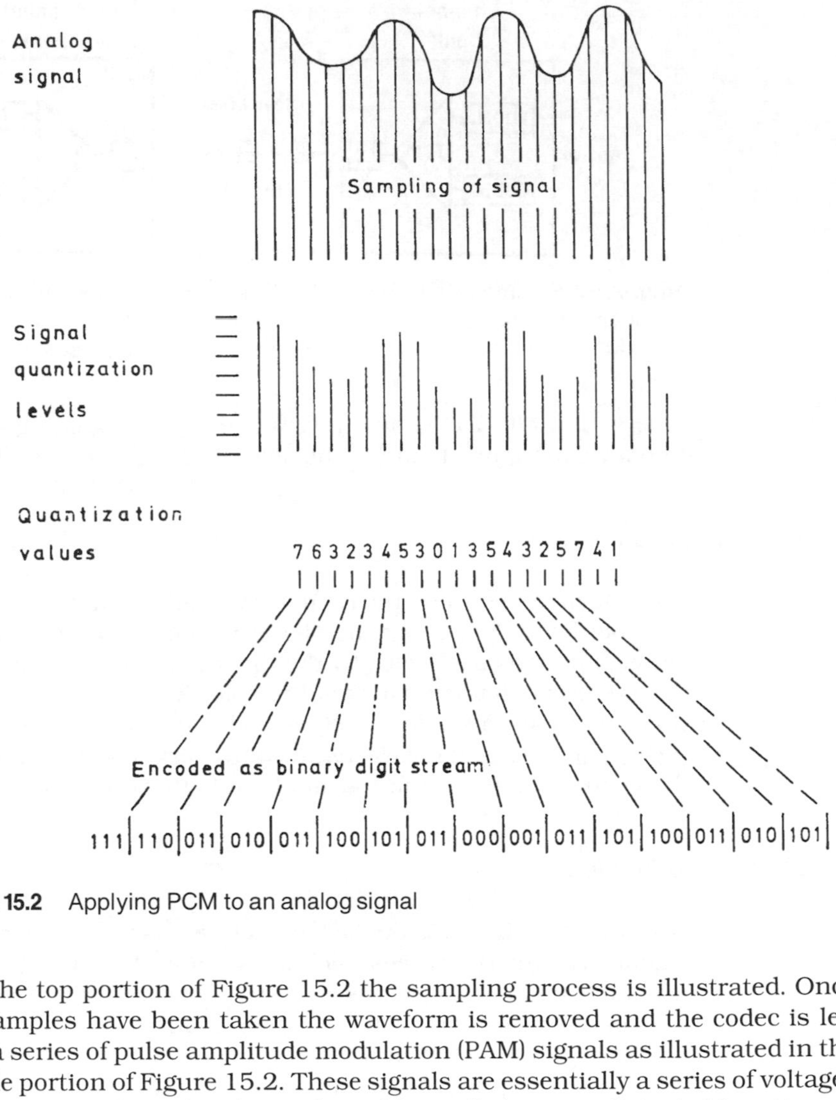

Figure 15.2 Applying PCM to an analog signal

In the top portion of Figure 15.2 the sampling process is illustrated. Once the samples have been taken the waveform is removed and the codec is left with a series of pulse amplitude modulation (PAM) signals as illustrated in the middle portion of Figure 15.2. These signals are essentially a series of voltages at the indicated amplitudes and would rapidly become distorted by attenuation and noise if transmitted in PAM form. Thus, a method is required to encode the PAM signals into digital form.

Since the PAM signals have an infinite number of amplitude levels, they must be coded into discrete steps. The process of reducing a PAM signal to a limited number of discrete amplitudes is called quantization.

Quantization

Experiments have shown that the use of 2048 uniform quantizing steps provides the ability to obtain a sufficient capability to reproduce a voice signal of high quality. For 2048 quantizing steps, an 11-element code (2^{11}) would be required. Using a sampling rate of 8000 samples per second, the data rate required to digitize a voice channel would become 88 000 bps. This data rate is reduced by quantizing the PAM signal.

To reduce the number of quantum steps two techniques are commonly used: non-uniform quantizing, and companding prior to quantizing followed by uniform quantizing.

Most PCM systems employ companding. The compression, as well as eventual expansion, are based upon logarithmic functions that follow one of two laws: the A law and the 'mu' (μ) law. Each law defines how many quantizing levels are used to describe a sample and how those levels are arranged. The μ law, used in North America, divides a quantization scale into 255 discrete units of two different sizes called chords and steps. The chords are spaced logarithmically from each other, with each higher chord spaced further apart from the preceding chord. Within each chord the 16 steps are spaced linearly. Thus, the step size is larger in the larger chord.

The encoding of a PAM signal into a binary string is a three-part process based on the use of 255 levels in the μ-law. Figure 15.3 illustrates the encoding scale used for the μ-law and its relation to the encoded PCM word. Since the 0 value is shared among positive and negative PAM signal heights, there are $16 \times 16 - 1$ or 255 levels available for use.

In the encoding process a 0 bit in bit position 1 indicates a negative polarity and a 1 bit indicates a positive polarity. Thus, the first bit isolates the PAM signal to above or below the zero reference level. The three bits used in the chord field define one of eight chords. Since there are 16 chords, 8 positive and 8 negative, the first bit in conjunction with the 3 chord bits defines one of 16 chords. The remaining 4 bits permit 1 of 16 steps within a chord to be coded.

In Europe, A-law companding is used in ITU-T systems. The A-law uses a different logarithmic scale and incorporates the use of 13 segments in place of 16 chords. Similar to μ-law coding, A-law coding uses an 8-bit word with 1 hit for polarity, 3 bits for the segment and 4 bits for the steps in the segment.

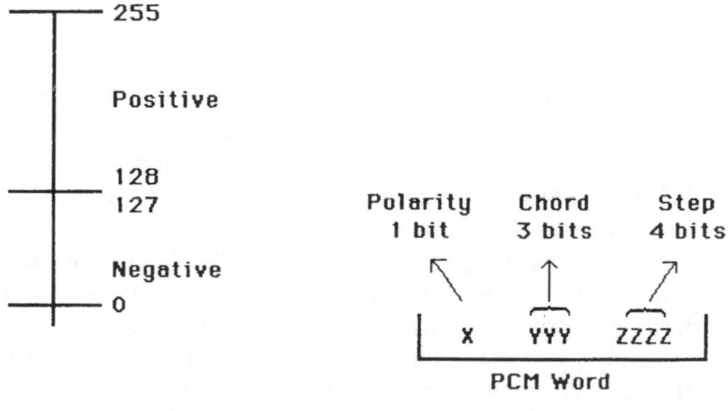

Figure 15.3 μ law encoding. Under μ-law encoding, 255 signal levels are definable. Each PAM is encoded into a PCM word consisting of one polarity bit, three chord bits and four bits used to represent the step in the chord

Irregardless of whether A-law or μ-law encoding is used, the end result is similar: the encoding of each PAM signal into an 8-bit PCM word. For the sake of simplicity the lower portion of Figure 15.2 illustrates encoding as a 3-bit word.

With 8 bits used to encode the amplitude of each sample the resulting data required to digitize voice is 8000 samples times 8 bits per sample per second or 64 kbps, which is also called a digital signal level zero or DS0 signal.

The 64 kbps data rate associated with PCM makes this voice digitization technique unsuitable for low-speed data networks which are based upon the utilization of analog 3002-type leased lines. This unsuitability is based on the maximum data rates obtainable on the previously mentioned facility, since, as an example, on 3002-typc lcased lines, a maximum data rate of 33.6 kbps has only recently been obtained through the use of trellis coded modulation modems. In addition, because a 64 kbps voice digitization rate equals the transmission rate on a DSO channel, you cannot multiplex multiple PCM encoded conversations on that channel. Due to its high data rate requirement, PCM is used to encode multiple voice conversations on T1 carriers which operate at 1.544 Mbps in North America and E1 carriers which operate at 2.048 Mbps in Europe.

TDM operations

The TDM in the channel bank scans the output of each codec in sequence, retrieving its 8-bit word and placing the word into the resulting multiplexed serial bit stream. In North America, the use of 24 codecs resulted in the multiplexing of 24 eight-bit words, one from each analog line input to the channel bank.

In Europe a different multiplexing scheme is employed, with 32 channels multiplexed, although one is normally used for synchronization and a second is used for transporting signaling information. Thus, a European 32 channel E1 circuit is normally restricted to carrying a maximum of 30 simultaneous voice conversations. Due to the differences in the composition of T1 and E1 circuits, we will examine TDM operations on each carrier as a separate entity.

North America

In North America the T1 circuit was developed to transport 24 digitized voice conversations, each operating at 64 kbps. Each voice conversation is sampled 8000 times per second using 8 bits per sample. Thus, one sample of all 24 channels results in a grouping of 24 PCM words which is referred to as a frame.

A single framing bit is inserted by the TDM to separate each frame from the next frame. Thus, the resulting frame has 193 bits and represents 24 voice digitized conversations. When transmitting on a T1 carrier this framing technique is called Digital Signal Level 1 (DS1) framing. Here DS1 represents digital signal level one whose resulting data rate of 1.544 Mbps in North America results from 24 voice signals being sampled 8000 times per second (193 bits $\times$ 8000). Figure 15.4 illustrates the composition of a DSI signal flowing on a span line.

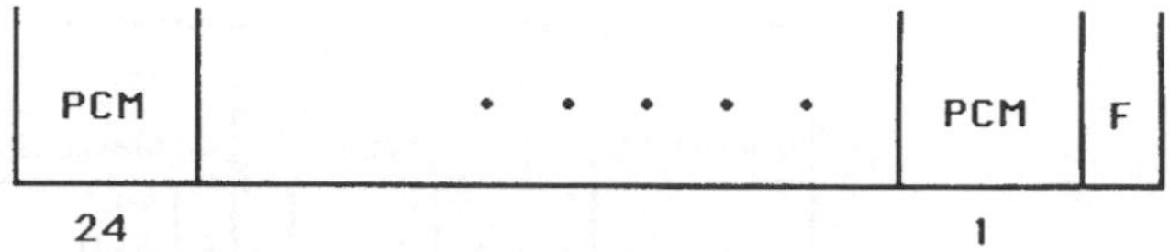

Figure 15.4 The DS1 frame. (F framing bit, PCM 8-bit word representing nearest digital value of the analog signal at the time of sampling)

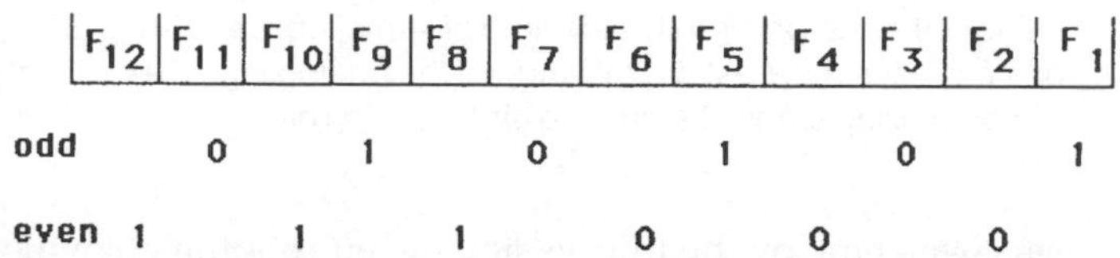

Figure 15.5 D4 superframe synchronization pattern. The framing bits from 12 frames under DS4 signaling form a distinct pattern that is repeated every 12 frames

Synchronization and D4 framing

For synchronization purposes, the basic frame of 193 bits was formed into a superframe of 12 frames. Here the framing bits are altered by the transmitting multiplexer to enable the receiving multiplexer to maintain synchronization with the received data.

The composition of framing bits within a superframe form a distinct pattern which is illustrated in Figure 15.5. This framing sequence, in which the 12-bit frame pattern continuously repeats itself, is called D4 framing. Note that the odd-numbered frames in Figure 15.5 alternate the sequence 1010..., and the even numbered frames alternate the sequence 000111... .

By counting the repeating pattern formed by the 193rd (framing) bit, the multiplexer in the channel bank always know where the receiving DS-0 signal is in the bit stream. To illustrate this, assume that the received framing bit pattern for the last four frames is 0010. From Figure 15.5, the next frame that will be received must be F_7, since the 0010 pattern only occurs in frame bits 3 through 6.

The odd framing bits in the D4 superframe are referred to as terminal framing (Ft) bits and are used to synchronize the bit stream. In comparison, the even framing bits are referred to as signal framing (Fs) bits and are used to define multiframe boundaries. In addition, the Fs bits are used to identify what is known as robbed bit signaling which occurs in frames 6 and 12 of the D4 superframe.

Signaling and robbed bits

With the DS1 signal representing 24 PCM encoded voice signals and the D4 frame bit pattern used for synchronization, both the signal and its frame pattern lack the capability to convey signaling information. This limitation

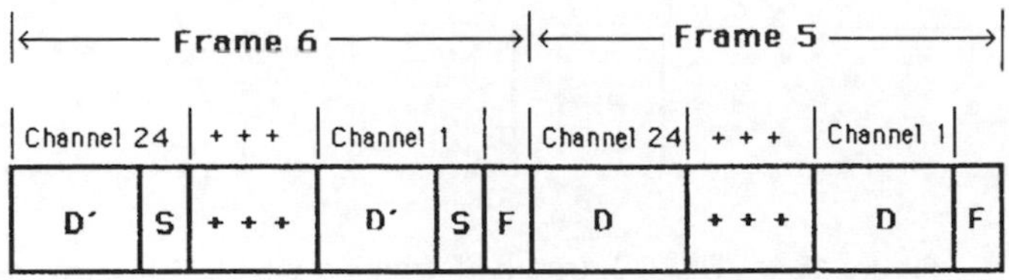

Figure 15.6 Bit robbing on a DS1 signal. Bit robbing uses the least significant bit of each PCM word in the sixth and twelfth frame for signaling. (D 8 data bits represent nearest digital value of the analog signal at the time of sampling, permitting 256 quantizing steps. D′ 7 data bits represent nearest digital value of the analog signal at the time of sampling, permitting 128 quantizing steps, S signaling bit, F framing bit)

was overcome by the impression of 'bit robbing', in which the least significant bit in the sixth and twelfth frame is 'stolen' for use as a signaling bit to convey such mechanical signaling information as 'on-hook' and 'off-hook' conditions.

The bit stolen from the sixth frame in a DSI signal is known as the 'A' bit, and the bit stolen from the twelfth frame is known as the 'B' bit. When E&M (ear and mouth) signaling is supported the A and B bits are used to indicate the status of the T1 circuit, with 00 used to denote an idle circuit condition. Here the term circuit refers to each of the 24 PCM channels that could carry a voice conversation. Figure 15.6 illustrates DS1 signaling based upon the use of robbed bits.

The bit robbing process has essentially no effect upon human hearing due to the manner in which it is effected. First, the use of the least significant bit in frames 6 and 12 means that the PCM word is only one step removed from the actual sample value. Secondly, since the frames are transmitted at 125 μs intervals, the 7-bit quantized value in frames 6 and 12 are rapidly replaced by an 8-bit quantized value from frames 1 and 7 just 125 μs later, with the 7-bit quantized value analog reconstruction thus basically imperceptible to the human ear.

Although bit robbing has no effect on voice transmission it obviously has a significant effect upon data transmission. This explains why AT&T's Dataphone Digital Service (DDS) and digital offerings by other communications carriers that flow over a T1 carrier using a DS1 signal are limited to a maximum data rate of 56 kbps instead of the DS-0 signal rate of 64 kbps. In Europe where signaling occurs on a separate channel outside of the voice digitized channels, digital service offerings equivalent to DDS have a maximum data rate of 64 kbps.

Extended superframe format

D4 framing only provides an indirect measurement of line quality through the monitoring of frame bits. Another limitation of the D4 format is the fact that obtaining a communications capability between devices on a T1 circuit required the use of a DSO time slot. To alleviate these problems, as well as to provide the T1 user with additional capability, AT&T introduced an extended superframe format in early 1985. Although this new framing format required

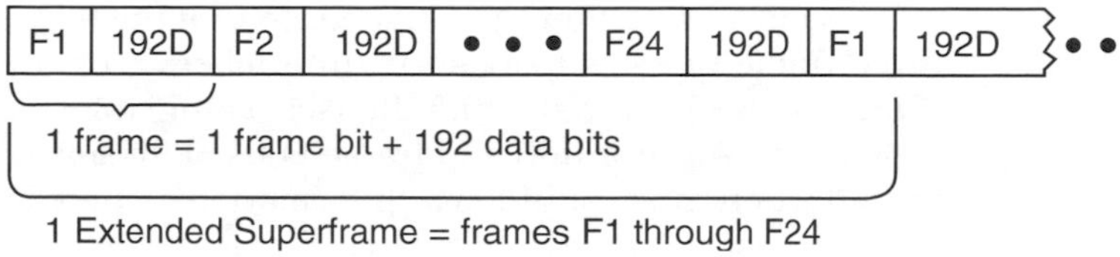

Figure 15.7 The extended superframe

the installation of equipment that supports the frame format, within a few years a majority of T1 circuits in North America conformed to it. Eventually, the extended superframe format can be expected to replace completely D4 framing.

Denoted as Fe and ESF, the extended superframe format extends D4 framing to 24 consecutive frame bits (F1 through F2), as illustrated in Figure 15.7.

Unlike D4 framing in which the 12 framing bits form a specific repeating pattern, the ESF pattern can vary. ESF consists of three types of frame bits.

ESF 'd' bits

The ESF 'd' bits, which represent a derived data link, are used by the telephone company to perform such functions as network monitoring to include error performance, alarm generation and reconfiguration to be passed over a T1 link. The 'd' bits appear in the odd frame positions, e.g. 1, 3, ..., 21, 23. Since they are used by 12 of the 24 framing bits, the 'd' bits represent a 4 kbps data link.

The data link formed by the 12 'd' frame bits is coded into higher-level data link control (HDLC) protocol format known as BX.25. Figure 15.8 illustrates the data link format carried by the 'd' bits.

The flag byte consists of the 8-bit sequence 01111110 and initiates and terminates each frame. The address field is used to identify a frame as either a command or response. A command frame contains the address of the device to which the command is being transmitted, and a response frame contains the address of the device sending the frame.

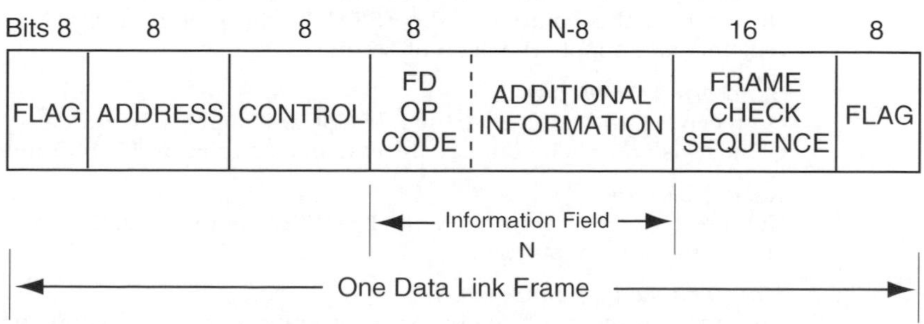

Figure 15.8 ESF data link format

The control field identifies the purpose of the frame and can indicate one of three frame types: supervisory, unnumbered, or information. A supervisory frame is used for data link housekeeping information, such as acknowledgements. An unnumbered frame is used for major system commands, line initialization and shutdown information, whereas an information frame contains user data. The frame check sequence (FCS) is a 16-bit CRC check used to insure the integrity of the data link.

One of the primary goals in the development of the BX.25 protocol was to provide a mechanism to extract performance information from ESF compatible CSUs. Doing so allows circuit quality monitoring without taking the circuit out of service and is a major advantage of ESF over D4 framing. Standard maintenance messages that are defined in AT&T's publication 54016 include messages that can return performance data concerning the number of errored seconds (ES), severely errored seconds (SES), and failed seconds (FS). An errored second is a second that contains one or more bit errors, and a severely errored second is considered to be a second with 320 or more bits in error. If ten consecutive severely errored seconds occur, this condition is considered as a failed signal state. Then each signal in a failed

Table 15.1 ESF data link maintenance messages

Send one hour performance data
On receiving this command the ESF CSU will supply the following: current status, elapsed time of current interval, ES and FS in the current 15 minute interval, number of valid intervals, count of ES and FS in 24 hour register, ES and FS during the previous four 15 minute intervals.

Send 24 hour ES performance data
On receiving this command, the ESF CSU will supply the following: current status, elapsed time of current interval, ESs and FSs in current 15 minute interval, number of valid intervals, count of ESs and FSs in 24 hour register, and ESs during previous ninety-six 15 minute intervals.

Send 24 hour FS performance data
On receiving this command, the ESF CSU will supply the following: current status, elapsed time of current interval, ESs and FSs in the current 15 minute interval, number of valid intervals, count of ESs and FSs in 24 hour register, and FSs during previous ninety-six 15 minute intervals.

Reset performance monitoring counters
On receiving this command, the ESF CSU will reset all interval times and ES and FS registers and supply the current status.

Send errored ESF data
On receiving this command, the ESF CSU will supply current data present in ESF error event registers. Each count represents one error event (65535 maximum).

Reset ESF register
Upon receiving this command, the ESF CSU will reset the ESF error event register and supply the current status.

Maintenance loop-back (DLB)
Energizes on receiving the proper code embedded in the 4 kbps data link. This loop-back loops through the entire CSU.

signal state is considered to be a failed second. Table 15.1 summarizes the standard maintenance messages transmitted on the ESF data link.

CRC-6 code bits

Frame bits 2, 6, 10, 14, 18 and 22 are used for a CRC-6 code. The 6-bit cyclic redundancy check sum is used by the receiving equipment to measure the circuit's bit error rate and represents 2 kbps of the 8 kbps framing rate. The CRC employs a mathematical algorithm which is used to check all 4632 bits in the ESF. Mathematically, the CRC check bit generation is performed by the use of a fixed polynomial whose composition is $X_6 + X + 1$, or 1000001 1. The data block is first multiplied by X_6 or 1000000 and then divided by the polynomial. The remainder is then transmitted in the six ESF CRC bit positions. At the receiver, a similar operation is performed on the received data block using the same polynomial and the locally generated check bit sequence. The CRC-6 code yields an accuracy of 98.4%, and the occurrence of a mismatch between the locally generated check bit sequence and the received check bit sequence indicates that one or more bits in the extended superframe are in error.

To conform with ESF CRC-6 coding and reporting requirements, the use of an ESF compatible channel service unit is required. This CSU not only generates the CRC-6 but must also be capable of detecting CRC errors and storing a CRC error count over a 24 hour period. ESF compatible CSUs contain buffer storage which enables the device to store current line status information, including all error events and errored and failed seconds for the current 15 minute period and the previous ninety-six 15 minute periods that represent the prior 24 hour period. To enable the carrier to retrieve these data, as well as to reset any or all counters and activate or deactivate loop-back testing on the local span line, the CSU must also have the ability to respond to network commands. Thus, an ESF-compatible CSU must have the capability to send and receive data based on the BX.25 formation via the 4 kbps data link.

A competing standard from the T1E1 committee requires the CSU to broadcast this information all of the time instead of on request from the carrier. This standard is designed to overcome the potential problem resulting from an IEC resetting CSU registers prior to an LEC reading the CSU registers.

Framing pattern

The third type of frame bits is used to generate the framing pattern. Here, frame bits 4, 8, 12, 16, 20 and 24 are used to generate the Fe framing pattern whose composition is 001011. These six bits result in a 2 kbps framing pattern. Another difference between the ESF and D4 frame format is in the area of signaling. ESF has added two additional signaling bits, C and D, in frames 18 and 24. Thus, in ESF signaling data is accommodated by using bit robbing in frame 6 (A-bit), frame 12 (B-bit), frame 18 (C-bit) and frame 24 (D-bit). In Table 15.2 you will find a summary of the ESF framing pattern.

Table 15.2 ESF framing pattern

Frame	Bit composition	Frame	Bit composition
1	d	13	d
2	C1	14	C4
3	d	15	d
4	0	16	0
5	d	17	d
6	C2	18	C5
7	d	19	d
8	0	20	1
9	d	21	d
10	C3	22	C6
11	d	23	d
12	1	24	1

Clocking

Clocking, or timing between channel banks, is primarily accomplished by having one end of a span line function as a master, while the opposite end functions as a slave.

Under the master–slave relationship, the slave takes its timing from the data transmitted by the master side as illustrated in Figure 15.9. This method of timing is also known as loop timing, since the slave recovers clocking from the T1 loop.

The master channel bank can receive its timing from one of several sources: from another channel bank, from a carrier's central office switch or from a master reference clock. The latter is called the Basic System Reference Frequency (BSRF) and is provided by an atomic clock located in Hillsboro, MO, the geographic center of the United States. The BSRF clock's location provides a uniform propagation delay due to its geographical location and it is known as a Stratum 1 clock. This clock is used to supply a hierarchy of timing to all carrier offices that use channel banks. Figure 15.10 illustrates the hierarchy of timing supplied from AT&T's master reference clock. As shown, the resulting subsystem is a tree-like network containing no closed loops.

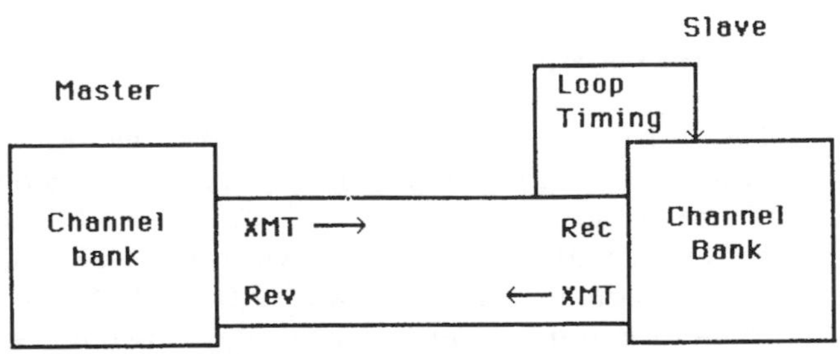

Figure 15.9 Loop timing. In loop timing the slave channel bank maintains timing from the received data stream

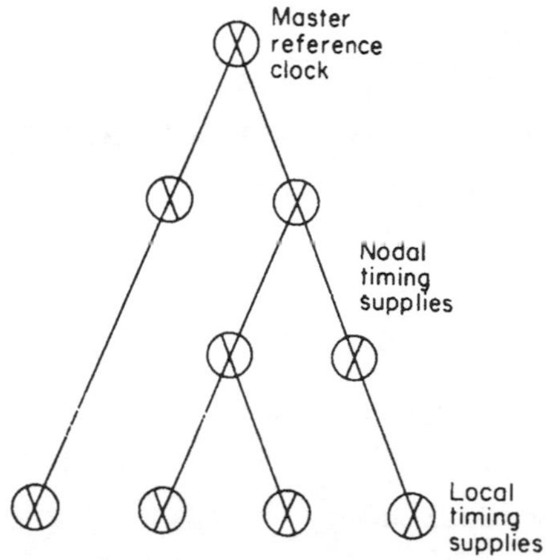

Figure 15.10 Channel bank timing. A hierarchy of timing provides synchronization between channel banks

Bipolar violations and timing

Repeaters are placed approximately every 6000 feet on span lines to regenerate digital pulses. The repeater uses incoming marks (1s) to maintain timing, since each mark represents a distinct voltage from which the repeater can take its timing.

If too many binary zeros occur in sequence the repeater's timing recovery circuits can become confused, resulting in a loss of timing. Although clocks could be built into each repeater, the cost of doing so would most likely be prohibitive. Thus, a more practical solution to maintaining repeater timing is to provide an adequate number of ones for timing recovery.

The maximum number of consecutive zeros that repeaters can tolerate prior to causing a timing problem is 15. In addition, at least one mark must be present in every eight bits. To meet this 'ones density' requirement channel banks were initially designed to invert the least significant bit of a PCM word which resided in a string of 15 zeros. Figure 15.11 illustrates this technique which minimizes voice distortion since, like E & M signaling, the least significant bit is used.

In the mid-1980s AT&T began implementing a new signaling method called bipolar 8 zero substitution (B8ZS). B8ZS signaling provides a significant improvement in providing a required ones density as it results in an intentional bipolar violation that is correctable. Hence it does not result in the altering of data.

Under B8ZS coding each eight consecutive 0s in a byte are removed and replaced by a B8ZS code. If the pulse preceding an all-zero byte is positive, the

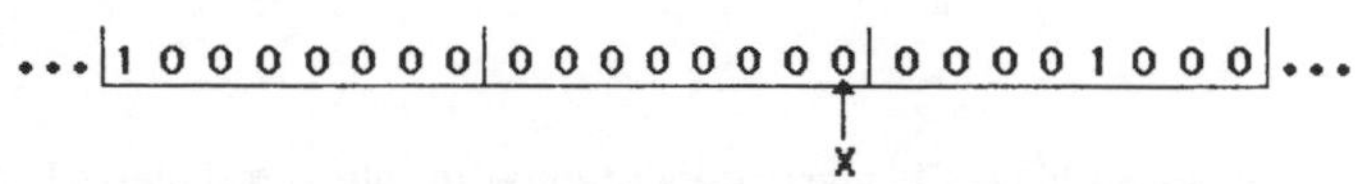

Figure 15.11 Ensuring ones density. By changing bit *X*, which is the least significant bit position from a zero to one, ones density is assured when a string of 15 zeros is encountered

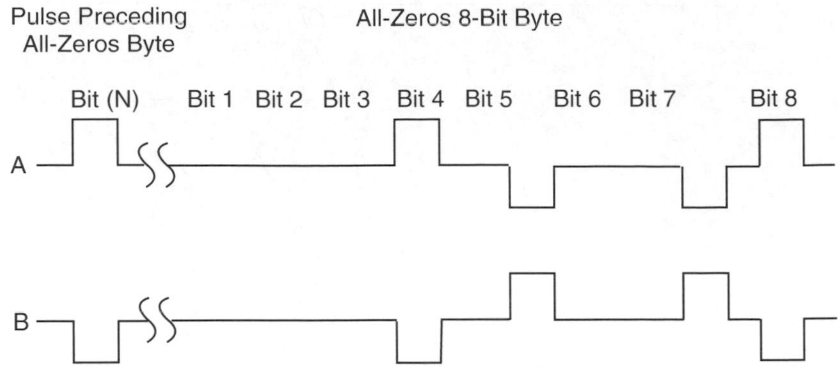

Figure 15.12 B8ZS coding

inserted code is 000+−0−+. If the pulse preceding an all-zero byte is negative, the inserted code is 000−+0−+. Figure 15.12 illustrates the use of B8ZS coding in which an all-zeros byte is replaced by one of two binary codes, with the actual code used based on whether the pulse preceding the all-zeros byte was positive or negative.

Both examples result in bipolar violations occurring in the fourth and seventh bit positions. Both carrier and customer equipment must recognize these codes as legitimate signals, and not as bipolar violations or errors, for B8ZS to work to enable a receiver to recognize the code and restore the original eight zeros.

T1 alarms and error conditions

There are several alarms and error conditions that are monitored and reported under the T1 D4 and ESF formats. Principal T1 alarms include a red alarm which is produced by a receiver to indicate that it has lost frame alignment and a yellow alarm which is returned to a transmitting terminal to report a loss of frame alignment at the receiving terminal. Normally, a T1 terminal will use the receiver's red alarm to request that a yellow alarm be transmitted.

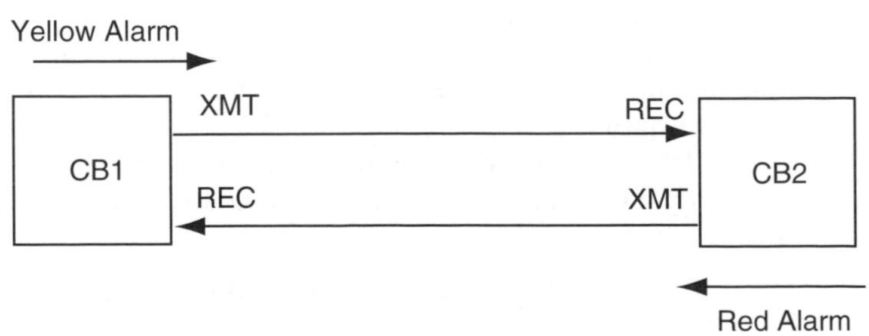

Figure 15.13 T1 alarm generation example. A red alarm is produced by a receiver to indicate it has lost frame alignment, and a yellow alarm is returned to a transmitting terminal to report a loss of some alignment at the receiving terminal

Table 15.3 T1 alarm formats

Mode	Format
Transmitted red alarm	
T1 D4	bit 2 = 0 in all data channels and Fs = 1 in frame 12
T1 ESF	repeated pattern of 8 zeros, 8 ones on data link
Yellow alarm generated at receiver	
T1 D4	bit 2 = 0 for 255 consecutive channels and Fs = 1 in frame 12
T1 ESF	16 patterns of 8 zeros, 8 ones on data link

To illustrate the operation of T1 alarms, consider the configuration illustrated in Figure 15.13 that shows two channel banks connected together via the use of a four-wire T1 circuit. If the second channel bank (CB2) loses synchronization with the first channel bank (CB1) or completely loses the signal it then transmits a red alarm to CB1. The transmission of the red alarm is accomplished by CB2 forcing bit position 2 in each PCM word to zero and the frame signaling bit in frame 12 to a binary 1 under the D4 format. If ESF framing is used, CB2 sends a repeated pattern of eight zeros and eight ones on the data link.

When CB1 recognizes the red alarm, it transmits a yellow alarm. This alarm, in effect, says 'I'm sending data but the other end is not receiving the transmitted information and the problem is elsewhere'. Under D4 framing the yellow alarm is generated at the receiver by setting bit 2 to zero for 255 consecutive channels and the frame alignment signal (Fs) to one in frame 12. Under ESF a pattern of eight zeros and eight ones repeated 16 times is used to indicate a yellow alarm.

A red alarm is generated when either the network or a DTE senses an error in the framing bits for either two out of four or five framing bits and this condition persists for more than 2.5 seconds. Table 15.3 summarizes the method of transmitting alarms on D4 and ESF T1 circuits.

In addition to red and yellow alarms, a third type of alarm that warrants attention is a blue alarm. This alarm, which is also known as an alarm indicating signal (AIS), is generated by a higher-order system (HOS), such as a T1C system operating at 3.152 Mbps. Figure 15.14 illustrates the generation of blue alarms when a failure between higher-order systems occurs. This alarm is generated by each HOS system to channel banks or multiplexers and,

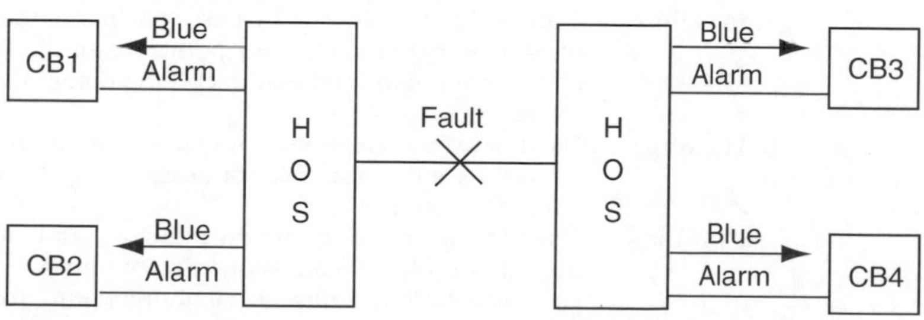

Figure 15.14 Blue alarm generation: HOS = higher order system, CB = channel bank

in effect, tells each lower-ordered system (DTE) that the problem is between higher ordered systems. Thus, this alarm is primarily used by communications carriers and avoids the dispatching of maintenance personnel to lower-order system facilities. Normally, a blue alarm is generated after 150 ms of loss of an incoming signal. This alarm is produced by transmitting a continuous ones pattern across all 24 channels.

Digital signal levels

In the United States, the telephone network's digital hierarchy contains several digital signals (DS) levels. With the exception of the first level known as DS0 signaling, each succeeding level is made up of a number of lower level signals.

Table 15.4 lists the four most widely used DS levels, their data rates, the numbers of voice channels they carry and use in a telephone network. In such networks, 24 DS0 signals form one DS1 signal, while four DS1s make up a DS2 and seven DS2s for one DS3 signal.

If you multiply the data rate of the lower level signal by the number of low level signals used to make up a higher level signal, the sum will be slightly less that the data rate of the higher level signal. This difference is used to frame the lower level signals onto the higher level signal and is not employed for actual data transmission. Thus multiplying the data rate of a lower level signal by the number of signals that make up a higher level signal can be used to obtain the actual data carrying rate of a DS level while the data rate in Table 15.4 represents the signaling rate of data and framing information.

In Section 15.3 when we examine the T3 carrier we will also examine the DS3 signal. In doing so we will focus our attention upon both framing bits and stuffing bits, with the latter used to compensate for clocking differences between digital signals during their multiplexing.

Table 15.4 Digital transmission hierarchy

Digital signal level	Data rate	Telephone company network use	Number of voice channels
DS0	64 kbps	Basic voice bandwidth data channel encoded via PCM; digital data service (DDS), analog/digital channel bank inputs	1
DS1	1.544 Mbps	This is the well known T1 carrier which consists of 24 DS0 signals. Used for point-to-point communications between telephone company offices	24
DS2	6.312 Mbps	Used between telephone company central offices as well as inter- and intra-building communications	96
DS3	44.736 Mbps	Used in high-capacity digital radio, coaxial cable and fiber-optic transmission systems for communications between telephone company offices	672

European E1 facilities

In Europe, a 32-channel system was developed to encode and multiplex voice signals in comparison to the 24-channel system used in the United States and other North American countries. Under the 32-channel system, 30 channels digitize voice signals resulting from incoming telephone lines, and the remaining two channels are used to provide signaling and synchronization information.

Frame composition

The PCM format used for time division multiplexing of 30 voice or data circuits onto a single twisted pair cable in Europe is referred to as CEPT PCM-30, whereas the European frame structure is standardized under ITU-T Recommendations G703/732. Each CEPT PCM-30 frame consists of 32 time slots to include 30 voice, one alignment and one signaling, with each time slot represented by 8 bits. Since each PCM channel is sampled 8000 times per second, the standard CEPT-30 data rate is $32 \times 8 \times 8000$, or 2.048 Mbps.

An alignment signal (0011011) is transmitted in bit positions 2 to 8 of time slot 0 in alternating frames. This signal is used to enable each channel to be distinguished at the receiver. Bit position 1 in time slot 0 carries the international bit, and frames not containing the frame alignment signal are used to carry national and international signaling and alarm indication for loss of frame alignment. Figure 15.15 illustrates the composition of the CEPT-30 frame and multiframe, where the multiframe consists of 16 frames, numbered from frame 0 to frame 15.

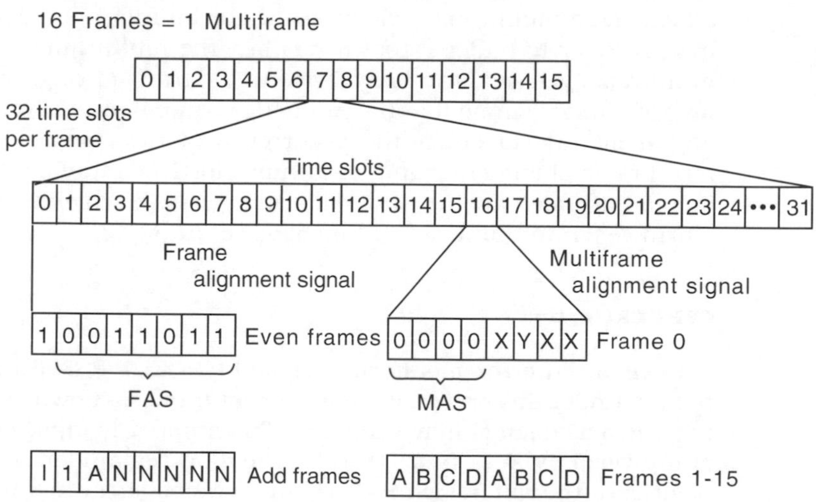

Figure 15.15 CEPT PCM-30 frame and multiframe composition: I – international bit, N = national bit, A = alarm indication signal, FAS = frame alignment signal, ABCD = ABCD signaling bits, X = extra bit for signaling, Y = loss of multiframe alignment, MA = multiframe alignment signal

To avoid imitation of the frame alignment signal, alternating frames fix bit 2 to a 1 in time slot 0, which is the reason why a 1 is entered into that bit position for odd time slot 0 frames.

Time slot 16 in each CEPT-30 frame is used to transmit such signaling data as on-hook and off-hook conditions, dialing digits and call progress. This is indicated by the characters ABCD for frames 1 to 15 in time slot 16 illustrated in Figure 15.15. Since a common channel is dedicated for the signaling data of all-voice circuits, this method of signaling is referred to as common channel signaling and enables each time slot to operate at 64 kbps. In comparison, T1 uses bit robbing to pass signaling information which reduces the effective data rate of time slots to 56 kbps. To enable each frame in a multiframe to be distinguished at the receiver for the recovery of ABCD signaling CEPT-30 uses a multiframe alignment signal. This signal, denoted by the symbol MAS in Figure 15.15, is transmitted in bit positions 1 through 4 of time slot 0 of frame 0.

CEPT alarms and error conditions

The principal alarms defined by the CEPT PCM-30 format include a red alarm which is produced by a receiver to indicate that it has lost frame alignment and a yellow alarm which is returned to the transmitting terminal to report a loss of frame alignment at the receiving terminal. These alarms function in the same way, as previously described in the section covering the T1 carrier.

Both red and yellow alarms are generated through the use of the alarm indication signal bit (bit 3) in time slot 0 (TS0) of odd frames. A red alarm is generated by setting bit $3 = 1$ in TS0 of non-frame alignment frames. A receiver then indicates the reception of a red alarm by generating a yellow alarm by setting bit $3 = 1$ in TS0 of non-alignment frames.

Two additional alarms generated by CEPT PCM-30 include a multiframe red alarm and a multiframe yellow alarm. The multiframe red alarm is produced by a receiver to indicate that it has lost the multiframe alignment, and the multiframe yellow alarm is returned to the transmitting terminal to report a loss of frame alignment at the receiving terminal. A receiver loses multiframe alignment due to either the occurrence of two consecutive errors in the multiframe alignment signal or, if time slot 16 contains all zeros for at least one multiframe, causing the red alarm to go high, which is coupled to a yellow alarm generator (bit $6 = 1$ in time slot 16, frame 0).

CEPT CRC option

For enhanced error monitoring capability, CEPT PCM-30 includes a CRC-4 option. Under this option, a group of eight frames, known as a submultiframe, is treated as a long binary number. This number is multiplied by X^4 (10000) and divided by $X^4 + X + 1$ (10011). The 4-bit remainder is transmitted in bit position 1 (the I bit in Figure 15.15) in time slot 0 in even frames which contain the frame alignment signal. After the receiver computes its own CRC-4 check, it uses bit position 1 in time slot 0 of frames 13 and 15 for CRC error performance reporting. Table 15.5 summarizes how these bits are used for CRC error performance reporting purposes.

Table 15.5 CEPT PCM-30 CRC error performance reporting

Bit 1 Frame 13	Bit 1 Frame 15	
1	1	CRC for SMF I, II error-free
1	0	SMF II in error, SMF I error free
0	1	SMF II error-free, SMF I in error
0	0	Both SMF I and II in error

The principal CEPT PCM-30 error conditions include the occurrence of bipolar violations, frame alignment errors and multiframe alignment errors. A bipolar violation is a failure to meet the AMI CEPT PCM-30 line code, where marks alternate as positive and negative pulses and spaces are represented by a zero voltage. A frame alignment error is a failure to synchronize on the frame alignment signal (0011011) contained in time slot 0 of alternating frames, whereas a multiframe alignment error is a failure to synchronize on the multiframe pattern (0000) contained in bits 1–4 of time slot 16 of frame 0.

Zero suppression

In Europe the high density bipolar three-zero maximum (HDB3) coding is used by CEPT PCM-30 to obtain a minimum ones density for clock recovery from received data. Under HDB3, the data stream to be transmitted is monitored for any group of four consecutive zeros. A four-zero group is then replaced with an HDB3 code. Two different HDB3 codes are used to ensure that the bipolar violation pulses from adjacent four-zero groups are of opposite polarity as indicated in Figure 15.16. The selection of the HDB3 code is based on whether there was an odd or even number of ones since the last bipolar violation (BV) occurred. If an odd number of ones has occurred since the previous bipolar violation, the coding method in Figure 15.16(a) is used to replace a sequence of four zeros. If an even number of ones occurred since the previous bipolar violation, the coding method in Figure 15.16(b) is used to replace a sequence of four zeros.

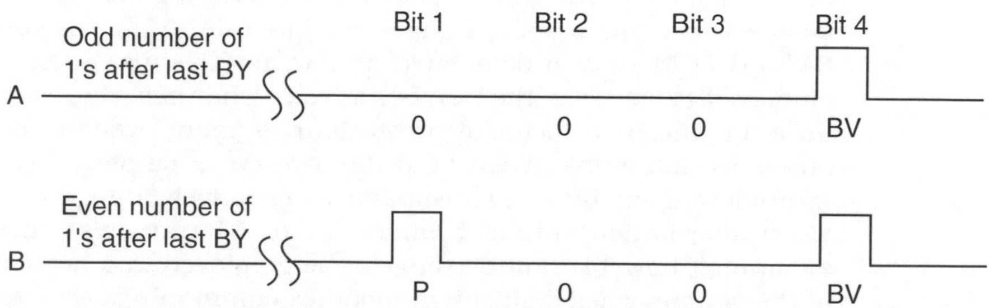

Figure 15.16 HDB3 coding: P = polarity bit, BV = bipolar violation

Table 15.6 European G703/732 frame structure

Time slot	Type of information
0	Synchronization
1–15	PCM encoded speech
16	Signaling
17–31	PCM encoded speech

Table 15.5 lists the composition of the 32 channels that are used to establish the G703/732 frame structure. In comparing the North American T1 frame structure to the European G703/732 structure, there are several key differences between the two in addition to the different data rates. First, North American T1 systems derive the 1.544 Mbps data rate from the use of 24 channels, whereas the European G703/732 system uses 30 voice channels plus separate synchronization and signaling channels, with each channel operating at 64 kbps to produce a 2.048 Mbps data rate. Secondly, the North American T1 system uses the 193rd bit in each frame for synchronization, where the G703/732 system provides a separate 64 kbps channel for this function. Finally, the T1 system uses the eighth bit in every sixth frame for signaling, whereas the G703/732 system uses a separate 64 kbps channel for this function. Table 15.6 summarizes the signaling characteristics of the G703/732 frame structure.

15.2 T1 MULTIPLEXERS

When T1 became available for subscriber use in 1980, it was initially relegated to use in voice networks. This limited utilization was based upon the lack of equipment to efficiently combine voice and data on one T1 facility as well as the availability of telephone company equipment designed exclusively for the combination of 24 voice channels on to one T1 circuit. Because the cost of less than 20 analog voice grade circuits equals the cost of one T1 facility while providing approximately 6% of the capacity of a T1 facility, many communications equipment manufacturers developed T1 multiplexers designed to economically multiplex both voice and data on to a composite T1 channel.

In addition to performing data multiplexing, most T1 multiplexer manufacturers offer users a variety of optional voice digitization modules that can be used to digitize voice signals at data rates ranging from the standard 64 kbps PCM rate to data rates as low as 9.6 kbps. Such modules can be employed to increase the number of voice channels that can be transmitted on a T1 line by a factor of two to four or more over the normal 24 or 30 channels obtainable when PCM digitization is employed. The use of these multiplexers enables organisations to effectively integrate voice, data and video information onto a common T1 or E1 transmission facility. Prior to examining how T1 multiplexers can be employed in a network, a discussion of the various voice digitization modules commonly available is warranted to obtain an understanding of the benefits and limitations of these modules.

15.2.1 Waveform-based voice digitization modules

In a conventional PCM digitization process the height of the analog signal is converted into an 8-bit word which represents the analog signal at the time sampling occurred. Since sampling occurs 8000 times per second, an analog voice signal is converted into a 64 kbps digital data stream, resulting in a maximum of 24 voice channels that can be carried on a North American T1 facility when PCM is employed. To increase the number of voice signals that can be carried on this facility, a variety of voice digitization techniques were developed, including Adaptive Differential Pulse Code Modulation (ADPCM) and Continuous Variable Slope Delta (CVSD) Modulation.

The previously mentioned voice digitization techniques are referred to as waveform encoding methods. Each of those techniques involves sampling the analog signal which can be considered to form a wave and its encoding into a digital value. Two other voice digitization techniques that warrant notation are vocoding and hybrid coding. Vocoding, which represents a contraction of the terms 'voice' and 'coding', refers to a series of coding methods based upon the modeling of human speech. As you might expect, hybrid coding represents a combination of waveform and vocoding.

During the 1980s just about all T1 multiplexer voice digitization modules were based upon waveform coding techniques. During the latter part of the 1980s vendors experimented with vocoding methods as their low bit rate enabled a T1 line to transport up to approximately 1000 voice conversations. Unfortunately, vocoding methods did not result in a high quality of reproduced speech and never gained popularity. However, during the 1990s hybrid coding that combined waveform and vocoding resulted in a family of low data rate, high quality voice reproduction digitization techniques. Some of these techniques are now used by T1 multiplexer manufacturers while other techniques, are used by router vendors to support voice over IP and voice over frame relay. In addition, several hybrid voice digitization techniques are used in many Internet telephony products. Because waveform encoding techniques are naturally associated with T1 multiplexers, this author elected to discuss all three categories of voice digitization in this section.

Adaptive Differential Pulse Code Modulation (ADPCM)

When Adaptive Differential Pulse Code Modulation (ADPCM) is employed, a transcoder is utilized to reduce the 8-bit samples normally associated with PCM into 4-bit words, retaining the 800 samples-per-second PCM sampling rate. This technique results in a voice digitization rate of 32 kbps, which is one-half the PCM voice digitization data rate.

Under the ADPCM technique, the use of 4-bit words permits only 15 quantizing levels; however, instead of representing the height of the analog signal each word contains information required to reconstruct the signal. This information is obtained by circuitry in the transcoder which adaptively predicts the value of the next signal based on the signal level of the previous sample. This technique is known as adaptive prediction, and its accuracy is based on the fact that the human voice does, not significantly change from one sampling interval to the next.

There are several versions of ADPCM standardized by the ITU-T. In addition to ADPCM at 32 kbps other versions of this voice digitization method operate at 24 kbps and 40 kbps.

Continuously Variable Slope Delta (CVSD) Modulation

In the Continuously Variable Slope Delta (CVSD) Modulation digitization technique, the analog input voltage is compared to a reference voltage. If the input is greater than the reference a binary '1' is encoded, whereas a binary '0' is encoded if the input voltage is less than the reference level. This permits a 1-bit data word to represent each sample.

At the receiver, the incoming bit stream represents changes to the reference voltage and is used to reconstruct the original analog signal. Each '1' bit causes the receiver to add height to the reconstructed analog signal, while each '0' bit causes the receiver to decrease the analog signal by a set amount. If the reconstructed signal is plotted, the incremental increases and decreases in the height of the signal will result in a series of changing slopes, resulting in the naming of this technique: continuously variable slope delta modulation.

Since only changes in the slope or steepness of the analog signal are transmitted, a sampling rate higher than the PCM sampling rate is required to recognize rapidly changing signals. Typically, CVSD samples the analog input at 16 000 or 32 000 times per second. With a 1-bit word transmitted for each sample, the CVSD data rate normally is 16 or 32 kbps. Other CVSD data rates are obtainable by varying the sampling rate. Some T1 multiplexer manufacturers offer a CVSD option which permits sampling rates from 9600 to 64 000 samples per second, resulting in a CVSD data rate ranging from 9.6 to 64 kbps, with the lower sampling rates reducing the quantity of the reconstructed voice signal. Normally, voice signals are well recognizable at 16 kbps and above, whereas a data rate of 9.6 kbps will result in a marginally recognizable reconstructed voice signal.

15.2.2 Vocoding

The term vocoding, which represents a contraction of the terms 'voice' and 'coding', references a series of voice digitization methods based upon the development of different models of human speech. Most vocoding methods are based upon the assumption that speech can be generated by exciting a linear system by a series of periodic pulses if sound is voiced, or the use of noise to represent the vocal track if sound is unvoiced. Vocoders perform an analysis of speech, extracting voice parameters that will be transmitted instead of waveform samples used by waveform coding techniques. At the receiver, the transmitted parameters are used to reconstruct voice via a synthesis process. Although vocoding methods result in a low bit rate, with a voice conversation reduced to between 2400 bps and 4800 bps, their extraction of speech parameters is performed without considering whether or not the synthesis process will generate a waveform resembling the original signal. This in turn often results in the reconstructed synthesized voice sounding a bit 'metallic'. In addition, the analysis performed at the transmitter often has difficulty in separating background noise from the actual voice conversation, reducing the clarity of the resulting synthesized voice reconstruction process. In spite of these problems several types of vocoders were commonly employed on international circuits by corporations seeking to reduce the cost of their global communications bill during the 1980s. One of the most popular types of vocoding is the linear predictive vocoder, part of whose historical evolution involves a child's toy that was popular during the 1970s.

Linear predictive coding

During the 1970s a toy in the shape of an owl was introduced that represented one of the earliest commercial applications for voice synthesis. The toy, called 'Speak and Spell', included a keyboard, speaker, and red and green colored owl eyes. A child (and many adults) would press a button and the owl would randomly select a word from its memory and pronounce the word using voice synthesis. The toy operator would then use the keyboard to spell the word just pronounced. If they spelled the world correctly, the owl's green colored light would illuminate; otherwise the red colored light would illuminate. In addition to becoming a popular toy, Speak and Spell also illustrated the practicality of voice synthesis, especially when data storage was limited which in effect is equivalent to requiring a low bit rate.

The basis behind the operation of linear predictive coding (LPC) is the analysis it performs on analog speech. Figure 15.17 illustrates the speech-producing elements of the human vocal tract. During LCP encoding the analog voice input is analyzed and then converted into a set of digital parameters for transmission. At the receiver, a synthesizer recreates an analog voice output based upon the received set of digital parameters. By limiting the analysis of the voice signal to four sets of voice parameters, a very low data rate can be used to transmit voice data in digital form.

Operation

In linear predictive coding, the voice signal is first sampled by a 12-bit analog-to-digital converter. The output of the converter is then used as input to four parametric detectors. A pitch detector analyzes the data to obtain the fundamental pitch frequency at which vocal cords vibrate. Next, a voice/unvoiced detector senses whether sound is caused by the vibration of vocal cords (voice) or by sounds such as 'shhh' (unvoiced) that do not vibrate. A power detector then determines the amplitude (volume or loudness) of the sound, while a spectral data decoder models the resonant cavity formed by the throat and the mouth.

Since LPC sends speech parameters rather than the amplitude of waveforms, it is actually a method of speech synthesis. During the 1980s LPC was

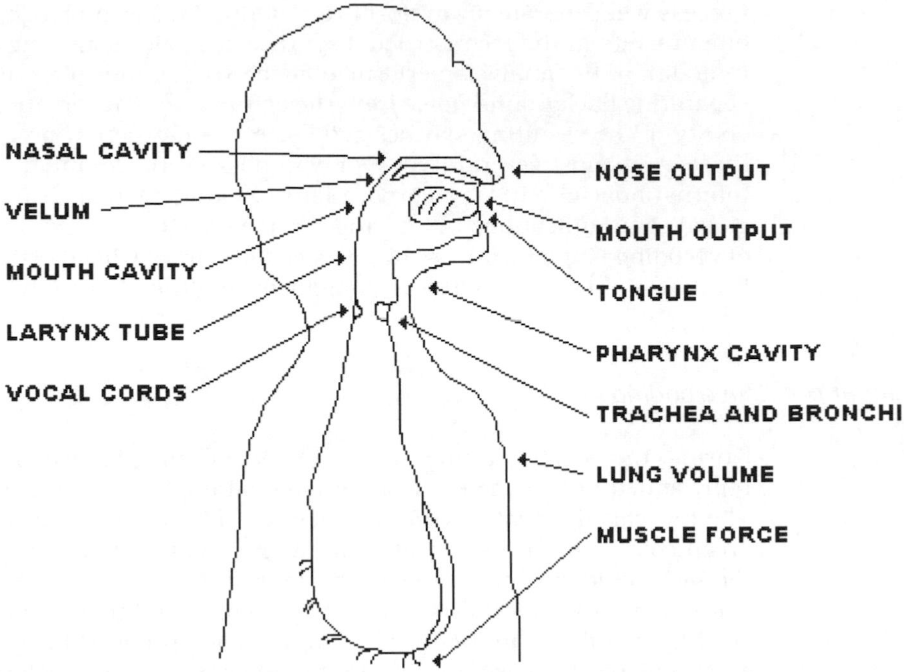

Figure 15.17 Speech-producing elements of the human vocal tract. These are analyzed by a linear predictive encoder to synthesize a voice conversation

integrated into several devices collectively known as voice digitizers that can operate at 2400 or 4800 bps. Typically, data from the four LPC detectors is stored in a 54-bit buffer that is released every 22.5 milliseconds. This results in 44.444 samples per second which, when multiplexed by 54 bits per sample, permits an analog voice conversation to be digitized at 2400 bps.

Constraints to consider

The primary constraints associated with the use of devices that employ linear predictive encoding are the cost of equipment and the fidelity of the reconstructed signal. When voice digitizers were introduced their cost commonly exceeded $7000 per unit, making their use economically feasible only for international long distance. Although voice digitizer prices have significantly fallen, most linear predictive coding methods have been superseded by hybrid coding built into multiplexers and Frame Relay Access Devices (FRADs).

With respect to the fidelity of reconstructed voice, LPC synthesizes both the speaker's conversation and any background noise. Due to the method used to synthesize voice conversations, background noise is accentuated, which can result in some disturbance to the reconstructed analog signal. Thus, the use of LPC is more suited to conversations originating in an office environment than to conversations on a factory floor where there may be a large amount of background noise.

15.2.3 Hybrid coding

Hybrid coding represents a combination of waveform coding and vocoding that results in a high quality of reconstructed speech transmitted at relatively low bit rates. Under hybrid coding speech is passed through a vocal tract predictor and a pitch predictor. The output of the prediction process, which represents synthesized speech, is used to generate the sampled waveform. The predictors of the generated waveform are then compared to the original speech parameters. Any variations within certain defined tolerances are considered as acceptable, while variations beyond those tolerances are adjusted by revising the synthesis parameters. Thus, a popular term associated with hybrid coding is synthesis by analysis. Two of the more popular implementations of hybrid coding are GSM and a family of codebook excited linear prediction (CELP) methods.

Global System for Mobile (GSM)

One of the more popular examples of hybrid coding is the Global System for Mobile (GSM) communications. GSM is employed for digital cellular operations in Europe and uses a hybrid coding method known as Regular Pulse Excited (RPE) coding which enables voice conversations to be transmitted at 13 kbps. Under RPE coding voice samples are split into frames 20 ms in length. For each frame a set of eight short-term predictor coefficients are obtained. Each frame is then further divided into four 5 ms subframes. The encoder determines a delay and gain for the long-term predictor for each subframe. A total of 40 samples is divided into three possible excitation sequences of 13 samples, and the sequence with the highest energy level is then selected as the best representation of the 20 ms sample.

Code Excited Linear Prediction (CELP)

Although Code Excited Linear Prediction (CELP) dates from the 1980s, it was not until the early 1990s that digital speech processors (DSPs) with sufficient processing capability were developed to implement theory. The development of 200–300 MHz DSPs was accompanied by the development of several versions of CELP.

As its name implies, CELP is based upon the construction of a codebook of excitations. The codebook is adaptively constructed and synthesis parameters resulting from voice samples are matched against entries in the adaptively constructed table. This enables the index of the code to be transmitted instead of actual speech parameters, significantly reducing the quantity of transmission. The original version of CELP, which operates at 16 kbps, was standardized by the ITU in 1992. Although CELP was a considerable improvement in bandwidth utilization over PCM and ADPCM, the size of the speech period and the method used to create a codebook resulted in a relatively long delay, which made this technique unsuitable for use on packet networks whose routing delays introduce additional latency. Recognizing this problem resulted in the development of a low-delay version of CELP referred to as Low-Delay Code Excited Linear Prediction (LD-CELP).

Low-Delay Code Excited Linear Prediction (LD-CELP)

Low-Delay Code Excited Linear Prediction (LD-CELP) provides a high quality of speech reproduction using a 16 kbps encoding rate and a relatively low encoding delay. To accomplish this, LD-CELP uses a 10 ms frame length that minimizes the delay time. LD-CELP was standardized by the ITU as recommendation G.728.

Conjugate-Structure Algebraic Code Excited Linear Prediction (CS-ACELP)

Conjugate-Structure Algebraic Code Excited Linear Prediction (CS-ACELP) provides a high quality of speech reproduction through the use of a series of coding methods. Those methods include interfame correlation preselection of the codebook structure, and the use of a conjugate structure. CS-ACELP was standardized by the ITU as recommendation G.729 and provides an 8 kbps voice digitization operating rate.

Other versions of CELP

In addition to LD-CELP and CS-ACELP, there are several other versions of CELP that warrant discussion. One recently promulgated ITU standard, Recommendation G.723.1, defines a low bit-rate voice compression method that includes a G3 fax transmission capability. Referred to as Multi-Pulse Maximum Likelihood Quantization (MP-MLQ), this speech compression method is combined with ACELP to provide selectable voice digitization rates of 5.3 and 6.3 kbps. By the late 1990s G.723.1 was being implemented by a number of vendors for inclusion in FRADs and routers to provide a voice over Frame Relay and voice over IP networking capability. When used to transmit digitized voice over packet networks, the voice digitization rate of 5.3 and 6.3 kbps results in packet headers being added to transport digitized speech. Depending upon the method used by FRADs and routers to fragment portions of speech into small segments for transport over the packet network, protocol overhead can result in the actual bandwidth consumed increasing by 10–25%. When ACELP/MP-MLQ compression is employed by multiplexers, each voice channel is commonly transported in an 8 kbps multiplexer time slot between geographically separated locations connected by a private network. Another version of CELP, referred to as Enhanced CELP (E-CELP), was being marketed by one vendor as a proprietary version of CELP during 1999. This version of CELP operates at 2400 and 4800 bps.

15.2.4 T1 multiplexer employment

Modern T1 multiplexers are microprocessor-based time division multiplexers designed to combine the inputs from a variety of data, voice and video sources on to a single communications circuit that operates at 1.544 Mbps in North

Table 15.7 Typical T1 multiplexer channel rates

Type	Data rates (bps)
Asynchronous	110; 300; 600; 1200; 1800; 2400; 3600; 4800; 7200; 9600; 19 200
Synchronous	2400; 4800; 7200; 9600; 14 400; 16 000; 19 200; 32 000; 38 400; 40 800; 48 000; 50 000; 56 000; 64 000; 112 000; 115 200; 128 000; 230 400; 256 000; 460 800; 700 000; 750 000
Voice	8000; 16 000; 32 000; 48 000; 64 000

America and 2.04 Mbps in Europe. Table 15.7 lists the typical input channel rates accepted by most T1 multiplexers. It should be noted that, although digitized voice is treated as synchronous input, its digitized data rate can vary considerably based on the type of optional voice digitization modules offered by the T1 multiplexer manufacturer. Table 15.6 compares the signaling characteristics of T1 and G703/732 systems.

In most applications, input to the T1 multiplexer's voice channels results from an interface to an organization's private branch exchange (PBX), resulting in one or more tie lines being obtained through the use of two T1 multiplexers and a T1 carrier facility. Figure 15.18 illustrates a typical T1 multiplexer application where voice, video and data are combined on to one T1 carrier facility. In this example, it was assumed that PCM digitization channel modules were selected for use in the T1 multiplexer, resulting in the ten voice channels on the PCM interface using 640 kbps of the available 1.544 Mbps T1 operating rate.

Since digitized video normally requires a data rate of 256 kbps to be effective, it was assumed that the organization using the T1 multiplexer has a conference room to connect to a distant location for video conferencing. Thus, 256 kbps

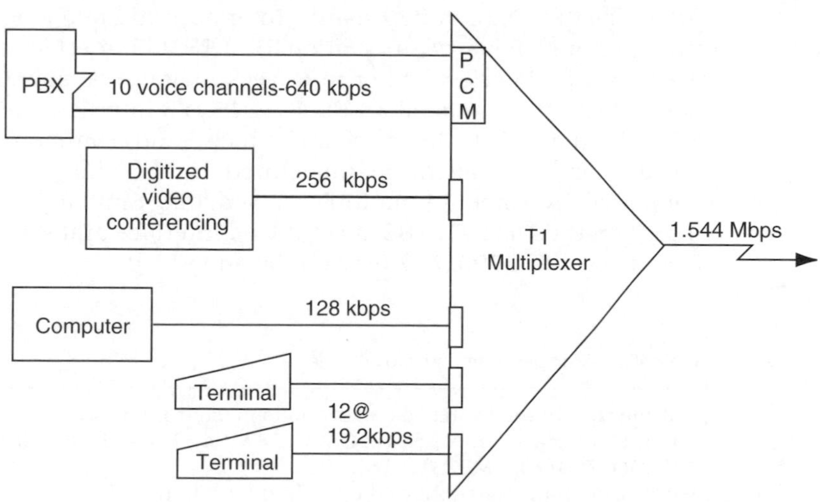

Figure 15.18 Typical T1 multiplexer network application

Table 15.8 Signaling characteristics comparison

	T1	G703/732
Composite data rate (Mbps)	1.544	2.048
Number of channels	24	32
Channel data rate (kbps)	64	64
Synchronization	Frame bit	Channel 0
CRC	CRC-6	CRC-4
Zero suppression	B8ZS coding	HDB3 coding
Signaling	Eighth bit in sixth frame	Channel 15

input to the T1 multiplexer in Figure 15.18 represents a digitized video conferencing signal. Similarly, it is assumed that the organization has two data centers, one at each location where there is also a PBX. This permits computer-to-computer transmission to occur at 128 kbps. Finally, it was assumed that 12 data terminals, each operating at 19.2 kbps at one site, required access to the computer located at the other end of the T1 link. Since the cost of lO tie lines and few leased lines to support the data terminal would normally equal the cost of a T1 carrier facility, in effect, the employment of T1 multiplexers provides the organization with no cost for the bandwidth required for video conferencing and wideband computer-to-computer transmission.

Features to consider

Table 15.9 lists some of the more important features that users contemplating the acquisition of a T1 multiplexer should consider. As previously indicated in Table 15.7 the data rates supported for asynchronous, synchronous and voice transmission can vary considerably from multiplexer to multiplexer and must be examined to ensure that the proposed equipment can support the user's requirements. Concerning channel interfaces, although most vendors support the Electronic Industry Association (EIA) RS-232 interface as well as the North American T1 interface, other interfaces may not be supported by some vendors. Thus, the more modern RS-422, RS-423 interface which is equivalent to ITU-T V.10 and V.11 interface and which is not commonly included in communications equipment manufactured in the United States may not be supported by some T1 multiplexer vendors. Similarly, the V.35 wideband interface, the MIL-STD-188 current loop interface and the European T1 interface known as G703/732 may not be supported.

Table 15.9 Multiplexer features to consider

Multiplexer channel rates: asynchronous; synchronous; voice
Channel interfaces: EIA RS-232-C (CCITT V.24); RS-422, RS-423 (CCITT V.10 V.11); V.35; G703/732; MIL-STD-188; T1
Number of channels supported and channel mixture
Multinode capability

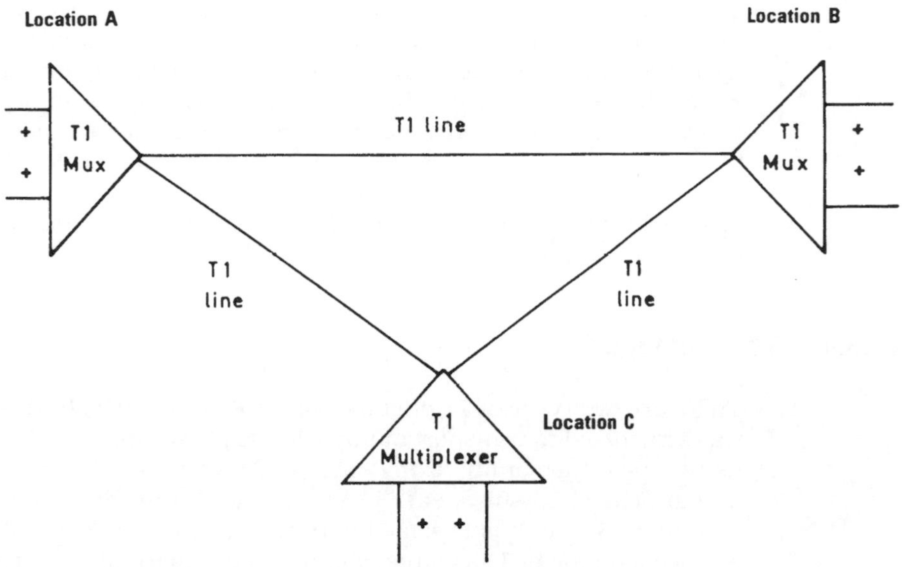

Figure 15.19 Multinodal T1 multiplexers can be networked

Even if all the user's interface requirements are supported, one must determine the total number of channels supported by the T1 multiplexer as well as the manufacturer's constraints regarding the mixture of such channels. Some T1 multiplexers cannot support more than 48 independent channels whereas other T1 multiplexers can support up to 512 or 1024 channels. Similarly, some T1 multiplexers are designed so that asynchronous data channels actually use a 64 kbps synchronous channel time slot on the aggregate T1 link regardless of the asynchronous data rate. In comparison, other T1 multiplexers may more efficiently service asynchronous data traffic, permitting a larger number of channels to be supported by the multiplexer.

Concerning multinode capability, some T1 multiplexers offering this feature permit users to network the multiplexers to service several locations. Figure 15.19 illustrates an example where three T1 multinodal multiplexers are employed to interconnect three distinct locations via three T1 lines. In comparison, uninodal T1 multiplexers can only be employed on point-to-point lines and would not provide the user with the routing flexibility illustrated in Figure 15.19.

15.3 THE T3 CARRIER

Although many organizations think of the T3 carrier as a modern digital transport system, it represents an AT&T developed standard that has been in use for approximately 30 years. Until the mid-1990s the relatively large transmission capacity of the T3 carrier as well as its cost placed it beyond the practical use of most organizations. The growth in the use of the Internet resulted in many popular Web sites experiencing millions of user accesses per day, which created a demand for transmission bandwidth that could only be satisfied through the use of a T3 transmission facility. In addition, Internet Service Providers as well as commercial organizations and government agencies found the use of T3 carriers ideal for consolidating transmission, enhancing its popularity. The growth in the construction of fiber optic long-distance transmission facilities resulted in a significant amount of bandwidth becoming available for use. This resulted in competitive pressure among long-distance communications carriers to fill up available bandwidth. To do so they began to offer significant discounts and price reductions that further enhanced the popularity of T3 circuits.

15.3.1 T3 circuit types

There are two types of T3 circuits available for use. The first type of T3 circuit was structured to transport multiple levels of multiplexed digital voice signals using a two-stage multiplexing scheme. This type of T3 circuit is referred to as a channelized or subrated T3. As we will shortly note, the channelized or subrated T3 circuit is designed to transport the equivalent of 28 T1 circuits.

A second type of T3 is a non-channelized T3 circuit. This type of T3 circuit does not support multiple levels of multiplexed digitized voice signals. However, it uses every 85th bit as a framing bit, resulting in a 44.210 Mbps transmission capacity on the 44.736 Mbps signal.

15.3.2 Evolution

Although the 1.544 Mbps T1 circuit significantly reduced cable congestion in urban areas, its operating rate was not sufficient for certain high volume locations. In 1972 AT&T implemented testing of what is now the second level of the digital transmission hierarchy in North America which is referred to as a DS2 signal.

The DS2 signal

The DS2 signal is formed through the use of an M12 multiplexer which combines four asynchronous DS1 signals (96 DSOs) for transmission over a synchronous link. Since the DS1 signals are asynchronous, the multiplexer required a method to compensate for the variations in the clock rates among the DS1 signals. The method of clock compensation selected by AT&T is referred to as pulse stuffing, which is effected by the addition of another layer of framing beyond DS1 framing to the DS2 signal. This additional layer of framing provides for the correct alignment of bytes in a manner similar to DS1 framing; however, it also enables variations in the clocking of DS1 signals through the use of stuffing bits. Such stuffing bits are added as needed to ensure that each DS1 signal has an identical bit rate prior to the M12 multiplexer bit interleaving the signals. A relatively 'fast' signal will receive less stuffing, while a relatively 'slow' signal will receive more stuffing. At the far end of the transmission path the stuffing bits are removed.

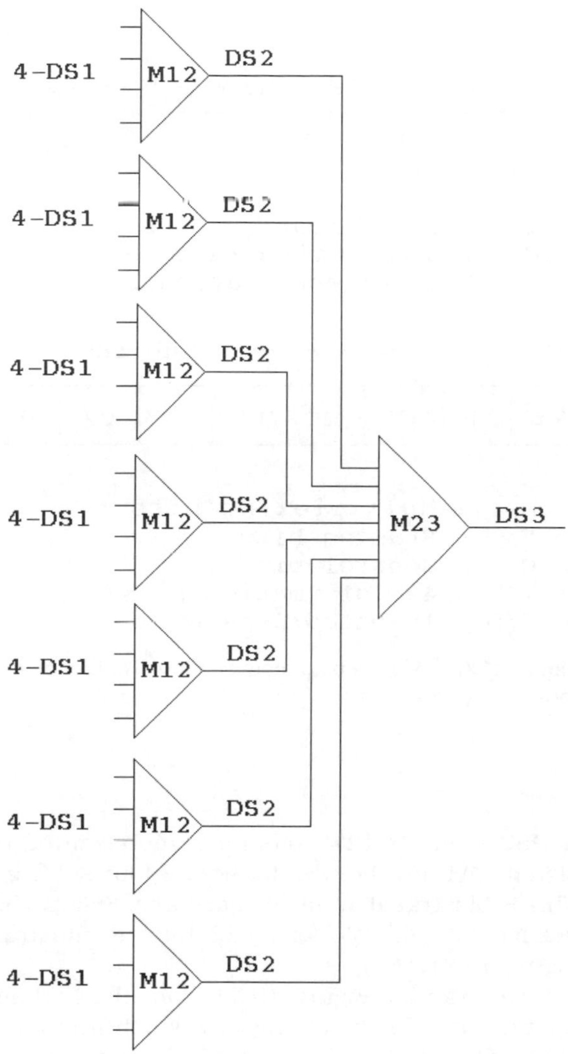

Figure 15.20 The formation of a T3 channelized carrier is accomplished via a two-stage multiplexing process

Once the DS2 hierarchy was in operation its 6.312 Mbps proved insufficient for many applications. A further development resulted in the output of seven M12 multiplexers being re-multiplexed into one composite high-speed circuit referred to as a T3 carrier. This second-stage level of multiplexing is performed by an M23 multiplexer, whose acronym references layer 2 in the digital hierarchy input to provide a layer 3 signal.

Figure 15.20 illustrates the construction of a DS3 signal as a result of a two-stage multiplexing process. Since an understanding of T3 framing requires an understanding of T2 or DS2 framing, let's first focus our attention on that topic.

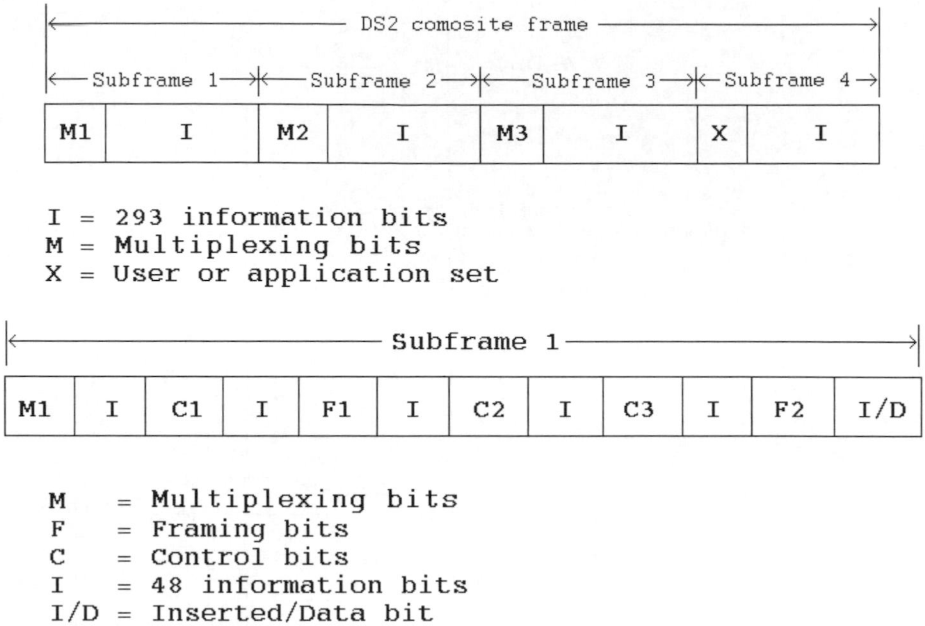

Figure 15.21 DS2 framing results in the creation of four subframes that are multiplexed together by time

DS2 framing

A DS2 frame is 1272 bits long, representing the multiplexing of four DS1 signals. Within the DS2 frame are four subframes, each 318 bits in length. This is illustrated in the top portion of Figure 15.21. Each subframe is further divided into six blocks of 53 bits as illustrated in the lower portion of Figure 15.21.

If we examine Figure 15.21 from the bottom up we can obtain a better appreciation for the manner by which framing is accomplished. At the first stage of framing an alternating 01 bit pattern is formed by the F bits that occur every 159 bits in the composite DS2 frame. The second framing stage results from the use of Multiplexing (M) bits which reside in the first bit position of each subframe. The sequence of M bits form the pattern 011X, with the 01 transition used to identify the beginning of the DS2 master frame. The X bit position can be set to either a 0 or 1 and reflects a user or application value.

Pulse stuffing

As previously noted, stuffing bits are used to synchronize four DS1 signals to compensate for clocking differences between each signal. To identify the fact that stuffing bits are employed resulted in the requirement to use control bits. Thus, the C bits in the subframe are used to identify the contents of the stuffing bits. That is, if a majority of the C bits (two out of three) are set to ‘1’, the stuffing bit is holding an inserted (I) bit. Otherwise, the stuffing bit contains data. Since the stuffing bit may or may not contain data, each subframe can contain either 311 or 312 data bits.

Maintaining ones density

Earlier in this chapter we noted that one common method used to maintain a ones density on a T1 line is B8Zs. A DS2 signal uses a similar technique, referred to as Bipolar with Six Zero Substitution (B6ZS). Under B6ZS up to a maximum of five consecutive zeros are permitted prior to a six-bit byte to include intentional bipolar violations being substituted for a sequence of six consecutive zeros.

15.3.3 T3 framing

As previously indicated, the formation of a T3 channelized carrier is accomplished via a two-stage multiplexing process. In the first stage four DS1 signals are combined into a DS2 signal. During the second stage seven DS2 signals are multiplexed to form a DS3 signal. Although Figure 15.20 illustrated two separate multiplexers as being used to form the two-stage multiplexing process, that process can also occur as the result of a single multiplexer. That multiplexer is referred to as an M13 device as it takes 28 DS1 signals, and although it internally first forms seven DS2 signals, it multiplexes those signals within the device to form one DS3 signal. Regardless of the number of separate multiplexers used to form a DS3, the resulting signal has a specific framing format. In actuality, although there is only one framing format, the use of certain bits falls into one of two camps that result in two standards. The first standard for T3 framing assumes that bit stuffing occurs at each stage in the multiplexing process and results in the M13 format. Since bit stuffing from T2 to T3 is unnecessary when one device performs T1 to T3 multiplexing, there is no need to use certain bits for bit stuffing. Recognition of this as well as the need for an enhanced maintenance capability resulted in the development of a new T3 framing format. That format is known as the C-bit parity format; it was proposed by AT&T and ratified as ANSI standard T1.107A in 1990.

The M13 framing format

A DS3 signal is divided into frames 4760 bits in length that are referred to as M-frames. Since each M-frame represents seven multiplexed DS2 signals, it is no surprise that each M-frame also consists of seven M-subframes. However, these subframes do not represent individual DS2 signals, as the DS3 signal is formed by bit-interleaving multiplexing of the seven DS2 signals. Since the M-frame is 4760 bits in length, each M-subframe is 680 bits in length.

Figure 15.22 illustrates the T3 M-frame and an M-subframe. Note that each M-subframe is further subdivided into eight blocks, each consisting of an overhead bit followed by 84 data bits.

A. The T3 M-frame

Subframe 1		Subframe 2		Subframe 3		Subframe 4		Subframe 5		Subframe 6		Subframe 7	
XI	I	X2	I	P1	I	P2	I	M1	I	M2	I	M3	I

X = user or application value
I = 679 information bits
P1 = Modulo 2 sum of all information bits in previous frame
M = Multiframe Alignment; 010

B. The T3 M-subframe

X1	I	F1	I	C1	I	F2	I	C2	I	F3	I	C3	I	F4	D/I

X = originally user or application value, now commonly loss of signal, loss of frame indicator
F = subframe alignment; 1001
C = indicates if last bit carries data or stuffing

Figure 15.22 The T3 M-frame and M-subframe

In examining the M-frame shown in Figure 15.22 note that the X bits were originally used to establish a signaling channel and prefix user information on the first two subframes used to form the M-frame. Today the X bits are used to indicate loss of signal and loss of frame. The P bits, which prefix the information bits in subframes 3 and 4, contain the Modulo 2 sum of data bits in the previous frame. The X bits and P bits are controlled by transmitting equipment to provide basic alarm and error indication information across a T3 line. The M bits that prefix the information bits in subframes 5 through 7 identify the relative subframe position. To do so they transport the bit sequence 010. Because the M bits provide a frame alignment capability they are also cumulatively referred to as the MultiFrame Alignment signal.

The lower portion of Figure 15.22 illustrates the general composition of an M-subframe. The four F bits form a sub-frame alignment signal that has the repeating pattern '1001'. Similar to the DS2 frame format, the three C bits indicate whether the last bit position in the subframe carries payload data or stuffing.

To facilitate visual observation of the M13 framing process, Figure 15.23 illustrates the M13 framing bit pattern. Note that the F bits are fixed in value to either 1 (F1) or 0 (F0). Similarly, the M bits are also fixed in value to either 1 (M1) or 0 (M0). Note that the control bits (C1, C2, C3) define the use of the last bit position in each subframe. That is, when two or more bits are set, the last bit in the subframe is a stuffed bit; otherwise the last bit transports data.

Subframe	Block #1							Block #8
1	X1	F1	C1	F0	C2	F0	C3	F1
2	X2	F1	C1	F0	C2	F0	C3	F1
3	P1	F1	C1	F0	C2	F0	C3	F1
4	P2	F1	C1	F0	C2	F0	C3	F1
5	M0	F1	C1	F0	C2	F0	C3	F1
6	M1	F1	C1	F0	C2	F0	C3	F1
7	M0	F1	C1	F0	C2	F0	C3	F1
	Block #49							Block #56

F = Subframe alignment; 1001
C = Stuffing indicator codes
M = Multiframe alignment; 010
X = Loss of frame, loss of signal indicators

Figure 15.23 The M13 framing bit pattern

The C-bit parity format

As previously noted in our examination of the M13 framing format, each DS3 signal is developed by combining seven subframes, each containing eight overhead bits. This results in each DS3 M-frame containing 58 overhead bits, of which 21 were used to indicate bit stuffing. The ability to use the 21 C bits for other purposes resulted in the development of the C-bit parity format. This framing format results in the use of the C bits for performance monitoring and maintenance instead of as a mechanism to define whether or not the last bit in each subframe is carrying data or is a stuffing bit.

Figure 15.24 illustrates the DS3 C-bit parity framing format. Note that only the bits in blocks 3, 5 and 7 are changed from their use from the DS3 M13 frame format previously illustrated in Figure 15.23.

The Application Identification Channel indicates whether or not C-bit parity is being used. If so, framing bit 3 in subframe 1 is set to 1. The Far-End Alarm and Control Channel (FEAC) bit is used for transmitting performance data as well as loopback commands on an end-to-end basis. Thus, the FEAC provides the ability to automatically place a T3 circuit into a loopback mode of operation.

The three CP bits represent path parity bits that are calculated by the transmitter. The receiver uses these C-bit parity bits to perform a parity check for transmission errors, enabling line quality statistics to be computed without having to take the line out of service. Information about received parity errors is returned to the transmitter via the use of three Far-End Block Error (FEBE) bits.

One additional change resulting from the C-bit parity framing format concerns the use of the first framing bit in the first two subframes. Those X bits, which were used to indicate loss of frame or loss of signal under the M13 format, are used to indicate a degraded second under the C-bit parity framing format. When used in this manner, a degraded second simply denotes a second in which a Loss of Frame or Loss of Signal occurred.

Subframe	Block number 1	2	3	4	5	6	7	8
1	X	F1	AIC	F0	N	F0	FEAC	F1
2	X	F1	DL_1	F0	DL	F0	DL	F1
3	P	F1	CP	F0	CP	F0	CP	F1
4	P	F1	FTBE	F0	FEBE	F0	FEBE	F1
5	M0	F1	DL_2	F0	DL	F0	DL	F1
6	M1	F1	DL	F0	DL	F0	DL	F1
7	M0	F1	DL	F0	DL	F0	DL	F1

AIC = Application Identification Channel = 1

N = Network Application bit (reserved)

FEAC = Far-End Alarm and Control Channel

DL = Data Link

CP = C-bit Parity

FEBE = Far-End Block Error

X = degraded second

Figure 15.24 The C-bit parity framing format

Ones density

Similar to other types of T-carrier transmission systems, the DS3 signal needs to provide a minimum ones density to enable repeaters to obtain clocking from pulses on the line. To accomplish this, Binary 3 Zero Substitution (B3ZS) coding is employed. Another difference between T3 and T1 and T2 is in their transmission media. While T1 and T2 are transported over twisted-pair cable, T3 is commonly transported over 75-ohm coaxial cable to the carrier's equipment. From the network interface, T3 is commonly transmitted over fiber optic cable or via microwave radio.

Access and network utilization

Unlike lower speed T-carriers, the use of a T3 transmission system can require a considerable degree of planning due to the special interfaces needed to support relatively high transmission rates. Although access to a T3 circuit is similar to the manner by which a T1 is accessed, the DSU/CSU used with a T3 needs to support a high-speed interface.

Data rate interface

For data rates up to 6 Mbps you can use V.35, RS449 and EIA 530 interfaces. However, this means that unless you have multiple data sources and the DSU/CSU performs multiplexing, you will not be able to utilize the full transmission capacity of the T3 system. As an alternative or supplement to the previously mentioned data interfaces, vendors offer one or more High Speed Serial Interface (HSSI) connections. The HSSI supports data rates from 3 to 52 Mbps and can permit a more effective use of a T3 transmission system. In fact, many routers are now manufactured to support an optional HSSI data interface.

The top part of Figure 15.25 illustrates the use of a multiport T3 DSU/CSU to provide access to a T3 transmission system, while the lower portion of that figure indicates the use of a T3 multiplexer. Similar to a T1 DSU/CSU, the T3 CSU provides the electrical interface, signal formatting, performance monitoring, alarm signal generation and loopback operations while the DSU provides the data interface. The network interface side of the CSU must support an operating rate of 44.736 Mbps. Due to the popularity of M13 and C-bit parity framing formats, most CSUs have the ability to be set to support either framing format. Table 15.10 provides a list of seven common T3 CSU network interface parameters.

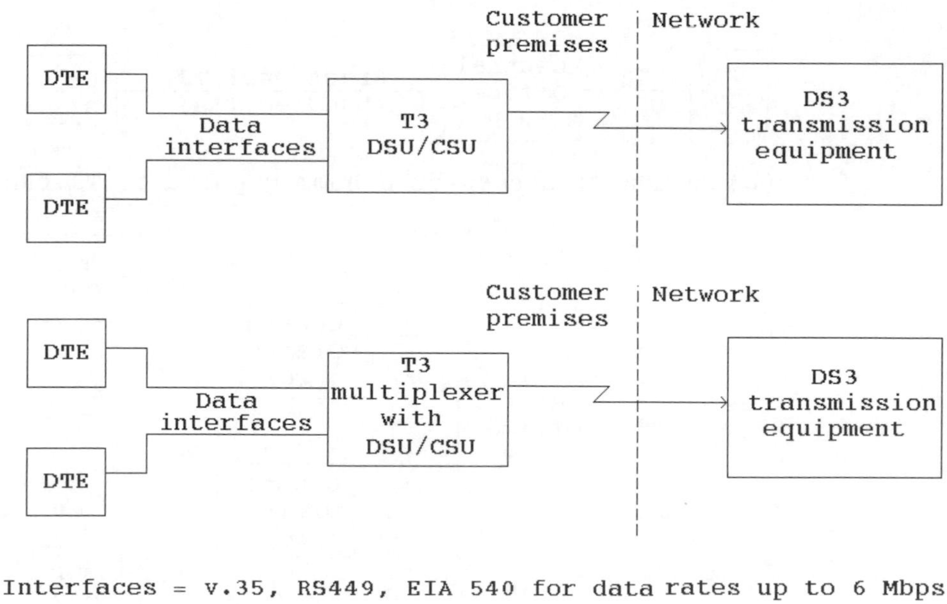

Figure 15.25 Common T3 network access methods

Table 15.10 CSU T3 network interface parameters

Line operating rate	DS3 44.736 Mbps
Framing format	M13 or C-bit parity
Input signal	+62 dBm to −11.7 dBm
Output signal	DS3
Impedance	75 ohm coax
Connector	BNC socket
Timing	Looped or internal clock

Network access

Although most ISPs use full T3 circuits on a point-to-point basis to connect to an Internet homing location, other organizations typically use T3 circuits either as access to a fractional T3 transmission facility or as a homing point for individual T1 circuits. When used as an access mechanism to a fractional T3 transmission service, a full T3 circuit is installed between the customer premises and the central office point of presence. The T3 access line is configured as a combination of T1s in the same manner that a T1 represents a series of DS0s used to provide access to a fractional T1 service. This type of T3 is referred to as a subrated T3 and is illustrated in the top part of Figure 15.26.

When a T3 circuit is used as a homing point for multiple T1s routed to the same or different geographical areas, the T3 access line is also divided into

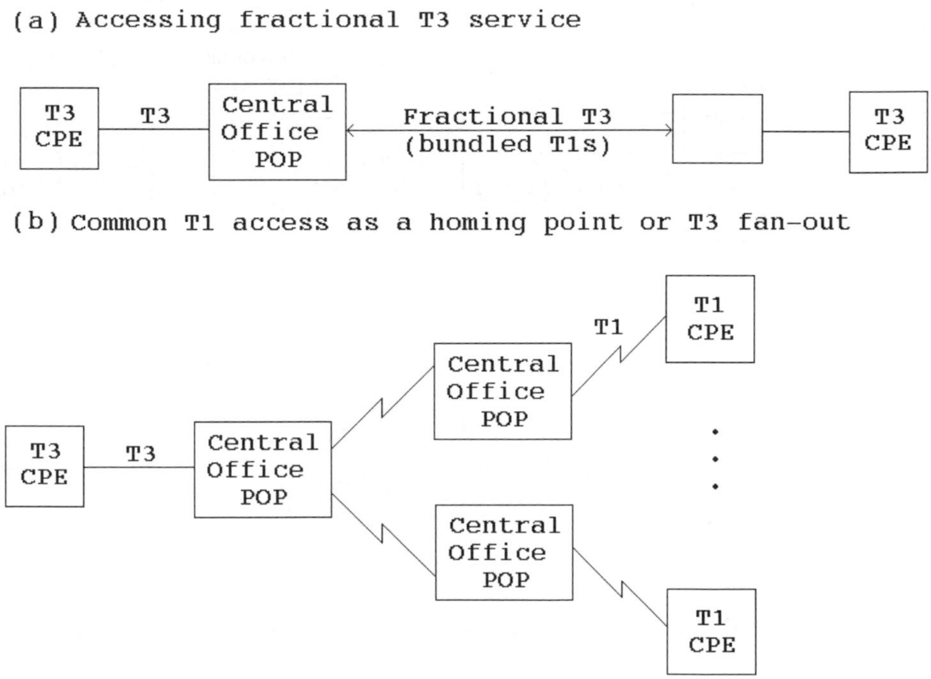

Figure 15.26 Common T3 utilization methods

combinations of T1s. Thus, the T3 access line is also a subrated transmission facility. Since the use of a single T3 access line to provide access to multiple T1 lines forms a fan-out network configuration, this network configuration is also commonly referred to as a fan-out.

When a T3 is used to provide access to fractional T3 service or as a homing point for multiple T1s, it is configured as a channelized or subrated signal. This means that the C-bit framing format cannot be maintained between customer premises; however, it can be used by the network operator. Thus, although you may not be able to obtain end-to-end performance measurements, the network operator can act upon your behalf and note the condition of this important transmission facility.

15.4 DDS, ASDS AND KILOSTREAM FACILITIES

At the beginning of the 1970s, communications carriers began offering communication systems designed exclusively for the transmission of digital data. Specialized carriers, including the now-defunct DATRAN, performed a considerable service to the information-processing community through their pioneering efforts in developing digital networks. Without their advancements, major communications carriers may have delayed the intro-duction of an all-digital service.

In December 1974, the FCC approved the Bell System's DATAPHONE© digital service (DDS), which was shortly thereafter established between five major cities. Since then the service has been rapidly expanded to the point where, by the early 1990s, more than 100 cities had been added to the DDS network. Western Union International set another milestone in February 1975, by applying to the FCC for authority to offer their International Digital Data Service (IDDS) from New York to Austria, France, Italy and Spain. Digital data transmission by major carriers had become a reality.

In Europe, several countries implemented digital networks, with British Telecom's KiloStream service currently being the most comprehensive network Equipment was placed into production in the United Kingdom in September of 1982, and the first customer was commercially connected to KiloStream in January, 1983. By 1995, over 600 exchanges in the United Kingdom were equipped to offer KiloStream services.

AT&T's DDS is strictly a synchronous facility providing full-duplex, point-to-point and multipoint service limited to speeds of 2.4, 4.8, 9.6, 19.2, and 56 kbps. In addition to these leased-line services, AT&T introduced a switched 56 kbps digital service in 1985. Access to AT&T's Switched 56 service is obtained by dialing a '700' number, which was available in over 100 cities in early 1999.

In the United Kingdom and other European countries, digital service similar to AT&T's DDS is available at many locations. The primary difference between European offerings and DDS is the ability of the former to provide subscribers with a full 64 kbps data rate in comparison to DDS's maximum rate of 56 kbps. European offerings can achieve a 64 kbps data rate since the T1 facilities they use carry signaling information in a separate channel, whereas North American T1 facilities use a bit robbing process that limits the maximum achievable data rate to 56 kbps on a DS0 channel.

Rates for leased line digital! services are based primarily on distance and transmission speeds. This type of digital service is normally cost-effective for high volume users that could justify the expense associated with a dedicated communications facility. In comparison, pricing for Switched 56 kbps service is usage-sensitive, based both on connection duration and distance between calling and called parties. Due to the cost of this service essentially corresponding to usage, it is attractive for such applications as the backup of critical DDS and T1 lines, peak-time overload; usage to eliminate the necessity of installing additional leased digital circuits as well as for infrequent activities that may require a high data rate, such as still-frame video conferencing and facsimile transmission.

Terminal access to the DDS network is accomplished by means of a data service unit which alters serial unipolar signals into a form of modified bipolar signals for transmission and returns them to serial unipolar signals at the receiving end. The various types of service units were previously discussed in detail in Chapter 7.

15.4.1 Applications

AT&T's Dataphone Digital Service leased line offerings (as well as competitive offerings from MCI, Sprint, and other communications carriers), although they are important steps in increasing communications performance, represent dated technology.

DDS was based on the construction of a separate network that only permitted the transmission of seven data bits per eight bit byte, reducing the maximum data rate to 56 kbps. In addition DDS is based on a multiplexing arrangement in which customer data operating at or below 19.2 kbps are multiplexed twice at a carrier's central office, the first time to generate a DSO operating at 64 kbps and the second time to take 24 DSOs and place them onto a T1 line operating at 1.544 Mbps.

15.4.2 ASDS

Recognizing the inefficiencies inherent in DDS, AT&T introduced its ACCUNET Spectrum of Digital Services (ASDS) which significantly expanded its digital transmission offerings. Under AT&T's ACCUNET Spectrum of Digital Services, users can select digital circuits operating in 56/64 kbps multiples to obtain a fractional T1 (FT1) capability.

Access to a communications carrier's FT1 service is obtained by the installation of a full T1 line from the customer's premises to the central office serving the customer. At that office the number of DSOs representing 56/64 kbps increments of service selected by the customer are removed from the local T1 line and placed into predefined time slots on the long distance carrier's T1 line. The process of removing time slots from one T1 line and placing them on another line is referred to as Digital Access and Cross Connect, abbreviated as DACS where the S references the system that performs the cross-connect operation.

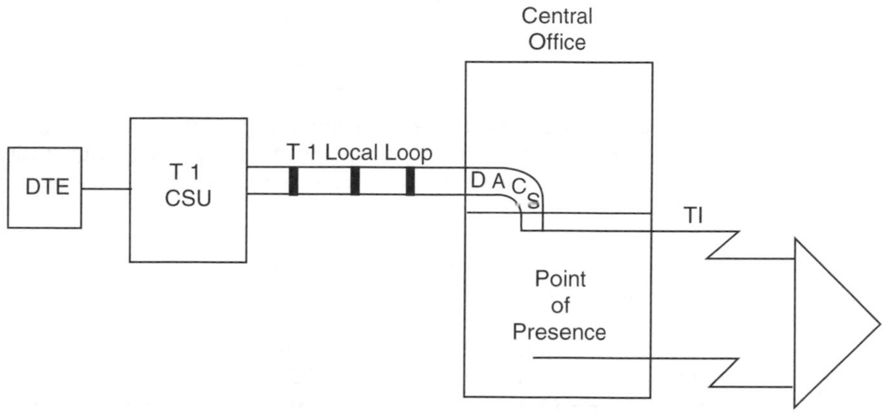

Figure 15.27 Fractional T1 access

Figure 15.27 illustrates the manner by which an end-user accesses a fractional T1 service. The local loop routed between the central office of the serving telephone company and the customer is a full T1 line which is terminated through the use of an intelligent T1 CSU. That CSU is configured to place data from the DTE on predefined DSO 64 kbps increments of bandwidth which flows to the central office. At the central office a DACS moves the predefined DSOs onto a long distance carrier's T1 line which will more than likely be multiplexed onto a T3 carrier for transmission between cities.

The requirement to obtain access to FT1 service via a full T1, even if only a fraction of the T1 local loop circuit is used, results from the necessity of DACS to be able to extract predefined DSO time slots. To do so requires the DACS to receive a formatted and recognized framing signal, such as a D4 or EFS T1 frame which denotes the placement of each DSO in the frame.

15.4.3 KiloStream service

Although British Telecom's KiloStream service is similar to DDS, there are several significant differences that warrant discussion.

The British Telecom customer is provided with an interface device which is called a network terminating unit (NTU), which is similar to a DSU. The NTU provides CCITT interface for customer data at 2.4, 4.8, 9.6, or 48 kbps to include performing data control and supervision, which is known as structured data. At 64 kbps, the NTU provides a CCITT interface for customer

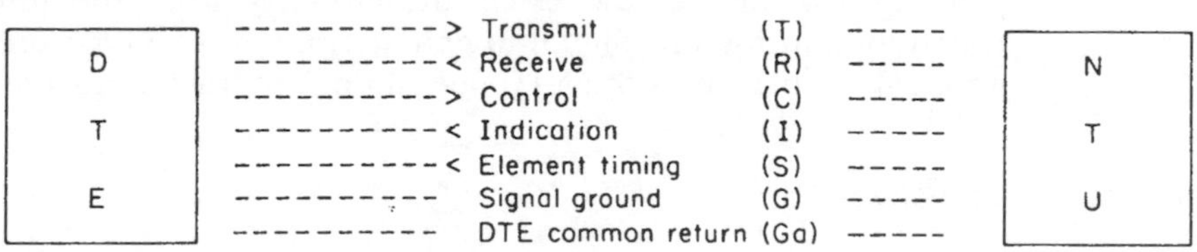

Figure 15.28 CCITT X.21 interface circuits

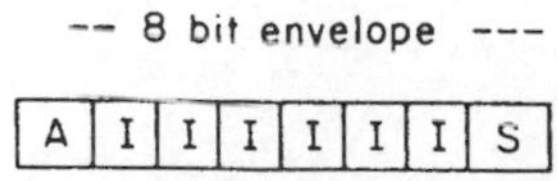

Figure 15.29 KiloStream envelope encoding. (A, alignment bit, alternates between '1' and '0' in successive envelopes to indicate the start and stop of each 8-bit envelope. S, status bit, is set or reset by the control circuit and checked by the indicator circuit. I, information bits)

Table 15.11 KiloStream NTU operational characteristics

Customer data rate (kbps)	DTE/NTU interface	Line data rate (kbps)	NTU operation
2.4	X.21	12.8	6 + 2 envelope encoding
4.8	X.21	12.8	6 + 2 envelope encoding
9.6	X.21	12.8	6 + 2 envelope encoding
48	X.21	64	6 + 2 envelope encoding
64	X.21	64	No envelope encoding
48	X.21 bis/V.35	64	6 + 2 envelope encoding
2.4	X.21/V.24	12.8	6 + 2 envelope encoding
4.8	X.21/V.24	12.8	6 + 2 envelope encoding
9.6	X.21/V.24	12.8	6 + 2 envelope encoding

data without performing data control and supervision, which is known as unstructured data.

The NTU controls the interface via CCITT recommendation X.21, which is the standard interface for synchronous operation on public data networks. An optional V.24 interface is available at 2.4, 4.8, and 9.6 kbps and an optional V.35 interface can be obtained at 48 kbps. The X.21 interface is illustrated in Figure 15.28. Here the control circuit (C) indicates the status of the transmitted information: data or signaling, while the indication circuit (I) signals the status of information received from the line. The control and indication circuits control or check the status bit of an 8-bit envelope used to frame six information bits.

Customer data are placed into a 6 + 2 format to provide the signaling and controlling information required. This is known as envelope encoding and it is illustrated in Figure 15.29.

The NTU performs signal conversion, changing unipolar non-return to zero signals from the V.21 interface into a di-phase WAL 2 encoding format. This ensures that there is no d.c. content in the signal transmitted to the line, provides isolation of the electronic circuitry from the line, and provides transitions in the line signal to enable timing to be recovered at the distant end. Table 15.11 lists the NTU operational characteristics of KiloStream.

The KiloStream network

In the KiloStream network, the NTUs on a customer's premises are routed via a digital local line to a multiplexer operating at 2.048 Mbps. This data rate is

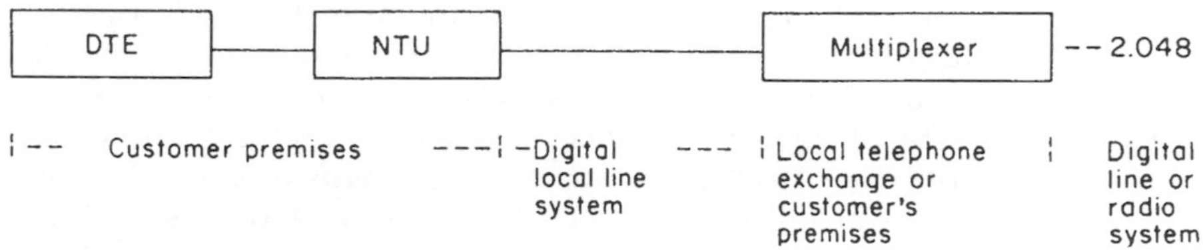

Figure 15.30 The KiloStream connection

the European equivalent of the T1 line in the United States that operates at 1.544 Mbps. The multiplexer can support up to 31 data sources and may be located at the local telephone exchange or on the customer's premises if traffic justifies. It is connected via digital line or a radio system into the British Telecom KiloStream network, as illustrated in Figure 15.30.

15.5 INTEGRATED SERVICES DIGITAL NETWORK (ISDN)

No discussion of digital networks would be complete without discussing the evolving future. In this section we will examine the future of both data and voice communications in the form of an Integrated Services Digital Network (ISDN), which can be expected to eventually replace most, if not all existing analog networks.

ISDN offers the potential for the development of a universal international digital network, with a series of standard interfaces that will facilitate the connection of a wide variety of telecommunications equipment to the network. Although the transition to ISDN may require several decades and some ISDN functions may never be offered in certain locations, its potential cannot be overlooked. Since many ISDN features offer a radical departure from existing services and current methods of communications, we will review the concept behind ISDN and projected features and new services that may result from its implementation in this section. This information is included to provide you with a familiarity of ISDN that will be used to illustrate how to effectively utilize this digital service.

15.5.1 Concept behind ISDN

The original requirement to transmit human speech over long distances resulted in the development of telephone systems designed for the transmission of analog data. Although such systems satisfied the basic requirement to transport human speech, the development of computer systems and the introduction of remote processing required a conversion of digital signals into an analog format. This conversion was required to enable computers and business machines to use existing telephone company facilities for the transmission of digital data. Not only was this conversion awkward and expensive due to the requirement to employ modems, but, in addition, the analog facilities of telephone systems limited the data transmission rate obtainable when such facilities were used.

The evolution of digital processing and the rapid decrease in the cost of semiconductors resulted in the application of digital technology to telephone systems. By the late 1960s, telephone companies began to replace their electromechanical switches in their central offices with digital switches; and by the early 1970s, several communications carriers were offering end-to-end digital transmission services. By the mid-1980s, a significant portion of the transmission facilities of most telephone systems was digital. On such systems, human speech is encoded into digital format for transmission over the backbone network of the telephone system. At the local loop of the network, digitized speech is reconverted into its original analog format and then transmitted to the subscriber's telephone.

Based on the preceding, ISDN can be viewed as an evolutionary progression in the conversion of analog telephone systems into an eventual all-digital network, with both voice and data to be carried end-to-end in digital form.

While the concept behind ISDN received the support of numerous communications carriers, competitive technologies such as Digital Subscriber Line modems, cable modems and satellite and wireless transmission techniques make it doubtful if ISDN will realize its intended potential. However, during the late 1990s its availability was significantly increased and the elimination of its per minute charge for local calls increased its use as a mechanism for high-speed Internet access.

15.5.2 ISDN architecture

With the evolving ISDN architecture, access to this digital network will result from one of two major connection methods: basic access and primary access.

Basic access

Basic access defines a multiple channel connection derived by multiplexing data on twisted-pair wiring. This multiple channel connection is between an end-user terminal device and a telephone company office or a local private automated branch exchange (PABX). The ISDN basic access channel format is illustrated in Figure 15.31.

As indicated in Figure 15.31, basic access consists of two bearer (B) channels and a data (D) channel that are multiplexed by time onto a common twisted-pair wiring media. Each bearer channel can carry one pulse code modulation (PCM) voice conversation or data at a transmission rate of 64 kbps. This enables basic access to provide the end-user with the capability to simultaneously transmit data and conduct a voice conversation on one telephone line or to be in conversation with one person and receive a second

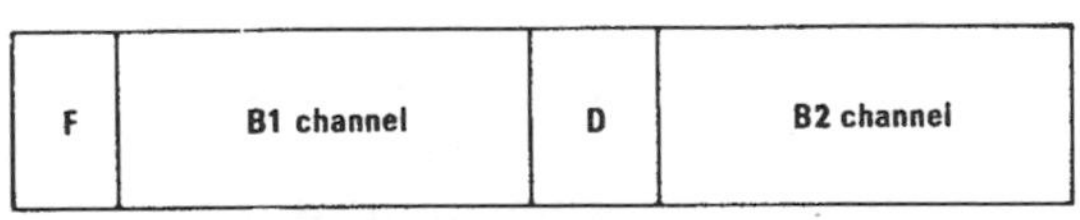

Figure 15.31 ISDN basic access channel format

telephone call. In the case of the latter situation, assuming that the end-user has an appropriate telephone instrument, he or she could place one person on hold and answer the second call.

The D channel was designed to both control the B channels through the sharing of network signaling functions on this channel, and transmit packet switched data. Concerning the transmission of packet switched data, the D channel provides the capability for a number of new applications, including monitoring home alarm systems and the reading of utility meters on demand. Since these types of application have minimum data transmission requirements, the D channel can be expected to be used for a variety of applications in addition to providing the signaling required to set up calls on the B channels. In fact, a recent innovation with respect to ISDN is an 'always-on' capability. When offered, ISDN always-on means that the D channel is always active. This would allow users to receive email as it occurs rather than have to connect to a service to retrieve electronic mail and represents one of many applications an ISDN always-on capability could support.

One of the problems associated with ISDN is terminology. For example, basic access which consists of a 2B+D channel is referred to as single line ISDN service by US West. The reason for this name is the fact that the same pair of telephone wires that delivers one call at a time basic telephone service, when converted to ISDN service, can provide two primary 64 kbps channels and one D channel. In this section we will use standardized notation to reference ISDN technology.

Primary access

Primary access can be considered as a multiplexing arrangement whereby a grouping of basic access users shares a common line facility. Typically, primary access will be employed to directly connect a PABX to the ISDN network. This access method is designed to eliminate the necessity of providing individual basic access lines when a group of terminal devices shares a common PABX which could be directly connected to an ISDN network via a single high-speed line. Due to the different types of T1 network facilities in North America and Europe, two primary access standards have been developed.

In North America, primary access consists of a grouping of 23 B channels and one D channel to produce a 1.544 Mbps composite data rate, which is the standard T1 carrier data rate. In Europe, primary access consists of a grouping of 30 B channels plus one D channel to produce a 2.048 Mbps data rate, which is the T1 carrier transmission rate in Europe.

15.5.3 Network characteristics

Four of the major characteristics of an ISDN network are listed in Table 15.12. These characteristics can also be considered as driving forces for the implementation of the network by communications carriers.

Due to the digital nature of ISDN, voice, data and video services can be integrated, alleviating the necessity of end-users obtaining separate facilities

Table 15.12 ISDN characteristics

Integrates voice, data and video services
Digital end-to-end connection resulting in high transmission quality
Improved and expanded services due to B and D channel data rates
Greater efficiency and productivity resulting from the ability to have several simultaneous calls occur on one line

for each service. Since the network is designed to provide end-to-end digital transmission, pulses can be easily regenerated throughout the network, resulting in the generation of new pulses to replace distorted pulses. In comparison, analog transmission facilities employ amplifiers to boost the strength of transmission signals, which also increases any impairments in the signal. As a result of regeneration being superior to amplification, digital transmission has a lower error rate and provides a higher transmission signal quality than an equivalent analog transmission facility.

Due to basic access in effect providing three signal paths on a common line, ISDN offers the possibilities of both improvements to existing services and an expansion of services to the end-user. Concerning existing services, current analog telephone line bandwidth limitations normally preclude data transmission rates over 33.6 kbps occurring on the switched telephone network. In comparison, under ISDN each B channel can support a 64 kbps transmission rate and the D channel will operate at 16 kbps. In fact, if both B channels and the D channel were in simultaneous operation, a data rate of 144 kbps would be obtainable on a basic access ISDN circuit, which would exceed current analog circuit data rates by a factor of 4.

Concerning expanded services, ISDN's D channel provides a transmission signaling path that can be used for many types of supplementary service. These services range from called number identification, display of charge of service during a call and call diversion, to reverse charging and closed user group identification. Called number identification not only provides subscribers with the identity of the calling party, but substantially reduces malicious calls because the calling party is identified.

Since each basic access channel in effect consists of three multiplexed channels, different operations can occur simultaneously without requiring an end-user to acquire separate multiplexing equipment. Thus, the end-user could receive a call from one person, transmit data to a computer and have the utility company read his or her electric meter at a particular point in time. Here, the ability to conduct up to three simultaneous operations on one ISDN line should result in both greater efficiency and productivity. Efficiency should increase because one line can now support several simultaneous operations, whereas the productivity of the end-user can increase due to the ability to receive telephone calls and then conduct a conversation while transmitting data.

Terminal equipment and network interface

One of the key elements of ISDN is a small set of compatible multipurpose user–network interfaces that were developed to support a wide range of

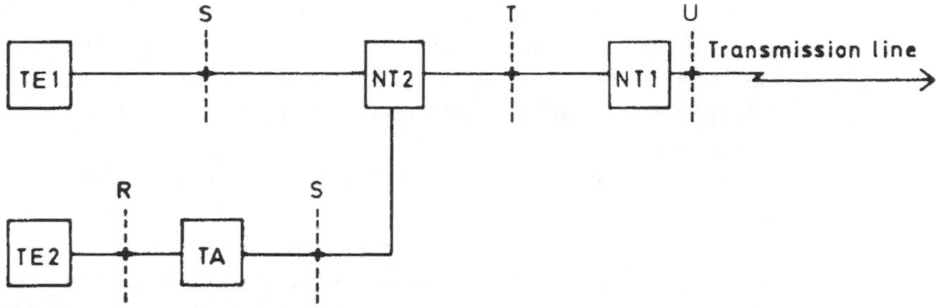

Figure 15.32 ISDN reference points and network interfaces. (TE1 terminal equipment 1) type devices comply with the ISDN network interface. TE2 (terminal equipment 2) type devices do not have an ISDN interface and must be connected through a TA (terminal adapter) functional grouping. NT2 (network termination 2) includes switching and concentration equipment which perform functions equivalent to layers 1 through 3 of the OSI reference model. NT1 (network termination 1) includes functions equivalent to layer 1 of the OSI reference model)

applications. These network interfaces are based upon the concept of a series of reference points for different user terminal arrangements which is then used to define these interfaces. Figure 15.32 illustrates the relationship between ISDN reference points and network interfaces.

The ISDN reference configuration consists of functional groupings and reference points at which physical interfaces may exist. The functional groupings are sets of functions that may be required at an interface, and reference points are employed to divide the functional groups into distinct entities.

The TE (terminal equipment) functional grouping is composed of TE1 and TE2 type equipment. Examples of TE equipment include digital telephones, conventional data terminals and integrated voice/data workstations. TE1 type equipment complies with the ISDN user–network interface and permits such equipment to be directly connected to an ISDN 'S' type interface which supports multiple B and D channels.

TE2 type equipment are devices with non-ISDN interfaces, such as RS-232 or the CCITT X or V series interfaces. This type of equipment must be connected through a TA (terminal adapter) functional grouping, which in effect converts a non-ISDN interface (R) into an ISDN Sending interface (S), performing both a physical interface conversion and protocol conversion to permit a TE2 terminal to operate on ISDN.

The NT2 (network termination 2) functional group includes devices that perform switching and data concentration functions equivalent to the first three layers of the OSI Reference Model. Typical NT2 equipment can include PABXs, terminal controllers, concentrators, and multiplexers.

The NT1 (network termination 1) functional group is the ISDN digital interface point and is equivalent to layer 1 of the OSI Reference Model. Functions of NT1 include the physical and electrical termination of the loop, line monitoring, timing and bit multiplexing. In Europe, where most communications carriers are government-owned monopolies, NT1 and NT2 functions may be combined into a common device, such as a PABX. In such situations, the equipment serves as an NT12 functional group. In comparison, in the United States the communications carrier may provide only the NT1, and third-party vendors can provide NT2 equipment. In such situations, the third-party equipment would connect to the communications carrier equipment at the T interface.

In North America the telephone company provides its basic access customers with a U interface. The U interface represents a two-wire (single pair) interface from a central office telephone switch that supports full-duplex data transfer. In other non-North American locations where NT1 and NT2 are combined, the customer is provided with an S/T interface. Unlike the U interface that is limited to supporting a single device, the S/T interface represents a four-wire, full-duplex interface that has separate pairs for transmit and receive, allowing up to seven devices to be placed on the S/T bus. Because of the ability of the S/T bus to support multiple devices, many equipment manufacturers now incorporate an NT1 containing an S/T bus in their products.

Terminal adapters

The terminal adapter (TA) will be a key device in achieving ISDN connectivity well into the 21st century. This is because the TA provides the large base of installed non- ISDN devices with the ability to be connected to the ISDN during the evolutionary process of vendors manufacturing ISDN compatible equipment which will gradually replace non-ISDN equipment.

Key functions performed by TAs include the conversion of electrical, mechanical, functional and procedural characteristics of non-ISDN equipment interfaces to those required by ISDN, mapping of network layer data to enable a signaling terminal to be 'understood' by ISDN equipment, and bit-rate adaptation.

Rate adaptation is the process in which the data rate of slow-speed devices is increased to the 64 kbps synchronous data rate of an ISDN B channel. During the rate adaptation process the data stream produced by a non-ISDN device is padded with dummy bits by the terminal adapter and clocked at a 64 kbps data rate. Rate adaptation schemes currently employed are based upon ITU-TV.14, V.110 and V.120 standards.

Although the actual rate adaptation process under each standard is different, each process is similar with respect to functionality. That is, the rate adaptation process involves inserting a received data stream into a synchronous frame structure that contains additional bits that allows an operating rate below 64 kbps to be transported as a 64 kbps ISDN B channel.

The V.14 standard results in the support of asynchronous rate adaptation up to 57.6 kbps. The V.110 standard that was developed prior to the V.120 standard and is popularly employed in Europe provides an adaptation capability up to 38.4 kbps. The third rate adaptation technique is the V.120 standard. The V.120 standard permits rate adaptation over a 1B or 2B ISDN facility. Although both the V.110 and V.120 standards support rate adaptation for synchronous and asynchronous data streams, in actuality many equipment vendors support only asynchronous rate adaptation.

Table 15.13 ISDN 2B1Q encoding

Bit pairs	Quaternary symbol	Line voltage level
00	−3	−2.5
01	−1	−0.833
10	+3	+2.5
11	+1	+0.833

15.5.4 ISDN layers

ISDN can be considered to represent the first three layers of the ISO Reference Model, with its physical, data link and network layers standardized by the ITU-T.

Physical layer

For basic access ISDN uses a data encoding scheme referred to as 2B1Q at the U interface in North America. The reason for the selection of 2B1Q involves the data rate of a basic access connection. If we examine Figure 15.31 we note that each ISDN basic access frame consists of frame overhead, one D channel and two B channels. The B channels each operate at 64 kbps while the D channel operates at 16 kbps. Because frame overhead is also 16 kbps, this results in a data rate of 160 kbps.

When designing a signaling rate it was important to select a method that would allow extended transmission distances from a telephone company switch to the subscriber. This resulted in the selection of 2 Binary 1 Quaternary (2B1Q) encoding in which two bits are encoded into one signal and which reduces the signal rate to 80 kbaud.

Under 2B1Q encoding four voltage levels define four quaternary symbols. Thus, each pair of bits are encoded into one of four symbols and the applicable voltage is placed on the line. Table 15.13 summarizes the relationship between bit pairs, their quaternary symbol value, and resulting line voltage level.

At the physical layer each U interface frame is 240 bytes in length. This results in the frame consisting of frame overhead, two B channels and one D channel having a duration of 1.5 ms at a data rate of 160 kbps.

The beginning of each physical layer frame consists of 18 synchronization bits transmitted as a sequence of nine quaternaries (+3 +3 −3 −3 −3 +3 +3 +3 −3). The synchronization bits are followed by 12 sequences of B channel and D channel data. Here 8 bits from the first and second B channel followed by 2 bits of D channel data are grouped into 12 sequences and result in 216 bits of data as shown below:

SYNC 18 bits	12 * ($B_1 + B_2 + D$) 216 bits	Maintenance 6 bits

Maintenance data in the form of 6 bits for CRC information, block error detection and loopback testing complete the physical layer frame. Finally, eight frames are used to form a superframe consisting of eight 240-bit frames for a total of 1920 bits. To distinguish the beginning of the superframe, its synchronization field is inverted, producing the quaternary pattern −3 −3 +3 +3 +3 −3 −3 −3 +3.

Data link layer

At the data link layer ISDN uses an HDLC-like protocol on the D channel referred to as Link Access Protocol – D channel (LAP-D). Under LAP-D the frame used for signaling has six fields and is illustrated at the top of Figure 15.33. Similar to HDLC, the Flag field always consists of the bit sequence 01111110. The address field, which is illustrated in the lower portion of Figure 15.33, consists of two bytes that contain five fields. The Service Access Point Identifier (SAPI) represents a 6-bit field that identifies the location where Layer-2 provides a service to Layer-3. Table 15.14 lists the SAPI values and their current utilization.

The seven-bit field labeled TEI in Figure 15.33 represents a Terminal Endpoint Identifier. TEIs are unique Ids that are assigned to each terminal equipment (TE) on an ISDN S/T bus. The assignment of the TEI can occur either statically when the TE is installed or on a dynamic basis when the TE is activated. Concerning the latter, dynamic TEI assignments are effected by a telephone company switch. Table 15.15 lists presently defined ISDN TEI values.

Similar to an HDLC exchange, the ISDN data link layer uses LAP-D frames to establish a connection to a switch. Under the ISDN Layer-2 establishment

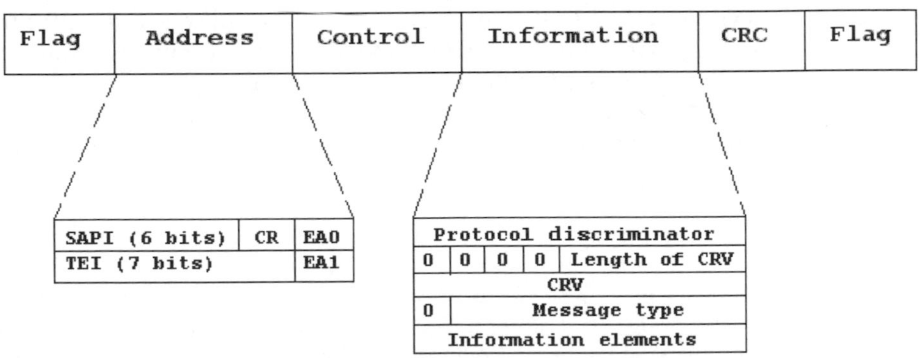

Legend:

SAPI	Service Access Point Identifier
C/R	Command/Response
EA0	Address extension
	0 = not final byte
	1 = final byte
TEI	Terminal Endpoint Identifier
CRV	Call Reference Value

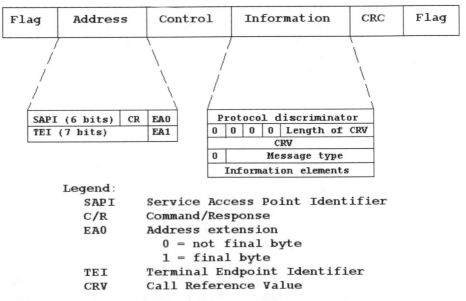

Figure 15.33 ISDN Link Access Protocol – D Channel (LAP-D) frame format

Table 15.14 ISDN Service Access Point Identifiers (SAPI)

SAPI value	Description
0	Call control procedures
1	Packet Mode using Q.931
?	Call control procedures
16	Packet Mode communications procedures
32–47	Reserved for national use
63	Management procedures
Others	Reserved for future use

Table 15.15 ISDN Terminal Endpoint Identifiers (TEI)

TEI value	Description
0–63	Fixed TEI assignment
64–126	Dynamic TEI assignment
127	Broadcast to all devices

process the TE and the network exchange Receiver Ready (RR) frames. Next, the TE transmits an Unnumbered Information (UI) frame with a SAPI of 63 (management procedure for network query) and a TEI of 127 (broadcast). The network switch assigns an available TEI from the range of 64 through 126. This is followed by the TE transmitting a Set Asynchronous Balanced Mode (SABME) frame. That frame has a SAPI value of 0 (call control) and a TEI value that was assigned by the network. The network then responds to the TE with an Unnumbered Acknowledgement (UA) frame that has a SAPI value of 0 and a TEI value that was previously selected by the network. At this point in time the connection is ready for a Layer-3 setup.

Network layer

The Information field within the LAP-D frame includes several fields that govern the services and features provided by the telephone switch to the subscriber's ISDN device. The general format of this variable field was previously illustrated in the lower right portion of Figure 15.33.

The Protocol Discriminator field identifies the Layer 3 protocol. The Call Reference Value (CRV) is used to uniquely identify each call on the user–network interface. This value can be carried in either one or two bytes and is assigned at the beginning of a call. When the call is cleared the CRV becomes available for another call.

The Message Type field identifies the type of message, such as SETUP, CONNECT, ALERTING and so on. The last field in the Information field, Information Elements, contains options that are set based upon the value of the Message Type field. The ISDN network layer is specified by a series of ITU-T Q-series documents that defines messages used to establish, maintain and terminate logical network connections between two devices.

At the beginning of a connection a Service Profile ID (SPID) is used to identify the services and features the telephone company switch will provide to the attached ISDN device at call setup time. The SPID is the subscriber's ten-digit telephone number usually followed by a suffix of four 1s that describes a specific ISDN device to the network. SPIDs are used since it is technically possible to attach up to eight different devices with up to 64 telephone numbers to a single ISDN telephone line. Thus, the SPID provides a mechanism to identify the capability of the line.

15.6 REVIEW QUESTIONS

1. List the three major components of a channel bank. What function does each component perform?

2. What is the rationale for using bipolar signaling for digital transmission?

3. What four functions does a codec perform?

4. How was a sampling rate of 8000 per second selected for PCM?

5. Why is it necessary to quantize a PAM signal?

6. What is the difference between a chord and a step with respect to their spacing?

7. Under μ-law encoding what would the PCM word 01011011 signify with respect to a PAM sample?

8. What does a basic T1 frame consisting of 193 bits represent?

9. How many framing bits per second flow over a DS1 line?

10. Assume that the D4 framing pattern ...111 was received. What frame number should the receiving multiplexer expect next?

11. Why is the maximum data rate 56 kbps on a DDS circuit in North America?

12. What sources of clocking can a channel bank use?

13. Describe the 'ones density' requirement for a DS1 line. Why does a repeater require a minimum separation between marks on a DS1 line?

14. What information is conveyed in robbed bits?

15. Describe the difference between D4 and ESF framing.

16. What is the primary benefit of B8ZS coding?

17. Assume that the last one bit was negative and a byte has all zero bits. What is the inserted B8ZS code?

18. Discuss the relationship between a red and yellow alarm.

19. What is the purpose of a blue alarm?

20. What is the primary advantage of ESF framing over D4 framing?

21. Where does framing take place on a CEPT PCM-30 frame?

22. What framing signal uses a CRC-6?

23. What framing signal uses a CRC-4?

24. What is the advantage of acquiring T1 multiplexers with voice digitization modules that operate at 32 kbps instead of PCMs 64 kbps data rate?

25. Describe an example of vocoding.

26. Describe two methods of hybrid coding.

27. Assume you selected a voice digitization module that operates at 16 kbps. How many voice conversations could be carried on a T1 line?

28. Describe the two types of T3 circuits. Which type of T3 circuit would you use for a high-speed Internet connection?

29. What does a DS2 signal represent? How many DS2 signals are multiplexed to form a DS3 signal?

30. Describe the use of the C-bits in a DS3 signal.

31. Describe the use of ones density methods for DS1, DS2 and DS3 signals.

32. What interface would you use to obtain the ability to utilize the full transmission capability of a T3 circuit?

33. Describe the use of a T3 circuit as a homing point in a network.

34. Why can you expect transmission quality on ISDN facilities to be superior to existing analog facilities?

35. Discuss the data transmission rate differences between a basic access ISDN circuit and that obtainable on the switched telephone network.

36. What is an advantage of 2B1Q line coding?

37. What protocol does ISDN's LAP-D resemble?

38. Describe an ISDN physical layer superframe.

39. What is the function of an ISDN Service Access Point Identifier?

40. Why is a Service Profile ID necessary?

16

NETWORK ARCHITECTURE

To satisfy the requirements of customers for remote computing capability, mainframe computer manufacturers developed a variety of network architectures. Such architectures define the interrelationshlp of a particular vendor's hardware and software products necessary to permit communications to flow through a network to the manufacturer's mainframe computer. Other sources of network architecture include the efforts of research laboratories, private organizations and government agencies.

Most network architectures are layered structures and bear resemblance to the ISO reference model. Although there are many important network architectures, no one book, nor possibly even a series of books, could provide more than a detailed overview of the numerous network architectures developed since the dawn of the computer age. Recognizing space limitations, we will focus our attention upon two key network architectures. One architecture we will examine represents a proprietary vendor product that can be considered past its prime, although it is still employed by most organizations with mainframe computers. The second network architecture represents a public architecture whose utilization is increasing at what Mr Spock would refer to as warp speed.

The proprietary network architecture we will examine is IBM's System Network Architecture (SNA). Through the mid-1990s SNA transported the vast majority of network-based traffic, representing the primary method of communications by banks, insurance companies and businesses. Although the Transmission Control Protocol (TCP/IP) network architecture was originally developed around the same time as SNA, TCP/IP was at that time primarily used by academic institutions. The development of the Web browser during the mid-1990s can be considered a milestone in the growth in the use of the TCP/IP protocol suite. Within a few years of its development tens of millions of persons were using Web browsers and the use of the Internet dramatically increased. In fact, by 1998 Internet traffic far surpassed SNA traffic and many organizations were gradually converting proprietary SNA networks to private but standardized TCP/IP networks. Such networks are referred to as intranets and represent the use of the TCP/IP protocol suite for private use. In addition, the downsizing of mainframes to client–server activity over LANs further reduced the use of SNA as a network architecture. However, because mainframes are great repositories for holding large quantities of data, their use has actually increased and the demise of both 'big iron', a term used to reference mainframes, and SNA traffic has failed to reach the level predicted by industry analysts.

The differences between SNA and TCP/IP are profound, and for many organizations that require the continued use of mainframes they represent a challenge when attempting to construct a common network based on the use of modern router technology. Recognizing some of the problems resulting from the original architecture of SNA, IBM introduced a more modern version of its network architecture, which changes the orientation of SNA networks from mainframe centric to peer-to-peer. This version of SNA is known as Advanced Peer-to-Peer Networking (APPN).

In this chapter we will first focus our attention on SNA, reviewing its concepts, examining the structure of an SNA network to include its major components and obtain an understanding of its major limitations in an era of client–server computing. Once this has been accomplished we will focus our attention on APPN concepts and then examine TCP/IP. In concluding this chapter we will use the previously presented information to discuss the integration of SNA and TCP/IP, which for many organizations remains a key issue.

16.1 SNA OVERVIEW

IBM's system network architecture (SNA) is a very complex and sophisticated network architecture which defines the rules, procedures and structure of communications from the input–output statements of an application program to the screen display on a user's personal computer or terminal. SNA consists of protocols, formats and operational sequences which govern the flow of information within a data communications network linking IBM mainframe computers, minicomputers, terminal controllers, communications controllers, personal computers and terminals.

When originally introduced by IBM in 1974, SNA was very limited with respect to functionality and capability in comparison to today. Its main purpose was to provide a well structured architecture for growth and development and in this area has achieved considerable success.

The first release of SNA included three major products: VTAM, NCP and cluster controllers. The first two products were obtained primarily from software, and the third product was essentially a combination of firmware and hardware.

VTAM, an acronym for Virtual Telecommunications Access Method, operates on the mainframe computer and provides the access method for communications between that computer and terminal devices in an SNA network. This software product provides a common interface to network communications facilities for all application programs operating under the computer's operating system. This enables application programmers to restrict communications operations to relatively simply input and output statements that are acted on by VTAM. In comparison, without VTAM, each application programmer would have to write his or her own routines to support communications for each application.

NCP, an acronym for Network Control Program, operates in the communications controller connected to the mainframe. This software product consists of coding that defines the cluster controllers attached to each communications controller line interface as well as the terminals connected to each cluster controller port. Figure 16.1 illustrates an elementary SNA configuration showing the placement of hardware and software. When originally announced VTAM ran on an IBM System/370 whereas the NCP operated on an IBM 3705 communications controller. Today, VTAM operates on a variety of IBM mainframes including the 30XX and 43XX series as well as the more modern Enterprise systems, while the NCP operates on IBM 3720, 3725 and 3745 communications controllers.

Since its original announcement SNA has dramatically expanded in terms of its architecture, evolving from the basic tree-like structure illustrated in Figure 16.1 to support a multitude of structures to meet the communications requirements of organizations that have multiple mainframe computers located in dispersed geographical areas. As approximately 70% of the mainframe computer market belongs to IBM, SNA can be expected to remain as a connectivity platform for the foreseeable future. This means that a large majority of the connections of local area networks to mainframe computers will require the use of gateways that support SNA operation.

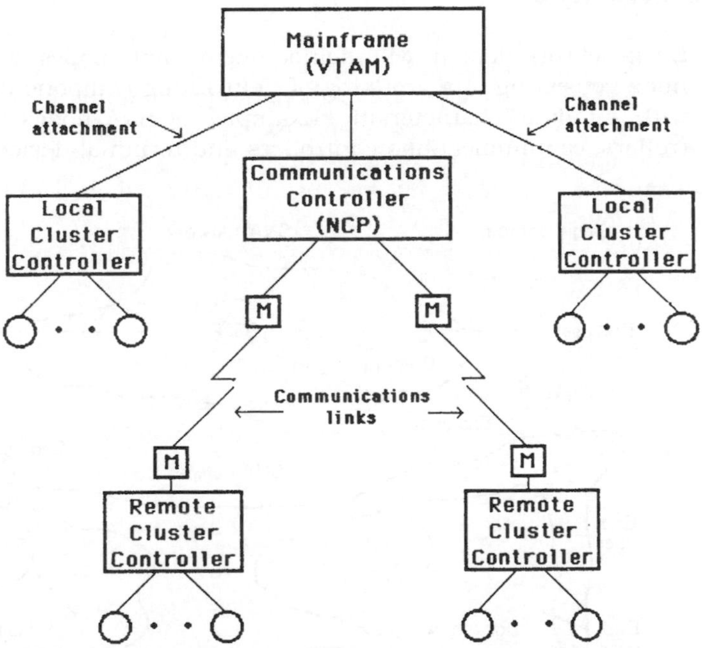

Figure 16.1 Early SNA network configuration. VTAM operates on the mainframe and provides a common interface between application programs and communication facilities. The NCP resides in the communications controller and defines the operational parameters of remote cluster controllers and terminals attached to those controllers

16.1.1 SNA elements

An SNA network consists of one or more domain, where a domain refers to all of the logical and physical components that are connected to and controlled by one common point in the network. This common point of control is called the System Service Control Point, which is commonly known by its abbreviation as the SSCP. There are three types of network addressable units (NAUs) in an SNA network: SSCPs, physical units and logical units.

16.1.2 System Services Control Point (SSCP)

The SSCP resides in the communications access method operating in an IBM mainframe computer, such as Virtual Telecommunications Access Method (VTAM), operating in a System/360, System/370, 43XX, 308X, 309X or Enterprise series computer, or in the system control program of an IBM minicomputer, such as a System/3 or an AS/400. The SSCP contains the network's address tables, routing tables and translation tables which it uses to establish connections between nodes in the network as well as to control the flow of information in an SNA network. Figure 16.2 illustrates single and multiple domain SNA networks.

16.1.3 Network nodes

Each network domain will include one or more nodes, with an SNA network node consisting of a grouping of networking components which provides it with a unique characteristic. Examples of SNA nodes include cluster controllers, communications controllers and terminal devices, with the address

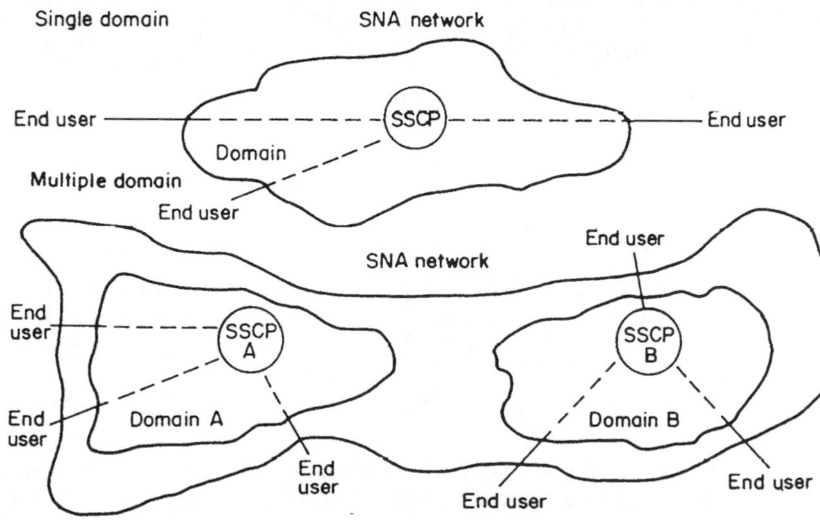

Figure 16.2 Single and multiple domain SNA networks

of each device in the network providing its unique characteristic in comparison to a similar device contained in the network.

16.1.4 The physical unit

Each node in an SNA network contains a physical unit (PU) which controls the other resources contained in the node. The PU is not a physical device as its name appear to suggest, but rather a set of SNA components which provide services used to control terminals, controllers, processors and data links in the network. In programmable devices, such as mainframe computers and communications controllers, the PU is normally implemented in software. In less intelligent devices, such as cluster controllers and terminals the PU is typically implemented in read-only memory, commonly referred to as firmware. In an SNA network each PU operates under the control of an SSCP. The PU can be considered to function as an entry point between the network and one or more logical units.

16.1.5 The logical unit

The third type of network addressable unit in an SNA network is the logical unit, known by its abbreviation as the LU. The LU is the interface or point of access between the end-user and an SNA network. Through the LU an end-user gains access to network resources and transmits and receives data over the network. Thus, all communications between SNA users must flow through at least two logical units. Each PU can have one or more LUs, with each LU having a distinct address.

16.1.6 SNA network structure

The structure of an SNA network can be considered to represent a hierarchy in which each device controls a specific part of the network and operates under the control of a device at the next higher level. The highest level in an SNA network is represented by a host or mainframe computer that executes a software module known as a communications access method. At the next lower level are one or more communications controllers, IBM's term for a front-end processor. Each communications controller executes a Network Control Program (NCP) which defines the operation of devices connected to the controllers, their PUs and LUs, operating rate, data code and other communications-related functions such as the maximum packet size that can be transmitted. Connected to communications controllers are cluster controllers, IBM's term for a control unit. Thus, the third level in an SNA network can be considered to be represented by cluster controllers.

The cluster controllers support the attachment of terminals and printers that represent the lowest hierarchy of an SNA network. Figure 16.3 illustrates the SNA hierarchy and the Network Addressable Units (NAUs) associated with each hardware component used to construct an SNA network. Note that

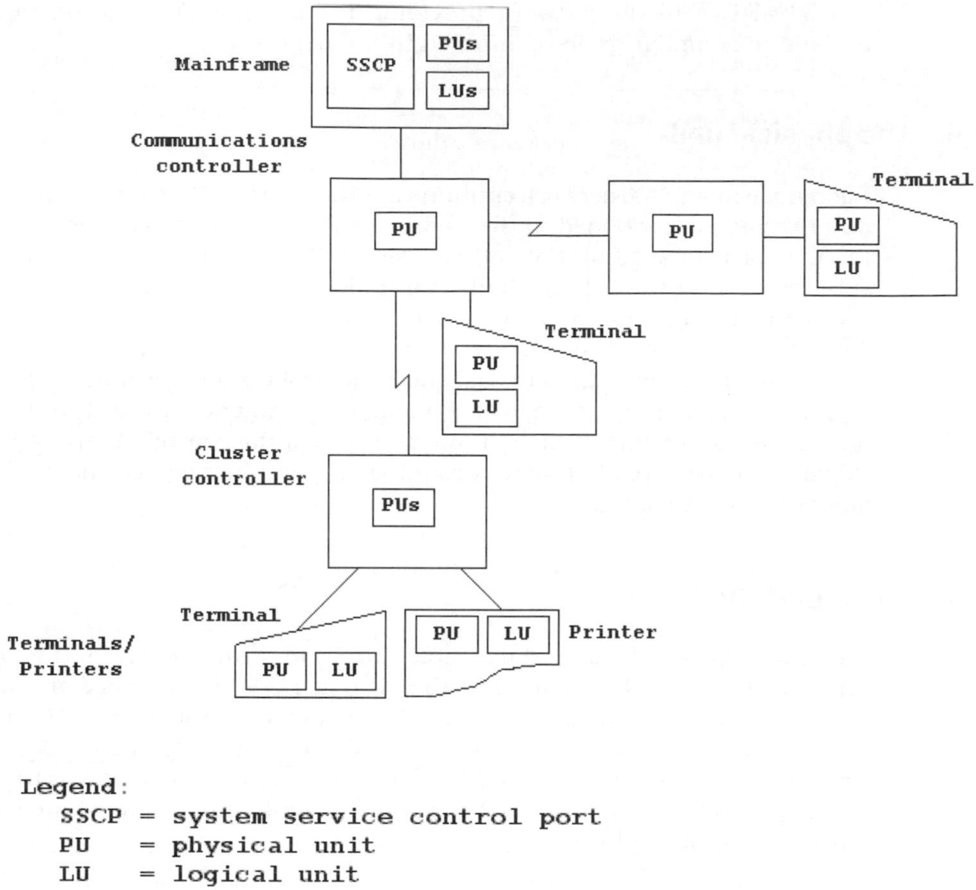

Figure 16.3 SNA hierarchical network structure. the structure of an SNA network is built upon a hierarchy of equipment, with mainframes connected to communications controllers and communications controllers connected to cluster controllers

NAUs include lines connecting mainframes to communications controllers to cluster controllers and connecting cluster controllers to terminals and printers. In addition, NAUs also define application programs that reside in the mainframe. Thus, NAUs provide the mechanism for terminals to access specific programs via a routing through hardware and transmission facilities that are explicitly identified.

As an example of the communications capability of SNA, consider an end-user with an IBM PC and an SDLC communications adapter who establishes a connection to an IBM mainframe computer. The IBM PC is a PU, with its display and printer considered to be LUs. After communication is established, the PC user could direct a file to his or her printer by establishing an LU-to-LU session between the mainframe and printer while using the PC as an interactive terminal running an application program as a second LU-to-LU session. Thus, the transfer of data between PUs can represent a series of multiplexed LU-to-LU sessions, enabling multiple activities to occur concurrently.

Table 16.1 SNA PU summary

PU Type	Node	Representative hardware
PU type 5	Mainframe	S/370, 43XX, 308X
PU type 4	Communications controller	3705, 3725, 3720
PU type 3	Not currently defined	n/a
PU type 2	Cluster controller	3274, 3276, 3174
PU type 1	Terminal	3180, PC with SNA adapter

16.1.7 Types of physical unit

Table 16.1 lists five types of physical units in an SNA network and their corresponding node type. In addition, this table contains representative examples of hardware devices that can operate as a specific type of PU. As indicated in Table 16.1, the different types of PUs form a hierarchy of hardware classifications. At the lowest level, PU type 1 is a single terminal. PU type 2 is a cluster controller which is used to connect many SNA devices on to a common communications circuit. PU type 4 is a communications controller which is also known as a front-end processor. This device provides communications support for up to several hundred line terminations, where individual lines in turn can be connected to cluster controllers. At the top of the hardware hierarchy, PU type 5 is a mainframe computer.

The communications controller is also commonly referred to as a front-end processor. This device relieves the mainframe of most communications processing functions by performing such activities as sampling attached communications lines for data, buffering the data and passing it to the mainframe as well as performing error detection and correction procedures. The cluster controller functions similarly to a multiplexer or data concentrator by enabling a mixture of up to 32 terminals and low-speed printers to share a common communications line routed to a communications controller or directly to the mainframe computer.

16.1.8 Multiple domains

Figure 16.4 illustrates a two-domain SNA network. By establishing a physical connection between the communications controller in each domain and coding appropriate software for operation on each controller, cross-domain data flow becomes possible. When cross-domain data flow has been, terminal devices connected to one mainframe gain the capability to access applications operating on the other mainframe computer.

SNA was originally implemented as a networking architecture in which users establish sessions with application programs that operate on a mainframe computer within the network. Once a session has been established a network control program (NCP) operating on an IBM communications controller, which in turn is connected to the IBM mainframe, works in conjunction with the access method operating on the mainframe, such as VTAM, to control the information flow between the user and the applications program. With the

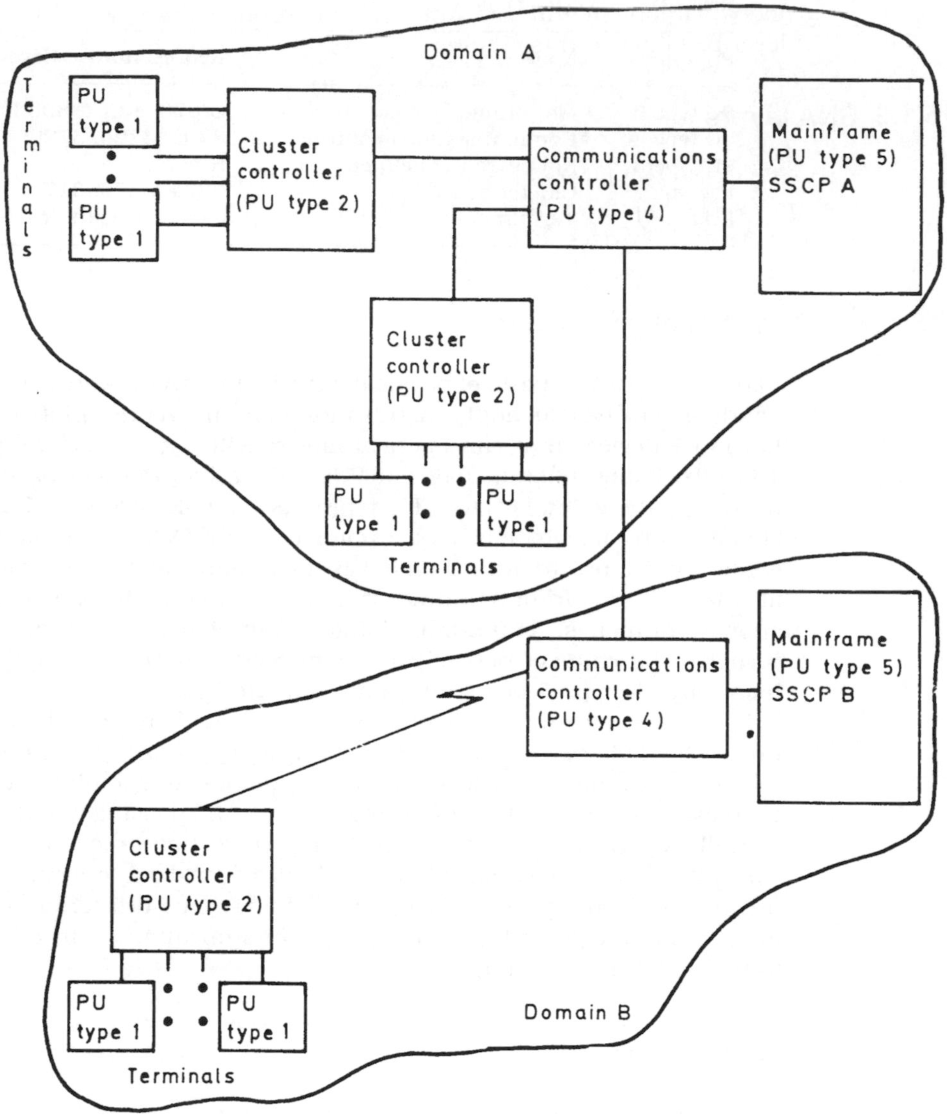

Figure 16.4 Two-domain SNA network

growth in personal computing many users no longer required access to a mainframe to obtain connectivity to another personal computer connected to the network. Thus, IBM modified SNA to permit peer-to-peer communications capability in which two devices on the network with appropriate hardware and software could communicate with one another without requiring access through a mainframe computer. In doing so, IBM introduced a PU2.1 node in 1987 in recognition of the growing requirements of its customers for peer-to-peer networking capability. Unfortunately, it was not until the early 1990s that its Advanced Peer-to-Peer Networking (APPN) became available and provided this networking capability, enabling communications between PU2.1

nodes without mainframe intervention through the use of LU6.2 sessions described later in this chapter.

16.1.9 SNA layers

IBM's SNA is a layered protocol that provides seven layers of control for every message that flows through the network. Figure 16.5 illustrates the SNA layers and provides a comparison to the seven-layer OSI Reference Model.

Physical and data link control

Similar to the OSI physical layer, SNA's physical control layer is concerned with the electrical, mechanical and procedural characteristics of the physical media and interfaces to the physical media. SNA's data link control layer is also quite similar to OSI's data link layer. Protocols defined by SNA include SDLC, System/370 channel, Token-Ring and X.25; however, only SDLC is used on a communications link in which master or primary stations communicate with secondary or slave stations. Some implementations of SNA using special software modules can also support Bisynchronous (BSC) communications, an IBM pre-SNA protocol still widely used, as well as asynchronous communications.

Path control layer

This layer is responsible for the creation of a virtual route through an SNA network that ties the message source to its destination. This activity is primarily accomplished by two of the major functions of the path control layer: routing and flow control. Concerning routing, since there can be many

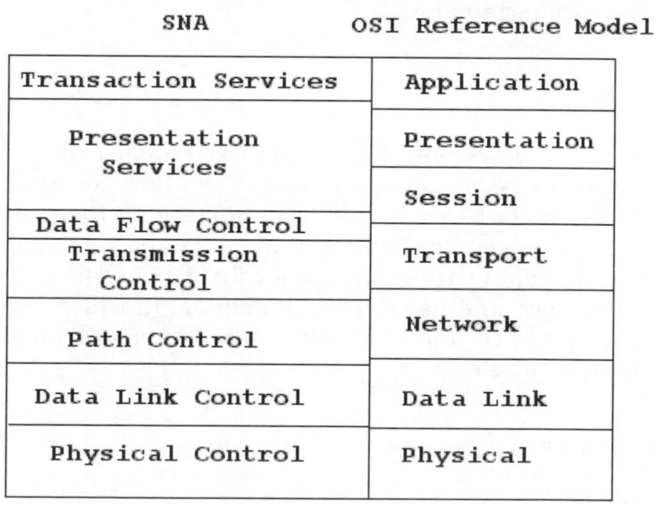

Figure 16.5 SNA and the OSI Reference Model

data links connected to a node, path control is responsible for ensuring that data is correctly passed through intermediate nodes as it flows from source to destination. At the beginning of an SNA session, both sending and receiving nodes as well as all nodes between those nodes cooperate to select the most efficient route for the session. Since this route is established only for the duration of the session, it is known as a virtual route. To increase the efficiency of transmission in an SNA network, the path control layer at each node through which the virtual route is established has the ability to divide long messages into shorter segments for transmission by the data link layer. Similarly, path control may block short messages into larger data blocks for transmission by the data link layer. This enables the efficiency of SNA's transmission facility to be independent of the length of messages flowing on the network.

Transmission control

The SNA transmission control layer provides a reliable end-to-end connection service, similar to the OSI Reference Model Transport layer. Other transmission control layer functions include session level pacing as well as encryption and decryption of data when so requested by a session. Here, pacing ensures that a transmitting device does not send more data than a receiving device can accept during a given period of time. Pacing can be viewed as similar to the flow control of data in a network, however, unlike flow control, which is essentially uncontrolled, NAUs negotiate and control pacing. To accomplish this the two NAUs at session end points negotiate the largest number of messages, known as a pacing group, that a sending NAU can transmit prior to receiving a pacing response from a receiving NAU. Here the pacing response enables the transmitting NAU to resume transmission. Session level pacing occurs in two stages along a session's route in an SNA network. One stage of pacing is between the mainframe NAU and the communications controller, while the second stage occurs between the communications controller and an attached terminal NAU.

Data flow control

The data flow control services layer handles the order of communications within a session for error control and flow control. Here, the order of communications is set by the layer controlling the transmission mode. Transmission modes available include full-duplex, which permits each device to transmit at any time, half-duplex flip-flop, in which devices can only transmit alternately, and half-duplex contention, in which one device is considered a master device and the slave cannot transmit until the master completes its transmission.

Presentation services

The SNA presentation services layer is responsible for the translation of data from one format to another. This layer also performs the connection and disconnection of sessions as well as updating the network configuration and performing network management functions. At this layer, the network addressable unit (NAU) services manager is responsible for formatting of data from an application to match the display or printer that is communicating with the application. Other functions performed at this layer include the compression and decompression of data to increase the efficiency of transmission on an SNA network.

The highest layer in SNA is the transaction services layer. This layer is responsible for application programs that implement distributed processing and management services, such as distributed databases and document interchange as well as the control of LU-to-LU session limits.

16.1.10 SNA developments

The most significant developments in SNA can be considered the addition of new LU and PU subtypes to support, what is known as Advanced Peer-to-Peer Networking (APPN). Previously, LU types used to define an LU-to-LU session were restricted to application-to-device and program-to-program sessions. LU1 through LU4 and LU7 are application-to-device sessions as indicated in Table 16.2 where LU4 and LU6 are program-to-program sessions.

The addition of LU6.2 which operates in conjunction with PU2.1 to support LU6.2 connections permits devices supporting this new LU to transfer data to any other device also supporting this LU without first sending the data through a mainframe computer. Other significant advantages of LU6.2 include the ability of distributed applications to communicate with one another as well as to negotiate each session.

16.1.11 SNA sessions

All communications in SNA occur within sessions between NAUs. Here a session can be defined and a logical connection established between two NAUs over a specific route for a specific period of time, with the connection and disconnection of a session controlled by the SSCP. SNA defines four types

Table 16.2 SNA LU session types

LU Type	Session type
LU1	Host application and a remote batch terminal
LU2	Host application and a 3270 display terminal
LU3	Host application and a 3270 printer
LU4	Host application and SNA word processor or between two terminals via mainframe
LU6	Between applications programs typically residing on different mainframe computers
LU6.2	Peer-to-peer
LU7	Host application and a 5250 terminal

of sessions: SSCP-to-PU, SSCP-to-LU, SSCP-to-SSCP and LU-to-LU. The first two types of sessions are used to request or exchange diagnostic and status information. The third type of session enables SSCPs in the same or different domains to exchange information. The LU-to-LU session can be considered as the core type of SNA session since all end-user communications take place over LU-to-LU sessions.

LU-to-LU sessions

In an LU-to-LU session one logical unit known as the Primary LU (PLU) becomes responsible for error recovery. The other LU which normally has less processing power becomes the secondary LU (SLU).

An LU-to-LU session is initiated by the transmission of a message from the PLU to the SLU. That message is known as a bind and contains information stored in the communications access method (known as VTAM) tables on the mainframe which identifies the type of hardware devices with respect to screen size, printer type, etc., configured in the VTAM table. This information enables a session to occur with supported hardware. Otherwise, the SLU will reject the bind and the session will not start.

Addressing

Previously we discussed the concept of a domain that consists of an SSCP and the network resources it controls. Within a domain a set of smaller network units exist that are known as subareas. In SNA terminology each host is a subarea as well as each communications controller and its peripheral nodes. The identification of a NAU in an SNA network consists of a subarea address and an element address within the subarea. Here the subarea can be considered as being similar to an area code, as it identifies a portion of the network. Figure 16.6 illustrates the relationship between a domain and three subareas residing in that particular domain.

In SNA a subarea address is 8 bits in length, while the element address within a subarea is restricted to 15 bits. This limits the number of subareas within a domain to 255 and restricts the number of PUs and LUs within a subarea to 255.

Each subarea address is shared by an SSCP and all of its LUs and PUs and represents a unique address within a domain. In comparison, element addresses are unique only within a subarea and can be duplicated. A third component of SNA addressing is a character-coded network name that is assigned to each component. Each name must be unique within a domain and SSCPs maintain tables that map names to addresses.

The routing of packets between SNA subarea nodes occurs through the use of a sequence of message units created at different layers in the protocol stack. An application on an SNA node generates a Request Header (RH) which prefixes user data to form a Basic Transmission Unit (BTU) as shown at the top of Figure 16.7. At the path control layer, a Transmission Header (TH) is added as a prefix to the BTU to form a Basic Information Unit (BIU). The TH contains source and destination addresses that represent the sender and

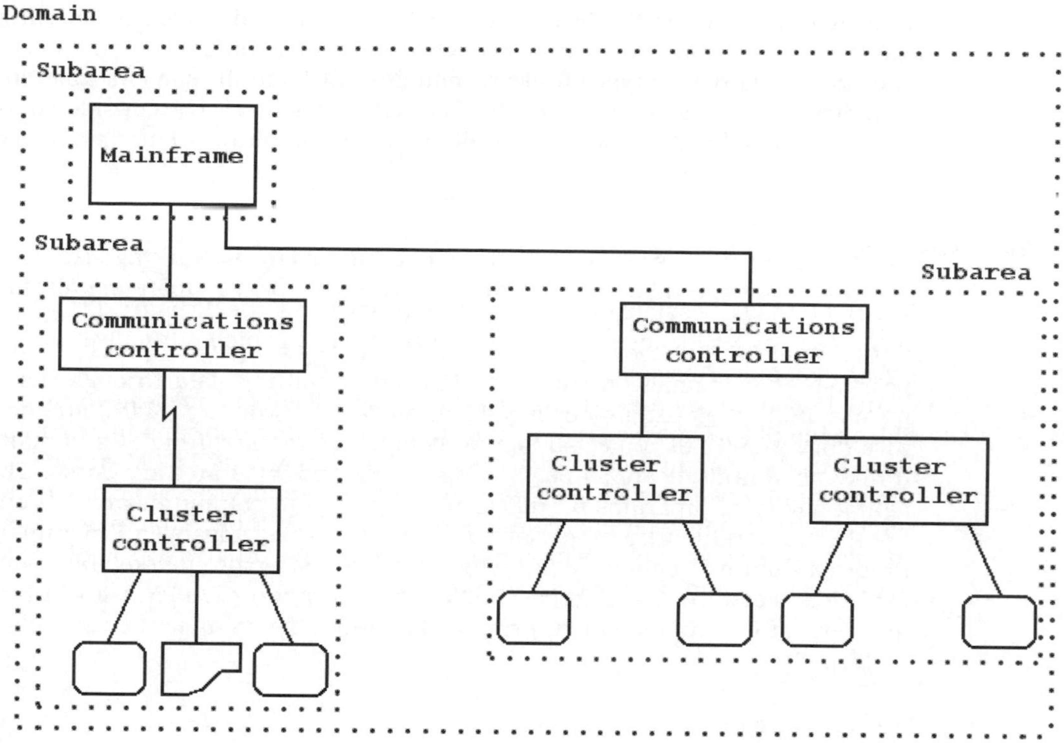

Figure 16.6 Relationship between a domain and its subareas. A subarea is a host or a communications controller and its peripheral nodes

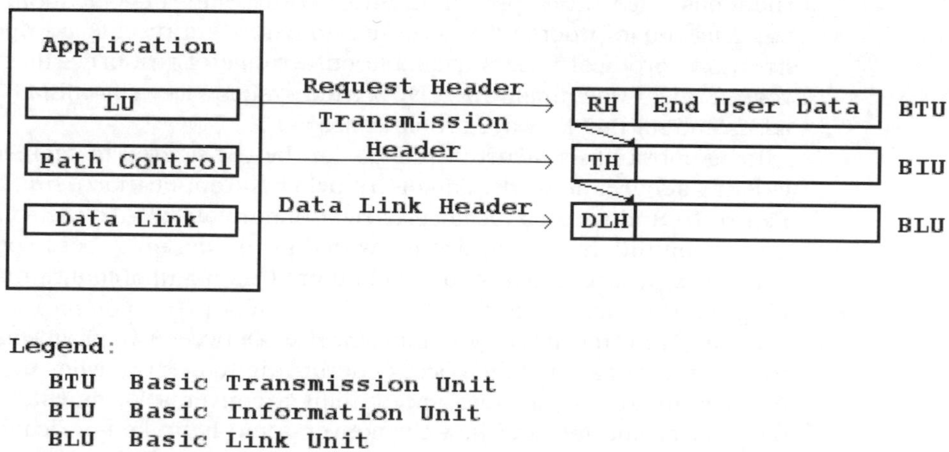

Figure 16.7 SNA routing based upon the headers in different message units

receiver of the packet, with subarea nodes examining the destination address within the TH of the BIU to make forwarding decisions.

Once a packet arrives at its destination subarea, routing must occur between the subarea and peripheral nodes in the destination subarea. That routing is based upon the Data Link Header (DLC) added to the BIU to form a Basic Link Unit (BLU), shown at the bottom of Figure 16.7.

Now that we have a basic understanding of SNA, let us turn our attention to APPN.

16.2 ADVANCED PEER-TO-PEER NETWORKING (APPN)

Although SNA represents one of the most successful networking strategies developed by a vendor, its centralized structure based upon mainframe-centric computing became dated in an evolving era of client–server distributive computing. Recognizing the requirements of organizations to obtain peer-to-peer transmission capability instead of routing data through mainframes, IBM developed its Advanced Peer-to-Peer Networking (APPN) architecture during 1992 as a mechanism for computers ranging from PCs to mainframes to communicate as peers across local and wide area networks. The actual ability of programs on different computers to communicate with one another is obtained from special software known as Advanced Program-to-Program Communication (APPC) which represents a more marketable name for LU6.2 software. Since APPC enables the operation of APPN, we will first focus our attention upon APPC prior to examining the architecture associated with APPN.

16.2.1 APPC concepts

APPC represents a software interface between programs requiring communications with other programs and the network to which the computers running those programs are connected. APPC represents an open communications protocol that is available on a range of platforms to include PCs, mainframes, Macintosh, and UNIX-based systems as well as IBM 3174 control units and its 6611 Nways router series.

In its most basic structure, APPC can be considered to represent a stack existing above a network adapter but below the application using the adapter. Figure 16.8 illustrates the general relationship of APPC to the software stack on two computers communicating with one another on a peer-to-peer basis. In this example, program A on computer 1 is shown communicating with program B on computer 2.

Under APPC terminology, communication between two programs is referred to as a conversation which occurs according to a set of rules defined by the APPC protocol. Those rules specify how a conversation is established, how data is transmitted, and how the conversation is broken or deallocated.

Similar to most modern programming concepts, APPC supports a series of verbs that provide an application programming interface (API) between transaction programs and APPC software residing on a host. For example, the APPC verb ALLOCATE is used to initiate a conversation with another transaction program, the verb SEND_DATA enables the program to initiate the data transfer to its partner program, and the verb DEALLOCATE would be used to inform APPC to terminate the conversation previously established.

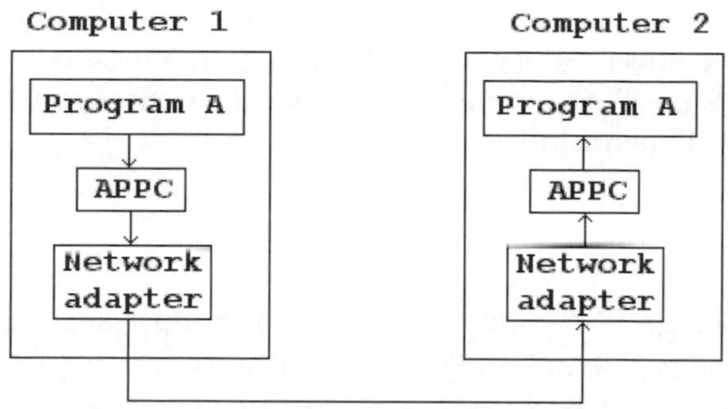

Figure 16.8 APPC software provides an interface between application programs and the network used

Now that we have a general appreciation for the software that enables program-to-program communications, let us turn our attention to the architecture of APPN.

16.2.2 APPN architecture

APPN is a platform-independent network architecture which consists of three types of computers – low-entry networking (LEN) nodes, end nodes (ENs) and network nodes (NNs). Similar to SNA, APPN applications use network resources via logical unit (LU) software. LUs reside on each type of APPN node and application-to-application sessions occur between LUs, which in the case of APPN are LU6.2 LUs. APPN nodes can support a nearly unlimited number of LUs in comparison to the 255 SNA supports. In addition, APPN LUs can support multiple users, significantly increasing its flexibility over SNA.

When an application program on one host requires communication with an application on another host, the first host tells its local LU to find a partner LU. The location of the partner LU is accomplished by a process in which several types of searches occur through an APPN network. Since those searches depend upon knowledge of the characteristics of the three types of APPN nodes, let us first turn our attention to the features and functions of those nodes.

LEN nodes

Low-entry networking (LEN) nodes date to the early 1980s when they were introduced as SNA Type 2.1 nodes. LEN nodes can be considered to represent the most basic subset of APPN functionality and have the ability to communicate with applications on other LEN nodes, end nodes or network nodes.

APPN includes a distributed directory mechanism that enables routes to be dynamically established through an APPN network. However, unlike IP networks that use 32-bit addresses, APPN uses alphanumeric names.

LEN nodes are manually configured with a limited set of LUs. Thus, to use APPN directory services a LEN node requires the assistance of an adjacent APPN node, where adjacency is obtained through a LAN connection or a direct point-to-point link.

End nodes

End nodes (ENs) can be viewed as a more sophisticated type of LEN. In addition to supporting all of the functions of LEN nodes, end nodes know how to use APPN services, such as its directory services. To learn how to use such services, an end node identifies itself to the network when it is initially brought up. This identification process is accomplished by the end node registering its LUs with a network node server. Here the NN represents the third type of APPN node and is discussed in the next section. In comparison, LEN nodes do not perform this activity.

Network nodes

Network nodes (NNs) are the third component of an APPN network. NNs provide all of the functions associated with end nodes as well as routing and partner LU location services. Concerning routing, NNs work together to route information between such nodes, in effect providing a backbone transmission capability.

The partner LU location service depends upon network node searches when the partner is not registered by the NN serving the requestor. In such situations, the NN server will broadcast a search request to adjacent network nodes, requesting the location of the partner LU. This broadcasting will continue until the partner LU is located and a path or route is returned. Since broadcast searches are bandwidth intensive, NNs place directory entries they locate into cache memory which serves to limit broadcasts being propagated through an APPN network.

16.2.3 Operation

To illustrate the operation of an APPN network, let us examine a small network in which two end nodes are connected by a network node. Figure 16.9 illustrates an example of this network structure.

When the links between EN1 and NN1 and EN2 and NN1 are activated, the computers on each link automatically inform each other of their capabilities, to include whether they are an end node or a network node, and ENs will register their capabilities with NNs. Thus, the NN will know the location and capability of both EN1 and EN2. When an application on EN1 needs to locate an LU in the network, such as LUX, it sends a request to its network node server, in this case NN1. Since NN1 is the server for EN1, both nodes establish a pair of control-point sessions to exchange APPN control information and EN1 registers its APPC LUs with NN1. Similarly, EN2 and NN1 establish a pair

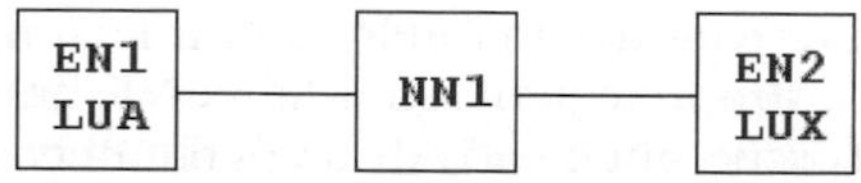

Legend:
EN End node
NN Network node
LU Logical Unit

Figure 16.9 An APPN network consisting of two end nodes and a network node

of control-point sessions when the link between those two nodes is brought up. Thus, NN1 knows how to get to EN1 and EN2 and which LUs are located at each node.

When EN1 asks NN1 to find LUX and determine a path through the network, NN1 checks its cache memory and notes that the only path available is 'EN1 to NN1 to EN2'. NN1 passes this path information back to EN1, which enables the application operating on EN1 to establish an APPN session to LUX and initiate the exchange of information.

Now that we have an appreciation for basic APPN routing, let us examine a more complex example in which originating and destination LUs reside on end nodes separated from one another by multiple NNs. Figure 16.10 illustrates this more complex APPN network, consisting of four end nodes and three network nodes grouped together in a topology which allows multiple path routing between certain nodes.

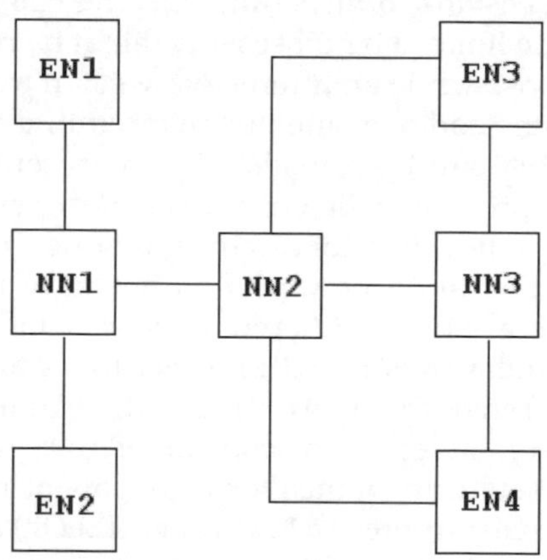

Legend:
EN End node
NN Network node

Figure 16.10 A more complex APPN network with multiple paths to some nodes

Let us assume an APPN application on EN1 wants to initiate a conversation with an application on EN4. EN1 first requests NN1 to locate EN4 and determine which path through the network should be used. Since NN1 is not EN4's network node server, it will initially have no knowledge of EN4's location. Thus, NN1 will transmit a request to each adjacent network node in its quest to locate EN4. Since NN2 is the only network node adjacent to NN1, it passes the request to its adjacent nodes. Based upon the configuration shown in Figure 16.10, there is only one adjacent network node, NN3. Although EN4 is connected to both NN2 and NN3, an end node can have only one network node server. Thus, if we assume NN3 is the server for EN4, then NN2 has no knowledge of EN4 and does not respond on its behalf.

Next, upon locating EN4, NN3 queries EN4 to determine its existing communications links. Upon receipt of information from EN4 that has links to both NN2 and NN3, NN2 passes the information about EN4 to NN2, which passes the information back to NN1. NN1 uses the information received to determine which route to EN4 is best, selects an appropriate route, and passes the selected route back to EN1, allowing that end node to establish an APPC session to EN4.

Route selection

Under APPN, routing is based upon network nodes maintaining a 'route addition resistance' value set up by network administrators and a 'class of service' of data to be routed. Class-of-service routing enables different types of data to be routed via paths optimized for batch, interactive, batch-secure and interactive-secure. APPN uses eight values that are defined for each network link that are used in conjunction with the class of service to select an appropriate route. Values defined for each link include propagation delay, cost per byte, cost for connect time, effective capacity and security. Using values defined for the links, a batch session might be routed on a path with high capacity and low cost, while an interactive session would probably be placed on a terrestrial link instead of a satellite link to minimize propagation delay.

APPN can be considered to represent a considerable enhancement to SNA as it provides efficient routing services that bypass the requirement of SNA data to flow in a hierarchical manner. However, APPN is similar to SNA in the fact that such networks do not have true network addresses, making pure routing between different networks difficult. In addition, the structure of APPN is similar to SNA with respect to its basic network layer operations that are illustrated in Figure 16.11. In examining Figure 16.11 note that APPN's network layer can be considered to represent APPC which converts LU service requests into frames for transport at the data link layer. Although SNA was originally limited to LLC Type 2 (LLC2) and SDLC transmission via a variety of physical layer interfaces, a number of conversion devices have been developed which extend both SNA and APPN transmission to Frame Relay.

In addition, other products enable SNA and APPN data to be transported under a different network layer, a technique referred to as encapsulation or tunneling. In fact, one common method referred to as Data Link Switching (DLSw) (see Section 16.4.3 below) enables SNA and APPN to be transported over an IP network.

<table>
<tr><td rowspan="2">Application</td><td colspan="4">Application Program</td></tr>
<tr><td colspan="4">LU Services</td></tr>
<tr><td>Network Layer</td><td colspan="4">APPC Path Control</td></tr>
<tr><td>Data Link Layer</td><td>LLC
SNAP
Ethernet 2</td><td colspan="2">LLC2
SNAP</td><td>Frame Relay, SDLC</td></tr>
<tr><td>Physical Layer</td><td>Ethernet/
802.5</td><td>Token-
Ring</td><td>FDDI</td><td>V.24, V.35, RS232,
T1, E1, HSSI</td></tr>
</table>

Figure 16.11 The general structure of APPN

16.3 TCP/IP

The Transmission Control Protocol/Internet Protocol (TCP/IP) represents two specific protocols within a protocol suite commonly referenced by its acronym TCP/IP. This protocol suite has its roots in the development of the Department of Defense Advanced Research Projects Agency (ARPA) network called ARPANET, a research network which was among the first to provide a reliable method of data interchange among computers manufactured by different vendors.

A bit of history

During the 1960s the United States Department of Defense sponsored research to interconnect a number of data centers located in laboratories and educational institutions. The initial goal of this networking research was to obtain a communications network that would provide the ability for data centers to exchange information even if certain network locations lost their operational capability due to a natural man-made catastrophe, such as a nuclear war. Communications research funds were provided by the Defense Advanced Research Projects Agency (DARPA), with the resulting network known by the mnemonic ARPANET.

ARPANET represents one of the first layered communications networks, preceding the development of the seven-layer ISO Reference Model by approximately a decade. The hardware and software developed to support ARPANET resulted in the development of a number of key networking concepts that are widely used in modern communications systems. Those concepts include packet switching and peer-to-peer communications.

Protocol development

The work involved in establishing ARPANET resulted in the development of three specific protocols for the transmission of information – the Transmission Control Protocol (TCP), the Internet Protocol (IP) and the User Datagram Protocol (UDP). Both TCP and UDP represent transport layer protocols. TCP is a transport layer protocol that provides end-to-end reliable transmission, while UDP represents a connectionless-mode layer 4 transport protocol. TCP includes such functions as flow control, error control and the exchange of status information, and is based upon a connection between source and destination being established prior to the exchange of information. Thus, TCP provides an orderly and error-free mechanism for the exchange of information. In comparison, UDP is well suited for transaction-based applications that can work on a best-effort delivery scheme, such as the transmission of network management information where transmission efficiency is more important than reliability. In addition, because the retransmission of packets transporting real-time voice would result in awkward delays, almost all Internet telephony products use UDP to transmit digitized voice. However, because it is important to ensure that a call is established to its appropriate destination, call control messages are commonly transported via TCP.

At the network layer, the IP protocol was developed as a mechanism to route messages between networks. To do so, IP was developed as a connectionless mode network layer protocol, and includes the capability to segment or fragment and reassemble messages that must be routed between networks that support different packet sizes than the size supported by the source and/or destination networks. The set of networking standards developed by DARPA was officially named the TCP/IP protocol suite after its two main standards.

16.3.1 The rise of the Internet

Although ARPANET was established as a research network, by the early 1980s DARPA began to convert other computers to use TCP/IP protocols. Gradually other networks were interconnected to the ARPANET, which by default became the backbone of a network of interconnected networks that was referred to as the Internet. In January 1983 the US Office of the Secretary of Defense mandated that all computers connected to long-distance networks should use the TCP/IP protocol suite for communications. At the same time, the US Government's Defense Communications Agency (DCA) that was responsible for ARPANET split that network into two separate entities. One of the two networks retained the name ARPANET and continued in its role as a research network. The second network, which was for military use, was appropriately named MILNET.

Since a considerable amount of communications research was being performed by university computer science departments, DARPA made the TCP/IP protocol stack available at a low cost that encouraged its adoption. Since most university computer science departments were using a version of the UNIX operating system, a large number of UNIX applications were initially developed for use with TCP/IP. Applications such as electronic mail, remote login, and file transfer were developed, resulting in other groups beginning to adopt TCP/IP. In 1986 the National Science Foundation established a network that interconnected its six supercomputer centers. Known as the NSFNET, this network was connected to the ARPANET. At approximately the same time, computer networks at universities, government agencies, and corporate research laboratories located throughout the world were being connected to the evolving Internet. To provide a mechanism for the adoption of standards necessary to maintain communications interoperability, DARPA formed the Internet Activities Board (IAB) in 1983. Recognizing that the expansion of the Internet included communications activities that now included basic file transfer and electronic mail necessary for daily business activities, the IAB was reorganized in 1989. That reorganization resulted in the formation of two major groups – the Internet Research Task Force (IRTF) and the Internet Engineering Task Force (IETF).

The IETF and IRTF

The IETF is responsible for working on medium- and short-term engineering problems and includes eight areas, with working groups formed under each area to work on specific problems. The IRTF is responsible for research activities related to the TCP/IP and internetwork architecture issues.

The IRSG

Similar to the IETF, the IRSG consists of a series of research groups responsible for performing research activities. To coordinate research activities performed by the research groups, a small group known as the Internet Research Steering Group (IRSG) was formed as part of the IRTF. That group also was given responsibility for coordinating priorities among various research groups in the IRTF.

RFCs

To facilitate the development of standards, a process was established that enables standards-related proposals to move from a draft state to implementation. This process involves the generation of technical reports referred to as Requests For Comments, or RFCs. RFCs can cover a variety of topics, ranging from proposals for new protocols and addressing schemes to the manner by which applications interact with transmission protocols.

RFCs are initially published as a draft by the IAB and a period of time is provided for comments and review. Each new or revised RFC is assigned a numeric identifier and after an approval process the draft can become an Internet standard. An organization currently known as the Internet Network Information Center (InterNIC) stores copies of RFCs so they are electronically available for retrieval.

16.3.2 The TCP/IP protocol suite

As a layered communications protocol, TCP/IP groups functions into defined network layers. Figure 16.12 illustrates a portion of the TCP/IP protocol suite and indicates the relationship between TCP/IP protocols and the seven-layer

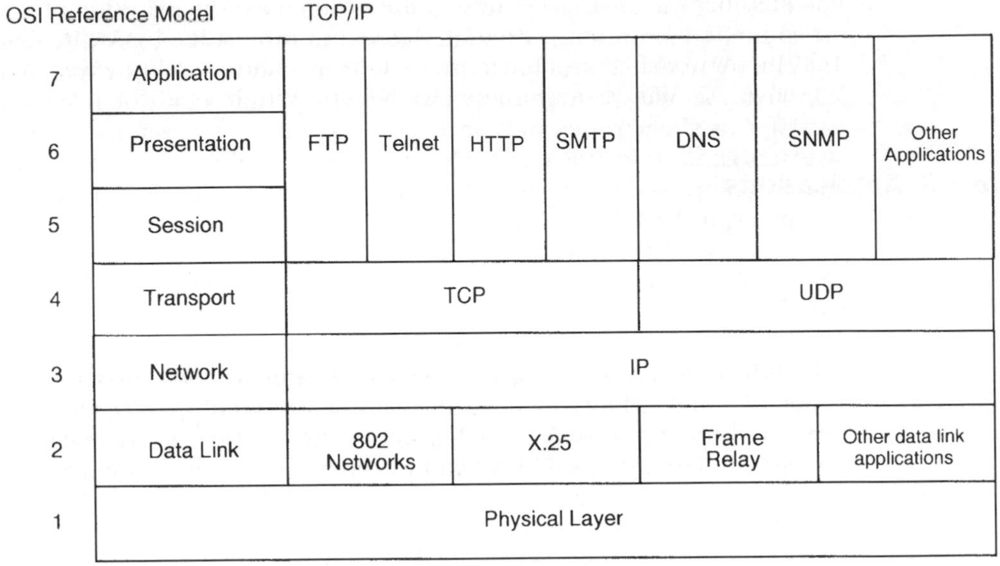

Figure 16.12 Comparing the TCP/IP protocol suite to the OSI Reference Model

OSI Reference Model. The left portion of Figure 16.12 indicates the seven layers of the ISO's Open System Interconnection Reference Model.

One of the key reasons for the dramatic growth in the use of the TCP/IP protocol suite can be traced to the development and structure of the TCP/IP. Because TCP/IP was developed using taxpayer funds, its specifications were placed in the public domain, and they are available for use royalty-free for vendors to develop 'protocol stacks' in software or firmware to implement TCP/IP's operational features. Another key reason for the growth in the use of TCP/IP for both private corporate networks as well as the interconnection of academic, commercial and private networks to the Internet, is its structure, which makes it suitable for both LAN and WAN operations.

At the Data Link Layer shown in Figure 16.12, TCP/IP can be transported within Ethernet, Token-Ring, FDDI, or another type of local area network frame. Because a considerable amount of effort was expended in developing LAN adapter cards to support the bus structure used in Apple Macintosh, IBM PCs and compatible computers, SUN Microsystem's workstations, and even IBM mainframes, the development of software-based protocol stacks to facilitate the transmission of TCP/IP on LANs provides the capability to interconnect LAN-based computers to one another, whether they are on the same network and merely require the transmission of frames on a common cable, or are located on networks separated thousands of miles from one another. Concerning the latter, through the use of transport layer and network layer services, TCP/IP can transport data via simple to complex meshed wide area network structures, providing a mechanism to satisfy academic, commercial and governmental communications requirements from LAN to LAN on both an intra- and inter-LAN basis. Based upon the preceding, the use of TCP/IP has significantly increased and its importance as a networking protocol has for some time been recognized by Microsoft Corporation by the inclusion of a TCP/IP networking capability in various versions of Windows, to include Windows 95, Windows 98, Windows NT and Windows 2000.

16.3.3 Applications

In examining Figure 16.12 you will note that in general TCP/IP applications can be equated to the upper three layers of the OSI Reference Model. Because the TCP/IP protocol suite predates the OSI Reference Model, most TCP/IP applications are considered to simply reside on top of the transport layer. The six applications listed in Figure 16.12 represent a small portion of over 100 applications supported by the protocol suite. Although the HyperText Transmission Protocol (HTTP) by far dominates current Internet usage as the transport for Web browsing, a significant amount of electronic mail is moved via SMTP, and many persons periodically use FTP and Telnet on a daily basis. In addition, without the Domain Name Service (DNS) we would be forced to enter numeric addresses instead of easy to remember host names, while the Simple Network Management Protocol (SNMP) provides us with a mechanism to monitor the status of our network. Although there are over 100 TCP/IP applications, we will focus our attention upon six in this section to obtain an appreciation of the diversity of the applications supported by the protocol suite.

File Transfer Protocol (FTP)

FTP supports the bulk transfer of data between hosts via TCP. Included within the FTP application are provisions for initiating bidirectional file transfers, displaying file directories, renaming and deleting files.

An FTP session requires the initiation of two connections between hosts. One connection is used to convey commands and status information, and the second connection is used for actual file transfers. Those connections occur on a common transmission facility by the use of two 'port' numbers. The function of TCP/IP ports is discussed later in this chapter.

FTP was originally a command-driven application, requiring the use of such commands as GET and PUT to perform file transfer operations. In fact, the latest versions of Microsoft's Windows operating systems limit built-in FTP support to the use of commands issued from the MS-DOS prompt compatibility box. An example of this command-based FTP support is illustrated in Figure 16.13 where the author first used the FTP command to connect to a server, entered his user identification and password, and then entered the question mark (?) at the prompt 'ftp>' to display a list of ftp commands.

In examining the screen shown in Figure 16.13 note that the command ftp www2.opm.gov informs the local ftp client that we wish to make a connection to the server whose host address is www2.opm.gov. Because all routing occurs via the use of IP addresses, the host address must be resolved into an IP address. This process is referred to as address resolution and is performed by

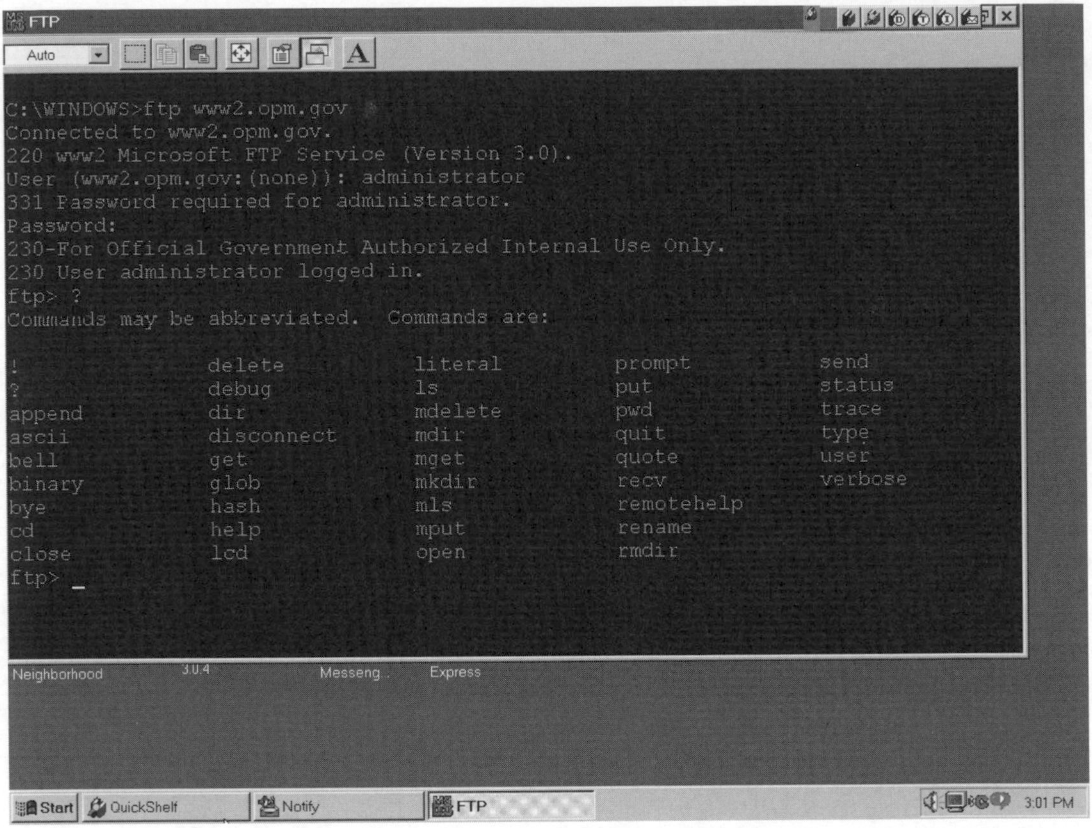
FTP
Auto
A
C:\WINDOWS>ftp www2.opm.gov
Connected to www2.opm.gov.
220 www2 Microsoft FTP Service (Version 3.0).
User (www2.opm.gov:(none)): administrator
331 Password required for administrator.
Password:
230-For Official Government Authorized Internal Use Only.
230 User administrator logged in.
ftp> ?
Commands may be abbreviated. Commands are:
! delete literal prompt send
? debug ls put status
append dir mdelete pwd trace
ascii disconnect mdir quit type
bell get mget quote user
binary glob mkdir recv verbose
bye hash mls remotehelp
cd help mput rename
close lcd open rmdir
ftp> _
Neighborhood
3.0.4
Messeng...
Express
Start
QuickShelf
Notify
FTP
3:01 PM

Figure 16.13 Connecting to an FTP server via the use of Microsoft's Windows operating system ftp command-based client program

the Domain Name Service (DNS), which is described later in this chapter. Readers familiar with host addresses will note that the host address previously mentioned (www2.opm.gov) represents a Web server address. This illustrates another important concept concerning the TCP/IP protocol suite. That is, one computer can support multiple applications, with each application transported in a packet identified by the port number of the TCP or UDP header contained in the packet. Thus, a series of packets flowing to the same host address can transport different applications, assuming the distant server is configured to support multiple applications.

In Figure 16.14 an example of an FTP command-based file transfer is shown. In this example we first used the CD (change directory) command to change to the Y2K directory. Once this was accomplished we used the DIR command to display a directory listing. This was followed by the use of the FTP GET command to transfer the contents of the file index.htm from the server to the local computer operating the FTP client program. Because an HTM file is encoded in ASCII we did not have to issue a specific ASCII mode command. If we required the transfer of a binary file we would have to enter the command binary prior to initiating the get command.

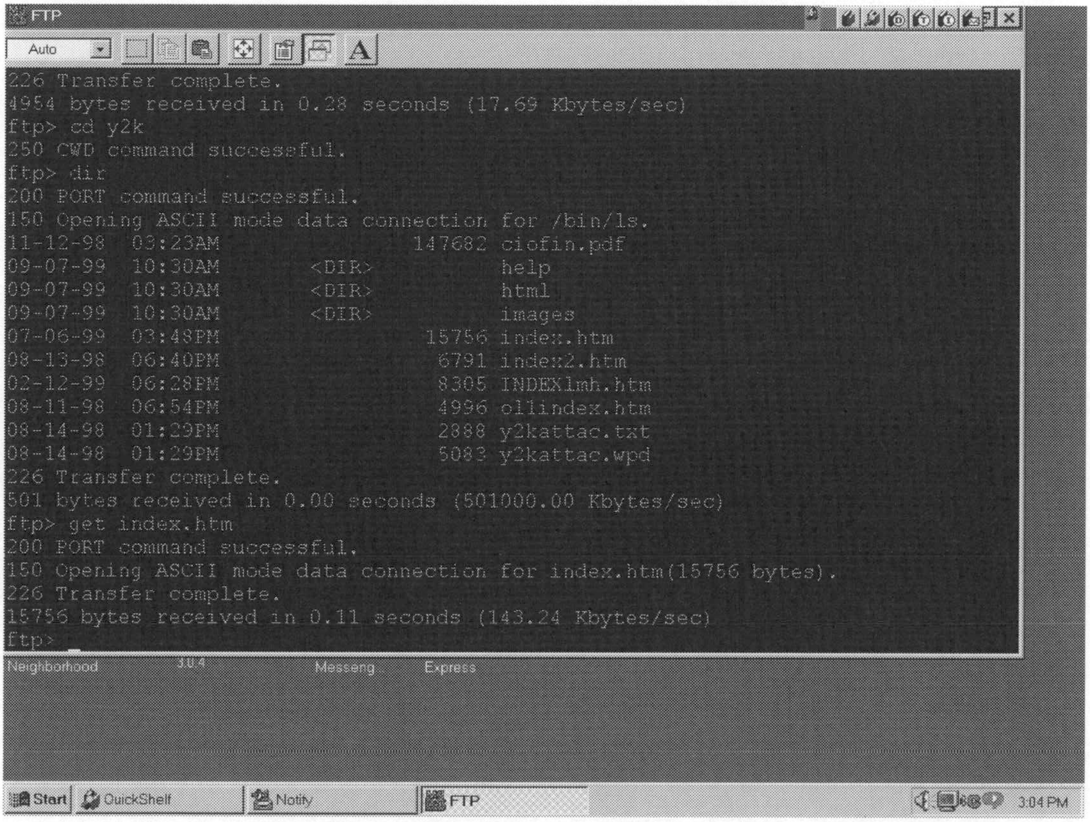

Figure 16.14 Using the Microsoft Windows ftp command-based program to transfer an ASCII file

Since the original development of FTP as a command-driven application, numerous vendors have introduced graphic user interface (GUI) versions of FTP and other TCP/IP applications. Figure 16.15 illustrates the NetManage Chameleon FTP application which essentially eliminates the requirement for users to know FTP commands. Instead, users can simply click on buttons to change directories, copy files and perform other FTP-related functions.

Telnet

Telnet represents an interactive remote access terminal protocol developed to enable users to log into a remote computer as if their terminal was directly connected to the distant computer. Several flavors of telnet have been developed, including TN3270 which is designed to support telnet access to IBM mainframes. TN3270 primarily differs from telnet in that it recognizes the screen control codes generated by IBM mainframes. Otherwise, the use of telnet to access an IBM mainframe would more than likely result in the display of what appears to be garbage on the terminal device. Telnet and TN3270 use a common TCP connection to transmit both data and control information.

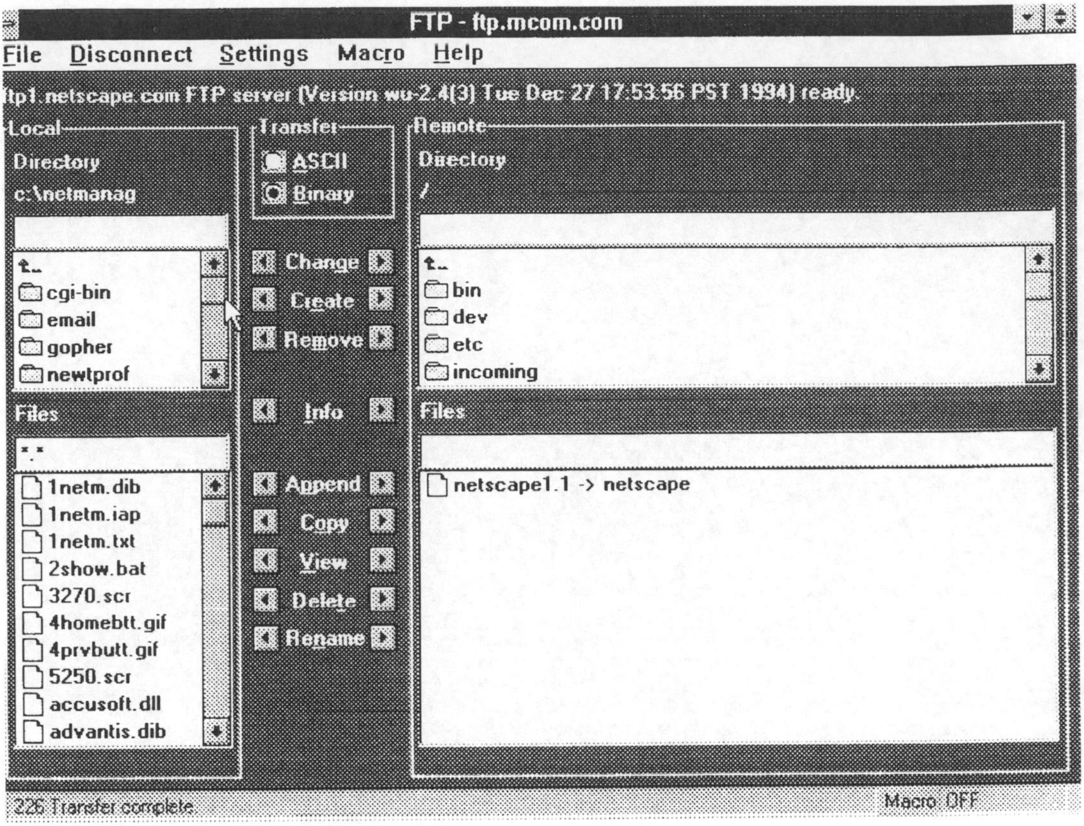

Figure 16.15 The NetManage Chameleon FTP application

Figure 16.16 illustrates the NetManage Chameleon Telnet application after the application's Connect menu has been selected. The resulting dialog box labeled Connection Dialog enables a user to enter the host address that they wish to establish a telnet session with. In addition, this GUI version of telnet supports eight terminal types which are selected through the Emulate pull-down menu. In Figure 16.16 port number 23 was selected, which is the default port used for telnet communications.

The Chameleon suite of TCP/IP applications to include FTP and Telnet were originally developed for Windows 3.1, which did not include a TCP/IP protocol stack. When Microsoft added support for TCP/IP in Windows 95 and succeeding versions of the Windows operating system it also included several TCP/IP applications. One such application is FTP, which was previously described. Another application is Telnet. Unlike FTP, Microsoft included a GUI-based Telnet application in all releases of Windows after version 3.1.

Figure 16.17 illustrates the Microsoft Telnet application after the Remote System entry from its Connect menu was selected. Similar to the Chameleon Telnet application, you specify a host name, port and terminal type, with a list of terminal types supported by the Microsoft version of Telnet pulled down in the illustration. Note that the Microsoft version of Telnet supports five

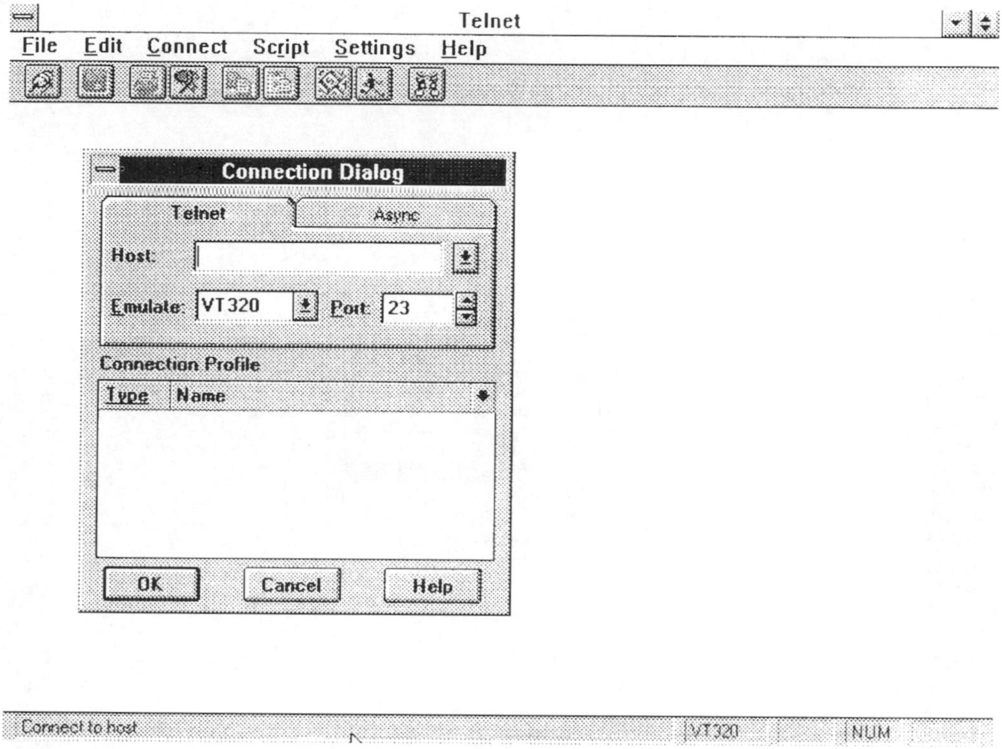

Figure 16.16 Preparing to initiate a telnet session using the NetManage Chameleon Telnet application

terminals and its port configuration is limited to predefined mnemonics, such as Telnet, which results in port 23 being used. Other ports supported by the Microsoft version of Telnet include daytime, echo, gold and chargen.

Simple Mail Transport Protocol (SMTP)

SMTP represents the Internet standard for transmitting electronic mail. This application permits the creation of a header containing such elements as the date, subject, to, cc and from, as well as a free-form ASCII text body. Extensions to SMTP now permit the attachment of binary files.

HyperText Transfer Protocol (HTTP)

The HyperText Transfer Protocol (HTTP) is a request/response client–server protocol developed to support the transfer of multimedia data. Since its beginnings in the early 1990s, HTTP has evolved to support the transfer of text, graphics and even full-motion video.

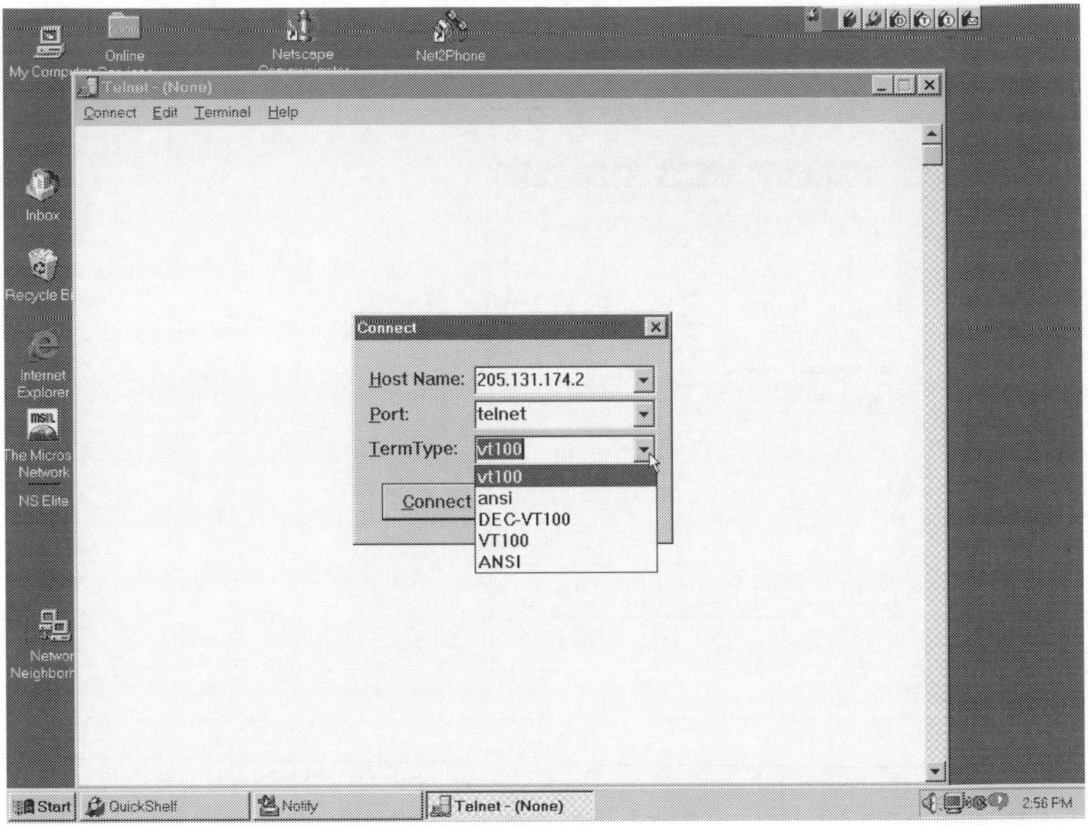

Figure 16.17 The Microsoft Windows Telnet client supports a limited number of predefined ports, such as Telnet for port 23 use

Web browsers, such as Netscape and Microsoft's Internet Explorer, use HTTP as the mechanism for transferring client requests for Web page displays to servers and displaying the response to those requests. HTTP communications use the default TCP port 80.

URLs

The HTTP protocol also uses the uniform resource locator (URL) whose format is as follows:

http://host[:port][absolute_path]

where host represents a legal Internet host name or IP address, while port represents the port communications occur on. If omitted, a default value of 80 is assumed. You can also specify an absolute path to a particular file on the destination server through the use of an optional path shown in brackets in the preceding format.

A URL can be viewed as a network extension of the standard filename concept. For example, http://www.opm/gov/index.htm represents the address of the file named index.htm located on the http server whose host address is *www.opm.gov*. Because almost all Web servers are configured to have a default home page, you can simply enter the URL as http://www.opm.gov to access the home page file of the HTTP server at *www.opm.gov*. If the HTTP server was configured to operate on a different port other than the normal default of 80, you would have to enter the appropriate port number in the URL. For example, if the network port was 123 you would enter the URL as http://www.opm.gov:123/index.htm or as http://www.opm.gov:123.

Figure 16.18 illustrates the use of the Netscape Navigator browser to access the URL http://www.altavista.com. That URL represents the home page of the Alta Vista Internet guide, one of several popular search engines you can use to rapidly locate specific information from the billions of Web pages residing on millions of servers connected to the Internet. In Figure 16.18 the author entered the query 'lexus + airbag + failures' in an attempt to locate information to ascertain if the failure of the side airbag on his wife's Lexus during a serious accident was an isolated incident or had been reported by other vehicle owners.

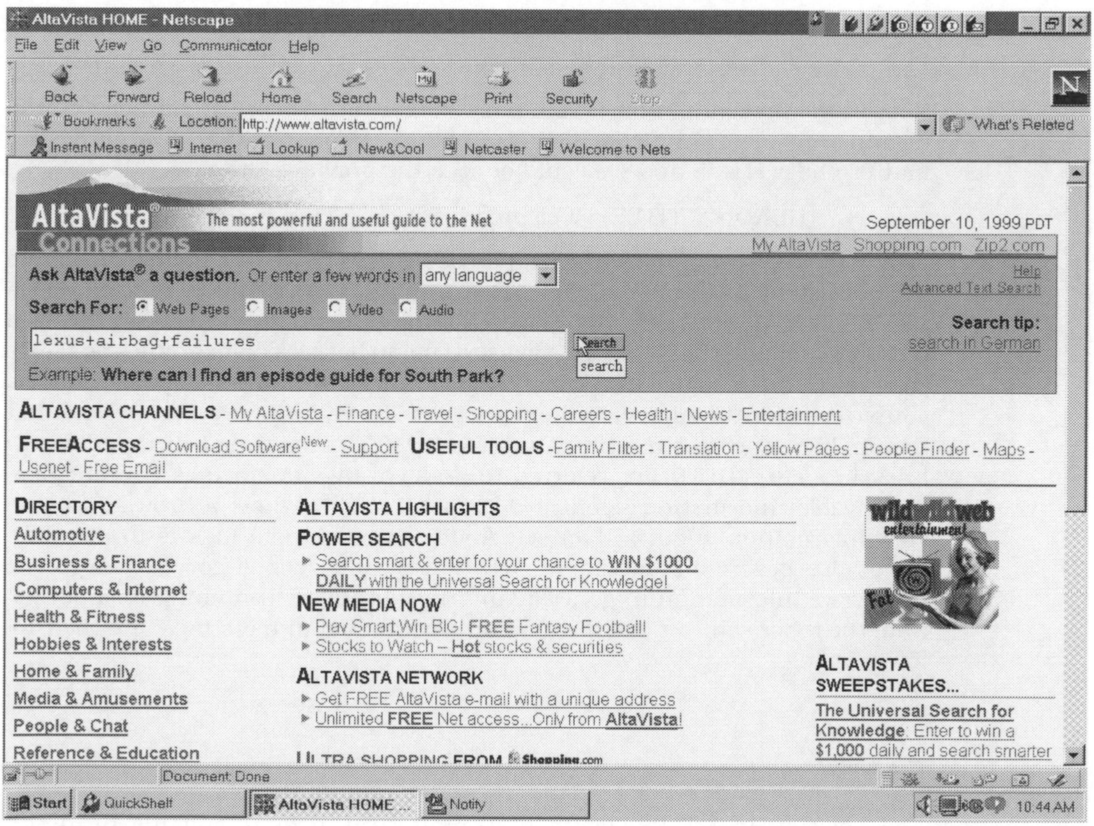

Figure 16.18 Creating a search on the home page of the Alta Vista Internet guide

HTML

The coding that makes the display of Web pages possible is referred to as HyperText Markup Language (HTML). An HTML coded document, which we normally refer to as a Web page, consists of a head and body, similar to a letter. However, instead of using English, French or another language in a letter, we use HTML code to develop the manner by which a Web page will be displayed.

The head

The head of an HTML document represents an unordered collection of information about the document. In HTML coding information to be displayed in some manner is commonly contained within a pair of tags which in turn are encased in brackets. For example, assume we wish to display the title of our document as 'Welcome to XYZ' on the top line of a browser when a user 'hits' our home page. Because the title tag resides in the head our coding would be as follows:

```
<HEAD>
<TITLE>Welcome to XYZ</TITLE>
</HEAD>
```

Note in the preceding that HTML uses the forward slash (/) character in the terminating tag to indicate its termination. Also note that spaces do not matter under HTML and we could encode the previous title as:

```
<HEAD> <TITLE>Welcome to XYZ</TITLE> </HEAD>
```

In addition, we can use either uppercase or lowercase in a tag as long as we are consistent in the use of case. Returning to Figure 16.18, note the term AltaVista HOME at the top left portion of the screen. If we select Source from the View menu in the Netscape Navigator menu bar we can examine the HTML code used to construct the page displayed in Figure 16.18. A portion of the source code for the Alta Vista home page is shown in Figure 16.19. Note that the top line generates the title AltaVista HOME. The second, third and fourth lines in Figure 16.19 are referred to as meta tags. Meta tags contain non-displayable information about a document, such as how a browser should view information about a data set. Another use of meta tags is to hide keywords that describe a document. When a so-called 'robot' goes to a Web site to retrieve information to generate an index for users searching for information, the robot will retrieve the keywords contained in meta tags and the URL for indexing.

The body

The body of an HTML document contains the text, to include headings, paragraphs, tables and similar information. In Figure 16.19 you will note

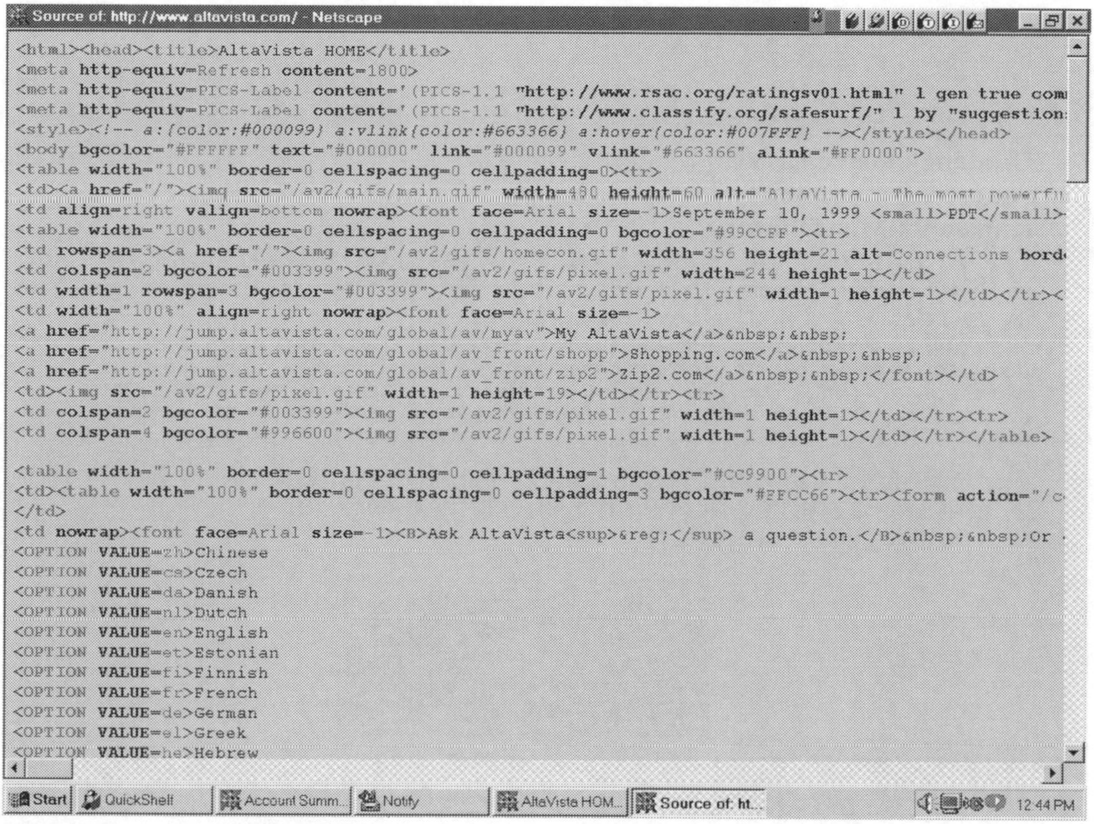

Figure 16.19 A portion of the source code used to construct the Alta Vista home page previously shown in Figure 16.18

that the body tag used also specifies the background color as well as other parameters.

If you look carefully through the source listing in Figure 16.19 you will note the tag <img src="/av2/gifs/main.gif" which indicates one method for placing a GIF image onto a Web page.

Web page creation

While the source coding shown in Figure 16.19 may appear intimidating, there are many tools available to use in creating Web pages. Figure 16.20 illustrates the use of Netscape Composer to create a portion of a Web page describing the failure of airbags on the Lexus RX300 used by the author's wife during a serious accident. By simply selecting different menu entries you can easily select font and type, text color, the use of bold and underlining, as well as the inclusion of images within a Web page.

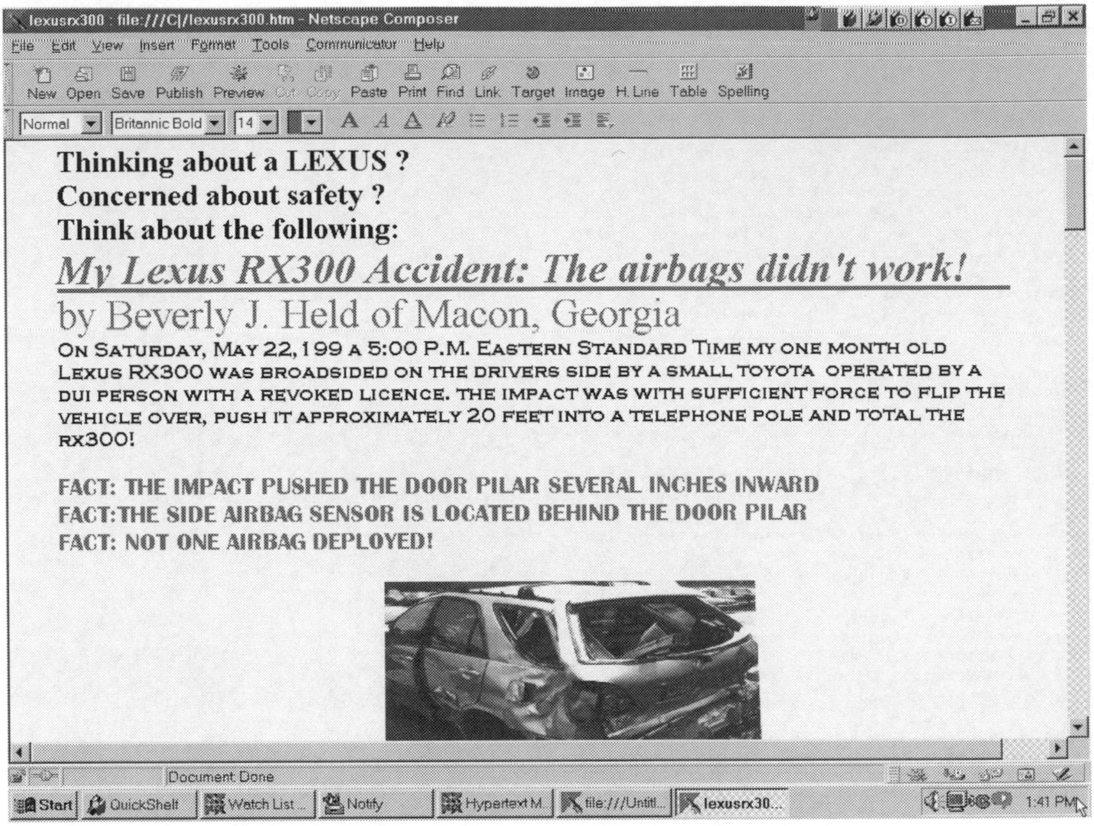

Figure 16.20 Creating a Web page through the use of Netscape Composer

Domain Name Service (DNS)

Actual addressing on a TCP/IP network occurs through the use of four decimal numbers ranging from 0 to 255 which are separated from one another by a dot. This dotted decimal notation represents a 32-bit address. Since numerical addresses are difficult to work with, TCP/IP also supports a naming convention based on mnemonics which are described later in this section. The conversion of mnemonic names to IP addresses represents one of two key functions performed by DNS. The second key function is to provide a pointer to a higher authority beyond a local network when a destination address is not on the local network. The method by which this is accomplished is also described later in this section.

Simple Network Management Protocol (SNMP)

The Simple Network Management Protocol provides a mechanism for obtaining information about the operating characteristics of networks as well as for changing the parameters of certain network such as routers and gateways. There are three key parts to SNMP: a manager, an agent and a Management Information Base (MIB).

The manager operates application software which uses SNMP to communicate with one or more agents. Agents respond to SNMP commands and return information about the state of the network they are connected to, or change the parameters of managed devices. Agents also have the capability to recognize the occurrence of predefined conditions and alert the manager that such conditions have occurred. The MIB can be considered to represent a database which is structured so that management information can be uniquely defined. Through the use of a structured MIB, third party vendors can develop communications products that can be controlled by management stations developed by other vendors.

16.3.4 TCP/IP communications

Two methods are supported by the TCP/IP protocol suite for thc routing of data between network nodes, a Layer 3 function. Those methods include the establishment of a virtual circuit and the use of datagrams.

Virtual circuit transmission

In a virtual circuit transmission environment network nodes establish a fixed path between the originating and destination locations. To accomplish this, network nodes must maintain a table of addresses and destination routes to permit a path to be established and, once established, to enable information to flow on the previously established path. The fixed path which is established for the duration of the transmission session is referred to as a virtual circuit. Once a session has been completed the previously established path is relinquished.

Although the establishment of a virtual circuit does not enable the use of an alternate route if a break in the path of the circuit occurs, it provides two advantages which facilitate its operation. First, information always flows on the same path, which precludes a data sequencing problem from occurring. Secondly, since only one transmission path is used the possibility of duplicate frames or packets of data occurring is reduced, providing an easier mechanism for the management of the flow of data.

Datagram transmission

A second mechanism that is used for the transmission of data at the network layer avoids the use of a fixed path. Referred to as datagram transmission, frames are broken into units of data known as datagrams which are broadcast onto every port other than the receiving port of each node in a network. Although this technique results in duplicate traffic occurring on some paths, it can considerably simplify network routing. In addition, since there are no

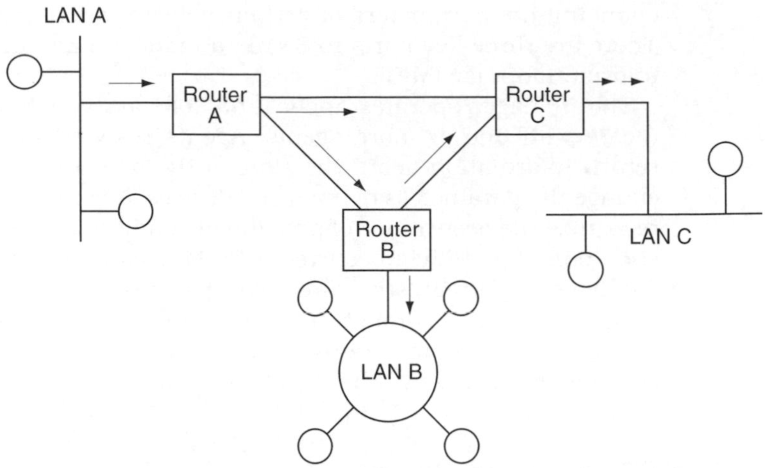

Figure 16.21 Depending on your network topology, datagram transmission can result in duplicate packets arriving at certain network nodes

fixed paths between nodes, no recovery method is required to re-establish a path if a line or intermediate node should fail.

One of the problems associated with datagram transmission is the fact that, depending on network topology, duplicate datagrams can arrive at a destination location. For example, consider Figure 16.21 which illustrates the transmission of a datagram from the router connected to LAN A to the router connected to LAN C. One copy of the datagram is transmitted on path A–C, whereas a second copy of the datagram arrives on LAN C via the paths A–B and B–C.

Since a Layer 3 protocol is responsible for the routing of information, the removal of duplicate datagrams becomes a Layer 4 or end-to-end responsibility. However, due to the overhead associated with keeping track of duplicate packets and the backwidth they consume, datagram transmission was essentially replaced by virtual circuit transmission.

16.3.5 The Internet Protocol (IP)

The Internet Protocol (IP) is a network layer protocol which supports a datagram gateway service between subnetworks, enabling hosts on one network to communicate with hosts on another network. To accomplish this IP will fragment large datagrams so they can be transferred over networks that support small maximum packet sizes, as well as perform the reassembly of those datagrams. Note that the term datagram references a physical quantity of bytes and not the data transmission mechanism referred to as datagram transmission.

Although IP is responsible for the routing of information, end-to-end transmission is the responsibility of the Layer 4 protocol in the protocol suite. At Layer 4 the TCP/IP protocol suite supports two transport protocols, the Transport Control Protocol (TCP) and the User Datagram Protocol (UDP). TCP provides a reliable, connection-oriented service which supports end-to-end transmission reliability. To accomplish this, TCP supports error detection and correction as well as flow control to regulate the flow of datagrams. In comparison, UDP provides an unreliable, connectionless transport service. This requires the higher layer application to be responsible for ensuring that messages are properly delivered. HTTP, FTP, Telnet, SMTP and other applications that are connection-oriented run on top of TCP. This is because TCP is a connection-oriented reliable protocol which requires a session to be established prior to data transfer being permitted. SNMP represents one of several types of applications that run on top of UDP, which means that a session does not have to be established for network management information to be transmitted. Thus, the use of UDP provides a degree of flexibility for the transmission of network management information. This is because the transmission of management information does not require the rigid structure associated with other types of applications obtained through the connection setup process, status exchange, and end-to-end flow control. Instead, management information can be transported in datagrams on a best effort basis and retransmitted when required without adversely effecting the operation of a management station. In comparison, HTTP, FTP, Telnet and other applications that require the establishment of a session must use TCP.

IP operation

Figure 16.22 illustrates the fields of an IP datagram and the octet length of each field. Note that the first 20 octets represent an IP header followed by user data.

The version field consists of four bits and identifies the format of the IP header. Currently version 4 is the latest version of IP. The header length is also four bits in length, and it denotes the length of the header in 32-bit words. The minimum value of this field is 5.

The type of service field indicates the quality of service desired for the datagram. Values for this field can be used to specify precedence, delay, throughput and reliability.

The total length field consists of two octets which denote the length of the datagram. The value in this field specifies the length of the header plus user data.

The identification field contains a 16-bit value assigned by the originator of the datagram. If the datagram must be fragmented to flow on another network and the flags field permits fragmentation two or more fragments consisting of a portion of the user data from the original datagram will be created. Since each fragment contains a portion of user data, the fragment offset field is used to identify the position of the fragment in units of 8 octets, in the original datagram.

The time to live field was designed to indicate, in seconds, the time that the datagram can flow on an internet. Each time the datagram passes through the internet layer on any network device, the value of this field is decremented, being discarded when a value of zero is reached. In most implementations

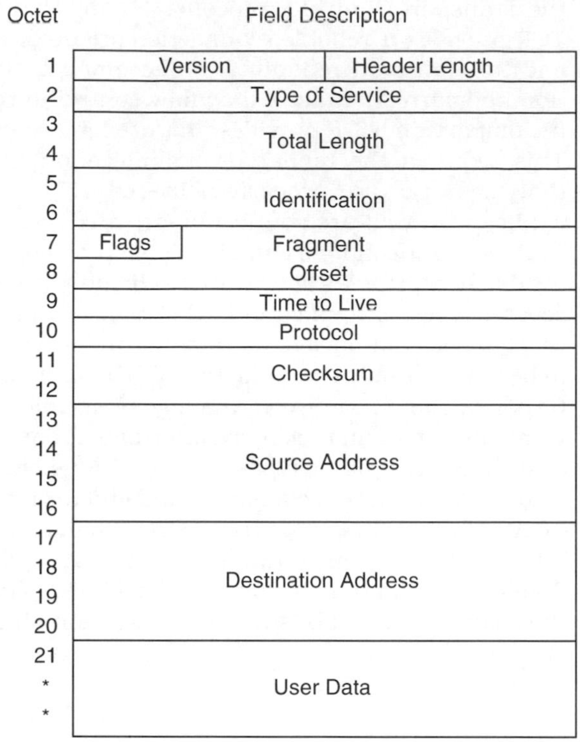

Figure 16.22 The IPv4 header

the time to live field is a hop-count field, with decrementation performed by each router.

The protocol field simply identifies the upper layer protocol using IP, such as TCP or UDP.

The two byte header checksum field contains a ones complement arithmetic sum which is computed over the header of the IP datagram. When the datagram is originally transmitted, as well as each time it is forwarded, the checksum is recalculated.

Both the source and destination addresses consist of 32 bits that represent the IP address of the originator and recipient, respectively. Because IP addressing is an important concept to understand, let us focus our attention on this topic.

IP addressing

In September 1981 the Internet Protocol (IP) was standardized. Included in the standard was a requirement for each host attached to an IP-based network to be assigned a unique, 32-bit address value for each network connection. This resulted in some networking devices, such as routers that have interfaces to more than one network, as well as host computers with multiple connections

Network prefix	Host address

Most IP addresses include a two-level hierarchy consisting of a network prefix and a host address.

Figure 16.23 The two-level IP addressing structure

to the same or different networks, being assigned a unique IP address for each network interface.

Basic address scheme

In most situations an IP address is used to identify a network as well as a host connected to the network. Thus, an IP address normally represents a two-level addressing hierarchy. This hierarchy is illustrated in Figure 16.23.

All hosts on the same network must be assigned the same network prefix but must have a unique host address to differentiate one host from another. Similarly, two hosts on different networks must be assigned a different network prefix; however, those hosts can have the same host address.

Address classes

When IP was standardized it was recognized that the use of a single method of subdivision of the 32-bit address into network and host portions would be wasteful with respect to the assignment of addresses. For example, if all addresses were split evenly, resulting in 16 bits for a network number and 16 bits for a host number, the result would allow a maximum of 65 534 ($2^{16} - 2$) networks with up to 65 534 hosts per network. Then, the assignment of a network number to an organization that had only 100 computers would result in a waste of 65 434 host addresses that could not be assigned to another organization. Recognizing this problem, the designers of IP decided to subdivide the 32-bit address space into different address classes, resulting in five address classes being defined. Those classes are referred to as Class A through Class E.

Figure 16.24 illustrates the five IP address formats to include the bit allocation of each 32-bit address class. In examining Figure 16.24, note that the address class can easily be determined through the examination of the values of one or more of the first 4 bits in the 32-bit address. Once an address class is identified, the subdivision of the remainder of the address into the network and host address portions is automatically noted. To obtain an appreciation for the use of each address class, let us examine the composition of the network and host portion of each address when applicable, as doing so will provide some basic information that can be used to indicate how such addresses are used.

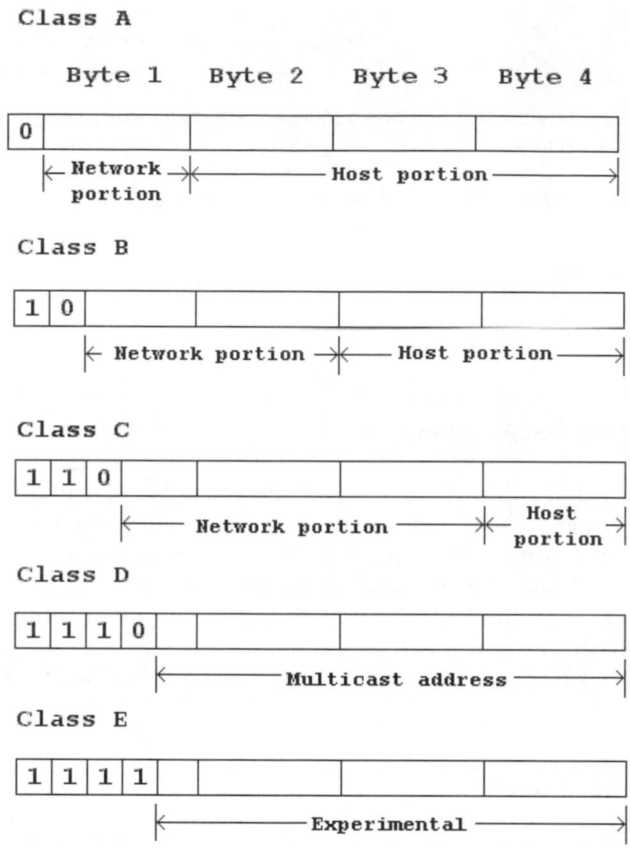

Figure 16.24 IP address formats

Class A

A Class A IP address is defined by a 0-bit value in the high-order bit position of the address. This class of addresses uses 7 bits for the network portion and 24 bits for the host portion of the address. As a result of this subdivision, 128 networks can be defined with approximately 16.78 million hosts capable of being addressed on each network. Due to the relatively small number of Class A networks that can be defined and the large number of hosts that can be supported per network, Class A addresses were commonly assigned to large organizations and countries that have national networks and are no longer available for new use.

Class B

A Class B network is defined by the setting of the two high-order bits of an IP address to 10. The network portion of a Class B address is 14 bits in width, while the host portion is 16 bits wide. This results in the ability of Class B addresses to be assigned to 16 384 networks, with each network having the ability to support up to 65 536 hosts. Due to the manner by which Class B addresses are subdivided into network and host portions, such addresses are normally assigned to relatively large organizations with tens of thousands of employees.

Class C

A Class C address is identified by the first 3 bits in the IP address being set to the value 110. This results in the network portion of the address having 21 bits while the host portion of the address is limited to 8 bit positions.

The use of 21 bits for a network address enables approximately 2 million distinct networks to be supported by the Class C address class. Since 8 bits are used for the host portion of a Class C address, this means that each Class C address can theoretically support up to 256 hosts; however, as we will soon note, addressing restrictions limit support to 254 hosts. Due to the subdivision of network and host portions of Class C addresses, they are primarily assigned for use by relatively small networks, such as an organizational LAN. Since it is quite common for academic institutions, government agencies and commercial organizations to have multiple LANs, it is also quite common for multiple Class C addresses to be assigned to organizations that require more than 256 host addresses but are not large enough to justify a Class B address.

Class D

A Class D IP address is defined by the assignment of the value 1110 to the first 4 bits in the address. The remaining bits are used to form what is referred to as a multicast address. Thus, the 28 bits used for that address enable approximately 268 million possible multicast addresses.

Multicast is an addressing technique which allows a source to send a single copy of a frame to a specific group through the use of a multicast address. Through a membership registration process hosts can dynamically enroll in multicast groups. Thus, the use of a Class D address enables up to 268 multicast sessions to occur simultaneously throughout the world.

Until recently the use of multicast addresses was relatively limited; however, their use can be expected to increase considerably as they provide a mechanism to conserve bandwidth, which is becoming a precious commodity.

To understand how Class D addressing conserves bandwidth, consider a digitized video presentation routed from the Internet onto a private network, which users working at five hosts on the network wish to review. Without a multicast transmission capability five separate video streams would be transmitted onto the private network, with each video stream consisting of frames with five distinct host destination addresses. In comparison, through the use of a multicast address one video stream would be routed to the private network.

Class E

The fifth address class defined by the IP address specification is a reserved address class known as Class E. A Class E address is defined by the first 4 bits in the 32-bit IP address having the value of 1111. This results in the remaining 28 bits being capable of supporting approximately 268.4 million addresses. Class E addresses are restricted for experimentation.

Dotted-decimal notation

Recognizing that the direct use of 32-bit binary addresses is both cumbersome and unwieldy, a technique more acceptable for human use was developed. That technique is referred to as dotted-decimal notation in recognition of the fact that the technique developed to express IP addresses occurs via the use of four decimal numbers separated from one another by decimal points.

Dotted-decimal notation divides the 32-bit Internet Protocol address into four 8-bit (one-byte) fields, with the value of each field specified as a decimal number. That number can range from 0 to 255 in bytes 2, 3 and 4. In the first byte of an IP address the setting of the first 4 bits in the byte used to denote the address class limits the range of decimal values that can be assigned to that byte. For example, from Figure 16.24 a Class A address is defined by the setting of the first bit position in the first byte to 0. Thus, the maximum value of the first byte in a Class A address is 128.

To illustrate the formation of a dotted-decimal number, let's first focus our attention upon the decimal relationship of the bit positions in a byte. Figure 16.25 indicates the decimal values of the bit positions within an 8-bit byte. Note that the decimal value of each bit position corresponds to 2^n, where n is the bit position in the byte. Using the decimal values of the bit positions shown in Figure 16.25, let's assume the first byte in an IP address has its bit positions set as 11000000. Then, the value of that byte expressed as a decimal number becomes $128 + 64$, or 192. Now let's assume that the second byte in the IP address has the bit values 01001000. From Figure 16.25 the decimal value of that binary byte is $64 + 8$, or 72. Let's further assume that the last two bytes in the IP address have the bit values 00101110 and 10000010. Then, the third byte would have the decimal value $32 + 8 + 4 + 2$, or 46, while the last byte would have the decimal value $128 + 2$, or 130.

Based upon the preceding, the dotted-decimal number 192.72.46.130 is equivalent to the binary number 11000000 01001000 00101110 10000010. Obviously, it is easier to work with and to remember four decimal numbers separated by dots than a string of 32 bits.

128	64	32	16	8	4	2	1

The decimal value of the bit positions in a byte correspond to 2^n where n is the bit position that ranges from 0 to 7.

Figure 16.25 Decimal values of bit positions in a byte

IP Networking basics

As previously noted, each IP network has a distinct network prefix and each host on a network has a distinct host address. When two networks are interconnected by the use of a router, each router port is assigned an IP address which reflects the network it is connected to. Figure 16.26 illustrates the connection of two networks via a router, indicating possible address assignments. Note that the first decimal number (192) of the four-byte dotted-decimal numbers associated with two hosts on the network on the left portion of Figure 16.26 denotes a Class C address. This is because 192 decimal is equivalent to 11000000 binary. Since the first two bits are set to the bit value 11, this indicates from Figure 16.24 a Class C address. Also note that the first three bytes of a Class C address indicate the network while the fourth byte indicates the host address. Thus, the network shown in the left portion of Figure 16.26 is denoted as 192.72.46, with device addresses that can range from 192.72.46.0 through 192.72.46.254. The reason a host address of 255 cannot be used is that it represents a setting of all ones, which denotes a broadcast address.

In the lower right portion of Figure 16.26 two hosts are shown connected to another network. Note that the first byte for the four-byte dotted-decimal number assigned to each host and the router port is decimal 226, which is equivalent to binary 11100010. Since the first two bits in the first byte are again set to 11, the second network also represents the use of a Class C address. Thus, the network address is 226.42.78, with device addresses on the network ranging from 226.42.78.0 through 226.42.78.254.

Host restrictions

Although it would appear that 256 devices could be supported on a Class C network (0 through 255 used for the host address), in actuality the host

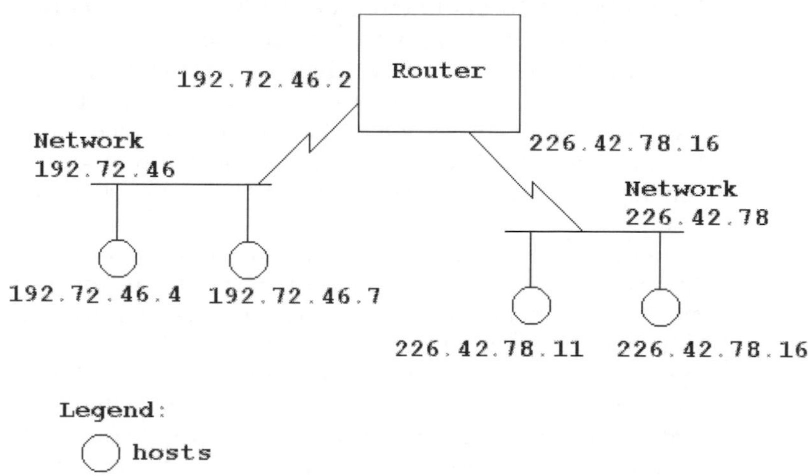

Figure 16.26 Router connections to networks require an IP address for each connection

portion field of an IP address has two restrictions. First, the host portion field cannot be set to all zero bits. This is because an all-0s host number is used to identify a base network or subnetwork number. Concerning the latter, we will shortly discuss subnetworking. Secondly, as briefly mentioned earlier, an all-1s host number represents the broadcast address for a network or subnetwork. Due to these restrictions, a maximum of 254 devices can be defined for use on a Class C network. Similarly, other network classes have the previously discussed addressing restrictions, which reduces the number of distinct addressable devices that can be connected to each type of IP network by two. Since, as previously explained, an all-0s host number identifies a base network, the two networks shown in Figure 16.26 would more commonly be shown numbered as 192.72.46.0 and 226.42.78.0.

Subnetting an IP address

The assignment of an address by the InterNIC to an organization provides the organization with a distinct identifier. Although that identifier is suitable for one physical network, from a practical perspective many organizations operate two or more physical networks. Rather than request an additional address from the InterNIC, which may not be granted due to the scarcity of IP addresses, organizations can subdivide their assigned IP address through a process referred to as subnetting. An important benefit of subnetting is that to the outside world the organization maintains a common network IP address. This permits, for example, a single router to be used for a corporate connection to the Internet, whereas two or more subnets could connect to the Internet via the router.

Subnetting represents a division of the host portion of an IP address into subnet and host entities. To the outside world information transmitted to the organization is addressed to the network portion of the address, whereas the host portion appears to represent a distinct host address. At the destination site the subnet field within the host portion of the address is used to identify a predefined physical network. Then, the remaining portion of the host portion is used to identify a host on the physical network. Thus, a second advantage associated with subnetting is the ability to reduce router table entries. This becomes possible because to the outside world an organization is viewed as having a single IP address, even if that address is subdivided into many subnets.

Figure 16.27 illustrates the subnetting of a Class C network address. Note that the subnet portion of the address can be one or more bit positions, with the length dependent on the number of physical networks that the organization has. For example, an organization with two networks at one location could assign one bit to the derived subnet portion of the IP address, resulting in seven bits being used for the host portion of the address. In this example each network could support up to 127 identifiable hosts. Before discussing how subnetting is accomplished let us first focus our attention on the routing of information on an IP network. Doing so will provide us with information necessary to understand the functionality of a process known as IP subnet

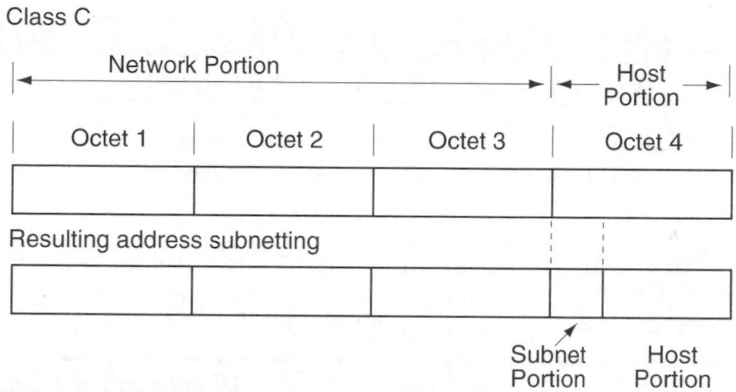

Figure 16.27 IP address subnetting

masking, which is the process which supports multiple networks via a common IP network address.

The transfer of information between two devices is based on an examination of the network portion of the IP address. If the destination address has the same network portion then the destination device must be on the same network. Thus, a router receiving a message with matching source and destination network addresses will not transmit the message off the local network. If the destination address contains a network identifier that is different from the existing network, the destination address must reside on a different network. In this situation the router would forward the message off the local network.

Since the subnet portion of an IP address represents an extension of the network address, a mechanism is required to inform each workstation as well as routers about this extension. The mechanism used to accomplish this is a subnet-mask. A subnet-mask is a 32-bit number for which the network identifier and subnet portions are set to 1s, whereas the host portion is set to 0s. The logical ANDing of the subnet-mask with an IP address results in the generation of the physical network address.

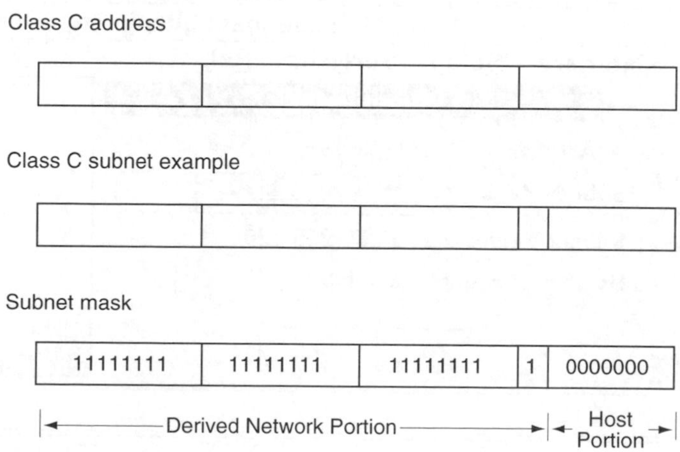

Figure 16.28 Creating a subnet mask

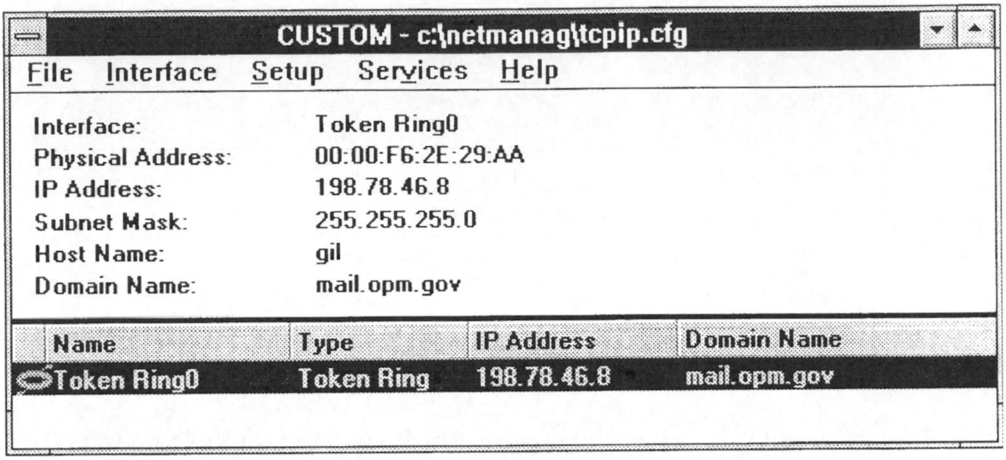

Figure 16.29 Viewing the current address settings via Chameleon's Custom application

Figure 16.28 illustrates the construction of a subnet-mask for the previously subnetted Class C address in which one bit position of the host portion of the IP address was used for the subnet portion of the address. By ANDing the 25 set bits in the subnet mask with each 32-bit IP address, an extended IP network portion is derived. This derived network address is used internally as a mechanism to determine whether or not a datagram should remain on the internal network.

Figures 16.29 and 16.30 illustrate the initial setting and modification of a subnet mask using the NetManage Chameleon TCP/IP application suite. Figure 16.29 illustrates the Chameleon custom application window which identifies current settings and provides the ability to modify those settings. In Figure 16.30 a subnet mask of 255.255.255.0 is shown, corresponding to setting every bit in the first three octets of a 32-bit mask. Figure 16.30 illustrates the display of a dialog box labeled Subnet Mask. This dialog box is

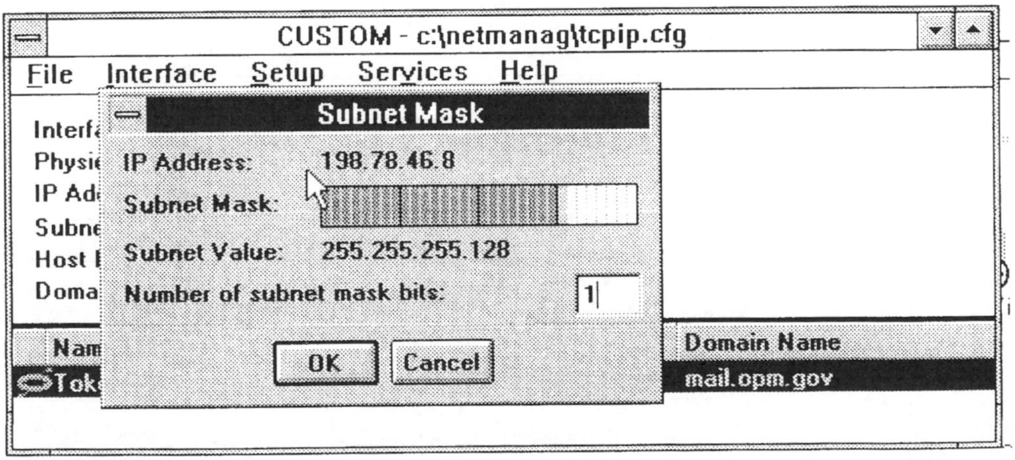

Figure 16.30 Changing the subnet mask

displayed by selecting a Subnet Mask entry from the Setup menu. In this example, setting the first bit in the fourth octet is shown, resulting in the value of the subnet mask changing to 255.255.255.128.

Because of the popularity of different versions of Microsoft's Windows operating system, a brief discussion of configuring the IP address and subnet mask under that operating system environment is warranted. From the Start menu you can select Control Panel and the network icon. If you previously installed TCP/IP, selecting Properties results in the display of a dialog box similar to the one illustrated in Figure 16.31.

In examining Figure 16.31 note the button associated with 'Obtain an IP address automatically'. Selecting that button allows your computer to be dynamically assigned an IP address by an Internet Service Provider or a Dynamic Host Configuration Protocol (DHCP) server located on an organization's network. The use of a DHCP server enables an organization to dynamically 'lease' IP addresses for predefined periods of time, allowing Internet access for more workstations than IP addresses assigned to the organization. Of course, the maximum number of stations that can simultaneously use TCP/IP is limited by the number of addresses assigned to the server for distribution.

If you focus attention upon the tabs located at the top of the dialog box shown in Figure 16.31, you will note one labeled DNS Configuration. This tab

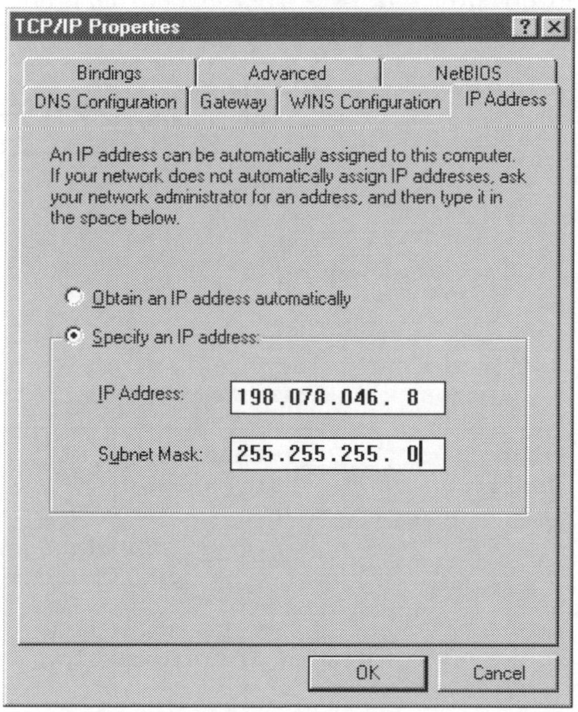

Figure 16.31 Using the Microsoft Windows TCP/IP properties dialog box to configure the IP address and subnet mask for a computer

allows you to assign a host name to your computer as well as to specify the address of one or more DNS servers for address resolution purposes. The tab labeled Gateway provides you with the ability to specify the IP address of a router on your network. Doing so provides TCP/IP with the ability to direct packets with a destination network address different from the local network address to the router for transmission off the network. The term gateway is used as the first series of devices used to transmit packets from one network to another were referred to by that term.

Address resolution

The physical address associated with a LAN workstation is referred to as its hardware address. For an Ethernet network, that address is 48 bits or six bytes in length. At the data link layer IP uses a 32-bit logical address. One common problem associated with the routing of an IP datagram to a particular workstation on a network is the delivery of the datagram to its correct destination.

Figure 16.32 illustrates the formation of a LAN frame transporting either a TCP or UDP application. Note that an IP header containing an IP destination address encapsulates the IP datagram and is transported in a LAN frame that contains a Layer 2 media access control (MAC) address. To correctly deliver the datagram requires knowledge of the relationship between data link and network layer addresses. This relationship is obtained by two protocols which

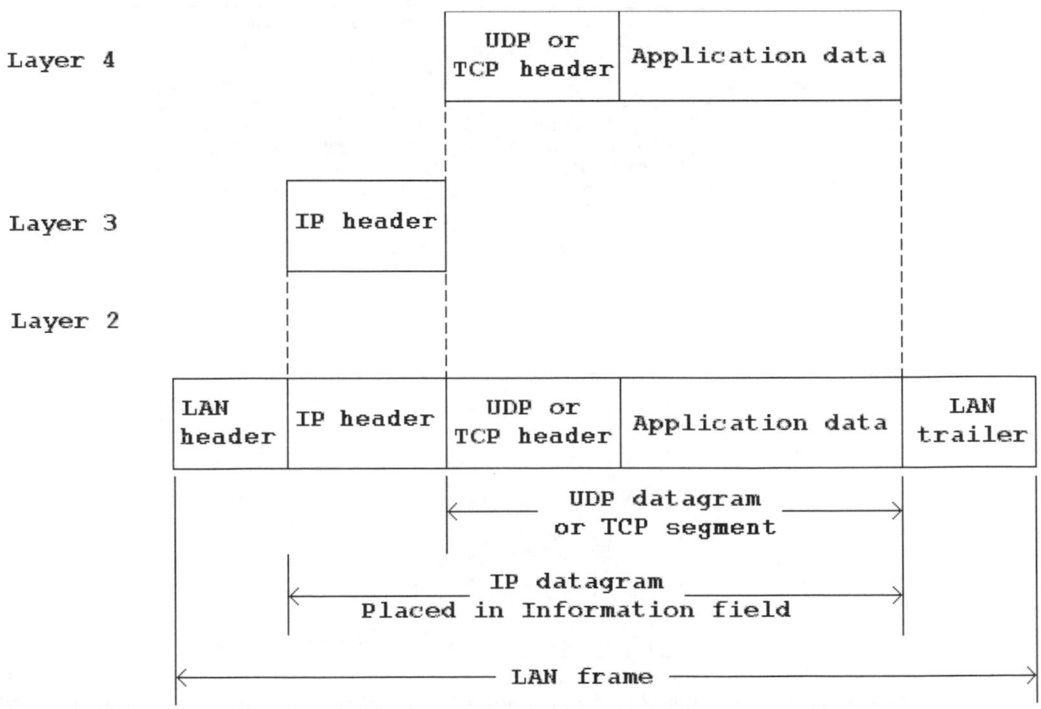

Figure 16.32 Forming a LAN frame

map the data link and network addresses to each other. One protocol, known as the Address Resolution Protocol (ARP), translates an IP address into a hardware address. The Reverse Address Resolution Protocol (RARP), as its name implies, performs a reverse mapping, converting a hardware address into an IP address.

To illustrate the use of ARP, assume one computer user wants to send a datagram to another computer and both computers are located on the same Ethernet network. The first computer would broadcast an ARP packet within an Ethernet frame to all devices on the LAN. That packet would contain the destination IP address, since it is known, setting the hardware address field to zeros as its value is unknown. Although each device on the LAN will read the ARP packet as it is transmitted as a broadcast packet, only the device that recognizes its own logical address will respond. When it does, it will transmit an ARP reply in which its physical address is inserted in the ARP address field previously set to zero. To reduce the necessity to constantly transmit ARP packets, the originator will record received information in a table known as an ARP cache, allowing subsequent datagrams to be quickly directed to the appropriate address on the LAN. Thus, ARP provides a well thought out methodology for equating physical hardware addresses to IP's network addresses and allows IP addressing at Layer 3 to occur independently from LAN addressing at Layer 2.

Now that we have an appreciation for IP and the method by which IP addresses are equated to Media Access Control Layer 2 hardware addresses, let's focus our attention on Layer 4 of the TCP/IP protocol suite.

TCP

The Transmission Control Protocol (TCP) represents a Layer 4 connection-oriented reliable protocol. TCP provides a virtual circuit connection mode service for applications that require connection setup and error detection and automatic retransmission.

Each unit of data carried by TCP is referred to as a segment. Segments are created by TCP subdividing the stream of data passed down by application layer protocols that use its services, with each segment identified by the use of a sequence number. This segment identification process enables a receiver, if required, to reassemble data segments into their correct order.

Figure 16.33 illustrates the format of the TCP protocol header. The source and destination ports are each 16 bits in length, and they identify a process or service at the host receiver. For example, port 23 is normally used for Telnet.

The sequence number is used to identify the data segment that is transported. The acknowledgement number interpretation depends on the setting of the ACK control flag (not directly shown in Figure 16.33). If the ACK control flag bit position is set, the acknowledgement field will contain the next sequence number that the sender expects to receive. Otherwise the field is ignored.

There are six control field flags that are used to establish, maintain and terminate connections. Those flags include URG (urgent), SYN, ACK, RST (reset), PSH (push) and FIN (finish).

2	Source Port
2	Destination Port
4	Sequence Number
4	Acknowledgement Number
2	Data Offset / Control Flags
2	Window
2	Checksum
2	Urgent Pointer
	Data

Figure 16.33 TCP protocol header

Setting URG = 1 indicates to the receiver that urgent data are arriving. The SYN flag is set to 1 as a connection request and thus serves to establish a connection. As previously discussed, the ACK flag when set indicates that the acknowledgement flag is relevant. The RST flag when set means that the connection should be reset, and the PSH flag tells the receiver to immediately deliver the data in the segment. Finally, the setting of the FIN flag indicates that the sender is finished and that the connection should be terminated.

The window field is used to convey the number of bytes that the sender can accept and functions as a flow control mechanism. The checksum provides error detection for the TCP header and data carried, and the urgent pointer field is used in conjunction with the URG flag. When the URG flag is set, the value in the urgent pointer field represents the last byte of urgent data.

When an application uses TCP, TCP breaks the stream of data provided by the application into segments and adds an appropriate TCP header. Next, an IP header is prefixed to the TCP header to transport the segment via the network layer. As data arrive at the destination network they are converted into a data link layer transport mechanism. For example, on a Token-Ring network TCP data would be transported within Token-Ring frames.

UDP

The User Datagram Protocol (UDP) is the second Layer 4 transport service supported by the TCP/IP protocol suite. UDP is a connectionless service,

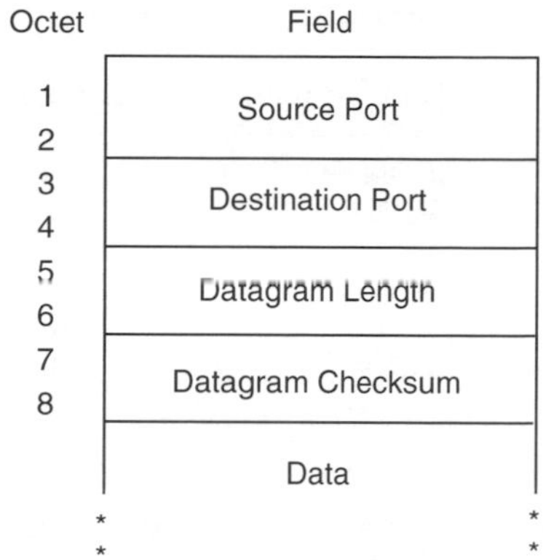

Figure 16.34 The UDP header

which means that the higher layer application is responsible for the reliable delivery of the transported message.

Figure 16.34 illustrates the composition of the UDP header. The Source and Destination port fields are each 16 bits in length and, as previously described for TCP, they identify the port number of the sending and receiving process, respectively. Here each port number process identifies an application running at the corresponding IP address in the IP header prefixed for the UDP header. The use of a port number provides a mechanism for identifying network services as they denote communications points where particular services can be accessed. For example, a value of 161 in a port field is used in UDP to identify SNMP.

The length field indicates the length of the UDP packets in octets, including the header and user data. The checksum, which is a ones complement arithmetic sum, is computed over a pseudo-header and the entire UDP packet. The pseudo-header is created by the conceptual prefix of 12 octets to the header illustrated in Figure 16.34. The first eight octets are used by source and destination IP addresses obtained from the IP packet. This is followed by a zero-filled octed and an octet which identifies the protocol. The last two octets in the pseudo-header denote the length of the UDP packet. By computing the UDP checksum over the pseudo-header and user data, a degree of additional data integrity is obtained.

16.3.6 Domain Name Service

The Domain Name Service (DNS) is the naming protocol used in the TCP/IP protocol suite which enables IP routing to occur indirectly through the use of names instead of IP addresses. To accomplish this DNS provides a domain name to IP address translation service.

A domain is a subdivision of a wide area network. When applied to the Internet, there are six top-level domain names which were specified by the Internet Network Information Center (InterNIC) at the time this book was being prepared. Those top level domains are listed in Table 16.3.

Table 16.3 Internet top-level domain names

Domain name	Assignment
.COM	Commercial organization
.EDU	Education organization
.GOV	Government agency
.MIL	Department of Defense
.NET	Networking organization
.ORG	Not for profit organization

Under each top level domain the InterNIC will register subdomains which are assigned an IP network address. An organization receiving an IP network address can further subdivide the domain into two or more subdomains. In addition, instead of using dotted decimal notation to describe the location of each host, they can assign names to hosts as long as they follow certain rules and install a name server which provides IP address translation between named hosts and their IP addresses.

To illustrate the operation of a name server, consider the network domain illustrated in Figure 16.35. In this example we will assume that a government agency has a local area network with several computers that will be connected to the Internet. Each host address will contain the specific name of the host plus the names of all of the subdomains and domains to which it belongs. Thus, the computer 'warrants' would have the official address

telnet.warrants.fbi.gov

Similarly, the computer 'cops' would have the address

ftp.cops.fbi.gov

In examining the domain naming structure illustrated in Figure 16.35, note that computers were placed in subdomains using common Internet application names, such as telnet and ftp. This is a common technique that organizations use to make it easier for network users within and outside of the organization to remember mnemonics that represent specific hosts on their network. Also note that since the TCP and UDP port numbers identify the application being transported, it is both possible and a common occurrence for one computer with a single IP address to support multiple applications. For example, many Web servers also function as FTP servers.

Although domain names provide a mechanism for identifying objects connected to wide area networks, hosts in a domain require network addresses to transfer information. Thus another host functioning as a name server is required to provide a name to address translation service.

Name server

The name server plays an important role in TCP/IP networks. In addition to providing a name-to-IP address translation service it must recognize that an

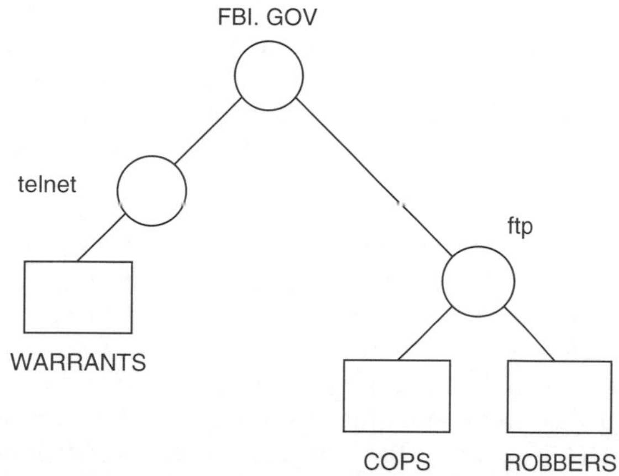

Figure 16.35 A domain-naming hierarchy

address is outside its administrative zone of authority. For example, assume that a host located on the domain illustrated in Figure 16.35 will use the address fred.microwear.com to transmit a message. The name server must recognize that that address does not reside in the domain, and must forward the address to another name server for translation into an appropriate IP address. Since most domains are connected to the Internet via an Internet service provider, the name server on the domain illustrated in Figure 16.35 would have a pointer to the name server of the Internet service provider and forward the query to that name server. The Internet service provider's name server will either have an entry in its table or forward the query to another name server. Eventually a name server will be reached that has administrative authority over the domain containing the host name to resolve, and will return an IP address through a reversed hierarchy to provide the originating name server with a response to its query. Most name servers cache the results of previous name queries which can considerably reduce off-domain or internet network traffic.

16.4 INTERNETWORKING

In this chapter we examined two network architectures, SNA and TCP/IP, as well as a variation of SNA designed to support peer-to-peer communications. SNA operates on a connection-oriented data link layer using end-to-end sequencing and error control with SDLC used as a transport mechanism. SDLC in turn has a set of rules concerning the number of frames that can be outstanding at one time, the time by which outstanding frames must be acknowledged and similar rules. In addition, SNA lacks the ability to directly associate an address with each network, in effect prohibiting routing at the network layer.

In a TCP/IP environment IP replaces SDLC, each host has a unique address and a methodology exists in the form of name servers which enable the translation of mnemonics to IP addresses. In addition, the domain naming structure enables IP to be a fully routable protocol.

Originally TCP/IP was developed in a research environment, whereas SNA has its routes in the commercial marketplace. As the need to interconnect computers from different organizations increased, the shortcomings associated with SNA resulted in an increase in the use of the TCP/IP protocol suite and its acceptance by the commercial community. However, many organizations still maintain large corporate networks based on the use of SNA, which has created a variety of problems, especially when attempting to integrate local area networks and mainframe-based SNA networks. A further complication is the fact that the extension of SNA onto LANs is accomplished via Logical Link Control 2 (LLC 2), a connection-oriented virtual circuit which requires stations at each end to negotiate a logical connection prior to transmitting data. Although LLC 2 provides for sequencing of frames, error control and flow control between end stations it is not routable. Since most routers use TCP/IP or UDP/IP, workstations on LANs accessing corporate SNA networks represent a routing problem. This problem is avoided when a 'pure' IBM SNA network is used; however, it becomes manifest when SNA is mixed with other protocols. Thus, in this section we will first examine how LAN workstations are connected to mainframes on a 'pure' IBM SNA network, after which we will examine the problems associated with providing a similar level of connectivity when two or more protocols are transported between network nodes.

16.4.1 SNA gateway operations

A variety of connectivity options are marketed which permits PCs on a local area network to be connected to IBM mainframes and minicomputers. The right portion of Figure 16.36 illustrates how a Token-Ring network can be connected to an IBM mainframe, such as a 43XX or 38XX computer system. In actuality, the MAU is cabled to the 3725 communications controller, with the communications controller requiring the installation of a Line Attachment Base (LAB) Type C and a Token-Ring Interface Coupler (TIC). Since up to four TICs can be installed in one LAB, up to four Token-Ring networks can be

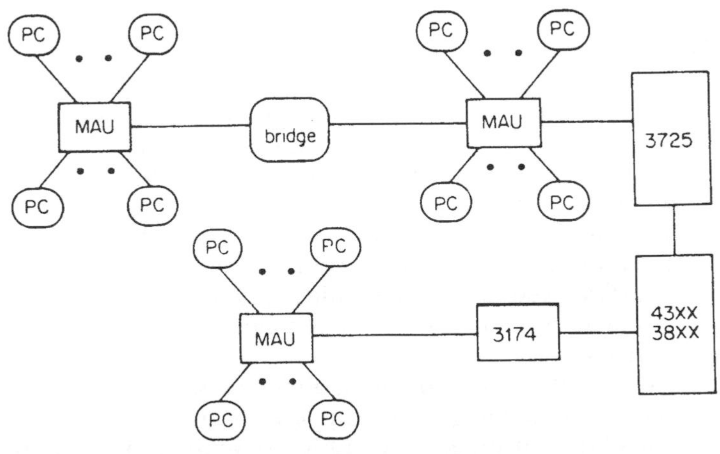

Figure 16.36 Token-Ring connection options

connected to one Type C LAB. Each personal computer on the Token-Ring network requiring access to the mainframe must operate the IBM PC 3270 Emulation Program Version 3.0, which results in the PC emulating a 3274 control unit with an attached 3278/9 display and 3287 printer. The actual transfer of data between the PC and the communications controller occurs using LLC 2, a time-sensitive protocol which is carried by SDLC.

In the lower portion of Figure 16.36, a Token-Ring network connection to an IBM mainframe through the use of an IBM 3174 controller is illustrated. Although several models of the 3174 controller were available only some of them offer Token-Ring support. Those controllers are designed to be channel attached to an IBM mainframe as illustrated in the lower portion of Figure 16.36 and provides a gateway to the mainframe for a 'local' Token-Ring local area network.

The key to the connection of workstations on a Token-Ring network to a mainframe is a combination of hardware and software that provides a gateway to the mainframe. IBM and other vendors market a series of products that provides a variety of methods to obtain gateway connectivity from a Token-Ring network to a main- frame. Due to space limitations, we will focus our attention upon IBM products; however, readers should note that many third-party vendors market similar functioning equipment and software to be discussed in this section.

IBM gateways are personal computers that can be connected to a mainframe via the use of one of three types of communications adapter cards: IBM 3278/9, IBM SDLC or an IBM Token-Ring Network adapter card. Due to the importance in understanding the constraints and limitations associated with each method of connection, we will cover each method in detail based on the type of adapter used in the gateway personal computer. For each of the connectivity methods discussed in this section, it is assumed that IBM's Workstation Emulation Program Version 3.0 or a later version of that program operates on the gateway computer.

SDLC adapter connectivity

The use of an SDLC adapter permits a gateway PC to be connected to an IBM communications controller, such as a 3720, 3725 or 3745 device. Since the SDLC adapter is used to obtain remote connectivity, the link between the gateway PC and the communications controller is obtained by the use of either a pair of modems or data service units. Modems are used if an analog communications facility is used, and data service units are used when a digital communications facility is employed. Figure 16.37 illustrates the use of an SDLC adapter in a gateway PC.

When a gateway PC uses an SDLC adapter, its data rate is limited to a maximum 56 kbps when communicating with the host. The gateway PC communicates with other workstations on the network via IBM's NETBIOS interface and a Token-Ring adapter board at 4 or 16 Mbps. One key limitation of this method of connectivity is thus the data rate between the gateway and mainframe, which can be a bottleneck with respect to the response time of other workstations on the network accessing mainframe applications.

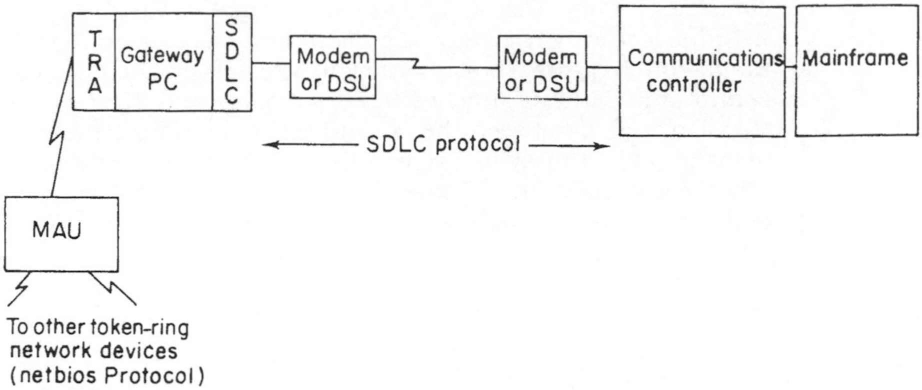

Figure 16.37 Using an SDLC adapter for gateway connectivity

When the PC is used as a gateway via an SDLC adapter, only the gateway PC is recognized as an SNA physical unit. Each of the workstation PCs on the network is designated as a logical unit (LU). Since a maximum of 32 LUs can be associated with a PU, a second limitation associated with this method of connectivity concerns the number of workstations supported by this type of gateway which is 32.

3278/9 adapter connectivity

The use of a 3278/9 adapter card requires a gateway PC to be cabled to a 3X74 control unit via the use of a coaxial cable. The control unit can be either locally attached to a host computer or remotely attached via the use of modems or DSUs. Figure 16.38 illustrates gateway connectivity obtained via the use of an IBM 3278/9 adapter.

A gateway using an IBM 3278/9 adapter is limited to supporting up to five concurrent network workstation communications sessions to the mainframe. This is because the port on the control unit must be configured as a distributed

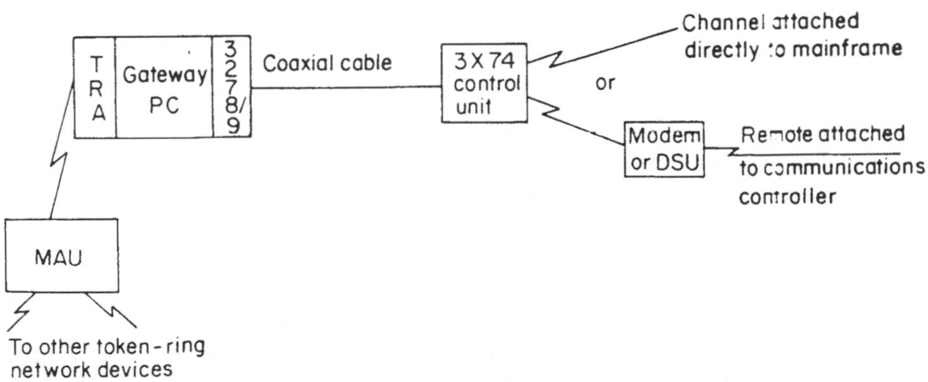

Figure 16.38 Using a 3278/9 adapter for gateway connectivity

function terminal (DFT) port, which is limited to supporting five concurrent sessions. Thus, seven gateways using 3278/9 adapter cards would be required to provide an equivalent level of concurrent sessions obtained through the use of a SDLC adapter card.

Similar to the use of an SDLC adapter, a throughput bottleneck can occur when a 3278/9 adapter card is used. This bottleneck resides in the 3X74 control unit to communications controller link which is limited to a maximum operating rate of 56 kbps.

Token-Ring hardware connectivity

Mainframe connectivity through the use of Token-Ring hardware is obtainable by two similar methods. Each method eliminates the use of a specific PC as a gateway, in effect, turning each workstation on the LAN into a combined workstation gateway.

The first method by which Token-Ring hardware can be used to provide mainframe connectivity is the use of a 3174 control unit equipped with a Token-Ring adapter. This hardware permits the control unit to become an active participant: on the LAN, operating at either 4, 16 or 100 Mbps. The second method involves the installation of a Token-Ring Interface Card (TIC) in an IBM communications controller, which also enables the controller to become an active participant on the LAN Figure 16.39 illustrates both methods of attachments.

Both techniques illustrated in Figure 16.39 extend IBM's SNA down to the individual workstation level on the LAN, with each workstation capable of being designated as a separate physical unit (PU). Since a more detailed level of control is required for SNA to be extended to each workstation, the IEEE 802.2 LLC 2 protocol is used on the LAN.

When a Token-Ring adapter on a 3174 control unit is used, the data rate bottleneck is between the control unit and communications controller. As previously discussed, this link is limited to 56 kbps, however, it extends support up to 256 workstations. When a TIC is used, the data rate on the

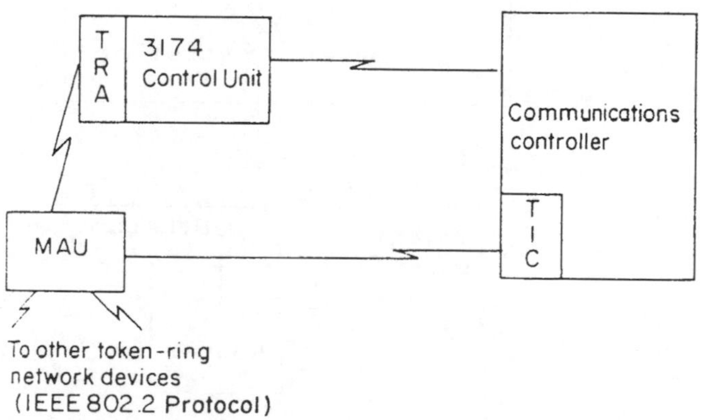

Figure 16.39 Using Token-Ring hardware for connectivity (TRA Token-Ring adapter, TIC Token-Ring interface card)

LAN is the rate data flowing to the communications controller, in effect, eliminating any data rate bottleneck that could affect response times when a significant number of workstation users access mainframe applications. Similar to the control unit TRA, the communications controller TIC supports up to 256 workstations on a LAN.

Using the 3172 Interconnect Controller

The IBM 3172 Interconnect Controller can be considered to represent the Swiss Army knife equivalent of gateway computers. The 3172 can be directly channel-attached to S/370 and S/390 computers via either a parallel or an ESCON channel, and supports the connection of a variety of LAN and WAN connections as well as a mixture of protocols that can run over those connections.

Figure 16.40 illustrates how a 3172 Interconnect Controller can be used as a gateway between an IBM host and the Internet. In this example, the 3172 Interconnect Controller enables TCP/IP communications to flow on the Token-Ring and Ethernet networks shown connected to the 3172. This in turn enables the router-based connections to the Internet to reach the mainframe via the 3172.

Software considerations

Although the 3172 and other gateway products provide users with the ability to route data between devices that communicate using different protocols, you must also consider the software on the end station accessing the host to

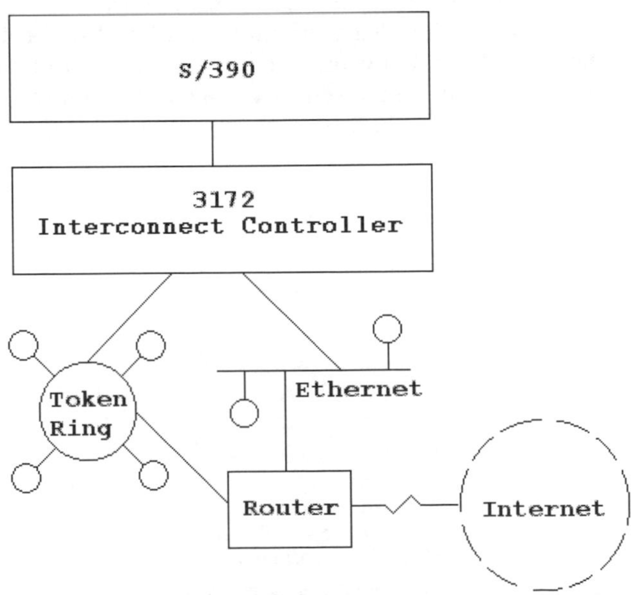

Figure 16.40 Using the 3172 Interconnect Controller

obtain a full gateway capability. IBM's 3270 Information Display system uses special codes to enable and disable a variety of terminal features associated with that vendor's original series of fixed logic terminal devices. As PCs began to replace dumb terminals, emulation programs were required to enable PC-based LAN workstations to obtain keyboard and screen operation compatibility with IBM hosts. One of the most popular IBM 3270 display station emulation programs is the PCOM/3270, an acronym for IBM's Personal Communications/3270 series of emulation programs.

If the 3172 can be considered as the Swiss Army knife of gateways, the PCOM/370 can be considered to represent the Swiss Army knife of emulation programs. PCOM/370 supports a variety of mainframe attachment methods, ranging from LAN adapters and coaxial cable to asynchronous and synchronous serial communications. Figure 16.41 illustrates the PCOM/3270 customize communications display screen which shows four options for communicating from a coaxial-based PC to a S/390 host. In examining Figure 16.41, note that the entry 'COAX' in the column labeled 'Adapter' is shown selected, which results in the display of four attachment methods shown in the middle of the illustration. In comparison, the use of a LAN adapter card

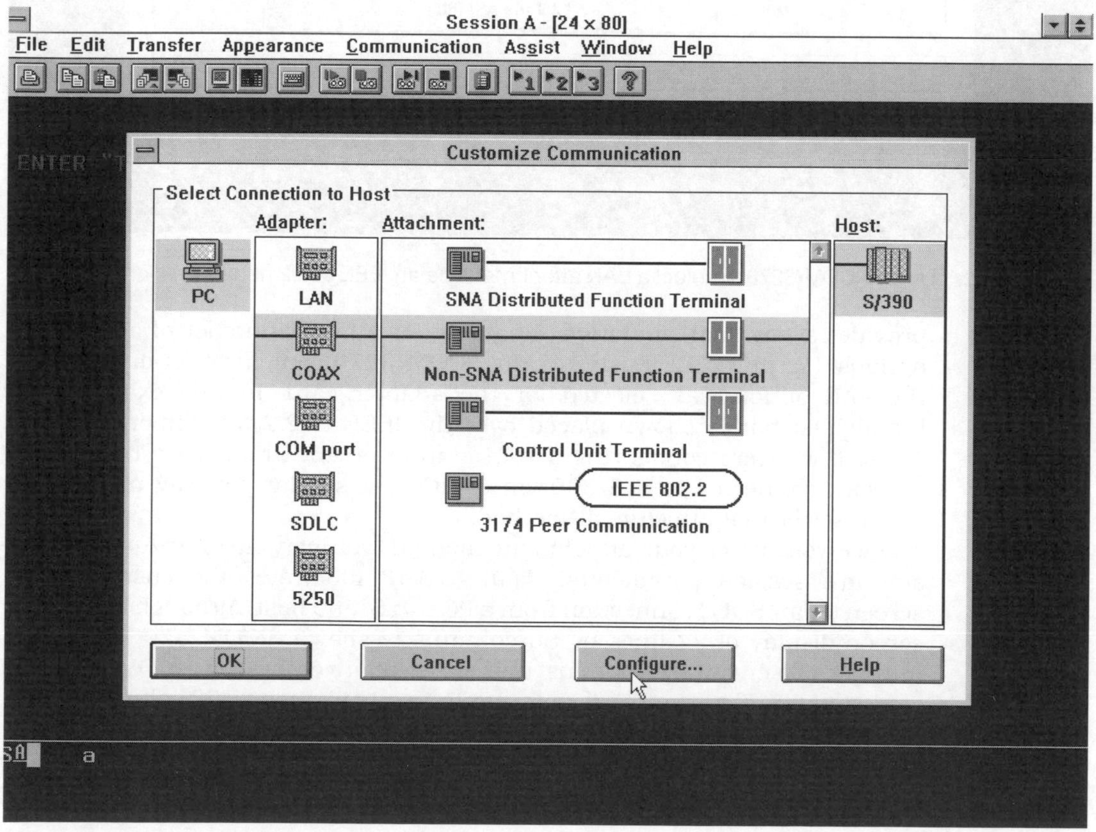

Figure 16.41 Using PCOM/3270 to select an appropriate attachment method when using a coaxial cable adapter for host communications.

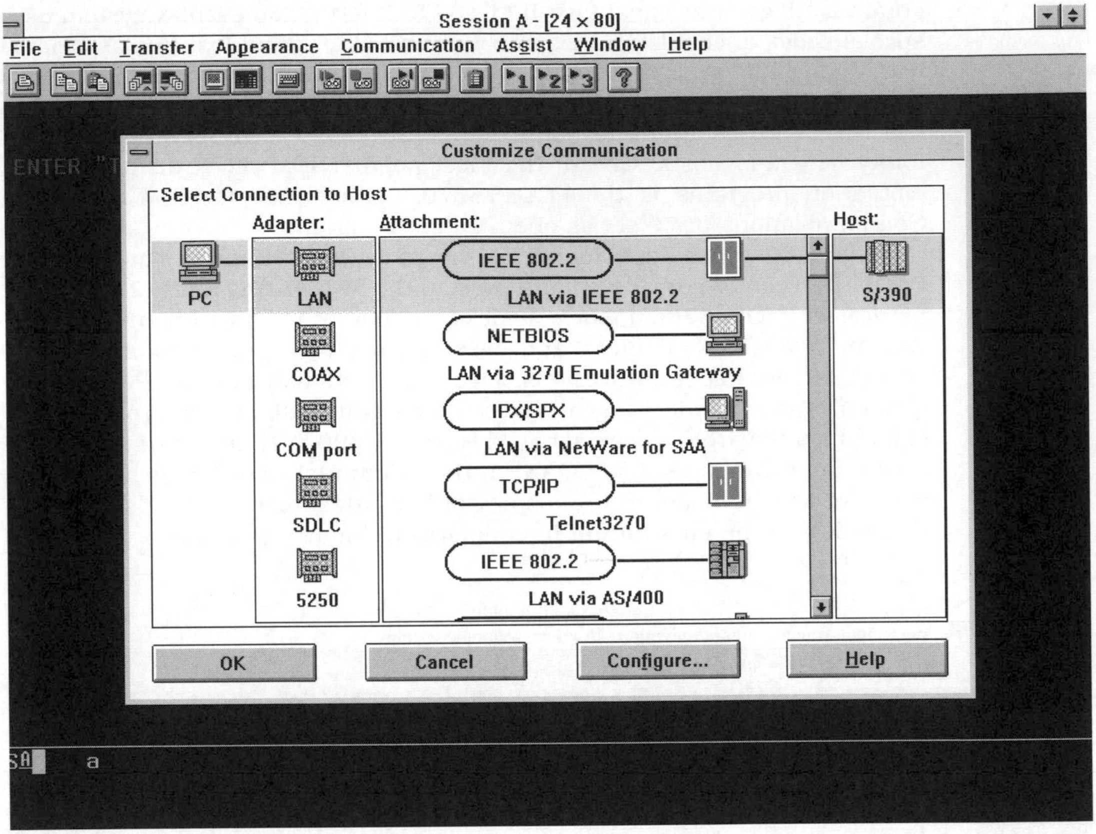

Figure 16.42 Using PCOM/3270 to select a LAN attachment via an IEEE 802.2 network.

provides a terminal emulation capability for a larger number of attachment options. Figure 16.42 illustrates the PCOM/3270 configuration display after the LAN option was selected for the adapter card. In this example, the highlighted bar is shown placed over the IEEE 802.2 attachment method. Thus, if we were configuring a LAN station or an Ethernet or Token Ring network connected to a S/390 via a TIC or a similar gateway method, we would select that attachment method.

Once you select your attachment method, you can customize your workstation's session parameters. Figure 16.43 illustrates the customization screen for an SDLC connection from a PC to a S/390 host. Although the default screen display of 24 lines by 80 columns is shown, PCOM/370 supports a number of screen size options that work relatively well with SVGA based workstations. For example, if your requirements include a 132 column display you can easily adjust the screen size by a simple point and click operation.

TN3270 operations

One of the more interesting methods for accessing S/390 hosts is via a special version of Telnet known as TN3270. Through the use of a TN3270 program

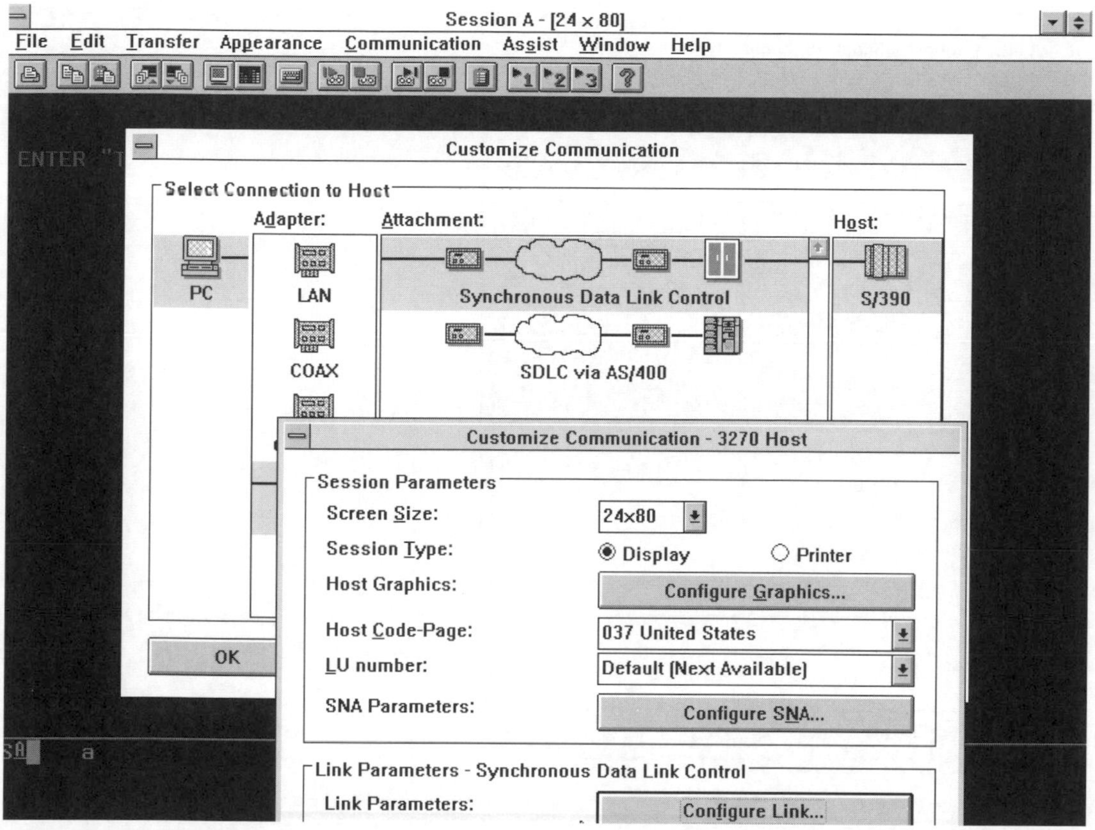

Figure 16.43 Using PCOM/3270 to adjust session parameters.

you can access IBM mainframes connected to the Internet via the Internet. This capability results from the fact that TN3270 represents a TCP/IP program that operates on top of a TCP/IP protocol stack.

Figure 16.44 illustrates the connection configuration of the NetManage Chameleon TN3270 terminal emulation program to initiate a session with a mainframe whose IP address is 192.76.46.1. Note that the session request will occur on TCP port 23, and the PC running TN3270 will operate as a 3270 model 2 display which uses a 24 by 80 character display. Figure 16.45 illustrates a TCP/IP connection to a S/390 using TN3270 via an Internet connection to a 3172 which in turn is connected to a S/390 mainframe. Although the TN3270 program used by the author uses a GUI interface, the actual display generated by the S/390 host is text based. This is shown in Figure 16.45 which illustrates the use of TN3270 to access the author's calendar stored on a mainframe calendar system.

Now that we have an appreciation for the methods by which workstations can obtain access to IBM mainframes, let us examine the problems that can occur when internetworking involves the transmission of multiple protocols.

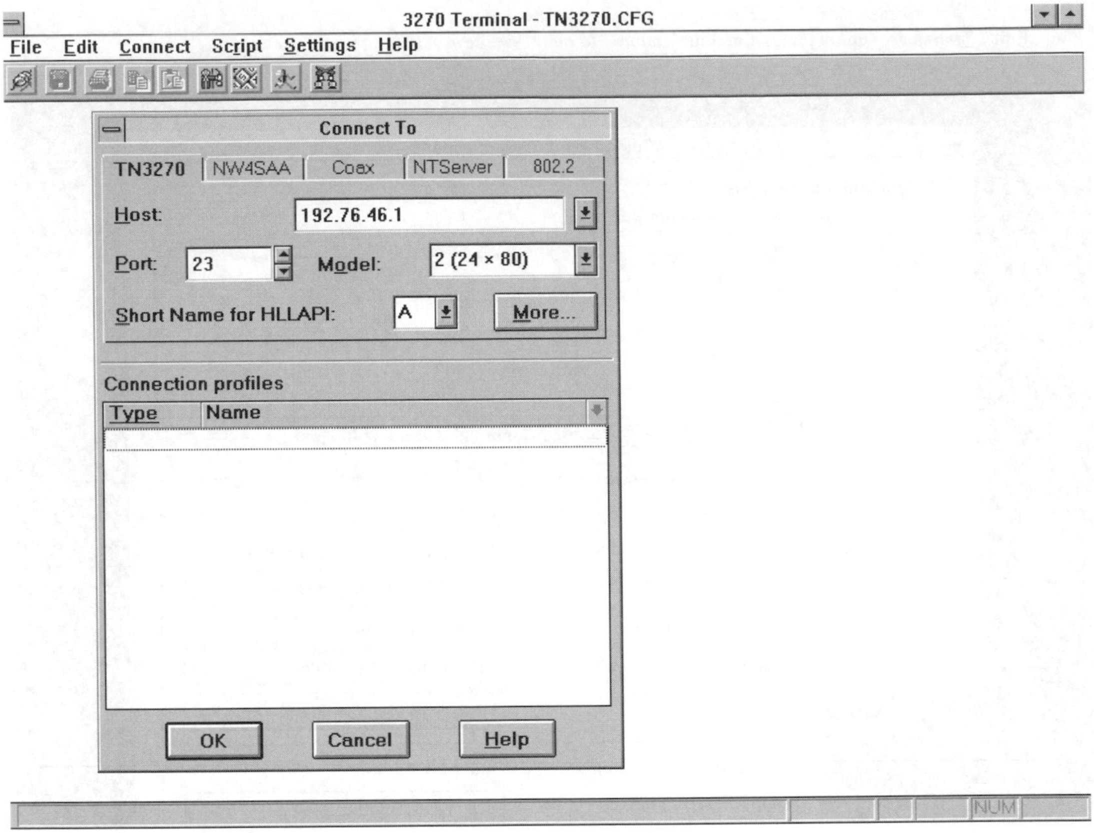

Figure 16.44 Using the NetManage Chameleon TN3270 terminal emulation program to obtain TCP/IP access to a #mainframe

16.4.2 Supporting multiple protocols

To illustrate the problems associated with supporting multiple protocols let us assume that an organization has a network consisting of LANs interconnected through the use of routers. Let us also assume that although the protocol used to interconnect LANs is TCP/IP, the organization operates an IBM mainframe and must provide workstations on each LAN with access to the mainframe. Figure 16.46 illustrates an example of the previously described router-based network.

Using a TIC connection to LAN A the communications controller can communicate with workstations on that local area network using LLC 2, a time-sensitive protocol that employs a series of timers which, when they expire, result in timeouts. Workstations on LAN B and LAN C that require access to the mainframe would also use LLC 2 as a transport mechanism, requiring the routers to function as bridges, because LLC 2 is not routable.

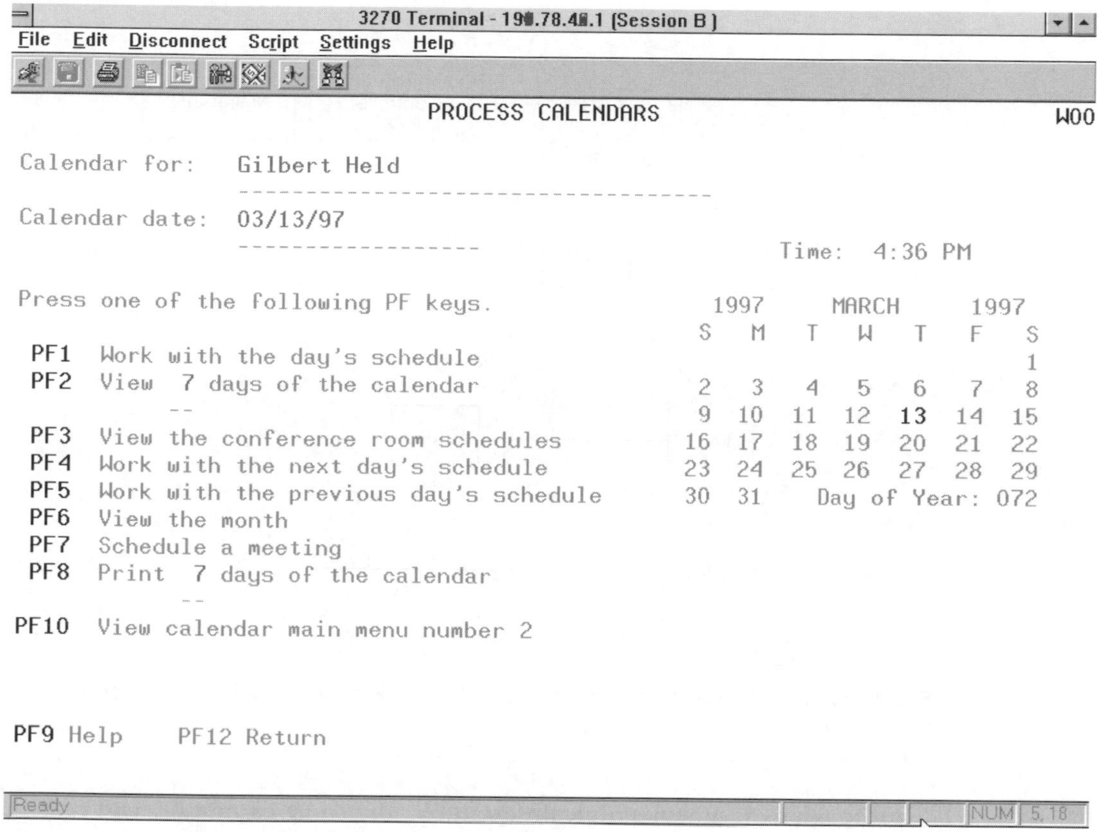

Figure 16.45 Using a TN3270 session to access a mainframe-based calendar system.

Session timeout problems

Now let us expand the use of the routers to provide LAN interconnectivity using TCP/IP. Suppose that a workstation on LAN B initiates a file transfer to a workstation on LAN A a moment prior to a workstation on LAN B attempting to respond to a mainframe request received via link A–B. Here the initial portion of the file transfer could delay the LLC session, resulting in a session timeout.

Encapsulation versus priority pass-thru

To prevent the previously described timeout situation from occurring, router vendors employ a variety of techniques. One common technique is to encapsulate SNA traffic within TCP/IP; however, doing so can result in an extended overhead due to the length of TCP and IP headers. A second method commonly employed is referred to as priority pass-thru. Under priority pass-thru, SNA to include LLC 2 traffic is transported as SNA SDLC between routers.

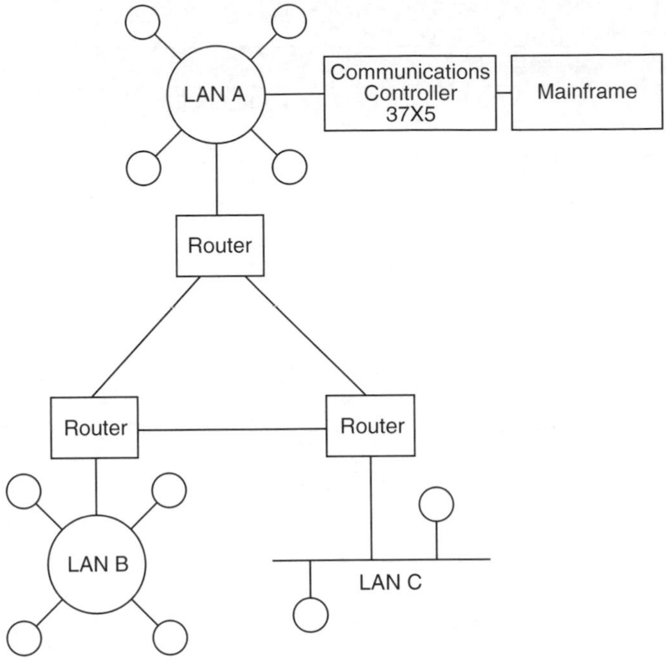

Figure 16.46 Using a router-based network to provide workstations with access to a mainframe

Although TCP/IP is dynamically routable, the same cannot be said of SNA traffic as it does not support network addresses. Thus, routers must either be manually configured to associate device addresses in the form of media access control (MAC) addresses to a specific network or support transparent bridging. Concerning the latter, the router would function as a bridge with respect to handling SNA LLC 2 traffic, constructing tables that associate MAC addresses with router ports. Since bridging cannot be prioritized this method could result in session timeouts when networks grow in complexity or other traffic carried between routers results in delays transporting bridged traffic. Thus, router vendors create a pass-thru segment of bandwidth which transports LLC 2 similar to the manner by which multiplexers use a bandpass channel. That is, routers reserve a portion of the bandwidth on the links connecting each other for the transmission of LLC 2 data, minimizing the potential for session timeouts. When a priority pass-thru scheme is used the routers dynamically adjust the size of the bandwidth allocated for LLC 2 transmission. Thus, a priority pass-thru mechanism represents a method whereby SNA and other protocols can coexist, so long as the other protocols are not also time-sensitive.

Now that we have an appreciation for APPN and SNA gateways and the use of emulation software, we can turn our attention to the integration of APPN and SNA traffic into a TCP/IP network. Since this capability can be considered to represent an SNA to TCP/IP gateway performed by Data Link Switching, we will next focus our attention upon this gateway technique.

16.4.3 Data Link Switching

In a quest to facilitate the use of a common network for transmitting both SNA and IP traffic, IBM developed a technique referred to as Data Link Switching (DLSw) whish was standardized by the IETF. Although this technique uses encapsulation and in effect tunnels SNA under IP, it uses several techniques that enhance the efficiency of the encapsulation process.

One of the major problems associated with SNA- and APPN-based networks is the fact that both are essentially proprietary. As the use of TCP/IP expanded for interconnecting LANs on private intranets as well as for communications with the Internet, many organizations were forced to maintain two separate networks, one for SNA and one for a second network protocol which is commonly TCP/IP.

In an effort to better support customer requirements for improving connectivity and lowering the cost associated with multiprotocol networking, IBM developed a tunneling mechanism referred to as Data Link Switching (DLSw). IBM first introduced DLSw in 1992 as a feature of its 6611 bridge-router series of products referred to as the Nways 6611 Network Processor. IBM submitted its effort to the Internet Engineering Task Force (IETF) which resulted in DLSw being standardized as RFC 1434. It was also standardized by an IBM-sponsored, multivendor forum known as the APPN Implementers' Workshop.

Overview

Data Link Switching was developed as a mechanism to support SNA and NetBIOS data traffic via both bridged and router based networks in a multiprotocol environment. Although DLSw is primarily used to tunnel SNA and NetBIOS under IP, it can also be used to tunnel other protocols. Since DLSw enables organizations to merge their SNA and APPN networks with their IP-based networks, many redundant communications links become candidates for removal, enabling organizations to operate a more efficient and less costly network infrastructure.

Operation

Under DLSw the connection-oriented protocols of SNA and NetBIOS in the form of Logical Link Control Type 2 (LLC2) and Synchronous Data Link Control (SDLC) packets are encapsulated into IP packets. Figure 16.47 illustrates an example of the manner by which point-to-point SDLC traffic and LAN-based LLC2 traffic are integrated into an IP router-based network. In examining Figure 16.47, note that the tunneling effort involves wrapping the IP header around SDLC or LLC2 data. Since such data then becomes encapsulated, as you might expect, another term used to reference the transport of SDLC and LLC2 data in IP packets is encapsulation.

Although the actual tunneling or encapsulation process appears to be relatively simple to accomplish, in actuality it is a complicated process. The complication results from the fact that SNA uses connection-oriented protocols, LLC2 and SDLC, which are based upon positive acknowledgment with

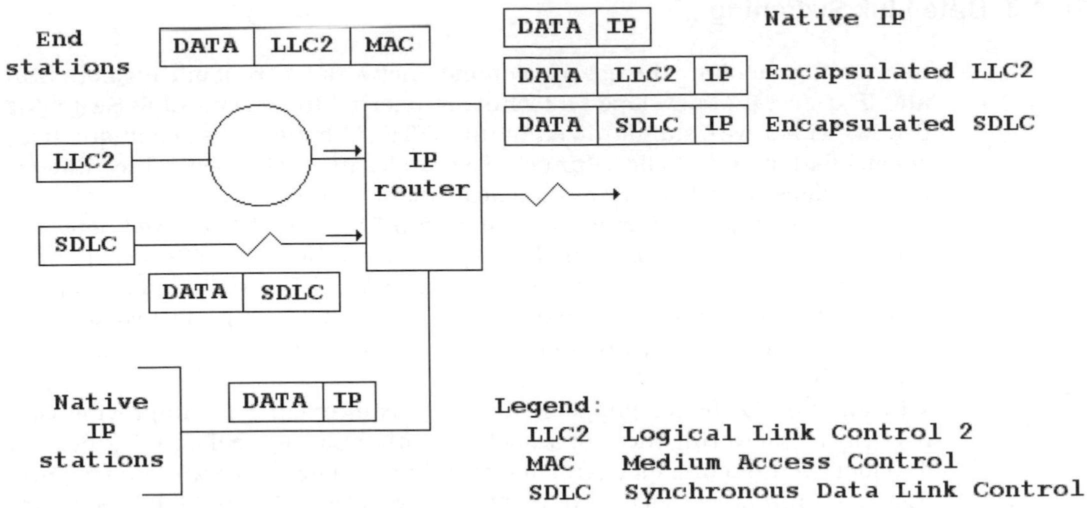

Figure 16.47 The DLSw tunneling process

retransmission (PAR). As such, if an ACK is not received within a period of time after a sequence of frames are transmitted, a timer will expire, resulting in the sending station retransmitting the data. If, due to network congestion, a circuit failure or other impairments, such retransmissions repeat, after a predefined number of repeats the connection will be terminated, resulting in the loss of any SNA and NetBIOS sessions in progress. Since IP is a connectionless protocol, the transport of SDLC and LLC2 data within IP would very likely result in periodic session timeouts as traffic density varies during the day. To prevent this situation from occurring, DLSw relies on spoofing, with the sending router acknowledging frames as they are received. The routers then use a reliable transport protocol, such as TCP, to ensure that data arrives at its intended destination. A slightly different procedure is used for SDLC that relies on the constant polling between primary and secondary SDLC stations. DLSw capable routers perform proxy polling. That is, the sending router intercepts polls from the SDLC primary station while another router polls the SDLC secondary station as if it was the SDLC primary station. Since polls are not passed onto the IP network, transmission efficiency is enhanced.

Although any encapsulation or tunneling method adds overhead in the form of an additional header, the use of spoofing and proxy polling can be considered as a significant counterbalance. Thus, in most cases the overhead associated with the use of additional headers will have a negligible effect upon the overall performance of DLSw.

16.5 REVIEW QUESTIONS

1. Why is SNA classified as an architecture?

2. What is the function of VTAM in an SNA network?

3. What is the function of the NCP in an SNA network?

4. Discuss the relationship between VTAM and the NCP in an SNA network with respect to their functions and location where they reside.

5. What are the three network addressable units in an SNA network? What is the function of each unit?

6. What is the relationship between peripheral nodes and subarea nodes in an SNA network?

7. Why is it difficult to interconnect separate SNA organizational networks?

8. What are two advantages obtained from the use of LU6.2 over LU2?

9. Describe the four types of SNA sessions.

10. What is an SNA subarea?

11. What is the function of Advanced Program to Program Communications (APPC)?

12. What is the relationship of APPN to APPC?

13. What are the three types of nodes in an APPN network?

14. Discuss the major functions of an APPN Network Node.

15. Describe the two variables that control APPN routing.

16. What is an RFC?

17. What is one of the key reasons for the dramatic growth in the use of the TCP/IP protocol suite?

18. What is the difference between FTP and Telnet with respect to the connections they require?

19. When using Microsoft Windows describe the general function performed by entering the following command:

ftp ftp www.irs.com

20. Discuss the flexibility associated with the use of the Microsoft Telnet application.

21. Describe the two key functions performed by the TCP/IP Domain Name Service (DNS).

22. What are the three major components of SNMP?

23. What is a Uniform Resource Locator (URL)?

24. What does the URL of http://www.irs.gov/refund.htm signify?

25. What is the difference between a datagram and a virtual circuit?

26. When UDP is used, where does the responsibility reside for ensuring that messages are properly delivered?

27. What are the two portions of an IP address?

28. Assume a dotted-decimal network IP address of 205.131.175.14. What is the class of the IP address?

29. Why would you want to subnet an IP address?

30. What is the function of a subnet mask?

31. When an application protocol uses TCP, where is the IP header placed with respect to the TCP header?

32. What is the function of a name server?

33. Discuss three key differences between SNA and TCP/IP.

34. What is the primary advantage obtained by the use of a communications controller TIC as opposed to the use of a 3174 Token-Ring Adapter for providing workstations access to an IBM mainframe?

35. Describe the problem associated with encapsulating SNA traffic within TCP/IP.

36. How does priority pass-thru operate?

37. Where would you use a TN3270 application?

17

PACKET NETWORKS

Until the early 1970s, access to mainframe computer systems was accomplished either through entry into a particular organization's data communications network or through the use of the public switched telephone network. Since then, a third method of accessing mainframe computers gained the acceptance of both the business and home computer user requiring access to a variety of computer facilities of independent organizations located at geographically dispersed locations. This method of communications is known as packet switching. For many persons, a packet network can be viewed as a transmission expressway, providing a fast, reliable and inexpensive method of communications to numerous locations throughout the United States and abroad. In fact, the so-called 'mother of all packet networks' is the Internet, which was covered in Chapter 16. While for many persons the Internet may appear to be the only packet network of interest, as we go about our daily lives we more than likely become users of two other packet networks that are the focus of this chapter – X.25 and frame relay.

The first widespread commercial packet network to be deployed is referred to as the X.25 network. Although this network was developed during an era of analog communications when digital transmission was a distant thought, it still represents a viable network over which a majority of credit card charges are verified.

The growth in client–server computing during the 1980s resulted in many organizations requiring a mechanism to effectively and efficiently interconnect geographically separated LANs. Although many organizations constructed internal networks through the use of leased lines and the acquisition of bridges and routers, the economics associated with sharing transmission facilities among many organizations resulted in the development of a new packet network technology designed to facilitate inter-LAN communications. That new packet network technology is known as frame relay, and it provides a high performance LAN interconnection capability in comparison with conventional packet switching technology.

In this chapter we will first focus our attention upon obtaining an overview of the basic technology associated with packet switching. This will be followed by examining how X.25 and frame relay networks operate, along with a comparison of the two technologies.

17.1 PACKET SWITCHING OVERVIEW

As previously noted in Chapter 16, the first packet-switching network was developed by the Advanced Research Projects Agency (ARPA) of the US Department of Defense. The initial goal of this network was to permit a wide variety of computer resources located throughout the United States to be simultaneously accessed via a common network by research personnel.

The resulting ARPAnet used the TCP/IP protocol suite and evolved into the modern Internet. While the Internet today achieves a high proportion of publicity, there are two other types of packet networks that are also used by businesses and individuals on both a direct and an indirect basis. As previously mentioned, a large portion of credit card transactions flow on the X.25 packet network. In comparison, the more modern frame relay packet network is not only extensively employed to interconnect LANs but in addition is commonly used to provide access to the Internet. In addition although Voice over IP (VoIP) currently receives a considerable amount of publicity as a replacement for conventional telephone service, most commercial transmission of voice over packet networks designed to transport data occurs over frame relay for reasons we will note later in this chapter.

Packet switching is a technique where data are broken into shorter sections known as packets. Each packet is forwarded over the network's communications facilities whenever it is ready to be transmitted, independent of the

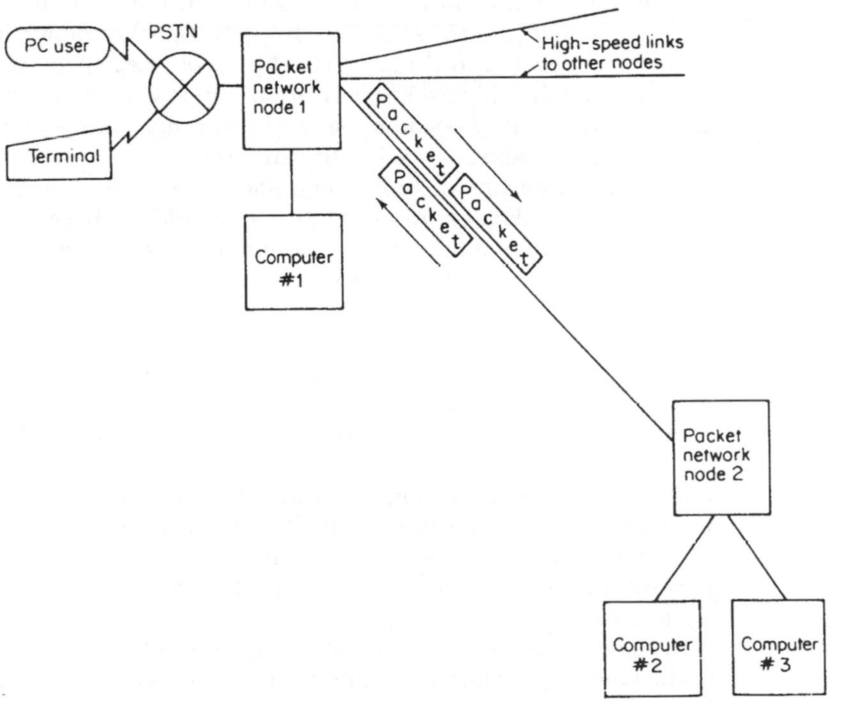

Figure 17.1 Simplified packet network. In a packet network, minicomputers at each node assemble data into packets for transmission over the network's facilities

remaining data. This technique then permits many devices to share a common communications facility by time as illustrated in Figure 17.1. The simplified packet network illustrated in Figure 17.1 shows two packet network nodal points, with three mainframe computer systems connected to the two nodes.

The packet network node is normally a minicomputer-based computer system which forms and routes packets to other destinations on the network. If access to a node is from a non-packet-forming device, such as an asynchronous terminal or personal computer, the node breaks the transmission into segments called packets. During the packet-assembly process, address and control information is appended to the data in the packet as illustrated in Figure 17.2. Similarly, when packetized data are received at its final destination, the header information is stripped from the packet prior to the node transmitting the data to the device connected to that node.

Packet network terminology varies based upon the type of packet network used. When an X.25 network is used, equipment that forms packets from data transmitted by non-compliant X.25 devices, such as PCs, is referred to as a packet assembler/disassembler (PAD). A PAD not only forms packets but, as its name implies, performs a reverse operation and disassembles received packets from a packet network into the serial protocol supported by a particular terminal device or computer. In a frame relay environment the conversion of a serial data stream produced by non-compliant frame relay devices into frames is performed by a frame relay access device (FRAD). Similar to a PAD, the FRAD also performs a reverse operation and converts frames received from a frame relay network into the protocol supported by a particular terminal device or computer. Thus, both PADs and FRADs can be viewed as protocol conversion devices.

Now that we have an appreciation for the general characteristics of a packet network and the basic role of protocol conversion devices, let's turn our attention to X.25 networks.

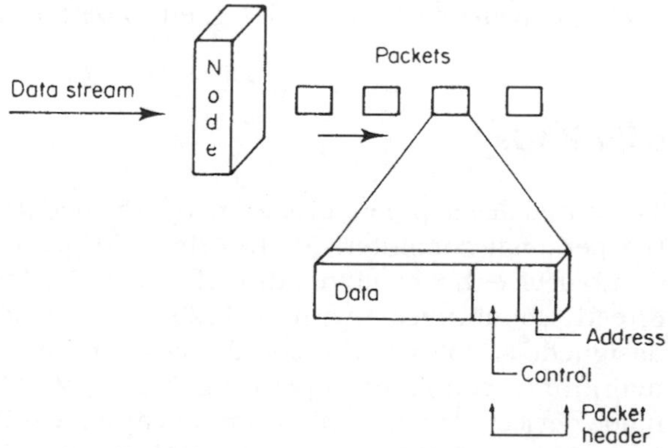

Figure 17.2 Packet assembly. During the packet assembly process, non-packetized data are packetized, with address and control information appended to each packet

17.2 X.25 NETWORKS

The ITU-T X.25 recommendation forms the basis of the flow of data within and to and from public packet data networks. The first draft of the X.25 recommendation was published in 1976, with major changes to the standard occurring in 1980 and 1984. Formally entitled 'Interface Between Data Terminal Equipment (DTE) and Data Circuit Terminating Equipment (DCE) for Terminals Operating in the Packet Mode on Public Data Network', the latest version of the X.25 recommendation provides a precise set of procedures for communications between DTE and DCE in a packet environment. Here, the DCE is a network node processor, normally a minicomputer operating specialized software. The DTE is a programmable device located at an end-user facility, such as a front-end processor, intelligent terminal or mainframe computer which supports the X.25 protocol.

17.2.1 Development period

The development of X.25 packet switching occurred during the period when high-speed data transmission was achieved primarily over analog wideband transmission facilities. Because the error rate of analog transmission facilities was relatively high, X.25 networks were designed to perform error checking at each node in the network. That is, when a packet arrives at an X.25 node the cyclic redundancy check (CRC) is recomputed and compared to the CRC contained in the packet. If the two match the packet is forwarded through the network. Otherwise the packet is dropped and the transmitting node is informed to retransmit the packet, with errors corrected by retransmission. Although error checking at each node results in an almost negligible probability of erroneous packets reaching their destination, the flow of packets is delayed at each node. If interactive query–response traffic is carried over more than a few nodes from source to destination, the nodal delays become very noticeable. As we will note later in this chapter, this built-in X.25 delay resulted in the development of frame relay as a more modern packet network technology designed to better support interactive traffic.

17.2.2 Need for PADs

The initial development of X.25 networks occurred prior to the introduction of the personal computer. At that time the primary business terminal was a dumb device that could not directly create packets, resulting in their ability to directly communicate with an X.25 network node. Similarly, most PCs were designed as 'character' mode devices, whereas network node processor and mainframe computers operating X.25 software communicate as 'packet' mode devices. Thus, dumb terminals and most PCs must communicate with a special network DTE known as a PAD, which converts the data stream of the terminal or personal computer into the X.25 protocol. The PAD can be located at the packet network node or at the end-user's facility.

To standardize the conversion of asynchronous to X.25 data flow, three ITU recommendations were developed: X.3, X.28 and X.29. The X.3 recommendation defines such parameters as data rate, the escape character and the flow-control technique, enabling the PAD to operate with a specific type of terminal, personal computer or mainframe computer. By the appropriate setting of these parameters, the PAD is able to correctly interpret the data stream received from the device communicating with it. Similarly, the device communicating with the PAD is able to understand the PAD when these parameters are set correctly.

Because the majority of access into an X.25 packet network occurs via the use of a PAD, let's turn our attention to the operational characteristics of this device prior to examining the flow of data through an X.25 network.

Packet Assembly/Disassembly (PAD)

The X.3 Packet Assembly/Disassembly (PAD) parameters which are used to control start–stop terminals within an X.25 network are listed in Table 17.1.

Each of the PAD parameters has two or more possible values. Ten of the more commonly altered parameters and their possible values are dealt with in the following paragraphs.

The 2:n echo parameter controls the echoing of characters on the user's screen as well as their forwarding to the remote DTE. Possible values are:

0 no echo
1 echo

Table 17.1 X.3 parameters

X.3 parameter number	X.3 meaning
1	PAD recall
2	Echo
3	Selection of data forwarding signal
4	Selection of idle timer delay
5	Ancillary device control
6	Control of PAD service signals
7	Procedure on receipt of break signal
8	Discard output
9	Padding after carriage return
10	Line folding
11	Binary speed
12	Flow control of the terminal pad
13	Line feed insertion after carriage return
14	Padding after line feed
15	Editing
16	Character delete
17	Line delete
18	Line display
19	Editing PAD service signals
20	Echo mask
21	Parity treatment
22	Page wait

The 3:n selection of data forwarding signal parameter defines a set of characters that act as data forwarding signals when they are entered by the user. Coding of this parameter can be a single function or the sum of any combination of the functions listed below. As an example, a 126 code represents the functions 2–64, which results in any character to include control characters being forwarded. Possible values of parameter 3 are:

0 no data forwarding character
1 alphanumeric characters
2 character CR
4 characters ESC, BEL, ENQ, ACK
8 characters DEL, CAN, DC2
16 characters ETX, EOT
32 characters HT, LF, VT, FF
64 all other characters: X′OO′ to X′IF′

The 4:n selection of idle timer delay parameter is used to specify the value of an idle timer used for data forwarding. Possible values of this parameter include:

0 no data forwarding on time-out
1 unit of 1/20 second, maximum 255

The 7:n procedure on receipt of break signal parameter specifies the operation to be performed upon entry of a break character. Possible values of this parameter include:

0 nothing
1 send an interrupt
2 reset
4 send an indication of break PAD
8 escape from data transfer state
16 discard output

Similarly to parameter 3, parameter 7 can be coded as a single function or as the sum of a combination of functions.

The 12:n flow control of the terminal PAD parameter permits flow control of received data using X-ON and X-OFF characters. Possible values are:

0 no flow control
1 flow control

The 13:n line feed insertion after carriage return parameter instructs the PAD to insert a line feed (LF) into the data stream following each carriage return (CR). Possible values are:

0 No LF insertion
1 Insert an LF after each CR in the received data stream
2 Insert an LF after each CR in the transmitted data stream
4 Insert an LF after each CR in the echo to the screen

The coding of this parameter can be as a single function or a combination of functions by summing the values of the desired options.

The 15 : n editing parameter permits the user to edit data locally or at the host. If local editing is enabled, the user can correct any data buffered locally, otherwise it must flow to the host for later correction. Possible values of this parameter include:

0 no editing in the data transfer state
1 editing in the data transfer state

The 16 : n character delete parameter permits the user to specify which character in the ASCII (International Alphabet Number 5) character set will be used to indicate that the previously typed character should be deleted from the buffer. Possible values for this parameter include:

0 no character delete
1–127 character delete character

The 17 : n line delete character is used to enable the user to specify which character in the character set denotes the previously entered line should be deleted. Possible values are:

0 no line delete
1–127 line delete character

The 18 : n display parameter enables the user to define the character which will cause a previously typed line to be redisplayed. Possible values are:

0 no line display
1–127 line display character

The 20 : n echo mask parameter is applicable only when parameter 2 is set to 1. When this occurs, parameter 20 permits the user to specify which characters will be echoed. Possible values include:

0 no echo mask (all characters echoed)
1 no echo of character CR
2 no echo of character LF
4 no echo of characters VT, HT, FF
8 no echo of characters BEL, BS
16 no echo of characters ESC, ENQ
32 no echo of characters ACK, NAK, STX, SOH, EOT, ETB and ETX
64 no echo of characters defined by parameters 16, 17 and 18
128 no echo of all other characters in columns 0 and 1 of International Alphabet Number 5 and the character DEL

The X.28 recommendation defines the commands and procedures for establishing (call request) and disconnecting calls (call clear). The X.29 recommendation defines the procedures for the exchange of PAD control information and how user data are transferred between a packet mode DTE and a PAD or between two PADs.

In essence, PADs are concentration devices which have the capability to multiplex between 2 and 64 or more character-mode devices onto an X.25 data link. Some PADs simply convert an asynchronous character stream into a synchronous X.25 data format, and other PADs are designed to provide 2780, 3780 and 3270 protocol conversion to X.25.

In actuality, when we refer to packet-switching standards we normally refer to the X.25 recommendation and related packet network standards. These related standards include the previously discussed X.3, X.28 and X.29 recommendations as well as the X.75 and X.121 recommendations.

X.75 describes the procedures for the interconnection of packet-switched networks. X.121 is the international numbering plan for public data networks. Together, these standards permit the user of one packet network, as an example, to access a computer connected to a different packet network.

PAD locations

The physical location of the PAD can be in one of three places. First, the PAD can reside on an X.25 network node, with terminals required to access the node via the PSTN or by leased lines. This is the most common PAD location, since the majority of persons accessing packet networks use asynchronous terminals and dial a node via the PSTN. Secondly, a stand-alone PAD can be installed at the end-user's terminal or host computer location as illustrated in Figure 17.3. The use of this type of PAD permits the multiplexed X.25 protocol to be routed on one circuit between the packet node and the end-user location, in effect serving to minimize the communications cost of the end-user. The third PAD location is inside a computer or terminal device. This type of PAD is normally a special adapter card that can be installed within a personal computer or on the channel adapter of a front-end processor. Most PADs built on an adapter card permit the personal computer user to establish simultaneous communications with two or more remote computers on one X.25 connection as illustrated in Figure 17.3. This capability permits the user to switch from one host session to another without requiring the log-off from one computer system prior to signing onto a second system.

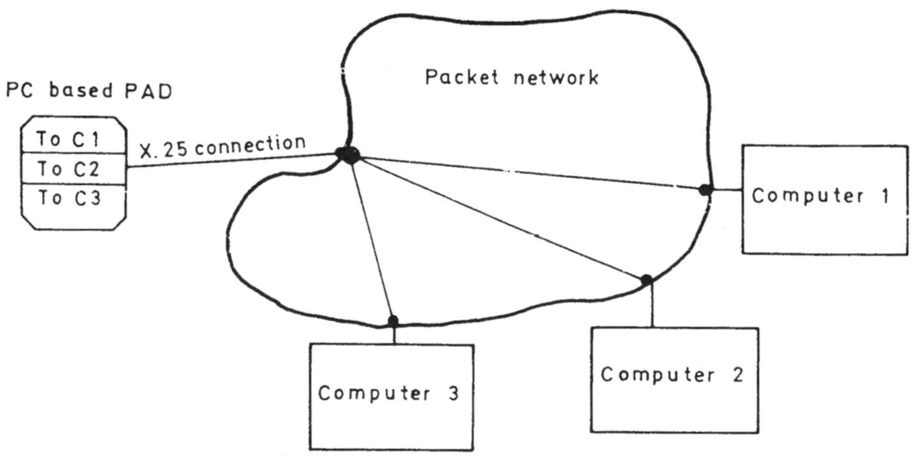

Figure 17.3 PC-based PAD utilization. The use of a PAD in a PC provides the personal computer with the ability to support multiple concurrent sessions through the packet network

17.2.3 X.25 layers

The X.25 recommendation defines three functional layers which correspond to layers 1–3 of the ISO Reference Model.

The X.25 physical layer (level 1) is defined by the X.21 recommendation. Unlike RS-232 or the V.24 interface that uses control signals on pin connectors, the X.21 interface employs the transmission of coded character strings across a 15-pin connector to define standard interface functions. Due to the limited availability of communications equipment with an X.21 interface, the ITU-T approved an interim interface, X.21 bis. X.21 bis is functionally equivalent to the RS-232 and V.24 interfaces in which a separate function is assigned to each pin in the 25-pin connector.

The X.25 link layer (level 2) is responsible for the transmission of packets of data between the user (DTE) and the node (DCE) without error. A data link control procedure very similar to HDLC is used, which adds control fields to the data packet as illustrated in Figure 17.4. The flag fields serve as the beginning and ending frame delimiters, and a 16-bit frame check sequence field provides the error-detection capability for the frame. The link layer is responsible for subdividing the user's data into packets which normally contain 128 characters. Some X.25 networks use 32, 256, 512 or 1024 character packets, with the packet size transparent to the user. When two packet networks are interconnected, X.75 defines the procedures for the orderly exchange of data between dissimilar sized packets.

The address field illustrated in Figure 17.4 identifies the station or stations for which command frames are intended o the station sending a response to a command in a response frame. The control field identifies the actual type of frame being transmitted, such as Information (I), Supervisory (S) or Unnumbered (U). In addition, this field supplies control information pertinent to the specific type of frame being transmitted, such as its transmitted sequence number $N(S)$, or sequence number expected to be received, $N(R)$, for an information frame. Figure 17.5 illustrates the X.25 command/response control field coding.

The original X.25 data link control procedure used a link access procedure (LAP) very similar to HDLC's asynchronous response mode (ARM). In 1984, the X.25 recommendation was revised to support link access procedure

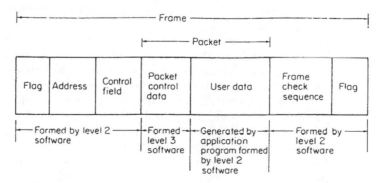

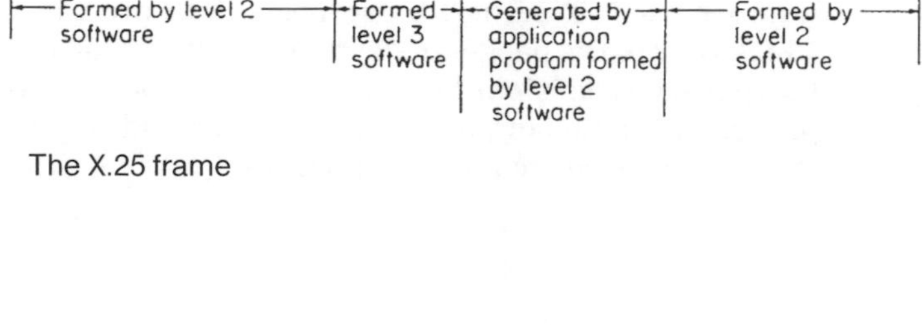

Figure 17.4 The X.25 frame

Format	Commands	Responses	Control Field Bit Positions 1 2 3 4	5	6 7 8
Information transfer	I		0 N(S)	P	N(R)
Supervisory	RR	RR	1 0 0 0	P/F	N(R)
	RNR	RNR	1 0 1 0	P/F	N(R)
	REJ	REJ	1 0 0 1	P/F	N(R)
Unnumbered	SARM	DM	1 1 1 1	P/F	0 0 0
	SABM		1 1 1 1	P	1 0 0
	DISC		1 1 0 0	P	0 1 0
		VA	1 1 0 0	F	1 1 0
		CMDR	1 1 1 0	F	0 0 1

Figure 17.5 X.25 command–response control field coding. (RR receiver ready, RNR receiver not ready, REJ reject, SARM set asynchronous response mode, SABM set asynchronous balanced mode, CMDR command reject, P/F poll/final, DM disconnect mode, DISC disconnect, UA unnumbered acknowledgement)

balanced (LAPB), which is similar to HDLC's asynchronous balanced mode (ABM). Under ABM, each side of a link consists of a combined primary/secondary station, permitting a station to originate as well as receive data.

Similar to HDLC, the control field of an I-frame keeps track of the data flow through the use of *N*(S) and *N*(R) values. *N*(S) contains the sequence number of the frame being sent from the transmitter to the originator, and the value of *N*(R) indicates the sequence number of the next frame the transmitter expects to receive from the receiver when that station becomes a transmitter. *N*(R) thus serves as an acknowledgement of all frames up to but not including that number. Finally, the poll/final bit indicates whether the receiver must acknowledge the transmission or when the final frame in a sequence has been transmitted.

The supervisory frame contains a poll/final bit, an *N*(R) value, and a four-bit command identifier. Supervisory commands acknowledge whether a receiver is or is not ready to receive transmission or is used to reject a frame that has a transmission error based on a frame check sequence computation.

The third X.25 layer is known as the packet layer. Level 3 defines the procedures for establishing and clearing calls and the exchange of packets between a DTE and DCE. Level 3 software prefixes the required packet control information to the user's data, after which it initiates level 2 software to create the X.25 frame. Included in the packet control information field are three subfields, known as octets since they consist of 8 bits. The first octet defines the logical channel group and the second octet defines the logical channel number used. Together, these two octets provide the routeing information which directs the packet to flow over a defined logical channel. The third octet, the packet type identifier, defines the type and resulting function of the packet. Examples of packet type include call request, DCE clear, DTE clear, DCE RR (receiver ready) and DTE RR.

To illustrate the use of packet control data, let us examine the sequence of events that occurs between a calling and called DTE on a packet network. This sequence of events is governed by X.25 level 3, which defines the procedures for call initiation, data transfer, call clear, and other functions between DTEs on the network.

During the call initiation process level 3 software in the calling DTE initiates transmission by sending a Call Request packet as illustrated at the top of Figure 17.6. The calling DTE's Call Request packet also contains the address of the called DTE as well as its own address, with the network using this information to establish a virtual circuit to link the DTEs together. The resulting virtual circuit represents a logical point-to-point connection established on a temporary basis to provide for data exchange. If the called DTE is available to accept the call, it will respond by transmitting a Call Accepted packet to the call originator. After the called DTE has received the

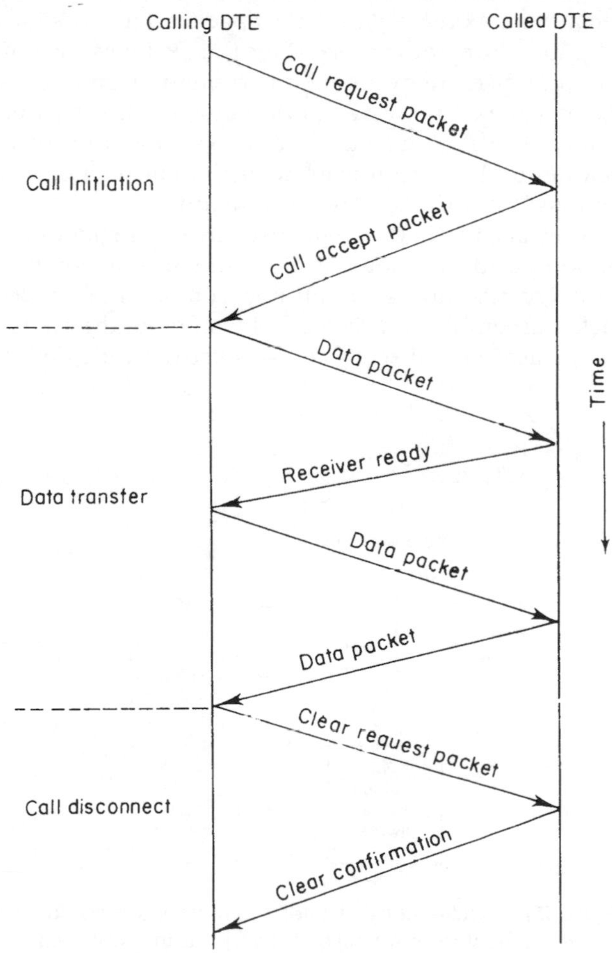

Figure 17.6 The X.25 packet network session activity

first data packet, it can authorize additional transmission by returning a Receiver Ready packet, or it can deny transmission by returning a Receiver Not Ready packet. Finally, when the calling DTE has completed its session, it transmits a Clear Request packet which is confirmed by the called DTE. At that time the virtual circuit is disassembled.

17.2.4 Methods of connection

You can connect a terminal, personal computer or a company's mainframe computer to a packet network in several ways. Figure 17.7 illustrates a few of the more common methods employed to interface a packet network. Note that a public port is a published telephone number that is available for all users to dial. In actuality, each node has a group of telephone numbers connected to a telephone rotary which causes an incoming call to be switched to the next available number if the dialed number is busy. To ensure that one never encounters a busy signal, many packet networks have private ports which are telephone numbers reserved for the exclusive use of a particular customer.

In an office environment where many terminals and PCs may require the use of a packet network, a PAD could be directly installed at the subscriber's facility. The PAD would directly assemble and disassemble packets at that location and communicate with the packet network node via a leased line obtained from the telephone company.

Two basic methods can be used by independent businesses as well as information utilities to connect their mainframe computers to a packet network. If an organization's mainframe computer as X.25 software that performs packet assembly and disassembly of data, they can connect their mainframe via a leased line directly to a packet network node This is illustrated in the

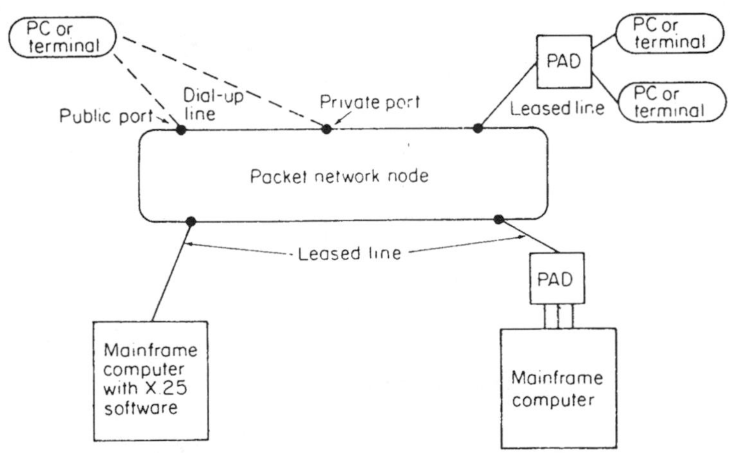

Figure 17.7 Packet network interface methods. Individual or grouped terminals and PCs can be connected to a packet network via the switched telephone network or by using leased lines. In comparison, mainframe computers are always connected to a packet network by a leased line

lower left portion of Figure 17.7. If such software does not operate on their mainframe, they can install a PAD at their location to perform the packet assembly and disassembly function as illustrated in the lower right part of Figure 17.7.

A packet-switched network consists of a large number of microcomputer-based nodes interconnected via high-speed data links. These data links resemble a mesh topology and are used by the minicomputers at the nodes to dynamically route packets of user data to their final destination. If traffic becomes too heavy on one link or a link failure occurs, the network control center in the packet network will sense such conditions and automatically inform the computer-based nodes in the network to perform alternative routing of data. Since hundreds or thousands of users can be simultaneously connected to the nodes of a packet network in larger cities, the utilization of the lines connecting most nodes in the packet network is high. In turn, this high utilization permits the cost per individual users of a packet network to result in an economical means of performing communications.

17.2.5 Utilization costs

As an example of the cost element associated with packet networks, we will first examine the cost of using the public switched telephone network. Here the cost of a long-distance call can vary based on the time the call was initiated, its duration, the distance between the called and calling parties and whether or not operator intervention was required. During business hours, a coast-to-coast call might cost 15¢ per minute while a call after 11 pm might be reduced to 5¢ per minute. Translated into an hourly rate, users can expect to pay between $3 and $9 per hour for such calls. In comparison, the cost of using a packet network is independent of the distance of the transmission. The principal cost elements are based on connect time and the amount of data transmitted, the latter charged on the basis of kilopackets or kilocharacters.

During prime time, which is usually 8 am to 5 pm, Monday through Friday, the average cost of accessing a mainframe computer via a packet network is approximately $3 per hour. During that hour you might transmit and receive 128 000 characters. If the packet network charges $1.00 per kilopacket, this adds an additional cost of $1.00 per connect hour. In actuality, for interactive applications most packets are transmitted only partially filled. Thus, responding to a mainframe prompt by transmitting 'RUN PROGRAM' results in the transmission of a packet consisting of 11 data characters and 117 pad characters which are inserted into the packet to fill it to 128 characters if that packet size is used by the packet network. Due to this a good 'rule of thumb' is to expect packet charges to equal 10% of connect charges. Even so, the cost associated with using a packet network can be significantly less than dialing long distance.

For non-prime time usage of packet networks, most networks have a flat rate per hour, typically ranging from $2 to $4 per connect-hour. This pricing mechanism is designed to encourage the use of such networks by the home computer hobbyist and businesses that can delay transmission until after normal working hours, representing the awareness of the packet networks that a good portion of their facilities is unused after prime time. Although a company using a packet network to access its mainframes would first establish an account with the network, users accessing most information utilities may never see a separate bill for the use of such facilities. This results from the fact that the cost of accessing the information utility via a packet network may be included in their general fee structure while other services may include a separate line item in their monthly bill to cover the cost of using the packet network.

Figure 17.8 illustrates a general comparison between the cost of using the PSTN, WATS, a leased line and a packet network. This comparison is for individually located terminals or personal computers accessing a common mainframe based on varying levels of communications ranging from 0 to 50 hours of usage per month. It was assumed that a terminal or personal computer accessing a mainframe was located at a distance of 650 miles from the mainframe. Thus, for comparison purposes the cost of a leased line was assumed to be $400 per month, and the cost of the use of the PSTN was assumed to be $10 per hour. For WATS, a cost of $7.50 per hour was assumed. In comparison, a packet network cost of $2.75 per hour, including both the correct time and character cost, was assumed.

A few words of caution are in order concerning the costs plotted in Figure 17.8. First, although the use of a packet network for an individually located terminal or personal computer appears to be the most economical, Figure 17.8 does not take into consideration the cost of connecting the mainframe computer to the packet network. This cost is based on the type of facility used to connect the mainframe to the packet network as illustrated in the lower portion of Figure 17.8, as well as the data rate used to connect the mainframe computer or PAD to the packet network. This latter cost is usually based on the data rate, because many packet networks furnish both a leased line and pair of modems for a set monthly fee to connect the end-user's mainframe computer to the packet network.

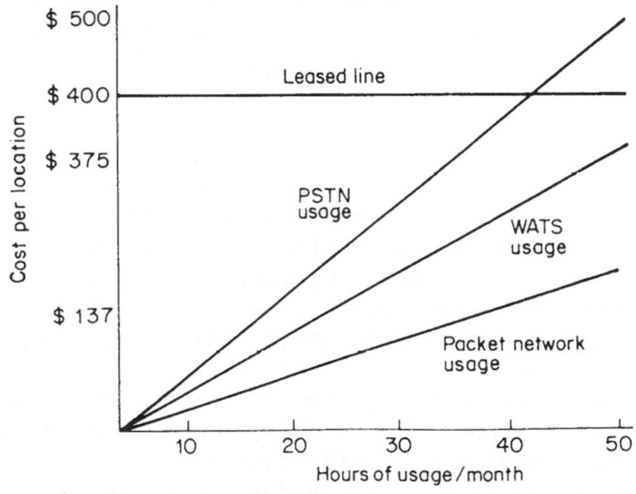

Figure 17.8 Cost comparison: PSTN, WATS, leased line and packet network usage

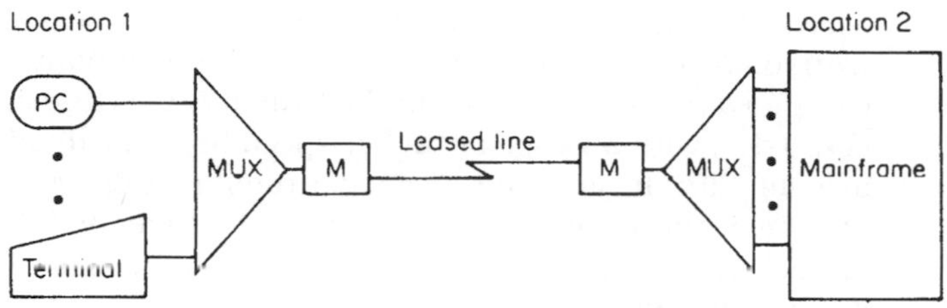

Figure 17.9 Using multiplexes. (MUX multiplexer, M high speed modem)

Assuming that the cost of a leased line, pair of modems and PAD is \$1000 per month, this cost must then be included in any analysis performed to determine the most economical data transportation facility. If there are 10 individual terminal or PC locations, the packet network usage cost illustrated in Figure 17.8 must be adjusted upward by \$100 per month. This figure represents the monthly mainframe connection cost (\$1000) divided by the number of individual terminal or personal computer locations (10). Thus, if a personal computer or terminal had no usage for the month the cost per location would be \$100.

A second consideration in evaluating the economics of using packet networks is the situation where many personal computers or terminals are collocated. Although you could use a PAD to connect to the packet network to reduce packet network costs, if there are more than three or four devices collocated and one's individual monthly usage is high, it is usually more economical to install a leased line to the mainframe and a pair of high-speed modems or DSUs/CSUs and multiplexers. Figure 17.9 illustrates the use of a pair of eight-channel multiplexes to permit up to eight collocated terminals and personal computers to share access to a mainframe computer.

Although the preceding cost comparisons were generalized, they do illustrate several key concepts concerning the economics associated with the use of packet networks. When only one or a few terminal devices are collocated and there are many such terminals distributed over a wide geographical area, the use of a packet network is normally the most economical data transportation facility. If there are only a few terminal devices and their connect time requirements per month are low, then the PSTN is usually the most economical data transportation facility. If there are many collocated terminals with large connect time requirements, then the installation of a leased line and multiplexers and high-speed modems is normally the most economical method to provide a data transportation facility.

17.2.6 Tymnet

In examining the method employed to access packet network facilities the use of Tymnet, a commercial packet network operated as a subsidiary of British Telecom, will be used for illustrative purposes. Tymnet was established in the early 1970s and its initial network served 30 cities in the United States. At that time, the company used 2400 bps lines to interconnect locations served in each city and its support of end-users was limited to asynchronous transmission. By the early 1990s, the company's packet network had expanded to over 600 cities in the United States and approximately 50 foreign countries, with 56 kbps or higher-speed lines used to interconnect equipment at each city. In addition to asynchronous support, Tymnet added a variety of synchronous protocol support, to include IBM 2780 and 3780 remote job entry, IBM 3270 interactive bisynchronous and SDLC support as well as other protocols.

Over the past decade the Tymnet network experienced several changes in ownership. After being purchased by McDonnell-Douglas during the 1980s, the Tymnet network was sold to British Telecom and became known as BT Tymnet. During the late 1990s British Telecom and MCI formed an alliance to provide X.25 communications on a global basis, which lasted until MCI was purchased by Worldcom. At the time this book revision was prepared AT&T was in talks with British Telecom concerning the formation of an alliance between the two telecommunications carriers. By late 1999 local modem dial-up access to the Tymnet Global Network was available in over 1000 cities within the United States and 1300 cities in over 110 countries around the globe.

One of the major reasons for the growth in the usage of packet networks is the ease with which such networks permit access to computer-based services connected to their facilities. For example, a person who wishes to use the Tymnet packet network would first dial the Tymnet access number nearest to his or her location. This call could be made to one of approximately 1000 locations throughout the United States or to over 1300 locations abroad, which almost assures persons of the ability to access a Tymnet node with a local telephone.

The telephone call is answered by a minicomputer at the Tymnet mode, which places a familiar high-pitched tone known as the carrier signal onto the line. When you have connected to the network, Tymnet displays a request for the user to enter his or her terminal identifier.

The Tymnet terminal identifier is a letter ranging from A to G as well as I and P which is used to enable the packet network to optimize its performance to the user's terminal. In general, the terminal identifier tells the network the code of the terminal, is data rate and the number of pad characters that should be sent from the packet network to the terminal to compensate for the line feed/carriage return delay built into most electromechanical devices. On Tymnet, the identifier A is used to indicate that there are no line feed or carriage return delays, and it is normally entered by users accessing this network with a personal computer or CRT terminal.

If access to Tymnet occurs at a data rate other than 300 bps, the network's 'Please type your terminal identifier' message appears garbled. Once the terminal identifier is entered, Tymnet displays the remote access node number the user is connected to followed by the node's port number. This information is displayed in the form -NNN-PPP-, with NNNN being the node number and PPP the port number. After these numbers have been displayed, Tymnet will display the message 'please log in:'. At this point, the end-user would enter the network address of the facility he or she wishes Tymnet to route them to.

Once a session has been completed, Tymnet transmits a 'disconnected' message in the form of a call cleared. This is a packet-switching message which indicates that the connection previously established between Tymnet and a computer has been cleared. Then, you receive the Tymnet message 'please log in:' again. At this point, you can enter a new address or hang up your telephone to disconnect from Tymnet.

17.2.7 Network information

Many packet networks have an information facility that users can access without cost to obtain such information as their dial-up access telephone numbers, international access cost and locations served as well as other relevant information. For Tymnet users, this packet nbetwork's Information Directory can be accessed by simply typing INFORMATION after the 'please log in:' message has been generated by the network. Figure 17.10 illustrates the topics that users can select from the Tymnet Information Directory.

From the Information Directory, users can obtain valuable information. As an example, a person traveling to Copenhagen might access the International Access Information portion of the directory and request information about Denmark. Doing so, he or she would obtain the point of contact and telephone number of the Danish Post Telegraph & Telephone (PTT) representative that coordinates packet network facilities, as well as a price list in Danish Kroner of different types of packet network services available in that country.

17.2.8 Features

Table 17.2 lists nine of the more common features associated with the use of packet networks. Due to the inclusion of these features or added values, most packet-switching networks are also known as value-added carriers.

Automatic baud detection is the ability of the network node to sample the width of the data pulses and adjust its speed to the device requiring access to the network. Most networks support 0–28 800 bps asynchronous dial-in data transmission.

```
1. HELP IN USING THE INFORMATION SERVICE
2. DIAL-UP ACCESS INFORMATION
3. PUBLIC DATA BASE AND TIMESHARING SERVICES AVAILABLE OVER TYMNET
4. INTERNATIONAL ACCESS INFORMATION
5. X.25 PRODUCTS CERTIFIED BY TYMNET
6. COMMUNICATIONS PRODUCTS VERIFIED BY TYMNET
7. HOST TYPES CURRENTLY INTERFACED ON TYMNET
8. TYMNET SALES OFFICE DIRECTORY
9. TYMNET TECHNICAL AND USER DOCUMENTATION
TYPE THE NUMBER OF THE DESIRED MENU ITEM FOLLOWED BY A CARRIAGE
RETURN:
```

Figure 17.10 Tymnet information directory

Table 17.2 Packet network features

Automatic baud detection
Automatic code detection
Low transmission error rates
Reliable service
Efficient handling of peak traffic loads
Electronic mail facilities
Usage and accounting reports
International access
Protocol conversion

By the use of a terminal identifier the packet network may perform code conversion, permitting, as an example, a PC using the ASCII character set to communicate with a mainframe using the EBCDIC code. Since data are transferred between network nodes using a sophisticated error detection and correction method, very low error rates occur once data reach a packet network node. In addition, the mesh structure of these networks permits highly reliable service, since nodes are usually inter-connected by at least two paths and packets are automatically routed around any link that may become inoperative. Similarly, the mesh structure permits packets to be routed over less heavily used circuits, to increase the level of performance on the overall network when traffic on any one link exceeds a certain level.

Usage and accounting reports provided by value-added carriers can include a variety of information which is basically dependent on the type of features users select during the month. As an example, basic access through a packet network might result in a listing of network nodes accessed, destination node and duration of each network session to include the characters transferred during the session. If users employed other features of the packet network, including sending electronic mail or obtaining international access, the usage and accounting report would identify this. In general, the communications manager of an organization using a packet network to provide access to his or her computational facilities receives this report on a monthly basis. When used correctly, this report can assist the communications manager in assessing whether to upgrade or contract communications facilities as well as to determine if packet network usage costs warrant the consideration of an alternative networking procedure.

Since many packet networks basically cover the entire United States and provide international access from a large number of foreign locations, a logical development was the offering of electronic mail facilities to corporate users. Several packet networks were among the first commercial services to offer electronic mail. Such systems provide subscribers with an electronic mailbox, which is their address for receiving electronic messages. In addition, many electronic mail systems interface AT&T Mail, MCI Mail and other mail systems, permitting hard-copy messages to be generated electronically to distant locations where they are then distributed on conventional media.

Although electronic mail was a popular offering of value added carriers during the 1980s, the growth of the Internet and availability of email from America OnLine, CompuServe, and other so-called information utility organizations substantially reduced the demand for value added carrier email.

Today a large proportion of traffic transported by X.25 packet networks is represented by credit card transactions protocol conversion.

17.2.9 Protocol conversion

Due to the large base of asynchronous terminals and personal computers, one of the earliest protocol conversion efforts involved the development of software by value-added carriers to enable asynchronous devices to access IBM 3270 mainframe applications. In effect, value-added carriers added protocol conversion software to the mini-computers they installed at a customer's location which previously provided a PAD facility between a user's mainframe computer and the packet network. Typically, this software converted the line-by-line transmission of the terminal or personal computer into a full-screen format, including character attributes and screen-positioning information acceptable to the mainframe, making the mainframe think it was communicating with an IBM 3270 type terminal. When the mainframe transmitted a full screen of information, the protocol conversion software would convert the character attributes and screen positioning information into the appropriate character codes that the terminal or personal computer was built to work with. Due to the large number of different types of personal computers, most protocol conversion assumes that the personal computer operates as a DEC VT100 device. This simplifies the protocol conversion software development process since that software only has to translate character attributes and screen positioning information into the appropriate VT100 character codes. Then, any VT100 emulator operating on an IBM PC, Apple Macintosh, or other type of personal computer will be able to correctly communicate in full-screen mode of operation with a mainframe that believes it is connected to a 3270 type of display station.

With the market success of asynchronous to 3270 protocol conversion, value-added carriers began to offer users many additional types of conversion software. Figure 17.11 illustrates Tymnet's protocol conversion services that were offered by that vendor when this book was written.

Terminal protocol \ Host protocol	ASYNC	3270 BISYNC	3270 SNA	X.25	SDLC	2780/ 3780 HASP	2946
ASYNC	A	A	A	A			A
3270 BISYNC	A	A	A	A			
3270 SNA	A	A	A	A			
X.25	A	A	A	A	A		
SDLC				A	A		
2780/3780 HASP	A			A		A	
2946	A			A			A

Figure 17.11 Tymnet protocol conversion services (A = available)

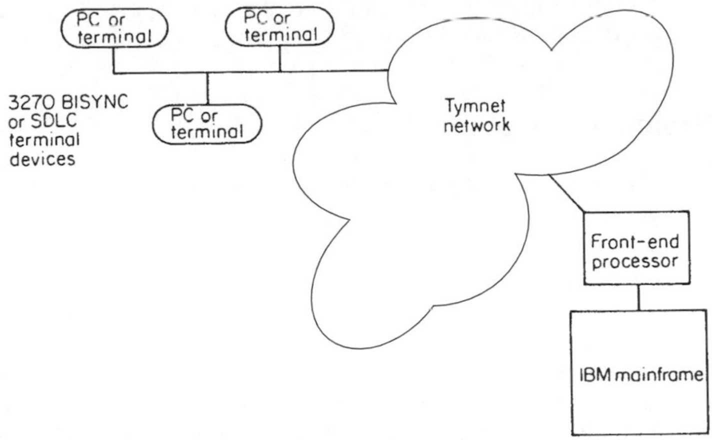

Figure 17.12 Multidrop network access through Tymnet

The Tymnet 3270 BISYNC and 3270 SNA host computer interfaces permit multidropped 3270 BSC and SDLC terminals to be connected to the user's mainframe via the Tymnet network. Figure 17.12 illustrates how a multidrop line connecting a series of BISYNC or SDLC terminal devices, including IBM PCs with SDLC adapter boards could be connected to a mainframe via Tymnet. One example of the economics associated with this networking strategy would be the situation where an organization has three terminals or personal computers located in three suburbs of Dallas that have to access the corporate mainframe in New York as 3270 devices. By routing the multidrop line from the three Dallas suburbs into a Tymnet node in Dallas, the cost of a leased line from Dallas to New York is eliminated. Although the user's organization will be billed for the interface at the mainframe and terminal node to support 3270 transmission, this cost may be less than the cost of a leased line between Dallas and New York.

Although the use of the Tymnet network shown in Figure 17.12 for multidrop access to an IBM mainframe appears to be simplistic, in actuality it involves both protocol conversion and protocol spoofing. Concerning the latter, the ability to support a multidrop line requires poll and select software support. However, the flow of polls through a packet network would result in a significant amount of wasted traffic, since the response to the vast majority of polls are 'no data to transmit'. Recognizing that most polls result in a negative response, Tymnet responds to polling at the node where the front-end resides, in effect spoofing the front end. Similarly, the node that connects to the multidrop line issues polls to each device on the line, removing polling packets from the network.

17.2.10 LAN interconnectivity

Although X.25 based packet switching services can be and are used to interconnect LANs their use can result in operational problems, especially when used to provide workstations with access to mainframes. Data flowing between packet network nodes are error-checked at each node, requiring the packet to be placed in memory so the cyclic redundancy check can be computed, and the status of the received packet, either error-free or containing one or more bit errors, is returned to the transmitting node. Error checking at each node adds a slight delay to the throughput of each packet through the packet network which cumulatively increases as packets are routed from node to node. Depending on the number of nodes traversed, the cumulative delay can result in session timeouts as well as adversely affecting internetworking performance.

X.25 packet network error checking on a node basis was a necessity when network infrastructures were based on the use of high speed analog leased lines to interconnect network nodes. The installation of tens of thousands of miles of optical fiber by communications carriers during the late 1980s and early 1990s facilitated a conversion of the infrastructure of packet networks to digital transmission facilities that have a significantly lower bit error rate than analog facilities. The resulting infrastructure made error checking on a node-by-node basis unnecessary and it resulted in the development and deployment of a new packet switching technology known as frame relay. Although many organizations migrated their LAN interconnections to frame relay, other organizations continue to use X.25 packet networks, because those networks continue to offer many features, such as protocol conversion, that are not obtainable from a frame relay provider.

17.3 FRAME RELAY

Frame relay represents a high-speed statistically multiplexed network interface standardized for the transmission of packets, unlike X.25 which operates at layers 1 through 3 of the OSI Reference Model. Frame relay only operates through layer 2 and can be considered to represent a streamlined version of X.25 packet switching.

17.3.1 Comparison to X.25

The key difference between X.25 and frame relay are in the areas of network access methods and operating rates, error handling, and throughput delays associated with packets flowing through each network.

Conventional X.25 packet networks support access via analog and digital transmission facilities to include dial and leased lines; however, network access operating rates are primarily limited to 56 kbps with some networks recently adding support for T1 access. In comparison, frame relay access is limited to digital leased lines, with network access operating rates primarily ranging from 56 kbps to 1.544 Mbps in North America and up to 2.048 Mbps in Europe.

The growth in the use of frame relay resulted in many organizations requiring network access at data rates beyond that offered by a single T1 line. Until 1999 frame relay subscribers requiring high speed network access above 1.544 Mbps would install multiple T1 lines from their office to a frame relay network node. During 1999 the Frame Relay Forum, an organization that promotes the use of frame relay, standardized an Implementation Agreement (IA) for the use of T3 lines to access frame relay networks. As frame relay network operators expand their support of T3 access, subscribers will obtain the ability to access such networks at a range of data rates from 56 kbps to approximately 45 kbps.

Through the elimination of layer 3 processing frame relay simply forwards packets on a hop-by-hop basis from the source node to the destination node. This significantly increases the throughput associated with frame relay networks, as well as explaining the naming of this type of network. That is, because frames are transported at layer 2 and they are relayed from node to node to their destination, the name frame relay as used as an identifier for this network technology.

Table 17.3 compares and contrasts TDM based circuit switching features with packet switching and frame relay. Although a comparison of several of the features listed in Table 17.3 are self-explanatory, some require a degree of elaboration. The virtual circuit capability feature refers to the ability of both X.25 packet switching and frame relay to establish temporary routes through their networks for the duration of a transmission session. Those logical circuits are established over existing physical circuits and numerous logical circuits can share the use of the bandwidth of each physical circuit. Under TDM circuit switching the bandwidth of circuits are switched to form a physical path of the duration of a transmission session and cannot be shared.

Under X.25 packets are error-checked at each node in the network. In addition, if too much data arrives at a switch and could cause congestion, the switch can regulate the flow of data via flow control. In comparison, TDM

Table 17.3 Comparing switching technologies

	Switching technology		
Feature	TDM circuit switching	X.25 packet switching	Frame relay
Access operating rate	any	<56 kbps	56 kbps–2.048 Mbps with T3 being introduced
Access method	switched or leased line	switched or leased line	leased line
Virtual circuit capability	no	yes	yes
OSI layer operation	n/a	1–3	1–2
Throughput	low to very high	low	high
Error checking at intermediate nodes	no	yes	no
Packet sequencing	n/a	yes	no
Flow control	n/a	yes	no, drop frames when congestion occurs
Switch delay	a few μs	10–40 ms	2–6 ms
One-way delay	a few μs	20–500 ms	40–150 ms
Suitability for voice and data	yes	no	yes

circuit switching establishes a fixed path transparent to data, avoiding the necessity of having to support both error detection and correction and flow control. In a frame relay network only the header is error checked at each node. If in error the frame is dropped. Under periods of congestion a network node switch will first warn devices in effect to 'slow down' via the setting of bit settings in frames. If congestion persists, the switch will then begin to drop frames. When a frame relay network drops frames it becomes the responsibility of higher layers in the application to recognize the absence of a query or response and retransmit a dropped frame.

Continuing our comparison TDM circuit switching results in circuits or portions of circuits representing fixed bandwidths of circuit capacity being interconnected to provide a transmission path. This results in a minimal delay time, as data are not examined for errors as they flow from node to node in a circuit switched network. In addition, the minimal delay associated with TDM circuit switching makes this technology very suitable for transporting voice and data. In comparison, the delays associated with X.25 packet networks make that technology totally unsuitable for carrying voice, as delays encountered by packets flowing through the network would distort a transported conversation. Although frame relay was not designed to transport digitized voice, several manufacturers of frame relay products introduced equipment to do so during 1994. During 1997 the Frame Relay Forum published its IA for Voice over Frame Relay (VoFR), which defines specifications necessary to enable products from different vendors to interoperate. By the new millennium VoFR represented a popular application for the use of frame relay networks in conjunction with traditional LAN interconnections.

17.3.2 Standards

Frame relay is based on a series of and ITU-T standards. The original formulation of the frame relay standard was based on efforts performed during the standardization of ISDN.

The communications protocol for frame relay is specified in ITU-T Recommendations I.122 and Q.922 which resulted in the addition of routing and relay functions to layer 2 of the OSI Reference Model. ANSI T1.606 describes the frame relay service and an addendum to that standard describes congestion management. ITU-T Recommendation I.233 is equivalent to ANSI T1.606 and ITU-T Recommendation I.370 is equivalent to the T1.606 addendum. There are several additional standards that define the operation of frame relay, to include Q.933 which is designed to support switched virtual circuits through a frame relay network, but which was implemented by only a handful of communications carriers when this book revision was prepared.

Although standards are very important, they are often not sufficient by themselves to obtain equipment interoperability. Recognizing this fact, the previously mentioned Frame Relay Forum periodically publishes Implementation Agreement (IA) documents which further define a particular facet of frame relay and enable vendors following the IA to have the equipment interoperate with other vendor products.

17.3.3 Network access

Figure 17.13 illustrates the attachment of two mainframes at one location and local area networks at two other locations to a frame relay network. The FRAD, an acronym for frame relay access device, functions similar to an X.25 PAD, converting a DTE data stream into frames capable of flowing on a frame relay network. The routers shown connecting the Token-Ring and Ethernet LANs to the frame relay network must be frame relay compatible. This is normally accomplished through the addition of an adapter card or the loading of software which provides the functionality of a FRAD within the router. Similar modular FRAD products enable bridges and even stand-alone personal computers to be directly connected to a frame relay network. In fact, IBM mainframes are normally connected to frame relay networks through frame relay compliant communications controllers that obtain the functionality associated with a FRAD via software.

The key difference between a FRAD and a frame relay compliant router concerns overall capability. While the FRAD is limited to assembling and disassembling frames, a frame relay compliant router has the capability to perform the functions of a FRAD as well as perform router functions. Due to this a router is usually but not always more expensive than a FRAD.

The actual line connections in Figure 17.13 linking the FRAD and routers to the frame relay network are digital circuits. Those circuits in North America can be 56 kbps DDS type facilities, ISDN lines, fractional T1, T1 or T3 lines. In Europe fractional T1 and T1 access lines would replace fractional E1 and E1 access. Although the use of DSUs and CSUs is not shown in Figure 17.13 for simplicity of illustration, they would be required to provide unipolar to bipolar conversion and digital line termination functions, as previously discussed in Chapter 8.

Now that we have an appreciation for the method by which data terminal equipment is attached to a frame relay network, let us turn our attention to how frame relay works.

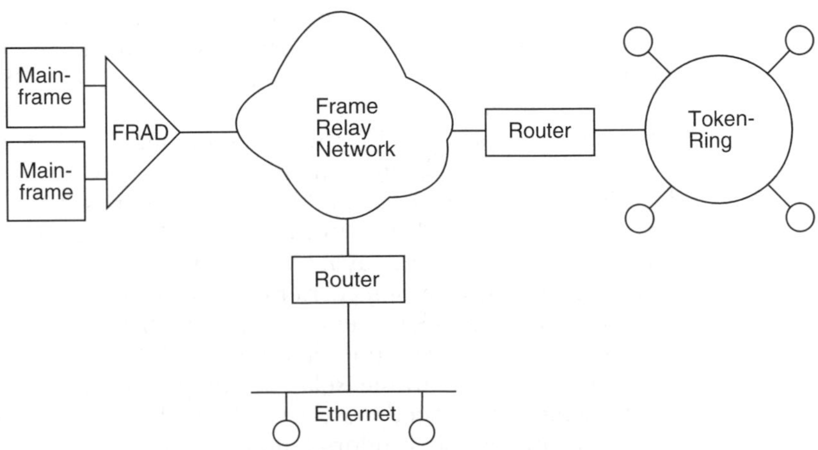

Figure 17.13 Accessing a frame relay network

17.3.4 Frame construction

The top portion of Figure 17.14 illustrates the construction of a frame relay frame with respect to the manner by which Layer 2 information is transported by the frame. The lower portion of Figure 17.14 illustrates the three types of frames that can be generated based upon the length of the frame Header field

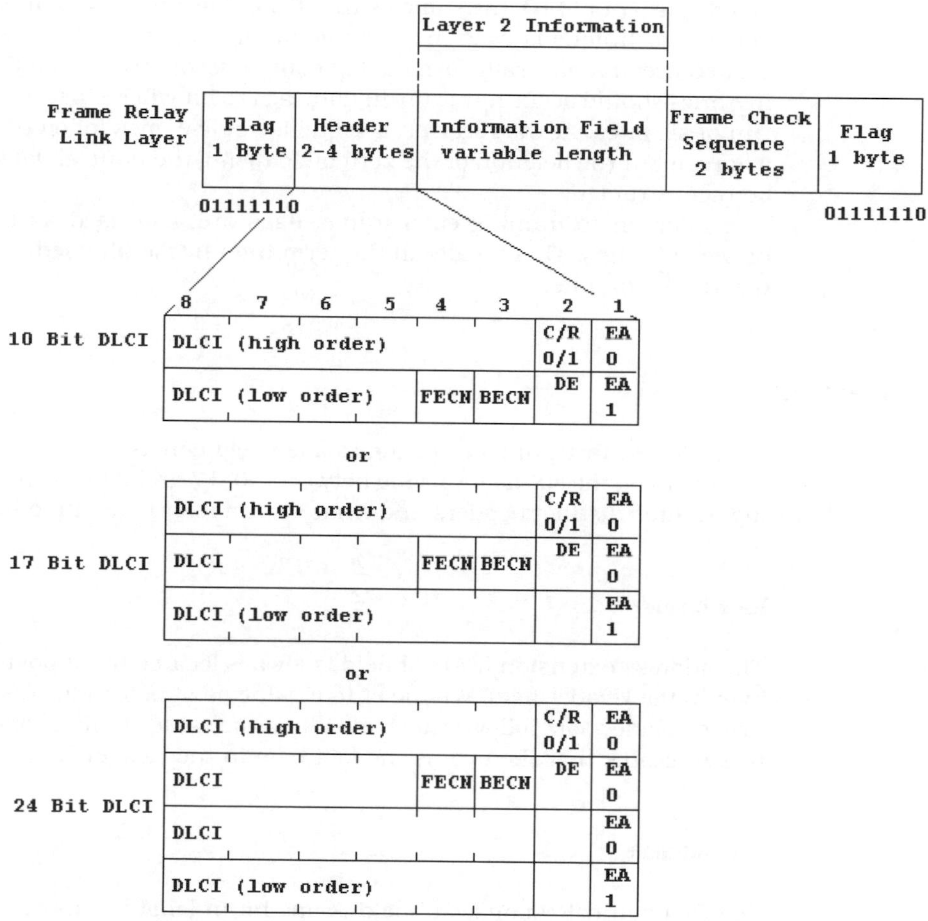

Legend:

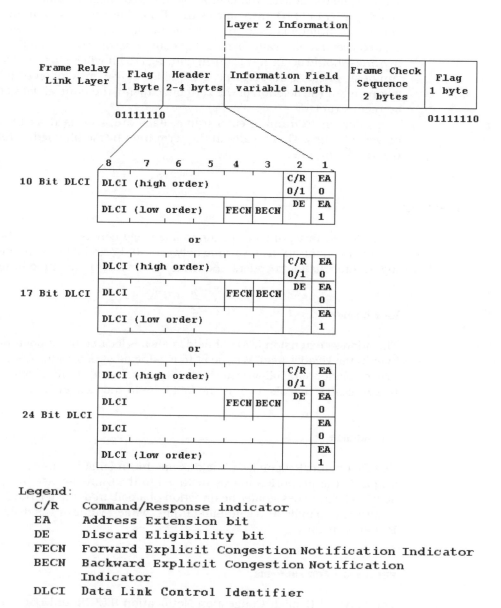

C/R	Command/Response indicator
EA	Address Extension bit
DE	Discard Eligibility bit
FECN	Forward Explicit Congestion Notification Indicator
BECN	Backward Explicit Congestion Notification Indicator
DLCI	Data Link Control Identifier

Figure 17.14 Frame relay construction and frame format

in bytes. To obtain an appreciation for the manner by which frames flow through a frame relay network, we will examine the use of each field as well as the composition of certain subfields and individual bit positions in the frame.

Flag fields

The Flag fields indicate the beginning and end of each frame. The bit pattern of a flag is 01111110, the same as the HDLC flag, and a zero-bit insertion and deletion technique is used in the same manner as for HDLC to eliminate the occurrence of a naturally formed flag being misconstrued as a flag. That is, if five ones should occur naturally in data, a zero bit will be inserted to prevent a run of six ones. At the receiver, a sequence of five ones followed by a zero bit will result in the deletion of the zero bit, causing the natural data sequence to be reconstructed.

In addition to framing each frame, flags are also used as fill characters between frames. This results in the term inter-frame fill used to reference its use in this manner.

Header field

As indicated in Figure 17.14, the Header field consists of six subfields, with the length of one subfield varying between 10, 17 and 24 bits in length, resulting in three distinct headers and what we can refer to as three frame formats.

EA subfield

The address extension (EA) subfield is always located in bit position 1 in each byte in the Header field. When set to a value of zero, it indicates that another byte of header data follows this byte. When the value of this subfield is set to 1, this indicates that the byte is the last byte in the Header field.

C/R subfield

The Command/Response subfield is one bit in length. This subfield is transparent to the network and was included in the frame for use by applications, such as SNA, to designate the direction of a poll. In doing so, the field is set to a value of 0 to indicate a command frame, while a value of 1 is used to denote a Response frame.

FECN and BECN subfields

The Forward Explicit Congestion Notification (FECN) and Backward Explicit Congestion Notification (BECN) subfields are each one bit in length. The FECN bit is set to a value of 1 by the network to inform the receiving station that the network is experiencing congestion. Similarly, the value of the BECN bit is also set to a value of 1 by the network. Here the network sets the BECN bit as a mechanism to inform the transmitting station that the network is congested in the opposite direction.

Congestion control

There are two main methods by which a frame relay network can control congestion. The first method involves switches monitoring their buffer occupancy or a similar metric, such as the ratio of inbound to outbound characters routed to another switch. When a predefined threshold is exceeded, the switch sets the FECN and BECN bits in the headers of frames passing through the switch. A second method by which a network can control congestion is for a switch experiencing congestion to notify switches at the edge of the network of this situation via the use of a Consolidated Link Layer Management (CLLM) message. A CLLM message is transmitted as a Layer 2 management connection using a DLCI value of 1007. The transmission of a CLLM message notifies edge switches in bulk of the fact that congestion has occurred, and the message contains a list of the virtual connections that are currently affected by congestion. Edge node switches will then directly set FECN and/or BECN bits as appropriate, or they can issue additional CLLM messages to the appropriate data terminal equipment to lower the input of frames with the affected DLCIs into the network.

Figure 17.15 illustrates the operation of a frame relay network with respect to the setting of FECN and BECN bits in frames as they flow through the network. In this example, network switch 2 is experiencing congestion and directly sets the bits. Here the cause of congestion could result from the data flow through switch 2 and from other network switches in the direction of the session shown between the sending and receiving stations flowing through switch 2. At a certain point in time the data flow through switch 2 in the direction of switch 3 might result in filling the buffers in switch 2. For example, assume the data flow from switch 1 to switch 2 destined to switch 3, as well as other network switches providing traffic towards switch 3, reaches 3 Mbps. If the connection between switches 2 and 3 is a T1 line operating at 1.544 Mbps, this means the buffer in switch 2 fills at the rate of approximately 1.5 Mbps. Thus, within a short period of time the buffer would overflow and frames would be lost. Recognizing this potential data loss, switch 2 carefully monitors its buffer occupancy and sets the FECN bit in frames being forwarded from that switch to their destination when the level of buffer occupancy reaches a predefined threshold. Similarly, switch 2 sets the BECN bit in frames transmitted by the receiving station to the sending station as a mechanism to inform the sending station that the network is congested in the opposite direction.

The criteria for the setting of FECN and BECN bits are left to the network operator. Thus, it is highly probable that different network operators will employ different criteria for the setting of those bits. Similarly, the response to those bits by data terminal equipment is left to the equipment developer.

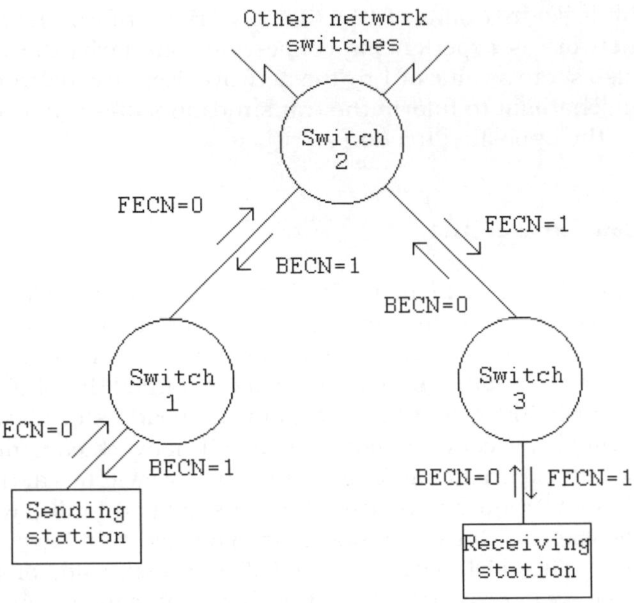

Figure 17.15 Network congestion and the use of FECN and BECN bits

In most situations, data terminal equipment representing the sending station will lower its transmission rate upon receipt of frames with their BECN bit set. At the receiving station the receipt of frames with their FECN bit set indicates to the receiving station that frames headed in its direction are experiencing or will experience congestion. Although the receiver cannot directly influence the flow of frames it is receiving, it can indirectly control the flow. To do so, the receiving station could slightly delay acknowledgments, which in turn would slightly delay the forward transmission rate towards the receiving device.

Under the second major method of congestion control, switch 2 would not directly set FECN and BECN bits in frames flowing through the switch. Instead, switch 2 would generate a CLLM message to edge switches, such as switches 1 and 3. Those switches would then either set FECN and BECN bits in frames with the appropriate DLCI, or issue CLLM messages to the appropriate data terminal equipment.

Differences between the two primary congestion control methods concern response time, data flow, and switch processing capability. When a switch sets the FECN and BECN bits directly, the sending and receiving stations obtain knowledge of congestion slightly faster than when CLLM messages are used. Concerning data flow, the ability to throttle the sending station directly from the edge of the network instead of possibly from within the network should theoretically improve network performance. However, since the throttling is slightly delayed, there may be no discernible improvement. Instead, by placing the effort of setting FECN and BECN bits on edge switches, the internal network switches may be relieved of some of the processing effort, which could improve packet flow through internal network switches.

DE subfield

The Discard Eligible (DE) bit provides a mechanism to indicate to the network a priority scheme for discarding frames during periods of congestion. That is, a DE bit value of 0 indicates to the network that the frame should not be discarded unless there is no alternative. In comparison, a DE bit value of 1 indicates that the frame can be discarded. To understand how and when the DE bit is set requires a detailed understanding of what is referred to as the Committed Information Rate (CIR). The CIR will be covered after we review the frame relay frame format. For now we will simply note that the DE bit is set by customer-provided equipment to identify frames generated by users that can be discarded before other frames if the network becomes overloaded. Thus, the DE bit enables the packet network to discard frames among all subscribers in an equitable manner by first operating upon those whose DE bit is set prior to randomly dropping traffic.

Although you might be tempted to think that frames with their DE bit set are similar to the 'mark of Cain', in actuality the vast majority of such frames flow through most frame relay networks. In fact, several network operators quote statistics which imply that over 99.9% of frames with their DE bit set flow through their network to their destination. Some frame relay service operators will provide service level guarantees to include guaranteeing that a certain percentage of all frames and those frames with their DE bit set will reach their destination.

In addition to discarding frames that have their DE bit set, switches will also discard erroneous and non-erroneous frames. Concerning erroneous frames, frames that are not bounded by opening and closing flags, frames that do not contain an information field, frames that have an invalid DLCI, frames that do not fall on a byte boundary, and frames that have an FCS error will be discarded. Non-erroneous frames are also subject to being discarded during periods of network congestion. However, non-erroneous frames whose DE bit is not set represent a worst case for frame discarding. That is, the network will first drop erroneous frames whether or not it is congested. Next, the network will drop frames whose DE bit is set during periods of network congestion. If the dropping of frames whose DE bit is set proves to be insufficient during periods of network congestion, then non-erroneous frames whose DE bit is not set will be randomly discarded.

DLCI subfield

The Data Link Control Identifier (DLCI) is used by the Frame Relay network to enable multiple applications to share the use of a common access line to the network. Although this function is similar to the X.25 LCN, unlike X.25 the DLCI does not specify a destination address. Instead, it simply represents a connection resulting from the establishment of a virtual circuit and has only local significance. Thus, the sending and receiving stations establishing a flow through a frame relay network for a common application may or may not use the same or different DLCIs.

The most commonly used DLCI consists of a 10-bit field which provides 2048 possible identifiers. Some DLCIs are reserved for specific management functions, such as CLLM messages that were previously discussed, as well as for other functions, a few of which are described later in this chapter. There are, however, approximately 995 DLCIs available for use by a frame relay compliant DTE, which should be sufficient for most organizations to identify the logical connections multiplexed onto the physical access channel.

DLCI utilization

To illustrate the basic use of DLCIs, consider Figure 17.16 which shows the connection of three FRADS to a frame relay network. In this example, let's assume a private virtual circuit (PVC) was established between Atlanta and New York, and a second PVC was established between Chicago and New York.

The PVC from Atlanta to Chicago is shown using DLCI = 16 to the serving network switch. The PVC is then shown using DLCI = 33 from switch 2 to the FRAD in Chicago. The PVC from Chicago to Atlanta uses DLCI = 34 in Chicago and DLCI = 61 in New York. It is important to note that each DLCI serves as a bidirectional PVC identifier. For example, DLCI = 16 would be used by the

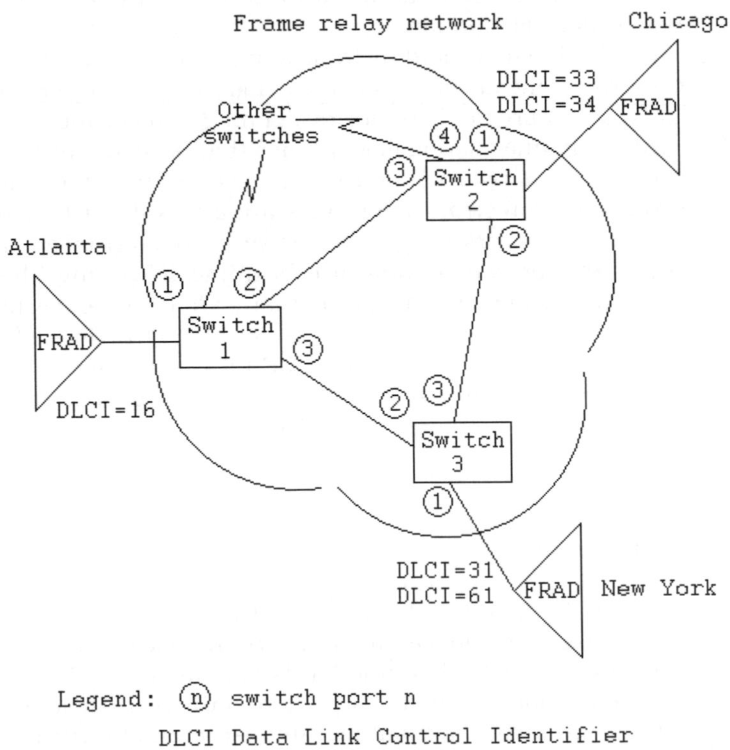

Figure 17.16 Sample frame relay network structure to illustrate DLCI assignments

FRAD in Atlanta as a mechanism to identify the local portion of the PVC routed through the network to the FRAD in New York. Similarly, the return PVC from New York to Atlanta would also use DLCI = 16 from switch 1 to the FRAD in Atlanta.

Switch operation

To illustrate how switches operate in a frame relay network, assume that the switch ports for the three network switches are numbered by the figures contained in circles in Figure 17.16. When a frame relay switch receives a frame, it performs a series of predefined operations. First, it recomputes the frame check sequence contained in the trailer. If the recomputed FCS does not match the FCS in the frame, the switch drops or discards the frame. At the same time the switch checks the DLCI for a valid value as well as the length of the frame. If the frame is too long, too short, falls on a boundary other than a byte boundary, or has an invalid DLCI, it is passed to the great bit bucket in the sky. Once this preliminary processing is completed, the switch examines its routing tables to determine which port the inbound frame will be transmitted on.

Figure 17.17 illustrates an example of possible routing table entries for switch 1 shown in Figure 17.16. Note that each switch routing table consists of a pair of port–DLCI values with each pair associated with input or output data flows. In the routing table shown in Figure 17.17, a frame received on port 1 that has a DLCI value of 16 will be transferred out of the switch via port 3 for routing to New York. The DLCI in the table is listed as 'xyz' to indicate that it is assigned by the service provider when the PVC was set up and may have no relationship to either of the local DLCIs in Atlanta and New York. Similarly, frames flowing into switch 1 from New York and destined to the FRAD shown connected to switch 1 in Atlanta in Figure 17.17 are input on port 3 and should be output on port 1. Since the service provider could use a different DLCI between switches 2 and 1, the entry in Figure 17.17 is shown as 'abc'.

The actual routing of frames via switches is a relatively fast operation. A switch simply uses the input port and DLCI as a mechanism to search its look-up table. Once an entry is found, the switch changes the DLCI in the

Switch 1 routing table

Input		Output	
Port	DLCI	Port	DLCI
1	16	3	xyz
3	abc	1	16
⋮	⋮	⋮	⋮

Figure 17.17 Frame relay switch routing table example

Table 17.4 DLCI assignments

DLCI number	DLCI assignment
0	Signaling
1–15	Reserved
16–1007	Assigned to user logical connections
1000–1018	Reserved
1019–1022	Reserved; used for some vendors for multicasting
1023	Identifies LMI control frame

frame to the value in the output portion of the table. However, since the DLCI is changed, the switch must recalculate the FCS and replace the prior FCS with the newly computed value.

DLCI assignments

Previously in this chapter we noted the use of a DLCI number for the transmission of a CLLM message. In actuality there are 1024 DLCI numbers available when a 10-bit DLCI is used, with DLCI numbers 16 through 1007 available for assignment to user logical connections. The remaining DLCIs are assigned as indicated in Table 17.4.

In examining Table 17.4, note that the Local Management Interface (LMI) in which all 10 DLCI bits are set to ones (1023 value) represents an extension to frame relay which permits multicasting, global addressing, unsolicited status updates by the network and other functions to be performed. Also note that multicasting is an optional LMI feature in which multicast groups are designated by four reserved DLCI values (1019 through 1022). Frames transmitted by one of those reserved DLCIs are replicated by the network and transmitted to all exit points in a designated set of points associated with the multicast group. Actual LMI multicast messages are used to notify user data terminal equipment of the addition, deletion and presence of multicast groups.

Information field

The Information field is the area of the frame that transports user data. As indicated in Figure 17.16, the length of this field is variable. Although some network providers support an Information field up to 4096 bytes, other network providers may support an Information field length of 512, 256 or even fewer bytes. The actual frame length is negotiated between a user's DTE and the network via the Local Management Interface. In addition to transporting user data, the use of a DLCI of 1023 to denote an LMI frame results in the Information field conveying LMI data. Thus, the actual type of data transported in the Information field, user or management data, depends upon the value of the DLCI in the frame.

FCS field

The frame check sequence (FCS) field contains a two-byte cyclic redundancy check (CRC) which represents the remainder formed by dividing the binary value of the frame by a fixed polynomial. Although each switch in a frame relay network will recompute the CRC and use it as a mechanism for deciding whether to forward or discard the frame, a CRC mismatch will not result in a switch issuing a request for retransmission. Thus, the burden of error recovery is placed upon end-user data terminal equipment, unlike X.25 which places the burden on both network switches and end-user data terminal equipment.

Now that we have an appreciation for the composition of the fields and subfields with the three types of frame relay frames and their use, let's turn our attention to the function of key frame relay service parameters. This will enable us to note how and when the DE bit in certain frames is set.

17.3.5 Service parameters

Frame relay standards specify a core set of key parameters which characterize the quality of service the provider supplies in response to the demand placed on the network by end-users. Key frame relay service parameters include the Committed Information Rate (CIR), the Committed Burst Size (Bc), the Excess Burst Size (Be), and the average time period (Tc) used for defining the CIR, Bc and Be. In this section, we will turn our attention to obtaining an understanding of each of these service parameters and their relationship to the setting of the DE bit in frames.

The CIR

The Committed Information Rate (CIR) denotes the minimum data transfer rate a subscriber is guaranteed upon the creation of a permanent virtual circuit established through the network. It is important to note that the CIR is not an instantaneous measurement of transmission but an average rate over time. For example, consider the use of a T1 line operating at 1.544 Mbps, used to provide access to a frame relay service. If the CIR is 256 kbps, this does not mean that the user cannot transmit more than 256 kbps. What it means is that the user will not transmit more than 256K bits in one second or 512K bits in two seconds. Although the distinction may appear trivial, it is extremely important and deserves a degree of elaboration, as CIRs can be assigned to each PVC and used to provide equitable access to the total bandwidth of a connection to a frame relay network.

To illustrate the use of CIRs, assume you are using a frame relay network to support inter-LAN and mainframe-to-mainframe communications. Let's further assume that you connect to a public frame relay network using a 512 kbps fractional T1 line and have the service provider supply each PVC, so that the mainframe-to-mainframe communication on one PVC has a CIR of 128 kbps and LAN-to-LAN communication on another PVC has a CIR of

384 kbps. This method of provisioning assigns a greater portion of the bandwidth to LAN-to-LAN communication; however, since that type of communication is bursty in nature it could shrink to 128 kbps, 56 kbps or even 0 bps, with the excess used by the mainframe-to-mainframe communications session. Thus, enabling another PVC to transmit at a rate above its CIR has a degree of merit, especially when the other PVCs are operating below their CIRs. Therefore, the mainframe-to-mainframe session can burst its transmission above its CIR when the LAN-to-LAN session falls below its CIR; however, excess transmission above the mainframe-to-mainframe CIR is not guaranteed.

Figure 17.18 illustrates how the bandwidth on the previously discussed access line could vary. The top portion of Figure 17.18 illustrates the guaranteed bandwidth for mainframe-to-mainframe and LAN-to-LAN communications based upon the previously noted CIR values. The middle portion of Figure 17.18 illustrates the data flow on the 512 kbps access line when there is a limited amount or no mainframe-to-mainframe communication and LAN-to-LAN communication bursts up to the full 512 kbps access line capacity. However, it should be noted that since the LAN-to-LAN PVC has a CIR of 384 kbps, the excess data rate above the contracted CIR, which can range up to 128 kbps, is subject to being discarded. Thus, the DE bit in frames representing a data flow exceeding 384 kbps will be set to indicate that the service

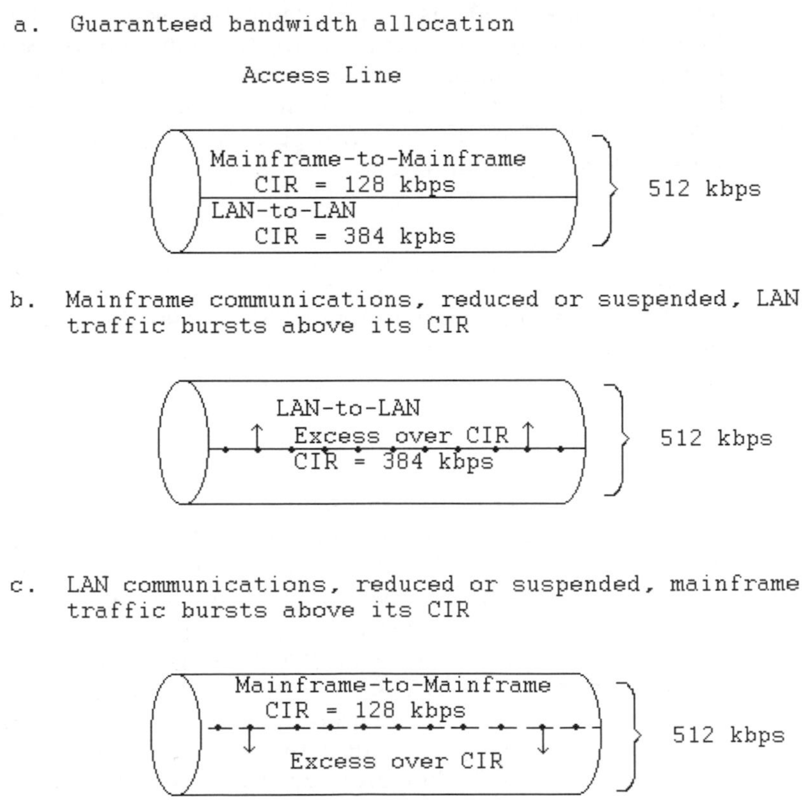

Figure 17.18 Varying bandwidth utilization on a frame relay access line

provider's network should consider those frames as eligible to be discarded during periods of network congestion. In actuality, the DE bit can be set either by the network or by the data terminal equipment used by a subscriber. The most common approach is for the network to set the DE bit in all frames that exceed the CIR of a PVC. However, subscribers can also indicate that frames from non-time-dependent data flows are discardable. This can be an important consideration when using a zero CIR whose use is discussed later in this section. The lower portion of Figure 17.18 illustrates the effect of mainframe-to-mainframe communication increasing over its CIR when LAN-to-LAN communication is reduced or suspended. Once again, when the data flow exceeds the CIR, which in this example is 128 kbps, all frames flowing on the PVC in excess of the CIR will have their DE bit set.

CIR computation

The CIR is computed based upon a measurement interval (Tc) and committed burst size (Bc). The measurement interval represents the period of time over which information transfer rates are computed and can vary from one frame relay service provider to another. The committed burst size (Bc) represents the maximum number of bits the network guarantees to deliver during the measurement interval (Tc). Thus, the CIR can be computed by dividing the committed burst size by the measurement interval, such that:

$$CIR = Bc/Tc$$

Carriers that use a small Tc value minimize the ability of users to burst transmission. In comparison, frame relay providers that use a relatively large value for Tc may be more suitable for users as they enable a higher degree of bandwidth aggregation over time by supporting longer bursts. Concerning transmission bursts, a third parameter known as the excess burst size (Be) is used by frame relay service providers to specify the maximum number of bits above the CIR that the network will attempt to deliver during the measurement interval, but for which delivery is not guaranteed.

Parameter relationships

Figure 17.19 illustrates the relationship between Tc, Bc and Be. Note that Bc + Be represents the maximum amount of information that can be transmitted during the time interval Tc. If a user transmits more than Bc + Be bits during the interval Tc, the network will immediately discard the excess frames. Many frame relay service providers set Be as the difference between Bc and the interface access speed to the network. Thus, this setting then results in the avoidance of frame discards at the entry node to the network, and results in (Bc + Be)/Tc becoming equal to the access speed.

In examining Figure 17.19, note that the actual information rate can vary over the measurement interval. As long as there is a sufficient amount of data to transmit, the actual information rate can never fall below the CIR and can burst above it to a maximum of Bc + Be.

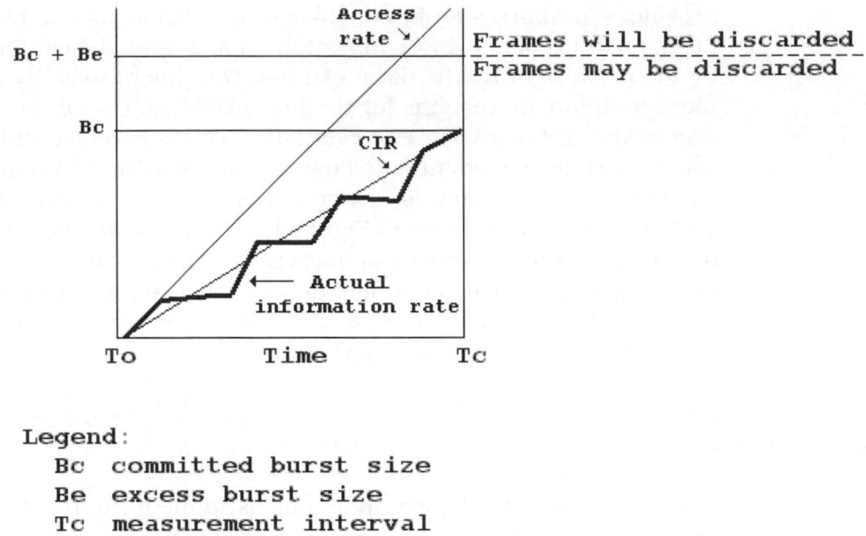

Figure 17.19 Committed Information Rate (CIR)

CIR assignment factors

There are several key factors that warrant consideration when selecting an appropriate CIR from a service provider. Those factors are listed in Table 17.5.

Service provider offerings

The first CIR assignment factor you should consider is the extent of service provider offerings. Although many service providers have similar CIR offerings, there are distinct differences between some providers. For example, some service providers offer CIRs from 0 in increments of 16 kbps up to certain fractional T1, full T1, or higher operating rates. In comparison, other service providers may not offer a zero (0) CIR and may provide CIRs in increments of 4 kbps or even 64 kbps. Since the cost per CIR is generally proportional to its operating rate, a slight difference in the CIR value selection factored over many PVCs can result in a considerable difference in the monthly cost associated with the use of a frame relay network.

Table 17.5 CIR assignment factors

Service provider offerings
Zero CIR selection
Aggregate CIR values to exceed the access line operating rate
Use asymmetrical CIRs
Use adjustable CIRs

Zero CIR selection

If you are setting up a PVC to transport only non-time-dependent data, you may wish to consider the use of a CIR of 0 if this option is supported by the network provider. The use of a CIR value of 0 means that the DE bit will be set in every frame flowing on the PVC. While this means that each frame is eligible for being discarded, it is important to realize that many network providers enable in excess of 99% of frames marked as discard eligible to flow end to end. Since the higher layer application will continue to retransmit discarded packets, data will not actually be lost. Instead, data may be slightly delayed until network congestion clears. With the monthly cost of a PVC with a zero CIR lower than the cost of a CIR with a higher value, you can reduce your organization's expenditure for frame relay service by selecting 0 CIRs for PVCs transporting non-time-critical data, such as file transfers and certain types of query–response applications.

Aggregate CIRs

One interesting technique that can be extremely beneficial when traffic bursts occur on different PVCs is to use a series of CIR values whose aggregated rate exceeds the line access rate. For example, assume a single access line operating at 256 kbps is used to support four PVCs, each with a CIR set to 64 kbps. This setup configuration allows each of the four PVCs to transmit up to the maximum rate of 256 kbps when the other PVCs are idle. However, when they do so the excess transfer over 64 kbps results in frames having their DE bit set. Now suppose you wish to support bursts to 128 kbps and feel that the probability of more than two PVCs operating at the same time is minimal. In this situation, especially if traffic is time critical, you could set the CIR of one or more PVCs to 128 kbps. Setting the CIR on each of the four PVCs to 128 kbps results in an aggregate CIR of 512 kbps while the access line operating rate remains fixed at 256 kbps. The ratio of the aggregated CIR to the line access rate is referred to as the subscription level. When the aggregated CIR value exceeds the line access rate, the term 'over-subscription' is used to reference this situation. Over-subscription can be extremely beneficial if data bursts can be expected to occur at different times during the day.

Asymmetrical CIRs

For interactive query–response traffic, such as clients on one LAN accessing servers connected to a distant network, the required bandwidth in each direction can be expected to differ. Thus, since some service providers support asymmetrical data rates, you may wish to adjust PVC CIRs to reflect the required bandwidth used in each transmission direction.

Adjustable CIRs

If your traffic is such that events can be preplanned, or you want to use frame relay for both production and backup purposes, you may wish to consider the use of adjustable CIRs. Some service providers allow subscribers to set their bandwidth requirements on a demand basis. This capability enables you to dynamically alter your organization's use of a frame relay service to satisfy changing network requirements. Of course, you can expect service providers to add a fee for this feature, so you should compare the fee against the flexibility it may provide.

Now that we have an appreciation for CIR assignment factors and the relationship between various frame relay metrics that define the operation of the network, let's turn our attention back to the FRAD. In doing so we will examine some of the key features of this communications product and discover how such features support different network techniques.

17.3.6 FRAD features

Today there are over 50 vendors producing FRADs. Thus, as you might surmise, the number of different features and the manner by which some of those features operate vary between vendor products as well as within the product line of a vendor. In this section we will focus our attention on key FRAD features you may wish to consider when evaluating different FRADs. In our examination of those features we will discuss and describe their use, providing information you can use to determine your organization's requirement for a specific feature.

Virtual circuit support

Until recently frame relay was restricted to the use of Permanent Virtual Circuits (PVCs) which represent logical, point-to-point connections established on a permanent basis through a service provider's network. PVCs are only permanent in the fact that they are not set up and torn down with each session. Instead, they remain in effect until a subscriber requests a change in the origination or destination of a circuit or terminates their contract. Recently, frame relay networks began to support Switched Virtual Circuits (SVCs). SVCs can be considered similar to a telephone connection, established on demand between two locations for the duration of a communications session, after which the SVC is torn down. By supporting several simultaneous PVCs, a FRAD can be used to provide a connection to multiple locations via a single access line. Since a Data Link Control Identifier (DLCI) is assigned by the service provider to identify each PVC, it is important to know the number of PVCs or DLCIs a FRAD can support.

Protocol support

Due to a lack of standards concerning protocol support, this feature can differ significantly between vendor products as well as within the product line of a vendor. Protocols supported by FRADs range in scope from Token-Ring and Ethernet to the transportation of voice using different voice compression modules built into some FRADs. One exception to the lack of standards is the Internet Engineering Task Force (IETF) Request for Comment (RFC) number 1490 entitled 'Multiprotocol Interconnect over Frame Relay'. RFC 1490 defines how protocol frames are encapsulated and, when necessary, fragmented into multiple frame relay frames. The fact that two products support the RFC, however, does not guarantee they are compatible and are capable of interoperating.

The ability of two FRADs to interoperate when supporting a particular protocol depends upon the manner by which other functions are implemented. For example, IBM's System Network Architecture represents a polled network with data transported by SDLC. To reduce the effect of end-to-end polling upon network bandwidth utilization, some FRADs suppress Logical Link Control (LLC) 802.2 'keep-alive' messages. Instead of passing those messages through the network, the FRAD locally acknowledges each 'keep-alive' message. In addition to providing more bandwidth for user data, the local acknowledgment of 'keep-alive' messages ensures that SNA sessions do not terminate if an acknowledgment flowing through the network is dropped or delayed. If one FRAD is set to locally acknowledge 'keep-alive' messages while another does not support this capability, you will obviously have an interoperability problem.

Congestion control

Although switches are responsible for alerting FRADs of the occurrence of network congestion, the response of FRADs can differ. FRADs generally throttle back traffic to the CIR rate upon the receipt of frames having their BECN bit set. However, some FRADs permit burst rate parameters to be configured as another congestion control mechanism. In the opposite direction the setting of the FECN bit indicates to a FRAD that it can lower its rate of acknowledgments to the transmitting station as a mechanism to alleviate congestion. However, there is no standard that requires FRADs to throttle back their rate of acknowledgment.

Hardware

FRADs are manufactured in a variety of configurations ranging from desktop and rack-mount units to products in tower PC-type cases. Figure 17.20 illustrates the rear view of a typical desktop FRAD designed to support a LAN and two serial communications devices, such as a directly connected control unit or front-end processor port.

When examining the device support capability of a FRAD, it is important to note the type and number of LANs supported as well as the protocols and data rates of its serial port support. Most FRADs support synchronous data rates of 2.4, 4.8, 9.6, 14.4, 19.2, 28.8, 38.4, 48, 56 and 64 kbps. Other FRADs support higher speed serial communications, commonly in multiples of 56/64 kbps to the T1 operating rate of 1.544 Mbps or the E1 operating rate of 2.048 Mbps.

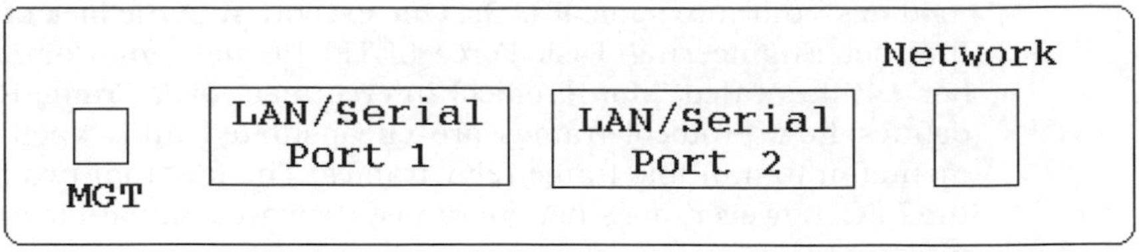

Figure 17.20 Common FRAD backplane

The network interface governs the type of frame relay access line connection supported by the FRAD. Some FRADs are limited to 64 kbps while other FRADs support T1 and E1. Some FRADs also include a high-speed interface which enables the connection of the FRAD to a T3 line operating at 45 Mbps, or enables an inverse multiplexer connected to two or more T1 lines to support data rates above 1.544 Mbps. In addition to switched 56 kbps and leased lines operating at 64 kbps and T1/E1 data rates, many FRADs also support ISDN connectivity via a Basic Rate Interface (BRI) capability.

Returning to Figure 17.20, the port labeled 'MGT' represents a management port by which the FRAD is configured. Most FRADs support the connection of an ASCII terminal or PC using a teletype emulator to issue configuration commands. Other FRADs support Telnet, enabling the FRAD to be configured via a workstation on a LAN.

Data compression

Until recently, the majority of data compression via FRAD transmission was accomplished by the use of routers, with a data compression feature either connected to a FRAD or directly connected to a frame relay network. Some routers incorporate hardware compression while other routers use software compression. For both methods data compression is proprietary to a particular vendor, which means that decompression at the other end of the network depends upon the use of an appropriately configured router. A second method of data compression that is also proprietary involves the use of stand-alone data compression units.

The use of stand-alone data compression units in a frame relay environment depends upon the protocols they support on both an input and an output basis. For example, if a FRAD supports SNA's SDLC protocol and a stand-alone compression unit operates on the information portion of SDLC frames, then its use is transparent to a FRAD.

A third method of compression is obtained by the use of FRADs that support the Frame Relay Forum (FRF) Implementation Agreement (IA) number 9. That IA defines how data compression should be performed to obtain a compatibility between devices implementing this function. FRADs that support FRF IA 9 can be expected to obtain a compression ratio between 2:1 and 4:1, with the actual compression ratio varying based upon the susceptibility of data to compression. Figure 17.21 illustrates the three previously described methods that could be used to compress data over a frame relay network.

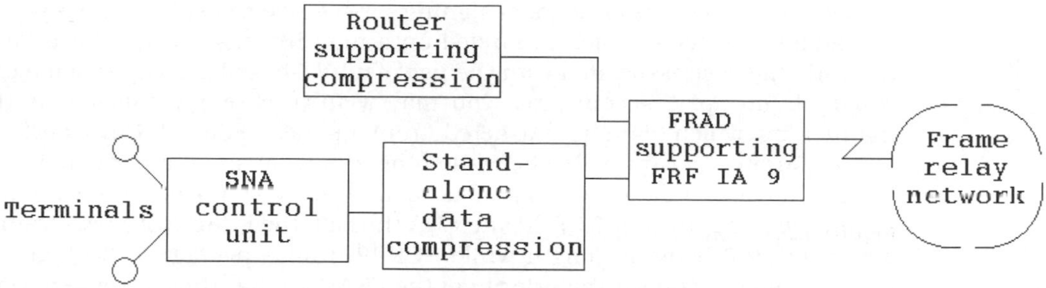

Figure 17.21 There are three basic methods that can be used to compress ata for transmission over a frame relay network. You can use a compression performing router, a compression performing FRAD, or a stand-alone data compression unit

Quality of Service

One common misunderstanding associated with frame relay is the fact that the network does not support Quality of Service (QoS). Instead, the ability to prioritize traffic is based upon proprietary schemes, ranging in scope from service providers offering prioritization mechanisms within their network to FRAD manufacturers incorporating priority schemes in their hardware products.

There are several schemes used by FRAD vendors to prioritize traffic. Each scheme is similar in that they are implemented by allocating memory buffers for defined data flows and then servicing the buffers in accordance with the priority scheme used. Since the buffers in effect are queues, we can say that FRAD QoS is based upon the assignment of data flows to defined queues and the extraction of data based upon a predefined criteria.

Common methods used to prioritize traffic include prioritization based upon port, IP address, PVC, protocol and frame length. Although not as common, some FRAD manufacturers provide an urgent queue capability. Here the use of an urgent queue results in traffic assigned to that queue always being serviced first regardless of the priorities assigned to other traffic. An urgent queue capability could effectively support real-time and video transported via frame relay since they cannot tolerate delays resulting from data being inserted ahead of their traffic.

The prioritization method provides a distinct number of service levels. Thus, it is important to note the number of service levels supported by a FRAD, and to compare that metric to your organization's requirements. Since a service level is implemented by the use of a buffer or queue, it is also important to recognize the buffering capability of a FRAD, and whether or not it can be expanded. Common FRAD levels of service for products offering a QoS range from 2 to 16, while buffering can range from kilobytes to megabytes. Thus, there is a considerable difference between certain vendor products with respect to the manner by which they implement a QoS. In addition, you may wish to carefully study their implementation method, as techniques that appear to be the same can have surprising results. For example, assume a FRAD vendor supports the reservation of traffic on PVCs, and you allocate

25% for video and 75% for data. Assigning service levels via PVCs forces you to use additional PVCs, which are billed on a monthly basis. Thus, for a large network, the cost associated with the use of a PVC-based priority scheme can rapidly build up. A second area you may wish to carefully consider is the manner by which data is extracted from queues. Some FRADs perform extraction on a per frame basis, while others may do so on a per byte basis. While this may not appear significant, the effect could be. For example, assume you are running TCP/IP and SNA through the FRAD, with SNA frames averaging 240 bytes in length while TCP/IP frames average 1600 bytes in length. Suppose you set the priority of the FRAD so that time-dependent SNA has a higher priority than TCP/IP. If the FRAD extracts data from queues on a frame basis, the result could be more time spent servicing TCP/IP frames even though you set the priority to favor SNA.

Packet fragmentation

Recognizing that not all data requires equal treatment, many FRADs include a packet fragmentation feature. This feature is normally associated with obtaining a Quality of Service by prioritizing certain predefined types of traffic and ensuring that when lower priority traffic is serviced, its frame length is adjusted so that it is sufficient to adversely affect time-dependent, high priority traffic.

Voice/fax support

A relatively recent addition to FRADs is the ability of some devices to support voice and fax transmission. To do so, the FRAD will commonly support the addition of different voice compression algorithms as well as priority and fragmentation techniques to enable voice and fax data to be prioritized over other data. In concluding this chapter we will examine the methods by which FRADs support voice and fax to include voice digitization techniques.

Management

There are several frame relay Management Information Bases (MIBs) supported by most FRADs. Most FRADs support SNMP and RMON. Through the use of MIB statistics you can manage both the use of FRADs and a frame relay network.

FRAD comparison shopping

To facilitate the comparison of FRADs, Table 17.6 summarizes the previously discussed FRAD features. In addition to listing those features that are beneficial for conducting an evaluation of vendor equipment, the table contains a column into which you can enter your organization's requirements for a

Table 17.6 FRAD features to evaluate

Feature	Requirement	Vendor A	Vendor B
Virtual circuit support			
PVC	_____	_____	_____
SVC	_____	_____	_____
Protocol support			
Token-Ring	_____	_____	_____
Ethernet	_____	_____	_____
SDLC	_____	_____	_____
Voice	_____	_____	_____
RCF 1490 compliance	_____	_____	_____
Other	_____	_____	_____
Congestion control			
BECN response	_____	_____	_____
FECN response	_____	_____	_____
Other	_____	_____	_____
Hardware			
Configuration	_____	_____	_____
Port support			
LANs	_____	_____	_____
Serial devices	_____	_____	_____
Network interface	_____	_____	_____
Management port	_____	_____	_____
Data compression	_____	_____	_____
Quality of Service			
Number of priority levels	_____	_____	_____
Buffer/queue extraction method	_____	_____	_____
Priority method			
Port	_____	_____	_____
IP address	_____	_____	_____
PVC	_____	_____	_____
Protocol	_____	_____	_____
Frame length	_____	_____	_____
Urgent queue	_____	_____	_____
Other	_____	_____	_____
Packet fragmentation	_____	_____	_____
Voice/fax support	_____	_____	_____
Management	_____	_____	_____

specific feature. Then you can use the columns labeled 'Vendor A' and 'Vendor B' to evaluate those two products against those requirements. If you need to evaluate more than two products, you can copy the page and overlap the right two columns to extend the number of vendor products to be evaluated.

Now that we have an appreciation for the features of a FRAD that should be examined during an equipment acquisition process, we will conclude our coverage of frame relay by turning our attention to the currently most talked about aspect of this network technology. The aspect of frame relay we will examine is the transmission of digitized voice over this network.

17.3.7 Voice over Frame Relay

Although not originally designed as a digitized voice transport mechanism, advances in voice digitization technology and the incorporation of prioritization techniques into vendor products resulted in many organizations turning to frame relay as a transmission scheme to support their voice, data and fax requirements. In this section we will focus our attention upon the rationale as well as the technology associated with the use of frame relay to transport voice along with data.

Rationale

Since many organizations have spare bandwidth on a PVC that connects geographically separated locations, the actual cost of implementing this technology can be limited to the cost of voice digitization equipment. Thus, it becomes highly probable for organizations to amortize the cost of equipment over several years, which can reduce the cost per call minute with sufficient traffic to below a penny a minute. Needless to say, this cost is quite enticing for many organizations and resulted in Voice over Frame Relay (VoF) becoming one of the more popular networking technologies being implemented by organizations during the latter part of the 1990s.

Overview

The ability to transmit voice over frame relay requires the use of equipment that recognizes the fact that voice must be treated differently from data. Such equipment can be in the form of a separate device capable of presenting a voice or fax digitized data stream to a FRAD, since many FRADS are limited to supporting data protocols. This conversion device commonly attaches to a PBX to accept an analog voice or fax or a digitized voice signal and compresses and digitizes the signal, generating a data flow in a protocol the FRAD supports.

Although many implementations of voice over frame relay can operate without modification to a FRAD or frame relay compliant router, this networking technique is not recommended unless your service provider and/or FRAD can perform priority queuing as well as packet fragmentation and several additional techniques described later in this section. Those techniques enable packets transporting voice to have priority over packets transporting data, to alleviate the potential for the transmission of data packets adversely delaying the transmission of a sequence of packets transporting digitized voice. This is necessary, as without priority queuing, fragmentation and several other techniques you can periodically expect to receive a degree of distortion at the receiver when the voice signal is reconstructed. Even with a queuing capability a FRAD may not be suitable for transporting digitized voice and data unless it has a frame fragmentation and jitter buffer, two techniques we will examine in detail later in this section when we discuss the operation of voice-compliant FRADs and routers. A fourth area that also requires consideration when transporting voice over frame relay is the voice digitization method used, since it significantly affects the bandwidth required to transport a human conversation. Concerning voice digitization, PCM produces a 64 kbps digital data stream which is more susceptible to delays through a frame relay network when carried in a large number of time-dependent frames than a lower digitization rate carried in fewer frames. Thus, many equipment developers now offer compression methods that can result in a voice conversation being carried by a 4 kbps to 16 kbps data stream.

In addition to stand-alone converters that provide an interface to FRADS, many vendors now market multi-functional FRADS and routers that perform analog to digital conversion and voice compression. Included in many products is a silence detection process that further enhances voice compression, and a frame sizing algorithm that limits the length of digitized voice, fax and data frames delivered to the network. This frame limiting algorithm minimizes end-to-end delay through the frame relay network being used and ensures that voice packets which are time sensitive are not unacceptably delayed behind fax and data packets that can tolerate a relatively lengthy delay.

Figure 17.22 illustrates the use of a multi-functional FRAD to fragment and prioritize the flow of voice, data and fax packets into a frame relay network. After a frame is created based upon a maximum length associated with the data it transports, it is sent into a priority queue (labeled Q in Figure 17.22). Since voice has the least tolerance for delay, frames carrying digitized voice would be placed at the top of the queue, while more delay-tolerant fax and data-carrying frames would be placed in lower priority areas in the queue.

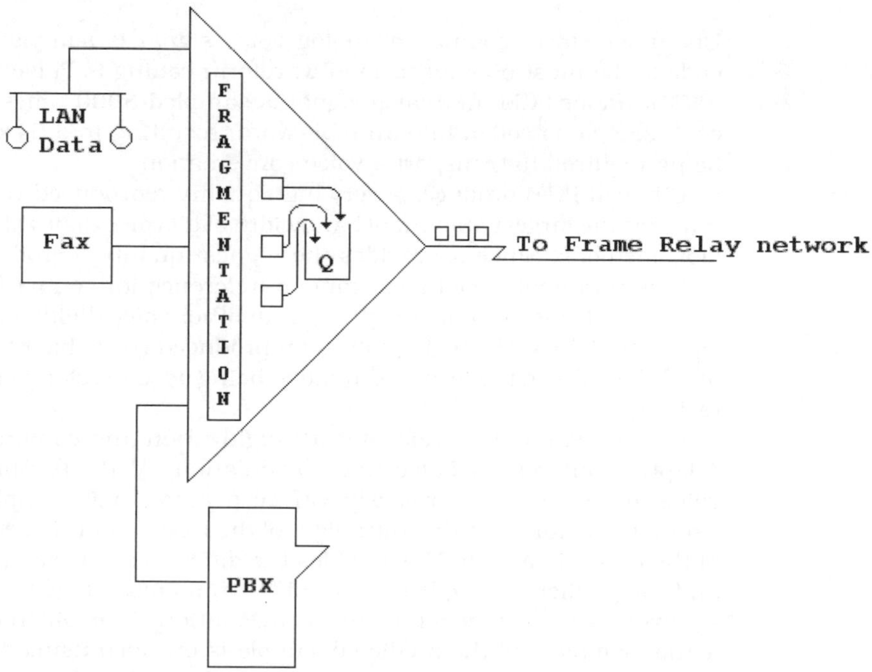

Figure 17.22 Using a multifunctional FRAD to support voice, data and fax

Now that we have an overview for the general technology required to support voice over frame relay, let's turn our attention to specifics and examine the technology in detail.

Technology

The ability to transport voice over frame relay requires the use of an efficient voice compression method as well as several FRAD or router features which facilitates the delivery of voice. In this section we will turn our attention to the technology required to transmit voice over frame relay effectively and efficiently, to include voice compression techniques and other features that are being incorporated into FRADs and frame relay-compliant routers.

Voice compression

The earliest method of voice digitization and compression was referred to as waveform coding. As we will shortly note, waveform coding produces the best quality of reproduced speech; however, it also requires the largest amount of bandwidth among all voice digitization and compression techniques.

Waveform coding

Under waveform coding an analog voice signal is sampled, quantized and coded. The most popular form of waveform coding is Pulse Code Modulation (PCM). Under PCM an analog signal is sampled 8000 times per second, and each sample is coded into an 8-bit word, resulting in a 64 kbps data stream being required to transport a voice conversation.

Although PCM produces a very high quality reproduced voice signal, it also requires the largest amount of bandwidth of all voice digitization and compression methods. Since it provides the highest quality reproduced voice signal, PCM is commonly used as a frame of reference for comparing the quality of reproduced voice signals resulting from other voice digitization and compression techniques. The high quality of reproduced voice that occurs from the use of PCM results in the term 'toll-quality' being used to reference this digitization technique.

Another common waveform voice digitization and compression method is Adaptive Differential Pulse Code Modulation (ADPCM). Since changes in a voice conversation occur slowly with respect to a high sampling rate, ADPCM uses a predictor to predict the height of the next sample based upon the height of the current sample. This enables the difference between the actual sample and the predicted sample to be used for transmission, with a predictor at the receiver used to regenerate the conversation. The difference between the actual sample and the predicted sample is encoded using 4 bits, which, at a sampling rate of 8000 samples per second, results in a transmission rate of 32 kbps.

Both PCM and ADPCM are commonly used by telephone companies and provide toll-quality voice reconstruction. Although support for both methods is incorporated into many FRADs and routers, their use requires a relatively large bandwidth. Since PVCs are priced based upon their CIR value, the use of PCM and ADPCM can be more expensive than other techniques that result in a smaller bandwidth being required to transport a digitized voice signal.

Linear predictive coding

A second category of voice compression and digitization techniques involves examining speech parameters, and coding those parameters for transmission. This method of voice compression is referred to as linear predictive coding (LPC). Although linear predictive coding techniques can result in a data rate as low as 2.4 kbps to transport a new voice conversation, the reproduction of the voice signal can result in a 'metallic' and unnatural sound.

Hybrid coding

The low digitization rates of LPC voice digitization methods were very appealing; however, the quality of reproduced speech was considerably below that of waveform coding techniques. Recognizing the strengths of each technique resulted in the development of a series of hybrid coding techniques that incorporate technology from both waveform coding and linear prediction. Hybrid coding uses speech analysis to develop a series of parameters, such as inflection, tone and energy, that represent a model of the voice conversation that can be used to predict speech in a manner similar to linear predictive coding. However, instead of transmitting those parameters directly as under LPC, hybrid coding techniques use those parameters to reconstruct speech at its origin and compare it to the recent sample. Then the hybrid coder will adjust one or more parameters so that there is a better fit between the samples that will be forwarded to the receiver and the actual sample. Although this introduces a slight delay, it significantly improves the quality of the reconstructed voice signal. Two of the more commonly used hybrid coding techniques are Code Excited Linear Prediction (CELP) and Algebraic Excited Linear Prediction (ACELP).

Several hybrid coding schemes have been standardized and are supported by many FRAD and router manufacturers. Those hybrid coding methods include Low Delay-CELP (LD-CELP) which provides a minimum degree of latency between encoded samples and transports a voice conversation at 16 kbps, and Conjunctive Structure ACELP (CS-ACELP). LD-CELP was standardized by the ITU as Recommendation G.728, and Conjunctive Structure ACELP (CS-ACELP), which transports a voice conversation at a data rate of 8 kbps, was standardized by the ITU as Recommendation G.729.

In addition to supporting standardized hybrid voice digitization techniques, some vendors support proprietary methods at data rates as low as 4.8 kbps. If you anticipate restricting equipment purchase to one vendor, then the use of a proprietary voice digitization technique can be considered. Low bandwidth, high toll-quality techniques can considerably facilitate transmission of voice over frame relay. However, if you anticipate obtaining a mixture of products from different vendors, you should avoid the use of proprietary techniques.

Silence detection and digital speech interpolation

Very rarely do two persons attempt to talk at once, and when they do the results are obviously less than satisfactory. Thus, voice communications are normally half-duplex. In addition, unless you speak like a spokesman did on a television commercial for an express delivery vendor several years ago, you use pauses between sentences. Recognizing these speech characteristics, many vendors use the silent period of one voice conversation for another active conversation. This technique is referred to as Digital Speech Interpolation (DSI) and its use can enhance bandwidth utilization by up to 50%. Thus, the use of a hybrid coding technique with DSI can result in an average bandwidth of 2.4 kbps per voice conversation, enabling up to six simultaneous voice conversations to be transported over a PVC with a 16 kbps CIR.

Now that we have an appreciation for voice digitization and compression techniques, let's turn our attention to other features that warrant consideration when evaluating the ability of a FRAD or router to transport voice over a frame relay network.

Frame fragmentation

As briefly discussed earlier in this section, it is important for a FRAD or router that transports voice to fragment all frames into relatively short segments. Fragmentation is necessary since a relatively long frame transporting data that falls between two frames transporting digitized voice could adversely delay the receipt of the second frame transporting voice. This in turn would result in a degree of distortion at the receiver, which would make the reconstructed conversation appear awkward.

Prioritization

When a FRAD or router accepts multiple inputs to include voice and data, it is important for the device to recognize that voice cannot tolerate any significant degree of delay. Thus, the ability to prioritize voice by placing frames transporting digitized voice in high-priority queues for servicing before frames transporting data that are placed into lower-priority queues is an important feature for FRAD and routers that will be used to transport voice over frame relay.

Jitter compensation

There are a variety of delays frames transporting voice will encounter as they flow from end to end. Some frames may be slightly delayed as they become interspersed behind other frames carrying either voice or data. Other frames may encounter a delay as they are serviced by one or more switches within a frame relay network. The result of these delays is that a uniform presentation of frames transporting digitized voice is received as a series of frames with random delays between each frame. If left unchecked, this could result in reconstructed voice sounding a bit awkward. Recognizing this problem, many vendors now incorporate jitter buffers into their FRADs and routers. The jitter buffer enables frames to be extracted in a continuous manner, resulting in a continuous and natural-sounding reconstructed conversation.

The VoFR IA

In May 1997, the Frame Relay Forum ratified its Voice over Frame Relay (VoFR) Implementation Agreement (IA) FRF.11. This IA defines standards for the transmission of voice over a public frame relay network. Included in the IA is the method by which compressed voice should be placed within the payload of a frame relay frame, how DTEs can signal each other concerning the use of a specific voice compression method, and a method for multiplexing up to 255 voice sub-channels on a single DLCI. As vendors develop FRF.11 compliant products, you can expect that additional voice transfer over frame relay will occur, to include the eventual use of frame relay in a manner similar to the public switched telephone network, with employees of one company contacting employees of another company via the use of frame relay.

17.4 REVIEW QUESTIONS

1. What key functions occur during the packet assembly process?

2. What is the primary function of the CCITT X.25 recommendation?

3. How does a character mode device communicate on a packet network?

4. Assume that the ASCII character] was assigned to X.3 parameter number 16. If you typed the message LOGOM]N, what would the contents of the PAD buffer be?

5. Describe the three locations where a PAD can be installed.

6 What is the primary advantage obtained by using a PAD manufactured as an adapter board in a PC to obtain a connection on a packet network?

7. Discuss the difference between functions performed by X.25 level 2 and level 3 software.

8. Suppose the *N*(R) field contains the value 6. What does this value indicate?

9. When referring to a packet network, what does the term public port signify?

10. Assume that the cost per minute of PSTN usage is 15 cents and ten terminals geographically dispersed from a computer center communicate with that center 2 hours per day, 20 days per month. If the monthly cost for connecting a mainframe to the packet network is $1000 and the cost per hour for packet network transmission is $3 including packet charges, prepare an analysis of the cost of using the PSTN as opposed to using the packet network. Which method of transmission is more economical? What happens if the number of terminals requiring access to the mainframe increases to 20?

11. Why do most packet networks that provide protocol conversion require personal computers to emulate a VT100 terminal?

12. Why is protocol spoofing important when a multidrop line is supported through an X.25 packet network?

13. Discuss the differences between X.25 and frame relay with respect to the network access methods and operating rates, error handling and throughput delays.

14. What is the general purpose or goal of a Frame Relay Forum Implementation Agreement (IA)?

15. Why are delays associated with a frame relay network lower than those associated with an X.25 packet switching network?

16. What is the purpose of a FRAD?

17. What is the main difference between a FRAD and a frame relay compliant router?

18. What is the function of the flag fields in a frame relay frame?

19. When is the address extension (EA) subfield in the frame relay header set to 1?

20. What is the purpose of the DLCI?

21. Who sets the FECN and BECN bits and in what situation are those bits set?

22. When is the Discard Eligible (DE) bit set in a frame relay frame?

23. What is the Committed Information Rate and how is it related to the line access rate?

24. Why would an organization select a Committed Information Rate of zero bps?

25. What is the relationship between the Committed Burst Size (Bc), the Committed Information Rate (CIR) and the measurement interval (Tc)?

26. Describe the three basic methods that can be used to compress data for transmission over a frame relay network.

27. Describe three methods that could be used by a FRAD to prioritize traffic.

28. What is the advantage associated with the use of hybrid coding for digitizing voice for transmission over a frame relay network?

29. Why is frame fragmentation necessary when transmitting voice over frame relay?

30. What is the purpose of a jitter buffer built into a FRAD?

18

COMMUNICATIONS SOFTWARE

The effect of personal computers on users requiring desktop computation is probably as pronounced as that of the telephone and photocopying machine on all business users. Today it is difficult to find a business, educational institution or government agency that does not have a large and growing base of personal computers. Accompanying this growing base of personal computers has been a corresponding requirement for communications software. In this chapter, we will first review the major communications software features that should be considered in selecting an appropriate program to satisfy different types of telecommunications requirements. Then, using the discussion of these parameters as a foundation of communications software knowledge, we will examine the operation of several commercial general-purpose communications programs as well as the rationale and utilization of programs that provide the user with specific terminal emulation capability.

In our examination of communications software we will focus our attention upon three specific types of communications programs. First, we will examine programs that enable a PC to function as an interactive remote terminal, a technique that was very popular during the 1980s and 1990s but has declined in use as the Internet and its World Wide Web (www) grew in popularity. However, even though the use of interactive terminal emulation programs has declined, they continue to be used by many businesses and individuals.

The second type of communications program we will examine represents a specific type of emulation program developed for providing access to mainframe computers.

In concluding our examination of communications programs, we will focus our attention upon the Web browser. Although its design is very user friendly, it includes many communications-related features that warrant our attention. Because those features differ from other communications programs, we will examine the browser as a separate entity in this chapter.

18.1 TERMINAL EMULATION SOFTWARE FEATURES

The variety of uses to which personal computers can be put is limited only by the user's imagination and by the machine's physical constraints. Although

Table 18.1 Common features of communication programs

	Utilization performance							
	Hardware	Software	Operational	Dialing	Transmission	Efficiency	Flexibility	Security
Automatic operations supported				•				
Buffer control					•			
Character-transfer support					•			
Clock display			•					
Communications port selection							•	
Control-character transfer						•		
Copy protection								•
Data compression						•		
Dialing directory				•				
Dialing from the keyboard				•				
Dialing method (tone-pulse)				•				
Disk-directory access						•		
Display-width selection							•	
Documentation			•					
Drive use and requirements	•		•					
Editing capability						•	•	
Encryption capability								•
Error detection and correction						•		•
Exit to OS command level							•	
File closing						•		

File-transfer capability					•		
File-viewing capability							•
Foreground–background operations						•	
Journalization							•
Line-feed control							•
Lowercase–uppercase conversion							•
Memory requirements	•						
Mid-session prompts–help capability			•				
Mode selection (menu–command)			•				
Modem compatibility	•	•					
OS command use		•					
OS version requirements		•					
Pacing capability					•		
Preconfigured dialing directory				•			
Printer control		•	•				
Programmable-key capability						•	
Programmable macros						•	•
Protocol setting							•
Screen-dump capability						•	•
Selectable data rate						•	•
Stripping–converting characters					•		
Switchable printer ports							•
Terminal emulation					•		
Transmission mode					•		
Unattended operation capability						•	

word processing and electronic spreadsheet programs are by far the most common applications, no less important is the machine's ability to communicate information. In fact, the range of personal computer applications that depend on data communications is so pervasive that it is hard to imagine any PC without the ability to communicate.

One of the main problems facing personal computer users is the selection of appropriate communications is software. Today, literally hundreds of such programs are marketed, ranging in scope from simple line-by-line TTY emulators to programs that mimic specific types of terminals. Each program includes such features as file transfer, unattended operation, and automatic sign-on and sign-off procedures for accessing bulletin boards and mainframes.

Different programs offer a variety of features that the user may wish to consider. Table 18.1 lists 45 of the most common features of terminal emulation communications programs in alphabetical order. This table can serve as a checklist in comparing specific programs. Alternatively, users with the opportunity to test or analyse programs prior to purchase may wish to rate the capability of each package according to those features deemed important.

Many of these features are interrelated in function. Some programs may not treat them separately but may incorporate them into another utility. Thus, the specifications and operational characteristics of each program should be investigated thoroughly.

18.1.1 Hardware utilization

The first set of features to be considered includes those that the user might first encounter when examining communications programs. An initial question is, or should be, 'Which program will work on the hardware available?' Memory, the number and type of disk drives and storage media, and the presence of a hard disk will affect how a communications program can be utilized.

Drive use and requirements

Text-based communications programs can operate on PCs with either a limited amount of hard drive storage capacity or, in some cases, antiquated computers limited to using diskette drives for data storage. In comparison, some modern graphic user interface based communications programs which include support for fax modem operations can require 10 to 15 Mbytes of data storage and can only operate on computers with a hard drive. and a significant amount of memory.

Memory requirements

Most communications programs have memory requirements ranging from 64K to 640K which, when the computer is only used on a single application at a time basis does not represent a problem. However, if the computer is connected to a LAN and uses an operating environment, such as Windows 3.X or Windows 95, the amount of conventional memory used by the communications program has a direct bearing on the ability of the computer to support multiple applications. This is because the first 640K of PC memory is used by many older types of software programs and LAN drivers. The usage of that memory area is physically limited to 640K, which means that at a given point in time conventional memory will become fully utilized, precluding its use by another program until the use of some programs is terminated.

Modem compatibility

Communications programs may support a specific type of intelligent modem, several such modems, or both intelligent and non-intelligent devices. Compatibility with an existing modem can be ensured by checking the command set supported by the communications program. In some cases the communications program will list a large number of specific modems supported. The command set of the user's modem should be checked against those supported by the program.

Printer control

The ability to turn a printer on and off is supplied by most communications programs. Since an attached letter-quality printer operates at a slower rate than a dot-matrix printer, such a device would probably not be able to keep pace with a long response or a long file transmitted by the distant machine. To overcome this limitation, some programs use a segment of memory as a printer spooler, enabling the simultaneous printing and down-line loading of data when the printer cannot keep up.

If the communications program does not have this capability, or implements flow control, an alternative is to transfer a file and then list its contents. Without a printer spooler, the user can only print and receive data simultaneously when the speed of the printer exceeds the data-transfer rate of the communications line. Otherwise, the receive buffer in the communications program will probably overflow, resulting in the loss of data. Data-flow overrun may even cause a program to terminate the session.

18.1.2 Software utilization

Clearly, the proper hardware is required to run a communications program on a personal computer. Just as important is the program's compatibility with the machine's software environment, in particular its operating system (OS).

OS version requirements

Many communications programs are restricted to operating under a specific computer operating system. For example, HyperAccess has separate program versions for use with DOS, Windows and OS/2. Although Windows 95 and Windows 98 have DOS 'compatibility', permitting you to operate DOS programs, on occasion performance can be degraded, especially if the communications program is used for high speed operations at 9600 bps or above. To alleviate this problem you may wish to consider either running the program directly under DOS or obtaining a version of the program developed specifically for the operating system you intend to use.

OS command use

Many communications programs let the user execute operating-system command through the communications program. Programs that have this feature allow users to delete and rename files, display disk directories, format a disk, or perform other operations without first leaving the communications program. With the ability to use operating-system commands, files can be deleted from a storage disk or a new disk formatted, without leaving the communications program.

18.1.3 Operational considerations

One communications program category of features frequently overlooked is the interaction of the program with the user. Many program features have a pronounced effect upon the efficient operation of the program by the user. In addition, a communications package must also be compatible with the user's level of expertise and need for convenience, clarity and ease of use.

Mode selection (menu/command)

For the first-time user, a menu-driven or GUI-based communications program makes the various options easier to understand and to use. Users may, however, reach a level of expertise where it is simpler to type commands than to go through a series of menus to accomplish a given task. A the opposite end of the spectrum are command-driven programs. These programs may be more difficult to master but, once learned, more efficient than their menu-drive counterparts.

To strike a balance through the advantages and disadvantages associated with menu- and command-driven software, some programs let the user select the mode of operation. For example, the user might switch from menu-driven to command-driven mode after obtaining a degree of familiarity with the program.

Mid-session prompts/help capability

Some communications programs employ extensive prompt messages that make it difficult for the user to set an option incorrectly. Others provide

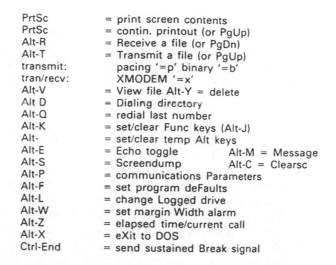

PrtSc	= print screen contents
PrtSc	= contin. printout (or PgUp)
Alt-R	= Receive a file (or PgDn)
Alt-T	= Transmit a file (or PgUp)
transmit:	pacing '=p' binary '=b'
tran/recv:	XMODEM '=x'
Alt-V	= View file Alt-Y = delete
Alt D	= Dialing directory
Alt-Q	= redial last number
Alt-K	= set/clear Func keys (Alt-J)
Alt-	= set/clear temp Alt keys
Alt-E	= Echo toggle Alt-M = Message
Alt-S	= Screendump Alt-C = Clearsc
Alt-P	= communications Parameters
Alt-F	= set program deFaults
Alt-L	= change Logged drive
Alt-W	= set margin Width alarm
Alt-Z	= elapsed time/current call
Alt-X	= eXit to DOS
Ctrl-End	= send sustained Break signal

Figure 18.1 PC-TALK III command summary

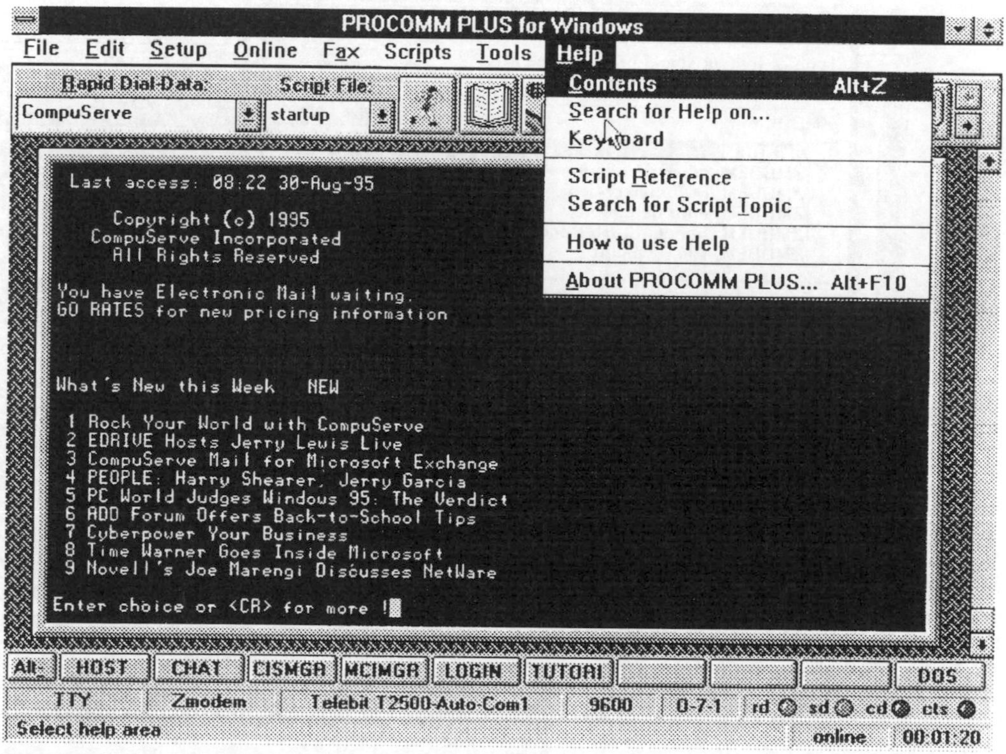

Figure 18.2 The Procomm Plus for Windows Help menu

limited prompting but may compensate for this lack with extensive help or assistance displays that appear in response to a predefined key sequence.

Figure 18.1 illustrates the PC-Talk III command summary that is displayed in the upper right-hand portion of the screen when the Home key is pressed on the keyboard. Since this program is command-driven, it is quite easy to forget many of the commands that are not used frequently. This help screen, while limited to a display of the available commands, may be sufficient for many users.

Other programs developed for use under a GUI environment provide either context-specific assistance that provides hints concerning operations that you are attempting to perform or permits you to select a specific topic. Once selected, the program will display a window containing information about the requested topic.

Figure 18.2 illustrates the ProcommPlus for Windows Help menu, and Figure 18.3 illustrates the dialog box resulting from selecting the 'Search for Help on' entry from the help menu and entering the word ZMODEM. Note that the program has a number of ZMODEM-related topics which are selectable for display. Once you select a topic and click on the Go to button, the program will display information about the selected topic.

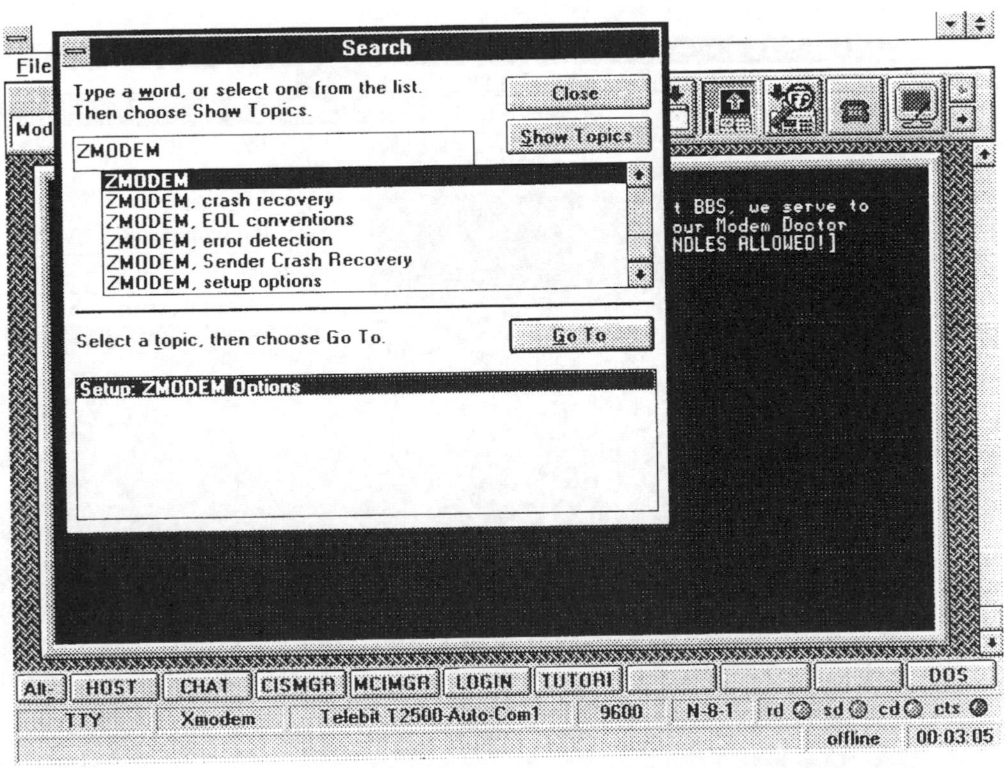

Figure 18.3 Using the Procomm Plus for Windows Help facility to obtain information about ZMODEM

18.1.4 Documentation

It is difficult to make full use of a sophisticated communications program without good documentation that indicates explicitly how each feature works. In addition, a good index is imperative, since thumbing through a manual wastes both time and communications dollars when the user is connected to a remote service.

Clock display

Like a fancy box of chocolates, interactive communications can easily become addictive. Calling long distance or using a value-added carrier that charges a fee based on the duration of the connection can quickly escalate communication costs. Search charges from information utilities or data-retrieval services may exceed $20 per connect-hour. Therefore, a mechanism to display the duration of a call can serve as a reminder that it may be time to terminate the session.

Many communications programs display the duration of a session. Some programs automatically display the elapsed time, and others display the time only when the user enters a predefined command or control code. Although this feature may be deemed frivolous by anyone with a wrist watch, once used, it can become like that indispensable box of chocolates.

An example of the clock display is shown in the lower right portion of Figure 18.2. At the time that the Help menu was selected this author had been on-line for 1 minute and 20 seconds.

18.1.5 Dialing

Once a software program has been found compatible with your requirements, it will almost certainly be used to set up a call over a switched public telephone network. Initiating a call involves dialing, which may be from a preconfigured or a user-developed dialing directory. Dialing has a number of aspects which are discussed in the following sections.

Automatic operations supported

A communications program may support a wide variety of automatic features. These can be as simple as control of an attached modem, or as sophisticated as automatic sign-on to an Internet Service Provider or information utility. Such sign-on tools are actually programmable macros, and will be considered in a separate category. Control of an attached modem, however, lets a program automatically dial, answer calls, redial numbers that are busy, and hang up the telephone when the session is over or when a call in progress ends due to loss of carrier.

Due to the popularity of the Hayes Smartmodem, most vendors marketing intelligent modems designed their products to be compatible with the Hayes command set. However, the Hayes compatibility actually means that only basic commands such as dialing, redialing and answering calls are supported. Many modems have proprietary 'extended' command sets to support certain functions, such as different error detection and correction and data compression methods. Thus, it is important to compare the modem support of the program that you are considering to the modem that you intend to use. Although most modern communications programs may support up to 100 or more different modems, that support may not include all of the features of each modem. Due to this, you may wish to use the modem manual and the manual setup feature of the communications program to enable specific modem features you may require.

Dialing method (tone/pulse)

Most intelligent modems work with both touch-tone and pulse telephone-dialing methods. Therefore, to make the most of an intelligent modem, the communication program may translate the dialing method requested into an appropriate modem command. Without this capability, it may be necessary to adjust a modem setting in order to change the dialing method.

Dialing directory

Programs can incorporate dialing directories ranging in scope from a simple listing of telephone numbers to a set of sophisticated entries, in which each

= = =Dailing directory 1= = =   Modem dialing command = ATDP
Long distance service +* =

Name	Phone *	-*= Comm Param	Echo	Mesg	Strip	Pace
1-FEINET	971-1001	300-E-7-1	N	N	N	N
2-FEINET *2	527-1800	300-E-7-1	N	N	N	N
3-UVA *1	924-0280	300-E-7-1	N	N	N	N
4-UVA *2	924-7401	300-E-7-1	N	N	N	N
5-MACON	2053	1200-E-7-1	Y	N	1	N
6-----------------	- --- --- ----	300-E-7-1	N	N	N	N
7-----------------	- --- --- ----	300-E-7-1	N	N	N	N
8-----------------	- --- --- ----	300-E-7-1	N	N	N	N
9-----------------	- --- --- ----	300-E-7-1	N	N	N	N
10----------------	- --- --- ----	300-E-7-1	N	N	N	N
11----------------	- --- --- ----	300-E-7-1	N	N	N	N
12----------------	- --- --- ----	300-E-7-1	N	N	N	N
13----------------	- --- --- ----	300-E-7-1	N	N	N	N
14----------------	- --- --- ----	300-E-7-1	N	N	N	N
15----------------	- --- --- ----	300-E-7-1	N	N	N	N

Dial Entry *: | or . . .   Enter: R to revise or add to directory
M for manual dialing
F/B to page through directory
X to exit to terminal
| For long distance service, precede entry * with +/−

Figure 18.4 PC-TALK III dialing directory

entry includes the number to be dialed, a text description of the destination and parameter settings. Figure 18.4 lists a partially completed dialing-directory screen from the PC-Talk III communications program. This program was among the first which provided communications for the original IBM PC, and its dialing directory structure formed the basis for similar structures used with many text-based programs developed during the 1980s. Up to four screens of 15 entries each, or a total of 60 dialing entries, can be maintained on a file created and accessed by the program.

Initially, the directory is empty. Default communications parameters are 300 bps, even parity, seven data bits and one stop bit, since this is a common setting for accessing mainframes and information utilities. As users add entries to the dialing directory, they can easily change the default parameters as well as the four rightmost columns, which affect the echoing of transmitted characters to the screen, letting a remote personal computer user know the status of file transfers (messaging), replacing of received characters (stripping), and how frequently new lines of data are sent (pacing). Once the parameters are set, selecting a particular entry tells the program to dial the number and to use the entry's parameters.

Figure 18.5 illustrates the Procomm Plus for Windows Dialing Directory. This GUI-based Windows program represents a significant evolution from the capabilities and functionality of PC TALK III. From the Procomm dialing

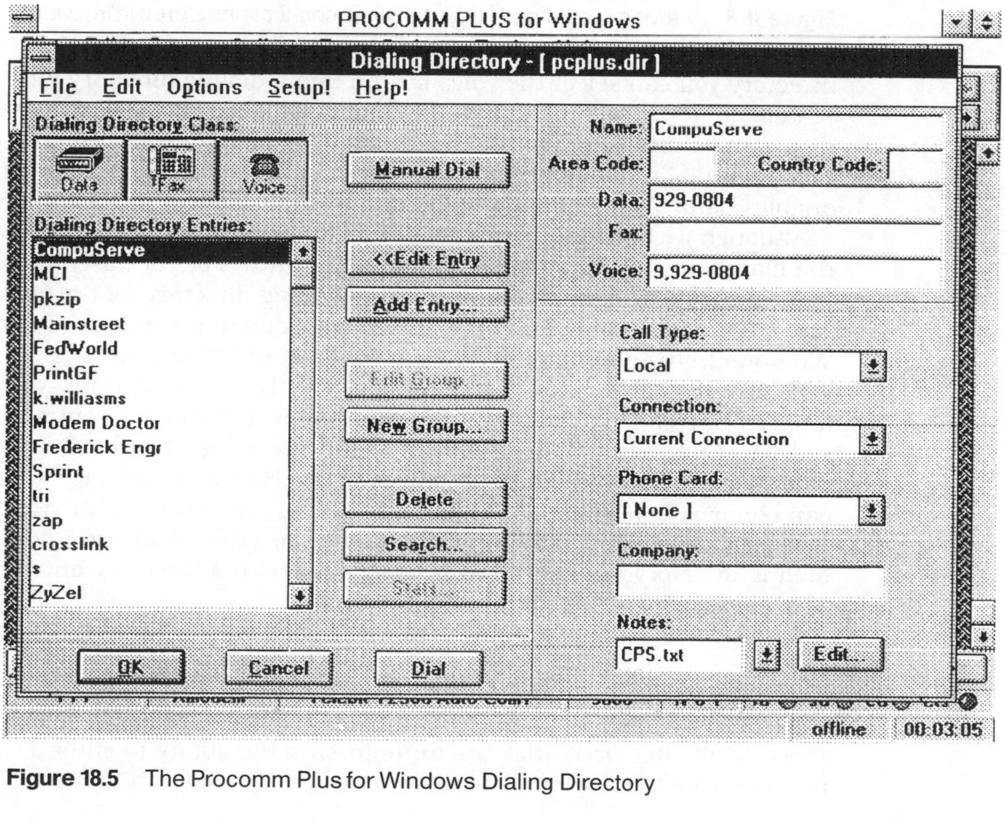

Figure 18.5 The Procomm Plus for Windows Dialing Directory

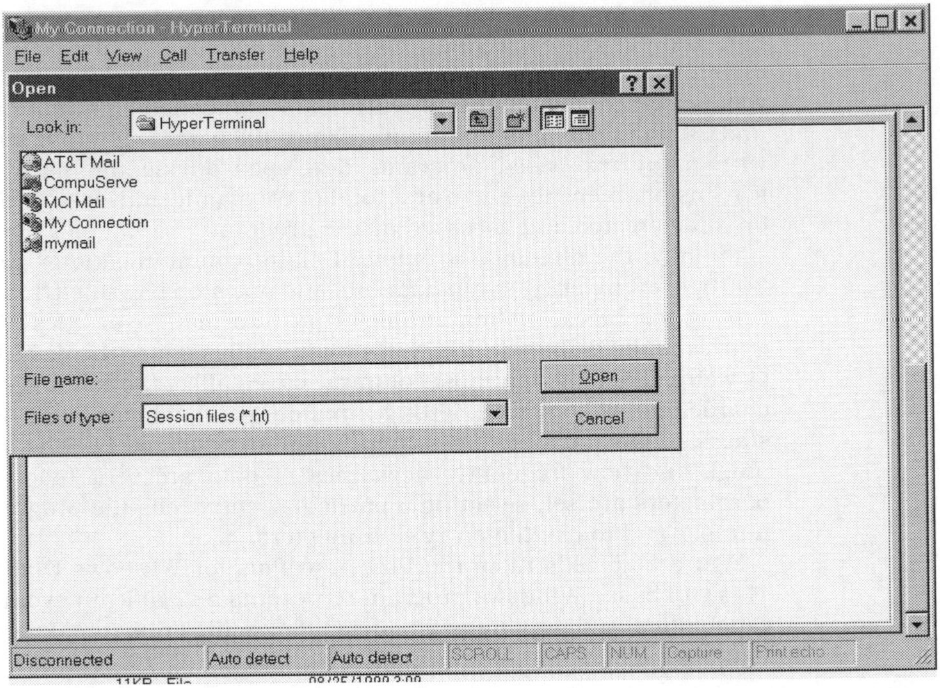

Figure 18.6 Viewing predefined HyperTerminal connections stored in individual .HT files

directory you can select the voice button and a dialing directory entry, which results in the program using the modem dialing the number and then displaying a message for you to lift the telephone handset. Also note that the Procomm dialing directory provides support for the use of phone cards and enables data, fax and voice numbers to be associated with each entry.

Although we like to think of modern technology providing improvements in the manner by which we work, this may not always be the case. In any event, you can be the judge by comparing the 'dialing directory' of the HyperTerminal communications program bundled with different versions of Microsoft's Windows to the dialing directories of PC-Talk and Procomm Plus.

Figure 18.6 illustrates the fact that HyperTerminal requires you to load a specific HyperTerminal (.HT) file to attempt to initiate a communications session. In Figure 18.7 you will note the resulting display of a previously configured MCIMail connection stored in an .HT file. Note that although you can change the telephone number and dialing properties within a file, HyperTerminal does not provide an easy method to view telephone numbers in a dialing directory nor the settings associated with a directory entry or series of entries.

Dialing from the keyboard

Since all dialing directories are finite in size, the ability to enter a telephone number from the keyboard can save the user a considerable amount of effort.

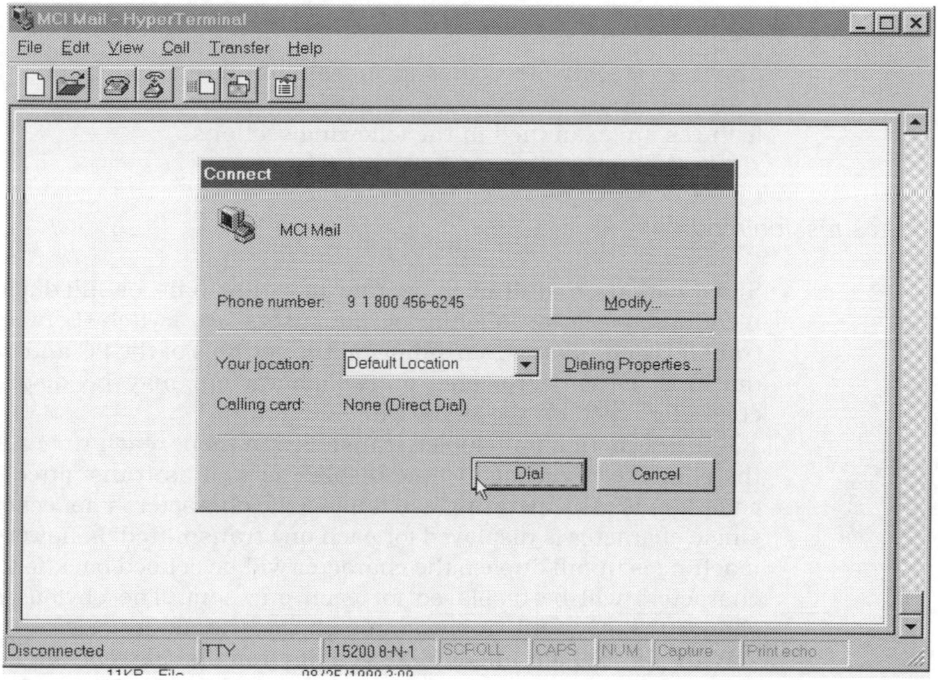

Figure 18.7 Displaying the basic parameters of the MCIMail HypterTerminal file

If the communications program permits dialing only through a directory entry and the directory is full, it may be necessary to remove an entry in order to dial. Figure 18.5 clicking on the Manual Dial entry permits you to enter a telephone number directly.

Preconfigured dialing directory

Some communications programs contain a dialing directory whose parameters are preconfigured for major information utility services, such as CompuServe and Dow Jones, among others. Normally, preconfigured dialing directories do not contain telephone numbers for the entries, since the program's purchaser could be located anywhere in the United States or abroad and any number included would probably not reflect the most economical telephone number to use.

The preconfiguration of transmission parameters may eliminate many minutes or hours of future effort; however, a word of caution is in order. At least one vendor lists 800 numbers in the dialing directory along with the parameter settings required to access several information utilities. Unfortunately, these 'toll-free' numbers are not free, since many information utilities add a surcharge to the user's connect time when access is via an 800 number. It is usually much less expensive to go through a value-added carrier that might be just a local call away.

18.1.6 Transmission

Another category of features includes those directly involved in the transmission, reception, and flow of data across a telecommunications link. These features are examined in the following sections.

Transmission mode

Since various mainframes operate in either half- or full-duplex, a communications program should permit users to switch between these two transmission modes. Depending on the settings of the PC and the destination machine, either zero, one, or two characters may be displayed for each character typed on the keyboard.

If the PC is in a half-duplex transmission mode, each time a key is pressed the character is echoed to the display as well as transmitted. If the remote computer is also operating half-duplex, the character is not echoed back and a single character is displayed for each one transmitted. If, however, the remote machine is in full-duplex, the character will be echoed back to the PC and two characters will be displayed for each one sent. The obvious solution is to change the PC to full-duplex to match the remote device.

With the PC in full-duplex and its remote partner in half-duplex, no character will be displayed when a key is pressed. This is because the PC transmits but does not echo the character to the screen while the remote unit receives the character and also does not echo it. Again, the solution is to change the transmission mode of the PC.

Terminal emulation

Most of the original communications programs first written for use with personal computers transmitted and received data on a line-by-line basis. These programs did not take advantage of the screen-display capability of the personal computer and simply emulated Teletype-style transmission.

More sophisticated terminal emulation can cause problems for the personal computer user. For instance, a terminal from Digital Equipment Corporation (DEC), which is now part of Compaq Computer Corporation, might not work with an IBM mainframe in full-screen mode, because it does not recognize many of the control codes used by the mainframe for cursor positioning, character attributes, and similar screen-control functions. To enable a personal computer to operate as a specific type of terminal, it became necessary to develop emulation software to convert these control codes into equivalents that the microcomputer recognizes. Similarly, the codes used to represent functions on the personal computer had to be converted into their mainframe equivalents. Thus, users wishing to use a PC or local processing and to act as a specific type of terminal must look for a package with the appropriate emulation capability.

Figure 18.8 illustrates the Procomm Plus for Windows Current Setup window after its Terminal Options button has been selected and the scrollable

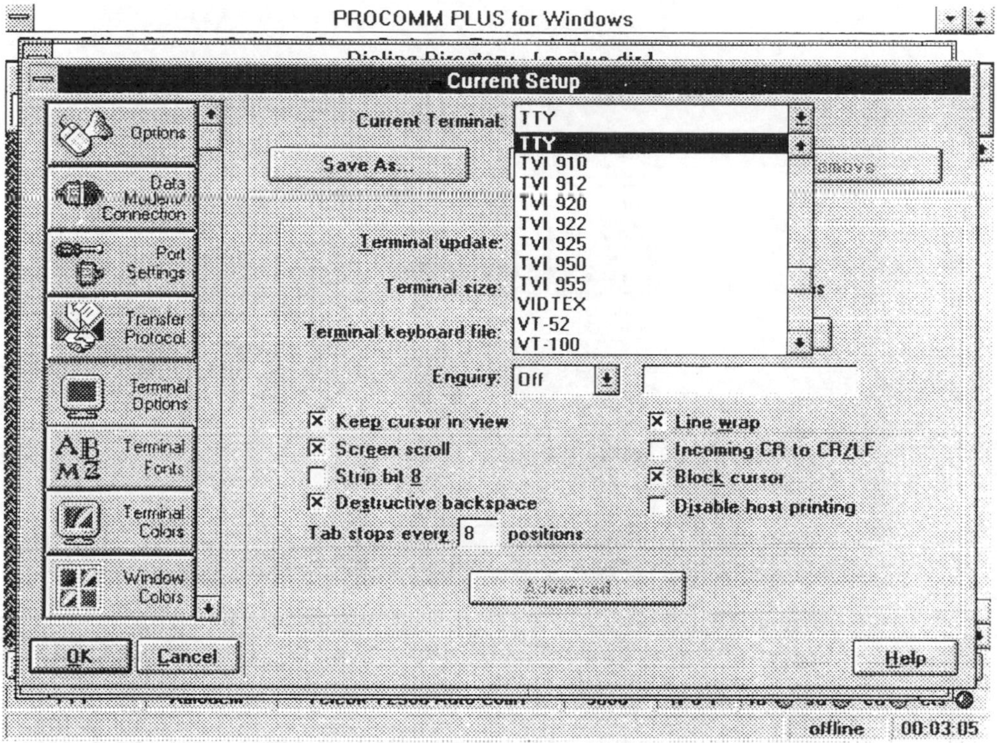

Figure 18.8 Selecting a terminal emulation using Procomm Plus for Windows

list of terminal emulations supported by the program were displayed within a scrollable dialog box. This program supports approximately 30 terminal emulations, which enable it to be used to access, on a full-screen basis, a variety of minicomputer and mainframes.

If you are using HyperTerminal bundled with different versions of Windows, you can also select from several types of terminal emulations. To do so you would click on the button labeled Modify and select the tab labeled Settings in the resulting display. Once the preceding is accomplished you would click on the down arrow associated with the box labeled Emulation to display a list of terminal emulationss supported by the HyperTerminal program. Figure 18.9 illustrates the terminal emulations supported by the HyperTerminal program.

Character transfer support

IBM PC applications involving communications ordinarily transmit data in seven-bit ASCII. Memory-image files can, however, be transferred only in an 8-bit representation. Also, the IBM PC and its compatibles use an extended ASCII character set, with 8-bit characters used to obtain ASCII characters from 128 to 255. Programs compiled, word processing files and most image files consist of data bytes that are represented using these extended-ASCII

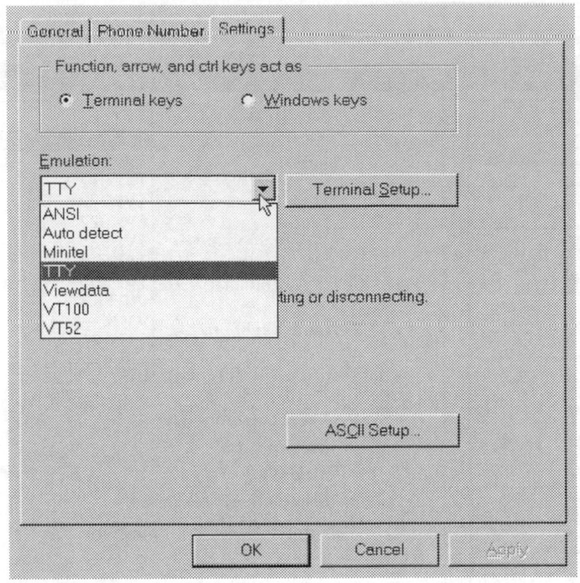

Figure 18.9 Viewing the terminal emulations supported by the HyperTerminal communications program bundled with Windows

characters, which would preclude file transfers between personal computers if the communications program did not support 8-bit ASCII transmission. Thus, the communications program should support 8-bit data transfer if the user wants to send binary data or programs that use extended ASCII characters.

File transfer capability

The transfer of ASCII or binary files requires the use of a file transfer protocol. Currently there are over 30 file transfer protocols in use, ranging from XMODEM and its derivatives to INDSFILE used to transfer data to and from IBM mainframes. Thus, it is important to consider the target computers that you anticipate communicating with and the file transfer protocol or protocols that they support. Doing so will enable you to compare the protocols supported by different programs against your specific file transfer requirements.

Stripping/converting characters

Some characters transmitted from a mainframe or information utility can interfere with the intended operation of a personal computer. Others can be a display nuisance. For example, an ASCII 127 character is sometimes used for timing delay. In such cases, the mainframe sends a variable number of these characters between lines, thinking that an attached electromechanical printer needs time to return its print-head to the home position. If the communications program does not automatically strip this character, it will appear on an IBM PC screen similar in size to a hotel from a Monopoly game. Not only does this clutter the display, but it also shifts each line a number of positions to the right. Since there are occasions where the user may wish to retain pad characters, some programs have an option for specifying which characters, if any, should be stripped from the incoming communications buffer.

Another character transmission area that can result in personal computer trouble is when a mainframe or information utility transmits an ASCII 126 character (control-Z) to mark the end of a response to a PC request. If received data were being logged to disk, the operating system might interpret the ASCII 126 as an end-of-file mark, which would cause any information in the file following this character to be inaccessible. The rest of the file would have been logged to disk, but the disk operating system would see an end-of-file mark. The only way to retrieve such a file would be with a special utility program.

To alleviate such misinterpretations, the communications program should contain a character-conversion feature. All ASCII 126 characters would be converted to ASCII 32 (space) characters. Similarly, a character-conversion feature could change XOFFs to spaces which could keep the printer running.

Buffer control

Communications programs use a buffer, which is a portion of random-access memory, for temporary storage when transmitting and receiving data. As the buffer fills, a flow-control procedure incorporated into most programs prevents the buffer from overflowing and thus keeps received data from being lost. The common method of implementing flow control is for the program to transmit an XOFF character when its receive buffer is between one-half and two-thirds full. If the transmitting device recognizes the XOFF, it will temporarily stop transmitting, so that the communications program can process the data in its receive buffer. Then, after the buffer has been emptied or reduced to a predefined value, the program transmits an XON character to tell the distant device that it may resume transmission.

Since some mainframes implement flow control with characters different from XON and XOFF, users who anticipate accessing a variety of mainframes may wish to consider a program that lets them select flow-control characters via a parameter table. Without such a feature, sending an XOFF will not halt the mainframe. Under certain situations, such as when receiving data and printing it concurrently, the buffer on the PC can overflow, resulting in data loss.

Pacing capability

Pacing control, intended to keep buffers from overflowing, can involve one of three methods. The host can slow down the transmission of files to a remote personal computer by pausing for a set amount of time between each transmitted line. Certain mainframes do not permit a connected device to send a line of data until the device receives a prompt from the host (typically an asterisk or question mark). Some communications programs may, therefore, wait for the receipt of the prompt before they transmit the next line. A third pacing mechanism is known as character-receipt delay. Here, the user specifies how many characters must be received from the mainframe before the program may transmit a new line of data.

Of the three pacing methods, transmission on prompt-character receipt is the most efficient, as it lets the user specify the exact character string that must be received from the mainframe before the program sends the next line of data. With time pacing, the user must specify an interval (in seconds or fractions of a second) as the worst-case time that it will take the mainframe to process each line of data.

A mainframe's response time is a function of its current load and of the activity to be performed. Pausing for a set interval between lines can, therefore, result in bloated connect time and transmission charges, because this method does not take advantage of the fact that some lines of data can be processed more quickly than others. Since impulse noise can destroy characters, specifying a number of characters that must be received from the mainframe before the microcomputer continues may cause an infinite wait. To avoid this situation, some communications programs that use this method permit manual user intervention or automatic transmission of the next line after a predefined time delay.

18.1.7 Performance efficiency

Of concern to most users is how a program functions once a connection to the target computer has been made and the program is mediating a session. One general set of requirements involves the software's efficiency. This element can show up in anything from compression, which hastens the transfer of file and reduces waiting time, to disk directory access, which can improve the user's productivity.

Selectable data rate

For remote communications, most programs let the user select a variety of data-transfer rates up to 33 600 or even 56 000 bps. This range is normally sufficient for many users. The introduction of modems that can operate at data rates up to 56 000 kbps means, however, that those operating rate settings may not be suitable for obtaining the full use of compression performing modems. For example, a 28 800 bps modem with a V.42 bis compression feature should have data transferred to the modem at three to four times its operating rate to effectively use is compression capability. This means the software should support a maximum selectable data rate of 86 400 to 115 200 bps.

Data compression

Several communications programs incorporate data-compression techniques. These range from simple blank suppression to sophisticated encoding schemes that reduce the actual quantity of data transferred by several hundred percent or more. Obviously, the computer at the other end of the data link must have the same communications program, since many software-based compression techniques are not standardized and are normally proprietary to each software vendor. The introduction of standardized compression routines built into modems significantly reduced the availability of this feature within communications programs. However, some programs that retain built-in compression do so because they provide a higher compression ratio than that obtainable through the use of a modem's built-in compression feature.

Error detection and correction

For file-transfer operations, the probability of an error occurring increases with the length of the file. The communications program should be able to detect the occurrence of an error, as well as to automatically correct the error without intervention. Typically, the program first detects that an error has occurred within a data block and then asks the other device to retransmit that block.

A wide variety of methods is used to detect and correct transmission errors. A simple checksum can be formed by adding up the ASCII values of the characters in a block and dividing this sum by a fixed number. More sophisticated algorithms use polynomial checking.

One of the more popular error detection and correction methods is incorporated into the XMODEM protocol created by Ward Christensen. In this protocol's error-handling technique, data are grouped into 28-byte blocks. A checksum for each block is formed by adding the ASCII values of all 128 characters and dividing the sum by 255. Then the quotient is discarded and the remainder is appended to the block as the checksum.

At the receiving device, a similar operation is performed on the received data. If the locally computed checksum matches the transmitted checksum, the data are considered to have been received correctly. If not, the receiver will ask the transmitting device to retransmit the block.

Although the above method of error detection and correction is extensively implemented by most non-commercial bulletin boards and in many commercial software programs, it does have several deficiencies. For one thing, it is a stop-and-wait protocol, where block number $N+1$ cannot be transmitted until block number N is acknowledged. Also, after nine negative acknowledgements are received the communications session will terminate and file transfer must restart from the beginning. Another limitation is the probability of receiving an undetected error, which is higher with the error-detection method used by the XMODEM protocol than with a cyclic redundancy-checking algorithm.

One or more of these limitations can be avoided by carefully selecting a communications program. The program should support a dozen or more file transfer protocols, to include the full-duplex ZMODEM protocol whose use in place of XMODEM can reduce file transfer time by 30% to 50% or more.

Foreground/background operations

Similarly to what is made possible by printer buffers, a few communications programs permit users to perform file-transfer operations while doing other work. This feature frees the microcomputer for other operations during lengthy file transfers and normally increases productivity. Other programs include the capability print information to include screen images and files at the same time the user performs other operations.

Programmable key capability

Some communications programs have a rigidly structured macro language. Others let the user assign a string of information for transmission to what is known as a programmable key. In this way, the user may define values associated with certain command-key combinations as well as function keys on their computer's keyboard. Pressing these keys during a session will cause the associated string to be transmitted. Both the number of keys that can be programmed and the length of the string that can be assigned to each key should be examined, because they vary considerably between programs.

Programmable macros

A macro is a command that can be stored and executed with a single keystroke. With these programmable macros, the user can store passwords and other log-on sequences and execute them with a single keystroke. This lets the user avoid the repetitive process of typing commands each time a mainframe computer or information utility is to be accessed. Since the number of macros varies between communications programs, this is another feature that should be investigated prior to selecting a communications program.

Control character transfer

The most common way to tell mainframes to terminate a previously initiated operation is to send a break signal. This signal should not be confused with, for example, the Ctrl + Break (control and break) key combination on the IBM PC keyboard, which terminates the current operation and return the user to command mode. Instead of transmitting a character, the break signal causes a stream of binary zeros to occur on a line for approximately one-tenth of a second, which tells the mainframe to stop what it is doing and return to its operating-system command level.

Normally, the personal computer's communications program will issue a break signal in response to a predefined character sequence being typed on the keyboard. Since the break signal is indispensable for host operations, micro-to-mainframes should learn how to tell the communications package to send one. Unfortunately, not all mainframes recognize a break signal, in which case an ability to transmit control characters from the keyboard becomes important. As an example, some mainframes recognize a control-C (ASCII 3) as a signal to terminate the current operation. Communications software that does not allow the transmission of control characters may force the user to resort to the in elegant solution of pulling the plug and then reconnecting to the mainframe.

Disk directory access

For file-transfer operations, an ability to check the contents of a floppy or hard disk without exiting to the operating system can be a valuable tool. Some communications programs not only let the user specify the disk drive and display files on the default drive (that is, the drive that the operating system will go to on a command that lacks a drive identifier) but also display the amount of storage available on the drive. This feature can be quite handy for down-line loading a file, since it indicates whether or not there is sufficient room to store the file.

For example, the communications program might indicate 10 240 available bytes of storage on the data disk, whereas the remote machine indicates that the file to be transferred contains 12 288 bytes. The disk obviously has insufficient storage. The user may then install a new disk, or perform an operating system command to delete an existing file, making room for the new one. (The ability to exit to the operating system command level will be examined as a flexibility issue.)

Editing capability

The ability to edit documents from within a communications program varies widely. Some packages provide only a backspace key. Others let the user perform full text editing of messages and files. To do this, the communications program stores data in an edit buffer and then transfers the edited data to a disk or to the communications port. Without this capability, users must store data on disk, leave the communications program, and then initiate a word processor or text editor to perform the desired edging.

Screen-dump capability

Members of the IBM PC family and Apple Macintosh series have a built-in command to send a dump of the screen image to an attached printer. Several communications programs have taken this feature one step further, permitting screen images to be stored to disk. Some will also time- and date-stamp each screen image as they are appended to a predefined file. If a program offers a 'view' option, the user can then look at selected screen images saved to disk during a prior communications session. Without this option, a user might have to end communications and run a word processor or text editor to view previously saved screen dumps.

File closing

Only a few programs close all open files to ensure that data are not lost if the transmission line drops during a file-transfer session. Some programs are quite sophisticated, not only closing and saving whatever information was received during a partial file transfer but also allowing the user to resume file transfer at the point where the disconnection occurred. This permits data to be appended to a file or to the portion of a file already received.

Unattended operation capability

Some communications programs, such as electronic bulletin-board programs, permit remote users to access and use all or a portion of the personal computer's hardware and software. This feature can be useful for unattended file transfer and electronic mail centralization. It does, however, require the habit of checking the mailbox on a periodic basis. Other programs operate in a master–slave relationship in which one module loaded on a PC at home allows access to another module operating on a PC at work as if the user was directly at the keyboard of the distant computer. This function is better referred to as a remote control feature, and it can be significantly helpful during the evening or on weekends to use the facilities of an office computer connected to the corporate LAN.

18.1.8 Performance flexibility

From within a communications session or normal command mode, a number of features can make a program more flexible. Users with non-standard screens will appreciate the ability to select display, whereas such features as editing capability benefit all users.

Protocol setting

This can mean as little as setting such parameters as the number of data bits, the type of parity, and the number of stop bits transmitted with each character or as much as specifying the type of error detection and correction to be used for transferring files. Programs offering a menu of schemes permit a wider scope of potential communications, since there is no universally standardized error-detection and correction mechanism used with PC operations.

Communications port selection

Expansion slots for adapter cards can support several asynchronous communications ports. Although most users have only a single modem, a serial printer could be run through the second communications port. Thus, being able to select which port is transmitting data provides some users with additional flexibility and may eliminate a time-consuming recabling process; if, for example, the printer is on port 1 with the modem on port 2, but the program inflexibly assumes that the modem will always be on port 1, the user must change cables.

Switchable printer ports

For users with both a letter-quality and a dot-matrix printer attached to their microcomputers, the ability to switch printer ports through the communications program can be a convenient feature. Users might select the letter-quality device to print electronic messages from the home office that are to be reproduced for distribution to other employees. The dot-matrix printer could rapidly list data that do not require high-quality printing.

Display-width selection

If the program emulates a 132-column display terminal and the hardware includes a special graphics-display board, the program can make full use of the hardware. Some programs emulate 'wide display' terminals by means of a horizontal scrolling mechanism. The program should be compatible with any specialized hardware used to display the full wide-display terminal screen.

Journalization

The journalization feature lets the user record communications sessions. Some programs can log all transactions to a disk file or printer, whereas other programs let the user journalize selectively (everything transmitted, everything received, or both).

For users of the IBM PC and compatible computers operating under DOS release 2.0 or later versions, a journalization feature is built into the operating system. Users can send a journal to the printer through DOS by pressing the Ctrl + PrtSc (control and print-screen) keys simultaneously.

File-viewing capability

This feature permits the user to view previously stored files. The rationale for having a file viewing capability while in a communications program are twofold. First, the user may wish to preview a program as it is down-line loaded from a mainframe or utility. Secondly, and perhaps more importantly, the user may want to up-line load a certain program but has to scan a few files first to make sure the right one is sent.

Exit to OS command level

Just about every communications program provides a simple mechanism to exit the program and re-enter the personal computer's operating system. If this feature is not offered, the operating system must be reinitialized and the program terminated each time another program is invoked, which may be awkward and inconvenient.

Lowercase/uppercase conversion

Some mainframe computers support only uppercase letters. To be able to communicate with such machines, a program must perform a lowercase-to-uppercase conversion prior to transmitting data. If the program does not include this feature, an alternative is to use the 'Caps Lock' key to ensure that conversational-mode communications are in uppercase.

Line-feed control

This option lets the user adjust the program's actions based on the presence or absence of line feeds after carriage returns. If neither the communications program nor the computer at the other end of the data link adds line feeds after each transmitted line during file transfer, the data received will be treated as one long, continuous line. It would thus not be listable or usable. In conversational mode, each line received overprints the previous line displayed on screen, which makes it difficult to conduct a dialogue with another device. If the communications program adds a line feed to each received line and the remote device adds one to each transmitted line, transferred data files and data received in conversational mode will appear double spaced. A communications program may include the option to enable or inhibit line feeds after carriage returns.

18.1.9 Security performance

Safeguarding user data is a familiar concern, but few users realize that the copy protection they can find so irritating is merely a security precaution taken by software vendors.

Encryption capability

A few communications programs incorporate a security-mode data-transfer feature. Typically, this feature uses modulo-2 addition, adding together the binary value of the data to be transferred and a binary key that the program generates in a byte-to-byte fashion. At the other end of the data link, modulo-2 subtraction is used to reproduce the original text. The following example illustrates modulo-2 addition and subtraction which forms the basis for the encryption of data.

Original character – A	01000001
Key	00001011
Modulo-2 addition	———
Transmitted data – F	01001010
Received data	01001010
Modulo-2 subtraction of key	00001011
	———
Reconstructed data	01000001

Since this basis of encryption obviously requires the same program at both ends of the data link, an alternative method is a stand-alone file-encryption program. With this method, data can be encoded prior to its transmission, independently of the communications program. The program must, however support 8-bit data transfers, since most file-encryption programs result in the conversion of 7-bit ASCII into a 8-bit binary character. So long as the recipient has a similar file-encryption program and knows the key used to encode the data, the original text can be reconstructed.

When the encryption is in effect, a key is added to each byte prior to its transmission. The receiver's program subtracts the key to restore the data to its original value.

Copy protection

To ensure that users obey their licence agreement, a few communications software vendors implemented copy protection schemes. Some suppliers let the user transfer the program to a hard disk, but protect their program by requiring that the original disk be inserted in a disk drive before the hard-disk copy is run. The program then checks periodically to make sure the 'master' disk (the one purchased by the user) is in the drive. Under such a scheme, the user can perform file transfers from disks with a configuration consisting of one disk drive and one hard disk. It is necessary, however, to first transfer all down-line loaded files to the hard disk and then, after leaving the communications session, copy the files from the hard disk to one or more floppy disks.

Another approach to copy protection is to embed a certain code, normally the serial number of the program, into the software. When one personal computer communicates with another using the same program, a check is initiated. If the two codes match, communications between the two computers is inhibited. Since this protection scheme permits the user to make an unlimited number of copies of the program as well as to transfer it to a hard disk and operate it without a master disk, it may be preferred to other schemes that limit operational flexibility.

18.2 TERMINAL EMULATION PROGRAM EXAMINATION

In Section 18.1, approximately 50 features common to communications software written for operation on personal computers were examined. Using the information as a base, this section will focus its attention upon the utilization of three popular communications programs, relating some of the features presented in the referenced section to the operational capabilities, constraints and limitations of these programs.

The programs examined in this section were selected based on the distinct market each program appears to be developed to satisfy. One program is Procomm Plus for Windows developed by DataStorm Technologies, which is one of the most popular asynchronous communications programs marketed during the mid- through late 1990s. The second program covered in this section is HyperTerminal, which is bundled with different versions of Windows. The third program is the IBM Personal Communications 13270 for Windows which provides a full 3270 emulation capability. This program enables a PC to access an IBM mainframe or AS/400 minicomputer via a coaxial cable, LAN connection and other methods, and it is commonly used in the business community to obtain full screen access to mainframe applications.

It should be noted that the programs reviewed in this section were selected for illustrative purposes only and should not be construed as a positive or

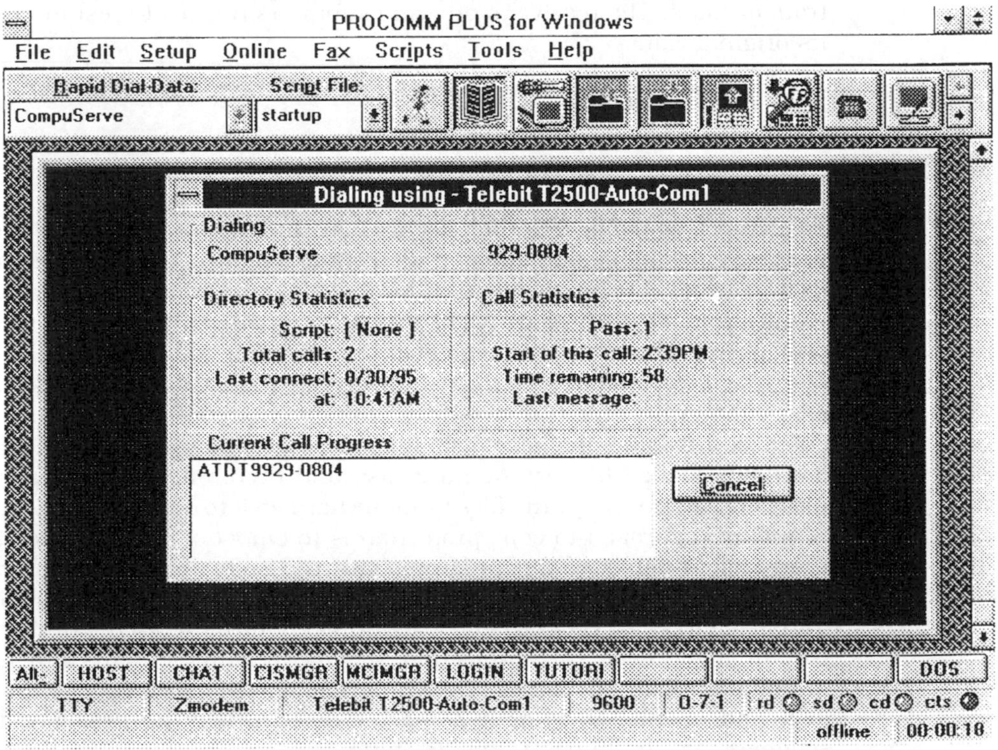

Figure 18.10 Initiating a connection to CompuServe via Procomm Plus for Windows

negative endorsement of any particular vendor product. In addition, readers are cautioned that it is the policy of many vendors to update their programs on a periodic basis, and a feature currently missing or lacking in capability on the version of software examined by the author may be added or revised in a later release.

18.2.1 Procomm Plus for Windows

Procomm Plus for Windows is a feature-packed communications program which provides an easy-to-use graphic user interface. Since a portion of the program's dialing directory was illustrated above, we will commence our brief examination of the operation of this program by selecting an entry from the program's Rapid Dial Data Box.

Figure 18.10 illustrates the dialing box displayed once a dialing directory entry has been selected. In this example the program was initiating a connection to the CompuServe information utility using a modem configuration selected to support a Telebit T2500 modem.

Figure 18.11 illustrates a most interesting feature associated with Procomm Plus for Windows: its scrolling capability. Through the use of the scrollbar, which is located on the right side of the window, you can view

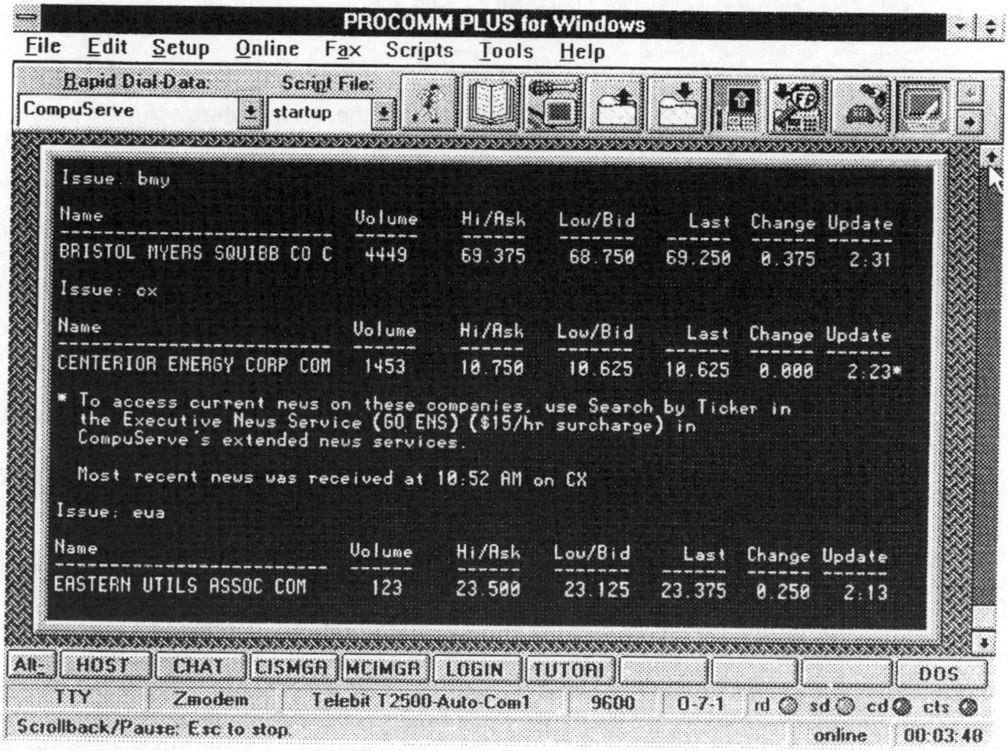

Figure 18.11 Using the Procomm for Windows scroll bar to view information previously scrolled off the display screen

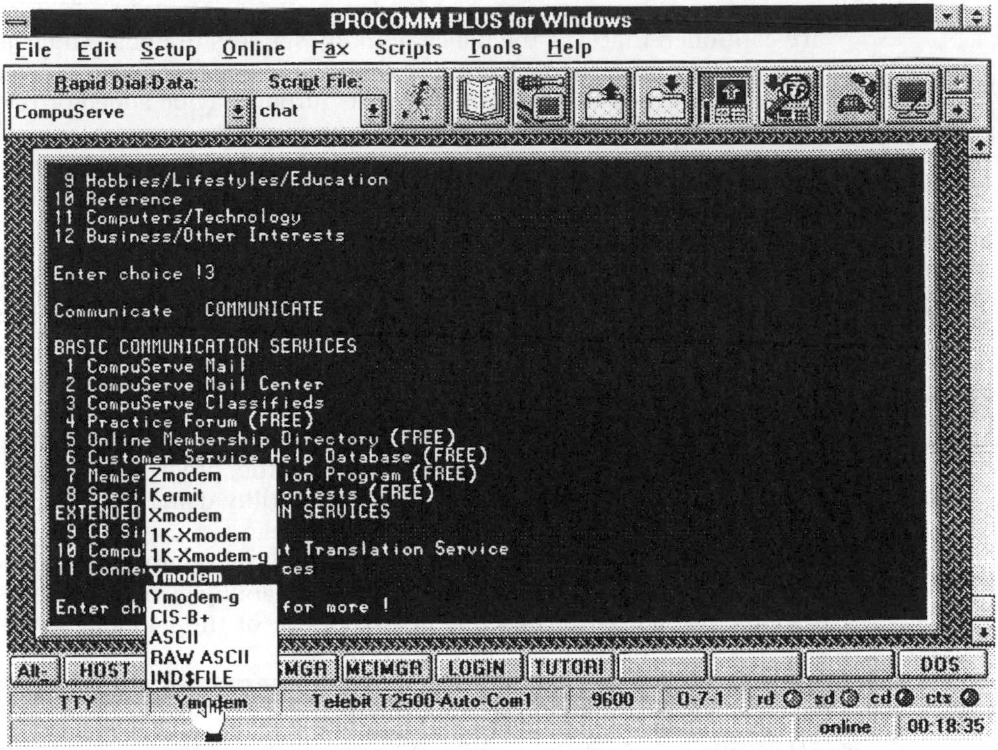

Figure 18.12 Clicking on the file transfer protocol button results in the display of a pop-up menu of selectable protocols

previously displayed information that scrolled off your display. This feature is extremely valuable when using a laptop or notebook computer that may not have a printer, as the scroll capability can be used as a substitute for logging portions of a communications session to disk.

The point and click operations associated with GUI programs provide an easy mechanism to control different communications options. Procomm Plus for Windows includes numerous point and click features which facilitate the use of the program. One such feature is illustrated in Figure 18.12, which shows the program's pop-up file transfer menu resulting from clicking on the file transfer button. In Figure 18.12 the YMODEM file transfer protocol is shown as being selected by the placement of the highlighted bar over that entry. Simply moving that bar over a different entry results in the selection of a different file transfer protocol.

Figure 18.13 illustrates the downloading of a file from CompuServe to the author's computer using the YMODEM protocol. This operation was initiated by clicking on the file folder icon that has a downward arrow pointing into the folder to represent the file download process. Note that Procomm displays a pop-up dialog box which provides information concerning the status of the file transfer as a percentage of completion as well as the number of bytes transferred. The program uses the current throughput, the file length obtained

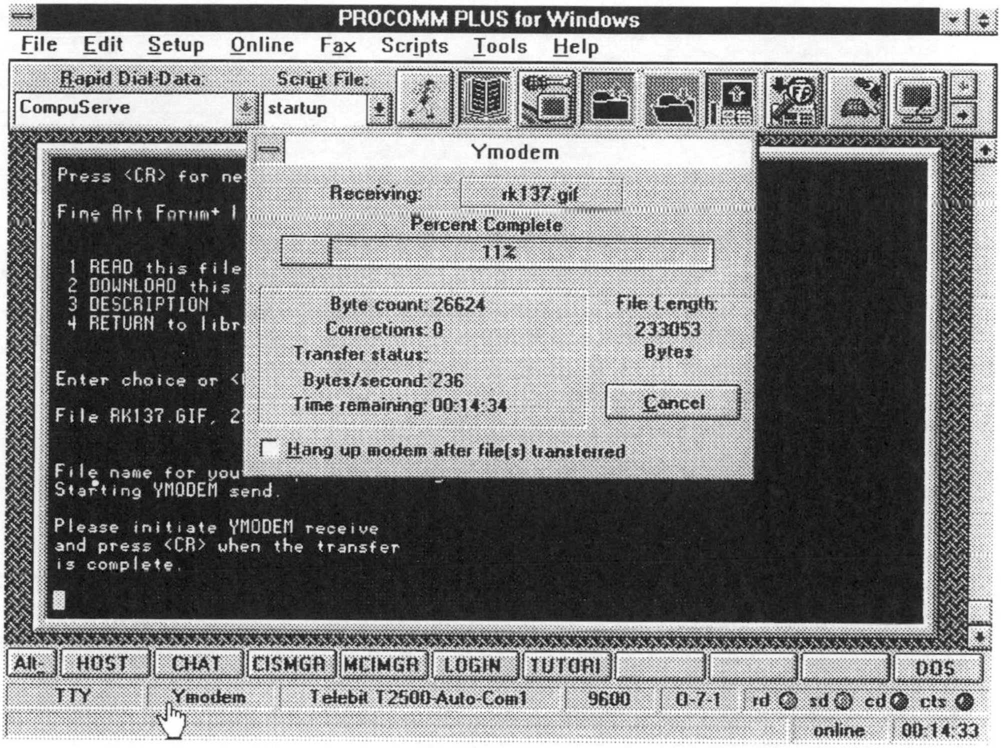

Figure 18.13 Using Procomm Plus for Windows to download a file from CompuServe

from the file header, and the byte count to project the time remaining to complete the file transfer.

In the example illustrated in Figure 18.13, the bytes/second display of 236 represents a throughput of 1888 bps, even though the operating rate of the modem is set to 9600 bps. This illustrates why XMODEM, YMODEM and other half-duplex transmission protocols are not wise selections if you can use a full-duplex protocol.

Another of the more interesting features built into Procomm Plus for Windows is a GIF file viewer. When you use the program to download a GIF file Procomm Plus for Windows, if setup to do so, will automatically display the image as it is being received. Although Procomm Plus for Windows and competitive products have literally hundreds of features, the previously presented examples of the operation of the program should provide an indication of the functionality associated with GUI based communications programs.

18.2.2 HyperTerminal

Because one of the best methods to illustrate the capabilities of a program is through its use, let us do so. In examining the use of HyperTerminal we will use the program to create a new connection name and associate an icon with that name. Then we will use the new connection to go online.

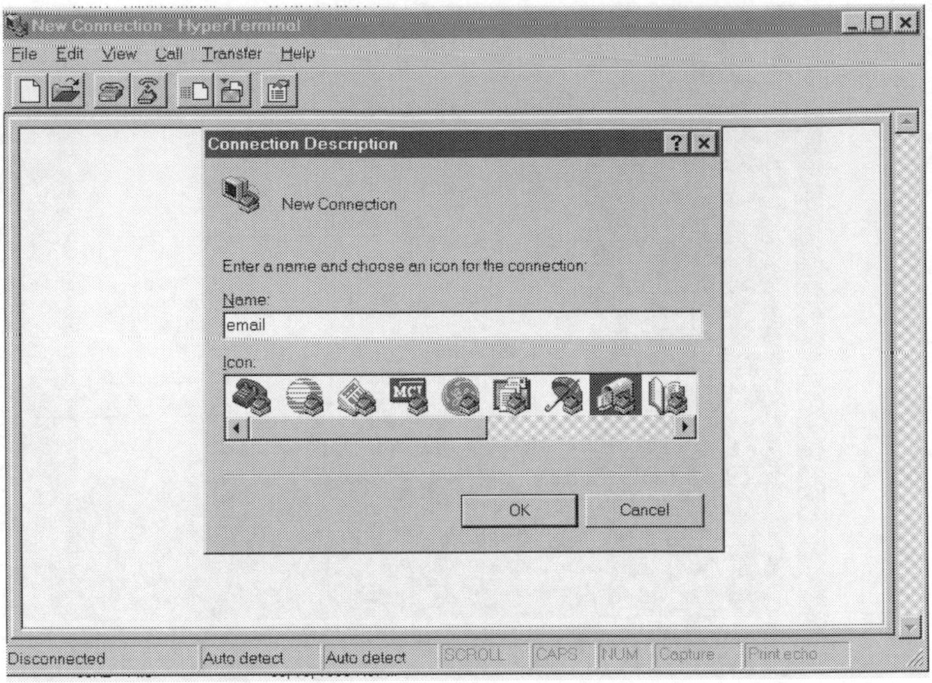

Figure 18.14 Assigning a name and icon when establishing a new HyperTerminal connection

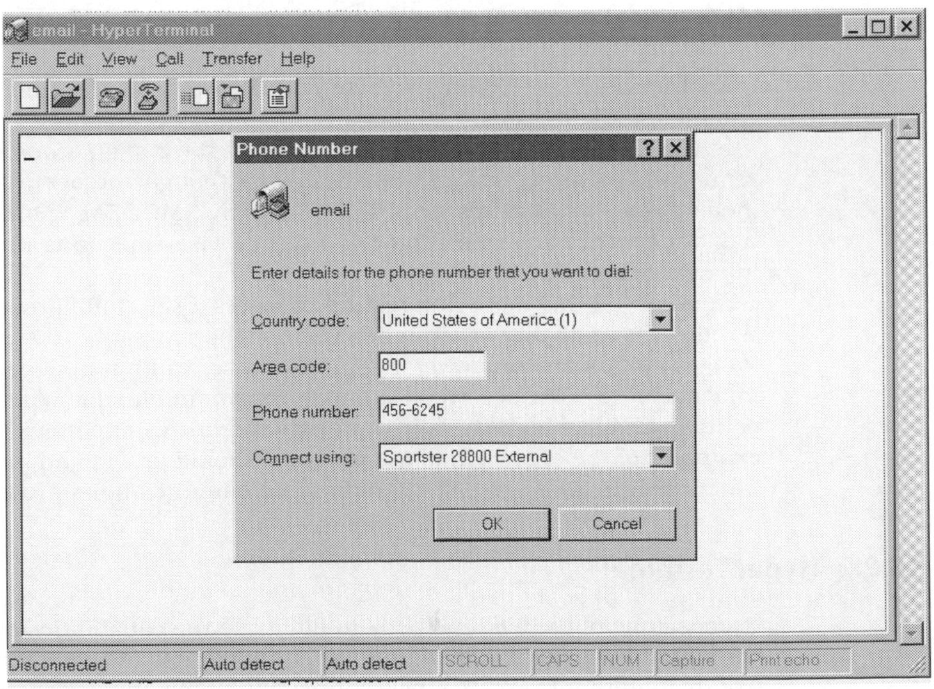

Figure 18.15 Entering the telephone number to be dialed for the HyperTerminal connection labeled email

Figure 18.14 illustrates the selection of HyperTerminal via the Windows Start button and the selection of the New Connection entry from the program's File menu. In this example we assigned the name email as the connection name and associated the icon that represents a mailbox with the just entered connection name. Once we complete the previously described assignment and icon selection, a new display appears that prompts us to enter the phone number to be dialed.

Figure 18.15 illustrates the completed display after this author entered the 800 toll-free telephone number required to access his email service.

Once you enter the applicable telephone number, a new dialog box labeled Connect will be displayed as illustrated in the left portion of Figure 18.16. At this time you can accept the program's dialing defaults and click on the button labeled Dial, or you can view and possibly change one or more defaults by clicking on the buttons labeled Modify and Dialing Properties. If you click on the button labeled Modify, a screen similar to the one previously illustrated in Figure 18.9 will be displayed, providing you with the ability to alter the terminal emulator as well as the echoing of characters and transmission of line feeds, with the latter two a few of several communications options accessible from the ASCII Setup button shown in Figure 18.9. If you select the

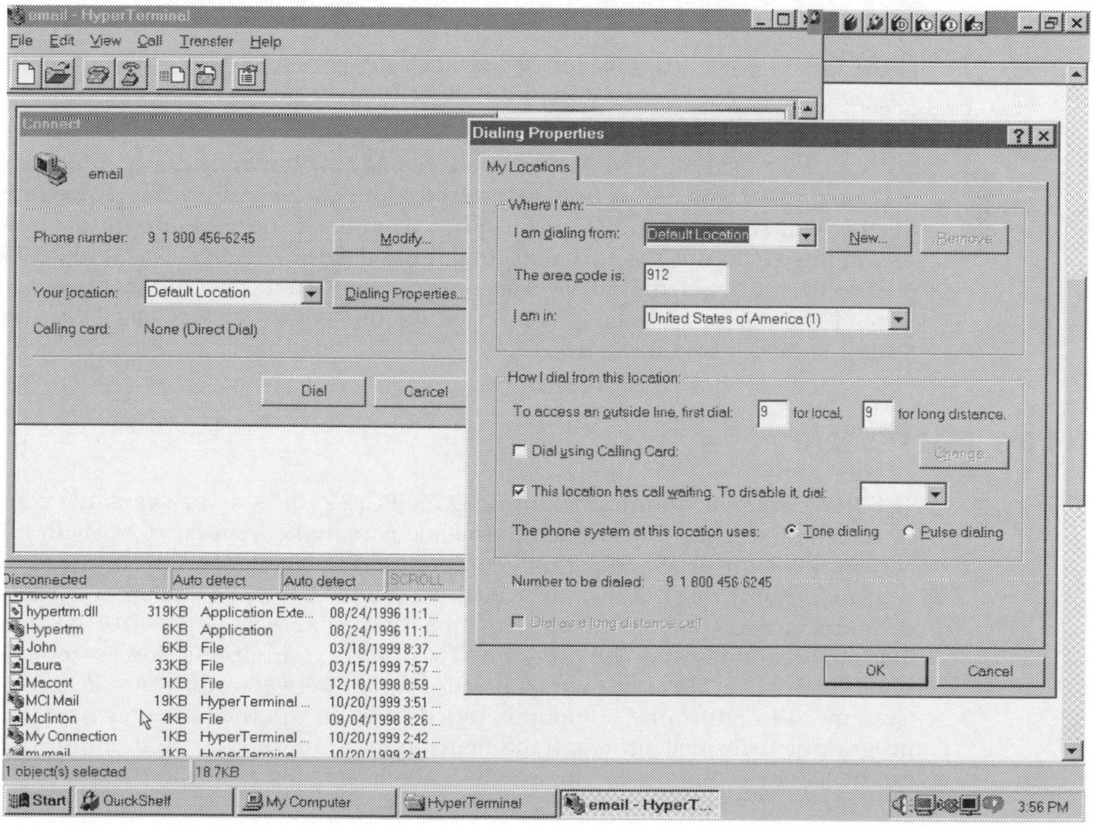

Figure 18.16 Viewing HyperTerminal dialing properties

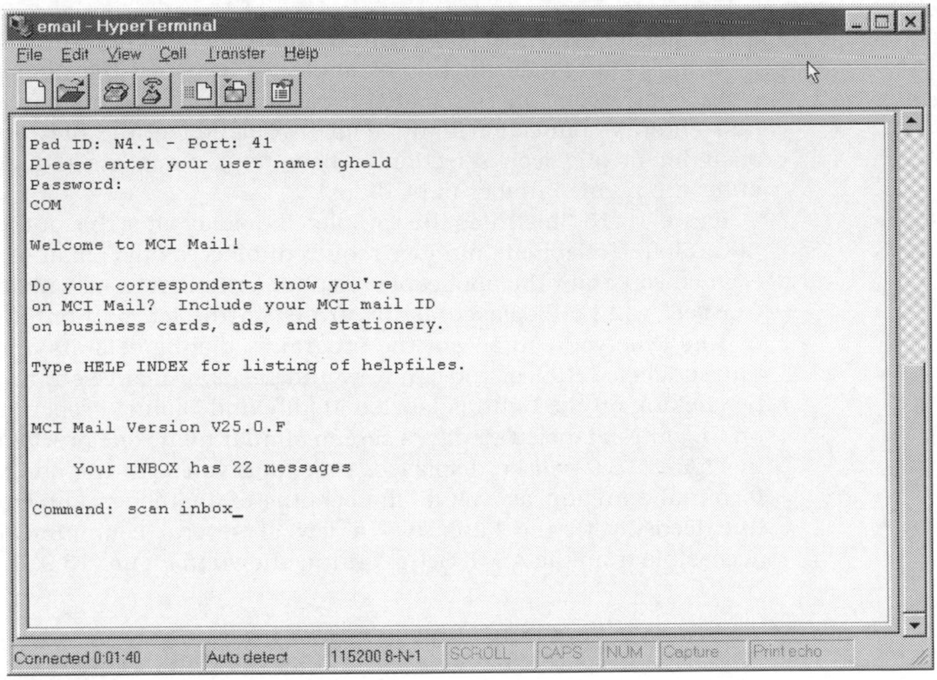

Figure 18.17 Establishing a HyperTerminal communications session

button labeled Dialing Properties whose display is shown in the right portion of Figure 18.16, you can define the prefix for long-distance and an outside line, disable call waiting, and perform other communications-related functions. Assuming we do not wish to alter any of the default settings, we would click on the button labeled Dial on the Connect dialog box. Doing so results in HyperTerminal providing a connection to the email system used by this author as illustrated in Figure 18.17.

18.2.3 IBM PC/3270

The IBM Personal Communications 3270 (PC/3270) is a very versatile communications program developed to provide personal computer users with full screen access to IBM mainframe computers. PC/3270 Version 4.0 used by this author provides mainframe access via coaxial cable, LAN adapter cards, a communications port, SDLC or through a 5250 adapter to an AS/400 system. Figure 18.18 illustrates the program's Customize Communication screen display which is set for a coaxial cable connection between the users PC and a System 390 mainframe computer. By clicking on another adapter icon, the program will display other attachment options for the selected connection method. For example, selecting the LAN adapter would result in the display of IEEE 802.2, NETBIOS, IPX/SPX, TCP/IP and similar LAN protocols that could be used to support transmission to a System 390's communications controller

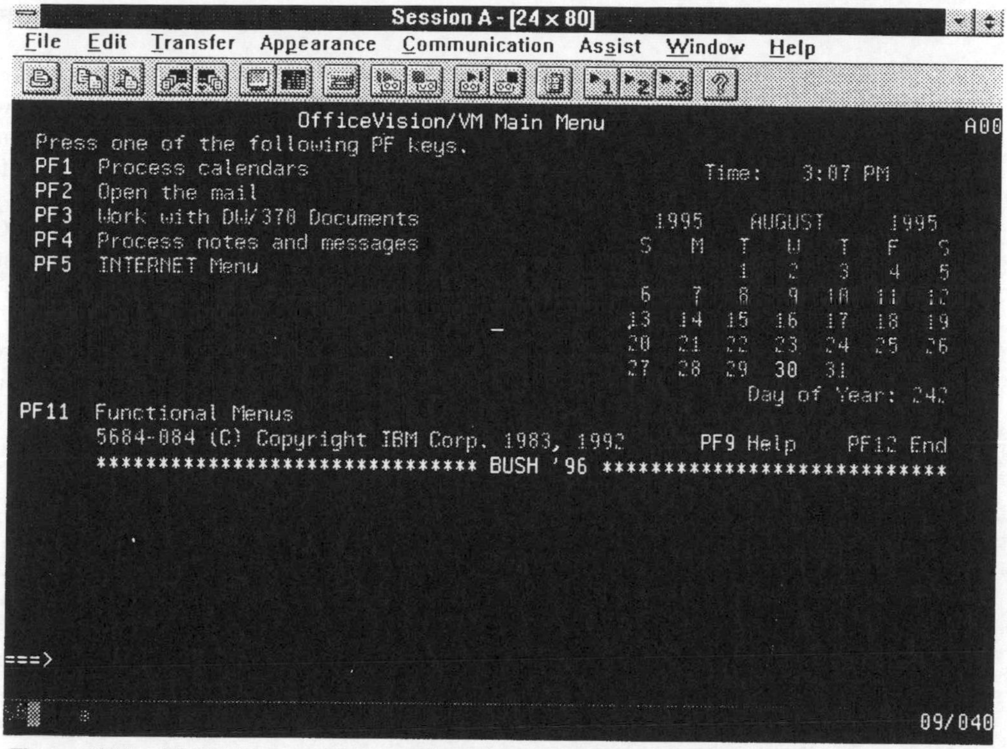

Figure 18.18 PC/3270 supports several methods whereby a PC can obtain connectivity to a host computer

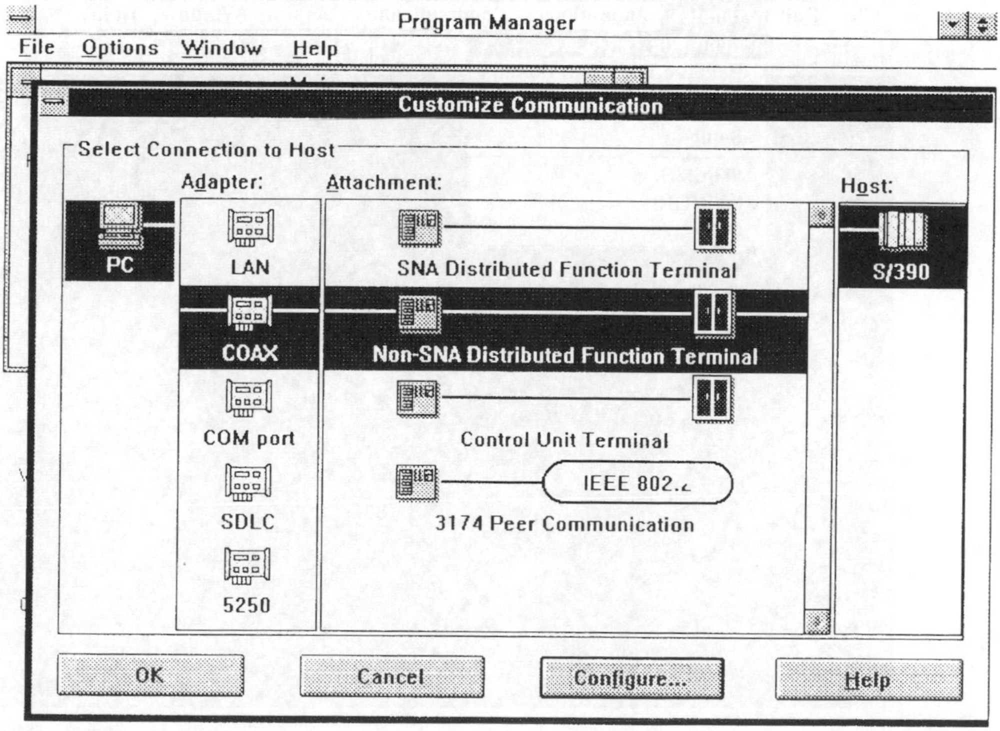

Figure 18.19 Using PC/3270 to access the Office Vision application running on a host

or another device connected to the mainframe and the local area network.

Figure 18.19 illustrates the use of the PC/3270 program to obtain access to IBM's OfficeVision main menu. When correctly configured, the PC's function keys are mapped to PF keys used by 3270 terminal keyboards to perform predefined operations. For example, pressing the F2 key on the PC would result in the generation of the code used to represent the PF2 key, instructing the application to display the mail menu.

To illustrate the ease by which files can be transferred between a PC and a host computer, Figure 18.20 shows the entries in the PC/3270 program's Transfer menu. To transmit a file from the PC to the mainframe you would first select the Send File to Host entry from the Transfer menu. This action would result in the subsequent display of a dialog box labeled Send Files to Host which is shown in Figure 18.21.

The Send Files to Host dialog box provides you with the ability to select one or more files from any directory on the PC to include LAN drives. In this example the file PCOM.TXT was selected and the Transfer Type button was pressed to illustrate the file transfer types supported by the program.

As indicated by this abbreviated example of the use of a GUI based terminal emulation program, most operations can be initiated by a point and click operation which significantly increases end-user productivity. However, a word of caution is in order concerning the effort required for program setup

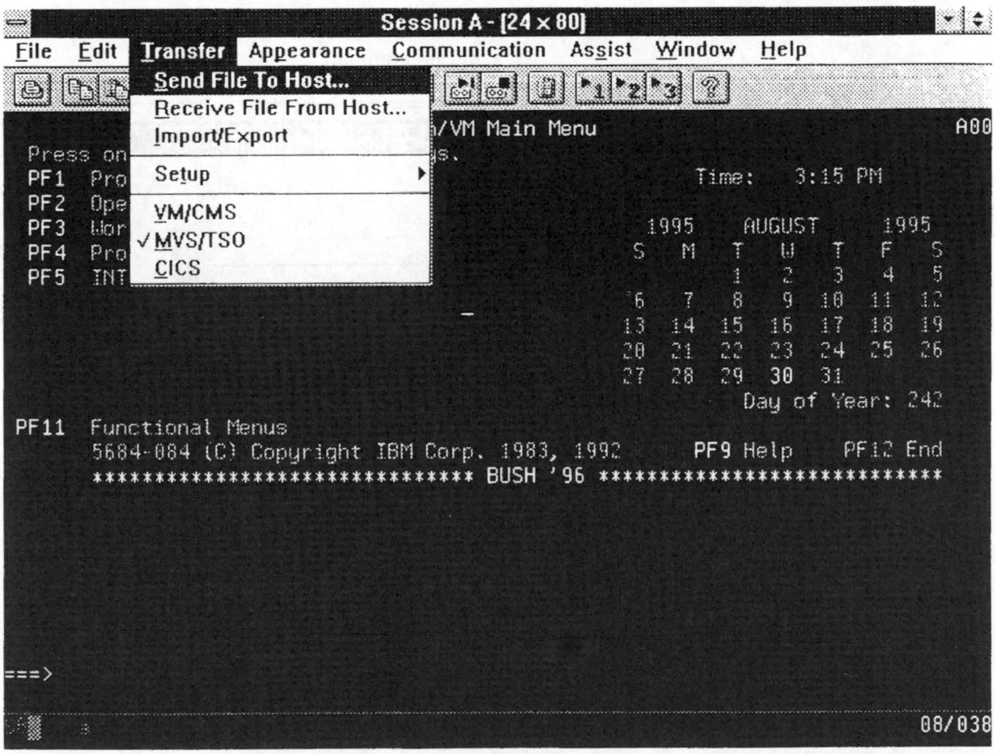

Figure 18.20 Preparing for a PC to host file transfer

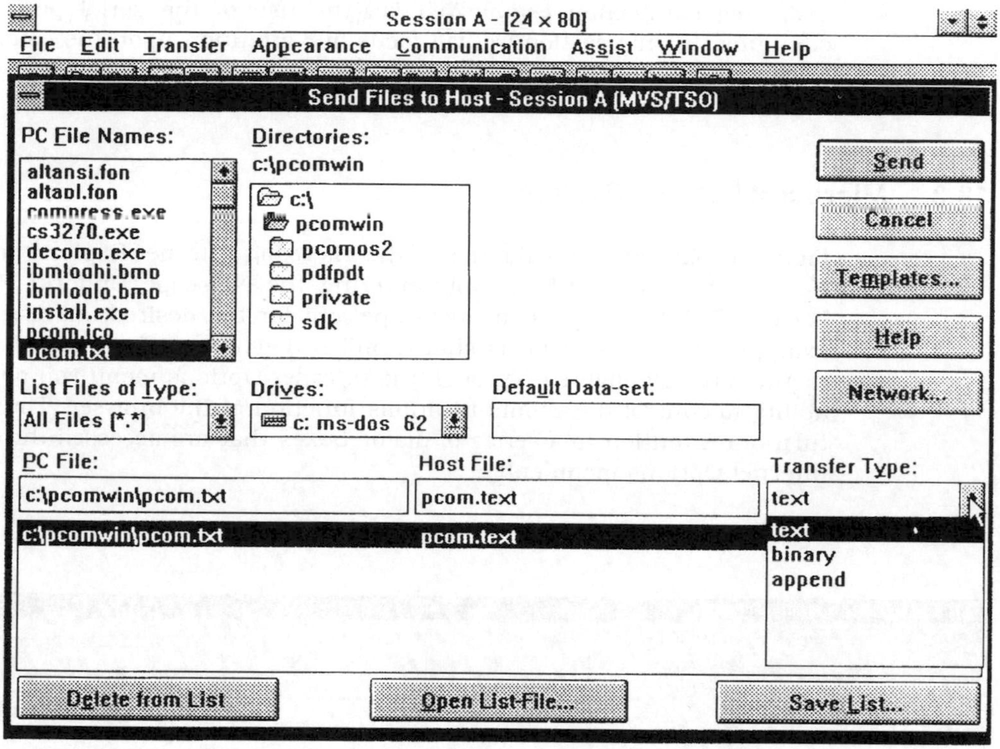

Figure 18.21 Selecting a file to transfer

operations. Some communications programs, especially when used on PCs connected to LANs using both Novell's IPX/SPX and TCP/IP can be extremely difficult to setup. This is because you must support dual network stacks, as well as setup the communications program.

18.3 WEB BROWSERS

An entirely new type of communications software program was developed during the late 1990s, collectively referred to as Web browsers. The first Web browser, which was called Mosaic, was developed at a government-funded supercomputer center. A group of programmers who had worked on the Mosaic browser left the government supercomputer center and with the backing of investors founded what became Netscape Communications, which is now part of America Online (AOL).

Although there are over a dozen browser products available for use, the browser market primarily represents a battle between competing products from Netscape and Microsoft. While each product supports an easy to use graphic user interface, behind that interface is a sophisticated program that allows you to browse the Internet either via a LAN connection or via a dial or dedicated connection established via the use of the serial port of your computer. In this section we will focus our attention upon the communications aspects of the Microsoft Internet Explorer browser.

18.3.1 Microsoft Internet Explorer

Figure 18.22 illustrates the use of the Microsoft Internet Explorer to access Netscape's Netcenter by simply entering the Netscape Uniform Resource Locator (URL) of http://home.netscape.com for the desired address. In this example, the Tools menu is shown pulled down, with the Internet Options menu entry highlighted. Through the Internet Options menu you obtain the ability to control the communications function of the browser. Thus, let us turn our attention to a series of dialog boxes that are accessed through the Internet Options menu entry.

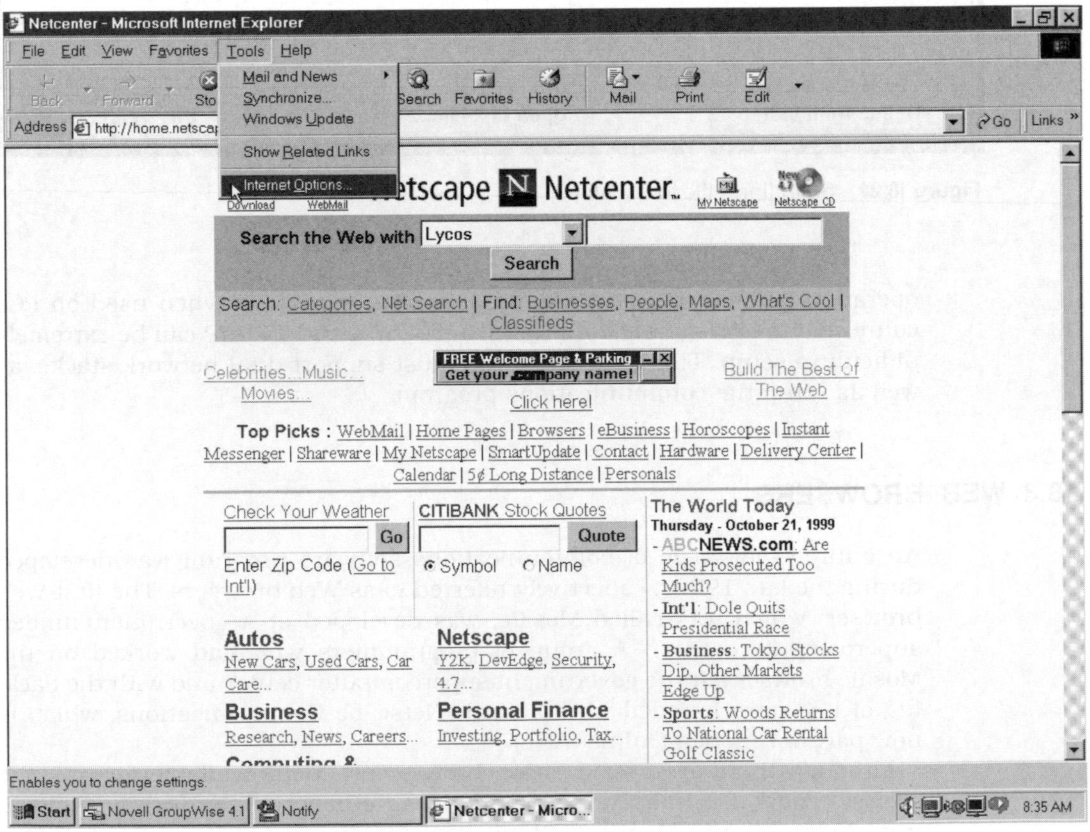

Figure 18.22 You can control the manner by which the Microsoft Internet Explorer communicates through the use of its Internet Options menu entry

Internet Options

If you examine the dialog box labeled Internet Options that is displayed in Figure 18.23, you will note it has six tabs, with the tab labeled General shown as the foreground display. The tab labeled Connections provides us with the ability to configure different communications options and we will shortly do so. However, there are a few options on the General tab display that warrant a brief discussion.

If you examine the address associated with the home page this entry provides you with a method to define the default location the browser will access upon startup. In Figure 18.23 this author configured the Microsoft browser to go to the Netscape Netcenter home page, which illustrates that business competitors can develop products that interoperate.

The second portion of the Internet Options General tab allows you to control the storage of Web pages you view. If your computer is short on storage space, you should click on the button labeled Settings and reduce the amount of disk space used for the storage of temporary files. The third portion of the General tab provides you with the ability to control the list of URLs visited. If you are a heavy surfer, the storage associated with tracking visited Web pages can quickly add up and you should consider periodically clearing the history or setting the number of days to a lower value.

Connections options

Selecting the tab labeled Connections from the Internet Options menu provides you with the ability to control the manner by which the Microsoft

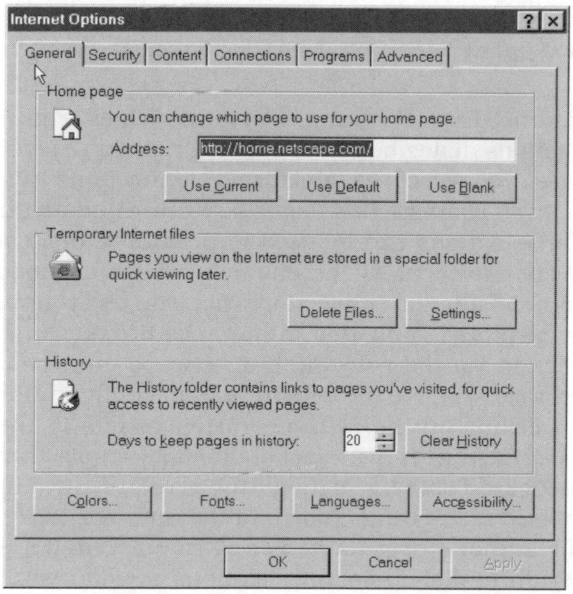

Figure 18.23 Through the Internet Options dialog box you can control the method of communications used by the browser as well as control other browser parameters

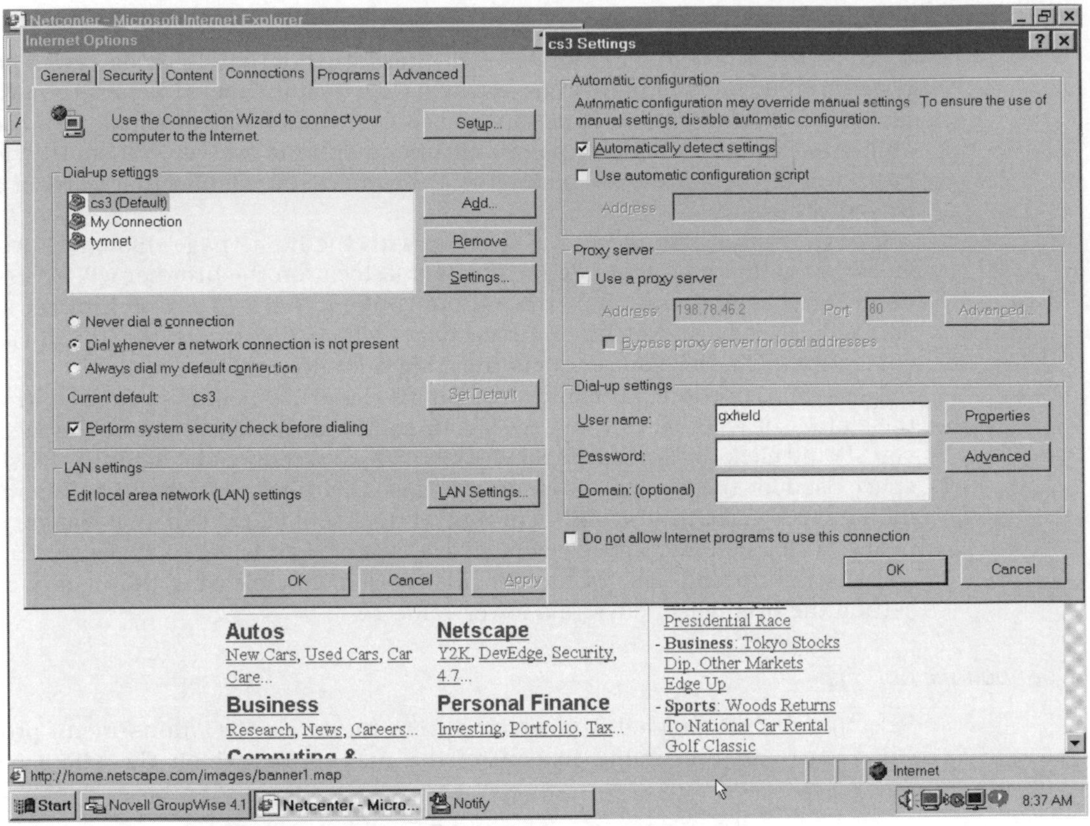

Figure 18.24 Viewing the Internet Options and dial-up settings dialog boxes

Internet Explorer browser communicates. Figure 18.24 illustrates two Connections dialog boxes. The dialog box on the left of Figure 18.24 is displayed when you select the tab labeled Connections from the Internet options menu.

If you examine this dialog box you will note three button selection options. Those options can be used to prohibit a dial connection, enable a dial connection to occur whenever a network connection is not present, or always use your default dial connection when you use your browser. The Connections tab is currently configured to have the browser first attempt to establish a connection via the network. If a network connection cannot be established the browser will use the default dial-up configuration specified.

Clicking on the Settings button results in the display of settings options associated with the selected dial-up settings entry. In the example illustrated in Figure 18.24 the highlighting of cs3 resulted in the display of the dialog box labeled cs3 Settings shown in the right portion of the display. In this example, the author's User Name is shown entered. If a password is also entered the resulting dial-up communications session will be established without user intervention, since a handshaking process between the author's computer and an Internet Service Provider (ISP) dialed by the cs3 settings will result in the User Name and password being communicated to the ISP.

If you click on the button labeled Properties shown in the right portion of Figure 18.24, a familiar dialog box will be displayed. That dialog box, which is shown in the left portion of Figure 18.25, is a modified version of the HyperTerminal communications program display previously discussed earlier in this chapter. From that dialog box you can change the telephone number to be dialed to include country code and area code, or even select a different modem for the connection. If you select the tab labeled Server Types the resulting display will appear similar to the dialog box located in the right portion of Figure 18.25. In actuality, the tab label is a bit misleading, as the dial-up server type really represents client software that is accessing the server. From this dialog box you can select one of several types of communications protocols, such as the Point-to-Point Protocol (PPP), CompuServe's version of PPP referred to as CISPPP, CSLIP and a Unix connection with IP header compression. If you click on the button labeled TCP/IP Settings, either you can have the Internet Service Provider provide a dynamic IP address or you can specify an IP address, in a manner similar to our prior examination of configuring a TCP/IP workstation in an earlier chapter in this book.

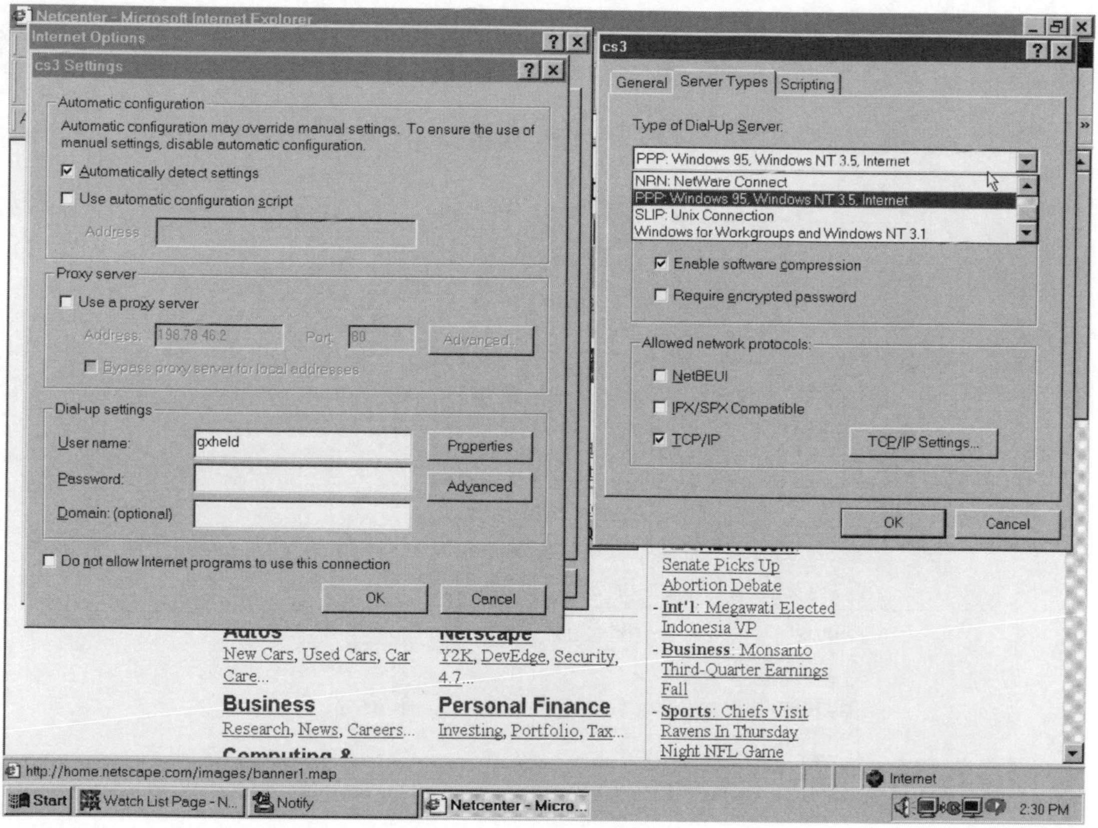

Figure 18.25 Viewing the settings of the default dial-up connection

18.3.2 LAN operation

Now that we have an appreciation for the applicable settings for dial-up communications, let us turn our attention to LAN settings.

If we turn our attention back to the dialog box labeled Internet Options that had its tab labeled Connections displayed, we will note a section of the display labeled LAN settings. That section, which was previously illustrated in the lower left portion of Figure 18.24, has a button labeled LAN Settings that you would invoke to configure certain local area network settings.

The left portion of Figure 18.26 illustrates the LAN Settings dialog box. Because a proxy firewall is in use which examines data sent from browsers on the network to the Internet and return traffic, this means it is not possible to use the gateway address in a normal TCP/IP protocol stack configuration. Instead, in the left portion of Figure 18.26 the box labeled Use a proxy server is checked and the address of the server and port number for HTTP traffic is entered.

If you click on the button labeled Advanced, you can specify proxy addresses for different applications. The display resulting from clicking on the button labeled Advanced is shown in the right portion of Figure 18.26. Because all

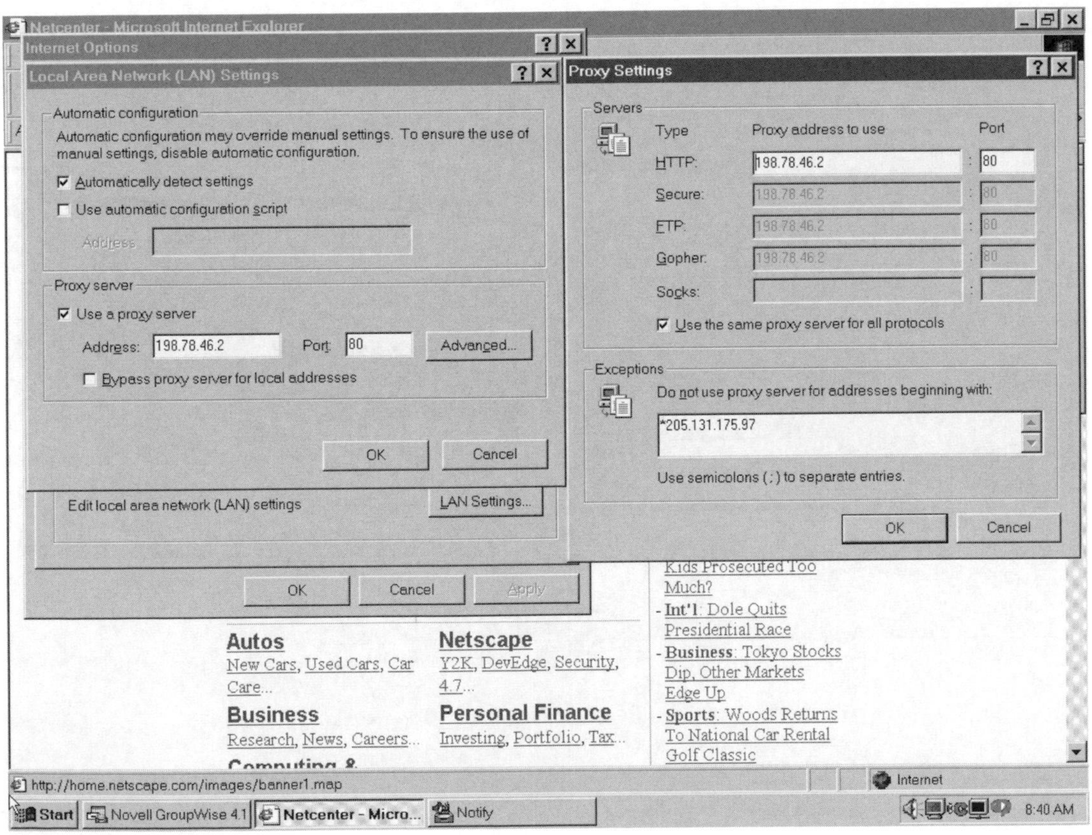

Figure 18.26 Configuring LAN settings when a proxy server is used

proxy services are performed by a common firewall the same proxy server address is used for all protocols and the applicable box is checked in the figure.

While it is a relatively easy process to configure applicable communications settings, it is also relatively easy to miss one or more. If you incorrectly configure a communications setting you may have a daunting task to locate the problem. For example, if your organization operates a proxy server and you either forget to check the applicable button or enter an incorrect address, the only message you will receive is an 'unable to display page' message from your browser. Thus, a careful examination of communications settings can alleviate potential problems.

18.4 REVIEW QUESTIONS

1. Discuss the relationship between the hardware of a personal computer and its ability to support the operation of communications software.

2. What are the advantages and disadvantages of menu-driven and command-driven software?

3. How can a clock display feature serve to control communication costs?

4. What is the function of a dialing directory in a communications program?

5. Assume that each time you press a key on your computer's keyboard two characters are displayed when you are using a communications software program. What is the most probable cause of the problem and how would you correct it?

6. What is the difference between unattended operations and remote control software?

7. Provide an example of the use of character conversion.

8. How can a data compression feature reduce the cost of communications?

9. What is the disadvantage of a stop-and-wait protocol?

10. What is the function of a break signal with respect to communications?

11. Under what circumstances would a line-feed control feature in a communications software program be of use?

12. If the bit composition of a data character is 01011010 and the encryption key for that character is 10100111, what would be the modulo-2 encrypted data character?

13. If you're using a 9600 bps modem and file transfer produces an average transfer rate of 2000 bps what do you attribute to the poor performance?

14. How does a communications program obtain information required to display the percentage of a file transfer in real time?

15. Describe a limitation of HyperTerminal in comparison to the dialing directories of other communications programs.

16. Describe the connectivity options of PC/3270.

17. How can you configure Microsoft's Internet Explorer to go to a predefined URL each time the browser is loaded?

18. Discuss two control options of Microsoft's Internet Explorer you can use to conserve computer storage.

19. Name two communications methods supported by Microsoft's Internet Explorer.

20. Describe what changes you should make to the configuration of a browser when your organization installs a firewall with proxy services.

19

FIBER-OPTIC, SATELLITE AND WIRELESS TERRESTRIAL COMMUNICATIONS

Both fiber-optic and satellite communications offer distinct advantages over conventional wire-based transmission. Although each method of communication is based on a different technological area, these two methods in many instances compete for a similar base of users. In fact, due to the widespread growth in the installed base of fiber by such vendors as MCI, Sprint, Frontier, Quest and AT&T in the United States, the new millennium may see a potential bandwidth glut, with the possibility that there will be more fiber capacity than commercial and residential customers can use.

In the area of satellite communications, the launch of numerous communication satellites during the 1980s and 1990s raises the potential for a different type of capacity glut. This glut is expected to arise from an increase in the number of communications satellites as well as an increase in the communications capacity of newly launched satellites in comparison with the capacity of earlier satellites.

In this chapter we will first focus our attention on fiber-optic transmission systems, examining their utilization with respect to their use within an organization as well as their use by communications carrier expanding network capacity. Next, we will examine the concepts and operation of communications satellites.

Although satellite communications represent a form of wireless, they are but one of several types of wireless communications. A second type that supports both voice and data transmission is terrestrial wireless communications, more commonly referred to as cellular communications, and is the focus of this chapter. In this section we will examine the basic operation of cellular communications as well as obtain an understanding of the different types of current and evolving cellular systems.

19.1 FIBER-OPTIC TRANSMISSION SYSTEMS

Although the majority of attention focused on fiber-optic transmission relates to its use by communications carriers, this technology can be directly applicable to many corporate networks. Due to the properties of light transmission, fiber-optic systems are well suited for many specialized applications to include the high-speed transmission of data between terminals and a computer or between computers located in the same building.

In this section we will first focus our attention on the components of a fiber-optic transmission system. After examining the advantages and limitations associated with such systems we will use the previous information presented as a foundation to analyze economics associated with cabling terminals to a computer, illustrating how this technology can be directly applicable to many computer installations. In doing so, we will examine the use of fiber-optic modems and multiplexers to denote the advantages and limitations associated with their use in a communications network. Finally, we will conclude this section by discussing the use of fiber networks by communications carriers and how this technology is being employed to provide communications to commercial and residential subscribers on a widespread scale.

Fiber optics is a rapidly evolving technology that has moved from a laboratory and consumer-product curiosity into a viable mechanism for low-cost, high data rate communications. Permitting the transmission of information by light, fiber optics is most familiar to individuals by the incorporation into contemporary desk lamps that were first marketed in the 1970s. In such lamps, bundles of individual fibers were attached to a common light source at the base of the lamp's neck. Radiating outward in a geometric pattern, the fibers were shaped to form a contemporary design. By polishing the end of each fiber, the common light source was reflected at the tip of each fiber, resulting in a very impressive visual display. Since the early 1970s a significant amount of fiber research has resulted in such fibers becoming available for practical data communications applications.

19.1.1 System components

Similar to conventional transmission systems, the major components of a fiber-optic system include a transmitter, transmission medium and receiver. The transmitter employed in a fiber-optic system is an electrical to optical

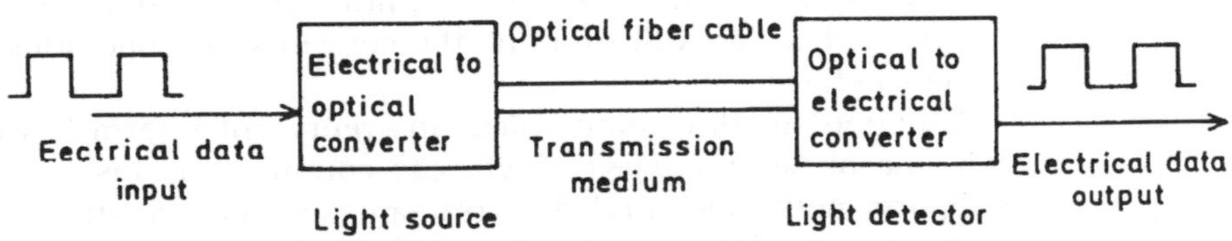

Figure 19.1 Fiber-optic system component relationship. The electrical to optical converter produces a light source which is transmitted over the optical fiber cable. At the opposite end of the cable an optical to electrical to electrical converter changes the light signal back into an electrical signal.

(E/O) converter. The E/O converter receives electronic signals and converts those signals into a series of light pulses. The transmission medium is an optical fiber cable which can be constructed out of plastic or glass material. The receiver used in fiber-optic systems is an optical to electric (O/E) converter. The O/E converter changes the received light signals into their equivalent electrical signals. The relationship of the three major fiber-optic system components is shown in Figure 19.1.

The light source and fiber transport

Currently two types of light sources dominate the electronic to optical conversion market: the light emitting diode (LED) and the laser diode (LD). Although both devices provide a mechanism for the conversion of electrically encoded information into light encoded information, their utilization criteria depends upon many factors. These factors include the type of fiber used to transport the light signal, response times, temperature sensitivity, power levels, system life, expected failure rate, and cost.

As previously noted in Chapter 3, there are two general types of optical fiber used to transport information: single mode and multimode. A single mode fiber has a core diameter of approximately 8 microns, which restricts the flow of light to one route in the fiber. This type of fiber is designed to transmit laser-generated light signals at distances of up to 20 to 30 miles, and is primarily used by communications carriers. In comparison, multimode fiber has a larger core diameter which permits light to travel on different paths, with each path considered a mode of propagation, giving the cable its name. Multimode cable can transport weaker light signals generated by LEDs over relatively short distances, typically under a few miles. LEDs using multimode cable are primarily used within buildings, such as to form an infrastructure for an optical LAN.

Two common types of multimode fiber are 50/125 μm and 62.5/125 μm. The numeric notation in the form of A/B signifies the diameter of the glass core through which the light travels (A) and the outside glass cladding diameter (B).

A LED is a PN junction semiconductor that emits light when biased in the forward direction; Typically a current between 25 and 100 mA is switched through the diode, with the wavelength of the emission a function of the material used in doping the diode.

The laser diode offers users a fast response time in converting electrically encoded information into light-encoded information; thereby, they can be used for very high data transfer rate applications. The laser diode can couple a high level of optical power into an optical fiber, resulting in a greater transmission distance than obtainable with light emitting diodes. Although they can transmit further at higher data rates than LEDs, laser diodes are more susceptible to temperature changes and their complex circuitry makes such devices more costly.

Optical cables

Many types of optical cable are marketed, ranging from simple one-fiber cables to complex 18-fiber, commonly jacketed, cables. In addition, a large

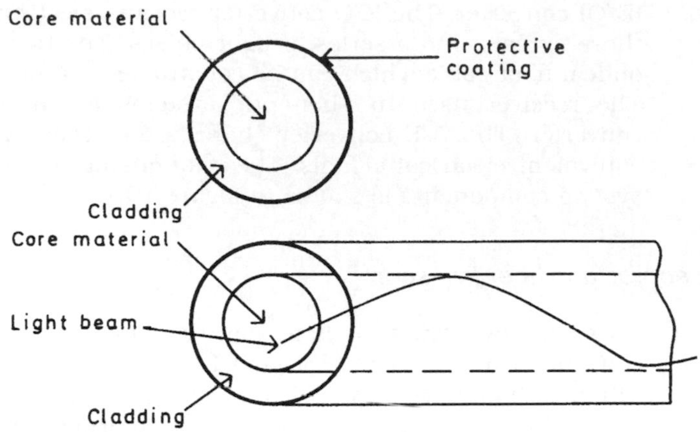

Figure 19.2 Optical fiber cable

variety of specially constructed cables can be obtained on a manufactured-to-order basis for vendors.

In its most common form an optical-fiber consists of a core area, cladding, and a protective coating. This is illustrated in the upper part of Figure 19.2. As a light beam travels through the core material the ratio of its speed in the core to the speed of light in free space is defined as the refractive index of the core. Thus

$$m = \frac{\text{speed of light in a vacuum}}{\text{speed of light in a medium}}$$

Typical values of the index of refraction for common materials are 1 for air, 1.33 for water, and 1.5 for glass.

A physical transmission property of light is that, while travelling in a medium of a certain refractive index, if it should strike another material of a lower refractive index, the light will bend towards the material containing the higher index. Since the core material of an optical fiber has a higher index of refraction than the cladding material, this index differential causes the transmitted light signal to reflect off the core–cladding junctions and propagate through the core. This is shown in the lower part of Figure 19.2

The core material of fiber cables can be constructed with either plastic or glass. Although plastic is more durable to bending, glass provides a lower attenuation of the transmitted signal. In addition, glass has a greater bandwidth, permitting higher data transfer rates when that material is used for fiber construction.

The capacity of a fiber-optic data link of a given distance depends on the numerical aperture (NA) of the cable as well as the core size, attenuation and pulse dispersion characteristics of the fiber. The NA value can be computed from the core and cladding refractive indices as follows:

$$\mathrm{NA} = \sqrt{m_1^2 - m_2^2}$$

where m_1 = core material refractive index, and m_2 = cladding material refractive index.

The numerical aperture value indicates the potential efficiency in the coupling of the light source to the fiber cable. Together with core material diameter, the numerical aperture indicates the level of optical power that can be transmitted into a fiber.

A function of both the fiber material and core–cladding imperfections, attenuation determines how much power can reach the far end of an unspliced link. When the numerical aperture, core diameter and attenuation are considered together one can determine the probable transmission power loss ratio.

Pulse dispersion is a measure of the widening of a light pulse as it travels down an optical fiber. Dispersion is a function of the cable's refractive index. Fibers with an appropriate refractive index permit an identifiable signal to reach the light detector at the far end of the cable.

Attenuation and bandwidth

Attenuation in fiber is similar to attenuation in a copper conductor. That is, it represents a reduction in the strength of a signal as it flows through a fiber. Fiber attenuation is measured in decibels per kilometer (dB/km) with a higher attenuation number meaning more loss and poorer performance.

Fiber bandwidth represents the capability of the fiber and is measured in units of MHz for a given distance in km, resulting in the use of the term MHz.km. Although the bandwidth value for a fiber is a constant, as the data rate increases the obtainable transmission distance decreases. Thus, the term MHz.km is used to reference both a data rate measured in MHz and a distance obtainable in km.

Types of fibers

Two types of optical fiber are available for use in cables: step index and graded index.

In a step index fiber an abrupt refractive index change exists between the core material and the cladding. In a graded index fiber the refractive index varies from the center of the core to the core–cladding junction. The gradual variation of the refractive index in this type of fiber serves to minimize the optical signal dispersion as light travels along the fiber core. The minimization of dispersion results from the light rays near the core–cladding junction traveling faster than those near the center of the core. The minimization of signal dispersion permits greater bandwidth, allowing transmission to occur over greater distances at higher data rates than available with a step index fiber. Thus, graded index fibers are commonly used for long haul, wide bandwidth communications applications. Conversely, step index fibers are used for communications applications requiring a narrow bandwidth and relatively low data rate requirement.

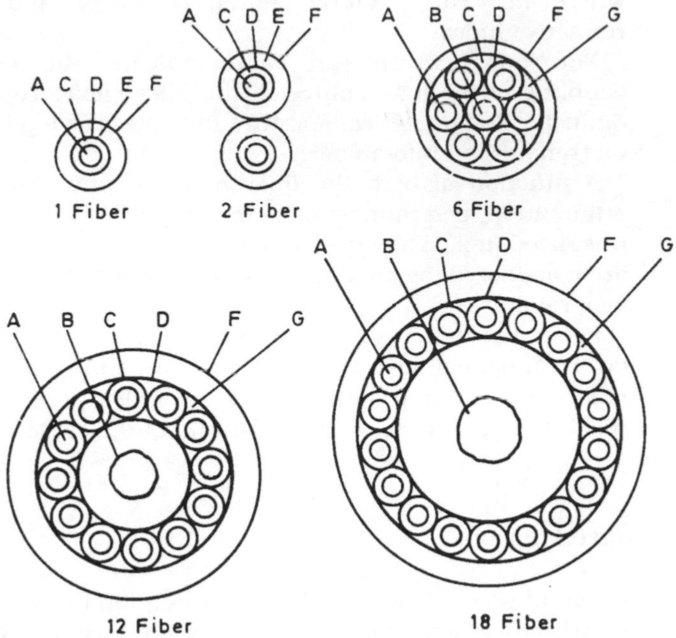

Figure 19.3 Common cable types (A optical fiber, B jacketed Kevlar strength member, C engineering plastic tubes, D plastic separator tape, E braided Kevlar strength member, FPVC jacket, Grip cord)

Common cable types

Five common types of optical cables are illustrated in Figure 19.3. Note that when a one-fiber cable has a transmitter connected to one cable and a receiver at the opposite end, it functions as a simplex transmission medium. As such, transmission is normally limited to one direction. The two-fiber cable can be considered a duplex transmission medium since each fiber permits transmission to occur in one direction. When transmission occurs on both fibers at the same time a full-duplex transmission sequence results. Thus, the 18-fiber cable is capable of providing nine duplex transmission paths.

Lightwave division multiplexing

Similarly to frequency division multiplexing (FDM) being used to permit multiple signals to be simultaneously transmitted on analog systems, a technique was developed to permit multiple transmissions to occur on an optical fiber. To accomplish this, multiple light signals at different wavelengths can be simultaneously transmitted, resulting in the technique referred to as lightwave division multiplexing (LDM). When applied to a one-fiber cable, this technique permits full-duplex transmission since different wavelengths are used for transmission in each direction.

During the late 1990s a relatively new technology referred to as Dense Wavelength Division Multiplexing (DWDM) evolved from increasing the capacity of LDM systems. In 1996 the first DWDM system enabled four wavelengths to be carried over a single fiber. Since then DWDM technology enabled eight, then doubled to 16, and further doubled, allowing 32 wavelengths of capacity to be obtained over a single fiber. In 1999 an 80-wavelength DWDM system was announced and some vendors anticipate an expansion of DWDM to 200 derived wavelengths of capacity.

The light detector

To convert the received light signal back into a corresponding electrical signal a photodetector and associated electronics are required. Photodetection devices currently available include a PIN photodiode, an avalanche photodiode, a phototransistor and a photomultiplier. Due to their efficiency, cost and light signal reception capabilities at red and near infrared (IR) wavelengths, PIN and avalanche photodiodes are most commonly employed as light detectors.

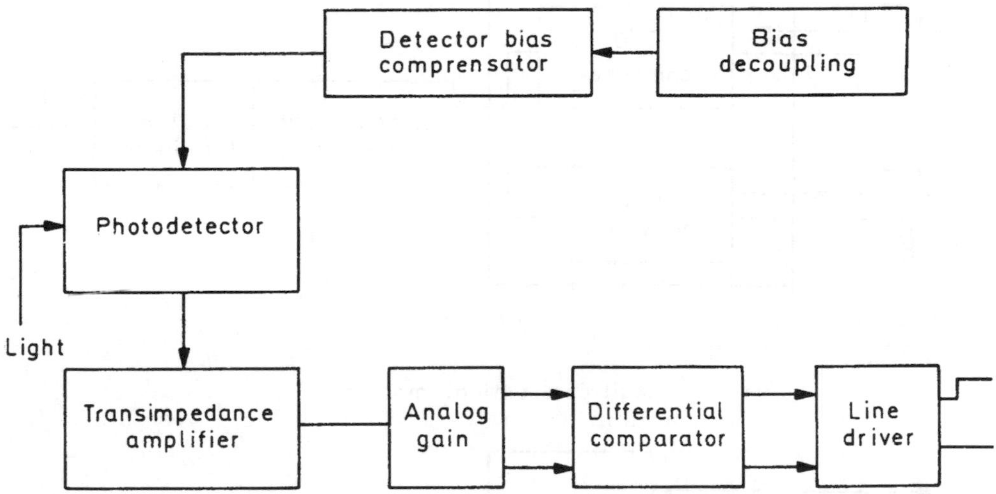

Figure 19.4 Light detector component. To filter d.c. input voltage, protect the photodiode and reduce the effect of electromagnetic interference, a light detector module has bias decoupling. Since an avalanche photodiode is a temperature-sensitive device, a detector-bias compensator is used to compensate for temperature changes that could affect the diode.

The photodetector converts the received optical signal into a low-level electrical signal. This detector can either be an avalanche or PIN diode, depending on the optical sensitivity requirements. The transimpedance amplifier is a low noise, current-to-voltage converter, whereas the analog gain element increases the voltage gain from the preceding amplifier to the level required for the decision circuitry. The differential comparator converts the analog signal into digital form by interpreting signals below a certain threshold as a '0' and above that threshold as a '1'. The line driver regenerates and drives the squared signal from the comparator for transmission over metallic cable

In comparison to PIN detectors, avalanche photodiode detectors offer greater receiver sensitivity. This increased light sensitivity results from their high signal-to-noise ratio, especially at high bit rates.

Since avalanche photodiode detectors require an auxiliary power supply which introduces noise, circuitry to limit such noise results in the device having a higher overall cost than a PIN photodiode. In addition, they are temperature-sensitive and their installation environment requires careful examination. A block diagram of the major components of a light detector is illustrated in Figure 19.4.

Other optical devices

Besides the optical devices previously discussed, two additional devices warrant coverage. The first device is an optical modem which performs 'optical' modulator and demodulation. The second device is an optical multiplexer which permits the transmission of many data sources over a single optical fiber.

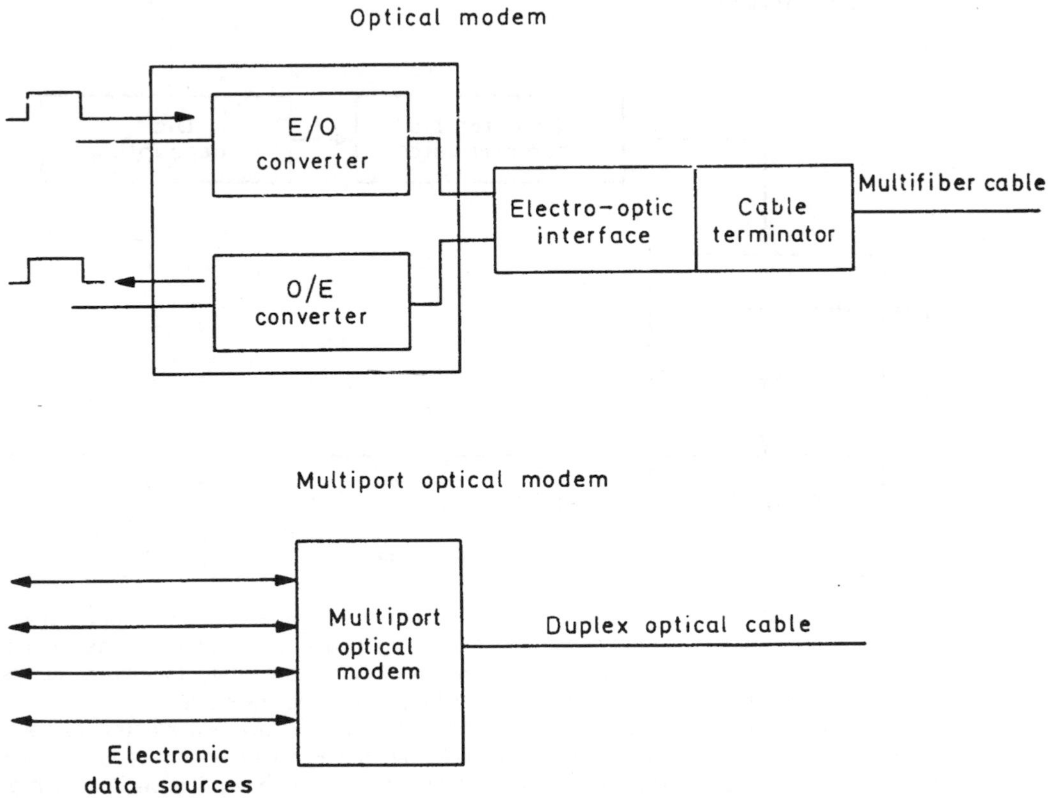

Figure 19.5 Optical modem. An optical modem can transmit and receive data over one multifiber cable, converting the electronic data source to light and the received light back into its corresponding electronic signal. Serving as a synchronous multiplexer, the multiport optical modem transmits data from up to four electronic sources as one optical signal

Optical modem

An optical modem is a device housing both an optical transmitter and an optical detector as shown in the top portion of Figure 19.5. Similarly to conventional modem development, a variety of optical modems have evolved. These variations range from single-channel stand-alone devices to multiport optical modems, the latter capable of functioning as a synchronous multiplexer and optical modem. The multiport optical modem permits conventional electronic bit streams from up to four data sources to be multiplexed and converted into one stream of light pulses for transmission on one optical fiber. In the lower part of Figure 19.5, a multiport optical modem servicing four data sources is shown.

Optical multiplexers

Functioning in a similar way to a conventional time division multiplexer, an optical multiplexer accepts the RS-232 electrical input of many data sources and multiplexes such signals for transmission over a single optical fiber. Included in the optical modem are a light source generator and detector. Thus, such devices are equivalent to a multiplexer with a built-in modem since they convert electronic multiplexed data into an optical signal and detect and convert light signals to their equivalent electronic signals prior to demultiplexing.

19.1.2 Transmission advantages

When used for data transmission, fiber-optic cables offer many advantages over cables with metallic electrical conductors. These advantages result from several distinct properties of the optical cables. The more common advantages associated with the utilization of a fiber-optic transmission system are listed in Table 19.1.

Table 19.1 Fiber-optic system advantages

Large bandwidth	Mixed voice, video, and data on one line
No electromagnetic interference (EMI)	No specially shielded conduits required Cable routing simplified Bit error rate improved
Low attenuation	Permits extended cable distances
No shock hazard, or short circuits	Can be used in dangerous atmospheres Common ground eliminated
High security	Transmission TEMPEST acceptable Tapping noticeable
Lightweight and small-size cable	Facilitates installation
Cable rugged and durable	Long cable life

Bandwidth

One of the advantages of fiber-optic cable over metallic conductors is the wide bandwidth of optical fibers. With potential information capacity directly proportional to transmission frequency, light transmission on fiber cable provides a transmission potential for very high data rates. Currently, data rates of up to 10^{14} bps have been achieved on fiber-optic links. When compared to the 56 kbps limitation of the public switched telephone network occurring over telephone wire pairs, fiber cable makes possible the merging of voice, video and data transmission on one conductor. In addition, the wide bandwidth of optical fiber provides an opportunity for the multiplexing of many channels of lower speed, but which are still significantly higher data rate channels than are transmittable on telephone systems.

Electromagnetic non-susceptibility

Since optical energy is not affected by electromagnetic radiation, optical-fiber cables can operate in a noisy electrical environment. This means that special conduits formerly required to shield metallic cables from radio interference produced by such sources as electronic motors and relays are not necessary.

Similarly, cable routing is easier because the rerouting necessary for metallic cables around fluorescent ballasts does not cause concern when routing fiber cables. Due to its electromagnetic interference immunity, fiber-optic transmission systems can be expected to have a lower bit error rate than corresponding metallic cable systems. By not generating electromagnetic radiation, fiber-optic cables do not generate crosstalk. This property permits multiple fibers to be routed in one common cable, simplifying the system design process.

Signal attenuation

The signal attenuation of optical fibers is relatively independent of frequency. In comparison, the signal attenuation of metallic cables increases with frequency. The lack of signal loss at frequencies up to 1 GHz permits fiber-optic systems to be expanded as equipment is moved to new locations. In comparison, conventional metallic cable systems may require the insertion of line drivers or other equipment to regenerate signals at various locations along the cable.

Electrical hazard

On fiber-optic systems light energy in place of electric voltage or current is used for the transfer of information. The light energy alleviates the potential of a shock hazard or short circuit condition. The absence of a potential spark makes fiber-optic transmission particularly well suited for such potentially dangerous industrial environment uses as petrochemical operations as well as refineries, chemical plants, and even grain elevators. A more practical benefit of optical fibers for most corporate networks is the complete electrical isolation that they afford between the transmitter and receiver. This results in the elimination of a common ground which is a requirement of metallic conductors. In addition, since no electrical energy is transmitted over the fiber, most building codes permit this type of cable to be installed without running the cable through a conduit. This can result in considerable savings when compared to the cost of installing a conduit required for conventional cables, whose cost can exceed $2500 for a 300 ft metal pipe.

Security

Concerning security, the absence of radiated signals makes the optical-fiber transmission TEMPEST acceptable. In comparison, metallic cables must often be shielded to obtain an acceptable TEMPEST level. Although fibers can be tapped like metallic cable, doing so would produce a light signal loss. Such a loss could be used to indicate to users a potential fiber tap condition.

Weight and size

Optical fibers are smaller and lighter than metallic cables of the same transmission capacity. As an example of the significant differences that can occur, consider an optical cable of 144 fibers with a capacity to carry approximately one million telephone conversations. The cable would be approximately one half inch in diameter and weigh about 3 ounces per foot. In comparison, the equivalent capacity copper coaxial cable would be about 3 inches in diameter and weigh about 10 pounds per foot. Thus, fiber-optic cables are normally easier to install than their equivalent metallic conductor cables.

Durability

Although commonly perceived as being weak, glass fibers have the same tensile strength as steel wire of the same diameter. In addition, cables containing optical fibers are reinforced with both a strengthening member inside and a protective jacket place around the outside of the cable. This permits optical cables to be pulled through openings in walls, floors, and the like without fear of damage to the cable.

With better corrosion resistance than that of copper wire, transmission loss at splice locations has a low probability of occurring when optical-fiber cable is used.

19.1.3 Limitations of use

As discussed, optical fiber cables offer many distinct advantages over conventional metallic cables. Unfortunately, they also have some distinct usage disadvantages. Two of the main limiting factors of fiber optic systems are cable splicing and system cost.

Cable splicing

When cable lengths of extended distances become necessary, optical fibers must be spliced together. To permit the transmission of a maximum amount of light between spliced fibers, precision alignment of each fiber end is required. This alignment is time-consuming, and depending upon the method used to splice the fibers, the installation cost can rapidly escalate.

Fibers may be spliced by welding, glueing, or through the utilization of mechanical connectors. All three methods result in some degree of signal loss between splice fibers.

Welding or the fusion of fibers results in the least loss of transmission between splice elements. The time required to clean each fiber end and then align and fuse the end with an electric arc does not make splicing easily suitable for field operations. A epoxy, or glueing, method of splicing requires the utilization of a bonding material that matches the refractive index of the core of the glass fiber. This method typically results in a higher loss than obtained with the welding process. Although mechanical connectors have the highest data loss among the three methods, this is by far the easiest method to employ. Although mechanical connectors considerably reduce splicing time requirements, the cost of good quality connectors can be relatively expensive. Currently, typical connectors cost approximately $10.

System cost

Good quality, low-loss single-channel fiber-optic cable costs between $0.50 and $1.00 per meter. A typical fiber-optic modem having a transmission range of 1 km costs approximately $300 to support a transmission rate of 10 to 100 Mbps. In comparison, conventional metallic cable for synchronous data transmission costs approximately 10 cents per foot while a line driver capable of regenerating digital pulses at data rates up to 56 kbps costs approximately $200.

Based on the preceding, the cost of a limited distance fiber-optic system at data rates up to 56 kbps will generally exceed that of an equivalent conventional metallic based transmission system. Only when high data rates are required or transmission distances expand beyond the capability of line drivers and, to some extent, limited distance modems, are fiber-optic systems economically viable.

19.1.4 Utilization economics

One of the most commonly used duplex fiber-optic cables costs $1.00 per meter which is equivalent to 33 cents per foot. An optical modem containing

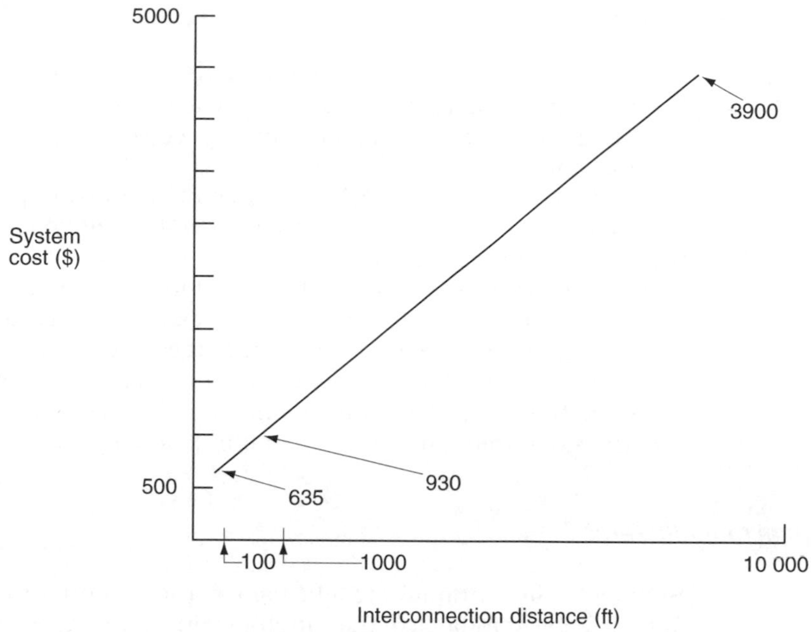

Figure 19.6 System cost varies with distance. Cost of a typical duplex optical-fiber cable and a pair of optical modems capable of transmitting up to 100 Mbps

an electric-to-optic and an optic-to-electric converter capable of transferring data at rates up to 10 Mbps costs approximately $300 per unit. The system cost of a pair of optical modems as a function of distance is illustrated in Figure 19.6. Note that as long as no splicing is required, costs are a linear function of distance.

Suppose that a requirement materializes which calls for the communications linkage of four high-speed digital devices located in one building of an industrial complex to a computer center located in a different building 10 000 feet away. What fiber-optic system can support this requirement and what are the economic ramifications of each configuration?

Dedicated cable system

Equivalent to individually connecting devices on metallic cables, four individual fiber-optic systems and four cables could be installed. Here eight optical modems would be required, resulting in the cost of the modems being $2700. With four cables being required, 40 000 ft of cable would cost $13 200, resulting in a total cost of $15 600 for this type of network configuration. In addition, a substantial amount of personnel time may be required to install four individual 10 000 ft cables.

Multichannel cable

A second method that can be employed to link multiple devices at one location is by the employment of multichannel cable. We can examine the economics associated with multichannel cable by considering the cost of an eight-channel cable.

A typical eight-channel cable capable of supporting four duplex transmissions costs approximately $5 per meter or $1.65 per foot. On a duplex channel basis, this represents a cost of 41 cents per foot per channel. This is approximately the same cost as an individual duplex cable. The use of most multichannel cable does not offer any appreciable savings over individual cable until ten or more channels are packaged together. Price to excluding the use of a multichannel cable, one should carefully consider cable installation costs since the time required to install one multichannel cable can be significantly less than the time required to install individual cables.

Optical multiplexers

Similar to conventional metallic cables, the potential installation of parallel optical cables indicates that multiplexing should be considered. Prior to deciding on the use of an optical multiport modem or optical multiplexer, you should examine the type of data to be transmitted as well as the data transfer rates required. If each data type is synchronous and no more than four data sources exist, the utilization of an optical multiport modem can be considered. If a mixture of asynchronous and synchronous data must be supported or if more than four data sources exist you should consider an optical multiplexer.

Currently, four-channel synchronous optical multiport modems cost approximately $500 per unit. Returning to our system requirement, a pair of optical multiport modems and one cable could support the four data terminals. Here the total system cost would be reduced to $4900, of which $100 would be for the pair of optical multiport modems and the remainder for the cable.

Suppose that one or more of the data terminals was an asynchronous device, or more than four terminals required communications support. For such situations an optical multiplexer should be considered. One optical multiplexer currently marketed costs $1500 and supports up to eight data channels at data rates up to 64 kbps. Since an optical receiver/transmitter is included in the multiplexer, only the cost of one duplex cable must be added to the cost of a pair of multiplexers. Doing so, we will obtain the cost of an optical transmission system capable of supporting up to eight data sources at data rates up to 64 kbps. For such a system the cost of 10 000 ft of cable and the two optical multiplexers would be $6900.

In Table 19.2 you will find a comparison of the four previously discussed network configurations. As indicated, both cost and expansion capacity varies widely between configurations. Although fiber-optic systems represent new technology, the financial aspects and expansion capability must be considered similar to the process involved with conventional metallic cable based transmission systems.

Table 19.2 Network configuration comparisons of four data sources and 10 000 ft transmission distance

Individual cable	Multichannel cable	Optical multiport modem	Optical multiplexer
System cost			
$15 600	$15 600	$4900	$6900
Expansion capability			
Extra cable and transmitter/receiver per data source	Requires more expensive cable and additional transmitters/receivers	Cannot support more than four channels	No cost to add support for up to four additional channels

19.1.5 Carrier utilization

The large bandwidth, immunity to noise and economics of installation were key driving forces that resulted in the installation of fiber-optic systems by communications carriers. Two of the earliest fiber-optic networks placed into commercial service by communications carriers were AT&T's Northeast Corridor and Pacific Telephone's North/South Lightwave Project, both of which were placed into service in 1983.

The Northeast Corridor system involved the installation of 78 000 fiber-km of lightwave circuits between Richmond, VA, and Boston, MA. In this system a cable containing 144 optical fibers has the capacity to simultaneously transmit over 240 000 voice conversations. AT&T's first standard digital lightwave system was known as Lightwave System FT3, which transmitted at the 44.7 Mbps T3 rate. The principal use of the Northeast Corridor system was for interoffice communications between metropolitan areas. A second Lightwave System standard was the Lightwave System FT3C. The FT3C system operated at 90.524 Mbps and provided support for up to 1344 voice conversations per fiber within a multi-fiber cable. The actual multi-fiber cable used in an FT3 system can contain up to 144 individual fibers packaged in ribbons of 12 fibers, providing a maximum capacity in excess of 80 000 voice conversations per multi-fiber cable to include backup strands in the cable that are used in the event that one or more primary individual fiber strands should fail.

The North/South Lightwave Project linked Sacramento to Los Angeles by over 515 miles of lightwave circuits. Since 1983 the growth in fiber-km of installed cable has been almost exponential, with several million fiber-km installed in the United States. In 1989, Sprint Communications completed the installation of their fiber network which spans the United States. The completion of this network allowed that communications carrier to remove its network of microwave towers that previously formed its backbone network. Since 1989 other long-distance communications carriers converted their infrastructure to a fiber optic based network. In addition to AT&T and MCI Worldcom, several relatively new communications carriers, such as Frontier Communications, Global Crossings and Quest Communications, created new fiber optic based networks that span the continental United States. In doing so most communications carriers created SONET rings which provide a high level of communications redundancy.

To obtain an appreciation for the operation of SONET rings requires knowledge of the operation of the Synchronized Optical Network which is the focus of the next section in this chapter.

19.1.6 SONET

One of the more significant developments in the use of fiber optic transmission systems occurred in 1985. In that year Bellcore proposed the Synchronized Optical NETwork (SONET) concept, an optical multiplexing hierarchy which defined a new series of optical carrier (OC) transmission rates, as well as a set of standard interfaces and signaling formats. The importance of SONET lies in the fact that the resulting standard defined by the ITU-T and ANSI addressed three key issues. Those issues were the definition of a multiplexing standard for transporting high speed digital signals, interoperability of the use of SONET by different vendors and a set of management functions.

Multiplexing

Until the introduction of SONET the multiplexing hierarchy for high speed transmission (T1 to T3 in North America and El to E3 in Europe) was based on asynchronous transmission with synchronization provided by framing bits. As the digital transmission hierarchy evolved based on asynchronous transmission, the resulting higher level signals required full demultiplexing for equipment to obtain the ability to extract individual DS0s. For example, in North America, four DS1 (Digital Signal Level 1) signals carried on individual T1 wire pairs were combined by a device known as an M12 multiplexer, resulting in a DS2 signal used exclusively by carriers for transmission on a T2 wire pair carrier system operating at 6.312 Mbps. A second multiplexer known as an M23 is then used to combine seven DS2 signals into a DS3 which is the T3 carrier operating at 44.736 Mbps. Thus, a T3 represents 28 T1s and 672 (28×24) individual DS0s.

Since the identification of individual DS0s is accomplished by examining framing bits of individual DS1 signals, the extraction of a single 64 kbps DS0 channel from a DS3 signal requires the DS3 to be demultiplexed into 28 individual DS1s. To do so requires multiplexing equipment at the location where one or more DS0s are to be dropped to demultiplex the DS3 signal, and then remultiplexing the DS1s. To accomplish this requires equipment costing $20 000 to $30 000, unless the carrier elects not to provide a DS0 dropping and inserting capability.

SONET addresses the previously described problem by establishing a standard multiplexing format using multiples of 51.84 Mbps signals as building blocks. The SONET specification defines a hierarchy of standardized digital data rates based on multiples of 51.84 Mbps, as noted in Table 19.3. To understand why the payload rate is lower than the data rate of the signal requires an examination of the SONET frame structure which is illustrated in Figure 19.7.

Table 19.3 SONET signal hierarchy

SONET designation	ITU-T designation	Data rate, Mbps	Payload rate, Mbps
STS-1	—	51.84	50.112
STS-3	STM-1	155.52	150.336
STS-9	STM-3	466.56	451.008
STS-12	STM-4	622.08	601.344
STS-18	STM-6	933.12	902.016
STS-24	STM-8	1244.16	1202.688
STS-36	STM-12	1866.24	1804.032
STS-48	STM-16	2488.32	2405.376

The SONET frame

The basic SONET frame illustrated in Figure 19.7 occurs 8000 times per second, or once per 125 μs. Each frame consists of 9 rows of 90 bytes (8100 bytes) of which three bytes per row or 27 bytes per frame represent line and section overhead. The first three bytes in the first three rows represent section overhead which conveys information about the span between two pieces of fiber equipment providing a point-to-point transmission path. Included in those 9 bytes are framing information, a parity count for span performance monitoring, a maintenance communications channel and data communications channels for remote monitoring and control functions. The next six rows of overhead consisting of three bytes per row represent line overhead. Line overhead contains parity checking on an office-to-office basis, interoffice data communications bytes, protection switching control information and payload pointers, the latter providing the

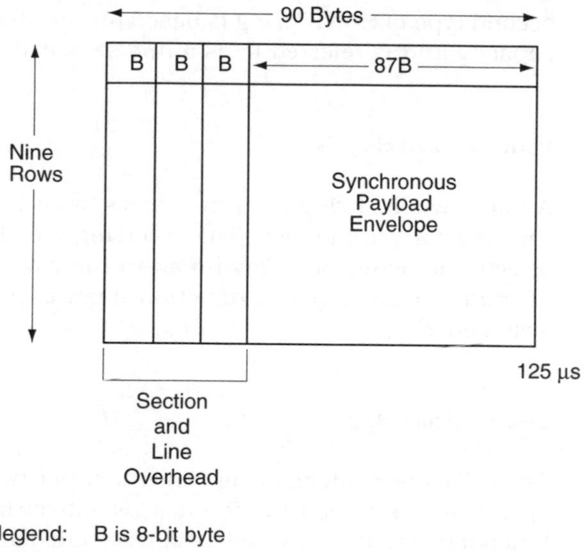

Figure 19.7 SONET frame structure

location where the payload begins within an STS-1 frame. The pointers enable equipment to locate and extract or insert DS0s without requiring the demultiplexing of the entire signal as is the case when asynchronous signaling is used

The frame illustrated in Figure 19.7 is more formally referred to as a Synchronous Transport Signal Level 1 (STS-1) frame. This frame format is used for all SONET transmission, with more copies of the STS-1 frame being transmitted per unit of time to create a higher level STS. For example, to create an STS-3 which represents an Optical Carrier (OC) 3 rate, three STS-1 frames are transmitted for each 25-μs time period. It should be noted that the 125-μs time period represents 1/8000 of a second and was selected as the duration of the STS-1 frame, since the PCM sampling rate is 8000 times per second.

SONET deployment

Although all communications carriers deploying SONET follow the same set of standards, differences exist in the network architecture over which SONET is deployed, Those differences are primarily between linear and ring-based network architectures which correspond to differences in the abilities of the architectures to survive cable cuts and other impairments that provide a near instantaneous restoral time.

SONET rings

Two types of SONET ring are used by some carriers, which provide both protection against cable cuts and other impairments, as well as quick restoral in the event that a portion of a ring is broken. One SONET ring is based on the switching of individual paths and is referred to as a path switched ring. The second type of SONET ring is based on the switching of the entire optical line capacity and is referred to as a line switched ring.

Path switched ring

A path switched ring uses two fibers, with traffic flowing both ways around the ring for redundancy. The receiving location monitors both signals and selects the better one. Any break in the ring is compensated for by the flow of traffic in the opposite direction. Figure 19.8(a) illustrates a SONET path switched ring.

Line switched ring

A SONET line switched ring can use either two or four fibers. During normal operating conditions traffic on a two-fiber ring flows only in one direction. When a fiber cut is detected, traffic is routed around the failure onto a 'protect' fiber in the opposite direction

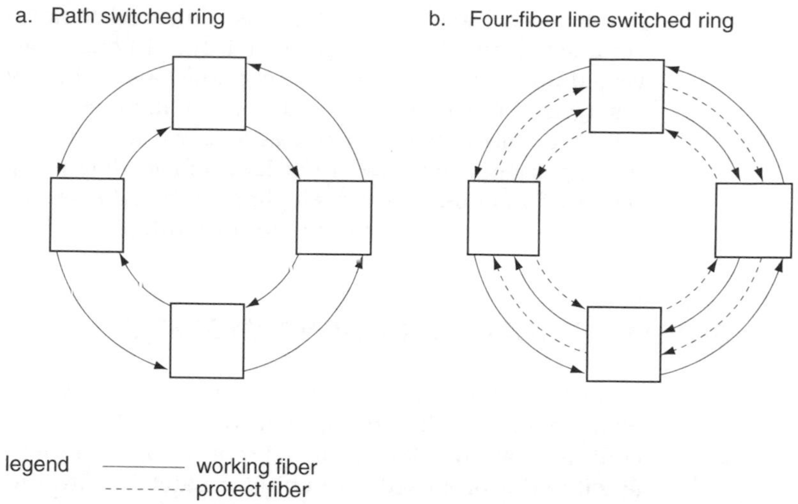

Figure 19.8 SONETring types

A four-fiber line switched ring provides two working fibers transmitting traffic in opposite directions and two 'protect' fibers used to reroute traffic in the event of a cable cut, as illustrated in Figure 19.8(b). This architecture enables a four-fiber line-switched ring architecture to share protection capacity, and it results in an increase in traffic capacity over a path switched ring.

One of the first communications carriers to deploy SONET rings on a nationwide basis was Sprint. By 1996, Sprint had approximately 30 SONET rings deployed across many states in the United States. Thosc rings, which use four-fiber lines, enable a cable cut to be compensated for in approximately 50 ms, which is unnoticeable for voice conversations and minimizes the loss of bits for users transmitting data. By 1998 most carriers converted their fiber infrastructure to SONET rings, due to the significant advantages they provide against cable cuts as well as in restored time when a cut has occured.

One of the key advantages of communications carrier fiber optic networks over conventional wire cable and microwave networks is the expansion capability of the former. In many instances the replacement of the transmitter and receiver is all that is required to increase the transmission capacity of the fiber cable. In other instances, when the physical capacity of a fiber is reached at one optical frequency, it becomes possible to use optical multiplexing to increase the capacity of the fiber. As previously discussed, under DWDM technology up to 200 derived wavelengths can be created through a common fiber. Other advantages of communications carrier fiber networks are similar to those advantages previously discussed for the use of fiber systems within an organization: immunity to noise, low attenuation, lightweight and small size cable, and high security.

The advantages of fiber-optic systems resulted in several demonstration projects that commenced operation in 1989. Each of these projects involves the wiring of fiber directly to residential and commercial users, with the bandwidth of fiber permitting voice, data and cable TV to reach the home and commercial user via a single local loop. Although regulatory approval and the vested interests of telephone companies and cable TV operators remain to be resolved, the use of fiber could eventually open a whole new realm of technology to subscribers. As an example, within a few years it may be possible to use a personal computer to order a movie, download the movie to your video recorder and talk on your telephone, with all communications occurring on a fiber cable linking your home to the world.

19.2 SATELLITE COMMUNICATIONS SYSTEMS

Arthur C. Clarke, the author of such well known books as *2001*, *2010* and *Rendezvous with Rama*, can be considered as the father of satellite communications. Several decades prior to the launch of the first geostationary satellite, he proposed launching a satellite into an Earth orbit that would result in the speed of the satellite being-matched to the rotation of the Earth. That orbit, approximately 22 300 miles above the equator, is known as a geostationary orbit. It is also referred to as the Clarke orbit in his honor.

19.2.1 Operation overview

When placed in a geostationary orbit a communications satellite has a fixed location in the sky with respect to ground stations, permitting the antenna of a ground station to be oriented to a fixed position. The satellite then functions as a line-of-sight microwave system, permitting one ground station to communicate with another ground station by 'bouncing a signal' off a satellite.

Each communications satellite has a number of transponders (transmitter–receiver) which function as a microwave relay in the sky. Ground stations transmit to the satellite on assigned frequencies on what is known as the uplink, whereas they receive information from the satellite on different frequencies, the latter referred to as the downlink. Figure 19.9 illustrates the relationship between a satellite and several ground stations and the transmission delay occurring from one satellite hop. Although propagation delay has no undesirable effect when signals, such as television and radio, are broadcast since they continuously flow in one direction and all signals are shifted in time, the same is not true with respect to data transmission. As previously discussed in Chapter 9, the approximate 250 ms delay can significantly decrease the throughput associated with stop-and-wait protocols. Even for voice communications, some persons will notice the quarter-second delay, which can become quite annoying.

19.2.2 Satellite access

Satellite technology involves the use of two methods to enable this communications facility to be shared among multiple users. These methods are

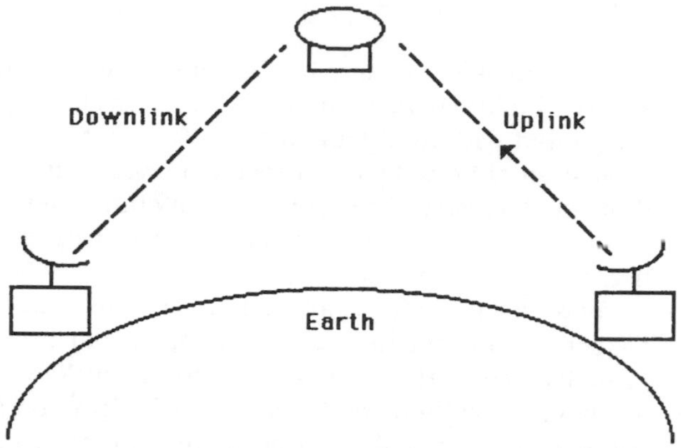

Figure 19.9 Round trip delay. For TV or radio, all signals experience a uniform delay. For voice or data, the round trip delay on a satellite link can reduce throughput (data) or be annoying (voice)

based upon multiplexing and are known as frequency division multiple access (FDMA) and time division multiple access (TDMA).

Frequency division multiple access (FDMA)

Frequency division multiple access involves the use of frequency division multiplexing to allocate portions of a satellite transponder to different users. As an example of the use of FDMA, assume that a satellite transponder supports a bandwidth of 500 MHz. If the bandwidth required for 128 voice-grade channels is 10 MHz, FDMA could be used to partition the satellite transponder into fifty 128 voice channel groupings. Then, each ground station would communicate with the satellite transponder using different frequency assignments to access a distinct group of 128 voice channels.

Time division multiple access (TDMA)

Time division multiple access results in the use of a satellite transponder being partitioned by time, similar to the method by which a multiplexer partitions access on to a high-speed data link. In the TDMA method of satellite access each ground station obtains the entire data rate capacity of the transponder for a very short period of time.

Because ground stations can be tuned to specific frequencies under FDMA, that method inherently ensures the correct ground station receives a designated transmission. Under the TDMA access method the position of data within a frame of information provides the mechanism for a receiving ground station to know that data are for that station. To minimize the effect of time delays causing the wrong ground station to receive data, TDMA also packetizes transmission into frames, incorporating an identifier into a header encoded into each slot.

Advantages of use

Four of the key advantages of satellite communications are network availability, distance-insensitive pricing, level of performance, and the ease associated with modifying a satellite network.

Most satellite networks provide a level of availability that at least matches that of terrestrial networks. Availability levels of 99.5% are currently achievable, with outages primarily occurring during violent thunderstorms, heavy rain, and periods of peak sunspot activity.

Concerning pricing, most satellite carriers have a distance-insensitive price schedule. This means, as an example, that a circuit from New York to San Francisco costs the same as a circuit from New York to Chicago because you are paying for the use of a portion of a transponder. Thus, for long distance transmission, satellite circuits are many times more economical than equivalent terrestrial facilities.

The quality of satellite communications is normally as bad as the best terrestrial network. Thus, the error rate for data transmission and the number of retransmissions on a satellite circuit are lower than on equivalent terrestrial circuits.

In the area of network expansion or modification, satellite communications in many instances may be vastly superior to terrestrial communications. As an example, consider a business that wants to implement an automated inventory system to connect 30 locations across the United States. Using conventional communications might require the installation of a multidrop network of leased lines requiring several months to complete. Using satellite communications would require the installation of ground stations at each location which can normally be set up in a few days.

19.2.3 Very small aperture terminal (VSAT)

Perhaps the highest growth area of satellite communications involves the use of very small aperture terminal (VSAT) ground stations. Since the mid-1980s, tens of thousands of retail stores, banks and other commercial organizations have installer VSATs to provide communications support for order entry, inventory control, financial transaction processing, and similar applications. Note that for each of these applications the delay associated with satellite communications has a minimal effect. As an example, if you drive into a drive-in bank the extra 1s delay associated with transmitting your transaction information by satellite is not noticeable. Thus, by focusing on delay-immune applications, satellite communications usage has begun to experience significant growth.

19.2.4 Low earth orbit satellites

One of the key advantages of a geosynchronous satellite is also a disadvantage. That is, although a geosynchronous satellite stays over a predefined location and constantly supports communications to the service area it covers, it cannot provide communications to other locations. Recognizing this limitation resulted in the development of low earth orbit satellites (LEOS).

The first generation of low earth orbit satellites was developed to perform military reconnaissance. Such satellites orbit the earth very quickly, completing an orbit in 90 minutes or less, and included thrusters to alter their orbit over special targets of interest. The orbits of the first generation of LEOS were rather low, resulting in an orbit lifetime measured in months compared to the decades associated with geosynchronous satellites. However, the relatively low orbit permits less complex and costly launch vehicles to be used. In addition, because the satellite orbits the earth in a low orbit, communications can be supported to ground stations that can use smaller antennas.

Low earth orbit satellites are used to support three main types of applications – navigation, messaging and voice. Unlike geosynchronous satellites that may result in awkward sounding voice due to the round trip delay associated with transmitting signals up to 26 000 miles for reflection back to earth, LEOS have a relatively low delay. Thus, they can be used to provide better support for voice calls. In addition, because they orbit the earth at a lower altitude than geosynchronous satellites, they can support the use of small antennas and less powerful handsets. In fact, it is difficult, if not impossible, for a geosynchronous satellite to support transmission to handsets. In comparison, several LEOS-based systems were designed to provide satellite-based communications to subscriber handsets similar to earth-based cellular handsets.

Iridium

One example of a commercial LEOS system used to support voice is Iridium, a name that represents the 77th element in the periodic table. Iridium was originally designed to use 77 satellites but was later scaled back to 66 satellites.

Iridium was designed by Motorola Corporation which remains a large backer; however, it is now a private company. One of the more interesting features of Iridium is the fact that instead of directly relaying communications back to an earth station it relays calls from one satellite to another (inter-satellite communications) prior to sending the call to an applicable earth station. Although Iridium launched its first satellite with great publicity in August 1997, its commercial success has not materialized. The high cost of its handsets that range between $1500 and $2000 and its call cost that can exceed $2.00 per minute limited its subscriber base to the wealthy.

Orbcomm

A second LEOS-based communications system that warrants attention is Orbcomm. Orbcomm is a satellite-based data communications network that supports messaging and not voice. The system is based upon the use of low frequency communications, with the 137–138 MHz band used for the down-link while the 148–149.9 MHz band is used for transmission on the uplink.

The first two of 36 planned Orbcomm satellites were launched in April 1995 and the remainder should be in orbit by the time you read this book.

19.3 WIRELESS TERRESTRIAL COMMUNICATIONS

There are numerous types of wireless terrestrial communications, ranging from methods employed to use industrial equipment to the remote control you operate your television with. In this section we will turn our attention to two specific types of wireless terrestrial communications – cellular communications, which allow persons to talk as well as send messages from both mobile and fixed locations without wires, and wireless LANs. Both types of communications are covered in this section, as they represent popular current and evolving methods of wireless terrestrial communications that either have or can be expected to have a considerable effect upon the manner by which we communicate.

19.3.1 Cellular communications

The first type of cellular communications system dates to work performed by Bell Telephone Laboratories during the 1970s. In order to support communications from mobile users, multiple fixed stations were located throughout a geographical area. Each area of coverage of a station is referred to as a cell. The frequency of use in a geographic area is structured so that no adjacent cells share common operational frequencies.

Figure 19.10 illustrates an example of the cell structure of a wireless communications system. In examining Figure 19.10 note that while certain frequencies can be simultaneously used in different cells those cells cannot be adjacent to one another. This restriction allows mobile users to move from one area to another and maintain communications by changing frequencies. This restriction also enables support for more simultaneous cellular users as it does not restrict the use of a particular frequency within the entire cellular system.

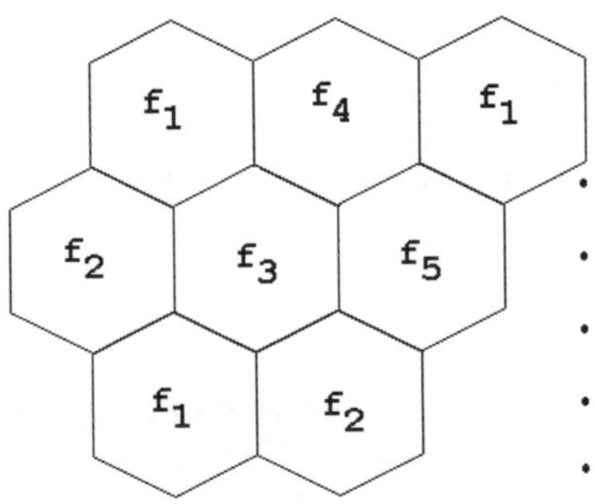

Figure 19.10 Adjacent cells within a geographic area support subscribers on different operating frequencies

As a mobile user begins to exit a cell area the cell's base station will note a loss of signal strength. This signal strength loss also tells the base station to send a message to the station that supports the cell the mobile user is entering, as well as to communicate with the mobile user on a special control channel. In doing so the base station in the cell the mobile user is leaving informs the new base station to take over, and switches the frequency of the mobile user to the frequency supported by the gaining station in the cell the user is entering. The actual switching process is performed by a mobile telephone switching office (MTSO) which, for economies of scale, typically is connected to numerous cell stations via land lines. In actuality, most cell stations are antennas and the MTSO switches calls among frequencies transmitted and received on those antennas. The previously described procedure is referred to as a 'handoff' and forms the basis for mobile cellular operations.

As indicated in the previous description of a mobile cellular system, there are three basic components required to establish the system. First, each subscriber requires a cellular phone that operates over a range of frequencies. Secondly, each cell has a station or antenna, with switching performed by the MTSO. Although cellular systems have considerably evolved and there are several types of systems in operation, each continues to be based upon the three basic components previously mentioned.

Now that we have an appreciation for the basic operation of a cellular system, let us turn our attention to the different types of cellular systems.

AMPS

The first commercial cellular system is referred to as the Advanced Mobile Phone Service (AMPS) or analog cellular and is both the oldest and currently the most widely used cellular service in the United States. AMPS is based on a Frequency Division Multiple Access (FDMA) technique under which the 800–900 MHz frequency spectrum is subdivided into 25 kHz channels, with subscribers switched from one 25 kHz channel to another as they move from one cell into another. Because the 800–900 MHz band is subdivided into channels by frequency and multiple users are supported by assigning each user to a different frequency, the term FDMA is used to reference this access technique.

Figure 19.11 provides a comparison of the three methods of cellular system access currently in use. The top portion of that illustration shows the FDMA access method used by AMPS, indicating that each conversation occurs on a distinct frequency.

During the 1980s AMPS evolved from a curiosity to a business tool, with the subscriber base of different analog systems rapidly expanding. Because the use of 25 kHz channels within the 800–900 MHz bandwidth limited the number of available channels per cell to approximately 400, the growth in the use of cell phones in urban areas soon resulted in many subscribers becoming unable to make a connection to their cellular service, a condition referred to as blocking. In an attempt to increase the capacity of cellular systems two new access methods evolved, TDMA and CDMA. Both represent digital cellular communications and differ in the manner by which subscribers obtain access to a cellular system.

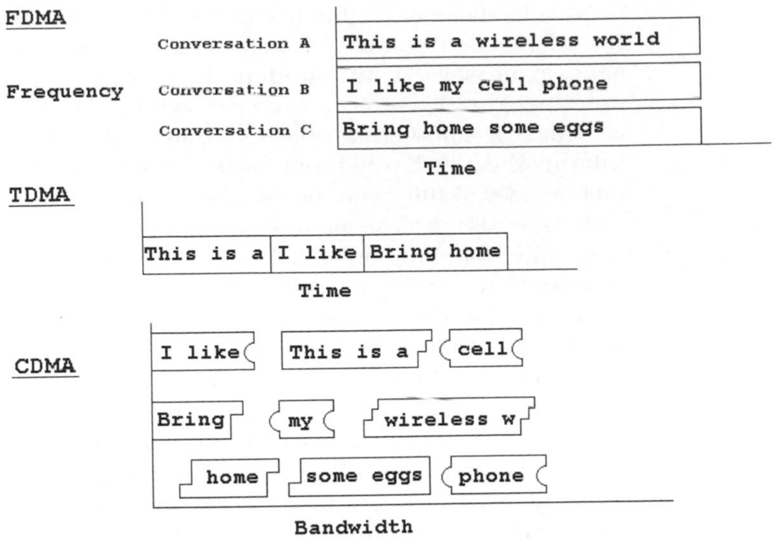

Figure 19.11 Cellular system access and transmission methods

TDMA

Time Division Multiple Access (TDMA) represents a digital cellular technique under which subscriber conversations are digitized and compressed and then placed into time slots assigned to a particular frequency.

The middle portion of Figure 19.11 illustrates the operation of TDMA with respect to FDMA. Note that portions of each of the three conversations occurring are multiplexed by time on the same frequency under TDMA. Because times are allocated as slots of one 150th of a second and assigned among three subscribers, it becomes possible for the TDMA access technique to support a tripling of the number of cell users over AMPS. Other benefits associated with TDMA and CDMA since both are digital techniques include the ability to authenticate calls, support for calling number ID, and voice privacy. Concerning the latter, unlike an AMPS conversation that can be easily scanned and listened to, a digitized call cannot be easily decoded by a scanner.

There are two operating ranges for TDMA based systems – 800 MHz and 1900 MHz. Digital cellular communications occurs at 800 MHz in North America while a new type of digital cellular system referred to as Personal Communications System (PCS) operates at 1900 MHz. Because the wavelength (λ) of a signal is inversely proportional to its frequency (f), i.e.,

$$\lambda \propto = 1/f$$

this results in PCS signals having a relatively short wavelength. Due to this, PCS cell sites are smaller than digital cellular cell sites and this explains why many PCS systems only provide coverage of the general area along an interstate highway instead of a wider area. One particular version of digital TDMA cellular that warrants discussion is Global System for Mobile Communications (GSM). GSM was originally developed in Europe and is a European standard that is also quite popular in Asia and the Middle East.

GSM was a narrowband version of TDMA that enables eight simultaneous communications on a single channel, considerably increasing the capacity of a cell. To do so GSM uses a voice compression technique that reduces the data rate required to transport the conversation to approximately 13 kbps.

A few years ago PCS operators in North America introduced GSM at 1900 MHz. Unfortunately, the European version operates at 900 MHz, which makes handsets incompatible between GSM in North America and other parts of the world. However, several vendors recently introduced dual-mode handsets that enable global GSM roaming. It should be noted that while dual-mode handsets supporting TDMA and FDMA have been available in the United States for several years, such handsets support North American digital frequencies and cannot be used in Europe. However, several vendors introduced 'universal' handsets during 1999, which, although currently relatively expensive, provide a global roaming capability.

CDMA

The third type of wireless access protocol is known as Code Division Multiple Access (CDMA). CDMA subdivides speech into small pieces that were numbered or coded and transmitted in both the time and frequency domain, as illustrated in the lower portion of Figure 19.11.

CDMA uses a relatively wide bandwidth and represents a form of spread-spectrum communications. Similar to TDMA, CDMA operates in both the 800 MHz and 1900 MHz bands in North America. The 1900 MHz band represents CDMA-based PCS wireless communications while 800 MHz band represents a conventional digital CDMA based wireless system. CDMA was pioneered by Qualcomm and provides between four and six times the capacity of TDMA-based systems. Table 19.4 provides a summary of the basic characteristics of the cellular wireless access methods supported in North America. Note that, as previously mentioned, GSM represents a narrowband version of TDMA.

In concluding our examination of current wireless cellular technology, we will turn our attention to the transmission of data. For many laptop, notebook and personal digital assistant (PDA) users, cellular communications provide a handy method to enter orders, retrieve email, and perform similar functions that overcome such problems as high hotel communications charges and hardwired telephones.

Table 19.4 North American wireless technologies

Technology	Classification	Frequency band
AMPS	Analog	800 MHz
CDMA	Digital cellular	800 MHz
	PCS	1900 MHz
TDMA	Digital cellular	800 MHz
	PCS	1900 MHz
GSM	PCS	1900 MHz

Wireless data

For several years, publications ran articles about the emerging ability of mobile business persons to relay email, purchase orders and reports via wireless communications as if they were in the office. Although it is possible to transmit data over different types of wireless networks, there are several constraints associated with existing networks that limit the use of current wireless networks for data transmission. Those constraints include the cost associated with using certain data-only wireless networks, achievable data transmission rates, and the cost of wireless modems.

Circuit switched networks

In general there are two types of wireless networks – circuit switched and packet switched. Circuit switched wireless networks are based upon the use of a circuit switched infrastructure under which a connection is established and bandwidth allocated between the originator and destination for the duration of the call. Examples of circuit switched wireless networks include AMPS, TDMA and CDMA digital networks. Charges for the use of circuit switched cellular networks are the same as when the network is used for voice, with usage billed based on connection time.

When AMPS is used you need an analog modem that supports an error control protocol suitable for mobile operations. Although the maximum data rate obtainable on AMPS is supposed to be 14.4 kbps, in most trials performed by the author of this book the highest data rate obtainable was 9.6 kbps.

Both TDMA and CDMA wireless are digital cellular systems where voice is digitized. To transmit data over a TDMA or CDMA a special modem is required. However, if a GSM wireless handset is used, you can use an adapter to directly connect your notebook or PDA directly to the handset and do not need to have a modem. Each type of digital network supports data transmission at 9.6 kbps.

Packet switched wireless

Similar to conventional packet networks, packet switched wireless networks are designed to break data into packets and route each packet through the network. One of the more popular wireless packet networks is called Cellular Digital Packet Data (CDPD). CDPD is constructed as an overlay to most cellular operators' existing digital networks and requires the use of a CDPD compatible modem to access the network. The maximum operating rate of a CDPD modem is 19.2 kbps, with many users reporting actual transmission rates closer to 14.4 kbps.

Because packet switching results in the statistical sharing of radio channels, CDPD makes more efficient use of cellular frequencies and allows their reuse by different subscribers. Several vendors, to include BellSouth Wireless Data, Ardis and RAM Mobile Data Services, provide nationwide or near-nationwide coverage in the United States. Unfortunately, current billing can be as high as $0.15 for 600 characters, or more than 1000 times the per-character cost for using a wired packet network, making the use of such networks more suitable for two-way paging-type messages rather than Internet surfing.

Now that we have an appreciation for existing cellular systems used for voice and data, let us conclude our coverage of this topic by turning our attention to the future. In doing so we will examine the next generation of wireless cellular service that within a few years will provide us with the ability to 'surf the net' and perform many activities that are impractical when using current technology.

The next generation

AMPS can be considered to represent the first generation of cellular communications, while current versions of TDMA and CDMA represent the second generation of cellular communications. While second generation cellular was being deployed, standardization bodies in the United States and Europe focused their attention upon developing a third generation (3G) wireless technology. Until early 1999 it appeared that separate and incompatible standards would be developed for North America and Europe, with the name cdma2000 used by the Telecommunications Industry Association (TIA) in the United States to specify a 3G wireless technology, while the ITU-T used the term International Mobile Telecommunications (IMT) 2000 for a standardized 3G technology that was incompatible with cdma2000. In June 1999 the ITU met in Beijing and endorsed a harmonization effort that will make cdma2000 compatible with the CDMA component of the IMT-2000 standard. Under this harmonization effort, the parameters for CDMA will enable the development of inexpensive, multimode handsets and infrastructure for global wireless operations.

Under the IMT-2000 specification both packet and circuit switched wireless communications will be supported. Both types of services support three types of mobility and the same data rates as indicated in Table 19.5. In examining Table 19.5 note that the minimum data rates under IMT-2000 could result in competition for the use of cable modems and digital subscriber lines (DSLs) as a mechanism to access the Internet. This is because there is no reason why service operators cannot use IMT-2000 technology to provide base stations in

Table 19.5 IMT-2000 data rates

Bearer service	Mobility requirements	Minimum data rate
Packet	Vehicular	144 kbps
	Outdoor to indoor and Pedestrian	384 kbps
	Indoor office	2 Mbps
Circuit switched	Vehicular	144 kbps
	Outdoor to indoor and Pedestrian	384 kbps
	Indoor office	2 Mbps

a neighborhood to support wireless communications from businesses and residential subscribers that do not roam.

19.3.2 Wireless LANs

Until 1997 wireless LANs represented proprietary technology. In 1997 the IEEE published its 802.11 standard for wireless LANs which allows vendors to develop products that are interoperable and can be expected to further boost an already growing industry.

The IEEE 802.11 standard governs the two lower layers of the ISO Reference Model, the physical layer and the medium access control (MAC) layer. Under the IEEE 802.11 standard three physical layers are specified. One is an optical-based physical layer that uses infrared light to transmit data. The other two physical layers represent two radio frequency (RF) types of spread spectrum radio communications. One RF type of spread spectrum radio communications is direct-sequence spread spectrum (DSSS), while the second type is frequency-hopping spread spectrum (FHSS).

Under DSSS the data stream is modulated with a spreading code into a broadened bandwidth, permitting the receiver to detect error-free data even if noise is encountered in portions of the transmission band. In comparison, FHSS results in the use of narrowband data transmission; however, the frequency of transmission is periodically changed at a fixed time interval around a spread using different center frequencies in a predefined sequence. Through the use of frequency hopping the FHHS avoids narrowband noise in portions of the transmission band. Under the IEEE 802.11 standard infrared transmission has a peak data rate of 1 Mbps while both types of RF communications operate at 2 Mbps. Both DSSS and FHSS operate in the 2.4 GHz Industrial, Scientific and Medical (ISM) band and provide a transmission distance that can cover a small campus area. In comparison, the infrared version is limited to line of sight and is most practical for use within a single office.

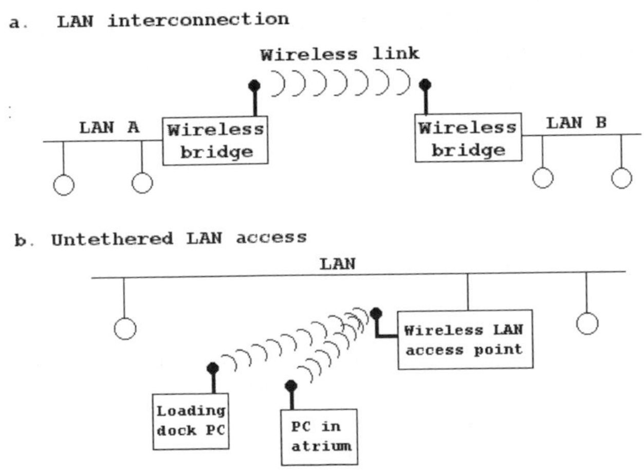

Figure 19.12 Wireless LAN applications

Applications

Wireless LANs open up a large area that until recently was difficult to support via conventional cable-based LANs. Those areas include the factory floor, warehouses, reception areas located in the center of a hotel atrium, and even loading docks. In addition, since wireless LAN components can allow connectivity to an existing LAN, the use of wireless provides a supplement to support devices that are normally untethered, such as laptops and PDAs.

Another potential application for wireless LANs is to bridge existing networks between buildings. The top portion of Figure 19.12 illustrates the use of several wireless LAN access points which allows stations not tethered to the LAN to operate via wireless communications as participants of the LAN.

19.4 REVIEW QUESTIONS

1. What types of light source are used with fiber systems?

2. What is the refractive index? What does its value indicate?

3. What is pulse dispersion?

4. How is light energy converted back into electrical energy?

5. Discuss the primary advantage associated with the use of lightwave division multiplexing.

6. What is an optical modem?

7. Discuss five advantages obtained from the use of a fiber-optic transmission system.

8. What are two limitations to the widespread use of fiber-optic systems?

9. Why is it costly to extract a DS0 signal from a DS3 signal on a T3 carrier?

10. Discuss why SONET enables individual-DS0s to be extracted from a high speed digital signal without demultiplexing.

11. What is the difference between a linear and ring-based SONET deployment with respect to reliability?

12. Discuss the operation of the two types of SONET rings.

13. Compare and contrast FDMA and TDMA.

14. Discuss four advantages obtained by the use of satellite communications.

15. What types of application are best suited for satellite communications?

16. Why are cellular systems structured to preclude the use of the same frequency in adjacent cells at the same time?

17. Discuss the cell capacity of AMPS, TDMA and CDMA wireless systems with respect to one another.

18. What are the three components of a wireless cellular system?

19. Why are more cell sites required to support PCS than a TDMA system within the same geographic area?

20. Describe three constraints associated with transmitting data over current wireless cellular systems.

21. Describe the three physical layer transmission methods supported by the IEEE 802.11 wireless LAN standard.

22. Describe two applications for wireless LANs.

20

EVOLVING TECHNOLOGIES

In concluding this book we will focus our attention upon two evolving communications technologies. Those technologies include Asynchronous Transfer Mode (ATM) and Virtual Private Networking (VPN).

ATM provides a mechanism for the integration of LANs and WANs as well as the ability to transport voice, data and video between private and public networks. VPNs enable you to use public networks, such as the Internet, to access computational facilities on a private network or to economically interconnect to private networks.

20.1 ATM

ATM represents a special type of packet switching network in which the packet length is fixed and referred to as a cell. While fixing the cell length may appear to represent a minor change from other packet network technologies, in actuality it provides several key benefits. First, it enables switching to be easily performed in hardware, which simplifies the cell switching process. In addition, the ability to perform switching in hardware permits very high switching rates to be achieved. A third benefit is the fact that the use of a fixed cell length minimizes delay when a few cells containing data are inserted between cells transporting digitized voice. This in turn provides ATM with a built-in design capability to transport voice, data and video between public and private networks.

20.1.1 Cell size

The ATM packet is relatively short, containing a 48 byte information field and a 5 byte header. This fixed length packet, which is illustrated in Figure 20.1, is called a cell and the umbrella technology for the movement of cells is referred to as cell relay. ATM represents a specific type of cell relay service which is defined under broadband ISDN (BISDN) standards.

The selection of a relatively short 53 byte ATM cell was based upon the necessity to minimize the effect of data transportation upon invoice. For

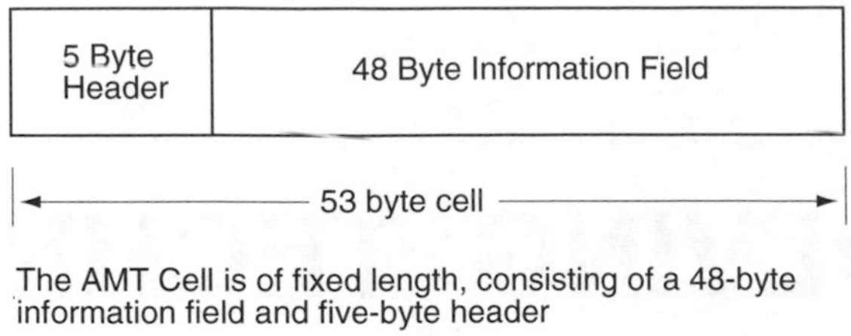
5 Byte Header
48 Byte Information Field
53 byte cell
The AMT Cell is of fixed length, consisting of a 48-byte information field and five-byte header

Figure 20.1 The ATM cell

example, under frame relay or X.25 packet switching, both of which support variable length information fields, a user could generate a burst of data which expands the frame length to a maximum value many times its minimum length. In doing so one user can effectively preclude other users from gaining access to the network until the first frame is completed. While this is a good design principle when all frames carry data, when digitized voice is transported the results can become awkward at the receiver. To illustrate this, assume your conversation is digitized and transported on a conventional packet switching network. As you say 'HELLO', 'HE' might be digitized and transported by a frame. Just prior to the letters 'LLO' leaving your mouth, assume a computer transmitted a burst of traffic that was transported by a long frame. Then, the digitization of 'LLO' would be stored temporarily until a succeeding frame could transport the data. At the receiver time gaps would appear, making the conversation sound awkward whenever bursts of data occurred. ATM is designed to eliminate this problem as cells are relatively short, resulting in cells transporting voice being able to arrive on a regular basis.

Another important advantage associated with the use of fixed length cells concerns the design of switching equipment. This design enables processing to occur from firmware instead of requiring more expensive software if variable length cells were supported.

In addition to its relatively short cell length facilitating the integration of voice and data, ATM provides three additional benefits. Those benefits are in the areas of scalability, transparency, and traffic classification.

20.1.2 Scalability

ATM cells can be transported on LANs and WANs at a variety of operating rates. This enables different hardware, such as LAN and WAN switches to support a common cell format, a feature lacking with other communications technologies. Within a few years, an ATM cell generated on a 25 Mbps LAN will be able to be transported from the LAN via a T1 line at 1.544 Mbps to a central office where it might be switched onto a 2.4 Gbps SONET network for transmission on the communications carrier infrastructure, with the message maintained in the same series of 53 byte cells, with only the operating rate scaled for a particular transport mechanism.

20.1.3 Transparency

The ATM cell is application transparent, enabling it to transport voice, data, images, and video. Due to its application transparency, ATM enables networks to be constructed to support any type of application or application mix instead of requiring organizations to establish separate networks for different applications.

20.1.4 Traffic classification

Five classes of traffic are supported by ATM to include one constant bit rate, three types of variable bit rates, and a user definable class. By associating such metrics as cell transit delay, cell loss ratio, and cell delay variation to a traffic class, it becomes possible to provide a guaranteed quality of service on a demand basis. This enables a traffic management mechanism to adjust network performance during periods of unexpected congestion to favor traffic classes based upon the metrics associated with each class.

20.2 THE ATM PROTOCOL STACK

Similar to other networking architectures ATM is a layered protocol. The ATM protocol stack is illustrated in Figure 20.2 and it consists of three layers: the ATM Adaptation Layer (AAL), the AEM Layer and the Physical Layer. Both AAL and Physical Layers are divided into two sublayers. Although the ATM protocol stack consists of three layers, as we will shortly note, those layers are essentially equivalent to the first two layers of the ISO Reference Model.

20.2.1 ATM Adaptation Layer

As illustrated in Figure 20.2, the ATM Adaptation Layer consists of two sublayers: a convergence sublayer and a segmentation and reassembly sublayer. The function of the AAL is to adapt higher level data into formats

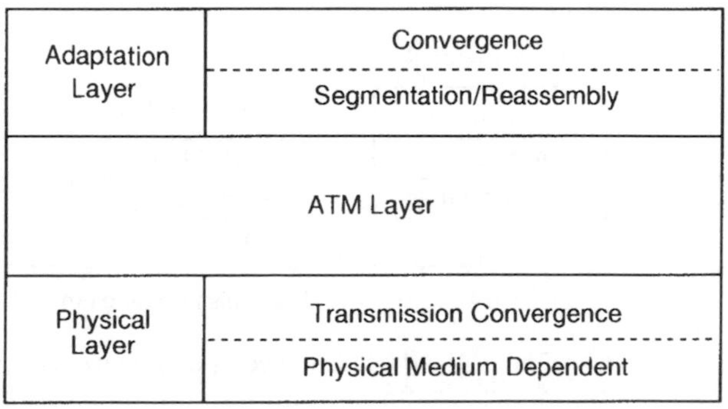

Figure 20.2 The ATM protocol stack

compatible with ATM Layer requirements. To accomplish this task the ATM Adaptation Layer subdivides user information into segments suitable for encapsulation into the 48 byte information fields of cells. The actual adaptation process depends on the type of traffic to be transmitted, although all traffic winds up in similar cells. Currently there are five different AALs defined, referred to as AA1 classes, which are described later in this chapter.

When receiving information, the ATM Adaptation Layer performs a reverse process. That is, it takes cells received from the network and reassembles them into a format that the higher layers in the protocol stack understand. This process is known as reassembly. Thus, the segmentation and reassembly processes result in the name of the sublayer that performs those processes.

20.2.2 The ATM Layer

As illustrated in Figure 20.2, the ATM Layer provides the interface between the AAL and the Physical Layer. The ATM Layer is responsible for relaying cells both from the AAL to the Physical Layer and to the AAL from the Physical Layer. The actual method by which the ATM Layer performs this function depends on its location within an ATM network. Since an ATM network consists of endpoint and switches, the ATM Layer can reside at either location. Similarly, a Physical Layer is required at both ATM endpoints and ATM switches.

Since a switch examines the information within an ATM cell to make switching decisions, it does not perform any adaptation functions. Thus, the ATM switch operates at layers 1 and 2, whereas ATM endpoints operate at layers 1 through 3 of the ATM protocol stack, as shown in Figure 20.3.

When the ATM Layer resides at an endpoint, it will generate idle or empty cells whenever there are no data to send, a function not performed by a switch. Instead, in the switch the ATM Layer is concerned with facilitating switching functions, examining cell header information which enables the switch to determine where each cell should be forwarded to. For both endpoints and switches, the ATM Layer performs a variety of traffic management functions, including buffering incoming and outgoing cells, as well as monitoring the transmission rate and conformance of transmission to service parameters

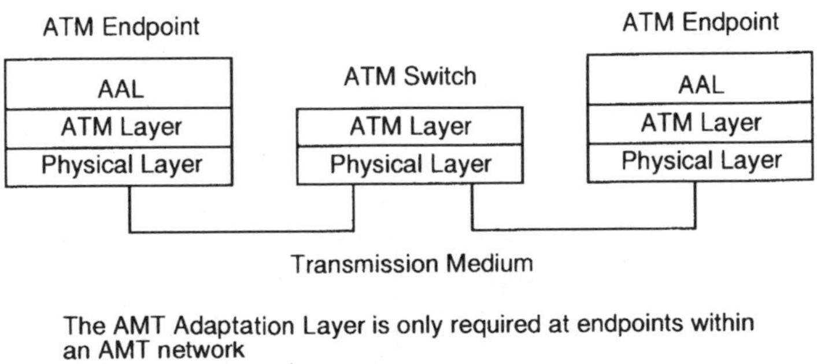

Figure 20.3 The ATM protocol stack within a network

that define a quality of service. At endpoints the ATM Layer also indicates to the AAL whether or not there was congestion during transmission, permitting higher layers to initiate congestion control.

20.2.3 Physical layer

Although Figures 20.2 and 20.3 illustrate an ATM Physical Layer, a specific physical layer is not defined within the protocol stack. Instead, ATM uses the interfaces to existing physical layers defined in other protocols which enable organizations to construct ATM networks on different types of physical interfaces, which in turn connect to different types of media. Thus, the omission of a formal physical layer specification results in a significant degree of flexibility which enhances the capability of ATM to operate on LANs and WANs.

20.3 ATM OPERATION

As previously discussed, ATM represents a cell-switching technology that can operate at speeds ranging from T1's 1.544 Mbps to the gigabit per second rate of SONET. In doing so the lack of a specific physical layer definition means that ATM can be used on many types of physical layers, which makes it a very versatile technology.

20.3.1 Components

ATM networks are constructed on the use of five main hardware components. Those components include ATM network interface cards, LAN switches, ATM routers, ATM WAN switches and ARM service processors.

ATM network interface cards

An ATM network interface card (NIC) is used to connect a LAN-based workstation to an ATM LAN switch. The NIC converts data generated by the workstation into cells that are transmitted to the ATM LAN switch and converts cells received from the switch into a data format recognizable by the workstation.

LAN switch

A LAN switch is a device used to provide interoperability between older LANs, such as Ethernet, Token-Ring or FDDI and ATM. To do so the LAN switch supports a minimum of two types of interface, with one being an ATM interface which enables the switch to be connected to an ATM switch that forms the backbone of the ATM infrastructure. The other interface or interfaces represent connections to older types of LAN.

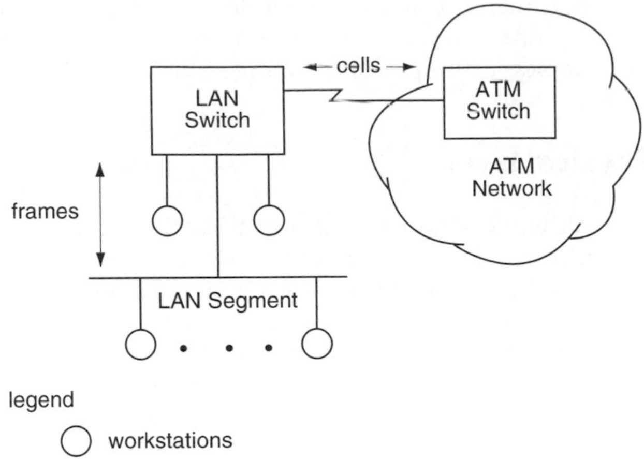

Figure 20.4 Using a LAN switch

The LAN switch functions as both a switch and protocol converter. Data received on one port destined to the ATM network are converted from frames to cells and transferred to the switch port providing a connection to the ATM switch. Since one LAN switch port can be capable of servicing a LAN segment, the use of a switch can minimize an organization's investment in ATM NICs. This is illustrated in Figure 20.4, which shows the use of a LAN switch with a single ATM port to provide access to an ATM network for individual workstations connected directly to individual switch ports as well as to a group of workstations on a LAN segment. Through the use of the LAN switch, an organization can selectively upgrade existing LANs to ATM while obtaining a connection to the ATM network.

ATM router

An ATM router is a router that contains one or more ATM NICs. As such, it can provide a direct or indirect-capability for LAN workstations to access an ATM network or for two ATM networks to be interconnected. For example, a network segment or individual workstations could be connected to a router, which in turn is connected to a LAN switch or directly to an ATM switch.

ATM switches

An ATM switch is a multi-port device which forms the basic infrastructure for an ATM network. Unlike a LAN switch, an ATM switch only permits a single end station to be connected to each switch port. By interconnecting ATM switches an ATM network can be constructed to span a building, city, country or the globe.

The basic operation of an ATM switch is to route cells from an input port onto an appropriate output port. To accomplish this the switch examines

Table 20.1 ATM communications operating rates

Operating rate (Mbps)	Transmission media
25–51	unshielded twisted pair category 3
100	multi-mode fiber
155	shielded twisted pair
622	single-mode fiber

fields within each cell header and uses that information in conjunction with table information maintained in the switch to route cells. Later in this chapter we will examine the composition of the ATM cell header in detail.

One of the key features of ATM switches reaching the market during the mid-1990s is their rate adaptation capability, which in general is a function of the transmission media used to connect endpoints and to connect switches to other switches. Table 20.1 lists some of the communication rates associated with different transmission media.

ATM service processor

An ATM service processor is a computer operating software which performs services required for ATM network operations. For example, an ATM network address can have one of three formats, with one format similar to a telephone number. In comparison, the NIC in a LAN has a hardware address burnt into the adapter. Stations on a LAN can register their addresses using the facilities of an Address Resolution Protocol (ARP) server. That server would then act as a translator between the burnt-in LAN specific hardware addresses and ATM public or private network addresses that could considerably differ from the LAN addressing scheme.

20.3.2 Network interfaces

ATM support two types of basic interfaces: a user-to-network interface (UNI) and a network-to-node interface (NNI).

User-to-network interface

The UNI represents the interface between an ATM switch and an ATM endpoint. Since the connection of a private network to a public network is also known as a UNI, the terms public and private UNI are used to differentiate between the two types of user-to-network interfaces. That is, a private UNI references the connection between an endpoint and switch on an internal, private ATM network, such as an organization's ATM based LAN. In comparison, a public UNI would reference the interface between either a customer's endpoint or switch and a public ATM network.

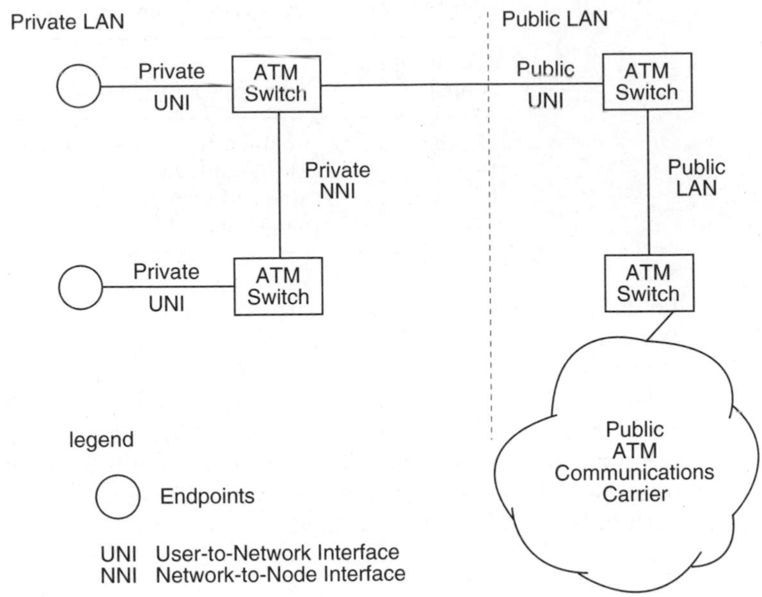

Figure 20.5 ATM network interfaces

Network-to-Node interface

The connection between an endpoint and switch is simpler than the connection between two switches. This results from the fact that switches communicate information concerning the utilization of their facilities, as well as pass setup information required to support endpoint network requests.

The interface between switches is known as a network-to-node or network-to-network interface (NNI). Similarly to the UNI, there are two types of NNIs. A private NNI describes the switch-to-switch interface on an internal network such as an organization's LAN. In comparison, a public NNI describes the interface between public ATM switches, such as those used by communications carriers. Figure 20.5 illustrates the four previously described ATM network interfaces.

20.3.3 The ATM cell header

The structure of the ATM cell is identical in both public and private ATM networks, with Figure 20.6 illustrating the fields within the five byte cell header. As we will soon note, although the cell header fields are identical throughout an ATM network, the use of certain fields depends on the interface or the presence or absence of data being transmitted by an endpoint.

Generic flow control field

The generic flow control (GFC) field consists of the first four bits of the first byte of the ATM cell header. This field is used to control the flow of traffic

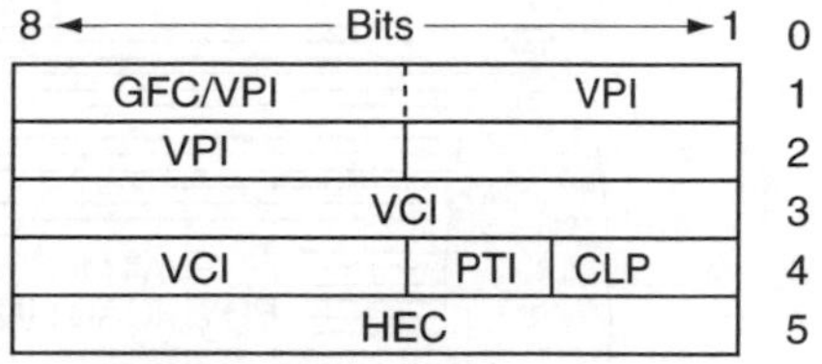

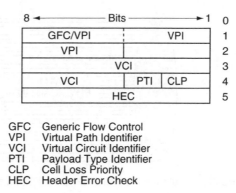

GFC Generic Flow Control
VPI Virtual Path Identifier
VCI Virtual Circuit Identifier
PTI Payload Type Identifier
CLP Cell Loss Priority
HEC Header Error Check

Figure 20.6 The ATM cell header

across the user-to-network interface (UNI) and is used only at the UNI. When cells are transmitted between switches, the four bits become an extension of the VPI field, permitting a larger VPI value to be carried in the cell header.

Virtual path identifier field

The virtual path identifier (VPI) identifies a path between two locations in an ATM network that provides transportation for a group of virtual channels, where a virtual channel represents a connection between two communicating ATM devices. When an endpoint has no data to transmit, the VPI field is set to all zeros to indicate an idle condition. As previously explained, when transmission occurs between switches, the GFC field is used to support an extended VPI value.

Virtual circuit identifier field

The virtual cell identifier (VCI) can be considered to represent the second part of the two-level routing hierarchy used by ATM, where a group of virtual channels are used to form a virtual path.

Figure 20.7 illustrates the relationship between virtual paths and virtual channels. Here the virtual channel represents a connection between two communicating ATM entities, such as an endpoint to a central office switch, or between two switches. The virtual channel can represent a single ATM link or a concatenation of two or more links, with communications on the channel occurring in cell sequence order at a predefined quality of service. In comparison, each virtual path (VP) represents a group of VCs transported between two points that can flow over one or more ATM links. Although VCs are associated with a VP, they are not unbundled nor processed. Thus, the purpose of a virtual path is to provide a mechanism for bundling traffic routed towards the same destination. This technique enables switches to examine the VPI field within the cell header to make a decision concerning the relaying of the cell instead of having to examine the entire three byte address formed by the

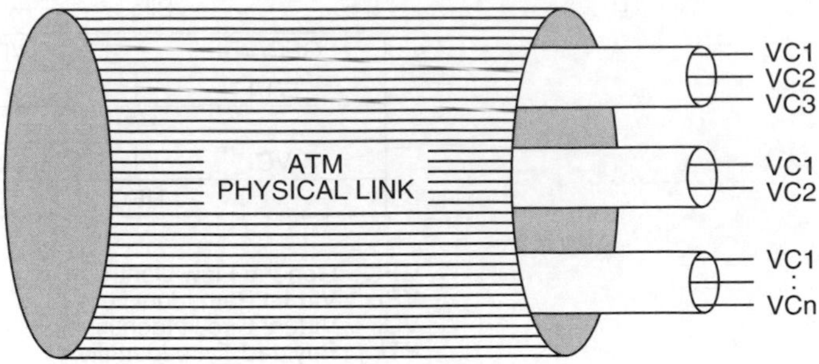

Figure 20.7 Relationship between virtual paths and virtual channels

VPI and the VCI. When an endpoint is in an idle condition, the VPI field is set to all zeros. Although the VCI field will also be set to all zeros to indicate the idle condition, other non-zero VCI values are reserved for use with a VPI zero value to indicate certain predefined conditions. For example, the VPI field value of zero when a VCI field value of 5 is used to transmit a signaling/ connection request.

Payload type identifier field

The payload type identifier (PTI) field consists of three bits in the fourth byte of the cell header. This field is used to identify the type of information carried by the cell. Values 0–3 are reserved to identify various types of user data, 4 and 5 denote management information, and 6 and 7 are reserved for future use.

Cell loss priority field

The last bit in the fourth byte of the cell header represents the cell loss priority (CLP) field. This bit is set by the AAL layer and is used by the ATM layer throughout an ATM network as an indicator of the importance of the cell. If the CLP is set to 1, this indicates that the cell can be discarded by a switch experiencing congestion. If the cell should not be discarded due to the necessity to support a predefined quality of service, the AAL layer will set the CTP bit to 0. The CLP bit can also be set by the ATM layer if a connection exceeds the quality of service level agreed to during the initial communications handshaking process when setup information is exchanged.

Header error check field

The last byte in the ATM cell header is the header error check (HEC). The purpose of the HEC is to provide protection for the first four bytes of the cell header against the misdelivery of cells due to errors affecting the addresses within the header. To accomplish this the HEC functions as an error detecting and correcting code. The HEC is capable of detecting all single and certain multiple bit errors as well as correcting single bit errors.

20.3.4 ATM connections and cell switching

Now that we have a basic understanding of the ATM cell header including the virtual path and virtual channel identifiers, we can turn our attention to the methods used to establish connections between endpoints as well as how connection identifiers are used for cell switching to route cells to their destination.

Connections

In comparison to most LANs that are connectionless, ATM is a connection-oriented communications technology. This means that a connection has to be established between two ATM endpoints prior to actual data being transmitted between the endpoints. The actual ATM connection can be established as a permanent virtual circuit (PVC) or as a switched virtual circuit (SVC).

A PVC can be considered as being similar to a leased line, with routing established for long term use. Once a PVC has been established, no further network intervention is required any time a user wishes to transfer data between endpoints connected via a PVC. In comparison, a SVC can be considered as being similar to a telephone call made on the switched telephone network. That is, the SVC requires network intervention to establish the path linking endpoints each time that a SVC occurs.

Both PVCs and SVCs obtain the 'V' as they represent virtual rather than permanent or dedicated connections. This means that through statistical multiplexing an endpoint can receive calls from one or more distant endpoints.

Cell switching

The VPI and VCI fields within a cell header can be used individually or collectively by a switch to determine the output port for relaying or transferring a cell. To determine the output port, the ATM switch first reads the incoming VPI, VCI, or both fields, with the field read dependent on the location of the switch in the network. Next, the switch will use the connection identifier information to perform a table lookup operation. That operation uses the current connection identifier as a match criterion to determine the output port that the cell will be routed onto as well as a new connection identifier to be placed into the cell header. The new connection identifier is then used for routing between the next pair of switches or from a switch to an endpoint.

Types of switches

There are two types of ATM switch, with the differences between each related to the type of header fields read for establishing cross-connections through the switch. A switch limited to reading and substituting VPI values is commonly referred to as a VP switch. This switch operates relatively fast. A switch that reads and substitutes both VPI and VCI values is commonly referred to as a virtual channel switch (VC Switch). A VC switch generally has a lower cell operating rate than a VP switch as it must examine additional information in each cell header. You can consider a VP switch as being similar to a central office switch, whereas a VC switch would be similar to end office switches.

Using connection identifiers

To illustrate the use of connection identifiers in cell switching, consider Figure 20.8, which illustrates a three switch ATM network with four endpoints. When switch 1 receives a cell on port 2 with VPI = 0, VCI = 10, it uses the VPI and VCI values to perform a table lookup, assigning VPI = 1, VCI = 12 for the cell header and switching the cell onto port 2. Similarly, when switch 1 receives a cell on port 3 with VPI = 0, VCI = 18 its table lookup operation results in the assignment of VPI = 1, VCI = 15 to the cell's header and the forwarding of the cell onto port 1. If we assume switch 2 is a VP switch, it only reads and modifies the VP, thus the VCI's are shown exiting the switch with the same values they had upon entering the switch. At switch 3, the VPC is broken down, with virtual channels assigned to route cells to endpoints C and D that were carried in a common virtual path from switch 1 to switch 3.

The assignment of VPI and VCI values is an arbitrary process which considers those already in use, with the lookup tables being created when a connection is established through the network. Concerning that connection, it results from an ATM endpoint requesting a connection setup via the User-Network Interface through the use of a signaling protocol which contains an address within the cell. That address can be in one of three formats. One known as E.164 is the same as used in public telephone networks, whereas the other two address formats include domain identifiers that allow address

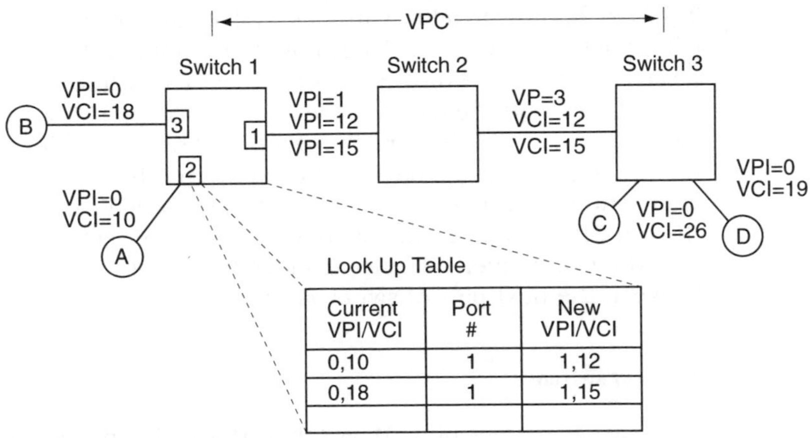

Figure 20.8 Cell switching example

Table 20.2 ATM traffic classes

Class	Name	Description
A	Constant bit rate (CBR)	Connection-oriented voice or video, such as DS0s
B	Variable bit rate (VBR)	Connection-oriented data services, such as packet video, requiring the transfer of timing information between end points
C	Connection-oriented (VBR)	Used for bursty data transfer for connection-oriented asynchronous traffic, such as IPX, X.25, Frame Relay
D	Connectionless VBR	Used for transmitting VBR data without requiring a previously established connection, i.e. SMDS
X	Simple and efficient adaptation layer (SEAL)	Similar to C but assumes higher layer process provides error recovery

fields to be assigned by different organizations. The actual signaling method is based on the signaling protocol used in ISDN and enables a quality of service to be negotiated and agreed during the connection setup process. The quality of service is based on metrics assigned to different traffic classes, permitting an endpoint to establish several virtual connections where each connection transports different types of data with different performance characteristics assigned to each connection. To obtain a better understanding of this important concept, let us first focus our attention on the traffic classes supported by ATM which are listed in Table 20.2. By using different qualities of service for different traffic classes, an ATM endpoint could prioritize Class B to correctly receive time-sensitive frames for a videoconference, while allowing relatively time-insensitive traffic, such as Class C, to be delayed or for cells to even be dropped during periods of congestion.

ATM is a complex evolving technology which provides the potential to integrate voice, data and video across LANs and WANs. Although it is still in a state of infancy, several communications carriers have gone on record to commit a considerable amount of money to implement the technology into their infrastructure for WAN operations. A similar commitment is not possible with respect to LANs as each organization is the ultimate controller of their destiny. Like all technologies, ATM will compete upon price and performance. Since such high speed technologies as Fast Ethernet offer 100 Mbps operations at a fraction of the current cost of ATM equipment, the adaptation of ATM on LANs may become more of a gradual evolutionary process than a technical revolution.

20.4 VIRTUAL PRIVATE NETWORKING

Virtual private networking (VPN) represents a communications technique that uses the transmission facilities of a public mesh structured packet network to access private networks. Such access can include mobile users connecting to a private network over a public packet network or connecting two or more private, geographically separated networks together via a public packet network. The public network could be any type of packet network, such as an X.25 or frame relay network; however, most references to virtual networking reference the use of the Internet as it can be accessed from thousands of cities on our globe. Because most public packet networks are considerably larger than private networks their use often provides a boundary-less networking capability. Since the mesh structure of a packet network provides more than one route between source and destination, the use of a packet network typically involves establishing a virtual route through the network. Thus, the use of a mesh structured packet network to access a private network is also referred to as virtual networking, a term popularly used to reference this communications method.

20.4.1 Rationale for use

There are two major reasons for high interest among businesses for virtual private networking. Those reasons are economics and reliability.

Economics

To illustrate the economics associated with the use of a VPN, let us examine the use of the Internet to interconnect several geographically separated LANs. Figure 20.9 compares the use of a private network and the Internet to interconnect three LAN locations.

If we assume the private network uses T1 lines the monthly cost of such lines is approximately \$3.00/mile. If we further assume that the distance from points A to B equals the distance from points A to C and each span is 500 miles in length, then the monthly networking cost for the private network to interconnect the three locations is \$3.00/mile × 1000 miles, or \$3000.

If we now turn our attention to the use of the Internet, let us assume that an Internet Service Provider (ISP) charges \$1000/month for a T1 connection in a

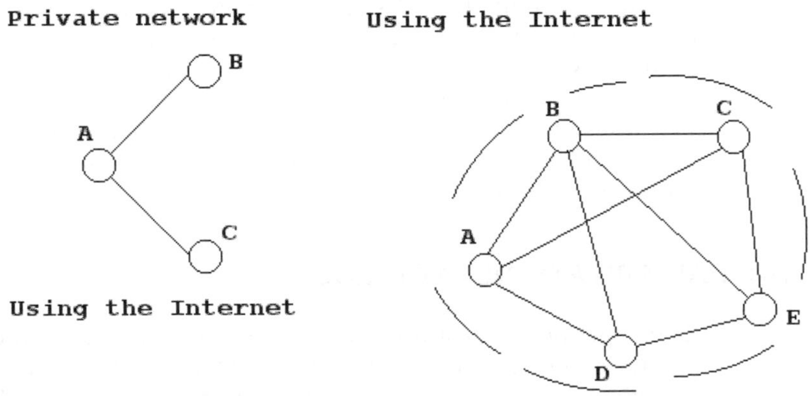

Figure 20.9 Comparing the use of a private network and the Internet

metropolitan area. Thus, connecting three locations via the Internet would result in a cost of \$3000 per month, which equals the cost associated with the use of the private network.

Now let us assume the three LAN locations to be interconnected are New York City, Los Angeles and Miami, increasing the total distance to approximately 3500 miles. The monthly cost of the private network now becomes \$3.00/mile $\times$ 3500 miles, or \$10 500. In comparison, the cost of using the Internet remains fixed at \$1000 per location, or \$3000 per month. Thus, as the distance between locations increases, the potential cost savings associated with the use of the Internet increase.

Another economic consideration associated with using a public network is the fact that only one router port is required to obtain access to multiple locations. In comparison, when private networking is used to interconnect geographically separated locations, as the network expands additional hardware in the form of router ports are required.

20.4.2 Reliability

The use of a private network can require multiple diversity routed circuits between interconnected locations if an organization is concerned about obtaining a very high level of reliability. In comparison, the mesh structure of the Internet as well as other packet networks provides a built-in alternate routing capability. Thus, the use of a mesh-structured network provides alternate routing which could be expensive to duplicate for a private network.

20.4.3 Problem areas

Although economics and reliability represent considerable advantages associated with VPNs they are not problem-free. Two key problems that require addressing include security and performance of the packet network used to interconnect geographically separated networks.

Security

The interconnection of private networks via the use of the Internet opens up a corporate network to potential attack from a virtually unlimited number of persons. For example, when the Internet is used to interconnect corporate networks it becomes possible for over 50 million persons who have access to the Internet to attempt to access the corporate network.

As a minimum most corporate networks connected to the Internet should have router access lists created that only permit traffic from addresses on interconnected networks to flow between networks. However, because it is a relatively easy process to spoof an IP address, most corporations will add a firewall with authentication and virus scanning capabilities. Thus, the added cost of security becomes an issue that must be considered.

Performance

A public packet network intermixes packets from different subscribers onto their backbone infrastructure. This means that relatively long packets transmitted by one organization performing a file transfer could be carried between short packets transmitting interactive queries or even digitized voice generated by another organization. The result of this means that latency delays are not normally predictable and can adversely affect time-dependent applications.

Applications that may not be correctly suitable for virtual networking via the Internet include digitized voice, interactive queries, and real-time control. However, a substantial effort is now underway to develop a mechanism to minimize delays and in effect provide a guaranteed bandwidth that would alleviate the previously described problem. Collectively the techniques being developed are referred to as providing a quality of service (QoS) and may further increase the use of the Internet over the next few years.

20.5 REVIEW QUESTIONS

1. Discuss two benefits obtained from the use of the relatively short fixed sized cell used in ATM.

2. When referring to ATM, what does the term scalability refer to?

3. What feature enables ATM to adjust network performance during unexpected congestion?

4. What is the function of the ATM Adaptation Layer?

5. Why is the ATM Adaptation Layer not required at an ATM switch?

6. What is the function of the ATM Layer?

7. Where in the ATM network are idle or empty cells generated?

8. What is the advantage associated with the absence of a specific Physical Layer being defined in the ATM protocol stack?

9. What are some of the functions of an ATM network interface card?

10. What are the two key functions performed by a LAN switch that has an ATM NIC?

11. Discuss the differences between a LAN switch and an ATM switch.

12. Discuss the difference between the user-to-network interface and the network-to-node interface.

13. What is the purpose of a virtual path?

14. What is the purpose of the cell loss priority field in the ATM cell header?

15. What is the purpose of the header error check field in the ATM cell header?

16. What are the two types of connections supported by ATM?

17. What is the difference between a VP switch and a VC switch with respect to their operating rate?

18. Describe two uses for VPNs.

19. What are the two major reasons behind organizations considering the use of the Internet to interconnect geographically separated LANs?

20. Describe two problem areas associated with the use of the Internet to interconnect private networks.

INDEX

BOOKS BY GILBERT HELD, PUBLISHED BY WILEY

DATA COMMUNICATIONS

UNDERSTANDING DATA COMMUNICATIONS
From Fundamentals to Networking
Third Edition

This comprehensive, introductory text offers a fundamental understanding of the role, operation and utilization of communication devices and facilities. Ideal for both the data communications student and the professional telecommunications engineer, this foundation text will appeal to anyone who wishes to gain a detailed knowledge of the evolution and future of communications.

2000 0 471 62745 3

DATA AND IMAGE COMPRESSION
Tools and Techniques
Fourth Edition

Considerably expanded, this fourth edition explains how practical data and image compression techniques are vital for efficient, low-cost transmission and data storage requirements. The book takes into account the advances in this area, including image and fax compression.

1996 0 471 95247 8

DATA COMMUNICATIONS NETWORKING DEVICES
Operation, Utilization and LAN and WAN Internetworking
Fourth Edition

This popular and authoritative text evaluates the networking products that can be used in the design, operation and optimization of a data communications network. Reflecting the many technological advances in the field, this unique book covers over 30 networking devices.

1998 0 471 97515 X

LAN MANAGEMENT WITH SNMP AND RMON

1996 0 471 14736 2

ETHERNET NETWORKS
From 10Base-T to Gigabit
Third Edition

1998 0 471 25310 3

REFERENCE

DICTIONARY OF COMMUNICATIONS TECHNOLOGY
Terms, Definitions and Abbreviations
Third Edition

1998 0 471 97516 8 (Cloth)
0 471 97517 6 (Paper)

THE COMPLETE MODEM REFERENCE
The Technician's Guide to Installation, Testing and Trouble-free Communications
Third Edition

1997 0 471 15457 1

ABOUT THE AUTHOR

Having gained a B.S.E.E. from Pennsylvania Military College, Gilbert Held majored in computer science for his M.S.E.E. from New York University. He also holds an M.B.A. and the M.S.T.M. degree from the American University.

Gilbert Held was the Chief of Data Communications for the United States Office of Personnel Management for 20 years. He serves as a consultant to a number of companies and teaches several college courses in LAN performance, Computers and Decision Theory, Management Information Systems and Data Communications. He is the author of over 50 books and 500 technical articles.

The only person to win the interface Karp award for excellence in technical writing twice, Gilbert Held is also the winner of the Association of American Publishers Professional and Scholarly Publishing Division award. He has been selected as one of the Federal Computer Week top 100 professionals in Government, Industry and Academia who have made an outstanding contribution in the field of computer science. He has also received several Government awards for exceptional performance.